BACTERIAL
PATHOGENOMICS

BACTERIAL
PATHOGENOMICS

Mark J. Pallen
Editor-in-chief
University of Birmingham
Birmingham, United Kingdom

Karen E. Nelson
Department of Biology
Howard University
Washington, DC
and
The Institute for Genomic Research
Rockville, Maryland

Gail M. Preston
Oxford University
Oxford, United Kingdom

ASM
PRESS

Washington, DC

Address editorial correspondence to ASM Press, 1752 N St. NW, Washington, DC 20036-2904, USA

Send orders to ASM Press, P.O. Box 605, Herndon, VA 20172, USA
Phone: (800) 546-2416 or (703) 661-1593
Fax: (703) 661-1501
E-mail: books@asmusa.org
Online: estore.asm.org

Library of Congress Cataloging-in-Publication Data

Bacterial pathogenomics / edited by Mark J. Pallen, Karen E. Nelson, and Gail M. Preston.
    p. ; cm.
   Includes bibliographical references and index.
   ISBN-13: 978-1-55581-451-9 (alk. paper)
   ISBN-10: 1-55581-451-4 (alk. paper)
  1. Bacterial genetics. 2. Bacterial genomes. 3. Pathogenic bacteria. 4. Genomics. I. Pallen,
Mark J. II. Nelson, Karen E. III. Preston, Gail M.
   [DNLM: 1. Bacteria—genetics. 2. Bacteria—pathogenicity. 3. Genome, Bacterial.
4. Sequence Analysis. QW 51 B1325 2007]
   QH434.B337 2007
   579.3165—dc22

2007011460

10 9 8 7 6 5 4 3 2 1

Cover: Circular view of the genome of the Sakai strain of *Escherichia coli* O157, colored according to GC content (red is high GC, yellow is average, and green is low). Taken from the coliBASE website (http://colibase.bham.ac.uk). Image courtesy of Chris Bailey, Roy Chaudhuri, and Mark Pallen.

CONTENTS

CONTRIBUTORS

Frederick M. Ausubel
Department of Genetics, Harvard Medical School, and Department of Molecular Biology, Massachusetts General Hospital, Boston, MA 02114

Roland Brosch
Unité de Génétique Moléculaire Bactérienne, 28 Rue du Dr. Roux, F-75724 Paris Cedex 15, France

Harald Brüssow
Nestlé Research Centre, Nutrition and Health Department. / Food and Health Microbiology, CH-1000 Lausanne 26, Vers-chez-les-Blanc, Switzerland

Carmen Buchrieser
Unité de Génomique des Microorganismes Pathogènes, Institut Pasteur, and CNRS URA 2171, F-75015 Paris, France

Roy R. Chaudhuri
Division of Immunity and Infection, IBR West, University of Birmingham, Edgbaston, Birmingham B15 2TT, United Kingdom

Stewart T. Cole
Unité de Génétique Moléculaire Bactérienne, 28 Rue du Dr. Roux, F-75724 Paris Cedex 15, France

Pascale Cossart
Unité de Interactions Bactéries-Cellules, Institut Pasteur; INSERM U604; and INRA, USC 2020, F-75015 Paris, France

Marien I. de Jonge
Unité de Génétique Moléculaire Bactérienne, 28 Rue du Dr. Roux, F-75724 Paris Cedex 15, France

Lynn G. Dover
School of Biosciences, University of Birmingham, Birmingham B15 2TT, United Kingdom

Andrea Dowling
Department of Biology, University of Bath, Bath BA2 7AY, United Kingdom

Derrick E. Fouts
Department of Microbial Genomics, The Institute for Genomic Research,
Rockville, MD 20850

Richard H. ffrench-Constant
Department of Biology, University of Bath, Bath BA2 7AY, United Kingdom

David S. Guttman
Department of Botany, Centre for the Analysis of Genome Evolution and Function,
University of Toronto, 25 Willcocks St., Toronto, ON M5S 3B3, Canada

Michelle Hares
Department of Biology, University of Bath, Bath BA2 7AY, United Kingdom

Matthew T. G. Holden
The Wellcome Trust Sanger Institute, Hinxton, Cambridge CB10 1SA, United Kingdom

Philip W. Jordan
The Sir William Dunn School of Pathology, University of Oxford, South Parks Road,
Oxford OX1 3RE, United Kingdom

Daniel G. Lee
Department of Genetics, Harvard Medical School, and Department of Molecular Biology,
Massachusetts General Hospital, Boston, MA 02114

Nicole T. Liberati
Department of Genetics, Harvard Medical School, and Department of Molecular Biology,
Massachusetts General Hospital, Boston, MA 02114

Jodi A. Lindsay
Department of Cellular & Molecular Medicine, St. George's, University of London,
Cranmer Terrace, London SW17 0RE, United Kingdom

Petra Matějková
Department of Biology, Faculty of Medicine, Masaryk University, 625 00 Brno,
Czech Republic

Dominique Missiakas
Department of Microbiology, University of Chicago, Chicago, IL 60637

Emmanuel F. Mongodin
Department of Microbial Genomics, The Institute for Genomic Research,
Rockville, MD 20850

Karen E. Nelson
Department of Microbial Genomics, The Institute for Genomic Research,
Rockville, MD 20850, and Department of Biology,
Howard University, Washington, DC 20059

Steven J. Norris
Department of Pathology and Laboratory Medicine, University of Texas Medical School at
Houston, Houston, TX 77030

Mark J. Pallen
Division of Immunity and Infection, Medical School, University of Birmingham,
Birmingham B15 2TT, United Kingdom

Timothy Palzkill
Department of Molecular Virology and Microbiology, Baylor College of Medicine,
1 Baylor Plaza, Houston, TX 77030

Gail M. Preston
Department of Plant Sciences, University of Oxford, South Parks Road,
Oxford OX1 3RB, United Kingdom

Timothy D. Read
Genomics Group, Biological Defense Research Directorate, Naval Medical Research
Center, Rockville, MD 20852

Nigel J. Saunders
The Sir William Dunn School of Pathology, University of Oxford, South Parks Road,
Oxford OX1 3RE, United Kingdom

Olaf Schneewind
Department of Microbiology, University of Chicago, Chicago, IL 60637

David Šmajs
Department of Biology, Faculty of Medicine, Masaryk University, 625 00
Brno, Czech Republic

Lori A. S. Snyder
The Sir William Dunn School of Pathology, University of Oxford, South Parks Road,
Oxford OX1 3RE, United Kingdom

Timothy P. Stinear
Department of Microbiology, Monash University, Wellington Road,
Clayton 3800, Australia

Gavin H. Thomas
Department of Biology, Area 10, University of York, P.O. Box 373, York YO10 5YW,
United Kingdom

Brendan Thomason
Genomics Group, Biological Defense Research Directorate, Naval Medical Research
Center, Rockville, MD 20852

Ian Toth
The Scottish Crop Research Institute, Invergowrie, Dundee DD2 5DA, United Kingdom

Jonathan M. Urbach
Department of Genetics, Harvard Medical School, and Department of Molecular Biology,
Massachusetts General Hospital, Boston, MA 02114

Nicholas Waterfield
Department of Biology, University of Bath, Bath BA2 7AY, United Kingdom

George M. Weinstock
Human Genome Sequencing Center, Baylor College of Medicine, 1 Baylor Plaza,
Houston, TX 77030

Jennifer J. Wernegreen

Josephine Bay Paul Center for Comparative Molecular Biology and Evolution, Marine Biological Laboratory, Woods Hole, MA 02543

Gang Wu

Department of Genetics, Harvard Medical School, and Department of Molecular Biology, Massachusetts General Hospital, Boston, MA 02114

Guowei Yang

Department of Biology, University of Bath, Bath BA2 7AY, United Kingdom

PREFACE

B liss was it in that dawn to be alive . . .
William Wordsworth, *French Revolution as it appeared to enthusiasts at its commencement*

The genomic era in bacteriology began when scientists at the Institute for Genomic Research in Maryland published the first two complete bacterial genome sequences in 1995. In the intervening 10 or more years, genome sequencing has exerted a stunning influence on bacteriology in general and the study of bacterial pathogenesis in particular. It has been a thrilling time to be in bacteriology. An apt analogy is with the space program, in particular, with the grand tour of the outer solar system during the 1970s and 1980s. Just as the two Voyager probes turned smudges glimpsed hazily through the telescope into newly mapped worlds, so bacterial genomics has allowed us to view the inner workings of our microscopic companions. Many of humankind's most fearful microbial adversaries—from the agents of the black death to the white plague—have now been captured in silico. We now grapple with them on our desktop computers, as well as in the laboratory. Where once we saw through a glass darkly, we now see face to face!

Indeed, now that we have genome sequences from almost every significant bacterial pathogen of humans, plants, and animals, it is scarcely possible to imagine the study of bacterial pathogenesis without the backcloth of genomics. The ready availability of genome sequences pervades every corner of our discipline and underpins the steady incremental accumulation of new information on each bacterial pathogen. This book stands as a monument to this stunning success of "bacterial pathogenomics." We have enlisted the contributions from over two dozen scientists from around the world to highlight the revolutionary contribution of genomics to the study of pathogenic bacteria and bacterial infection. We have adopted a twin-track approach: some chapters in this book survey the impact of genomics on our understanding of key taxonomic groups of pathogens, and others emphasize themes that cut across taxonomic boundaries,

integrating the impact of genomics on topics as diverse as bioterrorism, microbial ecology, bacterial evolution, bacterial protein secretion, and bacterial adaptations to pathogenic lifestyles. We hope that all readers, from student to profes-
sor, will gain new insights into pathogenomics from this book and will close its
pages with an enthusiasm for this subject that will endure well into the second
decade of bacteriology's postgenomic era!

MARK PALLEN
March 2007

INTRODUCTION TO PATHOGENOMICS

Mark J. Pallen

1

The genomic era in bacteriology appeared to begin abruptly in Maryland in 1995 with the completion of the first bacterial genome sequence, from *Haemophilus influenzae* Rd, followed quickly by a second genome sequence, from *Mycoplasma genitalium* (35, 39). Although in retrospect it is possible to identify a number of historical milestones en route to the first bacterial genome sequence (Table 1), none of these historical precedents adequately prepared the world for the revolution of 1995.

The prevailing thought before 1995 was that genomes should be sequenced by using a top-down approach, where detailed mapping and creation of an ordered library were essential prerequisites of any attempt at genome sequencing. The genius of Craig Venter, Hamilton Smith, and others involved in these first bacterial genome sequence projects was to exploit a "dumb," highly automated laboratory approach (whole-genome shotgun sequencing), followed by a "smart" bioinformatics approach (highly efficient genome assembly) that did not require onerous genome mapping. As proof of the revolutionary nature of this breakthrough, aside from a handful of projects already underway in 1995, virtually all bacterial genome-sequencing projects have capitalized on the speed, ease, and efficiency of whole-genome shotgun sequencing.

The choice of these two strains in the genomics revolution highlights the role of historical idiosyncrasy in science. The nonpathogenic unencapsulated Rd strain of *Haemophilus influenzae* was chosen for the first genome-sequencing project largely because it was the same strain in which Smith had discovered restriction enzymes in the late 1960s (a finding that led to Smith winning a Nobel prize) (62, 125). Similarly, the *Mycoplasma genitalium* genome was targeted because its small size made it easier for Venter and his colleagues to show that the project was repeatable.

WHY SEQUENCE BACTERIAL GENOMES?

Given the stunning success and pervasive impact of genome sequencing on bacteriology, it may now seem perverse even to ask: "why sequence bacterial genomes?" However, in the early years, there were many skeptics who saw genomic "stamp collecting" as a distraction of effort and funding away from "hypothesis-driven research" (91). Since then, bacterial genomics has amply validated its status as a platform for hypothesis generation. Furthermore,

Mark J. Pallen   Division of Immunity and Infection, Medical School, University of Birmingham, Birmingham B15 2TT, United Kingdom.

Bacterial Pathogenomics, Edited by M. J. Pallen et al.
© 2007 ASM Press, Washington, D.C.

TABLE 1 Milestones in bacterial genomics

Year	Milestone	Reference(s)
1977	Invention of dideoxy chain terminator sequencing ("Sanger sequencing")	118
1979	Sequencing of the 5.3-kilobase genome of bacteriophage phiX174	116
1981	First human mitochondrial genome sequence[a]	4
1982	Determination of the 48.5-kb genome sequence of bacteriophage lambda through first use of shotgun sequencing	117
1986	Development of automated fluorescent sequencing	126
1995	First complete genome sequences obtained of free-living bacteria (*Haemophilus influenzae* and *Mycoplasma genitalium*)	35, 39
1996	*Mycoplasma* becomes first bacterial genus that has completely sequenced genomes from two different species (*M. genitalium* and *M. pneumoniae*)	53, 54
1997	First genome sequences from *Escherichia coli* and *Bacillus subtilis*	12, 68
1998	First genome sequence from *Mycobacterium tuberculosis;* genome sequence from *Rickettsia prowazekii* provides first evidence of reductive evolution	5, 25
1999	*Helicobacter pylori* becomes the first species with completely sequenced genomes from two isolates	2
2000	Meningococcal genome sequence primes first application of reverse vaccinology	100
2001	Second *E. coli* genome sequences reveal unexpected level of horizontal gene transfer; genome sequence of *M. leprae* provides compelling evidence of bacterial pseudogenes and reductive evolution; first paper reporting genome sequences of two strains from one species (*Staphylococcus aureus*) in a single publication	26, 51, 70, 99
2002	Genome sequencing of multiple strains of *Bacillus anthracis* to provide markers for forensic epidemiology	108
2003	Genome sequencing of uncultivable *Tropheryma whipplei* leads to design of axenic growth medium	10, 106, 111
2004	Genome sequence of mimivirus blurs distinctions between bacteria and viruses	105
2005	Use of whole-genome sequencing used to identify target of new antituberculosis drug; *Mycoplasma genitalium* genome sequenced using pyrosequencing	6, 75
2006	Bacterial metagenomics survey of the Sargasso Sea yields >1 million new genes; metagenomic survey of human bowel microbiota	43, 136

[a]Now that genomics has validated the alpha-proteobacterial origin of mitochondria and cladistics suggests all taxonomy is phylogenetic, Fred Sanger can perhaps claim credit for sequencing the first bacterial genome through this project.

whole-genome sequencing in specialized centers has proven far more efficient than the competitive duplication of effort that preceded it, where multiple unspecialized laboratories often competed to sequence the same gene clusters.

For some pathogens, bacterial genome sequencing has been the only way to gain crucial information—for example, when appraising the metabolic capabilities of a bacterium that is resistant to in vitro culture (examples include *Tropheryma whipplei, Treponema pallidum,* and *Mycobacterium leprae* [10, 26, 40]). However, genomics has delivered new insights even when a species had already been subjected to intense study as a model organism—for example, it is now clear that much of the biology and evolutionary history of *Escherichia coli* was unknown to us before genome sequencing (chapter 2). Furthermore, in providing a genetic "parts list" of the organism under scrutiny, genomics can render otherwise open-ended problems finite. In many cases, all that remains is a sieving exercise, where researchers work through a few thousand protein coding sequences to the desired targets (an early example of this was the use of the genome sequence to unravel the complete lipopolysaccharide biosynthesis pathway in *H. influenzae* Rd [57]). On a more practical level, bacterial genomics has underpinned the search for new drug, vaccine, and diagnostic targets (for example, through so-called reverse vaccinology [100, 107, 121]). And finally, bacterial genomics has provided a test bed for more ambitious projects in human genomics, culminating in Celera's whole-genome shotgun sequence of the human genome (134, 135, 137). The two-way synergistic relationship between human and microbial genomics has clearly benefited bacteriology.

WHO IS SEQUENCING WHICH GENOMES?

At the time of writing (mid-2006), over 300 bacterial genomes have been sequenced, and nearly a thousand are still ongoing. However, these figures will almost surely have increased dramatically by the time the reader encounters this text—see the Genomes OnLine Database (http://www.genomesonline.org/) for up-to-date information (72).

At the present time, pathogens are clearly well represented among completed genome sequences, accounting for nearly 60% of the completed bacterial genome sequences and approximately 45% of ongoing projects. This emphasis on pathogens also skews the taxonomic distribution, with two bacterial phyla of clinical importance, the proteobacteria (159 genomes) and the firmicutes (75 genomes) accounting for over three-quarters of sequenced genomes. This taxonomic skewing is even more pronounced among ongoing projects, where proteobacteria account for over 500 projects and firmicutes for over 200.

As might be expected from its pioneering role in the field, The Institute for Genomic Research (TIGR) has sequenced more bacterial genomes than any other center (53, with 26 of them from pathogens), although the U.S. Department of Energy's Joint Genome Institute comes in a close second (50 genomes, 21 pathogens). However, it is worth stressing that TIGR's track record in sequencing to closure means that it remains preeminent in the field.

Over half the completed bacterial genome projects have been based in the United States. Outside the United States, major national players include Japan (34 genomes), France (19 genomes), Germany (18 genomes), and the United Kingdom's Wellcome Trust Sanger Institute (19 genomes, all but one of them from pathogens). Among newly industrializing nations, Brazil and China have both risen to prominence in bacterial genomics, with 10 and 11 completed genome projects, respectively. Curiously, India has yet to make an impact in this field.

WHAT HAPPENS IN A BACTERIAL GENOME-SEQUENCING PROJECT?

There must be a beginning of any great matter, but the continuing unto the end until it be thoroughly finished yields the true glory.

—Drake

Any bacterial genome-sequencing project comprises a number of steps (see below) and follows the 80:20 principle: that is, 80%, or more, of the work gets done in 20% of the time, while finishing the job consumes a disproportionate amount of effort. This is because the early stages can be automated (particularly the whole-genome shotgun sequencing and assembly), whereas the final stages (finishing, annotation, publication) require far more human intervention. This has thus led to a growing trend toward a "mix and match" approach, in which some genomes within a species are sequenced and annotated to completion, whereas others are left at the unannotated shotgun stage. Recent examples of this strategy include multistrain projects on *Streptococcus agalactiae* and *Campylobacter* (37, 131) (chapter 7). However, this approach has been criticized by those who prefer to see every genome closed, every base finished, and every gene annotated (95).

Strain Choice

The first stage in any bacterial genome project is to choose one or more strains to sequence. In the early years of bacterial genomics, when each project was treated as the genomic equivalent of an Apollo mission, this often proved to be a contentious issue. The choice was often presented in stark terms: should one sequence a well-characterized strain, for which there was a considerable body of prior knowledge, but which may be laboratory adapted and may have lost any pathogenic potential? Or should one instead go for a fresh, minimally passaged clinical isolate, but then risk finding that it proves genetically intractable in the laboratory?

It is clear that several genome-sequenced laboratory strains are compromised in their ability to survive in vivo and/or have undergone laboratory adaptation. Examples include *E. coli* K-12, the LT2 strain *of Salmonella enterica* serovar Typhimurium, the COL strain of *Staphylococcus aureus*, and the PAO1 strain of *Pseudomonas aeruginosa* (42, 138). In some cases, an apparently atypical phenotype of a genome-sequenced strain can be rapidly reversed; this occurred with the NCTC11168 strain of *Campylobacter jejuni*, which initially appeared to be a poor colonizer of chickens but was shown to regain this ability after longer maintenance in vivo (58). However, a more common approach to this problem, now that sequencing costs have dropped dramatically, is to sequence several strains from the same species, so that laboratory strains and fresh clinical isolates can be compared, often sampling a range of strains that vary in their pathogenic potential (Table 2).

From Shotgun to Finishing

The next stage in a genome-sequencing project is to grow enough bacterial cells to yield sufficient genomic DNA for the creation of a random shotgun library in *E. coli*. With most bacteria, this presents no problems. However, in some cases, sequencing centers may have to rely on external collaborators to supply the genomic DNA, particularly if it comes from a dangerous pathogen that can be propagated only under level 3 containment conditions. Organisms that grow extremely slowly in vitro or that can be grown only in vivo also present considerable challenges. In the case of the Sanger Institute's project on the Whipple's disease bacterium, *T. whipplei*, it took longer to

grow the organism (15 culture passages over 17 months) than it did to finish the genome sequence (10). Similar problems have arisen in other projects. For example, preparation of enough *Treponema pallidum* cells for genome sequencing required 11 days' growth in rabbit testis (40), while obtaining bacterial cells for the *M. leprae* genome-sequencing project required the organism's propagation in the nine-banded armadillo (26).

During the first bacterial genome-sequencing projects, great efforts were made to standardize insert size and prevent double inserts (35). However, more recent efforts have relaxed these constraints. Whole-genome shotgun sequencing now typically relies on obtaining paired end-reads from tens of thousands of inserts from small-insert plasmid libraries where average insert sizes are typically 1.5 to 6 kb (38). The plasmid libraries may also be supplemented with a few thousand reads from a library of much larger inserts (typically 10 to 40 kb), to provide a scaffold for assembly. High-throughput sequencing relies on the automated fluorescent Sanger dideoxy approach, with dye terminator chemistry; it is usually performed on automated capillary sequencers. Average read lengths are now typically in excess of 800 bp.

Shotgun sequencing continues until an acceptable coverage of the genome has been achieved. This is typically expressed as the "fold coverage" (amount of total sequence obtained/ estimated genome size). Typical coverage rates for bacterial genome-sequencing projects range from 5- to 10-fold. The reads from the shotgun-sequencing project are assembled into much longer contiguous sequences (contigs) by specialized assembly software, e.g., the TIGR assembler, the Celera assembler, or Phrap (http://www.phrap.org) (38, 101, 134). Typically, a few dozen to a few hundred contigs are obtained from a shotgun assembly, depending on how many repetitive sequences occur in the genome; the more repeats, the harder it is to assemble the sequence.

The final stage in obtaining a bacterial genome sequence is finishing, during which

TABLE 2 Bacterial genome sequencing centers and other resources

Country	Center or resource	URL
United States	Baylor College of Medicine Human Genome Sequencing Center	http://www.hgsc.bcm.tmc.edu
	DOE Joint Genome Institute	http://www.jgi.doe.gov/
	Stanford Genome Technology Center	http://sequence-www.stanford.edu
	The Institute for Genomic Research	http://www.tigr.org
	University of Minnesota Computational Biology Center	http://www.cbc.umn.edu/ResearchProjects/index.html
	University of Oklahoma's Advanced Center for Genome Technology	http://www.genome.ou.edu
	University of Washington Genome Center	http://www.genome.washington.edu
	University of Wisconsin *E. coli* Genome Center	http://www.genome.wisc.edu
	Washington University Genome Sequencing Center	http://genome.wustl.edu
United Kingdom	The Sanger Institute	http://www.sanger.ac.uk/Projects
	xBASE	http://xbase.bham.ac.uk/
France	Genoscope	http://www.genoscope.cns.fr/
Germany	Göttingen Genomics Laboratory	http://www.g2l.bio.uni-goettingen.de
Denmark	Genome Atlases	http://www.cbs.dtu.dk/services/GenomeAtlas/
Japan	National Institute of Technology and Evaluation, Biotechnology Center	http://www.bio.nite.go.jp/e/
	RIKEN Genomic Sciences Center	http://www.gsc.riken.go.jp
	Genome Information Research Center, Osaka University	http://www.gen-info.osaka-u.ac.jp/welcome_en.html

all gaps are closed and all ambiguities resolved. A number of strategies are used to close gaps, usually involving various forms of PCR (38). Indeed, sequencing of PCR-amplified DNA is essential if gaps are the result of toxicity of the DNA to the cloning host. If the sequence of a related strain is available, alignment of contigs to this genome can guide choice of primers in specific PCRs. Similarly, use of large insert libraries can also help order contigs. Software is available for checking the quality of the sequence (http://www.phrap.org). It may be necessary to improve the depth of coverage of some parts of the genome to remove all ambiguities. Finally, particular care has to be applied to repetitive regions. Thus, it is hardly surprising that the finishing stage is usually more time-consuming than the shotgun stage in a bacterial genome-sequencing project.

Annotation

Genome annotation does not usually begin until the genome sequence is complete, i.e., contiguous and free of ambiguities. A number of interesting features can be identified rapidly: repetitive sequences and homopolymeric tracts (often hallmarks of phase variation [97]); noncoding RNA genes (tRNAs, rRNAs, small regulatory RNAs); recently acquired DNA (through analysis of G+C content); leading and lagging strands and the origin of replication (through use of GC skew [85]). However, the major task once a genome sequence is complete remains the identification and annotation of protein-coding genes. Gene finding in bacterial genomes has progressed beyond the simple identification of open reading frames (ORFs) to the use of sophisticated methods, such as interpolated Markov models, to identify protein-coding sequences (or CDSs, a term now preferred to the less precise ORF). The industry standard in this regard is the program Glimmer (Gene Locator and Interpolated Markov Modeler), which uses interpolated Markov models to identify CDSs and distinguish them from noncoding DNA (115). A key feature of this approach is exploitation of a training set of trusted CDSs to generate a model that can then be used to predict all CDSs in the genome. This trusted set can be generated by the Long ORFs program, which identifies all ORFs over, say, 500 bp (i.e., long enough to be highly plausible) or by use of a set of CDSs that have been validated through BLAST hits at the protein level. Although Glimmer is over 98% accurate, problems can sometimes occur with CDSs with deviant GC content or with short CDSs.

Providing some kind of functional assignment for each CDS in a genome represents a formidable challenge. If done properly, this is a labor-intensive process that is a major bottleneck in genome-sequencing projects. In the worst cases, the annotation is superficial, misleading, or even plain wrong, particularly where fully automated annotation pipelines have been used. Here the annotation might simply reflect the title line of the highest-scoring BLAST hit. This can mean that a function of the database protein might be assigned to the search protein in error because the function resides in a part of the database protein sequence not represented in the BLAST result. The flagellar protein FlgJ provides a telling example here. In the conventional flagellar systems from *E. coli* and its close relatives, this is a two-domain protein, with one domain involved in rod assembly; the other plays an enzymatic role as a muramidase. However, in many other systems, FlgJ is a single-domain protein that lacks the enzymatic domain (94). Nonetheless, dozens of these single-domain FlgJ proteins that have been identified from genome sequencing have been annotated with terms like "peptidoglycan hydrolase" or "similar to flagellar muramidase protein," even though they lack the catalytic domain.

Another problem is that the annotation may blur the distinction between the molecular action of a protein or domain (e.g., iron-binding protein) and its physiological role or location (e.g., periplasmic) and inappropriately transfer functional assignments to a distant homolog embedded in an entirely different physiological context (e.g., in an organism lacking a periplasm—see Entrez protein entry NP_269229, or try searching the Entrez protein database with the terms "periplasmic" and "firmicutes"). Once made, such errors can be propagated through the databases. For this reason, one should never accept any functional assignment without researching the annotation and literature to find the original evidence that linked function to sequence, and then checking to ensure that this evidence is indeed pertinent to the sequence (89).

For the reasons already discussed, good-quality annotation relies on a human annotator scrutinizing each gene and each functional assignment. Simple arithmetic dictates how labor-intensive this task is—to spend just 5 min on each gene in a typical 2,000-gene genome would take up a month of human effort. The annotator is assisted in this task by automated homology searches of all CDSs against the current protein database and by searches of domain databases. However, the final decision as to how to annotate any given gene remains the prerogative of the annotator.

Another difficulty is that the very success of bacterial genomics undermines the utility of any annotation. Functional assignments based on homology (or lack of homology) are chasing a moving target in that what is true at the time of annotation ("closest homolog is X" or "has no known homologs") may well be untrue a year or two later. Similarly, the triumphs of high-throughput functional genomics approaches mean that new assignments can often be made on experimental grounds within a few years of a genome sequence appearing. To cope with these problems, there have been several attempts at reannotating genomes some years after initial publication (13, 17, 30, 31, 112). However, the safest advice is that whenever one is considering investigating any gene discovered through genome sequencing, one should repeat the homology searches and literature reviews for oneself before embarking on time-consuming and costly experimentation.

Policies for the release and acceptable use of genome sequence data have sometimes proved controversial (76, 77). Early in the genomic era, several major funding bodies agreed on an immediate data release policy for the human genome-sequencing project (the Bermuda accords) (102). This was soon applied to bacterial genome projects, so that sequences from the shotgun phase were made publicly available as soon as possible after they had been created. A decade later, this continues to be the policy of several major centers, including TIGR and the Sanger Institute. However, data release is contingent on acceptable use guidelines, which have been infringed in some cases (77). Furthermore, many sequencing centers have not signed onto the Bermuda accords. In such cases, the first that the scientific community knows of the project might be a journal publication and the release of a complete annotated genome sequence.

A genome sequence, even if well annotated, is of limited use to the average bacteriologist in the absence of additional visualization and analytical tools. Several major sequencing centers make genome viewers or browsers available through their own websites (Table 2). In addition, the Sanger Institute's Pathogen Sequencing Unit provides excellent platform-independent stand-alone tools, written in Java, for analyzing, visualizing, and annotating genomes (Artemis) and for bacterial genomic comparisons (the Artemis Comparison Tool, or ACT) (114). Several groups independent of major sequencing centers also maintain bacterial genome databases. Examples include the *coli*BASE and *x*BASE suite of databases (22) and the Genome Atlases (48).

WHAT HAS GENOMICS DONE FOR BACTERIOLOGY?

A Change in Perspective

One of the most important effects of the genomics revolution has been a decisive shift in perspective so that we now recognize that

what we knew in the pregenomic era represents just a small, often atypical section of a massively larger information landscape. In my experience alone, there are numerous examples of proteins and associated systems or mechanisms that previously seemed restricted to one small taxonomic group, but now, with the benefit of numerous genome sequences, can be seen to occur in all sorts of unanticipated contexts.

One important example is the ESAT-6 protein and its associated secretion system from *M. tuberculosis*. Although it was clear before genomics that this protein was a key antigen and an important virulence factor in the tubercle bacillus, it was only in the postgenomic era that it became obvious that there were homologs of ESAT-6 in many other gram-positive bacteria (90). Crucially, these sequence-based observations primed laboratory investigations that demonstrated a key role for an ESAT-6 homolog in the virulence of *S. aureus* (15).

Similarly, before genomics, the study of sortase-mediated sorting and attachment of LPXTG proteins to peptidoglycan was largely restricted to *S. aureus* (81). After the demonstration, through analysis of genome sequences (93), that sortase-like proteins and their substrates occurred in numerous other gram-positive bacteria (and even in some gram-negative bacteria and archaea), the field mushroomed, with over 160 PubMed citations of "sortase" or "LPXTG."

As a third example of this phenomenon, we can consider the protein EspA, which forms part of a type IIII secretion apparatus crucial to the virulence of selected pathovars of *E. coli*. When first discovered (66), this appeared to be one of a kind, absent from all other known type III secretion systems. Now, thanks to genomics, we know of many more examples of EspA-like proteins in a variety of bacteria occupying highly diverse niches (*Salmonella, Edwardsiella, Shewanella baltica, Chromobacterium violaceum, Yersinia frederiksenii, Yersinia bercovieri, Sodalis glossinidius*) (92).

All three examples illustrate an important consequence of the postgenomic perspective:

we can now more easily differentiate the specific from the general and the typical from the atypical, and in so doing challenge the anthropocentric view that bacterial virulence factors have evolved merely to cause disease in humans. Thus, for example, the discovery of ESAT-6 homologs in organisms not normally thought of as pathogens (e.g., *Bacillus subtilis, Mycobacterium smegmatis, Corynebacterium glutamicum*) hints at a broader function than just pathogenesis; it suggests that even "nonpathogens" may have developed mechanisms for subverting eukaryotic cells (90).

Genomics and the Species Concept in Bacteriology

As well as providing novel insights into virulence-related proteins or systems, genomics has even led us to question the nature of bacterial species and whether type strains can usually or even ever be considered "typical" of a species. *E. coli* provides a prime example of a highly diverse species, in that strains can vary by more than a megabase in size (chapter 2). This calls into question whether the commonly used model strain, K-12, should ever be accepted as being representative of the species as a whole (42). However, despite huge and largely unanticipated strain-to-strain differences in the mobile or variable component of the genome, and despite a key role for horizontal gene transfer in generating these differences, the nucleotide sequences of housekeeping genes within this species are still very tightly conserved (typically >98% identity at the nucleotide level), suggesting that the species concept can still be defended in the postgenomic era.

At the opposite extreme from the variability of *E. coli*, comparative genomics has shown that several species of bacterial pathogens are largely clonal, reflecting a recent origin and/or population bottleneck. Examples include the anthrax, leprosy, typhoid, and plague bacilli (1, 61, 63, 83, 96, 98, 127). This was also initially thought to be true for the tubercle bacillus (128). However, these earlier assumptions about the near-clonal origin of human tuberculosis have recently been overturned by the discovery of highly divergent relatives of *M. tuberculosis* in East Africa (47) (chapter 3).

Genome Evolution

Several additional key themes in the evolution of pathogens have emerged in the postgenomic era. One is that much strain-to-strain variation is due to prophages—indeed, in many cases, such phages encode crucial virulence factors (chapter 10). Another key theme is that loss of genes occurs through the process of reductive evolution, particularly when bacteria adopt a new pathogenic or symbiotic niche (chapter 8).

The pregenomic view of bacteria was that they represented the most efficient of life forms, honed to perfection by natural selection to carry none of the junk DNA that burdens eukaryotes. We now know that junk DNA, in the form of pseudogenes, is common in bacteria. In some recently evolved pathogens such as *Mycobacterium leprae*, pseudogenes can even account for a sizable fraction of the genome (chapter 3). Thus, a more accurate postgenomic view is that no bacterial genome is perfect. Each is a work in progress, complete with its own baggage of history; never a definitive manuscript, but always a palimpsest, carrying the marks of unceasing rounds of insertion, deletion, and rearrangement. In the most extreme cases, where a bacterium has jettisoned the ability for independent existence or has become genetically isolated from any source of new DNA, the erosive force of Muller's ratchet leads to a progressive shrinking of the genome. Thus, *Buchnera* represents an extremely cutdown version of *E. coli* (chapter 8). However, even in *E. coli* itself, by comparing strains, one can find snapshots of gene gain and gradualistic gene loss—for example, in the acquisition, then the slow progressive loss, of the ETT2 genomic island or in the gain, then the more sudden loss, of the Flag-2 gene cluster (109, 110).

Applied and Functional Genomics

Bacterial genomics has underpinned advances in drug and vaccine discovery and in pathogen detection and isolation. An early example of this approach was the use of "reverse vaccinology" to identify novel subunit vaccine candidates in the meningococcus (100). More recently, as an example of how genomics can influence our ability to culture pathogens, a genome-based reconstruction of the metabolic potential of the Whipple's bacillus (initially not culturable in the absence of human cells) led to the development of rationally designed medium, which allowed the organism to be maintained in axenic culture (111).

The ready availability of bacterial genome sequences has spawned a number of high-throughput postgenomic experimental approaches that together constitute the field of functional genomics (139). These approaches share a common goal of capturing global information in a single experimental program that covers all constituents or products of the genome, rather than focusing on one gene or protein at a time. One incidental consequence of the adoption of functional genomics by bacteriologists is that it has reunified our discipline with the rest of biology. Bacteriologists now share common ground with experts on yeast, fruit flies, or even cancer cells: all are trying to get the most out of microarray technology, proteomics, or global interaction studies. Furthermore, the application of multiple functional genomic approaches to the same organism means that, for the first time, we can hope to gain an integrated view of pathogen biology (139).

DNA microarrays have seen widespread use in bacteriology over the past decade. Space does not permit a full review of their uses and allows only the briefest description of methodology. Instead, I refer readers to several recent reviews on this subject for overviews (16, 28, 32, 55, 120) and to subsequent chapters in this volume for examples of applications. In short, a DNA microarray consists of an assembly of microscopic DNA spots attached to a solid surface, such as glass, plastic, or silicon. Thousands of spots can be contained on a single microarray, where they can be used as probes to detect genomic or cDNAs corresponding to every gene in the bacterial genome. Thus, microarrays can be used in comparative genomic hybridization to compare the gene content of various strains or in expression profiling to compare the repertoires of active genes in the same strain under different environmental conditions or in different strains (say a mutant and its parent strain) under the same conditions. Other potential applications of microarrays include resequencing of large numbers of genes or entire genomes; rapid surveys of tagged mutants (see below); or chromatin immunoprecipitation, where microarrays can be used to identify DNA fragments pulled down by a given DNA-binding protein. Table 3 provides a handful of recent examples of the use of microarrays in pathogenomics; the list is illustrative rather than exhaustive.

Mass mutagenesis represents another functional genomics technology that has been exploited in bacteriology (139). Here one aims to create a bank of mutants, with at least one deletion mutant for every gene in the genome, and then to assess them for survival under various conditions or for other interesting phenotypes. High throughput can be obtained if each mutant is tagged with a unique molecular bar code, mutants are tested in pools, and the fate of individual mutants is determined by amplification and/or hybridization of tags to a macro- or microarray. Although such approaches were developed in the pregenomic era, in the form of signature-tagged mutagenesis (52), the availability of whole-genome sequences has meant that rationally defined, comprehensive, nonredundant mutant pools can be created and exploited through an approach known as signature tagged allele replacement (140). A recent example of the power of such technologies was the microarray-based detection of *Salmonella* mutants that cannot survive in macrophages and mice (20).

Bacterial genomics has also catalyzed developments in the identification and investigation of the proteome (50)—that is, theoretically, the entirety of proteins in a bacterium throughout its life cycle, or, more practically, the entirety of proteins found in a particular

TABLE 3 Recent applications of microarrays in bacterial pathogenomics[a]

Application	Examples	Reference(s)
Comparative genomic hybridization	Assaying genomic diversity among isolates of *Streptococcus pneumoniae*, *S. pyogenes*, and methicillin-resistant *Staphylococcus aureus*	11, 67, 123
Pathogen detection and identification	Broad-spectrum respiratory tract pathogen identification	71
Resequencing/SNP detection	Probing genomic diversity and evolution of *Escherichia coli* O157	141
Transcriptional profiling	Identification of a naturally occurring mutation that modulates global phenotype and disease specificity in *S. pyogenes*	129
Strain/strain comparisons in vitro Mutant/parent comparisons in vitro	Analysis of two-component gene regulatory systems in *S. pyogenes*	124
Bacterial response in vivo	*Salmonella* within macrophages; intracellular replication of *Listeria*; *Mycobacterium tuberculosis* from clinical lung samples	34, 59, 103
Eukaryotic cell response to bacterial product	Effector expression in yeast allows identification of a bacterial G protein mimic	3
Eukaryotic cell response to bacterial infection	Gene expression in mouse cell line in response to infection by *Porphyromonas gingivalis*	88
Detection of molecular barcodes	Microarray-based detection of *Salmonella* mutants that cannot survive in macrophages and mice	20
Chromatin immunoprecipitation	Detection of genomic binding sites of the H-NS protein in *Salmonella*	87

[a]The list is illustrative rather than exhaustive. Examples are drawn from publications that appeared in early 2006.

cell type under particular conditions. Traditionally, bacterial proteomics has relied on the twin approaches of protein separation by two-dimensional gel electrophoresis followed by mass-spectrometric determination of protein identity. These approaches usually capture only a small fraction of the proteins that might be predicted from the genome sequence. However, recently, more powerful gel-free approaches have been developed that can identify more than half the predicted proteins in a bacterium (73). As with transcriptomics, proteomics has been used to explore variations in the same strain under different conditions or among different strains under the same conditions. A number of emerging subdisciplines within proteomics focus on global views of the contribution of various posttranslational modifications—and the enzymes that mediate them—to cellular function. Within the field of bacteriology, glycoproteomics is the most prominent of such subdisciplines (56); how-

ever, given that eukaryotic cell biologists are already working on the "kinomes" and "phosphatomes" of their favorite cells (24, 74), similar subdisciplines are soon likely to emerge in our field. Similarly, rather than focus on the whole proteome, many studies have focused on the proteomics of subcellular compartments, e.g., the periplasmic proteome, the surface proteome, and the secreted proteome, or secretome (44, 69, 132).

Allied to proteomics is structural genomics (sometimes also called structural proteomics) (21, 41, 79, 119). The goal of structural proteomics is to gain three-dimensional structure of all proteins of a given organism, using experimental methods (X-ray crystallography, nuclear magnetic resonance spectroscopy) or, at least, by homology modeling. A particular emphasis is on the development and exploitation of high-throughput methods for the determination of protein structures. One pathogen, *Mycobacterium tuberculosis*, is leading the way in the field of structural genomics, with over a hundred structures determined, thanks to the efforts of the TB Structural Genomics Consortium (7, 113, 130). One curious feature of the structural genomics approach is that determination of a protein structure often comes before anything is known about the protein's function. This raises new challenges in deducing protein function from structure (64).

Another functional genomics approach centered on the proteome encompasses the global study of protein-protein interactions (the interactome). Two complementary approaches have been applied to this problem: high-throughput two-hybrid approaches (as applied to *Helicobacter pylori* [104]) and large-scale pulldown studies (as applied to *E. coli* [9]). Although there are currently only a handful of such studies, this is clearly set to be a growth area in future years. In addition, there have been several smaller-scale studies concentrating on subcellular systems (e.g., bacteriophage, flagellar and nonflagellar type III secretion systems [9, 29, 78]).

Spanning the divide between the interactome and transcriptome, chromatin immuno-precipitation (ChIP) has recently surfaced as an exciting new approach to the global study of protein-DNA interactions in bacteriology (27, 45, 46), although the technique already has a long track record in eukaryotic biology (65). ChIP relies on the immunoprecipitation of a regulatory protein bound to DNA, followed by identification of the DNA fragments (often through a hybridization to a microarray, an approach termed "ChIP-chip"). ChIP provides a snapshot of where in the genome a DNA-binding protein sits under a given set of conditions. Unlike transcriptomics, it can separate direct from indirect effects of global regulators. It may also help identify the binding sites of uncharacterized "orphan" regulators. Although to date applications of ChIP in bacteriology have dealt with global regulators and/or model systems, studies are underway to exploit this promising new approach to the study of virulence gene regulation.

QUO VADIS: WHITHER BACTERIAL GENOMICS?

Prediction is very difficult, especially about the future.
—Niels Bohr

As the subsequent chapters in this volume testify, in little more than a decade, genomics has permeated every aspect of bacteriology and delivered genome sequences from almost every significant pathogen of humans, plants, and animals. What, then, can we expect from bacterial genomics in the next 10 years?

One clear trend is that costs of sequencing will continue to drop. This will be facilitated by the development of novel sequencing technologies that promise to provide better, faster, and cheaper genome sequencing (122). Examples include technologies from companies such as 454 Life Sciences, Agencourt Bioscience, Solexa, and VisiGen Biotechnologies. At present, it is unclear which of these emerging technologies will prove the most competitive. At the time of this writing, only 454 sequencing (pyrosequencing on a massively parallel bead array) has resulted in a peer-reviewed publication describing bacterial genome sequencing (75). However, if one extrapolates from the widely cited goal of the $1,000 human genome sequence (122), one might one day be able to sequence a bacterial genome for a dollar! Even if this rather fanciful goal is still a long way off, it is worth stressing that now that costs are in the $10,000 to $100,000 range, bacterial genome sequencing falls well within the remit of an average project grant.

One consequence of the already substantial improvements in the cost and speed of bacterial genome sequencing is that sequencing of individual bacterial genomes ceases to be a publishable research goal in its own right. Instead, genome-sequencing projects have to be embedded in a broader research program—at the very least, they will have to incorporate a comparative component. Several illustrations of this trend toward comparative genomics are already apparent, i.e., through genomics projects in which multiple strains within a species are targeted simultaneously (Table 4). Similarly, there have already been projects in which

TABLE 4 Sequencing of multiple strains from the same species[a]

Species	No. of completed genome sequences
Escherichia coli/Shigella	12
Staphylococcus aureus	9
Streptococcus pyogenes	7
Bacillus anthracis/cereus	6
Salmonella enterica	5
Chlamydophila pneumoniae	4
Brucella melitensis	4
Xanthomonas campestris	4
Yersinia pestis/pseudotuberculosis	4
Mycoplasma hyopneumoniae	3
Streptococcus agalactiae	3
Buchnera aphidicola	3
Legionella pneumophila	3
Pseudomonas syringae	3

[a]These figures from mid-2006 illustrate the growing trend toward sequencing multiple strains from the same species. See the Genomes OnLine Database (http://www.genomesonline.org/) for up-to-date information.

genome sequencing has been exploited in a very focused way to answer specific research questions—for example, the use of genome sequencing to identify the target of a new antituberculosis drug (6).

Ever-deeper coverage of the most important species of bacterial pathogen also allows us to estimate the size of the mobile gene pool within a species, how many more new genes we can expect to discover from each new genome sequence from within that species, and how many more genome sequences need to be obtained before we gain a near-exhaustive view of the gene pool for that species (23, 131). In addition, interspecies comparisons might also enable us to estimate the full extent of what has been called the "pathosphere" (14), namely, the global repertoire of bacterial virulence genes shared between all pathogens. However, this optimism has to be tempered by the realization that so-called virulence determinants may often be more accurately viewed as components of colonization or symbiosis systems, or even have conserved roles that have

nothing to do with bacterial-eukaryotic interactions (is the bacterial flagellum a "virulence factor"?).

Few bacteriologists are prepared for the "future shock" that will result from faster, better, cheaper genome sequences. How will we cope when we have not one, or even a dozen, E. coli sequences, but hundreds or thousands? One consequence will be the long overdue banishment of typological thinking from bacteriology (80); instead of saying that the E. coli genome was sequenced in 1997, we will be forced to admit that the first of several hundred E. coli genomes was published in 1997! Also, given that, even with emerging technologies, finishing remains a bottleneck, we will have to cope with the arrival of many incomplete, discontinuous genome sequences or develop much better methods for finishing genomes. In addition, methods for storing, analyzing, and visualizing comparative genomics data will need to be improved far beyond what we have today.

Imagine a world in which genome sequencing becomes the gold standard, or even routine approach, in bacterial epidemiology, replacing the current plethora of underpowered "proxy" techniques (pulsed-field gel electrophoresis, multilocus sequence typing, variable-length tandem repeat typing, etc.). Through genomic epidemiology, the spread of epidemic clones could then be tracked from ward to ward, hospital to hospital, and country to country, prospectively monitoring the emergence of mutations in real time, and accurately reconstructing the spread of pathogens (e.g., SNP1 was acquired by the pathogen in such-and-such hospital, SNP2 is specific to isolates from the renal unit). Precedents for the use of genome sequencing as an epidemiological tool already exist (108), driven by concerns over bioterrorism (chapter 10). However, it seems likely that routine sequencing of epidemic clones will become commonplace in the next decade.

Another emerging theme is the growth of metagenomics—that is, the sequencing of genomic material derived from complex mixed communities of bacteria. A single metage-nomics experiment, sequencing the prokary-otic content of the Sargasso Sea (136), nearly doubled the number of known microbial pro-tein-coding genes. A steady rise in the number of such projects over the coming decade will present a severe challenge to our ability to ex-ploit the coming deluge of bacterial sequence data. An interesting sideline will be the metage-nomics of the bacteriophage community, given both the numerical preponderance of phages in the biosphere and their well-established role in the evolution and dissemination of bacterial virulence factors (18, 19, 33, 49).

The remarkable success of functional ge-nomics in exploiting bacterial genome se-quences is set to continue over the coming decade; indeed, the pace of progress is set to in-crease with the continual emergence of new technologies (for example, nonplanar array for-mats, such as bead arrays [133]). Systems biol-ogy holds great promise in integrating "omics" data from varied sources into a unified model of bacterial cells, host-pathogen interactions, or subcellular processes (8, 36, 60, 84, 86). In addi-tion, the introduction of an evolutionary aspect to bacterial systems biology is likely to trans-form our understanding of the evolution of complex regulatory or protein interaction net-works (82). In closing, it is worth stressing that despite over a decade of stunning progress, one can expect bacterial pathogenomics to deliver many new, unexpected insights into bacterial virulence over the next 10 years. The adventure has only just begun!

REFERENCES

1. **Achtman, M., K. Zurth, G. Morelli, G. Tor-rea, A. Guiyoule, and E. Carniel.** 1999. *Yersinia pestis*, the cause of plague, is a recently emerged clone of *Yersinia pseudotuberculosis. Proc. Natl. Acad. Sci. USA* **96:**14043–14048.

2. **Alm, R. A., L. S. Ling, D. T. Moir, B. L. King, E. D. Brown, P. C. Doig, D. R. Smith, B. Noonan, B. C. Guild, B. L. deJonge, G. Carmel, P. J. Tummino, A. Caruso, M. Uria-Nickelsen, D. M. Mills, C. Ives, R. Gibson, D. Merberg, S. D. Mills, Q. Jiang, D. E. Tay-lor, G. F. Vovis, and T. J. Trust.** 1999. Genomic-sequence comparison of two unrelated isolates of the human gastric pathogen *Helicobac-ter pylori. Nature* **397:**176–180.

3. **Alto, N. M., F. Shao, C. S. Lazar, R. L. Brost, G. Chua, S. Mattoo, S. A. McMahon, P. Ghosh, T. R. Hughes, C. Boone, and J. E. Dixon.** 2006. Identification of a bacterial type III effector family with G protein mimicry functions. *Cell* **124:**133–145.

4. **Anderson, S., A. T. Bankier, B. G. Barrell, M. H. de Bruijn, A. R. Coulson, J. Drouin, I. C. Eperon, D. P. Nierlich, B. A. Roe, F. Sanger, P. H. Schreier, A. J. Smith, R. Staden, and I. G. Young.** 1981. Sequence and organization of the human mitochondrial genome. *Nature* **290:**457–465.

5. **Andersson, S. G., A. Zomorodipour, J. O. Andersson, T. Sicheritz-Ponten, U. C. Als-mark, R. M. Podowski, A. K. Naslund, A. S. Eriksson, H. H. Winkler, and C. G. Kurland.** 1998. The genome sequence of *Rickettsia prowazekii* and the origin of mitochondria. *Nature* **396:**133–140.

6. **Andries, K., P. Verhasselt, J. Guillemont, H. W. Gohlmann, J. M. Neefs, H. Winkler, J. Van Gestel, P. Timmerman, M. Zhu, E. Lee, P. Williams, D. de Chaffoy, E. Huitric, S. Hoffner, E. Cambau, C. Truffot-Pernot, N. Lounis, and V. Jarlier.** 2005. A diarylquinoline drug active on the ATP synthase of *Mycobacterium tuberculosis. Science* **307:**223–227.

7. **Arcus, V. L., J. S. Lott, J. M. Johnston, and E. N. Baker.** 2006. The potential impact of struc-tural genomics on tuberculosis drug discovery. *Drug Discov. Today* **11:**28–34.

8. **Baker, M. D., P. M. Wolanin, and J. B. Stock.** 2006. Systems biology of bacterial chemotaxis. *Curr. Opin. Microbiol.* **9:**187–192.

9. **Bartel, P. L., J. A. Roecklein, D. SenGupta, and S. Fields.** 1996. A protein linkage map of *Escherichia coli* bacteriophage T7. *Nat. Genet.* **12:**72–77.

10. **Bentley, S. D., M. Maiwald, L. D. Murphy, M. J. Pallen, C. A. Yeats, L. G. Dover, H. T. Norbertczak, G. S. Besra, M. A. Quail, D. E. Harris, A. von Herbay, A. Goble, S. Rutter, R. Squares, S. Squares, B. G. Barrell, J. Parkhill, and D. A. Relman.** 2003. Sequencing and analysis of the genome of the Whipple's dis-ease bacterium *Tropheryma whipplei. Lancet* **361:**637–644.

11. **Beres, S. B., E. W. Richter, M. J. Nagiec, P. Sumby, S. F. Porcella, F. R. DeLeo, and J. M. Musser.** 2006. Molecular genetic anatomy of inter- and intraserotype variation in the human bacterial pathogen group A *Streptococcus. Proc. Natl. Acad. Sci. USA* **103:**7059–7064.

12. Blattner, F. R., G. Plunkett III, C. A. Bloch, N. T. Perna, V. Burland, M. Riley, J. Collado-Vides, J. D. Glasner, C. K. Rode, G. F. Mayhew, J. Gregor, N. W. Davis, H. A. Kirkpatrick, M. A. Goeden, D. J. Rose, B. Mau, and Y. Shao. 1997. The complete genome sequence of *Escherichia coli* K-12. *Science* **277:**1453–1474.

13. Bocs, S., A. Danchin, and C. Medigue. 2002. Re-annotation of genome microbial coding-sequences: finding new genes and inaccurately annotated genes. *BMC Bioinformatics* **3:**5.

14. Burland, V., Y. Shao, N. T. Perna, G. Plunkett, H. J. Sofia, and F. R. Blattner. 1998. The complete DNA sequence and analysis of the large virulence plasmid of *Escherichia coli* O157:H7. *Nucleic Acids Res.* **26:**4196–4204.

15. Burts, M. L., W. A. Williams, K. DeBord, and D. M. Missiakas. 2005. EsxA and EsxB are secreted by an ESAT-6-like system that is required for the pathogenesis of *Staphylococcus aureus* infections. *Proc. Natl. Acad. Sci. USA* **102:**1169–1174.

16. Call, D. R. 2005. Challenges and opportunities for pathogen detection using DNA microarrays. *Crit. Rev. Microbiol.* **31:**91–99.

17. Camus, J. C., M. J. Pryor, C. Medigue, and S. T. Cole. 2002. Re-annotation of the genome sequence of *Mycobacterium tuberculosis* H37Rv. *Microbiology* **148:**2967–2973.

18. Casjens, S. 2003. Prophages and bacterial genomics: what have we learned so far? *Mol. Microbiol.* **49:**277–300.

19. Casjens, S. R. 2005. Comparative genomics and evolution of the tailed-bacteriophages. *Curr. Opin. Microbiol.* **8:**451–458.

20. Chan, K., C. C. Kim, and S. Falkow. 2005. Microarray-based detection of *Salmonella enterica* serovar Typhimurium transposon mutants that cannot survive in macrophages and mice. *Infect. Immun.* **73:**5438–5449.

21. Chance, M. R., A. R. Bresnick, S. K. Burley, J. S. Jiang, C. D. Lima, A. Sali, S. C. Almo, J. B. Bonanno, J. A. Buglino, S. Boulton, H. Chen, N. Eswar, G. He, R. Huang, V. Ilyin, L. McMahan, U. Pieper, S. Ray, M. Vidal, and L. K. Wang. 2002. Structural genomics: a pipeline for providing structures for the biologist. *Protein Sci.* **11:**723–738.

22. Chaudhuri, R. R., and M. J. Pallen. 2006. *x*BASE, a collection of online databases for bacterial comparative genomics. *Nucleic Acids Res.* **34:**D335–D337.

23. Chen, S. L., C. S. Hung, J. Xu, C. S. Reigstad, V. Magrini, A. Sabo, D. Blasiar, T. Bieri, R. R. Meyer, P. Ozersky, J. R. Armstrong, R. S. Fulton, J. P. Latreille, J. Spieth, T. M. Hooton, E. R. Mardis, S. J. Hultgren,

and J. I. Gordon. 2006. Identification of genes subject to positive selection in uropathogenic strains of *Escherichia coli*: a comparative genomics approach. *Proc. Natl. Acad. Sci. USA* **103:**5977–5982.

24. Coito, C., D. L. Diamond, P. Neddermann, M. J. Korth, and M. G. Katze. 2004. High-throughput screening of the yeast kinome: identification of human serine/threonine protein kinases that phosphorylate the hepatitis C virus NS5A protein. *J. Virol.* **78:**3502–3513.

25. Cole, S. T., R. Brosch, J. Parkhill, T. Garnier, C. Churcher, D. Harris, S. V. Gordon, K. Eiglmeier, S. Gas, C. E. Barry III, F. Tekaia, K. Badcock, D. Basham, D. Brown, T. Chillingworth, R. Connor, R. Davies, K. Devlin, T. Feltwell, S. Gentles, N. Hamlin, S. Holroyd, T. Hornsby, K. Jagels, A. Krogh, J. McLean, S. Moule, L. Murphy, K. Oliver, J. Osborne, M. A. Quail, M. A. Rajandream, J. Rogers, S. Rutter, K. Seeger, J. Skelton, R. Squares, S. Squares, J. E. Sulston, K. Taylor, S. Whitehead, and B. G. Barrell. 1998. Deciphering the biology of *Mycobacterium tuberculosis* from the complete genome sequence. *Nature* **393:**537–544.

26. Cole, S. T., K. Eiglmeier, J. Parkhill, K. D. James, N. R. Thomson, P. R. Wheeler, N. Honore, T. Garnier, C. Churcher, D. Harris, K. Mungall, D. Basham, D. Brown, T. Chillingworth, R. Connor, R. M. Davies, K. Devlin, S. Duthoy, T. Feltwell, A. Fraser, N. Hamlin, S. Holroyd, T. Hornsby, K. Jagels, C. Lacroix, J. Maclean, S. Moule, L. Murphy, K. Oliver, M. A. Quail, M. A. Rajandream, K. M. Rutherford, S. Rutter, K. Seeger, S. Simon, M. Simmonds, J. Skelton, R. Squares, S. Squares, K. Stevens, K. Taylor, S. Whitehead, J. R. Woodward, and B. G. Barrell. 2001. Massive gene decay in the leprosy bacillus. *Nature* **409:**1007–1011.

27. Constantinidou, C., J. L. Hobman, L. Griffiths, M. D. Patel, C. W. Penn, J. A. Cole, and T. W. Overton. 2006. A reassessment of the FNR regulon and transcriptomic analysis of the effects of nitrate, nitrite, NarXL, and NarQP as *Escherichia coli* K12 adapts from aerobic to anaerobic growth. *J. Biol. Chem.* **281:**4802–4815.

28. Conway, T., and G. K. Schoolnik. 2003. Microarray expression profiling: capturing a genome-wide portrait of the transcriptome. *Mol. Microbiol.* **47:**879–889.

29. Creasey, E. A., R. M. Delahay, S. J. Daniell, and G. Frankel. 2003. Yeast two-hybrid system survey of interactions between LEE-encoded proteins of enteropathogenic *Escherichia coli*. *Microbiology* **149:**2093–2106.

30. Dandekar, T., M. Huynen, J. T. Regula, B. Ueberle, C. U. Zimmermann, M. A. Andrade, T. Doerks, L. Sanchez-Pulido, B. Snel, M. Suyama, Y. P. Yuan, R. Herrmann, and P. Bork. 2000. Re-annotating the *Mycoplasma pneumoniae* genome sequence: adding value, function and reading frames. *Nucleic Acids Res.* **28:**3278–3288.

31. Daraselia, N., D. Dernovoy, Y. Tian, M. Borodovsky, R. Tatusov, and T. Tatusova. 2003. Reannotation of *Shewanella oneidensis* genome. *Omics* **7:**171–175.

32. Dharmadi, Y., and R. Gonzalez. 2004. DNA microarrays: experimental issues, data analysis, and application to bacterial systems. *Biotechnol. Prog.* **20:**1309–1324.

33. Edwards, R. A., and F. Rohwer. 2005. Viral metagenomics. *Nat. Rev. Microbiol.* **3:**504–510.

34. Faucher, S. P., S. Porwollik, C. M. Dozois, M. McClelland, and F. Daigle. 2006. Transcriptome of *Salmonella enterica* serovar Typhi within macrophages revealed through the selective capture of transcribed sequences. *Proc. Natl. Acad. Sci. USA* **103:**1906–1911.

35. Fleischmann, R. D., M. D. Adams, O. White, R. A. Clayton, E. F. Kirkness, A. R. Kerlavage, C. J. Bult, J. F. Tomb, B. A. Dougherty, J. M. Merrick, et al. 1995. Whole-genome random sequencing and assembly of *Haemophilus influenzae* Rd. *Science* **269:**496–512.

36. Forst, C. V. 2006. Host-pathogen systems biology. *Drug Discov. Today* **11:**220–227.

37. Fouts, D. E., E. F. Mongodin, R. E. Mandrell, W. G. Miller, D. A. Rasko, J. Ravel, L. M. Brinkac, R. T. DeBoy, C. T. Parker, S. C. Daugherty, R. J. Dodson, A. S. Durkin, R. Madupu, S. A. Sullivan, J. U. Shetty, M. A. Ayodeji, A. Shvartsbeyn, M. C. Schatz, J. H. Badger, C. M. Fraser, and K. E. Nelson. 2005. Major structural differences and novel potential virulence mechanisms from the genomes of multiple campylobacter species. *PLoS Biol.* **3:**e15.

38. Frangeul, L., K. E. Nelson, C. Buchrieser, A. Danchin, P. Glaser, and F. Kunst. 1999. Cloning and assembly strategies in microbial genome projects. *Microbiology* **145(Pt. 10):**2625–2634.

39. Fraser, C. M., J. D. Gocayne, O. White, M. D. Adams, R. A. Clayton, R. D. Fleischmann, C. J. Bult, A. R. Kerlavage, G. Sutton, J. M. Kelley, R. D. Fritchman, J. F. Weidman, K. V. Small, M. Sandusky, J. Fuhrmann, D. Nguyen, T. R. Utterback, D. M. Saudek, C. A. Phillips, J. M. Merrick, J. F. Tomb, B. A. Dougherty, K. F. Bott, P. C. Hu, T. S. Lucier, S. N. Peterson, H. O. Smith, C. A. Hutchison III, and J. C. Venter. 1995. The minimal gene complement of *Mycoplasma genitalium*. *Science* **270:**397–403.

40. Fraser, C. M., S. J. Norris, G. M. Weinstock, O. White, G. G. Sutton, R. Dodson, M. Gwinn, E. K. Hickey, R. Clayton, K. A. Ketchum, E. Sodergren, J. M. Hardham, M. P. McLeod, S. Salzberg, J. Peterson, H. Khalak, D. Richardson, J. K. Howell, M. Chidambaram, T. Utterback, L. McDonald, P. Artiach, C. Bowman, M. D. Cotton, C. Fujii, S. Garland, B. Hatch, K. Horst, K. Roberts, M. Sandusky, J. Weidman, H. O. Smith, and J. C. Venter. 1998. Complete genome sequence of *Treponema pallidum*, the syphilis spirochete. *Science* **281:**375–388.

41. Frishman, D. 2003. What we have learned about prokaryotes from structural genomics. *Omics* **7:** 211–224.

42. Fux, C. A., M. Shirtliff, P. Stoodley, and J. W. Costerton. 2005. Can laboratory reference strains mirror "real-world" pathogenesis? *Trends Microbiol.* **13:**58–63.

43. Gill, S. R., M. Pop, R. T. Deboy, P. B. Eckburg, P. J. Turnbaugh, B. S. Samuel, J. I. Gordon, D. A. Relman, C. M. Fraser-Liggett, and K. E. Nelson. 2006. Metagenomic analysis of the human distal gut microbiome. *Science* **312:**1355–1359.

44. Gohar, M., N. Gilois, R. Graveline, C. Garreau, V. Sanchis, and D. Lereclus. 2005. A comparative study of *Bacillus cereus*, *Bacillus thuringiensis* and *Bacillus anthracis* extracellular proteomes. *Proteomics* **5:**3696–3711.

45. Grainger, D. C., D. Hurd, M. Harrison, J. Holdstock, and S. J. Busby. 2005. Studies of the distribution of *Escherichia coli* cAMP-receptor protein and RNA polymerase along the *E. coli* chromosome. *Proc. Natl. Acad. Sci. USA* **102:** 17693–17698.

46. Grainger, D. C., T. W. Overton, N. Reppas, J. T. Wade, E. Tamai, J. L. Hobman, C. Constantinidou, K. Struhl, G. Church, and S. J. Busby. 2004. Genomic studies with *Escherichia coli* MelR protein: applications of chromatin immunoprecipitation and microarrays. *J. Bacteriol.* **186:**6938–6943.

47. Gutierrez, M. C., S. Brisse, R. Brosch, M. Fabre, B. Omais, M. Marmiesse, P. Supply, and V. Vincent. 2005. Ancient origin and gene mosaicism of the progenitor of *Mycobacterium tuberculosis*. *PLoS Pathogens* **1:**e5.

48. Hallin, P. F., and D. W. Ussery. 2004. CBS Genome Atlas Database: a dynamic storage for bioinformatic results and sequence data. *Bioinformatics* **20:**3682–3686.

49. **Hamilton, G.** 2006. Virology: the gene weavers. *Nature* **441**:683–685.

50. **Han, M. J., and S. Y. Lee.** 2006. The *Escherichia coli* proteome: past, present, and future prospects. *Microbiol. Mol. Biol. Rev.* **70**:362–439.

51. **Hayashi, T., K. Makino, M. Ohnishi, K. Kurokawa, K. Ishii, K. Yokoyama, C. G. Han, E. Ohtsubo, K. Nakayama, T. Murata, M. Tanaka, T. Tobe, T. Iida, H. Takami, T. Honda, C. Sasakawa, N. Ogasawara, T. Yasunaga, S. Kuhara, T. Shiba, M. Hattori, and H. Shinagawa.** 2001. Complete genome sequence of enterohemorrhagic *Escherichia coli* O157:H7 and genomic comparison with a laboratory strain K-12. *DNA Res.* **8**:11–22.

52. **Hensel, M., J. E. Shea, C. Gleeson, M. D. Jones, E. Dalton, and D. W. Holden.** 1995. Simultaneous identification of bacterial virulence genes by negative selection. *Science* **269**:400–403.

53. **Himmelreich, R., H. Hilbert, H. Plagens, E. Pirkl, B. C. Li, and R. Herrmann.** 1996. Complete sequence analysis of the genome of the bacterium *Mycoplasma pneumoniae*. *Nucleic Acids Res.* **24**:4420–4449.

54. **Himmelreich, R., H. Plagens, H. Hilbert, B. Reiner, and R. Herrmann.** 1997. Comparative analysis of the genomes of the bacteria *Mycoplasma pneumoniae* and *Mycoplasma genitalium*. *Nucleic Acids Res.* **25**:701–712.

55. **Hinton, J. C., I. Hautefort, S. Eriksson, A. Thompson, and M. Rhen.** 2004. Benefits and pitfalls of using microarrays to monitor bacterial gene expression during infection. *Curr. Opin. Microbiol.* **7**:277–282.

56. **Hitchen, P. G., and A. Dell.** 2006. Bacterial glycoproteomics. *Microbiology* **152**:1575–1580.

57. **Hood, D. W., M. E. Deadman, T. Allen, H. Masoud, A. Martin, J. R. Brisson, R. Fleischmann, J. C. Venter, J. C. Richards, and E. R. Moxon.** 1996. Use of the complete genome sequence information of *Haemophilus influenzae* strain Rd to investigate lipopolysaccharide biosynthesis. *Mol. Microbiol.* **22**:951–965.

58. **Jones, M. A., K. L. Marston, C. A. Woodall, D. J. Maskell, D. Linton, A. V. Karlyshev, N. Dorrell, B. W. Wren, and P. A. Barrow.** 2004. Adaptation of *Campylobacter jejuni* NCTC11168 to high-level colonization of the avian gastrointestinal tract. *Infect. Immun.* **72**:3769–3776.

59. **Joseph, B., K. Przybilla, C. Stuhler, K. Schauer, J. Slaghuis, T. M. Fuchs, and W. Goebel.** 2006. Identification of *Listeria monocytogenes* genes contributing to intracellular replication by expression profiling and mutant screening. *J. Bacteriol.* **188**:556–568.

60. **Kalir, S., and U. Alon.** 2004. Using a quantitative blueprint to reprogram the dynamics of the flagella gene network. *Cell* **117**:713–720.

61. **Keim, P., and K. L. Smith.** 2002. Bacillus anthracis evolution and epidemiology. *Curr. Top. Microbiol. Immunol.* **271**:21–32.

62. **Kelly, T. J., Jr., and H. O. Smith.** 1970. A restriction enzyme from *Hemophilus influenzae*. II. Base sequence of the recognition site. *J. Mol. Biol.* **51**:393–409.

63. **Kidgell, C., U. Reichard, J. Wain, B. Linz, M. Torpdahl, G. Dougan, and M. Achtman.** 2002. *Salmonella typhi*, the causative agent of typhoid fever, is approximately 50,000 years old. *Infect. Genet. Evol.* **2**:39–45.

64. **Kim, S. H., D. H. Shin, I. G. Choi, U. Schulze-Gahmen, S. Chen, and R. Kim.** 2003. Structure-based functional inference in structural genomics. *J. Struct. Funct. Genomics* **4**:129–135.

65. **Kim, T. H., and B. Ren.** 2006. Genome-wide analysis of protein-DNA interactions. *Annu. Rev. Genomics Hum. Genet.* **7**:81–102.

66. **Knutton, S., I. Rosenshine, M. J. Pallen, I. Nisan, B. C. Neves, C. Bain, C. Wolff, G. Dougan, and G. Frankel.** 1998. A novel EspA-associated surface organelle of enteropathogenic *Escherichia coli* involved in protein translocation into epithelial cells. *EMBO J.* **17**:2166–2176.

67. **Koessler, T., P. Francois, Y. Charbonnier, A. Huyghe, M. Bento, S. Dharan, G. Renzi, D. Lew, S. Harbarth, D. Pittet, and J. Schrenzel.** 2006. Use of oligoarrays for characterization of community-onset methicillin-resistant *Staphylococcus aureus*. *J. Clin. Microbiol.* **44**:1040–1048.

68. **Kunst, F., N. Ogasawara, I. Moszer, A. M. Albertini, G. Alloni, V. Azevedo, M. G. Bertero, P. Bessieres, A. Bolotin, S. Borchert, R. Borriss, L. Boursier, A. Brans, M. Braun, S. C. Brignell, S. Bron, S. Brouillet, C. V. Bruschi, B. Caldwell, V. Capuano, N. M. Carter, S. K. Choi, J. J. Codani, I. F. Connerton, A. Danchin, et al.** 1997. The complete genome sequence of the gram-positive bacterium *Bacillus subtilis*. *Nature* **390**:249–256.

69. **Kurian, D., K. Phadwal, and P. Maenpaa.** 2006. Proteomic characterization of acid stress response in *Synechocystis* sp. PCC 6803. *Proteomics* **6**:3614–3624.

70. **Kuroda, M., T. Ohta, I. Uchiyama, T. Baba, H. Yuzawa, I. Kobayashi, L. Cui, A. Oguchi, K. Aoki, Y. Nagai, J. Lian, T. Ito, M. Kanamori, H. Matsumaru, A. Maruyama, H. Murakami, A. Hosoyama, Y. Mizutani-Ui, N. K. Takahashi, T. Sawano, R. Inoue, C. Kaito, K. Sekimizu, H. Hirakawa, S. Kuhara, S. Goto, J. Yabuzaki, M. Kanehisa, A. Yamashita, K. Oshima, K. Furuya, C. Yoshino, T. Shiba, M. Hattori, N. Ogasawara, H. Hayashi, and K. Hiramatsu.** 2001. Whole

genome sequencing of meticillin-resistant *Staphylococcus aureus*. *Lancet* **357:**1225–1240.

71. **Lin, B., Z. Wang, G. J. Vora, J. A. Thornton, J. M. Schnur, D. C. Thach, K. M. Blaney, A. G. Ligler, A. P. Malanoski, J. Santiago, E. A. Walter, B. K. Agan, D. Metzgar, D. Seto, L. T. Daum, R. Kruzelock, R. K. Rowley, E. H. Hanson, C. Tibbetts, and D. A. Stenger.** 2006. Broad-spectrum respiratory tract pathogen identification using resequencing DNA microarrays. *Genome Res.* **16:**527–535.

72. **Liolios, K., N. Tavernarakis, P. Hugenholtz, and N. C. Kyrpides.** 2006. The Genomes On Line Database (GOLD) v.2: a monitor of genome projects worldwide. *Nucleic Acids Res.* **34:**D332–D334.

73. **Lipton, M. S., L. Pasa-Tolic, G. A. Anderson, D. J. Anderson, D. L. Auberry, J. R. Battista, M. J. Daly, J. Fredrickson, K. K. Hixson, H. Kostandarithes, C. Masselon, L. M. Markillie, R. J. Moore, M. F. Romine, Y. Shen, E. Stritmatter, N. Tolic, H. R. Udseth, A. Venkateswaran, K. K. Wong, R. Zhao, and R. D. Smith.** 2002. Global analysis of the *Deinococcus radiodurans* proteome by using accurate mass tags. *Proc. Natl. Acad. Sci. USA* **99:**11049–11054.

74. **Manning, G., D. B. Whyte, R. Martinez, T. Hunter, and S. Sudarsanam.** 2002. The protein kinase complement of the human genome. *Science* **298:**1912–1934.

75. **Margulies, M., M. Egholm, W. E. Altman, S. Attiya, J. S. Bader, L. A. Bemben, J. Berka, M. S. Braverman, Y. J. Chen, Z. Chen, S. B. Dewell, L. Du, J. M. Fierro, X. V. Gomes, B. C. Godwin, W. He, S. Helgesen, C. H. Ho, G. P. Irzyk, S. C. Jando, M. L. Alenquer, T. P. Jarvie, K. B. Jirage, J. B. Kim, J. R. Knight, J. R. Lanza, J. H. Leamon, S. M. Lefkowitz, M. Lei, J. Li, K. L. Lohman, H. Lu, V. B. Makhijani, K. E. McDade, M. P. McKenna, E. W. Myers, E. Nickerson, J. R. Nobile, R. Plant, B. P. Puc, M. T. Ronan, G. T. Roth, G. J. Sarkis, J. F. Simons, J. W. Simpson, M. Srinivasan, K. R. Tartaro, A. Tomasz, K. A. Vogt, G. A. Volkmer, S. H. Wang, Y. Wang, M. P. Weiner, P. Yu, R. F. Begley, and J. M. Rothberg.** 2005. Genome sequencing in microfabricated high-density picolitre reactors. *Nature* **437:**376–380.

76. **Marshall, E.** 2001. Bermuda rules: community spirit, with teeth. *Science* **291:**1192.

77. **Marshall, E.** 2002. Data sharing. DNA sequencer protests being scooped with his own data. *Science* **295:**1206–1207.

78. **Marykwas, D. L., S. A. Schmidt, and H. C. Berg.** 1996. Interacting components of the flagellar motor of *Escherichia coli* revealed by the two-hybrid system in yeast. *J. Mol. Biol.* **256:**564–576.

79. **Matte, A., J. Sivaraman, I. Ekiel, K. Gehring, Z. Jia, and M. Cygler.** 2003. Contribution of structural genomics to understanding the biology of *Escherichia coli*. *J. Bacteriol.* **185:**3994–4002.

80. **Mayr, E.** 2000. Darwin's influence on modern thought. *Sci. Am.* **283:**78–83.

81. **Mazmanian, S. K., G. Liu, H. Ton-That, and O. Schneewind.** 1999. *Staphylococcus aureus* sortase, an enzyme that anchors surface proteins to the cell wall. *Science* **285:**760–763.

82. **Medina, M.** 2005. Genomes, phylogeny, and evolutionary systems biology. *Proc. Natl. Acad. Sci. USA* **102** (Suppl. 1)**:**6630–6635.

83. **Monot, M., N. Honore, T. Garnier, R. Araoz, J. Y. Coppee, C. Lacroix, S. Sow, J. S. Spencer, R. W. Truman, D. L. Williams, R. Gelber, M. Virmond, B. Flageul, S. N. Cho, B. Ji, A. Paniz-Mondolfi, J. Convit, S. Young, P. E. Fine, V. Rasolofo, P. J. Brennan, and S. T. Cole.** 2005. On the origin of leprosy. *Science* **308:**1040–1042.

84. **Mori, H.** 2004. From the sequence to cell modeling: comprehensive functional genomics in *Escherichia coli*. *J. Biochem. Mol. Biol.* **37:**83–92.

85. **Mrazek, J., and S. Karlin.** 1998. Strand compositional asymmetry in bacterial and large viral genomes. *Proc. Natl. Acad. Sci. USA* **95:**3720–3725.

86. **Musser, J. M., and F. R. DeLeo.** 2005. Toward a genome-wide systems biology analysis of host-pathogen interactions in group A *Streptococcus*. *Am. J. Pathol.* **167:**1461–1472.

87. **Navarre, W. W., S. Porwollik, Y. Wang, M. McClelland, H. Rosen, S. J. Libby, and F. C. Fang.** 2006. Selective silencing of foreign DNA with low GC content by the H-NS protein in *Salmonella*. *Science* **313:**236–238.

88. **Ohno, T., N. Okahashi, S. Kawai, T. Kato, H. Inaba, Y. Shibata, I. Morisaki, Y. Abiko, and A. Amano.** 2006. Proinflammatory gene expression in mouse ST2 cell line in response to infection by *Porphyromonas gingivalis*. *Microbes Infect.* **8:**1025–1034.

89. **Pallen, M.** 2002. From sequence to consequence: in silico hypothesis generation and testing. *Meth. Microbiol.* **33:**27–48.

90. **Pallen, M. J.** 2002. The ESAT-6/WXG100 superfamily—and a new gram-positive secretion system? *Trends Microbiol.* **10:**209–212.

91. **Pallen, M. J.** 1999. Microbial genomes. *Mol. Microbiol.* **32:**907–912.

92. **Pallen, M. J., S. A. Beatson, and C. M. Bailey.** 2005. Bioinformatics analysis of the locus for enterocyte effacement provides novel insights into type-III secretion. *BMC Microbiol.* **5:**9.

93. **Pallen, M. J., A. C. Lam, M. Antonio, and K. Dunbar.** 2001. An embarrassment of sortases—a richness of substrates? *Trends Microbiol.* **9:**97–102.

94. Pallen, M. J., C. W. Penn, and R. R. Chaudhuri. 2005. Bacterial flagellar diversity in the post-genomic era. *Trends Microbiol.* **13**:143–149.

95. Parkhill, J. 2000. In defense of complete genomes. *Nat. Biotechnol.* **18**:493–494.

96. Parkhill, J., G. Dougan, K. D. James, N. R. Thomson, D. Pickard, J. Wain, C. Churcher, K. L. Mungall, S. D. Bentley, M. T. Holden, M. Sebaihia, S. Baker, D. Basham, K. Brooks, T. Chillingworth, P. Connerton, A. Cronin, P. Davis, R. M. Davies, L. Dowd, N. White, J. Farrar, T. Feltwell, N. Hamlin, A. Haque, T. T. Hien, S. Holroyd, K. Jagels, A. Krogh, T. S. Larsen, S. Leather, S. Moule, P. O'Gaora, C. Parry, M. Quail, K. Rutherford, M. Simmonds, J. Skelton, K. Stevens, S. Whitehead, and B. G. Barrell. 2001. Complete genome sequence of a multiple drug resistant *Salmonella enterica* serovar Typhi CT18. *Nature* **413**:848–852.

97. Parkhill, J., B. W. Wren, K. Mungall, J. M. Ketley, C. Churcher, D. Basham, T. Chillingworth, R. M. Davies, T. Feltwell, S. Holroyd, K. Jagels, A. V. Karlyshev, S. Moule, M. J. Pallen, C. W. Penn, M. A. Quail, M. A. Rajandream, K. M. Rutherford, A. H. van Vliet, S. Whitehead, and B. G. Barrell. 2000. The genome sequence of the food-borne pathogen *Campylobacter jejuni* reveals hypervariable sequences. *Nature* **403**:665–668.

98. Parkhill, J., B. W. Wren, N. R. Thomson, R. W. Titball, M. T. Holden, M. B. Prentice, M. Sebaihia, K. D. James, C. Churcher, K. L. Mungall, S. Baker, D. Basham, S. D. Bentley, K. Brooks, A. M. Cerdeno-Tarraga, T. Chillingworth, A. Cronin, R. M. Davies, P. Davis, G. Dougan, T. Feltwell, N. Hamlin, S. Holroyd, K. Jagels, A. V. Karlyshev, S. Leather, S. Moule, P. C. Oyston, M. Quail, K. Rutherford, M. Simmonds, J. Skelton, K. Stevens, S. Whitehead, and B. G. Barrell. 2001. Genome sequence of *Yersinia pestis*, the causative agent of plague. *Nature* **413**:523–527.

99. Perna, N. T., G. Plunkett III, V. Burland, B. Mau, J. D. Glasner, D. J. Rose, G. F. Mayhew, P. S. Evans, J. Gregor, H. A. Kirkpatrick, G. Posfai, J. Hackett, S. Klink, A. Boutin, Y. Shao, L. Miller, E. J. Grotbeck, N. W. Davis, A. Lim, E. T. Dimalanta, K. D. Potamousis, J. Apodaca, T. S. Anantharaman, J. Lin, G. Yen, D. C. Schwartz, R. A. Welch, and F. R. Blattner. 2001. Genome sequence of enterohaemorrhagic *Escherichia coli* O157:H7. *Nature* **409**:529–533.

100. Pizza, M., V. Scarlato, V. Masignani, M. M. Giuliani, B. Arico, M. Comanducci, G. T. Jennings, L. Baldi, E. Bartolini, B. Capecchi, C. L. Galeotti, E. Luzzi, R. Manetti, E. Marchetti, M. Mora, S. Nuti, G. Ratti, L. Santini, S. Savino, M. Scarselli, E. Storni, P. Zuo, M. Broeker, E. Hundt, B. Knapp, E. Blair, T. Mason, H. Tettelin, D. W. Hood, A. C. Jeffries, N. J. Saunders, D. M. Granoff, J. C. Venter, E. R. Moxon, G. Grandi, and R. Rappuoli. 2000. Identification of vaccine candidates against serogroup B meningococcus by whole-genome sequencing. *Science* **287**:1816–1820.

101. Pop, M., and D. Kosack. 2004. Using the TIGR assembler in shotgun sequencing projects. *Methods Mol. Biol.* **255**:279–294.

102. Powledge, T. 2003. Revisiting Bermuda. *Genome Biol.* **March 11**.

103. Rachman, H., M. Strong, T. Ulrichs, L. Grode, J. Schuchhardt, H. Mollenkopf, G. A. Kosmiadi, D. Eisenberg, and S. H. Kaufmann. 2006. Unique transcriptome signature of *Mycobacterium tuberculosis* in pulmonary tuberculosis. *Infect. Immun.* **74**:1233–1242.

104. Rain, J. C., L. Selig, H. De Reuse, V. Battaglia, C. Reverdy, S. Simon, G. Lenzen, F. Petel, J. Wojcik, V. Schachter, Y. Chemama, A. Labigne, and P. Legrain. 2001. The protein-protein interaction map of *Helicobacter pylori*. *Nature* **409**:211–215.

105. Raoult, D., S. Audic, C. Robert, C. Abergel, P. Renesto, H. Ogata, B. La Scola, M. Suzan, and J. M. Claverie. 2004. The 1.2-megabase genome sequence of *Mimivirus*. *Science* **306**:1344–1350.

106. Raoult, D., H. Ogata, S. Audic, C. Robert, K. Suhre, M. Drancourt, and J. M. Claverie. 2003. *Tropheryma whipplei* Twist: a human pathogenic *Actinobacteria* with a reduced genome. *Genome Res.* **13**:1800–1809.

107. Rappuoli, R. 2000. Reverse vaccinology. *Curr. Opin. Microbiol.* **3**:445–450.

108. Read, T. D., S. L. Salzberg, M. Pop, M. Shumway, L. Umayam, L. Jiang, E. Holtzapple, J. D. Busch, K. L. Smith, J. M. Schupp, D. Solomon, P. Keim, and C. M. Fraser. 2002. Comparative genome sequencing for discovery of novel polymorphisms in *Bacillus anthracis*. *Science* **296**:2028–2033.

109. Ren, C. P., S. A. Beatson, J. Parkhill, and M. J. Pallen. 2005. The Flag-2 locus, an ancestral gene cluster, is potentially associated with a novel flagellar system from *Escherichia coli*. *J. Bacteriol.* **187**:1430–1440.

110. Ren, C. P., R. R. Chaudhuri, A. Fivian, C. M. Bailey, M. Antonio, W. M. Barnes, and M. J. Pallen. 2004. The *ETT2* gene cluster, encoding a second type III secretion system from *Escherichia coli*, is present in the majority of

strains but has undergone widespread mutational attrition. *J. Bacteriol.* **186:**3547–3560.

111. **Renesto, P., N. Crapoulet, H. Ogata, B. La Scola, G. Vestris, J. M. Claverie, and D. Raoult.** 2003. Genome-based design of a cell-free culture medium for *Tropheryma whipplei. Lancet* **362:**447–449.

112. **Riley, M., T. Abe, M. B. Arnaud, M. K. Berlyn, F. R. Blattner, R. R. Chaudhuri, J. D. Glasner, T. Horiuchi, I. M. Keseler, T. Kosuge, H. Mori, N. T. Perna, G. Plunkett III, K. E. Rudd, M. H. Serres, G. H. Thomas, N. R. Thomson, D. Wishart, and B. L. Wanner.** 2006. *Escherichia coli* K-12: a cooperatively developed annotation snapshot—2005. *Nucleic Acids Res.* **34:**1–9.

113. **Rupp, B.** 2003. High-throughput crystallography at an affordable cost: the TB Structural Genomics Consortium Crystallization Facility. *Acc. Chem. Res.* **36:**173–181.

114. **Rutherford, K., J. Parkhill, J. Crook, T. Horsnell, P. Rice, M. A. Rajandream, and B. Barrell.** 2000. Artemis: sequence visualization and annotation. *Bioinformatics* **16:**944–945.

115. **Salzberg, S. L., A. L. Delcher, S. Kasif, and O. White.** 1998. Microbial gene identification using interpolated Markov models. *Nucleic Acids Res.* **26:**544–548.

116. **Sanger, F., G. M. Air, B. G. Barrell, N. L. Brown, A. R. Coulson, C. A. Fiddes, C. A. Hutchison, P. M. Slocombe, and M. Smith.** 1977. Nucleotide sequence of bacteriophage phi X174 DNA. *Nature* **265:**687–695.

117. **Sanger, F., A. R. Coulson, G. F. Hong, D. F. Hill, and G. B. Petersen.** 1982. Nucleotide sequence of bacteriophage lambda DNA. *J. Mol. Biol.* **162:**729–773.

118. **Sanger, F., S. Nicklen, and A. R. Coulson.** 1977. DNA sequencing with chain-terminating inhibitors. *Proc. Natl. Acad. Sci. USA* **74:**5463–5467.

119. **Schmid, M. B.** 2002. Structural proteomics: the potential of high-throughput structure determination. *Trends Microbiol.* **10:**S27–S31.

120. **Schoolnik, G. K.** 2002. Microarray analysis of bacterial pathogenicity. *Adv. Microb. Physiol.* **46:**1–45.

121. **Serruto, D., and R. Rappuoli.** 2006. Postgenomic vaccine development. *FEBS Lett.* **580:**2985–2992.

122. **Service, R. F.** 2006. Gene sequencing. The race for the $1000 genome. *Science* **311:**1544–1546.

123. **Silva, N. A., J. McCluskey, J. M. Jefferies, J. Hinds, A. Smith, S. C. Clarke, T. J. Mitchell, and G. K. Paterson.** 2006. Genomic diversity between strains of the same serotype and multi-

locus sequence type among pneumococcal clinical isolates. *Infect. Immun.* **74:**3513–3518.

124. **Sitkiewicz, I., and J. M. Musser.** 2006. Expression microarray and mouse virulence analysis of four conserved two-component gene regulatory systems in group a streptococcus. *Infect. Immun.* **74:**1339–1351.

125. **Smith, H. O., and K. W. Wilcox.** 1970. A restriction enzyme from *Hemophilus influenzae.* I. Purification and general properties. *J. Mol. Biol.* **51:**379–391.

126. **Smith, L. M., J. Z. Sanders, R. J. Kaiser, P. Hughes, C. Dodd, C. R. Connell, C. Heiner, S. B. Kent, and L. E. Hood.** 1986. Fluorescence detection in automated DNA sequence analysis. *Nature* **321:**674–679.

127. **Spratt, B. G.** 2004. Exploring the concept of clonality in bacteria. *Methods Mol. Biol.* **266:**323–352.

128. **Sreevatsan, S., X. Pan, K. E. Stockbauer, N. D. Connell, B. N. Kreiswirth, T. S. Whittam, and J. M. Musser.** 1997. Restricted structural gene polymorphism in the *Mycobacterium tuberculosis* complex indicates evolutionarily recent global dissemination. *Proc. Natl. Acad. Sci. USA* **94:**9869–9874.

129. **Sumby, P., A. R. Whitney, E. A. Graviss, F. R. DeLeo, and J. M. Musser.** 2006. Genome-wide analysis of group a streptococci reveals a mutation that modulates global phenotype and disease specificity. *PLoS Pathog.* **2:**e5.

130. **Terwilliger, T. C., M. S. Park, G. S. Waldo, J. Berendzen, L. W. Hung, C. Y. Kim, C. V. Smith, J. C. Sacchettini, M. Bellinzoni, R. Bossi, E. De Rossi, A. Mattevi, A. Milano, G. Riccardi, M. Rizzi, M. M. Roberts, A. R. Coker, G. Fossati, P. Mascagni, A. R. Coates, S. P. Wood, C. W. Goulding, M. I. Apostol, D. H. Anderson, H. S. Gill, D. S. Eisenberg, B. Taneja, S. Mande, E. Pohl, V. Lamzin, P. Tucker, M. Wilmanns, C. Colovos, W. Meyer-Klaucke, A. W. Munro, K. J. McLean, K. R. Marshall, D. Leys, J. K. Yang, H. J. Yoon, B. I. Lee, M. G. Lee, J. E. Kwak, B. W. Han, J. Y. Lee, S. H. Baek, S. W. Suh, M. M. Komen, V. L. Arcus, E. N. Baker, J. S. Lott, W. Jacobs, Jr., T. Alber, and B. Rupp.** 2003. The TB Structural Genomics Consortium: a resource for *Mycobacterium tuberculosis* biology. *Tuberculosis* (Edinb.) **83:**223–249.

131. **Tettelin, H., V. Masignani, M. J. Cieslewicz, C. Donati, D. Medini, N. L. Ward, S. V. Angiuoli, J. Crabtree, A. L. Jones, A. S. Durkin, R. T. Deboy, T. M. Davidsen, M. Mora, M. Scarselli, I. Margarit y Ros, J. D. Peterson, C. R. Hauser, J. P. Sundaram, W. C. Nelson, R. Madupu, L. M. Brinkac,**

R. J. Dodson, M. J. Rosovitz, S. A. Sullivan, S. C. Daugherty, D. H. Haft, J. Selengut, M. L. Gwinn, L. Zhou, N. Zafar, H. Khouri, D. Radune, G. Dimitrov, K. Watkins, K. J. O'Connor, S. Smith, T. R. Utterback, O. White, C. E. Rubens, G. Grandi, L. C. Madoff, D. L. Kasper, J. L. Telford, M. R. Wessels, R. Rappuoli, and C. M. Fraser. 2005. Genome analysis of multiple pathogenic isolates of *Streptococcus agalactiae*: implications for the microbial "pan-genome." *Proc. Natl. Acad. Sci. USA* **102**:13950–13955.

132. Tjalsma, H., H. Antelmann, J. D. Jongbloed, P. G. Braun, E. Darmon, R. Dorenbos, J. Y. Dubois, H. Westers, G. Zanen, W. J. Quax, O. P. Kuipers, S. Bron, M. Hecker, and J. M. van Dijl. 2004. Proteomics of protein secretion by *Bacillus subtilis*: separating the "secrets" of the secretome. *Microbiol. Mol. Biol. Rev.* **68**:207–233.

133. Venkatasubbarao, S. 2004. Microarrays—status and prospects. *Trends Biotechnol.* **22**:630–637.

134. Venter, J. C., M. D. Adams, E. W. Myers, P. W. Li, R. J. Mural, G. G. Sutton, H. O. Smith, M. Yandell, C. A. Evans, R. A. Holt, J. D. Gocayne, P. Amanatides, R. M. Ballew, D. H. Huson, J. R. Wortman, Q. Zhang, C. D. Kodira, X. H. Zheng, L. Chen, M. Skupski, G. Subramanian, P. D. Thomas, J. Zhang, G. L. Gabor Miklos, C. Nelson, S. Broder, A. G. Clark, J. Nadeau, V. A. McKusick, N. Zinder, A. J. Levine, R. J. Roberts, M. Simon, C. Slayman, M. Hunkapiller, R. Bolanos, A. Delcher, I. Dew, D. Fasulo, M. Flanigan, L. Florea, A. Halpern, S. Hannenhalli, S. Kravitz, S. Levy, C. Mobarry, K. Reinert, K. Remington, J. Abu-Threideh, E. Beasley, K. Biddick, V. Bonazzi, R. Brandon, M. Cargill, I. Chandramouliswaran, R. Charlab, K. Chaturvedi, Z. Deng, V. Di Francesco, P. Dunn, K. Eilbeck, C. Evangelista, A. E. Gabrielian, W. Gan, W. Ge, F. Gong, Z. Gu, P. Guan, T. J. Heiman, M. E.

Higgins, R. R. Ji, Z. Ke, K. A. Ketchum, Z. Lai, Y. Lei, Z. Li, J. Li, Y. Liang, X. Lin, F. Lu, G. V. Merkulov, N. Milshina, H. M. Moore, A. K. Naik, V. A. Narayan, B. Neelam, D. Nusskern, D. B. Rusch, S. Salzberg, W. Shao, B. Shue, J. Sun, Z. Wang, A. Wang, X. Wang, J. Wang, M. Wei, R. Wides, C. Xiao, C. Yan, et al. 2001. The sequence of the human genome. *Science* **291**:1304–1351.

135. Venter, J. C., M. D. Adams, G. G. Sutton, A. R. Kerlavage, H. O. Smith, and M. Hunkapiller. 1998. Shotgun sequencing of the human genome. *Science* **280**:1540–1542.

136. Venter, J. C., K. Remington, J. F. Heidelberg, A. L. Halpern, D. Rusch, J. A. Eisen, D. Wu, I. Paulsen, K. E. Nelson, W. Nelson, D. E. Fouts, S. Levy, A. H. Knap, M. W. Lomas, K. Nealson, O. White, J. Peterson, J. Hoffman, R. Parsons, H. Baden-Tillson, C. Pfannkoch, Y. H. Rogers, and H. O. Smith. 2004. Environmental genome shotgun sequencing of the Sargasso Sea. *Science* **304**:66–74.

137. Venter, J. C., H. O. Smith, and L. Hood. 1996. A new strategy for genome sequencing. *Nature* **381**:364–366.

138. Wilmes-Riesenberg, M. R., J. W. Foster, and R. Curtiss III. 1997. An altered rpoS allele contributes to the avirulence of *Salmonella typhimurium* LT2. *Infect. Immun.* **65**:203–210.

139. Wren, B. W. 2000. Microbial genome analysis: insights into virulence, host adaptation and evolution. *Nat. Rev. Genet.* **1**:30–39.

140. Wren, B. W., D. Linton, N. Dorrell, and A. V. Karlyshev. 2001. Post genome analysis of *Campylobacter jejuni*. *Symp. Ser. Soc. Appl. Microbiol.* **30**:36S–44S.

141. Zhang, W., W. Qi, T. J. Albert, A. S. Motiwala, D. Alland, E. K. Hyytia-Trees, E. M. Ribot, P. I. Fields, T. S. Whittam, and B. Swaminathan. 2006. Probing genomic diversity and evolution of *Escherichia coli* O157 by single nucleotide polymorphisms. *Genome Res.* **16**: 757–767.

UNDERSTANDING THE MODEL AND THE MENACE: A POSTGENOMIC VIEW OF *ESCHERICHIA COLI*

Roy R. Chaudhuri and Gavin H. Thomas

2

Escherichia coli, initially described (as *Bacterium coli commune*) by the German physician Theodor Escherich (35; for English translation, see reference 36), is an almost universal component of the lower-gut flora of humans and animals. It is usually a commensal organism, although there are a number of clinically important pathogenic strains that are associated with a range of diseases. To scientists, *E. coli* is familiar as the workhorse of molecular biology; it is used as a standard tool for genetic manipulation. This role is performed in laboratories across the world by using derivatives of a single *E. coli* strain, designated K-12. Here we discuss the insights into pathogenicity and the wider processes of *E. coli* genome evolution that have resulted from the sequencing of the *E. coli* K-12 genome, and, more recently, those from a number of pathogenic *E. coli* strains, including several from the related "genus" *Shigella*. We also discuss the available resources for *E. coli* genomics and the progress that has been made in recent years toward a complete understanding of *E. coli* biology.

Roy R. Chaudhuri   Division of Immunity and Infection, IBR West, University of Birmingham, Edgbaston, Birmingham B15 2TT, United Kingdom.   *Gavin H. Thomas*   Department of Biology, Area 10, University of York, P.O. Box 373, York YO10 5YW, United Kingdom.

THE *E. COLI* K-12 GENOME

E. coli K-12 was originally obtained from a convalescent diphtheria patient in 1922 and subsequently cultured in the strain collection at Stanford University for many years before its adoption as an experimental model by E. L. Tatum in the 1940s (71) as a result of its rapid growth rate and tractable nature. Derivatives of the strain were distributed around numerous laboratories (extensively reviewed in reference 7), and a snowball effect ensued; as the now-standard tools for genetic manipulation were being developed and knowledge about the strain accumulated, it became an increasingly attractive model system. *E. coli* K-12 is now more completely understood than any other organism and is unrivaled as the bacterial model of choice for molecular biology.

The ubiquity of *E. coli* K-12 as a model organism meant that it was the obvious candidate when the possibility of complete bacterial genome sequencing was first discussed (12). However, the process of sequencing was far from straightforward, being hindered by both technological and financial problems (93), and although it was the first bacterial genome project to be initiated, it was not the first to be completed. Two competing groups, one based in Japan and the other in the United States, instigated projects to sequence the genomes of

Bacterial Pathogenomics, Edited by M. J. Pallen et al.
© 2007 ASM Press, Washington, D.C.

two K-12 derivative strains. The U.S. group, led by Frederick Blattner at the University of Wisconsin, published the complete genome of E. coli K-12 strain MG1655 as the culmination of a 6-year project (14). As noted at the time of the publication, the completion of the genome represented both the end of an era and a new dawn for E. coli biology (52). The Japanese group published several large regions of the genome of E. coli K-12 W3110 (2, 40, 54, 90, 128, 133), but at the time, the project did not independently sequence the region of the chromosome from 70 to 100 min, which had been previously published (13, 17, 18, 27, 96, 114). The W3110 genome was subsequently completed and has been recently published (46).

The MG1655 genome sequence published in 1997 consists of 4,639,221 bp and the accompanying version m49 of the annotation identified 4,288 protein-coding genes (14). Of these, 1,632 (almost 40%) were annotated as being of unknown function, although this number has steadily diminished since 1997 as additional genes have been characterized by a variety of means (112, 117). Novel features identified thanks to the genome sequence included six new tRNA genes, additional copies of previously described repetitive elements, three cryptic phages together with several "phage remnants," a flagellar system similar to that previously characterized in *Salmonella*, and two new operons involved in the degradation of aromatic compounds (14). The identification of many insertion sequences (IS), Rhs elements, phage remnants, and regions of atypical base composition suggested that horizontal transfer may have played a significant role in the sculpting of the genome. Such fluid genome dynamics would provide a mechanism for the acquisition of pathogenicity determinants and explain the wide diversity of E. coli pathovars.

E. COLI GENOME DYNAMICS: ESTIMATING THE IMPACT OF HORIZONTAL TRANSFER

Several methods exist for the identification of genes that have been putatively acquired by lateral transfer. Many of these are based on the observation that genes of foreign origin exhibit properties such as base composition, codon usage, and dinucleotide frequencies that are atypical in comparison to the host "genome signature" (66). These variations are due to the presence of distinct mutational patterns in different genomes that are reflected in the base composition of long-term resident genes. Genes that have been recently acquired by lateral transfer from a donor species with an alternative mutational pattern can be identified because their base composition is typical of genes from the donor species, rather than the recipient. Genes that are acquired by horizontal transfer subsequently undergo a process of amelioration (66) as their base composition becomes acclimatized to the new host. Genes that were acquired in this way but that have been present in the new genome for a sufficient period will have completed this process and are impossible to identify as being of foreign origin by such methods. Genes that have been recently acquired from a donor species that has similar mutational patterns to the recipient will also be difficult to identify in this way (68). An initial application of such methods to the K-12 MG1655 genome identified 755 genes, equivalent to 18% of the genome, that appeared to be of foreign origin (67). This estimate has subsequently been questioned because of the limitations of the methods used (60, 122).

An alternative method for the detection of horizontal transfer is to identify genes that display phylogenetic relationships that significantly differ from the "species" tree, often assessed by 16S rRNA sequences (89). Alternatively, programs such as BLAST (3) can be used to identify unexpected similarities between genes in phylogenetically distinct species, although it should be noted that BLAST results do not necessarily correspond to those from more robust phylogenetic analyses (59). These methods will preferentially identify genes that have been acquired from distantly related organisms. A similar approach is to identify genes that are present in the genome of interest (E. coli K-12 MG1655), but that are absent from the genomes of closely related species such as

Salmonella enterica. However, this approach requires the inference of polarity—it is possible that genes are present only in K-12 because they have been lost from other lineages, rather than acquired laterally.

It is also possible to identify horizontally transferred genes by virtue of their position in the genome. For example, otherwise unremarkable genes that are located between genes that appear to be of phage origin are likely to have been acquired at the same time. A combinatorial approach that uses several of these methods is likely to be the most effective method of estimating the degree to which horizontal transfer has contributed to the composition of a genome. By adopting such an approach, Lawrence and Ochman have extended their estimate of the proportion of horizontally transferred genes in the MG1655 genome to 24.5%, acquired in at least 221 separate events (68).

PATHOGENIC *E. COLI*

Although biologists most readily associate *E. coli* with its role as a model system, the same is not true of the general public (32, 118). *E. coli* has an alter ego as a pathogen, responsible for a wide variety of diseases, and it is in this role that it is commonly mentioned in the media. Pathogenic *E. coli* are classified into a number of so-called pathovars on the basis of the symptoms they produce and their modes of infection; it should be noted that these classifications do not always correspond with the phylogenetic relationships between strains (101). There are three major classes of disease caused by *E. coli* pathotypes: diarrheal, urinary tract infections, and sepsis/meningitis. The pathovars are characterized by a number of virulence factors that allow them to invade niches not colonized by commensal *E. coli*. These virulence factors are often encoded on plasmids within "pathogenicity islands," which are large regions in the genome that are likely to have been acquired by horizontal transfer and may contain multiple virulence determinants. Below is a brief summary of the pathovars and their associated virulence factors. For an in-depth discussion of pathogenic *E. coli*,

readers are directed to the reviews by Kaper et al. (57) and Nataro and Kaper (82), and references therein.

There are six defined diarrheagenic *E. coli* pathovars (82), the most notorious of which are the enterohemorrhagic *E. coli* (EHEC). Initially identified as a human pathogen in 1982 (104), EHEC is associated with hemorrhagic colitis (bloody diarrhea) and hemolytic uremic syndrome, which in some cases can lead to kidney failure and death. EHEC strains are typified by a number of virulence factors, including a large virulence plasmid, a type III secretion system that mediates the formation of characteristic attaching and effacing (A/E) lesions in the colon, and a Shiga-like toxin (verotoxin) encoded within an integrated bacteriophage genome (57, 82). The type III secretion system is encoded by a well-characterized pathogenicity island known as the locus for enterocyte effacement (LEE; 75). Most research into EHEC has concentrated on strains of the O157:H7 serotype, which is the prevalent cause of hemolytic uremic syndrome in the United States, the United Kingdom, and Japan, although some non-O157 strains, particularly O111 and O26, are prevalent in other countries. The reservoir of EHEC is in cattle, and outbreaks of hemolytic uremic syndrome are often associated with the consumption of undercooked beef, leading to the condition being dubbed "hamburger disease" in North America. Outbreaks have also been associated with other foodstuffs, notably in 1996 in Sakai City, Japan, where a massive outbreak of O157:H7 infection among schoolchildren was linked to consumption of contaminated white radish sprouts (77).

Among the other diarrheagenic *E. coli* pathotypes, enteropathogenic *E. coli* (EPEC) are similar to EHEC in that they also harbor the LEE pathogenicity island and show a similar attaching and effacing phenotype, albeit in the small intestine. They do not produce a Shiga toxin, and they do not cause the bloody diarrhea typical of EHEC. Nevertheless, they are an important cause of infant diarrhea in developing countries. Enterotoxigenic *E. coli* (ETEC) is likewise a cause of diarrhea in the developing world among infants and visitors to the region; ETEC is the major cause of such "traveler's diarrhea." ETEC does not cause an attaching and effacing phenotype; it is characterized by the secretion of enterotoxins, which are heat labile, heat stable, or in some cases both (82).

Enteroaggregative *E. coli* (EAEC, previously EAggEC) are characterized by their adherence to surfaces and to one another in a stacked-brick configuration, often mediated by aggregative adherence fimbriae (AAF-I and AAF-II) encoded by members of the plasmid family pAA (88). EAEC secretes several toxins, including Pic, which is a member of the SPATE family (serine protease autotransporters of the *Enterobacteriaceae*; 48), EAST-1, a relative of the ETEC heat-stable toxins, and *Shigella* enterotoxin 1 (ShET1), which is encoded by a gene on the opposite strand at the *pic* gene locus. Related to the EAEC are the diffusely adherent *E. coli* (DAEC) (26), distinguished by their characteristic diffuse pattern of adherence on HEp-2 cells. It is unclear whether these are diarrheagenic, although there have been reports of the presence of a type III secretion system (8, 62).

Enteroinvasive *E. coli* (EIEC) have perhaps the most distinctive lifestyle of all the *E. coli* pathotypes. Unlike the other pathotypes, EIEC demonstrate a true intracellular lifestyle and are capable of surviving and replicating within the cytoplasm and vacuoles of host colonic epithelial cells. EIEC share many characteristics with members of the *Shigella* genus, which are the cause of bacillary dysentery or shigellosis. The *Shigella* genus was adopted in the 1940s, but it has since become clear that the four "species" that comprise the genus (*S. dysenteriae*, *S. flexneri*, *S. boydii*, and *S. sonnei*) are phylogenetically indistinguishable from *E. coli* (64, 65, 99). However, the dysentery caused by *Shigella* (shigellosis) is generally considered to be more severe than that caused by EIEC, with a smaller infectious dose required, and for this reason, the existing nomenclature has been retained. Phylogenetic analyses indicate that *Shigella* strains have evolved independently on

several occasions, and it has been suggested that EIEC strains may represent intermediate forms that have the potential to become full-blown *Shigella* and that the two should be considered members of the same pathovar (64). EIEC and *Shigella* share a 220-kb plasmid referred to as pINV, which is sufficient for the invasive phenotype. This plasmid encodes a type III secretion system (encoded by the *mxi* and *spa* genes) that is distinct from the LEE-encoded system found in EHEC and EPEC, and it secretes the Ipa invasins that mediate the invasion of epithelial cells (76).

E. coli is not only an intestinal pathogen. There are a number of extraintestinal pathogenic *E. coli* (ExPEC; 107) that are also of clinical importance. These have traditionally been subdivided on the basis of their clinical manifestation; common categories include uropathogenic *E. coli* (UPEC), which cause urinary tract infections, and neonatal meningitis *E. coli* (NMEC). In contrast to the intestinal pathogenic *E. coli*, ExPEC may stably colonize the host intestine, and invasion into an appropriate extraintestinal site is required for pathogenesis.

Of course, the above represents an anthropocentric view of *E. coli*, one focused on *E. coli* as a human pathogen and commensal. However, *E. coli* is also an important commensal and pathogen of a wide range of animals. Notably, avian pathogenic *E. coli* (APEC) are of economic importance as pathogens of poultry. They cause extraintestinal infections, initially in the respiratory tract but then often progressing to infections of the internal organs such as pericarditis, perihepatitis, peritonitis, and salpingitis (30, 120).

E. COLI DIVERSITY AND PHYLOGENETICS

To aid in the study of *E. coli* diversity, Ochman and Selander assembled a collection of standard reference strains from humans and 16 other mammalian species (85). These strains, referred to as the ECOR (*E. coli* Reference) collection, were chosen to represent the full genotypic diversity of the species on the basis of the results of multilocus enzyme electrophoresis (MLEE) studies (125). The strains were also selected so as to maximize the geographical distribution and host range, and to include both pathogenic and nonpathogenic varieties. A second collection, consisting of 78 diarrheagenic *E. coli* (DEC) strains representing the 15 major electrophoretic variants that were detected by MLEE (126), has also been assembled. Further information about the ECOR and DEC collections is available from Thomas Whittam's laboratory website (http://foodsafe.msu.edu/whittam).

Phylogenetic analysis of the ECOR collection, which is based on neighbor-joining analysis of MLEE distance data for 38 enzyme loci (49; Fig. 1), separates the strains into five major groups: A, B1, B2, D, and E (sometimes considered to be unclassified). These groups have been supported across several studies that use a variety of methods of phylogenetic reconstruction (5, 31, 37, 44, 70, 95), although there have been disagreements regarding the relationships between the groups. Midpoint rooting of the MLEE tree suggests that group A (which includes sequences closely related to K-12) is the deepest-branching group. However, a subsequent study that used a variety of data sources (MLEE, randomly amplified polymorphic DNA, *rrn* restriction fragment length polymorphism, and nucleotide sequence data) suggested that group B2 is the most basal (70). Interestingly, strains classified as group B2 are rarely found as human commensals (34), and the group includes many strains with ExPEC virulence determinants (11, 95). It is unclear whether this is due to some innate property of the B2 strains that makes them particularly receptive to the acquisition of virulence factors, or whether key virulence factors were obtained by an ancestor of the B2 group and subsequently lost in some strains (56, 95).

The phylogenetic relationships between pathogenic and commensal *E. coli* have long been of interest. Initially, *E. coli* were thought to be nonpathogenic, with all pathogenic strains classified within the genus *Shigella*. For diagnostic purposes, *Shigella* were differentiated from *E. coli* by an inability to ferment lactose and a lack of mobility (100). This simple classification was complicated in the 1940s when pathogenic strains were identified that shared the other characteristics of *E. coli*. Although it was clear that *Shigella* and *E. coli* were closely related (16), the distinction has been retained, largely because of the clinical severity of shigellosis. From the results of molecular phylogenetic analyses of chromosomal housekeeping genes, it became clear that *E. coli* and *Shigella* are essentially indistinguishable—some *Shigella* strains are more closely related to *E. coli* than they are to other *Shigella* (99, 100). This observation suggested that there had been multiple independent origins of *Shigella*, with convergent evolution of a similar phenotype. Expansion of this study to use both chromosomal and plasmid genes and to include EIEC strains suggested that EIEC had also evolved independently on multiple occasions (64). It has since been argued, on the basis of a correspondence between phylogenies derived from chromosomal and pINV plasmid genes, that there was a single acquisition of the common ancestor of pINV (38) by an ancestral *E. coli* strain. This hypothesis suggests that the chromosomal changes typical of *Shigella* strains (such as inactivation of the *lac* operon) were necessary to maintain pINV, which was subsequently lost in many strains (ECOR groups A and B1). This hypothesis does not appear to be compatible with the position of outlying *Shigella* strains deep within the trees generated in other studies (64, 99, 100), although it might explain the evolution of the *Shigella* and EIEC strains that group with the ECOR A/B1 lineage. A possible alternative explanation for the correspondence between plasmid and chromosomal trees is that plasmids preferentially transmit between closely related strains, as has been suggested previously (43).

A recent study (37) constructed a phylogenetic tree of ECOR strains, together with strains from the DEC collection; reference strains of *Shigella*, EIEC, and EAEC; and the EHEC and ExPEC strains for which the complete genome sequence has been determined (EDL933, Sakai, CFT073, and RS218;

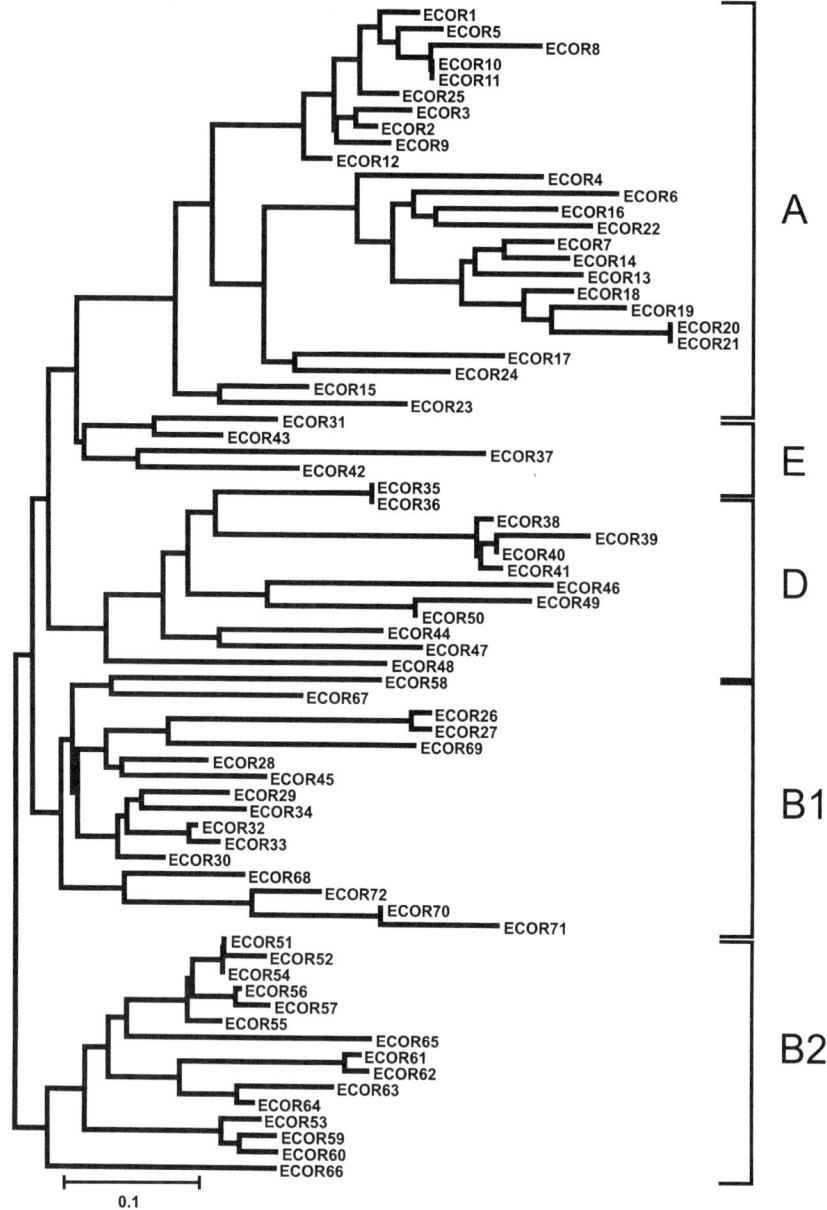

FIGURE 1 Phylogenetic relationships between the strains in the ECOR collection, as assessed by MLEE (123). Data were obtained from http://foodsafe.msu.edu/whittam, and a neighbor-joining tree was generated by neighbor, part of the PHYLIP package. The tree was rooted using group B2 as an outgroup (70).

see below). The tree was constructed on the basis of the partial sequences of six chromosomal genes (*trpA*, *trpB*, *pabB*, *putP*, *icd*, and *polB*), and analyzed by maximum parsimony, although the resultant topology was consis-

tent with that obtained when maximum-likelihood, neighbor-joining, and Bayesian methods were used. Sequence from *Escherichia fergusonii*, the most closely related species to *E. coli* (69), was used as an outgroup. This analysis represents the most comprehensive phylogenetic analysis of *E. coli* diversity published to date. The groups previously identified (A, B1, B2, D, and E) were supported, with an additional group, C, also present as a sister group of A and B1. The status of group B2 as the earliest-branching group was supported, with group D and then group E also branching off before the radiation of groups A, B1, and C (37, 38).

Several properties of *E. coli* genomes have been shown to correlate well with the phylogenetic structure of the ECOR collection. These include genome size, as estimated by the cumulative length of NotI and BlnI restriction fragments resolved by pulsed-field gel electrophoresis (10), and later by similar use of the restriction enzyme I-CeuI, which cleaves only at rRNA genes and enabled the evaluation of the relative contribution of chromosomal and plasmid DNA to the estimated genome size (9). In these studies, the *E. coli* chromosomes were found to vary in size from 4.5 to 5.5 Mb. The smallest genomes were found in group A, which includes the 4.6-Mb K-12 genome, with the larger genomes being derived from groups B2, D, and E. A correlation was found between the size of restriction fragments located on either side of the origin of replication, suggesting the possibility of selection to maintain the equidistant positions of the replication origin and terminus (9). Other properties that have been shown to be correlated to the phylogeny of the ECOR collection include the distribution of Rhs elements (50), the *Yersinia* high-pathogenicity island (25), the cryptic second type III secretion system ETT2 (103), and the virulence-associated *hly*, *kps*, *sfa*, and *pap* regions (15).

It has been observed that some virulence factors are found only within particular phylogenetic groups, despite their distribution being sporadic and consistent with horizontal transfer rather than vertical inheritance (37). This suggests that a specific genetic background is necessary for the expression and maintenance of such factors. The authors of the study suggest that this operates at two levels: an "ancestral" background common to all *E. coli* that is

sufficient for the expression of the virulence factors associated with mild and chronic diarrhea; and "derived" backgrounds necessary for more severe pathologies, such as those associated with EHEC, ETEC, and *Shigella*/EIEC. Diffusely adherent *E. coli* and EAEC appear to be found in all phylogenetic groups (with the exception of group E). Both EHEC and EPEC have been suggested to cluster in 2 groups (101). EHEC1 are found in group E, with the EHEC strains from group A referred to as EHEC2. Similarly, EPEC1 strains are in group B2, and EPEC2 strains are in group B1. Although such designations are useful, it should be noted that examples of EHEC and EPEC strains from outside these groupings have been identified (37). It is possible that even the strain collections used in this study do not adequately represent the full diversity of each of the pathovars. Indeed, recent multilocus sequence typing data indicate that ETEC diversity is comparable to that found in the ECOR collection (119).

It has recently been argued, on the basis of multilocus sequence typing data from an extensive collection of *E. coli* strains incorporating the ECOR collection, that, contrary to previous suggestion (101), the *E. coli* lineage is subject to extensive recombination and is hence not truly clonal (127). This casts doubt on the validity of phylogenetic approaches to the study of *E. coli* diversity because such methods will always reconstruct a single tree. This can be misleading if the data supplied incorporate sites with different evolutionary histories due to the effects of recombination. It was suggested that the major phylogenetic groupings of *E. coli* (A, B1, B2, and D) represent the descendants of four ancestral states before a population expansion. One-third of isolates were assigned as hybrid between these groups. Interestingly, pathogenic strains were overrepresented in the set of hybrid sequences, suggesting that recombination (and mutation) could be promoted as a result of the selective pressure imposed by host immune systems on pathogenic species. It is proposed that this effect is most pronounced in the "epidemic" pathogens EHEC, EIEC, and *Shigella*. It will be interesting to see whether the results of this multilocus sequence typing–based study are validated by the forthcoming availability of numerous complete genomes within the *E. coli* lineage (Table 1).

BEYOND K-12: OTHER *E. COLI* GENOMES

E. coli O157:H7

The development of shotgun sequencing strategies has allowed the sequencing of complete bacterial genomes to become increasingly routine. As an emergent pathogen of clinical importance, the EHEC strain O157:H7 was an obvious candidate to undergo this process. As with *E. coli* K-12, two rival groups simultaneously obtained the sequence of different but closely related strains. A Japanese group led by Tetsuya Hayashi was the first to publish a complete sequence (47), derived from the strain RIMD 0509952, which was isolated from a typical patient during the 1996 Sakai City outbreak of O157:H7 infection among schoolchildren (47; the sequence is commonly referred to as the Sakai genome). Around the same time, the Wisconsin group, who had been responsible for the K-12 MG1655 genome sequence (14), published an almost-complete genome of the O157:H7 strain EDL933 (94), which had been isolated from the 1982 American outbreak during which EHEC O157:H7 was first identified (104). This sequence was originally published with two gaps (94), but it has since been completed.

The 5.5-Mb O157:H7 genome shares a common backbone of around 4.1 Mb with the K-12 genome. This backbone is colinear, with the exception of a 422-kb inversion spanning the replication terminus of the EDL933 (but not Sakai) genome. The backbone is punctuated by a number of regions specific to O157, which were referred to as "S loops" (47) or "O islands," in an extension of the term "pathogenicity island" (94). A total of 177 O islands larger than 50 bp were identified, totaling 1.34 Mb (around a quarter of the genome). Perhaps surprisingly, 234 regions that were present in K-12 but absent from O157, dubbed "K islands" or "K loops," were also identified, totaling 0.53 Mb. This contradicted the implicit assumption that the genomes of pathogenic *E. coli* would consist of the K-12 genome plus a number of pathogenicity islands. Many of the O islands and K islands were at equivalent loci, suggesting the presence of hotspots for the insertion of foreign DNA.

It should be noted that most of the O islands had no obvious association with pathogenicity. Those with a known role in pathogenesis were two Shiga toxin (Stx) encoding phages, and the LEE region that encodes a type III secretion. A novel finding in the O157 genome was an island encoding a potential second type III secretion system, named ETT2. This was thought initially to be virulence associated (45, 47, 73), but it has since been shown to be incapable of producing a functional secretion apparatus (103). Other potential virulence factors located on O islands included an RTX toxin and associated transport system, fimbrial-biosynthesis operons, and nonfimbrial adhesins. Surprisingly, some genes located within K islands, such as those for ferric citrate utilization, might have been labeled as potential virulence factors had they been unique to a pathogenic *E. coli* strain. This raises the possibility that K-12 is derived from a pathogenic ancestor, is pathogenic for nonhuman hosts, or is on a path toward becoming a pathogen. Alternatively, these pathogenicity determinants may play a role in normal commensal colonization.

The two sequenced O157 strains both harbor a similar large (92 kb) virulence plasmid, referred to as pO157 (19, 72). This is related to the F plasmid and additionally encodes virulence determinants including a type II secretion system, the SPATE protein EspP, and the hemolysin HlyA (together with its associated secretion system, which consists of HlyC, HlyB, and HlyD). An unusually large

TABLE 1 *E. coli* genome projects completed and underway as of April 2006[a]

Genome	Type/pathovar	Status	Sequences	GenBank accession no.	Reference(s)	URL
E. coli K-12 MG1655	Laboratory	Complete	4.6-Mb chromosome	U00096	14, 46, 106	http://www.genome.wisc.edu/sequencing/k12.htm
E. coli O157:H7 RIMD 0509952 (Sakai)	EHEC	Complete	5.5-Mb chromosome 92-kb pO157 3.3-kb pOSAK1	BA000007 AB011549 AB011548	47, 72	http://genome.naist.jp/bacteria/o157
E. coli O157:H7 EDL933	EHEC	Complete	5.5-Mb chromosome 92-kb pO157	AE005174 AF074613	19, 94	http://www.genome.wisc.edu/sequencing/o157.htm
Shigella flexneri 2a 301	*Shigella*	Complete	4.6-Mb chromosome 221-kb pCP301	AE005674 AF386526	55	http://www.mgc.ac.cn/ShiBASE
E. coli CFT073	ExPEC (UPEC)	Complete	5.2-Mb chromosome	AE014075	124	http://www.genome.wisc.edu/sequencing/upec.htm
Shigella flexneri 2a 2457T	*Shigella*	Complete	4.6-Mb chromosome	AE014073	123	http://www.genome.wisc.edu/sequencing/sflex.htm
Shigella sonnei 046	*Shigella*	Complete	4.8-Mb chromosome 214-kb pSS	CP000038 CP000039	129	http://www.mgc.ac.cn/ShiBASE
Shigella boydii serotype 1 strain 227	*Shigella*	Complete	4.5-Mb chromosome 127-kb pSB4_227	CP000036 CP000037	129	http://www.mgc.ac.cn/ShiBASE
Shigella dysenteriae serotype 1 strain 197	*Shigella*	Complete	4.4-Mb chromosome 183-kb pSD1_197	CP000034 CP000035	129	http://www.mgc.ac.cn/ShiBASE
E. coli K-12 W3110	Laboratory	Complete	4.6-Mb chromosome	AP009048	46, 105	http://ecoli.aist-nara.ac.jp/gb5/Resources/G_seq/G_seq.html
E. coli UT189	ExPEC (UPEC)	Complete	5.1-Mb chromosome 114-kb pUT186	CP000243 CP000244	24	
E. coli 042	EAEC	Finished but unannotated	5.2-Mb chromosome 113-kb pAA		Unpublished	http://www.sanger.ac.uk/Projects/Escherichia_Shigella
Shigella sonnei 53G	*Shigella*	Finished but unannotated	5.0-Mb chromosome 216-kb plasmid A 9-kb plasmid B 5-kb plasmid C 2-kb plasmid D		Unpublished	http://www.sanger.ac.uk/Projects/Escherichia_Shigella

(Continued)

TABLE 1 *Continued*

Genome	Type/pathovar	Status	Sequences	GenBank accession no.	Reference(s)	URL
E. coli HS	Commensal	Complete but unannotated (automated CDD and COG-based annotation available in RefSeq)	4.6-Mb chromosome	AAJY00000000	Unpublished	http://msc.tigr.org/e_coli_and_shigella/escherichia_coli_hs
E. coli E23477A	ETEC	Complete but unannotated (automated CDD and COG-based annotation available in RefSeq)	5.0-Mb chromosome	AAJZ00000000	Unpublished	http://msc.tigr.org/e_coli_and_shigella/escherichia_coli_e24377a
E. coli K1 RS218	ExPEC (NMEC)	Complete but unpublished-assembly available for BLAST searches	5.1-Mb chromosome 114-kb pRS218 9.6-kb bacteriophage CUS-1		Unpublished	http://www.genome.wisc.edu/sequencing/rs218.htm
E. coli E2348/69	EPEC1	Preliminary assembly	12 contigs >1 kb		Unpublished	http://www.sanger.ac.uk/Projects/Escherichia_Shigella
Shigella dysenteriae M131649 (M131)	Shigella	Preliminary assembly	256 contigs >1 kb		Unpublished	http://www.sanger.ac.uk/Projects/Escherichia_Shigella
E. coli H10407	ETEC	Preliminary assembly	96 contigs >1 kb		Unpublished	http://www.sanger.ac.uk/Projects/E_coli_H10407
E. coli B171	EPEC	Preliminary	159 contigs	AAJX00000000	Unpublished	http://msc.tigr.org/e_coli_and_shigella/escherichia_coli_b171
E. coli 53638	EIEC	Preliminary assembly	119 contigs	AAKB00000000	Unpublished	http://msc.tigr.org/e_coli_and_shigella/escherichia_coli_53638
E. coli 101-1	EAEC	Preliminary assembly	70 contigs	AAMK00000000	Unpublished	http://msc.tigr.org/e_coli_and_shigella/escherichia_coli_101_1

E. coli E110019	EPEC	Preliminary assembly	115 contigs	AAJW00000000	Unpublished	http://msc.tigr.org/e_coli_and_shigella/escherichia_coli_e110019
E. coli B7A	ETEC	Preliminary assembly	198 contigs	AAJT00000000	Unpublished	http://msc.tigr.org/e_coli_and_shigella/escherichia_coli_b7a
E. coli F11	ExPEC (UPEC)	Preliminary assembly	88 contigs	AAJU00000000	Unpublished	http://msc.tigr.org/e_coli_and_shigella/escherichia_coli_f11
E. coli E22	EPEC	Preliminary assembly	109 contigs	AAJV00000000	Unpublished	http://msc.tigr.org/e_coli_and_shigella/escherichia_coli_e22
Shigella boydii BS512	Shigella	Preliminary assembly	79 contigs	AAKA00000000	Unpublished	http://msc.tigr.org/e_coli_and_shigella/shigella_boydii_bs512
Shigella dysenteriae 1012	Shigella	Preliminary assembly	42 contigs	AAMJ00000000	Unpublished	http://msc.tigr.org/e_coli_and_shigella/shigella_dysenteriae_1012
E. coli DH10B	Cloning	Preliminary assembly	399 contigs >2 kb		Unpublished	http://www.hgsc.bcm.tmc.edu/projects/microbial/EcoliDH10B

[a]Only projects for which sequence data are available are included. For an up-to-date list, see the NCBI Genome Project database (http://snipurl.com/ecoli) or GOLD (http://snipurl.com/genomes).

2.3-kb open reading frame is also present. This shows weak similarity to the toxin ToxB from *Clostridium difficile* and is also potentially involved in virulence. Additionally, the Sakai strain contains a 3.3-kb plasmid called pOSAK1, which is related to the drug-resistance plasmid NTP16 from *Salmonella enterica* serovar Typhimurium, and possibly represents the ancestral form of this plasmid because it lacks the transposons and antibiotic-resistance determinants found in NTP16.

The Sakai genome contains 18 prophage elements, of which 13 are lambda-like. All of these contain insertions and/or deletions that interrupt essential phage genes, suggesting that they are incapable of lytic growth (47). However, one lambdoid phage in the EDL933 genome, the Stx2-encoding BP-933W, has been shown to be capable of producing infectious particles (97). The Sakai genome also contains six prophage-like elements, which showed no homology to known phage genes with the exception of integrase-like proteins (47). Many of the prophage-like and prophage elements encode virulence-related proteins, including the two Stx toxins, superoxide dismutases, and the LEE-encoded type III secretion system (86). The lambda-like phages within the genome are very similar to each other, which suggested the possibility of duplication of and/or recombination between phages within the genome (47, 86). The observation that most of the lambda-like phages are present at the same loci but significantly different internally when comparing the Sakai and EDL933 genomes supported the possibility of extensive recombination between phage genomes (86). Subsequent analysis of eight diverse O157 genomes by long PCR genome scanning found that the prophage-like and prophage elements were diverse in both internal structure and chromosomal location (87). This suggested that prophages play a major role in *E. coli* strain diversity (86, 87). Genomic comparisons between the O157 and K-12 MG1655 genomes also enabled the identification of phage-like elements in K-12. There are 10 of these elements, plus the lambda prophage that was removed from the MG1655 strain, again demonstrating their role in *E. coli* diversity (47).

Shigella flexneri

The next *E. coli*–group genome sequence to be published was that of *Shigella flexneri* 2a strain 301, as determined by a Chinese group led by Qi Jin (55). A second *S. flexneri* 2a genome, from strain 2457T, was later obtained by the Wisconsin group (123). As with EHEC O157:H7, *S. flexneri* serotype 2a was an obvious target for whole-genome sequencing because of its importance as a pathogen. *Shigella* strains are human-specific intracellular pathogens, characterized by the presence of the 220-kb pINV virulence plasmid, together with a number of characteristic differences from *E. coli*. These differences include a loss of motility and an inability to utilize lactose, maltose, and xylose as carbon sources (55). It has been suggested that maintenance of the pINV plasmid requires these chromosomal changes (38).

The chromosomes of the two sequenced strains are approximately the same size as *E. coli* K-12 (4.6 Mb). The most striking feature of the genomes is the large number of IS elements present, 247 complete plus 67 partial in strain 301, compared with 35 complete and 45 partial in the Sakai strain. IS elements have been shown to mediate rearrangements including deletions and inversions (109), and the large number present in the *S. flexneri* genome seems to be responsible for a number of such events. Both genomes show large inversions relative to K-12 spanning the terminus of replication, with an additional large inversion surrounding the origin of replication in strain 2457T. Deletions include the disruption of the *cadA* gene. This gene encodes a lysine decarboxylase, which is responsible for the conversion of lysine to cadaverine, which has an inhibitory effect on *Shigella* enterotoxin activity and has been shown to adversely affect virulence (74). The *cad* operon has been found to be inactivated by a variety of methods in other *Shigella* and EIEC strains (21, 28), and is an example of a pathoadaptive mutation, where the inactivation of a gene pleiotropically promotes virulence. A second example of this phenomenon in the *Shigella* genome is the loss of *ompT*, the product of which attenuates *Shigella* virulence by degrading the pINV-encoded VirG, preventing the spreading of *Shigella* to adjacent epithelial cells (81).

The *S. flexneri* genomes also contain many pseudogenes (372 in strain 2457T), inactivated by a variety of mechanisms including IS element interruptions, premature stop codons, frameshifts, and truncations (55, 123). These account for many of the phenotypic differences between *Shigella* and *E. coli*, including the lack of motility and differences in carbohydrate metabolism. *Shigella* has a radically different lifestyle from noninvasive *E. coli*, and the pseudogenes could simply represent the decay of genes no longer required during the process of adaptation to an intracellular niche (65). It is interesting to compare in this light the genomes of EHEC O157:H7 and *S. flexneri* with the sequenced genomes of the closely related genus *Salmonella*. Both *S. flexneri* and *Salmonella enterica* serovar Typhi are human-specific pathogens, and this seems to be reflected in a large number of pseudogenes compared with the wide-host-range pathogens *Salmonella enterica* serovar Typhimurium and EHEC O157:H7 (55, 92).

Additional *Shigella* Genomes

A recent study by the Chinese group responsible for the first *S. flexneri* genome published in conjunction (as is increasingly the trend for genome-sequencing projects) three additional complete *Shigella* genomes, those of *S. sonnei* strain 046, *S. dysenteriae* serotype 1 strain 197, and *S. boydii* serotype 4 strain 227 (129). All consisted of a single chromosome together with a large virulence plasmid. Comparative genomics of these three genomes, together with the previously published *S. flexneri* genomes, revealed that the high prevalence of IS elements and pseudogenes was a theme conserved among all the *Shigella* strains. All the genomes demonstrated a number of chromosomal rearrangements facilitated by the IS elements, with the greatest deviation from colinearity with K-12 MG1655 found in the *S. boydii* and *S. dysenteriae* sequences. All strains had an inversion spanning the replication terminus relative to MG1655. A number of changes, such as the inactivation of *ompT* and *cadA*, had occurred by different mechanisms in the different strains, suggesting convergent evolution toward a *Shigella* phenotype. All the strains had acquired chromosomal virulence determinants through phage-mediated lateral transfer, and the diversity of these was implicated in the differential virulence of the different strains.

Uropathogenic *E. coli*

The only other *E. coli* genome sequences published to date are those of two uropathogenic strains. The strain CFT073 (124) was determined, like K-12 MG1655 and O157:H7 EDL933, by the Wisconsin group. The chromosome was 5.23 Mb in length; this represented the complete genome because no virulence plasmids were present. Pairwise comparison with the MG1655 genome indicated a similar pattern to the O157 genomes, with a conserved backbone punctuated by a number of strain-specific islands. These comprise around 25% of the genome and contain 2,004 genes, only 204 of which are also found in EDL933. The island genes were found to have significantly different codon usage for 52 of the 61 codons when compared with the backbone genes. By contrast, the backbone of EDL933 was shown to have codon use indistinguishable from the CFT073 backbone. This suggested that most strain-specific islands were of foreign origin. Island-encoded pathogenicity factors included P fimbriae encoded by two *pap* (pyelonephritis-associated pilus) operons, the *hlyCABD* genes, which encode a hemolysin and its secretion system (also present in EHEC on the plasmid pO157), and a number of autotransporters including homologues of Pic, Sat, and Vat, members of the SPATE family (91). A comparison with other UPEC strains indicated the presence of similar pathogenicity determinants, but variation in their chromosomal location and the islands on which they are present suggests multiple independent origins of uropathogenicity.

A second uropathogenic *E. coli* genome, that of strain UT189, has recently been published by a group led by Jeffrey Gordon at Washington University (24). The genome consists of a 5.1-Mb chromosome, together with the 114-kb plasmid pUT189. A single *pap* locus was present at a locus distinct from the *pap* loci on the CFT073 chromosome. The capsular biosynthesis locus was similar to that of CFT073, but the genomes differed in their polysialic synthase loci. Adjacent to the capsular biosynthesis locus of the UT189 genome was a locus that putatively encoded a type II secretion system that was mostly absent from the other published genomes. No type III or type IV secretion systems were present, suggesting that the type II system may be used to export virulence factors. Potential virulence determinants located on pUT189 included the *cjrABC* operon, apparently involved in iron uptake and conserved in the EIEC virulence plasmid O164. Also present is the gene *senB*, a putative secreted enterotoxin, the mutation of which has been shown to reduce the enterotoxicity of EIEC (83). The conjugative transfer and replication apparatus of the plasmid were similar to those of the F plasmid, and evidence of extensive plasmid recombination was found.

To further identify genes potentially involved in uropathogenicity, a phylogenetic approach was adopted. Maximum-likelihood methods (132) were used to identify genes that have apparently been subject to positive selection within the UPEC strains, but not in any of the other published *E. coli* genomes, on the basis of an increased ratio of the rates of nonsynonymous and synonymous substitution (dN/dS or Ka/Ks). Twenty-nine such genes were identified in this (relatively conservative) analysis. These included genes involved in DNA repair, which correlates well with the increased mutation rate typical of UPEC strains. This increased mutation rate is thought to confer a selective advantage, increasing persistence in the bladder and kidney (63). Other positively selected genes that were identified included genes involved in the production of surface structures, cell wall/membrane biogenesis, and iron acquisition.

E. coli Genomes in Progress

A number of additional *E. coli* genome sequencing projects are underway, with several of the sequences completed or in the finishing process (Table 1). The University of Wisconsin *E. coli* genome sequencing group has a project to sequence the K1 NMEC strain RS218. The sequence has been completed but has not yet been made available for download. BLAST searches of the sequence can be performed through the website, however (http://www.genome.wisc.edu/sequencing/rs218.htm). Göttingen Genomics Laboratory, in collaboration with Jörg Hacker, has sequenced the UPEC strain *E. coli* 536, with a publication currently in preparation (http://www.g2l.bio.unigoettingen.de/projects/c_proj_ec.html). The same group is also sequencing *E. coli* strain 789 AC/1 and the non-*coli Escherichia blattae*. The Wellcome Trust Sanger Institute has been funded to sequence five genomes: EAEC 042, EPEC E2348/69, *Shigella sonnei* 53G, *Shigella dysenteriae* M131649 (M131), and a non-K1 clinical isolate of NMEC. Of these, the EAEC 042 genome is finished and is currently in the process of annotation (gene predictions and a systematic nomenclature are already available), the *S. sonnei* sequence is complete but awaiting gene prediction, and the EPEC and *S. dysenteriae* strains are nearly complete and in the finishing process. The non-K1 NMEC strain that will be sequenced has not yet been determined. Further information and preliminary sequence data are available (http://www.sanger.ac.uk/Projects/Escherichia_Shigella). As a separate project, the Sanger Institute is also sequencing the genome of the ETEC strain H10407, which is currently virtually complete but awaiting finishing and annotation (http://www.sanger.ac.uk/Projects/E_coli_H10407).

Analyses of the EAEC 042 data before completion have revealed two findings of particular interest. The first of these is the *eip* island, which contains homologues of the *sip* type III secretion effector genes from *Salmonella*, and possibly has or once had a functional relationship with the ETT2 secretion system (103). Also of note is Flag2, a second flagellar gene cluster that appears to be evolutionarily ancient but almost completely deleted in most *E. coli* strains (102). EAEC 042 appears to encode a complete flagellar system, which shows homology to the lateral flagellar systems of *Aeromonas* and *Vibrio* at this locus. The system has been shown to be intact in around 20% of strains from the ECOR collection, but all that is found in the remainder, and the other complete or almost complete genome sequences, is a remnant of just two promoterless genes, *fhiA* and *mbhA*. This highlights the utility of sequencing and comparing multiple closely related genomes.

Recently, The Institute for Genome Research (TIGR) has deposited a number of *E. coli* genome sequences into the wgs (Whole Genome Shotgun) division of GenBank (http://snipurl.com/ecoli). These include two complete but unannotated genomes, the ETEC strain 23477A, and the commensal strain HS. These are accompanied by six strains that have a complete shotgun sequence and preliminary assembly but that have not as yet undergone finishing. These genomes are the additional ETEC strain B7A, the ExPEC strain F11, the EAEC strain 101-1, the EIEC strain 53638, and the EPEC strains E22, B171, and E110019. Also available are data from the *Shigella boydii* strain BS512 and *Shigella dysenteriae* 1012.

The probiotic *E. coli* strain Nissle 1917 is being sequenced by the German Research Centre for Biotechnology. This strain has been used for many years for the treatment of intestinal disorders. Although sequence data have not yet been released, an analysis of a preliminary assembly has been published (115). Comparison with the published *E. coli* genomes revealed that Nissle 1917 was closely related to the ExPEC strain CFT073. BLAST searches against all the *E. coli* sequences using a set of 3,239 proteins annotated as virulence factors found that 250 were present in at least one strain. A total of 70% of these were present in all the strains (including the nonpathogenic strains K-12 MG1655, K-12 W3110, and Nissle 1917), suggesting that their assignment as "virulence factors" may be misleading, and that they may have a more general contribution to fitness. Comparison of Nissle 1917 with CFT073 revealed that the former lacked 12 virulence determinants associated with renal pathogenesis, including proteins associated with the P fimbriae and alpha-hemolysin. These absences preclude the uropathogenicity of Nissle 1917 but may contribute to its probiotic properties.

Additional sequencing projects that are underway but for which sequence data are not yet available include a project by Genoscope in collaboration with a consortium of scientists led by Erick Denamur (http://www.genoscope.cns.fr/externe/English/Projets/Projet_MW/organisme_MW.html). This group is sequencing six *E. coli* strains, together with the genome of the most closely related non-*coli Escherichia, E. fergusonii*. The *E. coli* strains chosen are the commensal IAI1 and the EAEC 55989 from the phylogenetic group B1, the ExPECs UMN026 and IAI39 from group D, and the commensal ED1a and the ExPEC S88 from group B2. These strains have been strategically chosen to expand the representation of *E. coli* diversity among sequenced genomes. Also in progress (according to GOLD: http://www.genomesonline.org) are three *E. coli* genomes at Kitsato University: O26:H11, O111:H- and O103:H2, and *E. coli* B, at the Joint Genome Institute. GOLD can be searched for an up-to-date list of *E. coli* and *Shigella* genome projects (http://snipurl.com/genomes.).

E. COLI COMPARATIVE GENOMICS
The sequencing of MG1655, EDL933, and CFT073 provided an opportunity to perform a three-way comparison of the three genomes (124). This comparison showed that of the combined gene pool from the three genomes, only 39.2% were present in all three. This statistic is illustrative of the extraordinary degree of genomic variation between diverse members of this "species." It seems that the backbone genome is highly conserved and largely colin-

ear, with rearrangements unusual, notwithstanding the example of *Shigella*. However, precise definition of what constitutes backbone and what represents an island will depend on the sequencing of multiple genomes. Islands are often assumed to be the result of the acquisition of DNA by horizontal transfer, but they can equally be derived from deletions in one lineage. Comparisons between multiple genomes in a phylogenetic context can allow resolution of these possibilities. There are examples of genes in the *E. coli* K-12 genome that represent "evolutionary baggage," i.e., remnants of once functional gene clusters that are in the process of being eliminated from the genome. A comparative genomics approach can allow the function of these otherwise enigmatic genes to be determined.

Computational Resources for Whole-Genome Comparison
A number of developments in recent years have extended access to such genome comparisons. The programs blastn and tblastx from the BLAST suite (3) can be used to compare two genome sequences to highlight regions of difference. The Artemis Comparison Tool (ACT; 20) is a Java-based application that has been developed by the Sanger Institute to allow easy visualization of such BLAST comparisons on a genomic scale. Also available is WebACT (1), an online resource that uses ACT to display precalculated genome comparisons or alignments generated on the fly from user-supplied sequences. A Perl/Tk alternative to ACT/WebACT called GenomeComp is also available (131). This has a less flexible interface than ACT, but it has the facility to perform its own BLAST/megaBLAST comparisons. GenomeComp has been used as the basis for the *Shigella* comparative genomics website *Shi*BASE (130). Results from blastz, an independent implementation of BLAST designed specifically for genome alignments (110, 111), can be visualized by the EnteriX programs (39), which are Web based (http://bio.cse.psu.edu/). An alternative method of aligning genomes is to use the MUMmer suite of programs (29). Alignments

generated from the MUMmer programs can be visualized as dot plots from the TIGR Web pages (http://www.tigr.org/tigr-scripts/CMR2/webmum/mumplot), which can be useful for visualizing large-scale rearrangements. TIGR has also produced the Display-MUMs software (http://www.tigr.org/ software/displaymums), a Java-based application primarily used for displaying alignments of unfinished genomes to a completed reference genome. MUMmer is also used to generate the alignments for *coli*BASE, a comparative genomics database (22, 23). Examination of the distribution and genome context of a particular uncharacterized gene can reveal a lot about its function, and *coli*BASE is suitable for this kind of analysis. It also has many other useful features for examining and comparing genomes, including a whole-genome viewer and pattern-searching features.

Comparative Genome Hybridization

Although genome-sequencing efforts are ongoing, the currently available genomes are still some way short of fully representing the diversity present within *E. coli*. However, it is possible to perform comparative genome analyses by alternative means, albeit at a lower resolution. Microarray technology has been widely used for this purpose. Genomic DNA from a variety of test strains can be hybridized to an array of probes designed on the basis of the genome sequence of, for example, the K-12 MG1655 strain. Probes based on genes that are present in MG1655 but absent from the test strain will fail to hybridize the genomic DNA, thus allowing evaluation of the differences between strains. Note that this approach only identifies genes absent in the test genome; it will not provide any information about island genes that are absent from the reference genome. This approach has been applied in a number of studies (4, 33, 41, 84) by using a variety of pathogenic, laboratory, and ECOR strains hybridized to arrays of K-12 genes, in one case supplemented by a number of known virulence-associated genes (33). The *E. coli* backbone "core genome" was estimated to be

around 3,000 genes in these studies. An alternative approach involves the use of long-range PCR to scan along test genomes by using primers designed on the basis of a sequenced strain. Insertions and deletions can be inferred from variations in PCR product size or the failure to obtain a product. This has an advantage over the microarray strategy because it enables the identification of rearrangements and insertions as well as deletions. Such a strategy was applied to eight EHEC O157 strains using primers based on the Sakai genome sequence (87); it identified a high degree of genome diversity, largely attributable to variation within prophages.

THE EVOLUTION OF THE ANNOTATION OF *E. COLI* K-12

Since the completion of the K-12 MG1655 genome (14), many advances have been made in our understanding of *E. coli* biology from hundreds of research groups around the globe. The published version m49 of the annotation (GenBank accession no. U00096) has therefore become increasingly outdated. New genes have been identified that were not in the original annotation, and conversely, some regions that were originally predicted to be coding sequences are now not thought to be real. Also, a number of genes annotated with unknown functions have subsequently been characterized (112, 117). Annotation is not a static process, and even while the genome sequence was being released, efforts were being made to create bioinformatic tools to allow analysis and visualization of the genome sequence and its products. A summary of selected online resources for *E. coli* biology is shown in Table 2.

The ECDC database of Manfred Kröger was one of the first to collate and present information about the *E. coli* genome sequence and its products, and it was the first database to produce a useful, clickable map of the genome in 1995 (61, 121). This was followed shortly by the Colibri database, which was an *E. coli* implementation of a database initially constructed to view and manipulate the *Bacillus subtilis* genome (80). This database provided improved

databases. Relative to the U00096.2 release, there were 682 alterations to the start codon assignments of previously identified genes (based on experimental evidence), 31 old genes were eliminated, and 66 new genes were identified (including 17 pseudogenes and a single RNA). Other modifications included 2 gene fissions, 23 gene fusions, and a single "inversion" (where it was recognized that the previously annotated gene *phnQ* was incorrect, and the real coding sequence at that locus was on the opposite strand and designated *yjdP*). In addition, all gene assignments were reassessed on the basis of the published literature, annotations in existing *E. coli* databases, and homology searches of the most recent sequence data. The published snapshot therefore represented a state-of-the-art *E. coli* K-12 annotation as of March 2005, but by its very nature, it was immediately outdated.

The aim of the working group of *E. coli* bioinformaticians is to continue to update this annotation. As yet, no follow-up is planned, although funding by the National Institutes of Health for a dedicated *E. coli* K-12 community online information resource has recently been obtained by Barry Wanner at Purdue University. The update process may also be facilitated using the ASAP (a systematic annotation package) database (42), created by the group of Nicole Perna at the University of Wisconsin. ASAP contains a generic database schema for community-based annotation of genomes, and hence it has been chosen as the current repository for the m56 version and further versions of the MG1655 genome annotation. Any user can register with ASAP and suggest changes to any feature relating to the genome and its products. Guy Plunkett III, from the group of Fred Blattner, has a key role in curating these changes to the updated MG1655 genome sequence present in ASAP. The other *E. coli* genomes are also included in ASAP, but it is as yet unclear whether the annotation updates for K-12 will filter through to these other genomes. This highlights the use of genome comparisons for validating the existing annotation.

TOOLS FOR INTEGRATED BIOLOGY OF *E. COLI*

Analysis of the completed genome sequence has raised as many questions about the biology of *E. coli* as it solved. The K-12 genome clearly encodes around 4,200 different protein-encoding genes, but when the genome was realized, only about 2,000 of these had known functions. Although elucidation of the functions of 2,000 genes from one organism is still a remarkable feat achieved by *E. coli* biologists over many decades of research, the fact remains that around 50% of the genome had not been studied before 1997. The community has been quick to exploit this information, and around 200 uncharacterized genes have now been characterized at a functional level (112, 117). However, this still leaves many gene products without any known function, and one objective of using high-throughput global gene expression, proteomic, serial mutagenesis, and bioinformatic methods is to be able to collate information for these uncharacterized genes to enable the formation of testable hypotheses for their function. The ASAP database is one of the first to integrate data from high-throughput postgenomic experiments. It contains data from a large number of microarray and phenotype array studies. However, there are other attempts to integrate new experimental data onto the genome annotation such as the *Echo*BASE database, based out of the United Kingdom (78). This relational database is based on the m56 version of the genome and manually curates experiments selected from the published literature that provide evidence of the functions of previously uncharacterized gene products. This includes the full range of "omics" experiments, which produce data for large numbers of genes as well as information from smaller studies on particular gene products. Each experiment is linked to the genes for which it provides data, and in this way, by querying for a particular uncharacterized gene, the user can access curated experimental data that may provide a clue as to its function.

Other databases are evolving into highly integrative resources where the genome sequence and its genes are the simplest level of features. The most advanced and well-curated database relating to metabolism and gene regulation in *E. coli* K-12 is the EcoCyc database, pioneered by Peter Karp and colleagues (58). This database initially focused on gathering and representing information about metabolic pathways in *E. coli*, which are curated at a high level. The expansion of the contributors to include Ian Paulsen and Milton Saier has resulted in an excellent annotation of the membrane transport systems of this bacterium, and importing data from Julio Collado-Vides's RegulonDB database has added a vast collection of information on transcriptional regulation in *E. coli* (108). The EcoCyc template has been adapted to serve many other bacterial genomes, including *E. coli* O157:H7 EDL933 (EcoO157Cyc) and *Shigella flexneri* 2a strain 2457T (ShigellaCyc), but, again, these resources are not subject to the same degree of curation as the K-12–specific EcoCyc.

E. COLI IN THE POSTGENOMIC ERA

The year 2003 saw the launch of the International *E. coli* Alliance, an international group of major consortia aiming to coordinate efforts to model *E. coli* K-12 in silico. This highly ambitious project is being pursued by a number of active initiatives from Japan and Canada, and initiatives are planned in the United States and Europe. The Japanese effort, coordinated by Masura Tomita at Keio University, is producing data from bioinformatic, proteomic, and metabolomic studies for modeling efforts, as well as developing the software to hold and model this information. A similar process is being pursued by the Canadian CyberCell project; this consortium has a particular strength in quantitative proteomics. These data are collated in their CCDB database (108), and the proteomic data will be released to the community using this resource in the future.

To enable rapid functional genomic studies in *E. coli*, a number of resources have been constructed by the group of Hirotada Mori at the Nara Institute in Japan (reviewed in reference 79). These can be accessed through the Genobase database (http://ecoli.aist-nara.ac.jp). Most importantly, this group has created a set of knockout strains (the Keio collection) with single-gene deletions in 3,985 genes (6). These can be requested from the Mori group through Genobase and are an invaluable resource for the *E. coli* community. This group has also recently completed their analysis of the localization of a library of green fluorescent protein fusions to *E. coli* proteins and has released data from a large study of protein-protein interactions found in the *E. coli* proteome.

The continued annotation work by the bioinformatics group will also enhance the annotation for use in systems biology applications and will aim to contain much more information than the simple gene product and function lines currently found in the genome sequence file. For example, a complete mapping of terms for the gene ontology onto *E. coli* gene products is an important task that must be completed so that the genome can be accessed by many new bioinformatic tools that use this system.

CONCLUSIONS

E. coli, together with the now indisputably derivative and polyphyletic *Shigella* spp., represent an extraordinarily diverse group of pathogenic and commensal organisms. The recent and ongoing sequencing of multiple isolates continues to produce unexpected new findings, such as the second flagellar system present in *E. coli* 042 (102) or the novel type II secretion systems found in *S. dysenteriae* 1 strain 197 (129) and uropathogenic *E. coli* UT189 (24), indicating that genome sequencing of *E. coli* strains should continue for the foreseeable future; given the increasingly routine nature of the process, it is probably only a matter of time before a project to sequence the ECOR or DEC collections is feasible. A recent study that compared eight *Streptococcus agalactiae* genomes estimated that the pangenome (total gene reservoir of all strains) was huge and unlikely to be exhaustively characterized even if hundreds of complete genome sequences were obtained (116). Application of these methods to the published *E. coli* genomes (24) suggests a similar story, with each new genome predicted to add 441 genes to the pangenome. This figure is considerably larger than the ~30 genes expected to be added to the *Streptococcus* pangenome for each new genome, highlighting the remarkable diversity of the *E. coli* lineage.

The processes of *E. coli* genome evolution are clear from the genome sequences already available (and were even hinted at solely from analysis of the initial K-12 genome). A conserved core genome that is largely colinear (although subject to a number of IS-mediated rearrangements in *Shigella* spp.) is punctuated by a number of strain-specific "islands" that can represent either the acquisition or deletion of a region in one or multiple lineages. The current estimate for the size of the core genome is around 3 Mb and represents approximately two-thirds of the K-12 MG1655 chromosome (129). It is clear that the *E. coli* pathovars are not discrete units; they have generally arisen on multiple occasions as a result of convergent evolutionary processes, such as the horizontal acquisition of virulence plasmids or chromosomal pathogenicity determinants. However, they do not merely represent the K-12 genome with additional virulence factors. The genomes are in a constant state of flux, and many strain-specific islands will simply reflect this process and have no relationship to our anthropocentric definition of "virulence."

Much of the gene acquisition process appears to be mediated by bacteriophages (86), and deletion events are frequently facilitated by recombination between multiple copies of IS elements. Such deletions can, particularly in the case of *Shigella*, be selected for to remove from the proteome a product that interferes with a particular lifestyle (55, 123, 129). Alternatively, deletions may simply represent "mutational attrition," a cleaning up of regions that are no longer functional and merely represent evolutionary baggage. In the case of the ETT2 region (53, 103), a nonfunctional island can be seen at multiple stages of deletion from the case of O157:H7, where the island is intact save for a number of frameshifts in genes essential for function, through the increasingly large partial deletions seen in *E. coli* O78 strain 789, O111 strain B171, K-12, and *Shigella sonnei*, to the almost total absence seen in *Shigella flexneri*.

In many ways, the development of the *E. coli* research community has suffered from its ubiquity as a model system. To quote Fred Neidhardt, "All cell biologists have at least two cells of interest, the one they are studying and *E. coli*." Often *E. coli* researchers would not identify themselves as such, instead considering their field to be the particular system they were studying, be it recombination, replication, or transcription. Only in recent years, with the organization of the International *E. coli* Alliance and the development of community resources such as reannotation (105) and the Keio collection (6), has this situation started to be remedied and an *E. coli* community to rival those of comparable organisms such as yeast begun to develop. The focus remains on *E. coli* K-12, in its traditional role as a model for all cellular life (to quote Jacques Monod, "Tout ce qui est vrai pour le Colibacille est vrai pour l'éléphant" [all that is true for *E. coli* is true for the elephant]). This is no problem when considering individual systems, but it may be short-sighted when it comes to experiments on the level of the complete genome or proteome. The functions of many genes, such as the remnants of ETT2, may become apparent only when placed in their evolutionary context. Comparative genomic approaches are invaluable in this regard, and K-12 researchers should pay attention to the situation in other *E. coli* strains. The reverse is also true because the experimental characterization of gene products is far more commonplace in K-12 than the other *E. coli* and the annotation is more likely to be kept updated.

REFERENCES

1. **Abbott, J. C., D. M. Aanensen, K. Rutherford, S. Butcher, and B. G. Spratt.** 2005. WebACT—an online companion for the Artemis Comparison Tool. *Bioinformatics* **21:**3665–3666.

2. **Aiba, H., T. Baba, K. Hayashi, T. Inada, K. Isono, T. Itoh, H. Kasai, K. Kashimoto, S. Kimura, M. Kitakawa, M. Kitagawa, K. Makino, T. Miki, K. Mizobuchi, H. Mori, T. Mori, K. Motomera, S. Nakade, Y. Nakamura, H. Nashimoto, Y. Nishio, T. Oshima, N. Saito, G. Sampei, and T. Horiuchi.** 1996. A 570-kb DNA sequence of the *Escherichia coli* K-12 genome corresponding to the 28.0–40.1 min region on the linkage map (supplement). *DNA Res.* **3:**435–440.

3. **Altschul, S. F., T. L. Madden, A. A. Schaffer, J. Zhang, Z. Zhang, W. Miller, and D. J. Lipman.** 1997. Gapped BLAST and PSI-BLAST: a new generation of protein database search programs. *Nucleic Acids Res.* **25:**3389–3402.

4. **Anjum, M. F., S. Lucchini, A. Thompson, J. C. Hinton, and M. J. Woodward.** 2003. Comparative genomic indexing reveals the phylogenomics of *Escherichia coli* pathogens. *Infect. Immun.* **71:**4674–4683.

5. **Arnold, C., L. Metherell, G. Willshaw, A. Maggs, and J. Stanley.** 1999. Predictive fluorescent amplified-fragment length polymorphism analysis of *Escherichia coli*: high-resolution typing method with phylogenetic significance. *J. Clin. Microbiol.* **37:**1274–1279.

6. **Baba, T., T. Ara, M. Hasegawa, Y. Takai, Y. Okumura, M. Baba, K. A. Datsenko, M. Tomita, B. L. Wanner, and H. Mori.** 2006. Construction of *Escherichia coli* K-12 in-frame, single-gene knockout mutants: the Keio collection. *Mol. Syst. Biol.* **2:**2006.0008.

7. **Bachmann, B. J.** 1996. Derivations and genotypes of some mutant derivatives of *Escherichia coli* K-12, p. 2460–2488. *In* F. C. Neidhardt (ed.), Escherichia coli *and* Salmonella: *Cellular and Molecular Biology*, 2nd ed., vol. 2. ASM Press, Washington, DC.

8. **Beinke, C., S. Laarmann, C. Wachter, H. Karch, L. Greune, and M. A. Schmidt.** 1998. Diffusely adhering *Escherichia coli* strains induce attaching and effacing phenotypes and secrete homologs of Esp proteins. *Infect. Immun.* **66:**528–539.

9. **Bergthorsson, U., and H. Ochman.** 1998. Distribution of chromosome length variation in natural isolates of *Escherichia coli*. *Mol. Biol. Evol.* **15:**6–16.

10. **Bergthorsson, U., and H. Ochman.** 1995. Heterogeneity of genome sizes among natural isolates of *Escherichia coli*. *J. Bacteriol.* **177:**5784–5789.

11. **Bingen, E., B. Picard, N. Brahimi, S. Mathy, P. Desjardins, J. Elion, and E. Denamur.** 1998. Phylogenetic analysis of *Escherichia coli* strains causing neonatal meningitis suggests horizontal gene transfer from a predominant pool of highly virulent B2 group strains. *J. Infect. Dis.* **177:**642–650.

12. **Blattner, F. R.** 1983. Biological frontiers. *Science* **222:**719 720.

13. **Blattner, F. R., V. Burland, G. Plunkett III, H. J. Sofia, and D. L. Daniels.** 1993. Analysis of the *Escherichia coli* genome. IV. DNA sequence of the region from 89.2 to 92.8 minutes. *Nucleic Acids Res.* **21:**5408–5417.

14. **Blattner, F. R., G. Plunkett III, C. A. Bloch, N. T. Perna, V. Burland, M. Riley, J. Collado-Vides, J. D. Glasner, C. K. Rode, G. F. Mayhew, J. Gregor, N. W. Davis, H. A. Kirkpatrick, M. A. Goeden, D. J. Rose, B. Mau, and Y. Shao.** 1997. The complete genome sequence of *Escherichia coli* K-12. *Science* **277:**1453–1474.

15. **Boyd, E. F., and D. L. Hartl.** 1998. Chromosomal regions specific to pathogenic isolates of *Escherichia coli* have a phylogenetically clustered distribution. *J. Bacteriol.* **180:**1159–1165.

16. **Brenner, D. J., G. R. Fanning, F. J. Skerman, and S. Falkow.** 1972. Polynucleotide sequence divergence among strains of *Escherichia coli* and closely related organisms. *J. Bacteriol.* **109:**953–965.

17. **Burland, V., G. Plunkett III, D. L. Daniels, and F. R. Blattner.** 1993. DNA sequence and analysis of 136 kilobases of the *Escherichia coli* genome: organizational symmetry around the origin of replication. *Genomics* **16:**551–561.

18. **Burland, V., G. Plunkett III, H. J. Sofia, D. L. Daniels, and F. R. Blattner.** 1995. Analysis of the *Escherichia coli* genome VI: DNA sequence of the region from 92.8 through 100 minutes. *Nucleic Acids Res.* **23:**2105–2119.

19. **Burland, V., Y. Shao, N. T. Perna, G. Plunkett, H. J. Sofia, and F. R. Blattner.** 1998. The complete DNA sequence and analysis of the large virulence plasmid of *Escherichia coli* O157:H7. *Nucleic Acids Res.* **26:**4196–4204.

20. **Carver, T. J., K. M. Rutherford, M. Berriman, M. A. Rajandream, B. G. Barrell, and J. Parkhill.** 2005. ACT: the Artemis comparison tool. *Bioinformatics* **21:**3422–3423.

21. **Casalino, M., M. C. Latella, G. Prosseda, and B. Colonna.** 2003. CadC is the preferential target of a convergent evolution driving enteroinvasive *Escherichia coli* toward a lysine decarboxylase-defective phenotype. *Infect. Immun.* **71:**5472–5479.

22. **Chaudhuri, R. R., A. M. Khan, and M. J. Pallen.** 2004. *coli*BASE: an online database for *Escherichia coli*, *Shigella* and *Salmonella* comparative genomics. *Nucleic Acids Res.* **32** (Database issue)**:**D296–D299.

23. **Chaudhuri, R. R., and M. J. Pallen.** 2006. *x*BASE, a collection of online databases for bacterial comparative genomics. *Nucleic Acids Res.* **34:** D335–D337.

24. **Chen, S. L., C. S. Hung, J. Xu, C. S. Reigstad, V. Magrini, A. Sabo, D. Blasiar, T. Bieri, R. R. Meyer, P. Ozersky, J. R. Armstrong, R. S. Fulton, J. P. Latreille, J. Spieth, T. M. Hooton, E. R. Mardis, S. J. Hultgren, and J. I. Gordon.** 2006. Identification of genes subject to positive selection in uropathogenic strains of *Escherichia coli*: a comparative genomics approach. *Proc. Natl. Acad. Sci. USA* **103**:5977–5982.

25. **Clermont, O., S. Bonacorsi, and E. Bingen.** 2001. The *Yersinia* high-pathogenicity island is highly predominant in virulence-associated phylogenetic groups of *Escherichia coli*. *FEMS Microbiol. Lett.* **196**:153–157.

26. **Czeczulin, J. R., T. S. Whittam, I. R. Henderson, F. Navarro-Garcia, and J. P. Nataro.** 1999. Phylogenetic analysis of enteroaggregative and diffusely adherent *Escherichia coli*. *Infect. Immun.* **67**:2692–2699.

27. **Daniels, D. L., G. Plunkett III, V. Burland, and F. R. Blattner.** 1992. Analysis of the *Escherichia coli* genome: DNA sequence of the region from 84.5 to 86.5 minutes. *Science* **257**:771–778.

28. **Day, W. A., Jr., R. E. Fernandez, and A. T. Maurelli.** 2001. Pathoadaptive mutations that enhance virulence: genetic organization of the *cadA* regions of *Shigella* spp. *Infect. Immun.* **69**:7471–7480.

29. **Delcher, A. L., A. Phillippy, J. Carlton, and S. L. Salzberg.** 2002. Fast algorithms for large-scale genome alignment and comparison. *Nucleic Acids Res.* **30**:2478–2483.

30. **Dho-Moulin, M., and J. M. Fairbrother.** 1999. Avian pathogenic *Escherichia coli* (APEC). *Vet. Res.* **30**:299–316.

31. **Diamant, E., Y. Palti, R. Gur-Arie, H. Cohen, E. M. Hallerman, and Y. Kashi.** 2004. Phylogeny and strain typing of *Escherichia coli*, inferred from variation at mononucleotide repeat loci. *Appl. Environ. Microbiol.* **70**:2464–2473.

32. **Dixon, B.** 1998. *E. coli*'s double life. *ASM News* **64**:616–617.

33. **Dobrindt, U., F. Agerer, K. Michaelis, A. Janka, C. Buchrieser, M. Samuelson, C. Svanborg, G. Gottschalk, H. Karch, and J. Hacker.** 2003. Analysis of genome plasticity in pathogenic and commensal *Escherichia coli* isolates by use of DNA arrays. *J. Bacteriol.* **185**:1831–1840.

34. **Duriez, P., O. Clermont, S. Bonacorsi, E. Bingen, A. Chaventre, J. Elion, B. Picard, and E. Denamur.** 2001. Commensal *Escherichia coli* isolates are phylogenetically distributed among geographically distinct human populations. *Microbiology* **147**:1671–1676.

35. **Escherich, T.** 1885. Die Darmbakterien des Neugeborenen und Säuglings. *Fortsch. Med.* **3**: 515–522, 547–554.

36. **Escherich, T.** 1989. The intestinal bacteria of the neonate and breast-fed infant. 1885. *Rev. Infect. Dis.* **11**:352–356.

37. **Escobar-Paramo, P., O. Clermont, A. B. Blanc-Potard, H. Bui, C. Le Bouguenec, and E. Denamur.** 2004. A specific genetic background is required for acquisition and expression of virulence factors in *Escherichia coli*. *Mol. Biol. Evol.* **21**:1085–1094.

38. **Escobar-Paramo, P., C. Giudicelli, C. Parsot, and E. Denamur.** 2003. The evolutionary history of *Shigella* and enteroinvasive *Escherichia coli* revised. *J. Mol. Evol.* **57**:140–148.

39. **Florea, L., M. McClelland, C. Riemer, S. Schwartz, and W. Miller.** 2003. EnteriX 2003: visualization tools for genome alignments of Enterobacteriaceae. *Nucleic Acids Res.* **31**:3527–3532.

40. **Fujita, N., H. Mori, T. Yura, and A. Ishihama.** 1994. Systematic sequencing of the *Escherichia coli* genome: analysis of the 2.4–4.1 min (110,917–193,643 bp) region. *Nucleic Acids Res.* **22**:1637–1639.

41. **Fukiya, S., H. Mizoguchi, T. Tobe, and H. Mori.** 2004. Extensive genomic diversity in pathogenic *Escherichia coli* and *Shigella* strains revealed by comparative genomic hybridization microarray. *J. Bacteriol.* **186**:3911–3921.

42. **Glasner, J. D., P. Liss, G. Plunkett III, A. Darling, T. Prasad, M. Rusch, A. Byrnes, M. Gilson, B. Biehl, F. R. Blattner, and N. T. Perna.** 2003. ASAP, a systematic annotation package for community analysis of genomes. *Nucleic Acids Res.* **31**:147–151.

43. **Hall, B. G., L. L. Parker, P. W. Betts, R. F. DuBose, S. A. Sawyer, and D. L. Hartl.** 1989. IS103, a new insertion element in *Escherichia coli*: characterization and distribution in natural populations. *Genetics* **121**:423–431.

44. **Hall, B. G., and P. M. Sharp.** 1992. Molecular population genetics of *Escherichia coli*: DNA sequence diversity at the *celC*, *crr*, and *gutB* loci of natural isolates. *Mol. Biol. Evol.* **9**:654–665.

45. **Hartleib, S., R. Prager, I. Hedenstrom, S. Lofdahl, and H. Tschape.** 2003. Prevalence of the new, SPI1-like, pathogenicity island ETT2 among *Escherichia coli*. *Int. J. Med. Microbiol.* **292**:487–493.

46. **Hayashi, K., N. Morooka, Y. Yamamoto, K. Fujita, K. Isono, S. Choi, E. Ohtsubo, T. Baba, B. L. Wanner, H. Mori, and T. Horiuchi.** 2006. Highly accurate genome sequences of *Escherichia coli* K-12 strains MG1655 and W3110. *Mol. Sys. Biol.* **2**:E1–E5.

47. **Hayashi, T., K. Makino, M. Ohnishi, K. Kurokawa, K. Ishii, K. Yokoyama, C. G. Han, E. Ohtsubo, K. Nakayama, T. Murata, M. Tanaka, T. Tobe, T. Iida, H. Takami,**

T. Honda, C. Sasakawa, N. Ogasawara, T. Yasunaga, S. Kuhara, T. Shiba, M. Hattori, and H. Shinagawa. 2001. Complete genome sequence of enterohemorrhagic *Escherichia coli* O157:H7 and genomic comparison with a laboratory strain K-12. *DNA Res.* **8**:11–22.

48. Henderson, I. R., J. Czeczulin, C. Eslava, F. Noriega, and J. P. Nataro. 1999. Characterization of Pic, a secreted protease of *Shigella flexneri* and enteroaggregative *Escherichia coli. Infect. Immun.* **67**:5587–5596.

49. Herzer, P. J., S. Inouye, M. Inouye, and T. S. Whittam. 1990. Phylogenetic distribution of branched RNA-linked multicopy single-stranded DNA among natural isolates of *Escherichia coli. J. Bacteriol.* **172**:6175–6181.

50. Hill, C. W., G. Feulner, M. S. Brody, S. Zhao, A. B. Sadosky, and C. H. Sandt. 1995. Correlation of Rhs elements with *Escherichia coli* population structure. *Genetics* **141**:15–24.

51. Hill, C. W., and B. W. Harnish. 1981. Inversions between ribosomal RNA genes of *Escherichia coli. Proc. Natl. Acad. Sci. USA* **78**:7069–7072.

52. Hinton, J. C. 1997. The *Escherichia coli* genome sequence: the end of an era or the start of the FUN? *Mol. Microbiol.* **26**:417–422.

53. Ideses, D., U. Gophna, Y. Paitan, R. R. Chaudhuri, M. J. Pallen, and E. Z. Ron. 2005. A degenerate type III secretion system from septicemic *Escherichia coli* contributes to pathogenesis. *J. Bacteriol.* **187**:8164–8171.

54. Itoh, T., H. Aiba, T. Baba, K. Hayashi, T. Inada, K. Isono, H. Kasai, S. Kimura, M. Kitakawa, M. Kitagawa, K. Makino, T. Miki, K. Mizobuchi, H. Mori, T. Mori, K. Motomura, S. Nakade, Y. Nakamura, H. Nashimoto, Y. Nishio, T. Oshima, N. Saito, G. Sampei, Y. Seki, and T. Horiuchi. 1996. A 460-kb DNA sequence of the *Escherichia coli* K-12 genome corresponding to the 40.1–50.0 min region on the linkage map (supplement). *DNA Res.* **3**:441–445.

55. Jin, Q., Z. Yuan, J. Xu, Y. Wang, Y. Shen, W. Lu, J. Wang, H. Liu, J. Yang, F. Yang, X. Zhang, J. Zhang, G. Yang, H. Wu, D. Qu, J. Dong, L. Sun, Y. Xue, A. Zhao, Y. Gao, J. Zhu, B. Kan, K. Ding, S. Chen, H. Cheng, Z. Yao, B. He, R. Chen, D. Ma, B. Qiang, Y. Wen, Y. Hou, and J. Yu. 2002. Genome sequence of *Shigella flexneri* 2a: insights into pathogenicity through comparison with genomes of *Escherichia coli* K12 and O157. *Nucleic Acids Res.* **30**:4432–4441.

56. Johnson, J. R., and M. Kuskowski. 2000. Clonal origin, virulence factors, and virulence. *Infect. Immun.* **68**:424–425.

57. Kaper, J. B., J. P. Nataro, and H. L. Mobley. 2004. Pathogenic *Escherichia coli. Nat. Rev. Microbiol.* **2**:123–140.

58. Karp, P. D., M. Riley, M. Saier, I. T. Paulsen, J. Collado-Vides, S. M. Paley, A. Pellegrini-Toole, C. Bonavides, and S. Gama-Castro. 2002. The EcoCyc Database. *Nucleic Acids Res.* **30**:56–58.

59. Koski, L. B., and G. B. Golding. 2001. The closest BLAST hit is often not the nearest neighbor. *J. Mol. Evol.* **52**:540–542.

60. Koski, L. B., R. A. Morton, and G. B. Golding. 2001. Codon bias and base composition are poor indicators of horizontally transferred genes. *Mol. Biol. Evol.* **18**:404–412.

61. Kroger, M., and R. Wahl. 1998. Compilation of DNA sequences of *Escherichia coli* K12: description of the interactive databases ECD and ECDC. *Nucleic Acids Res.* **26**:46–49.

62. Kyaw, C. M., C. R. De Araujo, M. R. Lima, E. G. Gondim, M. M. Brigido, and L. G. Giugliano. 2003. Evidence for the presence of a type III secretion system in diffusely adhering *Escherichia coli* (DAEC). *Infect. Genet. Evol.* **3**:111–117.

63. Labat, F., O. Pradillon, L. Garry, M. Peuchmaur, B. Fantin, and E. Denamur. 2005. Mutator phenotype confers advantage in *Escherichia coli* chronic urinary tract infection pathogenesis. *FEMS Immunol. Med. Microbiol.* **44**:317–321.

64. Lan, R., M. C. Alles, K. Donohoe, M. B. Martinez, and P. R. Reeves. 2004. Molecular evolutionary relationships of enteroinvasive *Escherichia coli* and *Shigella* spp. *Infect. Immun.* **72**:5080–5088.

65. Lan, R., and P. R. Reeves. 2002. *Escherichia coli* in disguise: molecular origins of *Shigella. Microbes Infect.* **4**:1125–1132.

66. Lawrence, J. G., and H. Ochman. 1997. Amelioration of bacterial genomes: rates of change and exchange. *J. Mol. Evol.* **44**:383–397.

67. Lawrence, J. G., and H. Ochman. 1998. Molecular archaeology of the *Escherichia coli* genome. *Proc. Natl. Acad. Sci. USA* **95**:9413–9417.

68. Lawrence, J. G., and H. Ochman. 2002. Reconciling the many faces of lateral gene transfer. *Trends Microbiol.* **10**:1–4.

69. Lawrence, J. G., H. Ochman, and D. L. Hartl. 1991. Molecular and evolutionary relationships among enteric bacteria. *J. Gen. Microbiol.* **137(Pt. 8)**:1911–1921.

70. Lecointre, G., L. Rachdi, P. Darlu, and E. Denamur. 1998. *Escherichia coli* molecular phylogeny using the incongruence length difference test. *Mol. Biol. Evol.* **15**:1685–1695.

71. Lederberg, J. 2004. *E. coli* K-12. *Microbiology Today* **31**:116.

72. Makino, K., K. Ishii, T. Yasunaga, M. Hattori, K. Yokoyama, C. H. Yutsudo, Y. Kubota, Y. Yamaichi, T. Iida, K. Yamamoto, T. Honda,

C. G. Han, E. Ohtsubo, M. Kasamatsu, T. Hayashi, S. Kuhara, and H. Shinagawa. 1998. Complete nucleotide sequences of 93-kb and 3.3-kb plasmids of an enterohemorrhagic *Escherichia coli* O157:H7 derived from Sakai outbreak. *DNA Res.* **5:**1–9.

73. Makino, S., T. Tobe, H. Asakura, M. Watarai, T. Ikeda, K. Takeshi, and C. Sasakawa. 2003. Distribution of the secondary type III secretion system locus found in enterohemorrhagic *Escherichia coli* O157:H7 isolates among Shiga toxin–producing *E. coli* strains. *J. Clin. Microbiol.* **41:** 2341–2347.

74. Maurelli, A. T., R. E. Fernandez, C. A. Bloch, C. K. Rode, and A. Fasano. 1998. "Black holes" and bacterial pathogenicity: a large genomic deletion that enhances the virulence of *Shigella* spp. and enteroinvasive *Escherichia coli*. *Proc. Natl. Acad. Sci. USA* **95:**3943–3948.

75. McDaniel, T. K., K. G. Jarvis, M. S. Donnenberg, and J. B. Kaper. 1995. A genetic locus of enterocyte effacement conserved among diverse enterobacterial pathogens. *Proc. Natl. Acad. Sci. USA* **92:**1664–1668.

76. Menard, R., P. Sansonetti, and C. Parsot. 1994. The secretion of the *Shigella flexneri* Ipa invasins is activated by epithelial cells and controlled by IpaB and IpaD. *EMBO J.* **13:**5293–5302.

77. Michino, H., K. Araki, S. Minami, S. Takaya, N. Sakai, M. Miyazaki, A. Ono, and H. Yanagawa. 1999. Massive outbreak of *Escherichia coli* O157:H7 infection in schoolchildren in Sakai City, Japan, associated with consumption of white radish sprouts. *Am. J. Epidemiol.* **150:**787–796.

78. Misra, R. V., R. S. Horler, W. Reindl, I. I. Goryanin, and G. H. Thomas. 2005. EchoBASE: an integrated post-genomic database for *Escherichia coli*. *Nucleic Acids Res.* **33:**D329–D333.

79. Mori, H. 2004. From the sequence to cell modeling: comprehensive functional genomics in *Escherichia coli*. *J. Biochem. Mol. Biol.* **37:**83–92.

80. Moszer, I., P. Glaser, and A. Danchin. 1995. SubtiList: a relational database for the *Bacillus subtilis* genome. *Microbiology* **141(Pt. 2):**261–268.

81. Nakata, N., T. Tobe, I. Fukuda, T. Suzuki, K. Komatsu, M. Yoshikawa, and C. Sasakawa. 1993. The absence of a surface protease, OmpT, determines the intercellular spreading ability of *Shigella*: the relationship between the *ompT* and *kcpA* loci. *Mol. Microbiol.* **9:**459–468.

82. Nataro, J. P., and J. B. Kaper. 1998. Diarrheagenic *Escherichia coli*. *Clin. Microbiol. Rev.* **11:**142–201.

83. Nataro, J. P., J. Seriwatana, A. Fasano, D. R. Maneval, L. D. Guers, F. Noriega, F. Dubovsky, M. M. Levine, and J. G. Morris, Jr. 1995. Identification and cloning of a novel plasmid-encoded enterotoxin of enteroinvasive *Escherichia coli* and *Shigella* strains. *Infect. Immun.* **63:**4721–4728.

84. Ochman, H., J. G. Lawrence, and E. A. Groisman. 2000. Lateral gene transfer and the nature of bacterial innovation. *Nature* **405:**299–304.

85. Ochman, H., and R. K. Selander. 1984. Standard reference strains of *Escherichia coli* from natural populations. *J. Bacteriol.* **157:**690–693.

86. Ohnishi, M., K. Kurokawa, and T. Hayashi. 2001. Diversification of *Escherichia coli* genomes: are bacteriophages the major contributors? *Trends Microbiol.* **9:**481–485.

87. Ohnishi, M., J. Terajima, K. Kurokawa, K. Nakayama, T. Murata, K. Tamura, Y. Ogura, H. Watanabe, and T. Hayashi. 2002. Genomic diversity of enterohemorrhagic *Escherichia coli* O157 revealed by whole genome PCR scanning. *Proc. Natl. Acad. Sci. USA* **99:**17043–17048.

88. Okeke, I. N., A. Lamikanra, J. Czeczulin, F. Dubovsky, J. B. Kaper, and J. P. Nataro. 2000. Heterogeneous virulence of enteroaggregative *Escherichia coli* strains isolated from children in southwest Nigeria. *J. Infect. Dis.* **181:**252–260.

89. Olsen, G. J., and C. R. Woese. 1993. Ribosomal RNA: a key to phylogeny. *FASEB J.* **7:**113–123.

90. Oshima, T., H. Aiba, T. Baba, K. Fujita, K. Hayashi, A. Honjo, K. Ikemoto, T. Inada, T. Itoh, M. Kajihara, K. Kanai, K. Kashimoto, S. Kimura, M. Kitagawa, K. Makino, S. Masuda, T. Miki, K. Mizobuchi, H. Mori, K. Motomura, Y. Nakamura, H. Nashimoto, Y. Nishio, N. Saito, and T. Horiuchi. 1996. A 718-kb DNA sequence of the *Escherichia coli* K-12 genome corresponding to the 12.7–28.0 min region on the linkage map (supplement). *DNA Res.* **3:**211–223.

91. Parham, N. J., U. Srinivasan, M. Desvaux, B. Foxman, C. F. Marrs, and I. R. Henderson. 2004. PicU, a second serine protease autotransporter of uropathogenic *Escherichia coli*. *FEMS Microbiol. Lett.* **230:**73–83.

92. Parkhill, J., G. Dougan, K. D. James, N. R. Thomson, D. Pickard, J. Wain, C. Churcher, K. L. Mungall, S. D. Bentley, M. T. Holden, M. Sebaihia, S. Baker, D. Basham, K. Brooks, T. Chillingworth, P. Connerton, A. Cronin, P. Davis, R. M. Davies, L. Dowd, N. White, J. Farrar, T. Feltwell, N. Hamlin, A. Haque, T. T. Hien, S. Holroyd, K. Jagels, A. Krogh, T. S. Larsen, S. Leather, S. Moule, P. O'Gaora, C. Parry, M. Quail, K. Rutherford, M. Simmonds, J. Skelton, K. Stevens, S. Whitehead, and B. G. Barrell. 2001. Complete genome sequence of a multiple drug resistant

Salmonella enterica serovar Typhi CT18. *Nature* **413:**848–852.

93. **Pennisi, E.** 1997. Laboratory workhorse decoded. *Science* **277:**1432–1434.

94. **Perna, N. T., G. Plunkett III, V. Burland, B. Mau, J. D. Glasner, D. J. Rose, G. F. Mayhew, P. S. Evans, J. Gregor, H. A. Kirkpatrick, G. Posfai, J. Hackett, S. Klink, A. Boutin, Y. Shao, L. Miller, E. J. Grotbeck, N. W. Davis, A. Lim, E. T. Dimalanta, K. D. Potamousis, J. Apodaca, T. S. Anantharaman, J. Lin, G. Yen, D. C. Schwartz, R. A. Welch, and F. R. Blattner.** 2001. Genome sequence of enterohaemorrhagic *Escherichia coli* O157:H7. *Nature* **409:**529–533.

95. **Picard, B., J. S. Garcia, S. Gouriou, P. Duriez, N. Brahimi, E. Bingen, J. Elion, and E. Denamur.** 1999. The link between phylogeny and virulence in *Escherichia coli* extraintestinal infection. *Infect. Immun.* **67:**546–553.

96. **Plunkett, G., III, V. Burland, D. L. Daniels, and F. R. Blattner.** 1993. Analysis of the *Escherichia coli* genome. III. DNA sequence of the region from 87.2 to 89.2 minutes. *Nucleic Acids Res.* **21:**3391–3398.

97. **Plunkett, G., III, D. J. Rose, T. J. Durfee, and F. R. Blattner.** 1999. Sequence of Shiga toxin 2 phage 933W from *Escherichia coli* O157:H7: Shiga toxin as a phage late-gene product. *J. Bacteriol.* **181:**1767–1778.

98. **Pruitt, K. D., T. Tatusova, and D. R. Maglott.** 2005. NCBI Reference Sequence (RefSeq): a curated non-redundant sequence database of genomes, transcripts and proteins. *Nucleic Acids Res.* **33:**D501–D504.

99. **Pupo, G. M., D. K. Karaolis, R. Lan, and P. R. Reeves.** 1997. Evolutionary relationships among pathogenic and nonpathogenic *Escherichia coli* strains inferred from multilocus enzyme electrophoresis and *mdh* sequence studies. *Infect. Immun.* **65:**2685–2692.

100. **Pupo, G. M., R. Lan, and P. R. Reeves.** 2000. Multiple independent origins of *Shigella* clones of *Escherichia coli* and convergent evolution of many of their characteristics. *Proc. Natl. Acad. Sci. USA* **97:**10567–10572.

101. **Reid, S. D., C. J. Herbelin, A. C. Bumbaugh, R. K. Selander, and T. S. Whittam.** 2000. Parallel evolution of virulence in pathogenic *Escherichia coli*. *Nature* **406:**64–67.

102. **Ren, C. P., S. A. Beatson, J. Parkhill, and M. J. Pallen.** 2005. The Flag-2 locus, an ancestral gene cluster, is potentially associated with a novel flagellar system from *Escherichia coli*. *J. Bacteriol.* **187:**1430–1440.

103. **Ren, C. P., R. R. Chaudhuri, A. Fivian, C. M. Bailey, M. Antonio, W. M. Barnes, and M. J. Pallen.** 2004. The ETT2 gene cluster, encoding a second type III secretion system from *Escherichia coli*, is present in the majority of strains but has undergone widespread mutational attrition. *J. Bacteriol.* **186:**3547–3560.

104. **Riley, L. W., R. S. Remis, S. D. Helgerson, H. B. McGee, J. G. Wells, B. R. Davis, R. J. Hebert, E. S. Olcott, L. M. Johnson, N. T. Hargrett, P. A. Blake, and M. L. Cohen.** 1983. Hemorrhagic colitis associated with a rare *Escherichia coli* serotype. *N. Engl. J. Med.* **308:**681–685.

105. **Riley, M., T. Abe, M. B. Arnaud, M. K. B. Berlyn, F. R. Blattner, R. R. Chaudhuri, J. D. Glasner, T. Horiuchi, I. M. Keseler, T. Kosuge, H. Mori, N. T. Perna, G. Plunkett III, K. E. Rudd, M. H. Serres, G. H. Thomas, N. R. Thomson, D. Wishart, and B. L. Wanner.** 2006. *Escherichia coli* K-12: a cooperatively developed annotation snapshot—2005. *Nucleic Acids Res.* **34:**1–9.

106. **Rudd, K. E.** 2000. EcoGene: a genome sequence database for *Escherichia coli* K-12. *Nucleic Acids Res.* **28:**60–64.

107. **Russo, T. A., and J. R. Johnson.** 2000. Proposal for a new inclusive designation for extraintestinal pathogenic isolates of *Escherichia coli*: ExPEC. *J. Infect. Dis.* **181:**1753–1754.

108. **Salgado, H., S. Gama-Castro, M. Peralta-Gil, E. Diaz-Peredo, F. Sanchez-Solano, A. Santos-Zavaleta, I. Martinez-Flores, V. Jimenez-Jacinto, C. Bonavides-Martinez, J. Segura-Salazar, A. Martinez-Antonio, and J. Collado-Vides.** 2006. RegulonDB (version 5.0): *Escherichia coli* K-12 transcriptional regulatory network, operon organization, and growth conditions. *Nucleic Acids Res.* **34:**D394–D397.

109. **Schneider, D., E. Duperchy, E. Coursange, R. E. Lenski, and M. Blot.** 2000. Long-term experimental evolution in *Escherichia coli*. IX. Characterization of insertion sequence–mediated mutations and rearrangements. *Genetics* **156:**477–488.

110. **Schwartz, S., W. J. Kent, A. Smit, Z. Zhang, R. Baertsch, R. C. Hardison, D. Haussler, and W. Miller.** 2003. Human-mouse alignments with BLASTZ. *Genome Res.* **13:**103–107.

111. **Schwartz, S., Z. Zhang, K. A. Frazer, A. Smit, C. Riemer, J. Bouck, R. Gibbs, R. Hardison, and W. Miller.** 2000. PipMaker—a web server for aligning two genomic DNA sequences. *Genome Res.* **10:**577–586.

112. **Serres, M. H., S. Gopal, L. A. Nahum, P. Liang, T. Gaasterland, and M. Riley.** 2001. A functional update of the *Escherichia coli* K-12 genome. *Genome Biol.* **2:**RESEARCH0035.

113. **Serres, M. H., S. Goswami, and M. Riley.** 2004. GenProtEC: an updated and improved

analysis of functions of *Escherichia coli* K-12 proteins. *Nucleic Acids Res.* **32:**D300–D302.

114. **Sofia, H. J., V. Burland, D. L. Daniels, G. Plunkett III, and F. R. Blattner.** 1994. Analysis of the *Escherichia coli* genome. V. DNA sequence of the region from 76.0 to 81.5 minutes. *Nucleic Acids Res.* **22:**2576–2586.

115. **Sun, J., F. Gunzer, A. M. Westendorf, J. Buer, M. Scharfe, M. Jarek, F. Gossling, H. Blocker, and A. P. Zeng.** 2005. Genomic peculiarity of coding sequences and metabolic potential of probiotic *Escherichia coli* strain Nissle 1917 inferred from raw genome data. *J. Biotechnol.* **117:**147–161.

116. **Tettelin, H., V. Masignani, M. J. Cieslewicz, C. Donati, D. Medini, N. L. Ward, S. V. Angiuoli, J. Crabtree, A. L. Jones, A. S. Durkin, R. T. Deboy, T. M. Davidsen, M. Mora, M. Scarselli, I. Margarit y Ros, J. D. Peterson, C. R. Hauser, J. P. Sundaram, W. C. Nelson, R. Madupu, L. M. Brinkac, R. J. Dodson, M. J. Rosovitz, S. A. Sullivan, S. C. Daugherty, D. H. Haft, J. Selengut, M. L. Gwinn, L. Zhou, N. Zafar, H. Khouri, D. Radune, G. Dimitrov, K. Watkins, K. J. O'Connor, S. Smith, T. R. Utterback, O. White, C. E. Rubens, G. Grandi, L. C. Madoff, D. L. Kasper, J. L. Telford, M. R. Wessels, R. Rappuoli, and C. M. Fraser.** 2005. Genome analysis of multiple pathogenic isolates of *Streptococcus agalactiae*: implications for the microbial "pan-genome." *Proc. Natl. Acad. Sci. USA* **102:**13950–13955.

117. **Thomas, G. H.** 1999. Completing the *E. coli* proteome: a database of gene products characterised since the completion of the genome sequence. *Bioinformatics* **15:**860–861.

118. **Thomas, G. H., and K. A. Bettelheim.** 1998. *Escherichia coli* on the WWW. *Lett. Appl. Microbiol.* **27:**122–123.

119. **Turner, S. M., R. R. Chaudhuri, Z. D. Jiang, H. Dupont, C. Gyles, C. W. Penn, M. J. Pallen, and I. R. Henderson.** 2006. Phylogenetic comparisons reveal multiple acquisitions of the toxin genes by enterotoxigenic *Escherichia coli* strains of different evolutionary lineages. *J. Clin. Microbiol.* **44:**4528–4536.

120. **Vandemaele, F., D. Vandekerchove, M. Vereecken, J. Derijcke, M. DhoMoulin, and B. M. Goddeeris.** 2003. Sequence analysis demonstrates the conservation of *fimH* and variability of *fimA* throughout avian pathogenic *Escherichia coli* (APEC). *Vet. Res.* **34:**153–163.

121. **Wahl, R., and M. Kroger.** 1995. ECDC—a totally integrated and interactively usable genetic map of *Escherichia coli* K12. *Microbiol. Res.* **150:**7–61.

122. **Wang, H. C., J. Badger, P. Kearney, and M. Li.** 2001. Analysis of codon usage patterns of bacterial genomes using the self-organizing map. *Mol. Biol. Evol.* **18:**792–800.

123. **Wei, J., M. B. Goldberg, V. Burland, M. M. Venkatesan, W. Deng, G. Fournier, G. F. Mayhew, G. Plunkett III, D. J. Rose, A. Darling, B. Mau, N. T. Perna, S. M. Payne, L. J. Runyen-Janecky, S. Zhou, D. C. Schwartz, and F. R. Blattner.** 2003. Complete genome sequence and comparative genomics of *Shigella flexneri* serotype 2a strain 2457T. *Infect. Immun.* **71:**2775–2786.

124. **Welch, R. A., V. Burland, G. Plunkett III, P. Redford, P. Roesch, D. Rasko, E. L. Buckles, S. R. Liou, A. Boutin, J. Hackett, D. Stroud, G. F. Mayhew, D. J. Rose, S. Zhou, D. C. Schwartz, N. T. Perna, H. L. Mobley, M. S. Donnenberg, and F. R. Blattner.** 2002. Extensive mosaic structure revealed by the complete genome sequence of uropathogenic *Escherichia coli*. *Proc. Natl. Acad. Sci. USA* **99:**17020–17024.

125. **Whittam, T. S., H. Ochman, and R. K. Selander.** 1983. Multilocus genetic structure in natural populations of *Escherichia coli*. *Proc. Natl. Acad. Sci. USA* **80:**1751–1755.

126. **Whittam, T. S., M. L. Wolfe, I. K. Wachsmuth, F. Orskov, I. Orskov, and R. A. Wilson.** 1993. Clonal relationships among *Escherichia coli* strains that cause hemorrhagic colitis and infantile diarrhea. *Infect. Immun.* **61:**1619–1629.

127. **Wirth, T., D. Falush, R. Lan, F. Colles, P. Mensa, L. H. Wieler, H. Karch, P. R. Reeves, M. C. J. Maiden, H. Ochman, and M. Achtman.** 2006. Sex and virulence in *Escherichia coli*: an evolutionary perspective. *Mol. Microbiol.* **60:**1136–1151.

128. **Yamamoto, Y., H. Aiba, T. Baba, K. Hayashi, T. Inada, K. Isono, T. Itoh, S. Kimura, M. Kitagawa, K. Makino, T. Miki, N. Mitsuhashi, K. Mizobuchi, H. Mori, S. Nakade, Y. Nakamura, H. Nashimoto, T. Oshima, S. Oyama, N. Saito, G. Sampei, Y. Satoh, S. Sivasundaram, H. Tagami, and T. Horiuchi.** 1997. Construction of a contiguous 874-kb sequence of the *Escherichia coli*—K12 genome corresponding to 50.0-68.8 min on the linkage map and analysis of its sequence features. *DNA Res.* **4:**91–113.

129. **Yang, F., J. Yang, X. Zhang, L. Chen, Y. Jiang, Y. Yan, X. Tang, J. Wang, Z. Xiong, J. Dong, Y. Xue, Y. Zhu, X. Xu, L. Sun, S. Chen, H. Nie, J. Peng, J. Xu, Y. Wang, Z. Yuan, Y. Wen, Z. Yao, Y. Shen, B. Qiang, Y. Hou, J. Yu, and Q. Jin.** 2005. Genome dynamics and diversity of *Shigella* species, the etiologic

agents of bacillary dysentery. *Nucleic Acids Res.* **33:**6445–6458.

130. **Yang, J., L. Chen, J. Yu, L. Sun, and Q. Jin.** 2006. *Shi*BASE: an integrated database for comparative genomics of *Shigella*. *Nucleic Acids Res.* **34:**D398–D401.

131. **Yang, J., J. Wang, Z. J. Yao, Q. Jin, Y. Shen, and R. Chen.** 2003. GenomeComp: a visualization tool for microbial genome comparison. *J. Microbiol. Methods* **54:**423–426.

132. **Yang, Z., and R. Nielsen.** 2002. Codon-substitution models for detecting molecular adaptation at individual sites along specific lineages. *Mol. Biol. Evol.* **19:**908–917.

133. **Yura, T., H. Mori, H. Nagai, T. Nagata, A. Ishihama, N. Fujita, K. Isono, K. Mizobuchi, and A. Nakata.** 1992. Systematic sequencing of the *Escherichia coli* genome: analysis of the 0–2.4 min region. *Nucleic Acids Res.* **20:** 3305–3308.

THE MYCOBACTERIA: A POSTGENOMIC VIEW

Marien I. de Jonge, Timothy P. Stinear,
Stewart T. Cole, and Roland Brosch

3

The genus *Mycobacterium* comprises a high percentage of G+C actinobacteria that are characterized by a complex, lipid-rich cell wall. Whereas the great majority of the numerous species in the genus are harmless environmental saprophytes, this genus also harbors some of the best-known human pathogens, the etiological agents of tuberculosis (TB), leprosy, and Buruli ulcer. In spite of the availability of antituberculous drugs and the *Mycobacterium bovis* bacillus Calmette-Guérin (BCG) vaccine, approximately one-third of the world's population is infected with *Mycobacterium tuberculosis*, leading to a vicious infection cycle with 9 million new cases and 2 million deaths caused by the disease each year (75). The situation is worsening in many countries, particularly due to the human immunodeficiency virus epidemic and *M. tuberculosis* strains that have become multidrug resistant. The prevalence of leprosy has been reduced extensively in recent years, mainly by World Health Organization (WHO) multidrug therapy and vaccination with BCG (101); however, with more than 690,000 new *Mycobacterium leprae* infec-

tions annually, the incidence of the disease remains worrisome (7). TB and leprosy belong to the oldest known human diseases and are described in ancient historical records. In contrast, Buruli ulcer, a debilitating necrotic skin infection caused by *Mycobacterium ulcerans*, is recognized as an emerging infectious disease, whose prevalence in some West African countries exceeds that of leprosy and in some instances that of TB (127).

In this respect, there is an urgent need for new therapeutic agents and vaccines to control these major mycobacterial diseases. It was against this background that the genome-sequencing project of the paradigm strain of TB research, *M. tuberculosis* H37Rv, was first started in 1993 (50), followed by the genome-sequencing project on *M. leprae* (51) and most recently *M. ulcerans* (250a). *M. tuberculosis* is highly related to *Mycobacterium bovis,* the bovine tubercle bacillus, whose genome sequence was released in 2003 (91). With the pasteurization of milk, the incidence of *M. bovis* infections in humans has dramatically decreased, but the organism remains a primary cause of TB in a range of domesticated and wild mammals, responsible for annual losses of approximately $3 billion to world agriculture. *M. bovis* is a member of the *M. tuberculosis* complex, which includes mycobacteria that

Marien I. de Jonge, Stewart T. Cole, and Roland Brosch   Unité de Génétique Moléculaire Bactérienne, 28 Rue du Dr. Roux, 75724 Paris Cedex 15, France.   *Timothy P. Stinear*   Department of Microbiology, Monash University, Wellington Road, Clayton 3800, Australia.

Bacterial Pathogenomics, Edited by M. J. Pallen et al.
© 2007 ASM Press, Washington, D.C.

cause TB in different mammalian species and share more than 99.9% DNA sequence similarity. *M. tuberculosis* strain CDC1551 represents a recent clinical isolate of the *M. tuberculosis* complex whose complete genome sequence was published in 2002 (86). At the time of writing, sequencing projects of other subspecies of the *M. tuberculosis* complex, such as *Mycobacterium africanum*, BCG, *Mycobacterium microti*, or the more distantly related *Mycobacterium canettii* are underway (http://www.sanger.ac.uk/Projects/ Microbes/) or finished, in parallel with an initiative to sequence several clinical *M. tuberculosis* isolates (Broad Institute http:// www.broad .mit.edu/seq/ msc/). In addition, the genome sequence of *Mycobacterium avium paratuberculosis*, the etiological agent of ruminant Johne's disease, which is also suspected to be implicated in human Crohn's disease (192), has recently been accomplished (143). Various other genome-sequencing projects on environmental saprophytes or opportunistic human pathogens, such as *Mycobacterium smegmatis*, *Mycobacterium chelonae*, *Mycobacterium abscessus*, *Mycobacterium marinum*, *Mycobacterium flave-scens*, or members of the *Mycobacterium avium* complex, have been determined and are awaiting publication (http://www.genomesonline .org/) (Table 1).

The large number of mycobacterial sequencing projects reflects the extraordinary power of genomic analyses to unravel the genetic basis of the biology and evolution of a given organism. The insights from these studies are so important that biological and medical research without access to genomic data is nowadays unimaginable. Furthermore, comparison of genome data from virulent and nonvirulent organisms has the potential to identify genes that are involved in the infection cycle of the pathogenic variants. This approach of comparative genomics has been applied successfully to members of the *M. tuberculosis* complex, thereby identifying regions of difference (RD) that are implicated in the pathogenicity of *M. tuberculosis* (202). The distribution of the various RD in different members of the *M. tuberculosis* complex also allowed

the evolution of the *M. tuberculosis* complex to be elucidated (34). As another example, the comparison of the genomes of more distantly related species allows intragenus genome conservation to be studied (Fig. 1), while comparison of *M. tuberculosis* and *M. leprae* has revealed the extraordinary reductive evolution of the leprosy bacillus and has allowed the core genome of pathogenic mycobacteria to be identified (51).

In parallel with the progress in mycobacterial genome research, advances in mycobacterial genetics have enabled the inactivation and complementation of selected genes by means of conditionally replicating plasmid vectors or phages that mediate allelic exchange (123, 154, 184, 186). These techniques represent an immense step forward in the identification and confirmation of essential genes, virulence factors, and other important features of mycobacterial pathogens. Transposon mutagenesis has also been used to inactivate genes (141, 223), and variants such as signature-tagged mutagenesis (112) have been successfully applied to *M. tuberculosis* (38, 56) to identify genes that are essential in vivo. Another method for identifying genes implicated in the infection process was developed by Sassetti et al., who adapted the mariner transposon for high density mutagenesis in *M. tuberculosis* and named the method "transposon site hybridization" or "TraSH" (223). With this method, these authors first identified approximately 700 genes that are essential for growth of *M. tuberculosis* in vitro (224). Interestingly, the great majority of these genes are conserved in the degenerate genome of the leprosy bacillus, *M. leprae*, indicating that nonessential functions have been selectively lost since *M. leprae* diverged from other mycobacteria (51). In further studies, this technique was used to create pools of mutant strains whose in vivo growth characteristics were tested. Microarray hybridization served as the readout for the presence or absence of individual transposon mutants in the pool of bacteria after mouse passage (225). Together with results from all the other studies identifying virulence genes in *M. tuberculosis* (for an

TABLE 1 Mycobacterial strains that are subject of genome-sequencing projects

Species	Strain designation	Genome size	No. of ORFs	URL
Slow-growing mycobacteria				
M. tuberculosis	H37Rv	4,411 kb	4,402	http://genolist.pasteur.fr/TubercuList/
M. tuberculosis	CDC1551	4,403 kb	4,187	http://cmr.tigr.org/tigr-scripts/CMR/Genome Page.cgi?database=gmt
M. tuberculosis	Strain 210	Unfinished		http://tigrblast.tigr.org/ufmg/index.cgi? database=m_tuberculosis%7Cseq
M. tuberculosis	F11	Unfinished		http://www.broad.mit.edu/annotation/ microbes/mycobacterium_tuberculosis_f11/
M. tuberculosis	C	Unfinished		http://www.broad.mit.edu/annotation/microbes/ mycobacterium_tuberculosis_c/
M. tuberculosis	K	Unfinished		http://chimp.kribb.re.kr/~gsal/project/ Mycobacteriun1.php
M. bovis	AF2122/97	4,345 kb	3,955	http://genolist.pasteur.fr/BoviList/
M. bovis BCG	Pasteur 1173P2	4,375 kb		http://genolist.pasteur.fr/BCGList
M. microti	OV254	Unfinished		http://www.sanger.ac.uk/Projects/M_microti/
M. africanum	GM041182	Unfinished		http://www.sanger.ac.uk/sequencing/ Mycobacterium/africanum/
M. canettii	140010059	Unfinished		http://www.sanger.ac.uk/sequencing/ Mycobacterium/canettii/
M. marinum	M-strain	6,636 kb		http://www.sanger.ac.uk/Projects/M_marinum/
M. ulcerans	AGY99	5,805 kb		http://genolist.pasteur.fr/Burulist/
M. leprae	TN	3,268 kb	1,604	http://genolist.pasteur.fr/Leproma/
M. avium paratuberculosis	K-10	4,829 kb	4,350	http://www.cbc.umn.edu/ResearchProjects/ AGAC/Mptb/Mptbhome.html
M. avium avium	Strain 104	5,475 kb		http://tigrblast.tigr.org/ufmg/index.cgi?database= m_avium%7Cseq
Rapidly growing mycobacteria				
M. smegmatis	MC2 155	6,988 kb		http://tigrblast.tigr.org/ufmg/index.cgi?database= m_smegmatis%7Cseq
M. chelone	CIP 104535	~5 Mb		http://www.genoscope.cns.fr/externe/English/ Projets/Projet_LU/organisme_LU.html
M. abscessus	CIP 104536	~5 Mb		http://www.genoscope.cns.fr/externe/English/ Projets/Projet_LU/organisme_LU.html
M. flavescens	Pyr-GCK	Unfinished		http://genome.jgi-psf.org/draft_microbes/mycfl/ mycfl.home.html
M. vanbaalenii	Pyr-1	Unfinished		http://genome.jgi-psf.org/cgi-bin/runAlignment? db=mycva&advanced=1

overview, see Braunstein et al. [24]), this genome-wide approach gave an overview of the genetic "virulence potential" of *M. tuberculosis*, which concerns 194 genes (Table 2), i.e., approximately 5% of the total number of genes.

As a contribution for a book whose overall subject is pathogenomics, we shall focus in this chapter on insights gained from genome exploration and functional analyses and provide an overview of the evolution and of the infection strategies used by pathogenic my-

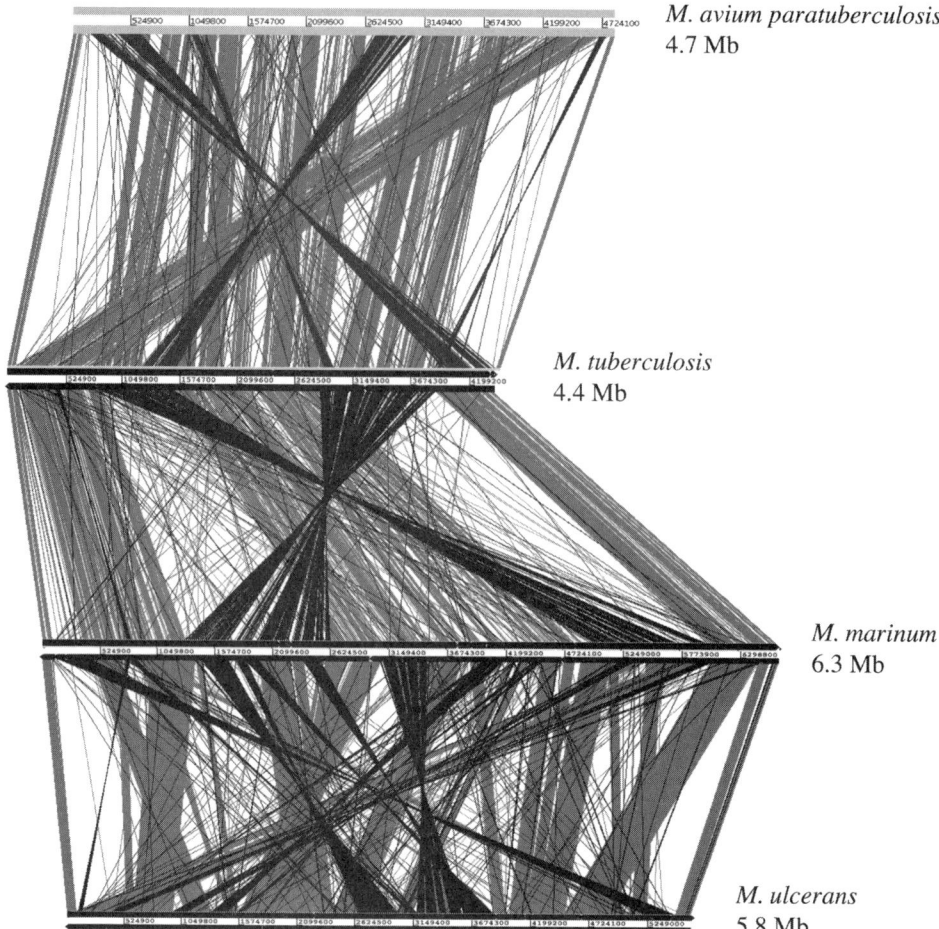

M. avium paratuberculosis
4.7 Mb

M. tuberculosis
4.4 Mb

M. marinum
6.3 Mb

M. ulcerans
5.8 Mb

FIGURE 1   DNA sequence comparison of *M. avium paratuberculosis* K-10, *M. tuberculosis* H37Rv, *M. marinum* M, and *M. ulcerans* Agy99 using the Artemis Comparison Tool (ACT) (43). Regions shown in light gray indicate similar genomic organization, whereas regions shown in dark gray represent genomic inversions.

cobacteria and how this information may be useful for developing new insights into disease control.

M. TUBERCULOSIS AND THE *M. TUBERCULOSIS* COMPLEX

Evolutionary Aspects

Since the publication of the *M. tuberculosis* H37Rv genome sequence (50), considerable insights have been gained into the evolution of

this important pathogen. This was enabled by comparative genomic analyses of members of the *M. tuberculosis* complex. As mentioned above, this complex represents a genetically very homogeneous group of bacteria that cause TB in different mammalian species and can be differentiated by few phenotypic characteristics, including different host preferences. However, recent genetic studies have identified numerous genetic markers that now enable us to identify and differentiate members

TABLE 2 Predicted functional classifications of genes identified by TraSH analysis (after references 224 and 225) as being essential for optimal growth of *M. tuberculosis* H37Rv under in vitro and in vivo conditions

Functional classification	In vivo		In vitro	
	No. of genes	Percent of category[a]	No. of genes	Percent of category[a]
Lipid metabolism	15	7.5	30	14.9
Carbohydrate transport and metabolism	9	8.4	24	22.4
Inorganic ion transport and metabolism	8	8.0	8	8.0
Cell envelope biogenesis, outer membrane	8	7.3	32	29.4
Amino acid transport and metabolism	8	4.3	80	43.0
Transcription	7	5.4	15	11.6
Coenzyme metabolism	7	6.0	38	32.8
DNA replication, recombination and repair	5	4.6	19	17.4
Translation, ribosomal structure	5	3.9	76	59.4
Secretion	3	13.6	8	36.4
Cell division and chromosome partitioning	2	8.7	8	34.8
Posttranslational modification, chaperones	2	2.5	27	34.2
Nucleotide transport and metabolism	1	1.5	25	38.5
Unknown	107	4.7	181	8.0
Total	194	5.0	614	15.7

[a]"Percent of category" refers to the fraction of genes of a particular functional class that are important for growth under each condi-

of the complex very efficiently. In this respect, the presence or absence of certain RDs has served to redefine the evolutionary pathway of the *M. tuberculosis* complex (34, 176), and this scheme has since then been completed by data on selected single nucleotide polymorphisms (SNPs) and microdeletions (83, 89, 91, 157). In contrast to previous hypotheses, these studies suggest that the agent of bovine TB, *M. bovis,* is not the ancestor of the human TB agent *M. tuberculosis*, as was thought for a long time (244). This perspective is further supported by recent results from paleomicrobiological investigations that demonstrate the presence of *M. tuberculosis* rather than *M. bovis* in ancient human remains (72). In one of these studies, genetic analyses of ancient mycobacterial DNA amplified from 4,000-year-old Egyptian mummies suggested that *M. tuberculosis* strains closely resembling "modern" *M. tuberculosis* of Sreevatsan's genetic group, genetic group 2 or 3 (242), were already common causative agents of human TB in ancient Egypt (281). Similarly, mycobacterial DNA amplified from a 2,000-

year-old mummy from England also showed characteristics of a "modern" *M. tuberculosis* strain (255). These studies give an idea of the genetic diversity of *M. tuberculosis* strains in ancient human populations.

However, modern molecular characterization methods are now also available to get an overview of the extant population of *M. tuberculosis* strains in the world, which is of importance to understanding the transmission patterns of *M. tuberculosis* better. One of these methods relies on the so-called direct repeat region, which harbors numerous direct repeat units of 36 bp that are interspersed with 27- to 41-bp segments of unique sequence. Recombination events between the direct repeats and/or copies of IS6110 elements inserted in this genomic region created polymorphisms that can be visualized and distinguished by a technique called spoligotyping (128). Comparative spoligotype analysis, together with analysis of RD, SNPs, and variation in mycobacterial interspersed repetitive units (252), has shown that combinations of these molecular markers appear to be strictly correlated in strains of the *M. tuberculosis* complex, allowing the development of rapid strain identification and typing strategies (12). Some of these data are now also available in large online databases (http://www.pasteur-guadeloupe.fr:8081/SITVITDemo/) that provide the opportunity to precisely compare strains isolated from different parts of the world (82).

The importance of molecular characterization of *M. tuberculosis* strains was also shown in a recent microarray analysis of a large number of epidemiologically well-defined *M. tuberculosis* isolates from San Francisco. In this study, Small and colleagues were able to demonstrate that strains show close association with their host populations over time, so much so that a patient's region of birth could be used as a predictor of strain carriage (89, 117). Their analyses confirmed the lack of any significant horizontal gene transfer in *M. tuberculosis*, a feature which was also pointed out by Supply and colleagues (253). In summary, all these data suggest that the population structure of present-day *M. tuberculosis* strains is highly clonal, with no significant horizontal gene transfer detected. However, coinfection with two or more different *M. tuberculosis* strains (21, 185) has been described, and it is possible that the very close genetic similarity between strains overall may mask occasional horizontal gene transfers. Such events may eventually be discovered when more genome sequences of closely related strains become available.

In contrast to the classical members of the *M. tuberculosis* complex, the situation with horizontal transfer is clearly different for tubercle bacilli, with its unusual smooth colony morphology. Such smooth tubercle bacilli were first described in the 1970s by Georges Canetti and later named *Mycobacterium canettii* (264). However, until recently, only a handful of *M. canettii* strains had been identified (264). In this respect, it was particularly interesting to investigate a collection of 37 strains with smooth colony morphology that were isolated from TB patients from Djibouti, East Africa (79). 16S rDNA sequences confirmed that

these isolates were tubercle bacilli, although minor variations at the beginning of the 16S rRNA gene were noted (106). On the basis of the molecular markers usually used for molecular typing of members of the *M. tuberculosis* complex and of the analysis of polymorphism within six housekeeping genes totaling more than 3 kb, it was obvious that some of these smooth tubercle bacilli did not correspond to *M. canettii* but to quite distant genetic clusters. In total, eight different lineages of smooth tubercle bacilli, four of them sharing many characteristics of *M. canettii*, were defined (106). The synonymous nucleotide variation within the smooth tubercle bacilli was much higher than in the *M. tuberculosis* complex alone.

Most interestingly, the distribution of SNPs in housekeeping genes displayed mosaic structures providing direct evidence of intragenic recombination within the smooth strains (106). As shown in Table 3, a substantially higher number of SNPs between *M. tuberculosis* and smooth tubercle bacilli than among classical members of the *M. tuberculosis* complex was also observed when gene cluster *rv0986* to *rv0988*, which apparently has been received by a progenitor of *M. tuberculosis* through horizontal transfer, was investigated in the various smooth tubercle bacilli and selected members of the *M. tuberculosis* complex (220). Together, these findings imply that the different lineages of smooth tubercle bacilli and the traditional members of the *M. tuberculosis* complex separated much earlier than the previously estimated age of 20,000 to 35,000 years for the etiological agent of human TB (122, 242) and that among the smooth lineages of the tubercle bacilli (proposed name *M. prototuberculosis*), interstrain recombination seems to have been a frequent event (106). It is also striking that the genetic diversity found in smooth strains, all originated from Djibouti, East Africa, and the otherwise high homogeneity within the globally spread tubercle bacilli is reminiscent of the distribution of human populations with the highest genetic divergence found in Africa. These findings suggest that tubercle bacilli have emerged in

TABLE 3 SNPs at genomic locus rv0986-88 that are specific for classical members of the *M. tuberculosis* complex (*M. tuberculosis* H37Rv, *M. tuberculosis* CDC1551, "ancestral" *M. tuberculosis* TbD1+, and *M. bovis* AF2122/97) and smooth tubercle bacilli (proposed name *M. prototuberculosis*) of lineages A to I[a]

SNP at the following genomic locus and position in gene

Strain	rv0986 (747 bp)					rv0987 (2,568 bp)																					rv0988 (1,161 bp)														
	15	233	261	354	682	247	428	429	525	562	808	855	867	876	999	1146	1271	1308	1347	1592	1640	1702	1770	2019	2149	2532	87	100	116	169	403	678	693	747	753	792	864	988	966	1093	1149
H37Rv	T	C	C	T	G	G	G	A	A	G	T	G	G	G	G	C	A	G	T	G	C	G	T	A	G	T	A	T	T	T	G	C	C	G	C	C	T	G	T	T	A
CDC1551	•	•	•	•	•	•	T	•	•	•	•	•	•	•	•	•	•	•	•	•	•	•	•	•	•	•	•	•	•	•	•	•	•	•	•	•	•	•	•	•	•
TbD1+	•	•	•	•	T	•	•	•	•	•	•	•	•	•	•	•	•	•	•	•	•	•	•	•	•	•	•	•	•	A	A	•	•	•	•	•	•	•	•	•	•
M. bovis	•	•	•	•	T	•	•	•	•	•	•	•	•	•	G	•	•	A	A	•	•	•	•	•	G	G	•	•	•	A	A	G	•	•	•	•	•	•	•	•	•
A	C	G	T	C	•	•	•	G	•	•	•	•	•	•	•	•	•	•	•	•	A	C	•	C	•	•	•	•	C	A	•	G	•	•	•	C	•	•	C	•	•
B	C	G	T	C	•	•	•	G	•	•	•	•	•	•	•	•	•	•	•	•	A	C	•	C	•	•	•	•	C	A	•	G	•	•	•	C	•	•	C	•	•
C/D	•	G	T	C	•	•	•	G	•	•	•	•	•	•	•	•	•	•	•	•	A	C	•	C	•	•	•	C	C	A	•	G	•	•	•	C	•	•	C	•	•
E	•	G	T	C	•	•	•	G	•	•	•	•	•	•	•	•	•	•	•	•	A	C	•	C	•	•	•	•	C	A	•	G	•	•	•	C	•	•	C	•	•
F	•	G	T	C	•	•	•	G	•	•	•	•	•	•	•	•	•	•	•	•	A	C	•	C	•	•	•	•	C	A	•	G	G	T	T	C	•	A	C	G	G
G	•	G	T	C	•	•	•	G	•	•	•	•	•	•	•	G	•	•	•	•	A	C	•	•	•	•	•	C	C	A	•	G	G	T	T	C	•	C	C	G	G
H	•	G	T	C	•	•	•	G	•	•	•	•	•	•	•	•	•	•	•	•	A	C	•	C	•	•	•	C	C	A	•	G	•	•	•	C	•	•	C	•	•
I	•	G	T	C	•	•	•	G	•	•	•	•	•	•	•	•	•	•	•	•	A	C	•	C	•	•	•	•	C	A	•	G	•	•	•	C	•	•	C	•	•
	s	N	s	s	N	s	N	N	s	N	s	s	s	s	s	s	N	N	s	s	N	N	s	s	s	s	s	s	s	N	N	s	s	s	s	s	s	N	N	s	s

[a] After references 220 and 106. Lineages A to D correspond to *M. canettii* strains. A dot indicates that the sequence at this locus is identical to that of *M. tuberculosis* H37Rv. s, synonymous mutation; N, nonsynonymous mutation.

55

Africa and that a dominant clone, consisting of the classical members of the *M. tuberculosis* complex (*M. tuberculosis, M. africanum, M. microti, M. bovis*), has then spread throughout the world. Among many other implications, such knowledge helps to explain the extraordinary adaptation of tubercle bacilli to their hosts.

Host–Related Factors

Before describing the genetic virulence determinants of *M. tuberculosis* in more detail, one should remember that the outcome of an infection with *M. tuberculosis* largely depends on the ability of the human host to mount an appropriate T-cell–mediated immune response (130). Critically, this response is usually only capable of containing the infection and rarely sterilizes the lungs. *M. tuberculosis* uses the respiratory tract as the principal portal of entry, and depending on the correct droplet size, it travels deep into the alveoli, where it is taken up by alveolar macrophages. Whereas these cells otherwise destroy most potential invaders, *M. tuberculosis* may persist and even replicate in this hostile environment. In the latter case, the macrophages are destroyed and the bacteria released and phagocytosed again by other macrophages and by dendritic cells. Bacteria ingested by dendritic cells are transported to draining lymph nodes, where antigen presentation to thymus-dependent lymphocytes takes place. After a period ranging from a few days to several weeks, these specific T lymphocytes leave the initial draining lymph node and migrate toward the sites of infection, where they release cytokines (e.g., interferon gamma [IFN-γ]) and cause activation of macrophages.

A local site of infection, also called a tubercle, contains macrophages, often fused to giant cells, bacteria, and lymphocytes. This tubercle may become a granuloma with central necrosis and fibrosis due to the release of various cytokines from the tightly interposed macrophages and other immune cells. In most situations, the development of specific cellular immunity in which various categories of T lymphocytes play a major role limits multiplication of bacilli, and

the infected person remains asymptomatic. This state is defined as TB infection, also called latent TB or primary infection, indicating a first contact with *M. tuberculosis*. It is characterized by a delayed-type hypersensitivity reaction to tuberculin or to partially purified proteins derived from tuberculin and is accompanied by a normal clinical status, normal chest X-ray, and negative bacteriology.

In some cases, multiplication of the bacilli is poorly controlled and active TB develops. This may happen soon after the initial infection or several years later. The active disease is accompanied by clinical and/or radiological signs and/or positive bacteriology. The final step in the infectious cycle of the tubercle bacillus is the development of tuberculous cavities in the lungs of the patient, which allows the efficient spread of high numbers of bacteria via aerosol to new hosts. From the perspective of the bacterium, this step is absolutely essential to maintain its so successful life cycle as an obligate pathogen, and it seems that this adaptation occurred during a long-lasting coevolution between the bacterium and its host. The shift from infection to disease depends on many factors. About 5% of immunocompetent individuals progress to tuberculous disease within 2 years of infection, and an additional 5% of infected individuals will develop the disease later in their lives. The risk of disease is increased in immunosuppressed people, particularly those infected with human immunodeficiency virus, when the number of CD4[+] lymphocytes decreases (239). Other key factors that confer susceptibility to mycobacterial diseases are genetic disorders that affect the function of the IFN-γ or interleukin-12 signaling cascade (2). Also, the use of tumor necrosis factor alpha blocking agents for the control of chronic inflammatory diseases, like rheumatoid arthritis, increases the risk of progression to active TB as a result of the disruption of the granuloma that normally compartmentalizes but does not kill *M. tuberculosis* in patients with latent TB (131). These examples underline the importance of the immune system in the control of TB.

General Features of the *M. tuberculosis* Genome

The genome of *M. tuberculosis* H37Rv consists of 4,411,532 bp containing approximately 4,000 genes encoding proteins and 50 genes encoding RNAs (40, 50) (http://genolist.pasteur.fr/TubercuList/). The original genome publication and subsequent reviews have dealt in detail with the initial findings (33, 35, 50, 99), and therefore in the following sections, selected features related to pathogenicity (Fig. 2) that have become the subject of recent functional studies will be discussed.

Lipid Metabolism

One of the key findings of the sequence analysis was that *M. tuberculosis* differs radically from other bacteria in that approximately 10% of its coding capacity is dedicated to genes encoding proteins that are involved in lipid metabolism, both biosynthetic activities and lipolytic functions. Whereas the first group of proteins is needed for the synthesis and modification of the various components of the extremely complex cell wall, the latter genes enable *M. tuberculosis* to degrade exogenous lipids and sterols from host tissues. Support for the hypothesis that *M. tuberculosis* may feed on lipids from host cells comes from an observation showing that mycobacteria recovered from infected tissues can degrade exogenous lipids (273). In addition to the classical β-oxidation cycle, catalyzed by the multifunctional FadA/FadB proteins, the genome sequence suggests that alternative lipid oxidation pathways may also exist in *M. tuberculosis*. Many of the genes involved in this process exist in multiple copies. In strain H37Rv, there are 36 *fadD* alleles encoding acyl-CoA synthase, 36 *fadE* genes encoding acyl-CoA dehydrogenase, and 21 *echA* genes for enoyl-CoA hydratase/isomerase (50). However, it is possible that the apparent redundancy may hide novel enzyme activities, as suggested by Gokhale and colleagues, who have recently shown that some of the *fadD* alleles do not encode fatty acyl-CoA ligases, but instead code for a new class of fatty acyl-AMP ligases that are linked to a proximal *pks* gene encoding a unique polyketide synthase (262).

A total of 15 *pks* genes were identified in the genome of strain H37Rv. Two of them, *pks1* and *pks15*, are located next to each other in the genomes of *M. tuberculosis* H37Rv and *M. tuberculosis* CDC1551. However, in certain other *M. tuberculosis* strains, *pks1* and *pks15* have fused. More detailed comparison revealed that a deletion of 7 bp is responsible for a frameshift, which separates *pks1* and *pks15* and has turned them into pseudogenes. It was shown that the intact *pks15/1* gene encodes an enzyme involved in the synthesis of a phenolic glycolipid (52), which is absent from strains H37Rv and CDC1551. So far, all tested *M. tuberculosis* strains of genetic groups 2 and 3 as defined by Sreevatsan et al. (242), including strain H37Rv (group 3) and strain CDC1551 (group 2), show this 7-bp deletion (157), suggesting that the deletion occurred in a common progenitor of the two strain lineages. By use of an *M. tuberculosis* strain of the Beijing lineage with an intact *pks15/1* gene together with a *pks15/1* knockout mutant of the same strain, Reed et al. found a correlation between synthesis of this particular phenolic glycolipid molecule and a decrease in the production of proinflammatory cytokines by infected host

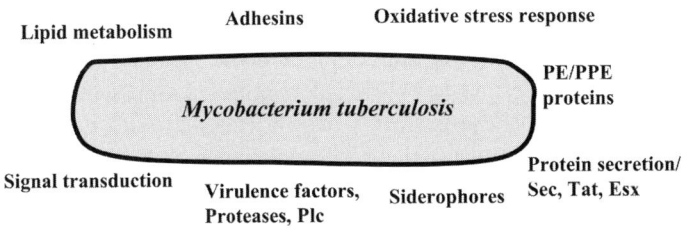

FIGURE 2 Bacterial factors described in this chapter that contribute to the pathogenicity of *M. tuberculosis*.

immune cells (211). This finding could explain in part the hypervirulent phenotype observed for strains from the Beijing family (147, 155). In *M. bovis* AF2122/97 and *M. bovis* BCG, a 6-bp deletion was observed at the same locus of the *pks15/1* gene (52). Because *M. bovis* and *M. bovis* BCG both produce phenolic glycolipid, it seems likely that this 6-bp deletion, which does not cause a frameshift in the *pks15/1* gene, has no influence on the enzymatic activity of the resulting gene product. However, because of mutations in a neighboring gene, encoding a glycosyl transferase, phenolic glycolipids produced by *M. bovis* and BCG are characterized by a shorter sugar moiety (187) and may therefore show different antigenic properties.

In the immediate proximity of the *pks15/1* genes is another important genomic locus for lipid metabolism, which contains genes that are involved in the synthesis and transport of a very abundant constituent of the cell wall, phthiocerol dimycocerosate. This genomic region was independently identified by two groups as being involved in the survival of *M. tuberculosis* in mice (38, 56). At least 13 genes of this 50-kb region are required for the synthesis or transport of phthiocerol dimycocerosate. The absence of these complex lipids affects the permeability and structure of the mycobacterial cell wall, which may explain the loss of virulence of *M. tuberculosis* mutants with transposon insertions in this locus. Detailed studies have also underlined the role of large transmembrane proteins, encoded by this region (e.g., MmpL7) in lipid transport (38, 56, 124). In a recent study it was found that besides MmpL7, MmpL4 is also required for the full virulent phenotype of *M. tuberculosis* (71).

As another example, the identification of the enzyme responsible for the final condensation step in mycolic acid biosynthesis should be mentioned here. This condensate, which remained unknown despite considerable efforts made by several laboratories during the last 40 years, was recently identified with the help of comparative genomics. In this case, comparison of the genome sequences of *M. tuberculosis* H37Rv and the more distantly related *Corynebacterium glutamicum* has shown that in contrast to *M. tuberculosis*, which harbors multiple *pks* genes, *C. glutamicum* carries only one *pks* gene, the orthologue of *pks13*. Because both species produce mycolic acids, Pks13, which contains the four catalytic domains theoretically required for the condensation reaction, was therefore selected as the candidate enzyme. Interestingly, in mycobacteria, *pks13* is an essential gene, whereas this is not the case for *C. glutamicum*, where a *pks13* knockout mutant was isolated that no longer produced mycolic acids (197). From a more practical perspective, some of the very specialized enzymes involved in lipid metabolism, encoded by genes that are unique and essential for *M. tuberculosis*, also represent important potential drug targets.

Cell Adhesion, Another Important Determinant of Virulence

Adhesion to host cells is a fundamental process in the life of pathogenic bacteria, and its specificity is determined by interactions between membrane and/or cell wall–attached bacterial and host-cell structures. In vitro studies have shown that *M. tuberculosis* is able to invade and multiply within epithelial cells (17, 232) even more efficiently than within macrophages (169, 226).

M. tuberculosis contains genes that are important for adhesion and entry into mammalian cells, as was first described by Arruda et al. in 1993 (9). Mycobacterial genome analysis revealed four copies of *mce* (which stands for *mammalian cell entry*) operons that are organized in the same manner and comprise eight genes each (50). At the 5′ end of each transcriptional unit are two genes, *yrbE1-4A* and *yrbE1-4B*, that have evolved from a tandem duplication, and whose products resemble YrbE, conserved hypothetical proteins found in *Escherichia coli* and *Haemophilus influenzae*. All of the YrbE proteins, including the eight from *M. tuberculosis*, are likely to be integral membrane proteins with six transmembrane alpha helices. The *mce* genes downstream of *yrbE* encode Mce proteins with hydrophobic stretches at the N terminus, probably representing signal or anchor sequences (50).

Bioinformatic analysis indicated that the Mce proteins are exposed on the cell surface, where they could associate with the integral membrane proteins YrbE1A and YrbE1B. The C-terminal domain is probably exposed to the exterior; this segment contains several copies of an extended motif that is predicted to adopt an amphiphilic alpha helical structure (256). The structure of Mce1A was modeled on the basis of the crystal structure of Colicin N of *E. coli* (62). Mce1A, as the model predicts, is a protein consisting of two major (alpha and beta) domains, connected by a long alpha helix. Mce1A was found to promote uptake into HeLa cells, while Mce2, which had the highest level of identity (67%) to Mce1, was unable to promote the association of microspheres with HeLa cells (47). Interestingly, Mce3A and Mce3B contain the RGD tripeptide Arg–Gly–Asp (98), a motif implicated in cell attachment (179, 212). The *mce3* operon however, is situated within the region of difference 7 (RD7), which is present in *M. tuberculosis* and many strains of *M. africanum,* but was found to be deleted from *M. bovis, M. bovis* BCG, and *M. microti* (98).

Results from different laboratories indicated that the different *mce* operons are regulated and expressed under in vivo conditions (1, 138, 222). Expression of the *mce3* operon is negatively regulated by Rv1963, located upstream of the *mce3* operon. Because no expression of *mce-3A* or *mce-4A* could be detected under in vitro oxygen-depleted conditions (107), Rv1963 might play a role in the negative regulation at the intracellular stage of infection. The expression of Mce3 proteins (Mce3A-F) is increased after uptake into macrophages (222) and was also detected during natural infections in humans (1). Others confirmed the expression of Mce1A, -3A, and -4A during infection of guinea pigs and rabbits (138). This shows that important progress has been made in unraveling expression of the *mce* operons and individual genes. However,

because of the potential functional overlap, a definite role of the Mce proteins in the infection process with *M. tuberculosis* has not yet been established.

Heparin-binding hemagglutinin (HBHA) was first described in 1996 (172), and by the use of knockout mutants, it was shown to be required for extrapulmonary dissemination of *M. tuberculosis* (189). Members of the *M. tuberculosis* complex but also other mycobacteria such as *M. leprae* and *M. avium* display this 28-kDa adhesin on their surface, where it plays an important role in binding to epithelial cells, but not to macrophages (170, 189). The major group of receptors targeted by HBHA are the heparan-sulfate-containing proteoglycans (190) that are recognized by the carboxy-terminal domain of HBHA (66), which contains Pro/Lys-rich repeats. Progressive removal of the repeats resulted in decreased binding to heparin. HBHA is posttranslationally modified by a carbohydrate moiety important for hemagglutination and protection against proteolytic cleavage that affects the heparin binding site (170). Another modification within the same domain, which also appeared to be important for the resistance against proteolysis, is methylation of the lysine residues in the Pro/Lys-rich repeat. The HBHA-dependent extrapulmonary dissemination was found to be mediated by receptor-dependent transcytosis not affecting the integrity of the tight junctions (171).

Apart from HBHA, other mycobacterial proteins might play a role in the interaction with epithelial cells (210) by binding to extracellular matrix protein, fibronectin. Well-studied examples are the fibronectin attachment protein from *M. avium avium* (229) and *M. avium paratuberculosis* (230), and proteins from the antigen 85 family from *M. tuberculosis* containing conserved surface-exposed residues that interact with human fibronectin (218).

The best-described receptor targeted by species from the *M. tuberculosis* complex is the C-type lectin DC-SIGN (dendritic-cell-specific intercellular molecule-3-grabbing nonintegrin)/CD209 (92, 254). The mycobacterium-specific lipoglycan mannose-capped lipoarabinomannan (ManLAM) was identified as a key ligand of DC-SIGN, a receptor expressed by dendritic cells but not by macrophages (254). Unlike macrophages, DCs are not permissive for the growth of *M. tuberculosis*, even though bacteria are not killed after uptake by these cells (254). Binding to DC-SIGN prevents DC maturation and induces the immunosuppressive cytokine interleukin-10, which might contribute to survival and persistence of *M. tuberculosis* (92). It was furthermore demonstrated that the human DC-SIGN homologue L-SIGN (CD209L), which has a different function and a different expression pattern (133), interacts with mycobacteria through ManLAM. It is in this context interesting to note that CD209L is expressed in human lung type II alveolar cells and endothelial cells (125). It was recently found that Man-LAM might not be the only ligand for DC-SIGN, because other ligands were also found to bind, like lipomannan and the mannose-containing 19- and 45-kDa glycoproteins (194). Furthermore, ManLAM was also found to be involved in the interaction with macrophages (227), as were other lipid molecules like phosphatidylinositol mannosides and polar glycopeptidolipids (163, 267).

Apart from the recently identified DC-SIGN, other receptors involved in adhesion, phagocytosis, and/or uptake of mycobacteria are the complement receptor 3 (191, 219, 228), the mannose receptor (10, 178, 267), and the scavenger receptor (109, 258).

The PE-PPE Family

One of the most surprising features of the *M. tuberculosis* sequence was the identification of two large gene families, the PE and PPE families, which occupy nearly 10% of the genome and comprise 100 and 67 members, respectively (50). Members of these families share a conserved N-terminal domain with the characteristic motifs ProGlu (PE) or ProProGlu (PPE), whereas the C-terminal segment varies in length and sequence. A particular subfamily of PE proteins contains multiple copies of polymorphic GC-rich repetitive sequences in

the C-terminal domains, and these proteins were therefore named PE-PGRS proteins. At first, no particular function could be attributed to PE and PPE proteins. However, more recent data suggest that these proteins are cell wall associated and surface exposed (13, 27). The two multigene families show significantly higher genetic variability compared with the genome as a whole (86, 91). They may represent an important source of antigenic variation and interfere with immune responses by inhibiting antigen processing (50). Disruption of the gene encoding a PE protein leads to a marked reduction of bacterial clumping, suggesting that this protein may mediate cell-cell adhesion and altered phagocytosis by macrophages (28). Another PE protein binds fibronectin and could mediate bacterial interaction with host cells (77, 235). Given the large number of genes encoding PE and PPE proteins in the genome of *M. tuberculosis*, it seems that these proteins fulfill a very important biological role. However, because of the extensive redundancy of PE and PPE proteins, which complicates gene knockout studies, the identification of the exact function of these proteins has not yet been fully elucidated.

Protection against Oxidative and Nitrosative Stress

Macrophages produce reactive oxygen and reactive nitrogen species that have potent antimicrobial activity. Protection against the toxic effects of exogenously generated reactive oxygen species is critical to mycobacterial virulence. One of the mechanisms of protection is the production of superoxide dismutases (SODs). SODs are widely distributed enzymes that are classified according to the metal cofactors involved in the redox reaction, which catalyze the disproportion of the superoxide radical (O_2^-) into molecular oxygen (O_2) and hydrogen peroxide (H_2O_2). Superoxide radicals are generated during aerobic respiration and can react with other cellular constituents to generate even more reactive oxygen species, such as hydroxyl radicals (OH^-). Aerobic bacteria, including saprophytic and pathogenic mycobacteria, degrade O_2^- to water and molecular oxygen by the sequential action of SOD and catalase.

M. tuberculosis sodA encodes an iron-containing SOD that is actively secreted through the alternative secA2 secretion pathway (26); it is thought to protect the bacterium from the oxidative attack of macrophages (139, 280). The structure of this enzyme is very similar to that of other iron- or manganese-containing bacterial SODs (54). Attempts to delete the *sodA* gene in *M. tuberculosis* have so far been unsuccessful, suggesting that SodA activity might be essential for bacterial survival. However, *M. tuberculosis* strains engineered to express reduced levels of SodA, using the antisense technology, were severely attenuated for virulence in mice (76).

SodC is a copper-zinc superoxide dismutase produced at much lower levels compared to SodA and is responsible for only a minor portion of the superoxide dismutase activity of *M. tuberculosis*. However, SodC has a lipoprotein binding motif, which suggests that it may be anchored in the membrane to protect *M. tuberculosis* from reactive oxygen intermediates at the bacterial surface (276). Deletion of the *sodC* gene led to sensitivity to O_2^- and H_2O_2 in vitro (74, 193) and to impaired survival in IFN-γ-activated murine peritoneal macrophages (193), but did not attenuate virulence in guinea pigs (74) or mice (193).

Another very important protein for the bacterial defense against oxidative stress is KatG, a 80-kDa hemoprotein that has catalase, peroxidase (116), and peroxynitritase activity (272). Although the exact role of KatG in virulence has not been elucidated, a number of different studies demonstrated that inactivation of *katG* results in various degrees of attenuation (115, 144, 204). It is very likely that KatG is required to catabolize the exogenous peroxides or peroxynitrites ($ONOO^-$) generated by the oxidative burst or by the reaction of O_2^- and NO, respectively (177). KatG might also be essential for detoxification of endogenous peroxides generated by respiration of nonfermentable substrates, such as fatty acids, men-

tioned above to be an important carbon source used by *M. tuberculosis* during persistent infection (165). All these data clearly show that KatG is an important virulence factor. However, apart from its physiological role in the infection process, KatG has attracted much attention because it is involved in the activation of the widely used anti-TB drug isonicotinic acid hydrazide (isoniazid or INH) (279). Formation of isonicotinoyl-NAD, the active form of the drug, involves removal of hydrazine from INH by KatG (238, 279). Naturally occurring deletions or mutations of *katG* are associated with INH resistance (114, 279). However, the most frequently selected mutations in *katG* in clinical isolates are those where Ser 315 is replaced by Thr (S315T), which allows KatG to retain a sufficient level of catalase activity for antioxidant defense but reduces its ability to activate INH (205).

The peroxiredoxin alkyl hydroperoxide reductase, AhpC, from *M. tuberculosis*, is the first element of an NADH-dependent peroxidase and peroxynitrite reductase system, functioning in direct reduction of peroxides and peroxynitrite (37). Oxidized AhpC is reduced by AhpD, which is regenerated by the reduction via the dihydrolipoamide acyltransferase (DlaT, Rv2215, originally annotated as *sucB*) by reduction of AhpD at its active-site cysteines (259). Finally, dihydrolipoamide dehydrogenase (Lpd) mediates the reduction of DlaT and completes the catalytic cycle (37, 259). In mycobacteria, AhpC can detoxify hydroperoxides and also affords protection to cells against reactive nitrogen intermediates (46, 161). The product of the *ahpC* gene of *M. tuberculosis* is thought to protect bacteria against both oxidative and nitrosative stress encountered within the infected macrophage (104, 161). However, other studies indicated that AhpC is of less or no importance for mycobacterial virulence (115, 241).

It was recently shown that *dlaT*, involved in the peroxynitrite reductase system, also encodes the E2 component of pyruvate dehydrogenase (259). Inactivation of *dlaT* in *M. tuberculosis* displayed phenotypes associated with DlaT's role in metabolism and in defense against nitrosative stress. The *M. tuberculosis* Δ*dlaT* strain showed retarded growth in vitro and was highly susceptible to killing by acidified sodium nitrite and mouse macrophages. It was therefore concluded that functional DlaT is required for optimal growth and protection against reactive nitrogen intermediates (RNIs) (233).

In mice, RNIs play an essential role in host defense against *M. tuberculosis* because inactivation of the gene coding for inducible nitric oxide synthase greatly increases susceptibility to TB (151). IFN-γ production, an important innate immune response against TB, triggers generation of nitric oxide by activating transcription of inducible nitric oxide synthase (45). In turn, nitric oxide acts as a precursor for the production of RNIs. The major action of RNIs is protein S-nitrosylation (251). A proteomic analysis of S-nitrosylation of mycobacterial proteins identified 29 S-nitroso proteins that all appeared to be enzymes involved in intermediary metabolism, lipid metabolism, and/or antioxidant defense (i.e., KatG, Lpd, and Mpa, described above).

The mycobacterial proteasome counters the destructive effects of RNI (59). Transposon-based mutagenesis, followed by a screen for variants hypersensitive to mildly acidified nitrite, identified five mutants with insertions in two proteasome-associated genes. One of the genes, called *mpa*, codes for an ATPase, required for unfolding of substrates about to be fed into the proteasome. The other (*rv2097c*), coding for a proteasome accessory factor (Paf), was already annotated as a putative component of the proteasome (50) because of its genetic association with *prcA* and *prcB*, coding for the α and β subunits of the 20S barrel-shaped core particle of the proteasome. The *M. tuberculosis* mutant deficient for the presumptive proteasomal ATPase was attenuated in mice, and exposure to proteasomal protease inhibitors markedly sensitized wild-type tubercle bacilli to RNI (60). It was therefore concluded that the mycobacterial proteasome serves as a defense system against nitrosative stress in mice. However, it remains unclear whether these findings also apply to human TB infections.

Protein Secretion and Virulence

The best-studied protein export pathway in gram-negative and gram-positive bacteria is the general secretion or Sec-dependent pathway, also referred to as the type II secretion system. Proteins secreted by this system are threaded in an unfolded state through an aqueous channel in the cytoplasmic membrane; they generally require N-terminal signal sequences, which are cleaved upon translocation (200). The Sec secretory machinery is highly conserved and composed of several proteins, with SecA, a major component of the system, generating the energy for secretion from the hydrolysis of ATP. Unusually, mycobacteria, along with several gram-positive bacteria (including *Staphylococcus*, *Streptococcus*, and *Listeria* species), have two *secA* genes (*secA1* and *secA2*). The *secA1* gene was found to be essential and encodes the housekeeping SecA, whereas *secA2* was not found to be essential but was clearly involved in protein export (25). Examples of proteins implicated in pathogenicity that are secreted by the SecA1-dependent pathway are Erp (exported repetitive protein), SA5K (secretion antigen, with an apparent molecular mass of 5 kD), or members of the antigen 85 (Ag85) complex. These proteins characteristically carry an N-terminal signal sequence. The Erp protein represents a surface-exposed protein produced by *M. tuberculosis* during phagosomal growth. Disruption of the gene in *M. tuberculosis* and in *M. bovis* BCG resulted in a marked attenuation of virulence, as judged from survival and multiplication in mouse lungs, spleen, or in bone marrow–derived macrophages. However, no differences in growth characteristics were observed in vitro between wild-type, mutant, and complemented H37Rv and BCG strains (18). The secretory antigen SA5K (or TB8.4) was described as being involved in intracellular survival (14, 20). Inactivation of *sa5k* in *M. bovis* BCG resulted in further attenuation and fewer and smaller granulomatous lesions compared with the parental strain. Proteins of the antigen 85 complex Ag85A (FbpA), Ag85B (FbpB), and Ag85C (FbpC) are the most abundant proteins in *M. tuberculosis* culture supernatants and are highly immunogenic. They act as mycolyl transferases required for the production of the cell wall constituents arabinogalactan-mycolate and trehalose dimycolate, and, as mentioned above, they also bind to fibronectin (58, 108, 218).

In contrast to the proteins secreted by the SecA1 machinery, proteins secreted by SecA2 do not contain signal sequences. Two prominent proteins, SodA and KatG, both implicated in intracellular survival of mycobacteria, were recently identified as being exported via the SecA2 pathway (26), and this also explains why Δ*secA2* mutants of *M. tuberculosis* are attenuated.

The twin-arginine translocation (Tat) system secretes fully folded proteins with a twin-arginine motif containing signal sequence across the cytoplasmic membrane of bacteria. Substrates of the Tat secretion system are synthesized as precursor proteins containing N-terminal signal sequences the characteristic RR motif in the charged region of the signal sequence (16). The *M. tuberculosis* genome contains 31 open reading frames with predicted Tat signal sequences, as identified by the TATFIND search algorithm (68). Among the predicted *M. tuberculosis* Tat substrates are the virulence-associated phospholipases C (PlcA, -B, and -C), described in more detail below, and the major β-lactamase BlaC (268). The genome of *M. tuberculosis* contains two distinct loci with open reading frames homologous to the bacterial Tat system. The proximity of *tatA* and *tatC* suggests that these genes are located in an operon, whereas the *tatB* gene is located about 1 Mbp away from *tatAC* (50). Inactivation of *tatA* and *tatC* in *M. smegmatis* causes a defect in β-lactamase secretion (164, 198).

Several small secreted proteins lack amino terminal signal sequences. Many of them belong to the 23-member protein family ESX, named after the 6-kDa early secreted antigenic target (ESAT-6), of *M. tuberculosis* (32, 240, 256). Both ESAT-6 and CFP-10 (culture filtrate protein, 10 kDa), the best-studied proteins of this family, are important virulence factors (29, 121, 142, 202, 246). The genes encoding ESAT-6 and CFP-10 are located in region of difference 1 (RD1) that was originally identified by subtractive hybridization as being absent from BCG but present in wild-type *M. bovis* strains (152). Because preliminary complementation experiments of BCG with the RD1 region from *M. bovis* did not result in increased virulence of the recombinant BCG strain in mice (152), the hypothesis that RD1 was involved in the attenuation of BCG remained unconfirmed for several years. The first experimental evidence that the loss of the RD1 locus did contribute to the attenuation of BCG was obtained only recently. A recombinant BCG::RD1, which contained the RD1 region and also large portions of its flanking regions, was more virulent in severe combined immunodeficient mice than the BCG-vector control strain. The same recombinant BCG strain was found to persist to a greater degree in the organs of immunocompetent mice but induced considerably less pathology than *M. tuberculosis* H37Rv (202). These data were then confirmed by RD1-knockout constructs of *M. tuberculosis* (121, 142). Overlapping portions of the RD1 region are also absent from other attenuated or avirulent members of the *M. tuberculosis* strains, like *M. microti* (30) or the Dassie bacillus (175). Genetic and biochemical approaches revealed that proteins encoded by the RD1 locus are part of a new secretion system, the ESAT-6 system 1 (ESX-1), required for the export of ESAT-6 and CFP-10 (105, 203, 243), and secretion does not seem to require host cell interaction, as was found for bacterial type III and IV secretion systems.

A region syntenous with the RD1 region of *M. tuberculosis* was identified in the nonpathogenic *M. smegmatis* (87, 157). The *M. smegmatis* RD1 proteins form a DNA-conjugation system, which suggests that pathogenic mycobacteria have adapted an ancestral conjugation system for protein secretion involved in virulence (87). There has been major progress in identifying RD1 genes required for ESAT-6 and CFP-10 secretion (31, 105, 203, 243). Inactivation or deletion of Rv3868 to Rv3871 and Rv3877 abolished ESAT-6 and CFP-10 export, which indicates that they are directly involved in secretion. Rv3868 and Rv3871 contain ATP binding sites; the former was predicted to function as an ATPase associated with various cellular activities (AAA). Rv3869, Rv3870, and Rv3877 are three putative membrane proteins with 1, 3, and 11 predicted transmembrane domains, respectively. Together with Rv3871, they are thought to form a membrane-bound ESX-1 secretory complex, with protein export driven by ATP hydrolysis. Inactivation of PE35 impaired the expression of ESAT-6 and CFP-10, suggesting a role in regulation (31).

Purified ESAT-6 and CFP-10 proteins have been described to form a stable 1:1 complex under in vitro conditions (168, 214), a finding confirmed by nuclear magnetic resonance structure analysis (213). In tubercle bacilli, copurification of ESAT-6 and CFP-10 was obtained from cytosolic fractions as well as from culture filtrates, suggesting that the proteins might be secreted as heterodimers (29). Alternatively, for an easier translocation, the two antigens might as well be separated with the help of the Rv3868 ATPase and re-form a heterodimer outside the bacterial cell. Work is ongoing to elucidate the mechanisms used during secretion.

Although ESX-1-like secretion systems occur in various mycobacterial species, they might function in different ways. This was indicated by the finding that the orthologue of Rv3868 was not found to be essential for secretion in *M. smegmatis* (53), whereas in the phylogenetically more closely related *M. marinum* and *M. tuberculosis*, this gene appeared to be indispensable for secretion of ESAT-6 and CFP-10 (31, 90). Whether proteins other than ESAT-6 and CFP-10 can be secreted by this system remains unclear; however, there is evidence that secretion of the proteins encoded by gene cluster *rv3614c* to *rv3616c*, situated in

a different chromosomal location than the genes for ESX-1, requires the ESX-1 secretion system. Mutants that lack these proteins failed to secrete ESAT-6 and CFP-10 and vice versa, which indicates that secretion of these proteins is mutually dependent (88, 150).

ABC Transporters

The genes encoding the ATP binding cassette (ABC) transporters occupy about 2.5% of the genome of *M. tuberculosis* (22, 50). These multisubunit permeases, which transport various molecules like ions, amino acids, peptides, antibiotics, polysaccharides, and proteins across biological membranes, are classified as importers and exporters. In contrast to *Bacillus subtilis* and *E. coli*, *M. tuberculosis* contains more exporters (21) than importers (16). The exporters are potentially implicated in the secretion of macrolides, antibiotics, and polysaccharides and of still-unknown substrates (22). One exporting system, which shows high homology to a similar system in *Agrobacterium tumefaciens* (162), is encoded by *rv0986* to *rv0988*. This system, which appears to have been acquired by a progenitor of *M. tuberculosis* by horizontal transfer, is required for the ability of *M. tuberculosis* to multiply in macrophages (220). Another export system, Rv3781 to Rv3783, was predicted to export polysaccharides (22), whereas a recently identified system containing two genes, *irtA* and *irtB*, strongly resembles the genes encoding the YbtP-YbtQ iron transport system in *Yersinia pestis* (81). Inactivation of the *irtAB* system decreases the ability of *M. tuberculosis* to survive under iron-deficient conditions and to multiply in human macrophages and mouse lungs (216). The *irtAB* genes are located in a chromosomal region previously shown to contain genes regulated by the iron-dependent regulator, IdeR (95, 217). Rodriguez and coworkers (95) postulate that IrtAB is a transporter of mycobacterial siderophore Fe-carboxymycobactin. The relevance of siderophore synthesis for iron acquisition by *M. tuberculosis*, and its implication in virulence is described below. The *M. tuberculosis* gene *rv1747* encodes a 92-kDa protein containing an ABC transporter domain. Although its function is unknown, it has been suggested that this gene may have the role of an exporter (22). It is one of two genes in a putative operon in which the first codes for a serine-threonine protein kinase, *rv1746 (pknF)* (50). The protein encoded by the neighboring gene, *rv1747,* was found to interact specifically with the serine-threonine protein kinase PknF in a kinase-dependent manner (57). A deletion mutant of *rv1747* is attenuated in mice, and growth of the mutant in mouse bone marrow–derived macrophages and dendritic cells is significantly impaired (57).

Siderophores: Synthesis and Regulation

Iron scavenging is an essential process in pathogenic bacteria because iron is important to maintain vital processes that depend on iron-containing cofactors, including electron transport, energy metabolism, and DNA synthesis, while free iron in the infected host is limited. Small molecules with high affinity for iron, also called siderophores, are used to acquire this essential element. There are two classes of mycobacterial siderophores produced by the different species of this genus. Exochelins are secreted peptide siderophores, which were isolated from *M. neoaurum* (231), and *M. smegmatis* (85). In contrast, the siderophores produced by pathogenic mycobacteria are named mycobactins and are more lipophilic. The core structure of these mycobactins is the hydroxyphenyloxazoline ring, which is derived from salicylate (65). The difference between mycobactins and carboxymycobactins is determined by the nature of the fatty acyl side chain linked to the ε amino group of the N-hydroxy lysine (65). Recently, Luo and coworkers found direct evidence that the lipophilic mycobactins efficiently extract intracellular macrophage iron. The metal-free siderophore diffusely associates with the macrophage membrane, ready for iron chelation, while the mycobactin-metal complex accumulates with high selectivity in macrophage lipid droplets, which are intracellular domains for lipid storage and lipid sorting.

These mycobactin-targeted lipid droplets contact phagosomes directly. The existence of this iron acquisition pathway indicates that mycobacteria have taken advantage of endogenous macrophage mechanisms for iron mobilization and lipid sorting for iron acquisition during infection (148). Annotation of the complete genome sequence of *M. tuberculosis* identified a cluster of 10 genes (*mbtA* to *mbtJ*), the *mbt* locus, which encodes the enzymes needed for the biosynthesis of mycobactin and the carboxymycobactin core (50, 206). Strikingly, *M. leprae* appears to have lost this gene cluster, although it still contains functional iron regulatory systems (51). Recently, a second *mbt* cluster was identified that contains four genes, *rv1347c, rv1344, fadD33,* and *fadE14,* annotated as *mbtK, mbtL, mbtM,* and *mbtN,* involved in the transfer of the alkyl substitutions that distinguish mycobactins and carboxymycobactins (137). The involvement of N-acyl transferase Rv1347c or MbtK in mycobactin biosynthesis was earlier shown by Card et al. (42) and was found to be essential for growth of *M. tuberculosis* (as was the hydroxylase, MbtG) in a genome-wide mutational analysis (224).

Members of four distinct families of metal sensing transcriptional regulators are found in the genome of *M. tuberculosis.* Among them were two members of the Fur family (FurA and FurB), two regulators of the DtxR family (IdeR and SirR), three regulators of the MerR family (Rv1674c, Rv1994c, and Rv3334) and seven regulators of the SmtB/ArsR family (Rv0324, Rv0576, Rv2034, Rv2358, Rv2640c, Rv2642, and Rv3744) (41) (http://genolist.pasteur.fr/TuberculList/). The best-characterized regulator is IdeR, which is found in all mycobacteria. The presence of iron causes a conformational change in the DNA-binding domains of IdeR (195). A 19-bp inverted repeat consensus sequence or iron box was identified as the IdeR binding site; these are found at the promoters of the *mbt* gene cluster and iron storage genes (*bfrA* and *bfrB*), encoding two ferritin-like proteins (50), repressing the former and positively regulating the latter (95). About 30% of all iron-regulated genes in *M. tuberculosis* are regulated by IdeR (217). MbtB is an essential gene for the mycobactin and carboxymycobactin biosynthesis as shown by mutation analysis (65), and inactivation of this gene strongly reduced replication in infected macrophages. It was furthermore shown that the expression of *mbtB* was induced after infection of macrophage cell cultures (95) and mice (260), while *bfrA* was found to be inhibited. These results confirmed that, in general, *M. tuberculosis* is exposed to low-iron conditions during infection that trigger siderophore biosynthesis.

Mycobacterial Proteases

The genome of H37Rv encodes over 30 proteases (50), but little is known about their role in virulence or turnover and modification of cellular proteins. Among the few proteases studied in pathogenic mycobacteria are ClpC identified in *M. leprae* (173), HtrA in *M. avium paratuberculosis* (39), and two secreted serine proteases Mtb32A (PepA) and Mtb32B (PepD) (237). The best-studied mycobacterial proteases are the transmembrane serine proteases of the subtilisin family, called mycosins. On the basis of their sequential identification, the proteins were designated mycosin 1 to 5 (Rv3883c, Rv3886c, Rv0291, Rv3449, and Rv1796, respectively) (36). Multiple genes encoding mycosins were found in other pathogenic bacteria, but just one variant was identified in *M. smegmatis* (53). Mycosin 1 was found to be expressed during infection of macrophages and was localized both in the membrane/cell wall fractions and in the supernatant of *M. tuberculosis* cultures (64). It was proposed that mycosins may intervene in the processing of secreted and/or extracellular proteins, such as the 19-kDa lipoprotein antigen, whose deglycosylation-dependent release from the cell surface is mediated by proteolytic cleavage (36, 113). This hypothesis was supported by the results of Converse and Cox, who showed that the protease mycosin 1 is essential for the secretion of ESAT-6 and CFP-10 in *M. smegmatis* (53).

Recently, a previously uncharacterized *M. tuberculosis* site two protease (S2P) homologue (Rv2869c) was identified that regulates *M. tu-*berculosis cell envelope composition, growth in vivo, and persistence in vivo (153). The protein contains the conserved HexxH and F/LDG motifs (36) and a transmembrane region, reminiscent of the intramembrane cleaving proteases (iCLIP) (271). The proteolytic activity of this iCLIP-like protease regulates cell envelope composition of pathogenic mycobacteria by either directly cleaving membrane-bound transcriptional regulators or through cleavage of other membrane proteins that control cell envelope composition. It was shown by microarray analysis that Rv2869c controls cell envelope composition through positive and negative transcriptional control of lipid anabolic and catabolic genes. Consistent with the comments made above about lipid metabolism, it seems that extractable lipids of the cell envelope, some of which are regulated by Rv2869c, act as direct effectors of pathogenesis and play an important role in the modulation of the host immune response (94, 208, 211). As such, the *M. tuberculosis* ΔRv2869c mutant showed fast clearance from the lungs and livers of mice, strongly suggesting that this protease is required for persistence of *M. tuberculosis* in its host.

Phospholipase C and the Significance of Genetic Polymorphisms

Four highly related genes encoding phospholipase C enzymes were identified in the *M. tuberculosis* genome (50). Three of these genes are organized in the *plcABC* locus (*plcA* and *plcB* are also known as *mpcA* and *mpcB*), while the fourth gene, *plcD*, is located in a different region. Phospholipase C enzymes have been described as important virulence factors in other pathogenic bacteria like *Pseudomonas aeruginosa*, *Bacillus cereus*, *Clostridium perfringens*, and *Listeria monocytogenes*, where they play a role in intracellular survival, cytolysis, and cell-to-cell spread (261). However, their function in the tubercle bacilli is less clear. In contrast to phospholipase D activity, which is detected in both virulent and saprophytic mycobacterial strains, phospholipase C activity is restricted to pathogenic mycobacteria (126, 247). The enzymes encoded by the four *plc* genes share about 70% amino acid identity (209) and have been shown to hydrolyze both phosphatidylcholine and sphingomyelin (126, 209, 274). Putative signal sequences for the Tat secretion system were identified in the mycobacterial Plc proteins; however, the four proteins from *M. tuberculosis* remain associated with the cell envelope (209), a finding that was earlier described for phospholipases of *M. microti* (274).

Raynaud and coworkers demonstrated with single knockout mutants that all four gene products contribute to the phospholipase C activity. They furthermore showed that *M. tuberculosis* triple ($\Delta plcABC$) and quadruple ($\Delta plcABCD$) knockout mutants were attenuated in the late phase of infection, as studied by bacterial growth kinetics in the mouse model. It has been proposed that intracellular phospholipase activity might be important for the release of fatty acids derived from host phospholipids, which the bacteria may use as a carbon source through the β-oxidation cycle and glyoxylate shunt (209).

The importance of the number of active phospholipase C enzymes remains enigmatic. This functional redundancy might be due to different substrate recognition or to different affinities. One of the regions of difference between *M. tuberculosis* and *M. bovis* BCG, RD5, contains the three clustered *plc* genes encoding PlcA, B, and C. This region was originally shown to be present in *M. tuberculosis* but absent from virulent *M. bovis*, attenuated *M. bovis* BCG, and some *M. microti* strains (30, 34, 98). *M. microti* is the infective agent of a TB-like disease in voles and is usually attenuated for humans. Nevertheless, in some cases, *M. microti* was detected as a causative agent of human disease. In contrast to the vole isolates in which the RD5 region was always deleted, all analyzed human isolates had retained the RD5 region (30). These results suggest that the number of functional *plc* genes necessary for the synthesis of an appropriate amount of phospholipase C may have an impact on the virulence and eventually might also influence the host specificity of the strain.

In the sequenced *M. tuberculosis* strain H37Rv, *plcD* is truncated by the insertion of a copy of IS*6110* (50) and by deletion of a genomic fragment designated RvD2 (34, 98). However, an intact version of *plcD* is present in *M. tuberculosis* strains CDC1551 (86) and most clinical isolates (98, 132, 265) and in *M. microti*, *M. bovis*, *M. microti*, and *M. bovis* BCG. The genetic analyses that focused on polymorphisms in the *plc* genes revealed significant numbers of clinical isolates with inactivated *plcA*, *plcB*, *plcC*, and *plcD*, most often caused by IS*6110* insertion due to the preferential integration site within these genes (132, 265). It seems that at least one of the four *plc* genes remained intact in each clinical isolate of *M. tuberculosis*. However, this does not seem to be the case for all virulent *M. bovis* isolates, which naturally lack *plcA-C*, and may still show IS*6110*-mediated inactivation of *plcD* (266). Thus, further conclusions about the contribution of each of the genes to pathogenesis for the moment remain speculative.

Protein Kinases and Phosphatases, Two-Component Systems, and Signal Transduction

Mycobacterial genome analysis revealed that in comparison to most other bacterial genome sequences, *M. tuberculosis* encodes an extraordinarily large number of Ser/Thr protein kinases named PknA,B and PknD-L, one Ser/Thr protein phosphatase (PstP), and two protein Tyr phosphatases (PtpA and PtpB) (50). Strikingly the eukaryotic and mycobacterial Ser/Thr kinases and phosphatases were found to be highly conserved (180, 277). However, some structural differences, which may have important functional consequences, were found between the human PP2Cα (61) and the phosphatase domain of PstP (4, 201). Ser/Thr protein kinases from prokaryotes are thought to be implicated in regulation of development, stress responses, interaction with the host cell, and pathogenicity. It is intriguing that *Legionella pneumophila*, a gram-negative bacterium that like *M. tuberculosis* can retard or inhibit the phagosome-lysosome fusion in infected macrophages, also encodes Ser/Thr protein kinases highly resembling the eukaryotic enzymes (44). In *M. tuberculosis, pknA* and *pknB,* localized near the origin of replication, are essential for growth in vitro according to Sassetti et al. (224) and reside within an operon together with *pstP,* encoding the Ser/Thr phosphatase that regulates kinase activity of PknA and PknB (19, 48). Protein kinases A and B are predominantly expressed during exponential growth and were also found to be upregulated during infection, whereas overexpression of these kinases slows down mycobacterial growth and alters cell morphology (129, 234). Little is known about the function of PknD, which contains a funnel-shaped β-propellor sensor domain which is thought to act as a receptor for multivalent ligands (97), and like PknB, PknE, and PknF, it is implicated in phosphorylation of forkhead-associated domains important for signal transduction (102). PknH is involved in the phosphate-dependent activation of EmbR, the putative transcriptional regulator of key cell wall arabinosyl transferases (EmbC, A, and B), which contains a forkhead-associated domain (3). The exact function of PknI, PknJ, and PknL kinases remains mainly undetermined. They are likely involved in signaling processes (11). The last two members of the mycobacterial Ser/Thr protein kinase family, PknG and PknK, have no apparent transmembrane regions and were therefore predicted to be soluble proteins. PknG was found to be secreted within the phagosomal lumen and cytosol of infected macrophages, inhibiting phagosome-lysosome fusion and mediating intracellular survival (269).

Genes *ptpA* (or *mptpA*) and *ptpB* (or *mptpB*) code for the two mycobacterial tyrosine phosphatases. PtpA is present in slow-growing mycobacterial species and in fast-growing saprophytes, whereas PtpB is restricted to members of the *M. tuberculosis* complex. Both phosphatases were present in whole-cell lysates and culture filtrates of *M. tuberculosis*, suggesting that these proteins are secreted into the extracellular medium (136). Tyrosine phosphatases are essential for the vir-

ulence of several pathogenic bacteria. The restricted distribution of *ptpB* suggested its involvement in virulence, which was later confirmed by Singh and colleagues. Disruption of the *ptpB* gene impairs the ability of the mutant strain to survive in activated macrophages and guinea pigs but not in resting macrophages, which suggests a role for PtpB in the host-pathogen interaction (236).

The *M. tuberculosis* genome contains a relatively low number of two-component systems, which are the molecular systems responsible for stimulus-response coupling in bacteria. Both *B. subtilis* and *E. coli* encode more than 30 members, whereas only 11 complete members have been identified in the mycobacterial genome (50). Some of these systems play a role in virulence, as has been described for PhoP-PhoR (96), PrrA-PrrB (78), SenX3-RegX3 (183), MprA-MprB (278), and DevS-DevR (63).

PhoP-PhoR, one of the best-described virulence-associated two-component systems, was first annotated as being involved in phosphate metabolism because of the observed similarity with counterpart genes in *Bacillus subtilis* (50). However, dramatic changes of the Δ*phoP* mutants during logarithmic phase and changes in the size and morphology of the colonies grown on solid media revealed the involvement of this two-component system in global regulatory circuits. The mutant showed impaired growth within mouse bone marrow–derived macrophages (188) and was also attenuated in immunocompetent mice. It was recently shown that the PhoP-PhoR system controls the biosynthesis of polyketide-derived lipids that are known to be restricted to pathogenic species of the *M. tuberculosis* complex (96). At present, it remains unknown which environmental signals are sensed by the sensor histidine kinase PhoR.

MprA-MprB (Mpr stands for mycobacterial persistence regulator), the other well-studied two-component system of *M. tuberculosis*, is required for the establishment and maintenance of persistent infection. The sensor kinase MprB autophosphorylates and donates a phosphate group to the response regulator, MprA (278).

MprA recognizes a 19-bp sequence comprising two loosely conserved 8-bp direct repeat subunits separated by three nucleotides, defined as the MprA box. Protein phosphorylation was not required for binding to this DNA sequence by MprA in vitro; however, phosphorylation enhanced DNA binding by MprA and was required for the regulation of *mprA* and *pepD* by MprA in vivo (111). Genes regulated by MprA in *M. tuberculosis* were identified by transcriptomics. Two important stress-responsive factors, SigB and SigE, which contain the conserved MprA box in their upstream regions, were found to be expressed in an MprA-dependent manner. The *mprA* gene itself was also found to be induced after exposure to stress factors, including ionic and nonionic detergents and alkaline pH (110). Deletion of 39 amino acids from the extracellular (putative sensor) domain of MprB blocked the transduction of the stress signals. Interestingly, these studies demonstrate a clear link between the detection of and the reaction to stress; they also demonstrated the importance of these systems for persistence of *M. tuberculosis* infection.

Another important regulation system in *M. tuberculosis*, the DosR/DosS/DosT regulon, is involved in the response to hypoxia and nitric oxide and may contribute to latency in vivo (215, 275). Up to 47 genes were found to be under the control of this system. The ability to respond to hypoxia is certainly a crucial step in the *M. tuberculosis* infection cycle. Although the tubercle bacilli are mainly considered as aerobes, the genome data revealed the potential for microaerophilic and anaerobic respiration. An operon, *narGHJI*, is present, encoding a nitrate reductase that allows utilization of nitrate as a terminal electron acceptor. Investigating the role of nitrate reductase, Bange and colleagues generated a *narG* mutant of *M. bovis* BCG (270) and found that immunodeficient mice infected with the *narG* mutant developed smaller granulomas than those infected with the wild-type strain. Furthermore, mice infected with the mutant presented no clinical signs of disease after more than 200 days. It therefore appears that the ability to respire

anaerobically contributes to virulence. It is also noteworthy that one of the classical microbiological methods to differentiate *M. bovis* from *M. tuberculosis* is based on nitrate reductase activity; *M. tuberculosis* reduces nitrate to nitrite, while *M. bovis* performs this reduction very poorly. In this respect, Bange and colleagues have shown that this defect in nitrate reductase activity is due to a point mutation in the promoter of the *M. bovis narGHIJ* cluster (245).

Anti-TB Drug Design

The current treatment, as recommended by the WHO, requires patients to adhere to a three- or four-drug regimen for a period of 6 to 8 months (67). Many patients fail to complete the treatment because of the side effects and the complicated treatment; this has resulted in drug resistance. The emergence of resistance to any anti-TB drug, together with an increase of multidrug-resistant TB, defined as resistance to at least isoniazid and rifampin, means that good TB control programs and the development of new drugs are crucial.

The availability of several mycobacterial genome sequences strongly stimulated the research on specific proteins and lipids that might serve as new drug targets. This approach was further encouraged by the development of genome-wide screens that allowed the essentiality of genes to be evaluated (141, 224). Confirmation that thereby identified essential genes that might be important drug targets was obtained by the discovery of a highly active anti-TB compound belonging to the diarylquinoline family. The C subunit of ATP synthase, AtpE, was identified as the target of this new compound (6). This essential enzyme uses the transmembrane proton-motive force to generate ATP for the cell.

Other prominent potential drug targets are enzymes that are involved in biosynthesis of essential cell wall lipids. For example, PimA is an essential α-mannosyltransferase responsible for the first mannosylation step in phosphatidylinositolmannoside synthesis (134), and many efforts are underway to determine the crystal structures of this protein (103). Acyl-CoA carboxylases commit acyl-CoAs to the biosynthesis of mycolic acids. Recently, the crystal structure of AccD5 has been determined and was used for extensive in silico screening, which resulted in the identification of a strong inhibitor (145).

The assembly of the arabinan portions of cell wall polysaccharides in mycobacteria involves a family of arabinosyltransferases (AraT) that promote the polymerization of decaprenolphosphoarabinose. Mycobacterial viability depends upon the ability of the organism to synthesize an intact arabinan, and thus compounds that inhibit these AraT are potential lead compounds. Some inhibiting compounds were found but still need to be optimized (49).

When KatG is mutated or deleted, as is often the case in INH-resistant strains, antioxidant protection is provided by enhanced expression of the peroxiredoxin AhpC, which is itself reduced by AhpD. Inhibition of AhpD might impair the antioxidant protection afforded by AhpC and make KatG-negative (INH resistant) strains more sensitive to oxidative stress. Computational studies provided support for a proposed AhpD substrate binding site and stimulated the search for potent inhibitors (135). Recently the structure and mechanism of AhpC were revealed, which might be exploited for the design of AhpC-specific inhibitors (104).

As described above, mycobactin has a critical role in iron acquisition inside host cells. Thus, mycobactin biosynthesis is an attractive target for the development of new anti-TB drugs that are effective under the iron-limiting conditions prevalent inside macrophages. Recently, it was shown that small-molecule inhibition of siderophore biosynthesis via an arylic acyl adenylate analogue inhibits growth of M. tuberculosis in an iron-dependent manner (80).

PtpA (or MptpA) is one of the two secreted phosphatases and is believed to be involved in survival of M. tuberculosis in host cells (236). Inhibitors of PtpA were recently identified on the basis of natural product analysis and fragment-based drug design (156).

Natural products are also the basis of pyranodibenzofuran derivatives, a novel class of recently identified potential inhibitors of M. tuberculosis (199).

As described above, the secreted serine-threonine PknG was found to promote survival of pathogenic mycobacteria within macrophages. A tetrahydrobenzothiophene was identified that specifically inhibited the kinase activity of PknG, thereby inhibiting the growth and/or survival of both M. bovis BCG and M. tuberculosis (269).

Development of New Diagnostics and Vaccines

The ability of M. tuberculosis to infect its host and remain latent for many years before reactivation is a key obstacle to the control of TB. Testing for latent TB has long been done in the form of tuberculin skin testing (Mantoux), but problems with cross-reactivity with environmental antigens and low sensitivity in immunosuppressed individuals, such as those infected with human immunodeficiency virus, underline the need for more specific and more sensitive tests (73, 182).

The enzyme-linked immunosorbent spot-forming cell (ELISPOT) assay measures IFN-γ production from whole blood or peripheral blood mononuclear cells by counting the number of producing cells that were stimulated either with purified protein derivative (PPD) or specific antigens (69, 140, 181).

The QuantiFERON-TB (QFT) test measures the amount of IFN-γ released by sensitized white blood cells after whole blood is incubated with PPD (181). An improved version of this test, QuantiFERON-TB Gold (QFT-G), recently gained U.S. Food and Drug Administration approval as an aid for diagnosing M. tuberculosis infections. This test detects the release of IFN-γ in freshly heparinized whole blood from sensitized persons when it is incubated with mixtures of synthetic peptides representing ESAT-6 and CFP-10. Because both ESAT-6 and CFP-10 are absent from all BCG vaccine strains and from commonly encountered nontuberculous mycobacteria except M. kansasii,

M. szulgai, and *M. marinum* (5), QFT-G is expected to be more specific in the detection of TB infections than tests based on PPD antigens.

New additional antigens to be used in diagnostic ELISPOT assays or eventually new QFT-like tests are still needed to improve the specificity. Apart from ESAT-6 and CFP-10, which have also been used in skin tests (196, 263) and ELISPOT (140), HBHA might be useful for diagnostic purposes. As described above, HBHA-mediated extrapulmonary dissemination was suggested to be one of the key steps in the development of latent TB (189, 257). The finding that latently infected human individuals mount a strong T-cell response to HBHA while patients with active disease do not suggests that HBHA could be used as a diagnostic marker (146).

Because prevention is better than cure, many efforts have been made and are still underway to develop improved vaccines against TB. The current BCG vaccine is efficient in the protection against TB meningitis, but it often provides only minor protection against pulmonary TB, the prevalent form in adults (84). In order to overcome this problem, several potential new vaccine candidates have recently been developed, some of which are presently entering clinical trials. They are either based on the secreted proteins ESAT-6 or members of the antigen 85 group or they make use of a fusion protein (Mtb72F) that is composed of a serine protease Rv0125 (Mtb32a) and PPE18, a member of the PPE protein family (mtb39, Rv1196). These experimental vaccines are administered either as subunit vaccines (23, 70, 166), as recombinant BCG (120), or by the use of viral vectors (167) or fused proteins in special adjuvant formulations (23, 70). In addition, attenuated *M. tuberculosis* (160, 221) and recombinant BCG strains that express proteins otherwise absent from BCG also represent promising candidates (100, 203).

It is clear that the attempts to discover new diagnostic markers, drugs, and vaccines against TB greatly benefit from the availability of genomic and postgenomic data on *M. tuberculosis*

and other members of the *M. tuberculosis* complex.

OTHER MAJOR MYCOBACTERIAL HUMAN PATHOGENS

M. leprae

Mycobacterium leprae, the infectious agent of leprosy, represents another classical major human pathogen among the slow-growing mycobacteria. As mentioned above, leprosy cases have been decreasing as a result of multidrug treatment campaigns introduced by the WHO, but the estimated annual incidence remains at more than 690,000 leprosy cases worldwide. In this respect, elucidating the genetic blueprint of this organism was particularly interesting because *M. leprae* does not grow on axenic medium. At 3,268,182 bp, the *M. leprae* genome is almost 1.2 Mb smaller than that of *M. tuberculosis*, and its GC content of ~58% also differs extensively (51). At one time, this exceptionally low GC value led to doubts as to whether the leprosy bacillus was a true member of the *Mycobacterium* genus, but the genome sequence has confirmed the identity of the leprosy bacillus as a true *Mycobacterium*. The difference in genome size results from loss of genetic information, with a coding capacity of less than 50% in *M. leprae* compared with 90.8% in *M. tuberculosis*. This reduction is due to both the deletion of chromosomal regions and the accumulation of mutations in genes resulting in pseudogenes. These decayed gene remnants are abundant and apparently represent the removal of functions no longer required for the highly adapted in vivo growth, where substrates may be provided by the host cell. Here it should also be mentioned that *M. leprae* has a temperature optimum of 31°C, and therefore it mainly affects parts of the colder sites of the human body.

One of the most striking differences in gene families between the leprosy and tubercle bacilli is the case of the PE and PPE proteins. Whereas these two families account for 167 genes in *M. tuberculosis*, only 9 intact PE or PPE genes could be identified in *M. leprae*, with complete loss of the PE-PGRS subfamily

(51). This most likely reflects both downsizing in *M. leprae* and expansion in *M. tuberculosis*. Interestingly, some of the intact PE and PPE genes in *M. leprae*, such as ML1828, ML1182, or ML0411, are in gene-poor regions or are surrounded by pseudogenes. Hence, while neighboring loci were deleted or accumulated mutations, these PE and PPE genes were maintained, underlining the importance of retaining a minimal set of functional PE and PPE proteins. In both *M. leprae* and *M. tuberculosis*, the single *rrn* operon is situated ~1.3 Mb from *oriC*, the chromosomal origin of replication. A copy of *rrn* is located adjacent to *oriC* in most bacterial genomes, and this proximity boosts rRNA production through increased gene dosage. It has been suggested that the atypical arrangement seen in mycobacterial pathogens may be related to their slow growth. However, the extremely reduced doubling time of 14 days of *M. leprae* and the failure to cultivate this bacterium on xenotic media are most probably caused by the loss of many functions related to its energy metabolism.

M. leprae also lacks a functional catalase-peroxidase because *katG* is a pseudogene. As mentioned above, in *M. tuberculosis* KatG is known to play two roles: the detoxification of oxygen radicals and the activation of the antimycobacterial drug isoniazid. Catalase-peroxidase has been clearly identified as a virulence factor in *M. tuberculosis* complex organisms, most likely functioning to protect the bacillus from the reactive oxygen species generated by the macrophage respiratory burst. Absence of this defense mechanism indicates either that the leprosy bacillus has some other means to withstand reactive oxygen or that it fails to trigger a respiratory burst. On the other hand, the absence of a functional KatG explains why *M. leprae* is resistant to isoniazid, and it makes the use of this efficient anti–TB drug inappropriate for the treatment of leprosy.

Until recently, because *M. leprae* strains cannot be cultivated on xenotic media, not much was known of the population structure of leprosy bacilli. This has greatly changed in the meantime as a result of the use of various comparative genomics techniques used on a large number of biopsy samples from leprosy patients from all over the world. This investigation has undoubtedly shown that extant *M. leprae* isolates show very little genetic variability among each other and represent a single clone that has spread efficiently around the world. However, a few SNPs have been identified that allowed the various *M. leprae* strains to be grouped according to their geographical origin and allowed an evolutionary scheme for the leprosy bacilli to be proposed (174). As with the tubercle bacilli (106), the results of the investigation of the population structure of *M. leprae* strains point to East Africa as the most probable geographical region from where leprosy bacilli emerged and were transmitted to other parts of the world (174).

M. ulcerans

M. ulcerans was first identified and shown to be the etiologic agent of Buruli ulcer in 1948 after the appearance of unusual skin ulcers in six patients from a remote farming community in southeastern Australia (149). Proximity to stagnant or slow-flowing water was a risk factor for contracting an ulcer, and there is now experimental evidence that *M. ulcerans* may be transmitted to mammalian hosts via aquatic insects (159).

M. ulcerans is an unusual mycobacterium because it produces a toxin, mycolactone. This polyketide-derived lipophilic molecule is responsible for the extensive subcutaneous necrosis seen in *M. ulcerans* infection that undermines the overlying epidermis and results in the typical pathology of Buruli ulcer. The structure of mycolactone is known (93), and its absolute configuration has been confirmed by total chemical synthesis (15). Mycolactone is a macrocyclic polyketide and is part of a large class of natural products typically produced by actinomycetes such as the *Streptomyces*. This class of compounds includes a number of extremely valuable drugs, such as the antibacterial erythromycin A, the antiparasitic avermectin, the antifungal amphotericin B, and the immunosuppressant and anticancer compound rapamycin. Injection of purified mycolactone replicates the human disease in rodent and cellular models (93). Mycolactone is cytotoxic, has antiphagocytic activity, induces apoptosis of antigen-presenting cells, and inhibits the proinflammatory cytokine response (55). From a relatively early stage in disease, *M. ulcerans* is found primarily in an extracellular location in mammalian tissues, in contrast to other pathogenic mycobacteria (55). Studies of mycolactone-deficient *M. ulcerans* mutants suggest that mycolactone is required for lysis of the plasmatocytes that transport the bacterium away from the coelomic cavity in *Naucoris cimicoides* (158), suggesting that mycolactone may have an important role in the adaptation of *M. ulcerans* to an arthropod niche.

In 1998, the WHO, alarmed at the dramatic increases of cases of Buruli ulcer across sub-Saharan Africa, launched the Global Buruli Ulcer Initiative (8). A principal aim of this initiative was to promote research to develop tools to diagnose, treat, and prevent the disease, particularly by genomics. In 2001, the Institut Pasteur launched a project to determine the complete genome sequence of a West African epidemic strain of *M. ulcerans*. The recently completed project (250a) has provided deep insight into the evolution of this pathogen and has revealed new leads for developing urgently needed diagnostic tests and therapeutics.

THE COMPLETE GENOME SEQUENCE OF *MYCOBACTERIUM ULCERANS* AGY99

The complete DNA sequence of the 5,805,761-bp genome of *M. ulcerans* Agy99 was obtained in 2005. The genome is composed of two circular replicons, a chromosome of 5,631,606 bp and a 174,155-bp plasmid named pMUM001. The chromosome has an average G+C content of 65.72%. It contains 4,281 protein-coding sequences (CDS). There were also 727 pseudogenes detected. The chromosome harbors two prophages and 302 insertion sequence elements (ISE), which in-

cludes 209 copies of IS*2404* and 83 copies of IS*2606*. The plasmid contains 81 CDS with an average G+C content of 62.5% and includes four copies of IS*2404* and eight copies of IS*2606* (247, 249, 250).

Comparisons with *M. tuberculosis* revealed the same general themes, such as an abundance of proteins involved in lipid metabolism and prominent PE and PPE families encoding the Gly-Ala-rich cell envelope proteins. The number of potential transport proteins and enzyme systems involved in carbon and energy metabolism was considerably higher in *M. ulcerans*, and regulatory proteins were also numerous.

INSIGHTS INTO THE EVOLUTION OF *M. ULCERANS*

At the nucleotide level, *M. ulcerans* is nearly indistinguishable from *Mycobacterium marinum*. *M. marinum* causes a tuberculoid-like disease in fish and frogs, and in contrast to *M. ulcerans*, it causes only a limited cutaneous infection in humans characterized pathologically by intracellular bacterial replication and a granulomatous host response. Unlike *M. ulcerans*, *M. marinum* grows relatively rapidly (4-h doubling time compared with >50 h) and produces light-inducible carotenoid pigments that protect the bacterium from UV-induced damage (207). Multilocus DNA sequence comparisons between these two species show greater than 98% sequence identity and have led to the hypothesis that *M. ulcerans* has recently diverged from a *M. marinum*–like progenitor (248). Data from the genome project confirmed this hypothesis and showed that *M. ulcerans* has evolved by acquisition of a plasmid, expansion of ISE, accumulation of pseudogenes, genome rearrangements, and genome reduction. These are all features consistent with a bacterium that has recently passed through an evolutionary bottleneck and is now adapting to a new niche environment.

The pMUM001 Virulence Plasmid

The discovery of pMUM001 was a key finding of the *M. ulcerans* genome project because the plasmid is essentially a blueprint for mycolactone production. It contains three very large genes (*mlsA1*, 50 kb; *mlsA2*, 7 kb; and *mlsB*, 40 kb) encoding giant type I polyketide synthases that span more than 60% of the total DNA sequence (249). The 12-membered macrocyclic core of mycolactone is synthesized by the polyketide synthase encoded by the *mlsA1* and *mlsA2* genes, and the side chain is produced through the action of a third polyketide synthase encoded by the *mlsB* gene. pMUM001 also contains genes required for postassembly mycolactone modification, including a cytochrome P450 (MUP053) and a putative type III ketosynthase, or joinase (MUP045), which may transfer the completed polyketide side chain onto the core of the toxin (249). High-performance liquid chromatography–linked sequential mass spectrometry has shown that *M. ulcerans* in culture produces a number of mycolactone structural variants (118). Most of them appear linked to the aberrant operation of the cytochrome P450 (247). Some strains of *M. ulcerans* produce variants that are due to an alteration of the *mls* DNA sequence within a specific functional domain (119). These alterations are thought to have arisen by recombination-mediated domain swapping because there is an extraordinarily high level of nucleotide homology between the many repeating functional domains of the *mls* genes. In some instances, the level of nucleotide identity is 100%, and there is less than 3% nucleotide variation for even the most variable functional domain, which is the ketosynthase domain present in 16 copies (249). In addition to facilitating "natural" combinatorial chemistry by domain swapping as described above, the high DNA homology also suggests that the entire region has recently evolved by a series of recombination-mediated in-frame duplication, deletion, and swapping events. Although all strains of *M. ulcerans* have been shown to contain a plasmid related to pMUM001 and to produce mycolactones, recombination also appears to mediate frequent deletion of functional domains, leading to the formation of spontaneous mycolactone–minus mutants among laboratory-passaged strains (247). These observations suggest that the evolution of the mycolactone locus is not only re-cent, but that there is a very strong selection process acting on the population to maintain those mycobacteria with intact *mls* genes and that produce mycolactone. The key questions surrounding the role and function of mycolactone remain to be determined.

Genome Reduction

Genome comparisons between *M. ulcerans* and the recently completed 6.6-Mb genome *M. marinum* M strain (http://www.sanger.ac.uk/Projects/M_marinum/) have shown how *M. ulcerans* has evolved from a *M. marinum*–like progenitor. Excluding ISE and PE/PPE genes, 80% of the *M. ulcerans* CDS have *M. marinum* orthologues with high synteny and an average DNA identity level of 98.8%. However, a key finding from these comparisons was that numerous chromosomal inversions and deletions had occurred in *M. ulcerans*, many of them mediated by the two high-copy-number ISE. There were 157 regions of difference or MURDs (*M. ulcerans* region of difference) greater than 1 kb, accounting for 1,232 kb present in *M. marinum* but absent from *M. ulcerans*. Approximately 170 kb of these MURDs represented insertions in *M. marinum*; however, the majority of the MURDs have arisen from the accumulated deletion of 1,100 kb of DNA from *M. ulcerans*. Like *M. leprae*, many paralogous gene families have been lost, notably several secondary metabolism loci and the PE and PPE genes, the latter accounting for 45% of all MURDs. The same pattern of loss was seen among the pseudogenes, suggesting that secondary metabolites, for example, carotenoid pigments, and PE and PPE proteins are no longer used by *M. ulcerans*. Interestingly, two of five ESX (ESAT-6) systems in *M. ulcerans* have been lost by mutation and deletion. One of these loci is orthologous to the ESX-1 locus (ESAT-6 system 1) in the tubercle bacillus (32). EspA family effector proteins secreted by ESX systems (88) have also been depleted. ESX systems promote granuloma formation by *M. marinum* and *M. tuberculosis*, and so their loss may partly explain the predominantly extracellular location of *M. ulcerans* in diseased tissue.

In addition to the 81 CDS predicted in pMUM001, 475 kb of DNA was identified in *M. ulcerans* Agy99 that is absent from the M strain of *M. marinum*. The very high number of copies of IS*2404* and IS*2606* account for 396 kb of this total. These elements have severely altered genome architecture by causing chromosome inversions, DNA deletions, and insertions within genes to cause 15% of the 727 pseudogenes. The remaining 79 kb consist of 143 regions, encompassing two prophages and extensive DNA sequence variation among the highly polymorphic 3′ sections of 97 PE and PPE genes and pseudogenes. There are also 32 other DNA segments that contain 11 predicted CDS, encoding hypothetical membrane proteins, hydrolytic enzymes, and lipases, some of which may play a role in pathogenesis or find application in the development of new diagnostic tools.

CONCLUDING REMARKS

The success of mycobacterial pathogens is undoubtedly a consequence of a finely tuned dialogue that has evolved over time between the pathogen and host. The availability of genome sequences from several mycobacterial species provides the principal resource for researchers to explore their pathogenic life cycle and evolution. This information will play a key role in the development of new therapeutic and preventive strategies to address the immense global burden of mycobacterial disease and relieve the human misery they cause.

ACKNOWLEDGMENTS

We are grateful to P. Brodin, D. Bottai, C. Demangel, T. Garnier, S. V. Gordon, and A. S. Pym for fruitful discussions and advice. This work was supported by the Institut Pasteur (PTR35, PTR110, GPH5), the European Union (QLK2-CT-2001-02018 and LSHG-CT-2003-503367), the Ministère de la Recherche et Nouvelles Technologies (ACI Microbiologie), and the Association Française Raoul Follereau.

REFERENCES

1. **Ahmad, S., S. El-Shazly, A. S. Mustafa, and R. Al-Attiyah.** 2004. Mammalian cell-entry proteins encoded by the *mce3* operon of *Mycobacterium tuberculosis* are expressed during natural infection in humans. *Scand. J. Immunol.* **60**:382–391.

2. **Alcais, A., C. Fieschi, L. Abel, and J. L. Casanova.** 2005. Tuberculosis in children and adults: two distinct genetic diseases. *J. Exp. Med.* **202**:1617–1621.

3. **Alderwick, L. J., V. Molle, L. Kremer, A. J. Cozzone, T. R. Dafforn, G. S. Besra, and K. Futterer.** 2006. Molecular structure of EmbR, a response element of Ser/Thr kinase signaling in *Mycobacterium tuberculosis*. *Proc. Natl. Acad. Sci. USA* **103**:2558–2563.

4. **Alzari, P. M.** 2004. First structural glimpse at a bacterial Ser/Thr protein phosphatase. *Structure* **12**:1923–1924.

5. **Andersen, P., M. E. Munk, J. M. Pollock, and T. M. Doherty.** 2000. Specific immune-based diagnosis of tuberculosis. *Lancet* **356**:1099–1104.

6. **Andries, K., P. Verhasselt, J. Guillemont, H. W. Gohlmann, J. M. Neefs, H. Winkler, J. Van Gestel, P. Timmerman, M. Zhu, E. Lee, P. Williams, D. de Chaffoy, E. Huitric, S. Hoffner, E. Cambau, C. Truffot-Pernot, N. Lounis, and V. Jarlier.** 2005. A diarylquinoline drug active on the ATP synthase of *Mycobacterium tuberculosis*. *Science* **307**:223–227.

7. **Anonymous.** 1998. Elimination of leprosy as a public health problem (update). *Wkly. Epidemiol. Rec.* **73**:308–312.

8. **Anonymous.** 1998. *Proceedings of the International Conference on Buruli Ulcer Control and Research.* Yamoussoukro, Cote d'Ivoire.

9. **Arruda, S., G. Bomfim, R. Knights, T. Huima-Byron, and L. W. Riley.** 1993. Cloning of an *M. tuberculosis* DNA fragment associated with entry and survival inside cells. *Science* **261**:1454–1457.

10. **Astarie-Dequeker, C., E. N. N'Diaye, V. Le Cabec, M. G. Rittig, J. Prandi, and I. Maridonneau-Parini.** 1999. The mannose receptor mediates uptake of pathogenic and nonpathogenic mycobacteria and bypasses bactericidal responses in human macrophages. *Infect. Immun.* **67**:469–477.

11. **Av-Gay, Y., and M. Everett.** 2000. The eukaryotic-like Ser/Thr protein kinases of *Mycobacterium tuberculosis*. *Trends Microbiol.* **8**:238–244.

12. **Banu, S., S. V. Gordon, S. Palmer, R. Islam, S. Ahmed, K. M. Alam, S. T. Cole, and R. Brosch.** 2004. Genotypic analysis of *Mycobacterium tuberculosis* in Bangladesh and prevalence of the Beijing strain. *J. Clin. Microbiol.* **42**:674–682.

13. **Banu, S., N. Honore, B. Saint-Joanis, D. Philpott, M. C. Prevost, and S. T. Cole.** 2002. Are the PE-PGRS proteins of *Mycobacterium tuberculosis* variable surface antigens? *Mol. Microbiol.* **44**:9–19.

14. **Batoni, G., D. Bottai, G. Maisetta, M. Pardini, A. Boschi, W. Florio, S. Esin, and M. Campa.** 2001. Involvement of the *Mycobacterium tuberculosis* secreted antigen SA-5K in intracellular survival of recombinant *Mycobacterium smegmatis*. *FEMS Microbiol. Lett.* **205**:125–129.

15. **Benowitz, A. B., S. Fidanze, P. L. Small, and Y. Kishi.** 2001. Stereochemistry of the core structure of the mycolactones. *J. Am. Chem. Soc.* **123:**5128–5129.

16. **Berks, B. C., F. Sargent, and T. Palmer.** 2000. The Tat protein export pathway. *Mol. Microbiol.* **35:**260–274.

17. **Bermudez, L. E., and J. Goodman.** 1996. *Mycobacterium tuberculosis* invades and replicates within type II alveolar cells. *Infect. Immun.* **64:** 1400–1406.

18. **Berthet, F. X., M. Lagranderie, P. Gounon, C. Laurent-Winter, D. Ensergueix, P. Chavarot, F. Thouron, E. Maranghi, V. Pelicic, D. Portnoi, G. Marchal, and B. Gicquel.** 1998. Attenuation of virulence by disruption of the *Mycobacterium tuberculosis* erp gene. *Science* **282:**759–762.

19. **Boitel, B., M. Ortiz-Lombardia, R. Duran, F. Pompeo, S. T. Cole, C. Cervenansky, and P. M. Alzari.** 2003. PknB kinase activity is regulated by phosphorylation in two Thr residues and dephosphorylation by PstP, the cognate phospho-Ser/Thr phosphatase, in *Mycobacterium tuberculosis*. *Mol. Microbiol.* **49:**1493–1508.

20. **Bottai, D., S. Esin, G. Batoni, M. Pardini, G. Maisetta, V. Donati, F. Favilli, W. Florio, and M. Campa.** 2005. Disruption of the gene encoding for secretion antigen SA5K affects growth of *Mycobacterium bovis* bacillus Calmette-Guerin in human macrophages and in mice. *Res. Microbiol.* **156:**393–402.

21. **Braden, C. R., G. P. Morlock, C. L. Woodley, K. R. Johnson, A. C. Colombel, M. D. Cave, Z. Yang, S. E. Valway, I. M. Onorato, and J. T. Crawford.** 2001. Simultaneous infection with multiple strains of *Mycobacterium tuberculosis*. *Clin. Infect. Dis.* **33:**e42–e47.

22. **Braibant, M., P. Gilot, and J. Content.** 2000. The ATP binding cassette (ABC) transport systems of *Mycobacterium tuberculosis*. *FEMS Microbiol. Rev.* **24:**449–467.

23. **Brandt, L., Y. A. Skeiky, M. R. Alderson, Y. Lobet, W. Dalemans, O. C. Turner, R. J. Basaraba, A. A. Izzo, T. M. Lasco, P. L. Chapman, S. G. Reed, and I. M. Orme.** 2004. The protective effect of the *Mycobacterium bovis* BCG vaccine is increased by coadministration with the *Mycobacterium tuberculosis* 72-kilodalton fusion polyprotein Mtb72F in *M. tuberculosis*–infected guinea pigs. *Infect. Immun.* **72:**6622–6632.

24. **Braunstein, M., S. S. Bardarov, and W. R. Jacobs, Jr.** 2002. Genetic methods for deciphering virulence determinants of *Mycobacterium tuberculosis*. *Methods Enzymol.* **358:**67–99.

25. **Braunstein, M., A. M. Brown, S. Kurtz, and W. R. Jacobs, Jr.** 2001. Two nonredundant SecA homologues function in mycobacteria. *J. Bacteriol.* **183:**6979–6990.

26. **Braunstein, M., B. J. Espinosa, J. Chan, J. T. Belisle, and W. R. Jacobs, Jr.** 2003. SecA2 functions in the secretion of superoxide dismutase A and in the virulence of *Mycobacterium tuberculosis*. *Mol. Microbiol.* **48:**453–464.

27. **Brennan, M. J., and G. Delogu.** 2002. The PE multigene family: a "molecular mantra" for mycobacteria. *Trends Microbiol.* **10:**246–249.

28. **Brennan, M. J., G. Delogu, Y. Chen, S. Bardarov, J. Kriakov, M. Alavi, and W. R. Jacobs, Jr.** 2001. Evidence that mycobacterial PE_PGRS proteins are cell surface constituents that influence interactions with other cells. *Infect. Immun.* **69:**7326–7333.

29. **Brodin, P., M. I. de Jonge, L. Majlessi, C. Leclerc, M. Nilges, S. T. Cole, and R. Brosch.** 2005. Functional analysis of early secreted antigenic target-6, the dominant T-cell antigen of *Mycobacterium tuberculosis*, reveals key residues involved in secretion, complex formation, virulence, and immunogenicity. *J. Biol. Chem.* **280:** 33953–33959.

30. **Brodin, P., K. Eiglmeier, M. Marmiesse, A. Billault, T. Garnier, S. Niemann, S. T. Cole, and R. Brosch.** 2002. Bacterial artificial chromosome-based comparative genomic analysis identifies *Mycobacterium microti* as a natural ESAT-6 deletion mutant. *Infect. Immun.* **70:**5568–5578.

31. **Brodin, P., L. Majlessi, L. Marsollier, M. I. de Jonge, D. Bottai, C. Demangel, J. Hinds, O. Neyrolles, P. D. Butcher, C. Leclerc, S. T. Cole, and R. Brosch.** 2006. Dissection of ESAT-6 system 1 of *Mycobacterium tuberculosis* and impact on immunogenicity and virulence. *Infect. Immun.* **74:**88–98.

32. **Brodin, P., I. Rosenkrands, P. Andersen, S. T. Cole, and R. Brosch.** 2004. ESAT-6 proteins: protective antigens and virulence factors? *Trends Microbiol.* **12:**500–508.

33. **Brosch, R., S. V. Gordon, K. Eiglmeier, T. Garnier, and S. T. Cole.** 2000. Comparative genomics of the leprosy and tubercle bacilli. *Res. Microbiol.* **151:**135–142.

34. **Brosch, R., S. V. Gordon, M. Marmiesse, P. Brodin, C. Buchrieser, K. Eiglmeier, T. Garnier, C. Gutierrez, G. Hewinson, K. Kremer, L. M. Parsons, A. S. Pym, S. Samper, D. van Soolingen, and S. T. Cole.** 2002. A new evolutionary scenario for the *Mycobacterium tuberculosis* complex. *Proc. Natl. Acad. Sci. USA* **99:**3684–3689.

35. **Brosch, R., S. V. Gordon, A. Pym, K. Eiglmeier, T. Garnier, and S. T. Cole.** 2000. Comparative genomics of the mycobacteria. *Int. J. Med. Microbiol.* **290:**143–152.

36. **Brown, G. D., J. A. Dave, N. C. Gey van Pittius, L. Stevens, M. R. Ehlers, and A. D. Beyers.** 2000. The mycosins of *Mycobacterium tuberculosis* H37Rv: a family of subtilisin-like serine proteases. *Gene* **254:**147–155.

37. **Bryk, R., C. D. Lima, H. Erdjument-Bromage, P. Tempst, and C. Nathan.** 2002. Metabolic enzymes of mycobacteria linked to antioxidant defense by a thioredoxin-like protein. *Science* **295:**1073–1077.

38. **Camacho, L. R., D. Ensergueix, E. Perez, B. Gicquel, and C. Guilhot.** 1999. Identification of a virulence gene cluster of *Mycobacterium tuberculosis* by signature-tagged transposon mutagenesis. *Mol. Microbiol.* **34:**257–267.

39. **Cameron, R. M., K. Stevenson, N. F. Inglis, J. Klausen, and J. M. Sharp.** 1994. Identification and characterization of a putative serine protease expressed in vivo by *Mycobacterium avium* subsp. *paratuberculosis. Microbiology* **140:**1977–1982.

40. **Camus, J. C., M. J. Pryor, C. Medigue, and S. T. Cole.** 2002. Re-annotation of the genome sequence of *Mycobacterium tuberculosis* H37Rv. *Microbiology* **148:**2967–2973.

41. **Canneva, F., M. Branzoni, G. Riccardi, R. Provvedi, and A. Milano.** 2005. Rv2358 and FurB: two transcriptional regulators from *Mycobacterium tuberculosis* which respond to zinc. *J. Bacteriol.* **187:**5837–5840.

42. **Card, G. L., N. A. Peterson, C. A. Smith, B. Rupp, B. M. Schick, and E. N. Baker.** 2005. The crystal structure of Rv1347c, a putative antibiotic resistance protein from *Mycobacterium tuberculosis*, reveals a GCN5-related fold and suggests an alternative function in siderophore biosynthesis. *J. Biol. Chem.* **280:**13978–13986.

43. **Carver, T. J., K. M. Rutherford, M. Berriman, M. A. Rajandream, B. G. Barrell, and J. Parkhill.** 2005. ACT: the Artemis Comparison Tool. *Bioinformatics* **21:**3422–3423.

44. **Cazalet, C., C. Rusniok, H. Bruggemann, N. Zidane, A. Magnier, L. Ma, M. Tichit, S. Jarraud, C. Bouchier, F. Vandenesch, F. Kunst, J. Etienne, P. Glaser, and C. Buchrieser.** 2004. Evidence in the *Legionella pneumophila* genome for exploitation of host cell functions and high genome plasticity. *Nat. Genet.* **36:**1165–1173.

45. **Chan, E. D., J. Chan, and N. W. Schluger.** 2001. What is the role of nitric oxide in murine and human host defense against tuberculosis? Current knowledge. *Am. J. Respir. Cell. Mol. Biol.* **25:**606–612.

46. **Chen, L., Q. W. Xie, and C. Nathan.** 1998. Alkyl hydroperoxide reductase subunit C (AhpC) protects bacterial and human cells against reactive nitrogen intermediates. *Mol. Cell* **1:**795–805.

47. **Chitale, S., S. Ehrt, I. Kawamura, T. Fujimura, N. Shimono, N. Anand, S. Lu, L. Cohen-Gould, and L. W. Riley.** 2001. Recombinant *Mycobacterium tuberculosis* protein associated with mammalian cell entry. *Cell Microbiol.* **3:**247–254.

48. **Chopra, P., B. Singh, R. Singh, R. Vohra, A. Koul, L. S. Meena, H. Koduri, M. Ghildiyal, P. Deol, T. K. Das, A. K. Tyagi, and Y. Singh.** 2003. Phosphoprotein phosphatase of *Mycobacterium tuberculosis* dephosphorylates serine-threonine kinases PknA and PknB. *Biochem. Biophys. Res. Commun.* **311:**112–120.

49. **Cociorva, O. M., S. S. Gurcha, G. S. Besra, and T. L. Lowary.** 2005. Oligosaccharides as inhibitors of mycobacterial arabinosyltransferases. Di- and trisaccharides containing C-3 modified arabinofuranosyl residues. *Bioorg. Med. Chem.* **13:**1369–1379.

50. **Cole, S. T., R. Brosch, J. Parkhill, T. Garnier, C. Churcher, D. Harris, S. V. Gordon, K. Eiglmeier, S. Gas, C. E. Barry III, F. Tekaia, K. Badcock, D. Basham, D. Brown, T. Chillingworth, R. Connor, R. Davies, K. Devlin, T. Feltwell, S. Gentles, N. Hamlin, S. Holroyd, T. Hornsby, K. Jagels, A. Krogh, A. McLean, S. Moule, L. Murphy, K. Oliver, J. Osborne, M. A. Quail, M.-A. Rajandream, J. Rogers, S. Rutter, K. Seeger, J. Skelton, R. Squares, S. Squares, J. E. Sulston, K. Taylor, S. Whitehead, and B. G. Barrell.** 1998. Deciphering the biology of *Mycobacterium tuberculosis* from the complete genome sequence. *Nature* **393:**537–544.

51. **Cole, S. T., K. Eiglmeier, J. Parkhill, K. D. James, N. R. Thomson, P. R. Wheeler, N. Honore, T. Garnier, C. Churcher, D. Harris, K. Mungall, D. Basham, D. Brown, T. Chillingworth, R. Connor, R. M. Davies, K. Devlin, S. Duthoy, T. Feltwell, A. Fraser, N. Hamlin, S. Holroyd, T. Hornsby, K. Jagels, C. Lacroix, J. Maclean, S. Moule, L. Murphy, K. Oliver, M. A. Quail, M. A. Rajandream, K. M. Rutherford, S. Rutter, K. Seeger, S. Simon, M. Simmonds, J. Skelton, R. Squares, S. Squares, K. Stevens, K. Taylor, S. Whitehead, J. R. Woodward, and B. G. Barrell.** 2001. Massive gene decay in the leprosy bacillus. *Nature* **409:**1007–1011.

52. **Constant, P., E. Perez, W. Malaga, M. A. Laneelle, O. Saurel, M. Daffe, and C. Guilhot.** 2002. Role of the *pks15/1* gene in the biosynthesis of phenolglycolipids in the *Mycobacterium tuberculosis* complex. Evidence that all strains synthesize glycosylated p-hydroxybenzoic methyl esters and that strains devoid of phenolglycolipids harbor a frameshift mutation in the pks15/1 gene. *J. Biol. Chem.* **277:**38148–38158.

53. **Converse, S. E., and J. S. Cox.** 2005. A protein secretion pathway critical for *Mycobacterium tuberculosis* virulence is conserved and functional in *Mycobacterium smegmatis. J. Bacteriol.* **187:**1238–1245.

54. **Cooper, J. B., K. McIntyre, M. O. Badasso, S. P. Wood, Y. Zhang, T. R. Garbe, and D. Young.** 1995. X-ray structure analysis of the iron-dependent superoxide dismutase from *Mycobacterium tuberculosis* at 2.0 Angstroms resolution reveals novel dimer-dimer interactions. *J. Mol. Biol.* **246:**531–544.

55. **Coutanceau, E., L. Marsollier, R. Brosch, E. Perret, P. Goossens, M. Tanguy, S. T. Cole, P. L. Small, and C. Demangel.** 2005. Modulation of the host immune response by a transient intracellular stage of *Mycobacterium ulcerans*: the contribution of endogenous mycolactone toxin. *Cell Microbiol.* **7:**1187–1196.

56. **Cox, J. S., B. Chen, M. McNeil, and W. R. Jacobs, Jr.** 1999. Complex lipid determines tissue-specific replication of *Mycobacterium tuberculosis* in mice. *Nature* **402:**79–83.

57. **Curry, J. M., R. Whalan, D. M. Hunt, K. Gohil, M. Strom, L. Rickman, M. J. Colston, S. J. Smerdon, and R. S. Buxton.** 2005. An ABC transporter containing a forkhead-associated domain interacts with a serine-threonine protein kinase and is required for growth of *Mycobacterium tuberculosis* in mice. *Infect. Immun.* **73:**4471–4477.

58. **Daffe, M.** 2000. The mycobacterial antigens 85 complex—from structure to function and beyond. *Trends Microbiol.* **8:**438–440.

59. **Darwin, K. H., S. Ehrt, J. C. Gutierrez-Ramos, N. Weich, and C. F. Nathan.** 2003. The proteasome of *Mycobacterium tuberculosis* is required for resistance to nitric oxide. *Science* **302:**1963–1966.

60. **Darwin, K. H., G. Lin, Z. Chen, H. Li, and C. F. Nathan.** 2005. Characterization of a *Mycobacterium tuberculosis* proteasomal ATPase homologue. *Mol. Microbiol.* **55:**561–571.

61. **Das, A. K., N. R. Helps, P. T. Cohen, and D. Barford.** 1996. Crystal structure of the protein serine/threonine phosphatase 2C at 2.0 Å resolution. *EMBO J.* **15:**6798–6809.

62. **Das, S., S. L. Chan, B. W. Allen, D. A. Mitchison, and D. B. Lowrie.** 1993. Application of DNA fingerprinting with IS986 to sequential mycobacterial isolates obtained from pulmonary tuberculosis patients in Hong-Kong before, during and after short-course chemotherapy. *Tuber. Lung Dis.* **74:**47–51.

63. **Dasgupta, N., V. Kapur, K. K. Singh, T. K. Das, S. Sachdeva, K. Jyothisri, and J. S. Tyagi.** 2000. Characterization of a two-component system, devR-devS, of *Mycobacterium tuberculosis. Tuber. Lung Dis.* **80:**141–159.

64. **Dave, J. A., N. C. Gey van Pittius, A. D. Beyers, M. R. Ehlers, and G. D. Brown.** 2002. Mycosin-1, a subtilisin-like serine protease of *Mycobacterium tuberculosis*, is cell wall–associated and expressed during infection of macrophages. *BMC Microbiol.* **2:**30.

65. **De Voss, J. J., K. Rutter, B. G. Schroeder, H. Su, Y. Zhu, and C. E. Barry III.** 2000. The salicylate-derived mycobactin siderophores of *Mycobacterium tuberculosis* are essential for growth in macrophages. *Proc. Natl. Acad. Sci. USA* **97:**1252–1257.

66. **Delogu, G., and M. J. Brennan.** 1999. Functional domains present in the mycobacterial hemagglutinin, HBHA. *J. Bacteriol.* **181:**7464–7469.

67. **DeRiemer, K., L. Garcia-Garcia, M. Bobadilla-del-Valle, M. Palacios-Martinez, A. Martinez-Gamboa, P. M. Small, J. Sifuentes-Osornio, and A. Ponce-de-Leon.** 2005. Does DOTS work in populations with drug-resistant tuberculosis? *Lancet* **365:**1239–1245.

68. **Dilks, K., R. W. Rose, E. Hartmann, and M. Pohlschroder.** 2003. Prokaryotic utilization of the twin-arginine translocation pathway: a genomic survey. *J. Bacteriol.* **185:**1478–1483.

69. **Doherty, T. M., A. Demissie, J. Olobo, D. Wolday, S. Britton, T. Eguale, P. Ravn, and P. Andersen.** 2002. Immune responses to the *Mycobacterium tuberculosis*–specific antigen ESAT-6 signal subclinical infection among contacts of tuberculosis patients. *J. Clin. Microbiol.* **40:**704–706.

70. **Doherty, T. M., A. W. Olsen, J. Weischenfeldt, K. Huygen, S. D'Souza, T. K. Kondratieva, V. V. Yeremeev, A. S. Apt, B. Raupach, L. Grode, S. Kaufmann, and P. Andersen.** 2004. Comparative analysis of different vaccine constructs expressing defined antigens from *Mycobacterium tuberculosis. J. Infect. Dis.* **190:**2146–2153.

71. **Domenech, P., M. B. Reed, and C. E. Barry III.** 2005. Contribution of the *Mycobacterium tuberculosis* MmpL protein family to virulence and drug resistance. *Infect. Immun.* **73:**3492–3501.

72. **Donoghue, H. D., M. Spigelman, C. L. Greenblatt, G. Lev-Maor, G. K. Bar-Gal, C. Matheson, K. Vernon, A. G. Nerlich, and A. R. Zink.** 2004. Tuberculosis: from prehistory to Robert Koch, as revealed by ancient DNA. *Lancet Infect. Dis.* **4:**584–592.

73. **Drobniewski, F. A., M. Caws, A. Gibson, and D. Young.** 2003. Modern laboratory diagnosis of tuberculosis. *Lancet Infect. Dis.* **3:**141–147.

74. **Dussurget, O., G. Stewart, O. Neyrolles, P. Pescher, D. Young, and G. Marchal.** 2001.

Role of *Mycobacterium tuberculosis* copper-zinc superoxide dismutase. *Infect. Immun.* **69:**529–533.

75. **Dye, C.** 2006. Global epidemiology of tuberculosis. *Lancet* **367:**938–940.

76. **Edwards, K. M., M. H. Cynamon, R. K. Voladri, C. C. Hager, M. S. DeStefano, K. T. Tham, D. L. Lakey, M. R. Bochan, and D. S. Kernodle.** 2001. Iron-cofactored superoxide dismutase inhibits host responses to *Mycobacterium tuberculosis. Am. J. Respir. Crit. Care Med.* **164:**2213–2219.

77. **Espitia, C., J. P. Laclette, M. Mondragon-Palomino, A. Amador, J. Campuzano, A. Martens, M. Singh, R. Cicero, Y. Zhang, and C. Moreno.** 1999. The PE-PGRS glycine-rich proteins of *Mycobacterium tuberculosis*: a new family of fibronectin-binding proteins? *Microbiology* **145:**3487–3495.

78. **Ewann, F., C. Locht, and P. Supply.** 2004. Intracellular autoregulation of the *Mycobacterium tuberculosis* PrrA response regulator. *Microbiology* **150:**241–246.

79. **Fabre, M., J. L. Koeck, P. Le Fleche, F. Simon, V. Herve, G. Vergnaud, and C. Pourcel.** 2004. High genetic diversity revealed by variable-number tandem repeat genotyping and analysis of hsp65 gene polymorphism in a large collection of "*Mycobacterium canettii*" strains indicates that the *M. tuberculosis* complex is a recently emerged clone of "*M. canettii." J. Clin. Microbiol.* **42:**3248–3255.

80. **Ferreras, J. A., J. S. Ryu, F. Di Lello, D. S. Tan, and L. E. Quadri.** 2005. Small-molecule inhibition of siderophore biosynthesis in *Mycobacterium tuberculosis* and *Yersinia pestis. Nat. Chem. Biol.* **1:**29–32.

81. **Fetherston, J. D., V. J. Bertolino, and R. D. Perry.** 1999. YbtP and YbtQ: two ABC transporters required for iron uptake in *Yersinia pestis. Mol. Microbiol.* **32:**289–299.

82. **Filliol, I., J. R. Driscoll, D. Van Soolingen, B. N. Kreiswirth, K. Kremer, G. Valetudie, D. D. Anh, R. Barlow, D. Banerjee, P. J. Bifani, K. Brudey, A. Cataldi, R. C. Cooksey, D. V. Cousins, J. W. Dale, O. A. Dellagostin, F. Drobniewski, G. Engelmann, S. Ferdinand, D. Gascoyne-Binzi, M. Gordon, M. C. Gutierrez, W. H. Haas, H. Heersma, G. Kallenius, E. Kassa-Kelembho, T. Koivula, H. M. Ly, A. Makristathis, C. Mammina, G. Martin, P. Mostrom, I. Mokrousov, V. Narbonne, O. Narvskaya, A. Nastasi, S. N. Niobe-Eyangoh, J. W. Pape, V. Rasolofo-Razanamparany, M. Ridell, M. L. Rossetti, F. Stauffer, P. N. Suffys, H. Takiff, J. Texier-Maugein, V. Vincent, J. H. De Waard, C. Sola, and N. Rastogi.** 2002. Global distribution of *Mycobacterium tuberculosis* spoligotypes. *Emerg. Infect. Dis.* **8:**1347–1349.

83. **Filliol, I., A. S. Motiwala, M. Cavatore, W. Qi, M. H. Hazbon, M. Bobadilla del Valle, J. Fyfe, L. Garcia-Garcia, N. Rastogi, C. Sola, T. Zozio, M. I. Guerrero, C. I. Leon, J. Crabtree, S. Angiuoli, K. D. Eisenach, R. Durmaz, M. L. Joloba, A. Rendon, J. Sifuentes-Osornio, A. Ponce de Leon, M. D. Cave, R. Fleischmann, T. S. Whittam, and D. Alland.** 2006. Global phylogeny of *Mycobacterium tuberculosis* based on single nucleotide polymorphism (SNP) analysis: insights into tuberculosis evolution, phylogenetic accuracy of other DNA fingerprinting systems, and recommendations for a minimal standard SNP set. *J. Bacteriol.* **188:**759–772.

84. **Fine, P. E. M., I. A. M. Carneiro, J. B. Milstein, and C. J. Clements.** 1999. *Issues Relating to the Use of BCG in Immunization Programmes. A Discussion Document.* Department of Vaccines and Biologicals. World Health Organization, Geneva, Switzerland.

85. **Fiss, E. H., S. Yu, and W. R. Jacobs, Jr.** 1994. Identification of genes involved in the sequestration of iron in mycobacteria: the ferric exochelin biosynthetic and uptake pathways. *Mol. Microbiol.* **14:**557–569.

86. **Fleischmann, R. D., D. Alland, J. A. Eisen, L. Carpenter, O. White, J. Peterson, R. DeBoy, R. Dodson, M. Gwinn, D. Haft, E. Hickey, J. F. Kolonay, W. C. Nelson, L. A. Umayam, M. Ermolaeva, S. L. Salzberg, A. Delcher, T. Utterback, J. Weidman, H. Khouri, J. Gill, A. Mikula, W. Bishai, W. R. Jacobs, Jr, J. C. Venter, and C. M. Fraser.** 2002. Whole-genome comparison of *Mycobacterium tuberculosis* clinical and laboratory strains. *J. Bacteriol.* **184:**5479–5490.

87. **Flint, J. L., J. C. Kowalski, P. K. Karnati, and K. M. Derbyshire.** 2004. The RD1 virulence locus of *Mycobacterium tuberculosis* regulates DNA transfer in *Mycobacterium smegmatis. Proc. Natl. Acad. Sci. USA* **101:**12598–12603.

88. **Fortune, S. M., A. Jaeger, D. A. Sarracino, M. R. Chase, C. M. Sassetti, D. R. Sherman, B. R. Bloom, and E. J. Rubin.** 2005. Mutually dependent secretion of proteins required for mycobacterial virulence. *Proc. Natl. Acad. Sci. USA* **102:**10676–10681.

89. **Gagneux, S., K. Deriemer, T. Van, M. Kato-Maeda, B. C. de Jong, S. Narayanan, M. Nicol, S. Niemann, K. Kremer, M. C. Gutierrez, M. Hilty, P. C. Hopewell, and P. M. Small.** 2006. Variable host-pathogen compatibility in *Mycobacterium tuberculosis. Proc. Natl. Acad. Sci. USA* **103:**2869–2873.

90. **Gao, L. Y., S. Guo, B. McLaughlin, H. Morisaki, J. N. Engel, and E. J. Brown.** 2004.

A mycobacterial virulence gene cluster extending RD1 is required for cytolysis, bacterial spreading and ESAT-6 secretion. *Mol. Microbiol.* **53:**1677–1693.

91. **Garnier, T., K. Eiglmeier, J. C. Camus, N. Medina, H. Mansoor, M. Pryor, S. Duthoy, S. Grondin, C. Lacroix, C. Monsempe, S. Simon, B. Harris, R. Atkin, J. Doggett, R. Mayes, L. Keating, P. R. Wheeler, J. Parkhill, B. G. Barrell, S. T. Cole, S. V. Gordon, and R. G. Hewinson.** 2003. The complete genome sequence of *Mycobacterium bovis. Proc. Natl. Acad. Sci. USA* **100:**7877–7882.

92. **Geijtenbeek, T. B., S. J. Van Vliet, E. A. Koppel, M. Sanchez-Hernandez, C. M. Vandenbroucke-Grauls, B. Appelmelk, and Y. Van Kooyk.** 2003. Mycobacteria target DC-SIGN to suppress dendritic cell function. *J. Exp. Med.* **197:**7–17.

93. **George, K. M., D. Chatterjee, G. Gunawardana, D. Welty, J. Hayman, R. Lee, and P. L. Small.** 1999. Mycolactone: a polyketide toxin from *Mycobacterium ulcerans* required for virulence. *Science* **283:**854-857.

94. **Glickman, M. S., J. S. Cox, and W. R. Jacobs, Jr.** 2000. A novel mycolic acid cyclopropane synthetase is required for cording, persistence, and virulence of *Mycobacterium tuberculosis. Mol. Cell* **5:**717–727.

95. **Gold, B., G. M. Rodriguez, S. A. Marras, M. Pentecost, and I. Smith.** 2001. The *Mycobacterium tuberculosis* IdeR is a dual functional regulator that controls transcription of genes involved in iron acquisition, iron storage and survival in macrophages. *Mol. Microbiol.* **42:**851–865.

96. **Gonzalo Asensio, J., C. Maia, N. L. Ferrer, N. Barilone, F. Laval, C. Y. Soto, N. Winter, M. Daffe, B. Gicquel, C. Martin, and M. Jackson.** 2006. The virulence-associated two-component PhoP-PhoR system controls the biosynthesis of polyketide-derived lipids in *Mycobacterium tuberculosis. J. Biol. Chem.* **281:**1313–1316.

97. **Good, M. C., A. E. Greenstein, T. A. Young, H. L. Ng, and T. Alber.** 2004. Sensor domain of the *Mycobacterium tuberculosis* receptor Ser/Thr protein kinase, PknD, forms a highly symmetric beta propeller. *J. Mol. Biol.* **339:**459–469.

98. **Gordon, S. V., R. Brosch, A. Billault, T. Garnier, K. Eiglmeier, and S. T. Cole.** 1999. Identification of variable regions in the genomes of tubercle bacilli using bacterial artificial chromosome arrays. *Mol. Microbiol.* **32:**643–656.

99. **Gordon, S. V., R. Brosch, K. Eiglmeier, T. Garnier, R. G. Hewinson, and S. T. Cole.** 2002. Royal Society of Tropical Medicine and Hygiene Meeting at Manson House, London, 18th January 2001. Pathogen genomes and human health. Mycobacterial genomics. *Trans R. Soc. Trop. Med. Hyg.* **96:**1–6.

100. **Grode, L., P. Seiler, S. Baumann, J. Hess, V. Brinkmann, A. Nasser Eddine, P. Mann, C. Goosmann, S. Bandermann, D. Smith, G. J. Bancroft, J. M. Reyrat, D. van Soolingen, B. Raupach, and S. H. Kaufmann.** 2005. Increased vaccine efficacy against tuberculosis of recombinant *Mycobacterium bovis* bacille Calmette-Guerin mutants that secrete listeriolysin. *J. Clin. Invest.* **115:**2472–2479.

101. **Group, K. P. T.** 1996. Randomised controlled trial of single BCG, repeated BCG, or combined BCG and killed *Mycobacterium leprae* vaccine for prevention of leprosy and tuberculosis in Malawi. Karonga Prevention Trial Group. *Lancet* **348:**17–24.

102. **Grundner, C., L. M. Gay, and T. Alber.** 2005. *Mycobacterium tuberculosis* serine/threonine kinases PknB, PknD, PknE, and PknF phosphorylate multiple FHA domains. *Protein Sci.* **14:**1918–1921.

103. **Guerin, M. E., A. Buschiazzo, J. Kordulakova, M. Jackson, and P. M. Alzari.** 2005. Crystallization and preliminary crystallographic analysis of PimA, an essential mannosyltransferase from *Mycobacterium smegmatis. Acta Crystallograph. Sect. F Struct. Biol. Cryst. Commun.* **61:**518–520.

104. **Guimaraes, B. G., H. Souchon, N. Honore, B. Saint-Joanis, R. Brosch, W. Shepard, S. T. Cole, and P. M. Alzari.** 2005. Structure and mechanism of the alkyl hydroperoxidase AhpC, a key element of the *Mycobacterium tuberculosis* defense system against oxidative stress. *J. Biol. Chem.* **280:**25735–25742.

105. **Guinn, K. I., M. J. Hickey, S. K. Mathur, K. L. Zakel, J. E. Grotzke, D. M. Lewinsohn, S. Smith, and D. R. Sherman.** 2004. Individual RD1-region genes are required for export of ESAT-6/CFP-10 and for virulence of *Mycobacterium tuberculosis. Mol. Microbiol.* **51:**359–370.

106. **Gutierrez, M. C., S. Brisse, R. Brosch, M. Fabre, B. Omais, M. Marmiesse, P. Supply, and V. Vincent.** 2005. Ancient origin and gene mosaicism of the progenitor of *Mycobacterium tuberculosis. PLoS Pathog.* **1:**e5.

107. **Haile, Y., D. A. Caugant, G. Bjune, and H. G. Wiker.** 2002. *Mycobacterium tuberculosis* mammalian cell entry operon (mce) homologs in *Mycobacterium* other than tuberculosis (MOTT). *FEMS Immunol. Med. Microbiol.* **33:**125–132.

108. **Harth, G., B. Y. Lee, J. Wang, D. L. Clemens, and M. A. Horwitz.** 1996. Novel insights into the genetics, biochemistry, and immunocytochemistry of the 30-kilodalton major extracellu-

lar protein of *Mycobacterium tuberculosis. Infect. Immun.* **64:**3038–3047.

109. Haworth, R., N. Platt, S. Keshav, D. Hughes, E. Darley, H. Suzuki, Y. Kurihara, T. Kodama, and S. Gordon. 1997. The macrophage scavenger receptor type A is expressed by activated macrophages and protects the host against lethal endotoxic shock. *J. Exp. Med.* **186:**1431–1439.

110. He, H., R. Hovey, J. Kane, V. Singh, and T. C. Zahrt. 2006. MprAB is a stress-responsive two-component system that directly regulates expression of sigma factors SigB and SigE in *Mycobacterium tuberculosis. J. Bacteriol.* **188:**2134–2143.

111. He, H., and T. C. Zahrt. 2005. Identification and characterization of a regulatory sequence recognized by *Mycobacterium tuberculosis* persistence regulator MprA. *J. Bacteriol.* **187:**202–212.

112. Hensel, M., J. E. Shea, C. Gleeson, M. D. Jones, E. Dalton, and D. W. Holden. 1995. Simultaneous identification of bacterial virulence genes by negative selection. *Science* **269:**400–403.

113. Herrmann, J. L., P. O'Gaora, A. Gallagher, J. E. Thole, and D. B. Young. 1996. Bacterial glycoproteins: a link between glycosylation and proteolytic cleavage of a 19 kDa antigen from *Mycobacterium tuberculosis. EMBO J.* **15:**3547–3554.

114. Heym, B., B. Saint-Joanis, and S. T. Cole. 1999. The molecular basis of isoniazid resistance in *Mycobacterium tuberculosis. Tuber. Lung Dis.* **79:**267–271.

115. Heym, B., E. Stavropoulos, N. Honore, P. Domenech, B. Saint-Joanis, T. M. Wilson, D. M. Collins, M. J. Colston, and S. T. Cole. 1997. Effects of overexpression of the alkyl hydroperoxide reductase AhpC on the virulence and isoniazid resistance of *Mycobacterium tuberculosis. Infect. Immun.* **65:**1395–1401.

116. Heym, B., Y. Zhang, S. Poulet, D. Young, and S. T. Cole. 1993. Characterization of the *katG* gene encoding a catalase-peroxidase required for the isoniazid susceptibility of *Mycobacterium tuberculosis. J. Bacteriol.* **175:**4255–4259.

117. Hirsh, A. E., A. G. Tsolaki, K. DeRiemer, M. W. Feldman, and P. M. Small. 2004. Stable association between strains of *Mycobacterium tuberculosis* and their human host populations. *Proc. Natl. Acad. Sci. USA* **101:**4871–4876.

118. Hong, H., P. J. Gates, J. Staunton, T. Stinear, S. T. Cole, P. F. Leadlay, and J. B. Spencer. 2003. Identification using LC-MSn of co-metabolites in the biosynthesis of the polyketide toxin mycolactone by a clinical isolate of *Mycobacterium ulcerans. Chem. Commun. (Camb.)* 21 Nov:2822–2823.

119. Hong, H., J. B. Spencer, J. L. Porter, P. F. Leadlay, and T. Stinear. 2005. A novel mycolactone from a clinical isolate of *Mycobacterium ulcerans* provides evidence for additional toxin heterogeneity as a result of specific changes in the modular polyketide synthase. *Chembiochem* **6:**643–648.

120. Horwitz, M. A., and G. Harth. 2003. A new vaccine against tuberculosis affords greater survival after challenge than the current vaccine in the guinea pig model of pulmonary tuberculosis. *Infect. Immun.* **71:**1672–1679.

121. Hsu, T., S. M. Hingley-Wilson, B. Chen, M. Chen, A. Z. Dai, P. M. Morin, C. B. Marks, J. Padiyar, C. Goulding, M. Gingery, D. Eisenberg, R. G. Russell, S. C. Derrick, F. M. Collins, S. L. Morris, C. H. King, and W. R. Jacobs, Jr. 2003. The primary mechanism of attenuation of bacillus Calmette-Guerin is a loss of secreted lytic function required for invasion of lung interstitial tissue. *Proc. Natl. Acad. Sci. USA* **100:**12420–12425.

122. Hughes, A. L., R. Friedman, and M. Murray. 2002. Genomewide pattern of synonymous nucleotide substitution in two complete genomes of *Mycobacterium tuberculosis. Emerg. Infect. Dis.* **8:**1342–1346.

123. Jacobs, W. R., Jr., G. V. Kalpana, J. D. Cirillo, L. Pascopella, S. B. Snapper, R. A. Udani, W. Jones, R. G. Barletta, and B. R. Bloom. 1991. Genetic systems for mycobacteria. *Methods Enzymol.* **204:**537–555.

124. Jain, M., and J. S. Cox. 2005. Interaction between polyketide synthase and transporter suggests coupled synthesis and export of virulence lipid in *M. tuberculosis. PLoS Pathog.* **1:**e2.

125. Jeffers, S. A., S. M. Tusell, L. Gillim-Ross, E. M. Hemmila, J. E. Achenbach, G. J. Babcock, W. D. Thomas, Jr., L. B. Thackray, M. D. Young, R. J. Mason, D. M. Ambrosino, D. E. Wentworth, J. C. Demartini, and K. V. Holmes. 2004. CD209L (L-SIGN) is a receptor for severe acute respiratory syndrome coronavirus. *Proc. Natl. Acad. Sci. USA* **101:**15748–15753.

126. Johansen, K. A., R. E. Gill, and M. L. Vasil. 1996. Biochemical and molecular analysis of phospholipase C and phospholipase D activity in mycobacteria. *Infect. Immun.* **64:**3259–3266.

127. Johnson, P. D., T. Stinear, P. L. Small, G. Pluschke, R. W. Merritt, F. Portaels, K. Huygen, J. A. Hayman, and K. Asiedu. 2005. Buruli ulcer (*M. ulcerans* infection): new insights, new hope for disease control. *PLoS Med.* **2:**e108.

128. **Kamerbeek, J., L. Schouls, A. Kolk, M. van Agterveld, D. van Soolingen, S. Kuijper, A. Bunschoten, H. Molhuizen, R. Shaw, M. Goyal, and J. van Embden.** 1997. Simultaneous detection and strain differentiation of *Mycobacterium tuberculosis* for diagnosis and epidemiology. *J. Clin. Microbiol.* **35:**907–914.

129. **Kang, C. M., D. W. Abbott, S. T. Park, C. C. Dascher, L. C. Cantley, and R. N. Husson.** 2005. The *Mycobacterium tuberculosis* serine/threonine kinases PknA and PknB: substrate identification and regulation of cell shape. *Genes Dev.* **19:**1692–1704.

130. **Kaufmann, S. H.** 2001. How can immunology contribute to the control of tuberculosis? *Nat. Rev. Immunol.* **1:**20–30.

131. **Keane, J.** 2005. TNF-blocking agents and tuberculosis: new drugs illuminate an old topic. *Rheumatology (Oxford)* **44:**714–720.

132. **Kong, Y., M. D. Cave, D. Yang, L. Zhang, C. F. Marrs, B. Foxman, J. H. Bates, F. Wilson, L. N. Mukasa, and Z. H. Yang.** 2005. Distribution of insertion- and deletion-associated genetic polymorphisms among four *Mycobacterium tuberculosis* phospholipase C genes and associations with extrathoracic tuberculosis: a population-based study. *J. Clin. Microbiol.* **43:**6048–6053.

133. **Koppel, E. A., K. P. van Gisbergen, T. B. Geijtenbeek, and Y. van Kooyk.** 2005. Distinct functions of DC-SIGN and its homologues L-SIGN (DC-SIGNR) and mSIGNR1 in pathogen recognition and immune regulation. *Cell. Microbiol.* **7:**157–165.

134. **Kordulakova, J., M. Gilleron, K. Mikusova, G. Puzo, P. J. Brennan, B. Gicquel, and M. Jackson.** 2002. Definition of the first mannosylation step in phosphatidylinositol mannoside synthesis. PimA is essential for growth of mycobacteria. *J. Biol. Chem.* **277:**31335–31344.

135. **Koshkin, A., X. T. Zhou, C. N. Kraus, J. M. Brenner, P. Bandyopadhyay, I. D. Kuntz, C. E. Barry III, and P. R. Ortiz de Montellano.** 2004. Inhibition of *Mycobacterium tuberculosis* AhpD, an element of the peroxiredoxin defense against oxidative stress. *Antimicrob. Agents Chemother.* **48:**2424–2430.

136. **Koul, A., A. Choidas, M. Treder, A. K. Tyagi, K. Drlica, Y. Singh, and A. Ullrich.** 2000. Cloning and characterization of secretory tyrosine phosphatases of *Mycobacterium tuberculosis.* *J. Bacteriol.* **182:**5425–5432.

137. **Krithika, R., U. Marathe, P. Saxena, M. Z. Ansari, D. Mohanty, and R. S. Gokhale.** 2006. A genetic locus required for iron acquisition in *Mycobacterium tuberculosis. Proc. Natl. Acad. Sci. USA* **103:**2069–2074.

138. **Kumar, A., M. Bose, and V. Brahmachari.** 2003. Analysis of expression profile of mammalian cell entry (mce) operons of *Mycobacterium tuberculosis. Infect. Immun.* **71:**6083–6087.

139. **Kusunose, E., K. Ichihara, Y. Noda, and M. Kusunose.** 1976. Superoxide dismutase from *Mycobacterium tuberculosis. J. Biochem. (Tokyo)* **80:**1343–1352.

140. **Lalvani, A., A. A. Pathan, H. Durkan, K. A. Wilkinson, A. Whelan, J. J. Deeks, W. H. Reece, M. Latif, G. Pasvol, and A. V. Hill.** 2001. Enhanced contact tracing and spatial tracking of *Mycobacterium tuberculosis* infection by enumeration of antigen-specific T cells. *Lancet* **357:**2017–2021.

141. **Lamichhane, G., M. Zignol, N. J. Blades, D. E. Geiman, A. Dougherty, J. Grosset, K. W. Broman, and W. R. Bishai.** 2003. A postgenomic method for predicting essential genes at subsaturation levels of mutagenesis: application to *Mycobacterium tuberculosis. Proc. Natl. Acad. Sci. USA* **100:**7213–7218.

142. **Lewis, K. N., R. Liao, K. M. Guinn, M. J. Hickey, S. Smith, M. A. Behr, and D. R. Sherman.** 2003. Deletion of RD1 from *Mycobacterium tuberculosis* mimics bacille Calmette-Guerin attenuation. *J. Infect. Dis.* **187:**117–123.

143. **Li, L., J. P. Bannantine, Q. Zhang, A. Amonsin, B. J. May, D. Alt, N. Banerji, S. Kanjilal, and V. Kapur.** 2005. The complete genome sequence of *Mycobacterium avium* subspecies paratuberculosis. *Proc. Natl. Acad. Sci. USA* **102:**12344–12349.

144. **Li, Z., C. Kelley, F. Collins, D. Rouse, and S. Morris.** 1998. Expression of katG in *Mycobacterium tuberculosis* is associated with its growth and persistence in mice and guinea pigs. *J. Infect. Dis.* **177:**1030–1035.

145. **Lin, T. W., M. M. Melgar, D. Kurth, S. J. Swamidass, J. Purdon, T. Tseng, G. Gago, P. Baldi, H. Gramajo, and S. C. Tsai.** 2006. Structure-based inhibitor design of AccD5, an essential acyl-CoA carboxylase carboxyltransferase domain of *Mycobacterium tuberculosis. Proc. Natl. Acad. Sci. USA* **103:**3072–3077.

146. **Locht, C., J. M. Hougardy, C. Rouanet, S. Place, and F. Mascart.** 2006. Heparin-binding hemagglutinin, from an extrapulmonary dissemination factor to a powerful diagnostic and protective antigen against tuberculosis. *Tuberculosis (Edinb.)* **86:**303–309.

147. **Lopez, B., D. Aguilar, H. Orozco, M. Burger, C. Espitia, V. Ritacco, L. Barrera, K. Kremer, R. Hernandez-Pando, K. Huygen, and D. van Soolingen.** 2003. A marked difference in pathogenesis and immune response in-

duced by different *Mycobacterium tuberculosis* genotypes. *Clin. Exp. Immunol.* **133**:30–37.

148. **Luo, M., E. A. Fadeev, and J. T. Groves.** 2005. Mycobactin-mediated iron acquisition within macrophages. *Nat. Chem. Biol.* **1**:149–153.

149. **MacCallum, P., J. Tolhurst, G. Buckle, and S. H.** 1948. A new mycobacterial infection in man. *J. Pathol. Bacteriol.* **60**:93–122.

150. **MacGurn, J. A., S. Raghavan, S. A. Stanley, and J. S. Cox.** 2005. A non-RD1 gene cluster is required for Snm secretion in Mycobacterium tuberculosis. *Mol. Microbiol.* **57**:1653–1663.

151. **MacMicking, J. D., R. J. North, R. La-Course, J. S. Mudgett, S. K. Shah, and C. F. Nathan.** 1997. Identification of nitric oxide synthase as a protective locus against tuberculosis. *Proc. Natl. Acad. Sci. USA* **94**:5243–5248.

152. **Mahairas, G. G., P. J. Sabo, M. J. Hickey, D. C. Singh, and C. K. Stover.** 1996. Molecular analysis of genetic differences between *Mycobacterium bovis* BCG and virulent *M. bovis.* *J. Bacteriol.* **178**:1274–1282.

153. **Makinoshima, H., and M. S. Glickman.** 2005. Regulation of *Mycobacterium tuberculosis* cell envelope composition and virulence by intramembrane proteolysis. *Nature* **436**:406–409.

154. **Malaga, W., E. Perez, and C. Guilhot.** 2003. Production of unmarked mutations in mycobacteria using site-specific recombination. *FEMS Microbiol. Lett.* **219**:261–268.

155. **Manca, C., L. Tsenova, A. Bergtold, S. Freeman, M. Tovey, J. M. Musser, C. E. Barry III, V. H. Freedman, and G. Kaplan.** 2001. Virulence of a *Mycobacterium tuberculosis* clinical isolate in mice is determined by failure to induce Th1 type immunity and is associated with induction of IFN-alpha/beta. *Proc. Natl. Acad. Sci. USA* **98**:5752–5757.

156. **Manger, M., M. Scheck, H. Prinz, J. P. von Kries, T. Langer, K. Saxena, H. Schwalbe, A. Furstner, J. Rademann, and H. Waldmann.** 2005. Discovery of *Mycobacterium tuberculosis* protein tyrosine phosphatase A (MptpA) inhibitors based on natural products and a fragment-based approach. *Chembiochem* **6**:1749–1753.

157. **Marmiesse, M., P. Brodin, C. Buchrieser, C. Gutierrez, N. Simoes, V. Vincent, P. Glaser, S. T. Cole, and R. Brosch.** 2004. Macro-array and bioinformatic analyses reveal mycobacterial "core" genes, variation in the ESAT-6 gene family and new phylogenetic markers for the *Mycobacterium tuberculosis* complex. *Microbiology* **150**:483–496.

158. **Marsollier, L., J. Aubry, E. Coutanceau, J. P. Andre, P. L. Small, G. Milon, P. Legras, S. Guadagnini, B. Carbonnelle, and S. T. Cole.** 2005. Colonization of the salivary glands of *Naucoris cimicoides* by *Mycobacterium ulcerans* requires host plasmatocytes and a macrolide toxin, mycolactone. *Cell. Microbiol.* **7**:935–943.

159. **Marsollier, L., R. Robert, J. Aubry, J. P. Saint Andre, H. Kouakou, P. Legras, A. L. Manceau, C. Mahaza, and B. Carbonnelle.** 2002. Aquatic insects as a vector for *Mycobacterium ulcerans. Appl. Environ. Microbiol.* **68**:4623–4628.

160. **Martin, C., A. Williams, R. Hernandez-Pando, P. J. Cardona, E. Gormley, Y. Bordat, C. Y. Soto, S. O. Clark, G. J. Hatch, D. Aguilar, V. Ausina, and B. Gicquel.** 2006. The live *Mycobacterium tuberculosis* phoP mutant strain is more attenuated than BCG and confers protective immunity against tuberculosis in mice and guinea pigs. *Vaccine* **24**:3408–3419.

161. **Master, S. S., B. Springer, P. Sander, E. C. Boettger, V. Deretic, and G. S. Timmins.** 2002. Oxidative stress response genes in *Mycobacterium tuberculosis*: role of ahpC in resistance to peroxynitrite and stage-specific survival in macrophages. *Microbiology* **148**:3139–3144.

162. **Matthysse, A. G., H. A. Yarnall, and N. Young.** 1996. Requirement for genes with homology to ABC transport systems for attachment and virulence of Agrobacterium tumefaciens. *J. Bacteriol.* **178**:5302–5308.

163. **McCarthy, T. R., J. B. Torrelles, A. S. MacFarlane, M. Katawczik, B. Kutzbach, L. E. Desjardin, S. Clegg, J. B. Goldberg, and L. S. Schlesinger.** 2005. Overexpression of *Mycobacterium tuberculosis* manB, a phosphomannomutase that increases phosphatidylinositol mannoside biosynthesis in *Mycobacterium smegmatis* and mycobacterial association with human macrophages. *Mol. Microbiol.* **58**:774–790.

164. **McDonough, J. A., K. E. Hacker, A. R. Flores, M. S. Pavelka, Jr., and M. Braunstein.** 2005. The twin-arginine translocation pathway of *Mycobacterium smegmatis* is functional and required for the export of mycobacterial beta-lactamases. *J. Bacteriol.* **187**:7667–7679.

165. **McKinney, J. D., K. Honer zu Bentrup, E. J. Munoz-Elias, A. Miczak, B. Chen, W. T. Chan, D. Swenson, J. C. Sacchettini, W. R. Jacobs, Jr., and D. G. Russell.** 2000. Persistence of *Mycobacterium tuberculosis* in macrophages and mice requires the glyoxylate shunt enzyme isocitrate lyase. *Nature* **406**:735–738.

166. **McShane, H.** 2004. Developing an improved vaccine against tuberculosis. *Expert Rev. Vaccines* **3**:299–306.

167. **McShane, H., and A. Hill.** 2005. Prime-boost immunisation strategies for tuberculosis. *Microbes Infect.* **7**:962–967.

168. **Meher, A. K., N. C. Bal, K. V. Chary, and A. Arora.** 2006. *Mycobacterium tuberculosis* H37Rv ESAT-6-CFP-10 complex formation confers thermodynamic and biochemical stability. *FEBS J.* **273:**1445–1462.

169. **Mehta, P. K., C. H. King, E. H. White, J. J. Murtagh, Jr., and F. D. Quinn.** 1996. Comparison of in vitro models for the study of *Mycobacterium tuberculosis* invasion and intracellular replication. *Infect. Immun.* **64:**2673–2679.

170. **Menozzi, F. D., R. Bischoff, E. Fort, M. J. Brennan, and C. Locht.** 1998. Molecular characterization of the mycobacterial heparin-binding hemagglutinin, a mycobacterial adhesin. *Proc. Natl. Acad. Sci. USA* **95:**12625–12630.

171. **Menozzi, F. D., V. M. Reddy, D. Cayet, D. Raze, A. S. Debrie, M. P. Dehouck, R. Cecchelli, and C. Locht.** 2006. *Mycobacterium tuberculosis* heparin-binding haemagglutinin adhesin (HBHA) triggers receptor-mediated transcytosis without altering the integrity of tight junctions. *Microbes Infect.* **8:**1–9.

172. **Menozzi, F. D., J. H. Rouse, M. Alavi, M. Laude-Sharp, J. Muller, R. Bischoff, M. J. Brennan, and C. Locht.** 1996. Identification of a heparin-binding hemagglutinin present in mycobacteria. *J. Exp. Med.* **184:**993–1001.

173. **Misra, N., S. Habib, A. Ranjan, S. E. Hasnain, and I. Nath.** 1996. Expression and functional characterisation of the *clpC* gene of *Mycobacterium leprae*: ClpC protein elicits human antibody response. *Gene* **172:**99–104.

174. **Monot, M., N. Honore, T. Garnier, R. Araoz, J. Y. Coppee, C. Lacroix, S. Sow, J. S. Spencer, R. W. Truman, D. L. Williams, R. Gelber, M. Virmond, B. Flageul, S. N. Cho, B. Ji, A. Paniz-Mondolfi, J. Convit, S. Young, P. E. Fine, V. Rasolofo, P. J. Brennan, and S. T. Cole.** 2005. On the origin of leprosy. *Science* **308:**1040–1042.

175. **Mostowy, S., D. Cousins, and M. A. Behr.** 2004. Genomic interrogation of the dassie bacillus reveals it as a unique RD1 mutant within the *Mycobacterium tuberculosis* complex. *J. Bacteriol.* **186:**104–109.

176. **Mostowy, S., D. Cousins, J. Brinkman, A. Aranaz, and M. A. Behr.** 2002. Genomic deletions suggest a phylogeny for the *Mycobacterium tuberculosis* complex. *J. Infect. Dis.* **186:**74–80.

177. **Ng, V. H., J. S. Cox, A. O. Sousa, J. D. MacMicking, and J. D. McKinney.** 2004. Role of KatG catalase-peroxidase in mycobacterial pathogenesis: countering the phagocyte oxidative burst. *Mol. Microbiol.* **52:**1291–1302.

178. **Nigou, J., A. Vercellone, and G. Puzo.** 2000. New structural insights into the molecular deciphering of mycobacterial lipoglycan binding to C-type lectins: lipoarabinomannan glycoform characterization and quantification by capillary electrophoresis at the subnanomole level. *J. Mol. Biol.* **299:**1353–1362.

179. **Ohno, S.** 1995. Active sites of ligands and their receptors are made of common peptides that are also found elsewhere. *J. Mol. Evol.* **40:**102–106.

180. **Ortiz-Lombardia, M., F. Pompeo, B. Boitel, and P. M. Alzari.** 2003. Crystal structure of the catalytic domain of the PknB serine/threonine kinase from *Mycobacterium tuberculosis. J. Biol. Chem.* **278:**13094–13100.

181. **Pai, M., L. W. Riley, and J. M. Colford, Jr.** 2004. Interferon-gamma assays in the immunodiagnosis of tuberculosis: a systematic review. *Lancet Infect. Dis.* **4:**761–776.

182. **Palomino, J. C.** 2005. Nonconventional and new methods in the diagnosis of tuberculosis: feasibility and applicability in the field. *Eur. Respir. J.* **26:**339–350.

183. **Parish, T., D. A. Smith, G. Roberts, J. Betts, and N. G. Stoker.** 2003. The senX3-regX3 two-component regulatory system of *Mycobacterium tuberculosis* is required for virulence. *Microbiology* **149:**1423–1435.

184. **Parish, T., and N. G. Stoker.** 2000. Use of a flexible cassette method to generate a double unmarked *Mycobacterium tuberculosis* tlyA plcABC mutant by gene replacement. *Microbiology* **146:**1969–1975.

185. **Pavlic, M., F. Allerberger, M. P. Dierich, and W. M. Prodinger.** 1999. Simultaneous infection with two drug-susceptible *Mycobacterium tuberculosis* strains in an immunocompetent host. *J. Clin. Microbiol.* **37:**4156–4157.

186. **Pelicic, V., M. Jackson, J. M. Reyrat, W. R. Jacobs, Jr., B. Gicquel, and C. Guilhot.** 1997. Efficient allelic exchange and transposon mutagenesis in *Mycobacterium tuberculosis. Proc. Natl. Acad. Sci. USA* **94:**10955–10960.

187. **Perez, E., P. Constant, A. Lemassu, F. Laval, M. Daffe, and C. Guilhot.** 2004. Characterization of three glycosyltransferases involved in the biosynthesis of the phenolic glycolipid antigens from the *Mycobacterium tuberculosis* complex. *J. Biol. Chem.* **279:**42574–42583.

188. **Perez, E., S. Samper, Y. Bordas, C. Guilhot, B. Gicquel, and C. Martin.** 2001. An essential role for phoP in *Mycobacterium tuberculosis* virulence. *Mol. Microbiol.* **41:**179–187.

189. **Pethe, K., S. Alonso, F. Biet, G. Delogu, M. J. Brennan, C. Locht, and F. D. Menozzi.** 2001. The heparin-binding haemagglutinin of *M. tuberculosis* is required for extrapulmonary dissemination. *Nature* **412:**190–194.

190. **Pethe, K., M. Aumercier, E. Fort, C. Gatot, C. Locht, and F. D. Menozzi.** 2000. Charac-

terization of the heparin-binding site of the mycobacterial heparin-binding hemagglutinin adhesin. *J. Biol. Chem.* **275:**14273–14280.

191. **Peyron, P., C. Bordier, E. N. N'Diaye, and I. Maridonneau-Parini.** 2000. Nonopsonic phagocytosis of *Mycobacterium kansasii* by human neutrophils depends on cholesterol and is mediated by CR3 associated with glycosylphosphatidylinositol-anchored proteins. *J. Immunol.* **165:**5186–5191.

192. **Pickup, R. W., G. Rhodes, S. Arnott, K. Sidi-Boumedine, T. J. Bull, A. Weightman, M. Hurley, and J. Hermon-Taylor.** 2005. *Mycobacterium avium* subsp. *paratuberculosis* in the catchment area and water of the River Taff in South Wales, United Kingdom, and its potential relationship to clustering of Crohn's disease cases in the city of Cardiff. *Appl. Environ. Microbiol.* **71:**2130–2139.

193. **Piddington, D. L., F. C. Fang, T. Laessig, A. M. Cooper, I. M. Orme, and N. A. Buchmeier.** 2001. Cu,Zn superoxide dismutase of *Mycobacterium tuberculosis* contributes to survival in activated macrophages that are generating an oxidative burst. *Infect. Immun.* **69:**4980–4987.

194. **Pitarque, S., J. L. Herrmann, J. L. Duteyrat, M. Jackson, G. R. Stewart, F. Lecointe, B. Payre, O. Schwartz, D. B. Young, G. Marchal, P. H. Lagrange, G. Puzo, B. Gicquel, J. Nigou, and O. Neyrolles.** 2005. Deciphering the molecular bases of *Mycobacterium tuberculosis* binding to the lectin DC-SIGN reveals an underestimated complexity. *Biochem. J.* **392:**615–624.

195. **Pohl, E., R. K. Holmes, and W. G. Hol.** 1999. Crystal structure of the iron-dependent regulator (IdeR) from *Mycobacterium tuberculosis* shows both metal binding sites fully occupied. *J. Mol. Biol.* **285:**1145–1156.

196. **Pollock, J. M., J. McNair, H. Bassett, J. P. Cassidy, E. Costello, H. Aggerbeck, I. Rosenkrands, and P. Andersen.** 2003. Specific delayed-type hypersensitivity responses to ESAT-6 identify tuberculosis-infected cattle. *J. Clin. Microbiol.* **41:**1856–1860.

197. **Portevin, D., C. De Sousa-D'Auria, C. Houssin, C. Grimaldi, M. Chami, M. Daffe, and C. Guilhot.** 2004. A polyketide synthase catalyzes the last condensation step of mycolic acid biosynthesis in mycobacteria and related organisms. *Proc. Natl. Acad. Sci. USA* **101:**314–319.

198. **Posey, J. E., T. M. Shinnick, and F. D. Quinn.** 2006. Characterization of the twin-arginine translocase secretion system of *Mycobacterium smegmatis. J. Bacteriol.* **188:**1332–1340.

199. **Prado, S., H. Ledeit, S. Michel, M. Koch, J. C. Darbord, S. T. Cole, F. Tillequin, and P. Brodin.** 2006. Benzofuro[3,2-f][1]benzopyrans: a new class of antitubercular agents. *Bioorg. Med. Chem.* **14:**5423–5428.

200. **Pugsley, A. P.** 1993. The complete general secretory pathway in gram-negative bacteria. *Microbiol. Rev.* **57:**50–108.

201. **Pullen, K. E., H. L. Ng, P. Y. Sung, M. C. Good, S. M. Smith, and T. Alber.** 2004. An alternate conformation and a third metal in PstP/Ppp, the *M. tuberculosis* PP2C-Family Ser/Thr protein phosphatase. *Structure* **12:**1947–1954.

202. **Pym, A. S., P. Brodin, R. Brosch, M. Huerre, and S. T. Cole.** 2002. Loss of RD1 contributed to the attenuation of the live tuberculosis vaccines *Mycobacterium bovis* BCG and *Mycobacterium microti. Mol. Microbiol.* **46:**709–717.

203. **Pym, A. S., P. Brodin, L. Majlessi, R. Brosch, C. Demangel, A. Williams, K. E. Griffiths, G. Marchal, C. Leclerc, and S. T. Cole.** 2003. Recombinant BCG exporting ESAT-6 confers enhanced protection against tuberculosis. *Nat. Med.* **9:**533–539.

204. **Pym, A. S., P. Domenech, N. Honore, J. Song, V. Deretic, and S. T. Cole.** 2001. Regulation of catalase-peroxidase (KatG) expression, isoniazid sensitivity and virulence by furA of *Mycobacterium tuberculosis. Mol. Microbiol.* **40:**879–889.

205. **Pym, A. S., B. Saint-Joanis, and S. T. Cole.** 2002. Effect of katG mutations on the virulence of *Mycobacterium tuberculosis* and the implication for transmission in humans. *Infect. Immun.* **70:** 4955–4960.

206. **Quadri, L. E., J. Sello, T. A. Keating, P. H. Weinreb, and C. T. Walsh.** 1998. Identification of a *Mycobacterium tuberculosis* gene cluster encoding the biosynthetic enzymes for assembly of the virulence-conferring siderophore mycobactin. *Chem. Biol.* **5:**631–645.

207. **Ramakrishnan, L., H. T. Tran, N. A. Federspiel, and S. Falkow.** 1997. A crtB homolog essential for photochromogenicity in *Mycobacterium marinum*: isolation, characterization, and gene disruption via homologous recombination. *J. Bacteriol.* **179:**5862–5868.

208. **Rao, V., N. Fujiwara, S. A. Porcelli, and M. S. Glickman.** 2005. *Mycobacterium tuberculosis* controls host innate immune activation through cyclopropane modification of a glycolipid effector molecule. *J. Exp. Med.* **201:**535–543.

209. **Raynaud, C., C. Guilhot, J. Rauzier, Y. Bordat, V. Pelicic, R. Manganelli, I. Smith, B. Gicquel, and M. Jackson.** 2002. Phospholipases C are involved in the virulence of *Mycobacterium tuberculosis. Mol. Microbiol.* **45:**203–217.

210. **Reddy, V. M., and D. A. Hayworth.** 2002. Interaction of *Mycobacterium tuberculosis* with human respiratory epithelial cells (HEp-2). *Tuberculosis (Edinb.)* **82:**31–36.

211. **Reed, M. B., P. Domenech, C. Manca, H. Su, A. K. Barczak, B. N. Kreiswirth, G. Kaplan, and C. E. Barry III.** 2004. A glycolipid of hypervirulent tuberculosis strains that inhibits the innate immune response. *Nature* **431:**84–87.

212. **Relman, D. A., M. Domenighini, E. Tuomanen, R. Rappuoli, and S. Falkow.** 1989. Filamentous hemagglutinin of *Bordetella pertussis*: nucleotide sequence and crucial role in adherence. *Proc. Natl. Acad. Sci. USA* **86:**2637–2641.

213. **Renshaw, P. S., K. L. Lightbody, V. Veverka, F. W. Muskett, G. Kelly, T. A. Frenkiel, S. V. Gordon, R. G. Hewinson, B. Burke, J. Norman, R. A. Williamson, and M. D. Carr.** 2005. Structure and function of the complex formed by the tuberculosis virulence factors CFP-10 and ESAT-6. *EMBO J.* **24:**2491–2498.

214. **Renshaw, P. S., P. Panagiotidou, A. Whelan, S. V. Gordon, R. G. Hewinson, R. A. Williamson, and M. D. Carr.** 2002. Conclusive evidence that the major T-cell antigens of the *Mycobacterium tuberculosis* complex ESAT-6 and CFP-10 form a tight, 1:1 complex and characterization of the structural properties of ESAT-6, CFP-10, and the ESAT-6★CFP-10 complex. Implications for pathogenesis and virulence. *J. Biol. Chem.* **277:**21598–21603.

215. **Roberts, D. M., R. P. Liao, G. Wisedchaisri, W. G. Hol, and D. R. Sherman.** 2004. Two sensor kinases contribute to the hypoxic response of *Mycobacterium tuberculosis*. *J. Biol. Chem.* **279:**23082–23087.

216. **Rodriguez, G. M., and I. Smith.** 2006. Identification of an ABC transporter required for iron acquisition and virulence in *Mycobacterium tuberculosis*. *J. Bacteriol.* **188:**424–430.

217. **Rodriguez, G. M., M. I. Voskuil, B. Gold, G. K. Schoolnik, and I. Smith.** 2002. ideR, An essential gene in *Mycobacterium tuberculosis*: role of IdeR in iron-dependent gene expression, iron metabolism, and oxidative stress response. *Infect. Immun.* **70:**3371–3381.

218. **Ronning, D. R., T. Klabunde, G. S. Besra, V. D. Vissa, J. T. Belisle, and J. C. Sacchettini.** 2000. Crystal structure of the secreted form of antigen 85C reveals potential targets for mycobacterial drugs and vaccines. *Nat. Struct. Biol.* **7:**141–146.

219. **Rooyakkers, A. W., and R. W. Stokes.** 2005. Absence of complement receptor 3 results in reduced binding and ingestion of *Mycobacterium tuberculosis* but has no significant effect on the induction of reactive oxygen and nitrogen intermediates or on the survival of the bacteria in resident and interferon-gamma activated macrophages. *Microb. Pathog.* **39:**57–67.

220. **Rosas-Magallanes, V., P. Deschavanne, L. Quintana-Murci, R. Brosch, B. Gicquel, and O. Neyrolles.** 2006. Horizontal transfer of a virulence operon to the ancestor of *Mycobacterium tuberculosis*. *Mol. Biol. Evol.* **23:**1129–1135.

221. **Sambandamurthy, V. K., X. Wang, B. Chen, R. G. Russell, S. Derrick, F. M. Collins, S. L. Morris, and W. R. Jacobs, Jr.** 2002. A pantothenate auxotroph of *Mycobacterium tuberculosis* is highly attenuated and protects mice against tuberculosis. *Nat. Med.* **8:**1171–1174.

222. **Santangelo, M. P., J. Goldstein, A. Alito, A. Gioffre, K. Caimi, O. Zabal, M. Zumarraga, M. I. Romano, A. A. Cataldi, and F. Bigi.** 2002. Negative transcriptional regulation of the mce3 operon in *Mycobacterium tuberculosis*. *Microbiology* **148:**2997–3006.

223. **Sassetti, C. M., D. H. Boyd, and E. J. Rubin.** 2001. Comprehensive identification of conditionally essential genes in mycobacteria. *Proc. Natl. Acad. Sci. USA* **98:**12712–12717.

224. **Sassetti, C. M., D. H. Boyd, and E. J. Rubin.** 2003. Genes required for mycobacterial growth defined by high density mutagenesis. *Mol. Microbiol.* **48:**77–84.

225. **Sassetti, C. M., and E. J. Rubin.** 2003. Genetic requirements for mycobacterial survival during infection. *Proc. Natl. Acad. Sci. USA* **100:**12989–12994.

226. **Sato, K., H. Tomioka, T. Akaki, and S. Kawahara.** 2000. Antimicrobial activities of levofloxacin, clarithromycin, and KRM-1648 against *Mycobacterium tuberculosis* and *Mycobacterium avium* complex replicating within Mono Mac 6 human macrophage and A-549 type II alveolar cell lines. *Int. J. Antimicrob. Agents* **16:**25–29.

227. **Schlesinger, L. S., S. R. Hull, and T. M. Kaufman.** 1994. Binding of the terminal mannosyl units of lipoarabinomannan from a virulent strain of *Mycobacterium tuberculosis* to human macrophages. *J. Immunol.* **152:**4070–4079.

228. **Schorey, J. S., M. C. Carroll, and E. J. Brown.** 1997. A macrophage invasion mechanism of pathogenic mycobacteria. *Science* **277:**1091–1093.

229. **Schorey, J. S., M. A. Holsti, T. L. Ratliff, P. M. Allen, and E. J. Brown.** 1996. Characterization of the fibronectin-attachment protein of *Mycobacterium avium* reveals a fibronectin-binding motif conserved among mycobacteria. *Mol. Microbiol.* **21:**321–329.

230. **Secott, T. E., T. L. Lin, and C. C. Wu.** 2004. *Mycobacterium avium* subsp. *paratuberculosis* fi-

bronectin attachment protein facilitates M-cell targeting and invasion through a fibronectin bridge with host integrins. *Infect. Immun.* **72:** 3724–3732.

231. **Sharman, G. J., D. H. Williams, D. F. Ewing, and C. Ratledge.** 1995. Determination of the structure of exochelin MN, the extracellular siderophore from *Mycobacterium neoaurum. Chem. Biol.* **2:**553–561.

232. **Shepard, C. C.** 1957. Use of HeLa cells infected with tubercle bacilli for the study of antituberculous drugs. *J. Bacteriol.* **73:**494–498.

233. **Shi, S., and S. Ehrt.** 2006. Dihydrolipoamide acyltransferase is critical for *Mycobacterium tuberculosis* pathogenesis. *Infect. Immun.* **74:**56–63.

234. **Singh, A., Y. Singh, R. Pine, L. Shi, R. Chandra, and K. Drlica.** 2006. Protein kinase I of *Mycobacterium tuberculosis*: cellular localization and expression during infection of macrophage-like cells. *Tuberculosis (Edinb.)* **86:**28–33.

235. **Singh, K. K., X. Zhang, A. S. Patibandla, P. Chien, Jr., and S. Laal.** 2001. Antigens of *Mycobacterium tuberculosis* expressed during preclinical tuberculosis: serological immunodominance of proteins with repetitive amino acid sequences. *Infect. Immun.* **69:**4185–4191.

236. **Singh, R., V. Rao, H. Shakila, R. Gupta, A. Khera, N. Dhar, A. Singh, A. Koul, Y. Singh, M. Naseema, P. R. Narayanan, C. N. Paramasivan, V. D. Ramanathan, and A. K. Tyagi.** 2003. Disruption of mptpB impairs the ability of *Mycobacterium tuberculosis* to survive in guinea pigs. *Mol. Microbiol.* **50:**751–762.

237. **Skeiky, Y. A., M. J. Lodes, J. A. Guderian, R. Mohamath, T. Bement, M. R. Alderson, and S. G. Reed.** 1999. Cloning, expression, and immunological evaluation of two putative secreted serine protease antigens of *Mycobacterium tuberculosis. Infect. Immun.* **67:**3998–4007.

238. **Slayden, R. A., and C. E. Barry III.** 2000. The genetics and biochemistry of isoniazid resistance in Mycobacterium tuberculosis. *Microbes Infect.* **2:**659–669.

239. **Sonnenberg, P., J. R. Glynn, K. Fielding, J. Murray, P. Godfrey-Faussett, and S. Shearer.** 2005. How soon after infection with HIV does the risk of tuberculosis start to increase? A retrospective cohort study in South African gold miners. *J. Infect. Dis.* **191:**150–158.

240. **Sorensen, A. L., S. Nagai, G. Houen, P. Andersen, and A. B. Andersen.** 1995. Purification and characterization of a low-molecular-mass T-cell antigen secreted by *Mycobacterium tuberculosis. Infect. Immun.* **63:**1710–1717.

241. **Springer, B., S. Master, P. Sander, T. Zahrt, M. McFalone, J. Song, K. G. Papavinasasundaram, M. J. Colston, E. Boettger, and V. Deretic.** 2001. Silencing of oxidative stress response in *Mycobacterium tuberculosis*: expression patterns of ahpC in virulent and avirulent strains and effect of ahpC inactivation. *Infect. Immun.* **69:**5967–5973.

242. **Sreevatsan, S., X. Pan, K. E. Stockbauer, N. D. Connell, B. N. Kreiswirth, T. S. Whittam, and J. M. Musser.** 1997. Restricted structural gene polymorphism in the *Mycobacterium tuberculosis* complex indicates evolutionarily recent global dissemination. *Proc. Natl. Acad. Sci. USA* **94:**9869–9874.

243. **Stanley, S. A., S. Raghavan, W. W. Hwang, and J. S. Cox.** 2003. Acute infection and macrophage subversion by *Mycobacterium tuberculosis* require a specialized secretion system. *Proc. Natl. Acad. Sci. USA* **100:**13001–13006.

244. **Stead, W. W., K. D. Eisenach, M. D. Cave, M. L. Beggs, G. L. Templeton, C. O. Thoen, and J. H. Bates.** 1995. When did *Mycobacterium tuberculosis* infection first occur in the New World? An important question with public health implications. *Am. J. Respir. Crit. Care Med.* **151:**1267–1268.

245. **Stermann, M., L. Sedlacek, S. Maass, and F. C. Bange.** 2004. A promoter mutation causes differential nitrate reductase activity of *Mycobacterium tuberculosis* and *Mycobacterium bovis. J. Bacteriol.* **186:**2856–2861.

246. **Stewart, G. R., J. Patel, B. D. Robertson, A. Rae, and D. B. Young.** 2005. Mycobacterial mutants with defective control of phagosomal acidification. *PLoS Pathog.* **1:**269–278.

247. **Stinear, T. P., H. Hong, W. Frigui, M. J. Pryor, R. Brosch, T. Garnier, P. F. Leadlay, and S. T. Cole.** 2005. Common evolutionary origin for the unstable virulence plasmid pMUM found in geographically diverse strains of *Mycobacterium ulcerans. J. Bacteriol.* **187:**1668–1676.

248. **Stinear, T. P., G. A. Jenkin, P. D. Johnson, and J. K. Davies.** 2000. Comparative genetic analysis of *Mycobacterium ulcerans* and *Mycobacterium marinum* reveals evidence of recent divergence. *J. Bacteriol.* **182:**6322–6330.

249. **Stinear, T. P., A. Mve-Obiang, P. L. Small, W. Frigui, M. J. Pryor, R. Brosch, G. A. Jenkin, P. D. Johnson, J. K. Davies, R. E. Lee, S. Adusumilli, T. Garnier, S. F. Haydock, P. F. Leadlay, and S. T. Cole.** 2004. Giant plasmid-encoded polyketide synthases produce the macrolide toxin of *Mycobacterium ulcerans. Proc. Natl. Acad. Sci. USA* **101:**1345–1349.

250. **Stinear, T. P., M. J. Pryor, J. L. Porter, and S. T. Cole.** 2005. Functional analysis and annotation of the virulence plasmid pMUM001 from *Mycobacterium ulcerans. Microbiology* **151:** 683–692.

250a. Stinear, T. P., T. Seemann, S. Pidot, W. Frigui, G. Reysset, T. Garnier, G. Meurice, D. Simon, C. Bouchier, L. Ma, N. Tichit, J. L. Porter, J. Ryan, P. D. Johnson, J. K. Davies, G. A. Jenkin, P. L. Small, L. M. Jones, F. Tekaia, F. Laval, M. Daffe, J. Parkhill, and S. T. Cole. 2007. Reductive evolution and niche adaptation inferred from the genome of *Mycobacterium ulcerans*, the causative agent of Buruli ulcer. *Genome Res.* **7:**192–200.

251. St. John, G., N. Brot, J. Ruan, H. Erdjument-Bromage, P. Tempst, H. Weissbach, and C. Nathan. 2001. Peptide methionine sulfoxide reductase from *Escherichia coli* and *Mycobacterium tuberculosis* protects bacteria against oxidative damage from reactive nitrogen intermediates. *Proc. Natl. Acad. Sci. USA* **98:**9901–9906.

252. Supply, P., S. Lesjean, E. Savine, K. Kremer, D. van Soolingen, and C. Locht. 2001. Automated high-throughput genotyping for study of global epidemiology of *Mycobacterium tuberculosis* based on mycobacterial interspersed repetitive units. *J. Clin. Microbiol.* **39:**3563–3571.

253. Supply, P., R. M. Warren, A. L. Banuls, S. Lesjean, G. D. Van Der Spuy, L. A. Lewis, M. Tibayrenc, P. D. Van Helden, and C. Locht. 2003. Linkage disequilibrium between minisatellite loci supports clonal evolution of *Mycobacterium tuberculosis* in a high tuberculosis incidence area. *Mol. Microbiol.* **47:**529–538.

254. Tailleux, L., N. Maeda, J. Nigou, B. Gicquel, and O. Neyrolles. 2003. How is the phagocyte lectin keyboard played? Master class lesson by *Mycobacterium tuberculosis*. *Trends Microbiol.* **11:**259–263.

255. Taylor, G. M., D. B. Young, and S. A. Mays. 2005. Genotypic analysis of the earliest known prehistoric case of tuberculosis in Britain. *J. Clin. Microbiol.* **43:**2236–2240.

256. Tekaia, F., S. V. Gordon, T. Garnier, R. Brosch, B. G. Barrell, and S. T. Cole. 1999. Analysis of the proteome of *Mycobacterium tuberculosis* in silico. *Tuber. Lung Dis.* **79:**329–342.

257. Temmerman, S. T., S. Place, A. S. Debrie, C. Locht, and F. Mascart. 2005. Effector functions of heparin-binding hemagglutinin-specific CD8$^+$ T lymphocytes in latent human tuberculosis. *J. Infect. Dis.* **192:**226–232.

258. Thorson, L. M., D. Doxsee, M. G. Scott, P. Wheeler, and R. W. Stokes. 2001. Effect of mycobacterial phospholipids on interaction of *Mycobacterium tuberculosis* with macrophages. *Infect. Immun.* **69:**2172–2179.

259. Tian, J., R. Bryk, S. Shi, H. Erdjument-Bromage, P. Tempst, and C. Nathan. 2005. *Mycobacterium tuberculosis* appears to lack alpha-ketoglutarate dehydrogenase and encodes pyruvate dehydrogenase in widely separated genes. *Mol. Microbiol.* **57:**859–868.

260. Timm, J., F. A. Post, L. G. Bekker, G. B. Walther, H. C. Wainwright, R. Manganelli, W. T. Chan, L. Tsenova, B. Gold, I. Smith, G. Kaplan, and J. D. McKinney. 2003. Differential expression of iron-, carbon-, and oxygen-responsive mycobacterial genes in the lungs of chronically infected mice and tuberculosis patients. *Proc. Natl. Acad. Sci. USA* **100:**14321–14326.

261. Titball, R. W. 1993. Bacterial phospholipases C. *Microbiol. Rev.* **57:**347–366.

262. Trivedi, O. A., P. Arora, V. Sridharan, R. Tickoo, D. Mohanty, and R. S. Gokhale. 2004. Enzymic activation and transfer of fatty acids as acyl-adenylates in mycobacteria. *Nature* **428:**441–445.

263. van Pinxteren, L. A., P. Ravn, E. M. Agger, J. Pollock, and P. Andersen. 2000. Diagnosis of tuberculosis based on the two specific antigens ESAT-6 and CFP10. *Clin. Diagn. Lab. Immunol.* **7:**155–160.

264. van Soolingen, D., T. Hoogenboezem, P. E. de Haas, P. W. Hermans, M. A. Koedam, K. S. Teppema, P. J. Brennan, G. S. Besra, F. Portaels, J. Top, L. M. Schouls, and J. D. van Embden. 1997. A novel pathogenic taxon of the *Mycobacterium tuberculosis* complex, Canetti: characterization of an exceptional isolate from Africa. *Int. J. Syst. Bacteriol.* **47:**1236–1245.

265. Viana-Niero, C., P. E. de Haas, D. van Soolingen, and S. C. Leao. 2004. Analysis of genetic polymorphisms affecting the four phospholipase C (plc) genes in *Mycobacterium tuberculosis* complex clinical isolates. *Microbiology* **150:**967–978.

266. Viana-Niero, C., C. A. Rodriguez, F. Bigi, M. S. Zanini, J. S. Ferreira-Neto, A. Cataldi, and S. C. Leao. 2006. Identification of an IS6110 insertion site in plcD, the unique phospholipase C gene of *Mycobacterium bovis*. *J. Med. Microbiol.* **55:**451–457.

267. Villeneuve, C., M. Gilleron, I. Maridonneau-Parini, M. Daffe, C. Astarie-Dequeker, and G. Etienne. 2005. Mycobacteria use their surface-exposed glycolipids to infect human macrophages through a receptor-dependent process. *J. Lipid Res.* **46:**475–483.

268. Voladri, R. K., D. L. Lakey, S. H. Hennigan, B. E. Menzies, K. M. Edwards, and D. S. Kernodle. 1998. Recombinant expression and characterization of the major beta-lactamase of *Mycobacterium tuberculosis*. *Antimicrob. Agents Chemother.* **42:**1375–1381.

269. Walburger, A., A. Koul, G. Ferrari, L. Nguyen, C. Prescianotto-Baschong, K. Huygen, B. Klebl, C. Thompson, G.

Bacher, and J. Pieters. 2004. Protein kinase G from pathogenic mycobacteria promotes survival within macrophages. *Science* **304:**1800–1804.

270. Weber, I., C. Fritz, S. Ruttkowski, A. Kreft, and F. C. Bange. 2000. Anaerobic nitrate reductase (narGHJI) activity of *Mycobacterium bovis* BCG in vitro and its contribution to virulence in immunodeficient mice. *Mol. Microbiol.* **35:**1017–1025.

271. Weihofen, A., and B. Martoglio. 2003. Intramembrane-cleaving proteases: controlled liberation of proteins and bioactive peptides. *Trends Cell Biol.* **13:**71–78.

272. Wengenack, N. L., M. P. Jensen, F. Rusnak, and M. K. Stern. 1999. *Mycobacterium tuberculosis* KatG is a peroxynitritase. *Biochem. Biophys. Res. Commun.* **256:**485–487.

273. Wheeler, P. R., K. Bulmer, and C. Ratledge. 1990. Enzymes for biosynthesis de novo and elongation of fatty acids in mycobacteria grown in host cells: is *Mycobacterium leprae* competent in fatty acid biosynthesis? *J. Gen. Microbiol.* **136:**211–217.

274. Wheeler, P. R., and C. Ratledge. 1992. Control and location of acyl-hydrolysing phospholipase activity in pathogenic mycobacteria. *J. Gen. Microbiol.* **138:**825–830.

275. Wisedchaisri, G., M. Wu, A. E. Rice, D. M. Roberts, D. R. Sherman, and W. G. Hol. 2005. Structures of *Mycobacterium tuberculosis* DosR and DosR-DNA complex involved in gene activation during adaptation to hypoxic latency. *J. Mol. Biol.* **354:**630–641.

276. Wu, C. H., J. J. Tsai-Wu, Y. T. Huang, C. Y. Lin, G. G. Lioua, and F. J. Lee. 1998. Identification and subcellular localization of a novel Cu,Zn superoxide dismutase of *Mycobacterium tuberculosis. FEBS Lett.* **439:**192–196.

277. Young, T. A., B. Delagoutte, J. A. Endrizzi, A. M. Falick, and T. Alber. 2003. Structure of *Mycobacterium tuberculosis* PknB supports a universal activation mechanism for Ser/Thr protein kinases. *Nat. Struct. Biol.* **10:**168–174.

278. Zahrt, T. C., C. Wozniak, D. Jones, and A. Trevett. 2003. Functional analysis of the *Mycobacterium tuberculosis* MprAB two-component signal transduction system. *Infect. Immun.* **71:**6962–6970.

279. Zhang, Y., B. Heym, B. Allen, D. Young, and S. Cole. 1992. The catalase-peroxidase gene and isoniazid resistance of *Mycobacterium tuberculosis. Nature (Lond.)* **358:**591–593.

280. Zhang, Y., R. Lathigra, T. Garbe, D. Catty, and D. Young. 1991. Genetic analysis of superoxide dismutase, the 23 kilodalton antigen of *Mycobacterium tuberculosis. Mol. Microbiol.* **5:**381–391.

281. Zink, A. R., C. Sola, U. Reischl, W. Grabner, N. Rastogi, H. Wolf, and A. G. Nerlich. 2003. Characterization of *Mycobacterium tuberculosis* complex DNAs from Egyptian mummies by spoligotyping. *J. Clin. Microbiol.* **41:**359–367.

NEISSERIA: A POSTGENOMIC VIEW

Lori A. S. Snyder, Philip W. Jordan, and Nigel J. Saunders

4

GENOME SEQUENCES OF THE *NEISSERIA* SPECIES

In 2000, two *Neisseria meningitidis* genome sequences were completed, published, and made publicly available (108, 152). These sequences have proven to be an invaluable resource for the neisserial research community and to those who have used bacterial genome sequences to draw conclusions about topics related to bacteria as a whole. As of the writing of this chapter, there were 404 publications that cited the *N. meningitidis* MC58 genome sequence (152) and 304 that cited the *N. meningitidis* 22491 genome sequence (108). Although many of these publications cite both genome sequences, there has still been a great deal of use made of these resources. In addition, some publications that have made use of these genome sequences have not cited the primary publications. Some groups made use of the incomplete sequence data available from these sequencing centers—The Institute for Genomic Research (TIGR) and the Wellcome Trust Sanger Institute, respectively—before their publication, which makes it difficult to identify studies using them, but shows the insight of those investigators

who realized the potential of whole genome sequence availability. In addition to these two genome sequences, there are three other unpublished neisserial genome sequences; however, as long as they remain unpublished, their use is nearly impossible to determine unless the relevant publications also cited the Tettelin et al. (152) or Parkhill et al. (108) publications. This chapter will therefore primarily focus on those studies that have made use of the two published *N. meningitidis* genome sequences (108, 152).

Investigators have used the genome sequences in a variety of ways. Some have made use of the genome sequence to verify the presence of a gene in *N. meningitidis* that has been sequenced from *N. gonorrhoeae*. Others have used it to flesh out the sequence of their region of interest, in which case the available complete genome sequence does away with the need to sequence the region. Although such simple operations are possible by using a complete genome sequence, the complete genome sequence was not required for these. The same conclusions could have been drawn from Southern blot tests or from screening a library by using the known region as a probe and extending the sequence of the region of interest. This chapter will not address this type of study but will instead focus on those inves-

Lori A. S. Snyder, Philip W. Jordan, and Nigel J. Saunders The Sir William Dunn School of Pathology, University of Oxford, South Parks Road, Oxford OX1 3RE, United Kingdom.

Bacterial Pathogenomics, Edited by M. J. Pallen et al.
© 2007 ASM Press, Washington, D.C.

tigations that were only made possible, or were greatly facilitated, by the availability of the meningococcal genome sequences.

FEATURES DESCRIBED BY THE *N. MENINGITIDIS* GENOME SEQUENCING PROJECTS

The Neisserial Silent Cassette Systems

The major pilin subunit, encoded by *pilE*, undergoes antigenic variation through recombination with silent *pilS* cassettes (50, 59, 60, 98, 108, 127, 152). The *pilS* sequences have homology with the *pilE* gene, but only the sequence in the *pilE* expression locus is capable of expressing a protein because it is the only site with a promoter and the 5′ end of the gene. The silent cassettes serve as a reservoir of variable sequences that generate change through homologous recombination with the expressed *pilE* site. Analysis of the *N. meningitidis* genome sequences revealed that there are probably additional expression/silent cassette systems in the meningococcus that have not yet been experimentally tested.

THE *mafB* SILENT CASSETTE SYSTEM

Potential silent cassettes were identified for *mafB* and *fhaB*, suggesting that the proteins encoded by the expression loci of these genes can undergo antigenic variation by a similar mechanism to *pilE*. For *mafB*, there are sequences in the *mafAB* region that have homology with *mafB* but lack initiation codons. This suggests that these are silent cassettes and that MafB undergoes antigenic variation. The *mafAB* system, encoding a potential lipoprotein (MafA) and predicted secreted protein (MafB), may be involved in adhesion to host cells (108).

THE *fhaB* SILENT CASSETTE SYSTEM

In the *N. meningitidis* Z2491 genome sequence, a gene with homology to the filamentous hemaglutinin precursor gene from *Bordetella pertussis*, *fhaB* (NMA0688), was identified. What are apparently silent cassettes of the *fhaB* sequence have also been identified, as has a gene with homology to *fhaA*, the accessory se-

cretion protein in the *B. pertussis* filamentous hemaglutinin system (69, 81).

Islands of Horizontal Transfer

The whole genome sequences also enabled predictions to be made about the recent horizontal acquisition of regions. On the basis of dinucleotide signatures and sequence comparisons, three islands of horizontal transfer (IHT) were identified in the *N. meningitidis* MC58 genome sequence (152). The first, IHT-A, has two subregions, with IHT-A1 (NMB0066 to NMB0074) containing the capsule locus and IHT-A2 (NMB0091 to NMB0100) containing an ABC transporter, a secreted protein, and eight hypothetical genes. The second region, IHT-B (NMB0498 to NMB0521), contains 24 hypothetical genes, and the third, IHT-C (NMB1746 to NMB1775), contains 30 genes, including a copy of *tspB* (NMB1747; paralogues are also located at NMB0976, NMB1548, NMB1628, and NMB0480, close to IHT-B), a *piv* homologue (NMB1750, other homologues are NMB1552 and NMB1625), and a putative hemagglutinin/hemolysin (NMB1768). The *N. meningitidis* Z2491 genome sequence does not contain IHT-A, IHT-B, or IHT-C (152). It is interesting to note that the two copies of *fhaAB* in *N. meningitidis* strain MC58 are located immediately adjacent to IHT-B (*fhaA* NMB0496 and *fhaB* NMB0497) and IHT-C (*fhaA* NMB1780 and *fhaB* NMB1779).

POSTGENOMIC INVESTIGATIONS USING NEISSERIAL GENOME SEQUENCES

Investigations of Genes, Pathways, and Processes of Interest

By conducting homology searches of the meningococcal genomes, homologues of genes known to be involved in a metabolic process, protein families, and biosynthetic pathways in other species can be identified and further investigated. Such searches for genes within the genomes can aid investigators in planning and interpreting the data from their experiments, particularly if these are conducted with the sequenced strains.

NITRIC OXIDE

Nitric oxide (NO) is released from macrophages in response to *N. meningitidis* lipopolysaccharide (LPS), and it is an intermediate in the denitrification of nitrite to nitrous oxide (11). Anjum et al. (4) investigated NO metabolism in the meningococcus. The *N. meningitidis* MC58 and *N. meningitidis* Z2491 genome sequences were searched for genes homologous to those involved in NO metabolism and reduction of nitrite to NO in other species (*Escherichia coli*, *Pseudomonas stutzeri*, *Pseudomonas aeruginosa*, *Achromobacter cycloclastes*, *Ralstonia eutropha*, and *Paracoccus denitrificans*). This identified two regions involved in NO metabolism and genes homologous to those responsible for reduction of nitrite to NO. The growth of the meningococcus microaerobically, the denitrification of nitrite to NO, and the anaerobic growth of the meningococcus in the presence of nitrite were investigated and interpreted on the basis of these genomic data of the available genes (4). It was concluded that *N. meningitidis* is therefore able to counteract exogenous NO, which may confer protection during infection. However, it is a recurrent theme of postgenomics studies conducted at a system level that a clear distinction frequently cannot be drawn between central physiology and virulence behaviors, and these differences may quite possibly be evolved and primarily adaptive to the use of nitrites in nonaerobic respiration rather than the attributed virulence-associated role, which may be secondary.

SUPEROXIDE

Host neutrophils and macrophages commonly use superoxide radicals to eradicate bacteria after invasion of the host. Radicals generate strand breaks in DNA and RNA and affect fatty acid structure in the cell membranes. Radicals can also be generated naturally as a result of aerobic respiration and the presence of transition metal ions within the cell. This means that during pathogenesis, production of proteins that prevent free radical damage is essential. Seib et al. (128) studied the Sco protein that is present in many prokaryotic and eukaryotic cells and that has been shown to be important for formation of the aa_3 cytochrome center commonly used as the terminal electron acceptor in oxidative phosphorylation (40). By searching the gonococcal strain FA1090 and meningococcal strain MC58 genome sequences with the yeast *sco* sequence, they found a *sco* homologue present in both genomes (128). *Neisseria* spp. have been shown to lack the genes encoding the aa_3 complex and instead use the cbb_3 cytochrome as the terminal acceptor. Seib et al. (128) showed that the neisserial gene was not important for aa_3 or for any other cytochrome production. The secondary structure prediction for Sco predicted similarities to a thioredoxin fold; therefore, the resistance to oxidative stress was assessed in an *sco* mutant and its parent strain by an oxidative killing assay, in which paraquat was used to generate superoxide radicals. It was found that the mutant was much more susceptible to paraquat-induced killing than the parent strain, suggesting this protein has an important role in resistance to oxidative damage (128). In this way, a genomics approach was able to correct and fill in important gaps in our understanding of this aspect of these species' physiology.

TRANSMEMBRANE TRANSPORT

The amino acid sequence of MsbA from *E. coli*, an LPS and phospholipids transporter (31), was used to search the *N. meningitidis* MC58 genome sequence, identifying NMB1919 (151). When this meningococcal gene was introduced into a temperature-sensitive *E. coli* *msbA* mutant, the *N. meningitidis* gene was able to complement the *E. coli* mutation, restoring growth at 44°C and the correct outer membrane localization of phospholipids and LPS (151). Mutants of *msbA* in *N. meningitidis* displayed differences in growth rate, colony side, colony opacity, and amount of LPS. Although the meningococcal MsbA is capable of complementing the defect in *E. coli*, it is not required for phospholipid transport in *N. meningitidis* (151). This study highlights the need to conduct experimental investigations of the actual functions of homologues within a species of interest and not to assume that the roles that these genes play in model species can be extrapolated without need for further investigation.

LPS MODIFICATION

The identification of the addition of phosphoethanolamine residues to LPS and phosphorylcholine to pilus has led to the use of the genome sequences to identify the presence of homologues of the protein families that catalyze these linkages. Cox et al. (21) searched the *N. meningitidis* strain MC58 genome sequence to find homologues of *lpt*-3, a neisserial gene known to add phosphoethanolamine residues to LPS (84). Although the function of *lpt*-3 and the linkage generated by this gene product were known, it was also known that other phosphoethanolamine substitutions occurred on the LPS of *N. meningitidis* (39, 71, 72) and that these were not mediated by *lpt*-3 (84). Two homologues of *lpt*-3 were identified in the strain MC58 genome sequence, NMB1638 and NMB0415 (21). PCR screening was conducted. The findings demonstrated the presence of both of these genes in most of the other meningococcal strains tested. Mutants were made, which showed that NMB0415 is not involved in modification of LPS. NMB1638, however, was determined to be a phosphoethanolamine transferase that modifies the lipid A region of LPS. The gene was therefore named *lptA* (21).

PILUS MODIFICATION

Warren and Jennings (158) also searched the *N. meningitidis* strain MC58 genome sequence to find homologues of *lpt*-3 and also to find homologues of genes involved in the addition of phosphorylcholine to pilin and in the biosynthesis and acquisition of phosphorylcholine. These searches identified three genes that were potentially involved in phosphorylcholine biosynthesis or uptake (NMB0939, NMB1270, and NMB1277) as well as the two *lpt*-3 homologues also identified by Cox et al. (21) (NMB1638 and NMB0415). Although the previous investigation of the *lpt*-3 homologues was looking for genes that modify LPS (21), this investigation was addressing those that modify pili. Therefore, NMB1638 mutants, of the LPS modification gene *lptA*, did not show altered pilus, whereas NMB0415 mutants, which showed no variation in their LPS modifications, were altered in their pili. NMB0415 was therefore named *pptA* for its pilin phosphorylcholine transferase activity (158).

Before either the LPS or pilus modification studies, NMB0415 was investigated by Snyder et al. (140) in both *N. meningitidis* and *N. gonorrhoeae*, in which the potential to phase vary NMB0415 in *N. meningitidis* was recognized. In addition, it was demonstrated that this gene, named *dca* by Snyder et al. (140) because of its association with the division cell wall synthesis cluster, was required for natural competence for transformation in *N. gonorrhoeae* strains FA19 and FA1090, but not of *N. meningitidis* strains NMB and O929.

RECOMBINATION

As mentioned above, the PilE undergoes antigenic variation through recombination with the silent *pilS* cassettes (50, 59, 60, 98, 108, 127, 152). Skaar et al. (133) sought to identify recombinases which might explain the recombination events that occur when *pilE* interchanges DNA with the *pilS* silent cassettes. Eight recombinases homologous to the *piv* genes present in *Moraxella* spp. (87) were identified as present in the *N. gonorrhoeae* FA1090 genome sequence (NGO0773, NGO1137, NGO1164, NGO1200, NGO1262, NGO1641, NGO1648, and NGO1703). However, inactivation of all of these genes had no effect on antigenic pilin variation. In addition to those intact genes identified by Skaar et al. (133), a *piv* homologue gene fragment is present between NGO0953 and NGO0955 in the *N. gonorrhoeae* strain FA1090 genome sequence (Snyder, personal observation). It is interesting to note that there are far more *piv* homologues in the gonococcal genome sequence than have been found in the meningococcal genome sequences; in *N. meningitidis* MC58, there are three (NMB1552, NMB1625, and NMB1750) plus two gene fragments (NMB0292 and NMB1232), and in *N. meningitidis* Z2491 there are two (NMA0772 and NMA1800) plus the same two gene fragments seen in MC58. The overall function of these genes is unknown although the presence of some of these genes only among the pathogenic members of the *Neisseria* spp. suggests that they may have some role in virulence (133). Indeed, the commensal *Neisseria* spp. tested by Snyder and Saunders (134) did not hybridize to the NMB1552/NMB1625 gene probe, but they did hybridize to the NMB1750 gene probe, suggesting that some alleles of these elements are absent from the nonpathogenic neisserial species.

PREPILIN PROTEINS
Before the publication of the neisserial genome sequences, Wolfgang et al. utilized the incomplete *N. gonorrhoeae* strain FA1090 genome sequence to investigate the potential presence of prepilin-like molecules other than PilE in gonococci (162). The N-terminal sequence of PilE was used in the search, identifying the *comP* gene that encodes a prepilin essential for natural transformation in *N. gonorrhoeae* (162). Upon publication of the meningococcal genome sequences, these completed sequences were used to identify another prepilin gene, *pilV*, which is essential for adherence of the gonococci to human epithelial cells (160). In another report, from 2005, a further search revealed a cluster of five additional prepilin genes that act in concert and may be involved in bidirectional pilin remodeling: *pilH*, *pilI*, *pilJ*, *pilK*, and *pilL* (161).

PENICILLIN-BINDING PROTEINS
Three penicillin-binding proteins (PBPs) were identified experimentally in the *N. gonorrhoeae* in 1981 (6) and were considered to be reasonably well established; however, a search of the *N. meningitidis* MC58 genome sequence and that of *N. gonorrhoeae* FA1090 identified a new PBP, which was called PBP 4 (144). It is likely that the expression of PBP 4 in cultures is too low to have been detected biochemically, making the availability of the genome sequence critical for its discovery. The investiga-

tors were then able to clone, express, and purify PBP 4 so that its enzymatic activity as a low-molecular-mass PBP could be studied.

TonB-DEPENDENT FAMILY PROTEINS
Iron has been shown to be essential for almost all bacterial species to survive, while its acquisition within the human host is made difficult by the presence of human iron-binding proteins that maintain free iron at a very low availability. Its importance for virulence has been observed, with the addition of iron supplements increasing the blood titer of bacteria present after infection (58), but it is important to be careful in the interpretation of such studies because more of the iron acquisition systems are species specific and therefore there are likely to be elements of comparing more normal growth with supplementation with poor growth and fitness during iron starvation. Conditions of iron starvation have been shown to enhance meningococcal capsule biosynthesis (94) and to increase meningococcal virulence in a mouse model (12, 58). The TonB-dependent family (Tdf) of outer-membrane receptors includes five receptors involved in the acquisition of iron from transferrin, lactoferrin, hemoglobin, and the siderophore ferric enterobactin (15, 18, 78, 113, 147). Searches of the meningococcal genome sequences, and the *N. gonorrhoeae* FA1090 genome sequence, were conducted with the amino acid sequences of *E. coli* FecA and FepA, *P. aeruginosa* FpvA, *Yersinia enterocolitica* HemR, and *Serratia marcescens* HasR. This revealed seven homologues of Tdf proteins, encoded by NMB1882 (named *tdfF*), NMB1497 (named *tdfH*), NMB0964, NMB1829, NMB0293, NMA1558, and one gene present only in the gonococcal genome, which was named *tdfG* (154). Three of these were examined. TdfF was found not to be expressed under the conditions tested. The gene that was present only in the gonococcal genome, *tdfG*, was found to be present in only 17% of the *N. gonorrhoeae* strains tested and none of the *N. meningitidis* strains. TdfH expression was not affected by iron availability, suggesting a different substrate

for this Tdf protein (154). The functions of these proteins therefore remain to be determined, but the genome-scale analysis indicates an important area of future studies to understand the roles of these genes.

TYPE V SECRETION SYSTEMS: AUTOTRANSPORTERS

In many bacterial infections, secreted proteins play a role in the pathogenesis of disease. The mechanisms of secretion of proteins from bacteria have been extensively studied in the postgenomic era, with five general types of secretion systems identified (77). Type V secretion systems include autotransporter proteins (55). In addition to the IgA protease common to all pathogenic *Neisseria* spp., van Ulsen et al. (156) were able to identify seven other putative autotransporters from a search of the *N. meningitidis* MC58 and Z2491 genome sequences. The amino acid sequences of 27 proteins were used to search the meningococcal genomes, with autotransporter proteins from *Bordetella pertussis*, *E. coli*, *Haemophilus influenzae*, *Helicobacter pylori*, *Helicobacter mustelae*, *Moraxella catarrhalis*, *Rickettsia* spp., *S. marcescens*, and *Shigella flexneri*. Each of the seven genes identified in the *N. meningitidis* genome sequences is homologous to autotransporters, has an N-terminal signal sequence, and has a C terminus containing a potential β-barrel domain and autotransporter signature sequence. The eight autotransporters of *N. meningitidis* are therefore: NMB0312 (*vapA/autA*, encoding a putative virulence-associated protein [2]), NMB0700 (*iga*, encoding IgA protease [115]), NMB0992 (*hsf/nhhA*, encoding a putative adhesin [110]), NMB1525 (*virG*, encoding an AIDA-related protein [109]), NMB1969 (*aspA/nalP*, encoding a putative serotype 1–specific antigen [153, 157]), NMB1985 (*app*, encoding a putative adhesion and penetration protein [51]), NMB1994 (*nadA*, encoding an adhesin [17]), and NMB1998 (encoding a putative serine-type peptidase). It is interesting to note that six of these genes are either known phase variable genes (*nadA*) (89) or are candidates for phase variable expression (*vapA*, *virG*, *aspA*, *app*, and NMB1998) (136).

TYPE I SECRETION SYSTEMS: RTX-LIKE PROTEIN TRANSPORT

Wooldridge et al. (163) hypothesized that the meningococcal homologues of the RTX (repeat in toxin)-like proteins might be secreted through a type I secretion system, as they are in other gram-negative species (9). The meningococcal genome sequences were searched by using the protein sequences of *E. coli* HlyB and HlyD, which are components of the *E. coli* hemolysin secretion system (85). The meningococcal homologues of HlyB and HlyD that were identified were shown to be functional components of a type I secretion system required for the export of the meningococcal RTX-like proteins (163).

RTX-LIKE PROTEINS

The genes encoding the RTX-like proteins are scattered across the chromosome (152) and are present in different allelic versions in different strains (106). Forman et al. (37) therefore took advantage of the availability of the complete meningococcal strain MC58 genome sequence to ensure that their RTX-like protein-deficient mutant had eliminated all copies of the gene. Six FrpC paralogues were identified: NMB0365, NMB0585, NMB1403, NMB1405, NMB1409, and NMB1415. It was determined that only two of these contained the nonapeptide repeats characteristic of RTX-like proteins; therefore, the RTX mutant only addressed the products of NMB0585 and NMB1415 (37), whereas the others were still present. This double mutant, in NMB0585 and NMB1415, was then tested in the infant rat model of meningococcal disease, in which the mutant was shown to be as virulent as the parent strain (37). This illustrates the important way in which genome sequence information can be used to facilitate the proper interpretation of experimental data.

VERIFICATION OF PRESENCE OF METABOLIC PATHWAY

Some investigators made use of the genome sequences to help them to understand the data obtained in their studies and to verify the presence of a particular metabolic pathway. The interpretation of the results of experiments investigating the function of the meningococcal γ-glutamyl aminopeptidase gene (*ggt*, NMB1057) (149) were made in light of the presence of the genes necessary for cysteine synthesis from glycine identifiable within the genome. It was therefore concluded that the differences in growth and phenotype observed on different media were not due to the lack of these enzymes, but may instead be due to the inefficiency of this pathway under nutrient-limiting conditions.

DEMONSTRATION OF ABSENCE OF A SYSTEM AND THUS A NEW MECHANISM

The genome sequences can also be used to show that something is not there, as was the case when Tzeng et al. (155) used the meningococcal genome sequences to show the absence of genes encoding proteins responsible for the biosynthesis and attachment of 4-amino-4-deoxy-L-arabinose (Ara4N) to the LPS core and lipid A regions, which mediates resistance to the polymyxin B (PxB) cationic antimicrobial peptide in other species. This led to the investigation of the specific mechanisms involved in cationic antimicrobial peptide resistance in *N. meningitidis* and the identification of the importance of genes involved in lipid A modification and efflux in polymyxin B resistance (155).

IDENTIFICATION OF EFFLUX PUMP SYSTEMS

At the mucosal surfaces, there are many antimicrobial compounds present, including fatty acids, bile salts, gonadal steroidal hormones (102), and antibacterial peptides (131), yet *Neisseria* spp. are able to colonize these surfaces. A number of these host-derived compounds are exported by the MtrCDE efflux pump system (53, 86, 107). However, there was evidence for an MtrCDE-independent mechanism underlying resistance to long-chain fatty acids (96).

The FarAB Efflux Pump System. The sequence of the neisserial genomes allowed for a search of homologues to known efflux pumps from other bacterial species. Lee and Shafer (76) searched the incomplete *N. gonorrhoeae* strain FA1090 genome sequence to identify a homologue of the *emrAB* genes from *E. coli*, which mediate energy-dependent transport of certain hydrophobic molecules out of the cell in *E. coli* (82). It was subsequently demonstrated experimentally that the neisserial homologues, named *farAB*, mediate resistance of *N. gonorrhoeae* to antibacterial long-chain fatty acids (76).

The NorM Efflux Pump System. By means of the same approach, a neisserial homologue of *norM* from *Vibrio parahaemolyticus* was discovered. This was previously implicated in quinolone resistance in *V. parahaemolyticus* and *E. coli* (101). By using *N. meningitidis* MC58 and Z2491 and *N. gonorrhoeae* FA1090 genome sequences, *norM* homologues were found among the neisserial genomes (119). It was shown to export molecules with a quarternary ammonium ion, including quinolones (119).

The MacAB Efflux Pump System. The *mtrCDE* operon encodes an efflux pump that removes macrolides from the cell; however, Rouquette-Loughlin et al. (121) show that the genome sequence of *N. gonorrhoeae* strain FA1090 also contains a homologue of *macAB* from *E. coli* (70). Much like MacAB in *E. coli*, the gonococcal gene product also acts as a macrolide pump. However, it is not as active as the MtrCDE system because of a poor -10 consensus sequence. Repair of the -10 to the consensus sequence results in strong expression with an appreciable effect on macrolide efflux (121).

IDENTIFICATION OF AN ADP-RIBOSYLTRANSFERASE

By means of a novel strategy, Masignani et al. (91) were able to predict the presence of an ADP-ribosyltransferase in *N. meningitidis* strain MC58. These molecules catalyze the transfer of ADP-ribose to protein targets. When expressed by bacteria, these molecules are usually toxic because they alter the normal functions of eukaryotic cells, e.g., cholera toxin (97). Usually the sequence of these genes is poorly conserved, but by using the conservation of catalytic amino acid residues and conserved secondary structure motifs to search the genome sequence, one possible homologue (NMB1343) was found. The predicted protein encoded by this gene was subsequently shown to have all the properties of this class of toxin. This gene is not present in either the *N. meningitidis* Z2491 or *N. meningitidis* FAM18 genome sequences (91).

Investigation of Regulatory Networks

The availability of the neisserial genome sequences meant the availability not only of the sequences of the genes, but also of the intergenic regions, including the promoters. The interactions between DNA-binding regulatory proteins and the sequences of these intergenic regions govern the regulation of gene expression.

REGULATORY PROTEINS

The MtrR Transcriptional Regulator. One of the most extensively investigated regulatory proteins of the *Neisseria* spp. is MtrR, which was first identified as a regulator of the expression of the MtrCDE (52, 53) and Far AB (76) efflux pumps in *N. gonorrhoeae*. In an attempt to identify other genes regulated by MtrR, a search of the *N. gonorrhoeae* strain FA1090 genome sequence was conducted using the known MtrR-binding site from the intergenic region between *mtrR* and *mtrC* (83), identifying one such sequence 5′ of a homologue of *E. coli yabB*, the first gene of the *E. coli dcw* cluster (140).

Although no specific binding of MtrR was detected in this region, this search did identify the neisserial *dcw* cluster, which includes three genes that are not normally found in this highly conserved cluster of essential genes: *dca*, which is involved in natural competence for transformation (140) and pilus modification (158); *dcaB* (141); and *dcaC*, which contains a

variable copy number of a 108-bp coding tandem repeat (65, 141).

The PtsN Regulator. As part of a structural proteomic approach to the investigation of the *Neisseria* spp., the known and putative regulatory proteins from the available genome sequences have been identified. These have then been systematically applied to high-throughput cloning, expression, crystallization, and structural determination. The first of these proteins to be described is the product of *ptsN* (118), which is part of the phosphoenolpyruvate–sugar phosphotransferase system in *E. coli* (116). The structure of the P_{II} protein, which is also involved in controlling the interactions between amino acid and sugar metabolism, has also been solved (Nichols et al., submitted). In an ongoing project, a further five neisserial regulatory proteins crystal structures have been solved, which means that even with its current progress, this project has made the regulators of the *Neisseria* spp. one of the best structurally characterized sets of regulatory proteins for any bacterial species.

PROMOTER-LOCATED MOTIFS AND SEQUENCES

The REP2 Repeats/CREN. Parkhill et al. (108) described 26 REP2 repeats, within the *N. meningitidis* Z2491 genome sequence. Although the function of these repeats was unknown at the time, the presence of a ribosome binding site at one end of this element suggested that they may be functionally important.

It was later shown that the REP2 repeat includes the CREN (contact regulatory element of *Neisseria*), which is important for cell contact response (23). After the identification of this motif within the promoter of *pilC1* (23), the complete genome sequences of *N. meningitidis* strains MC58 and Z2491 were searched for the CREN motif, identifying 12 CREN homologues within the predicted promoter regions of NMB1847 (*pilC1* itself), NMB1856 (*crgA*), NMB0607 (*secD* homologue), NMB0697 (*ksgA* homologue), NMB0776 (*gly*1), NMB0783, NMB1047, NMB1136, NMB1347 (a *suhB* homologue), NMB1508, NMB1664 (a peptidase homologue), and NMB2132 (a transferrin-binding protein homologue), which were all shown to be upregulated in their expression on contact with host cells (24). These genes were also investigated by PCR to determine the presence of these genes in commensal *Neisseria* spp. strains, which identified that NMB0697, NMB1347, NMB1508, NMB1847, and NMB1856 were absent in those strains tested. Snyder and Saunders (134), however, detected hybridization of strains from the commensal *Neisseria* spp. to gene probes for all of these genes, with the exception of *pilC*1.

Morelle et al. (100) also demonstrated the upregulation of the expression of REP2/CREN sequence-associated genes. In this case, 14 of 16 genes identified in the *N. meningitidis* Z2491 genome sequence were tested and shown to upregulate in their expression. The 16 genes tested included 8 of those from the Deghmane et al. study (24), but not NMB0783, NMB1047, NMB1347, and NMB2132. The remaining eight genes identified by Morelle et al. (100) were NMB0262 (*xseB* homologue), NMB0277 (*mviN* homologue), NMB0615 (*amtB* homologue), NMB0844, NMB0996, NMB1335, NMB1605 (*parC* homologue), and NMB2130. Three of these (NMB0277, NMB1605, and NMB2130) do not have a REP2/CREN sequence within their promoter regions in *N. meningitidis* strain MC58.

Subsequent work by Ieva et al. (61) did not find any evidence that the REP2/CREN repeat contained a promoter for the *crgA* gene and suggested that instead this may have a posttranscriptional effect on RNA stability and translation. REP repeats in *E. coli* have been shown to form secondary structures that protect against exonuclease digestion and so prolong the mRNA existence (104). Therefore, they may be involved in mRNA stability.

The Fur Box. Delany et al. (28) used the *E. coli* Fur box, which is the binding site for the Fur repressor (36), to search the *N. meningi-tidis* MC58 genome sequence. This identified four promoters containing homologues of the Fur box, which were investigated for binding and regulation by meningococcal Fur (28).

Microarrays

MICROARRAYS TO ASSESS TRANSCRIPTION PROFILES

Although cloned random libraries have been used to generate microarrays (32), these have often been low-cost or interim solutions used before the completion of a genome-sequencing project. Ideally, a microarray is designed against the complete genome sequence of the organism of interest or, in the case of the pan-*Neisseria* microarray (developed by Davies, Saunders, Shafer, Apicella, Seifert, and Dyer), against multiple genome sequences of the same species or closely related species. The pan-*Neisseria* microarray (138) was the first academically produced multispecies, multistrain microarray; it was completed in 2002. There are now several "pan" microarrays available from academic and commercial sources, and even other neisserial multistrain microarrays, from TIGR, Operon, and the Bacterial Microarray Group at St. George's Hospital Medical School (BμG@S). With the current availability of around 10 neisserial microarrays, the regulatory networks of the *Neisseria* spp. can now be specifically addressed by using this technology. This profusion of microarrays, which are based on different probe sets, means that much of the microarray data generated are not directly comparable between groups and publications. For maximum comparability and communication within the research community, it would be best if a limited number of these arrays were generally widely used.

Response to Autoinducer-2. The earliest neisserial microarray experiments used nylon membrane arrays produced by Eurogentec; one such investigation concluded that *luxS*-inactivated meningococci do not respond to autoinducer-2 (AI-2), unlike *E. coli* and other bacteria in which it has been studied

(33). A later proteomic analysis that used two-dimensional (2D) gels also showed a lack of global response of *N. meningitidis* to AI-2 (126).

Response to Iron Starvation. Because of the connection between the expression of virulence determinants, such as capsule (94), with iron availability, the influence of iron on the behavior of the *Neisseria* spp. has been one of the most intensively investigated expression responses to date. Grifantini et al. (46) observed the effect on transcription of the presence or absence of iron on *N. meningitidis* strain MC58 over the first 5 h of iron starvation or presence. They saw differential regulation of 235 genes. In agreement with its importance for virulence, the absence of iron led to the upregulation of many virulence-related genes, including *frpC* (NMB1415) and the *frpA*- and *frpC*-related genes (NMB0364, NMB0584, NMB1402, NMB1405, NMB1412, NMB1414), *comP* (NMB2016), and a putative bacteriocin resistance protein (NMB0103). Upregulated in the presence of iron were *aniA* and *norB*, involved in anaerobic respiration and NO metabolism, and *sodB*, important for preventing free radical damage. Clustering analysis performed on these genes identified one cluster that was in close agreement with those genes for which Fur, the ferric uptake regulator, binds to their promoters. In addition, the results of this microarray experiment were used to identify the function of hypothetical genes (45).

Ducey et al. (35) demonstrated that under iron starvation conditions, only a small proportion of the genes reported to be regulated in *N. meningitidis* (46) were similarly regulated in their study when they used *N. gonorrhoeae*. In the gonococcal study, a total of 109 upregulated and 94 downregulated genes were reported in iron-starved *N. gonorrhoeae* by using the pan-*Neisseria* microarray (35). Of these, approximately 30% are considered to be directly dependent on Fur, with additional control proposed to be mediated by a MerR-like regulator (NMB1303).

This pair of studies illustrates an important point about the use of functional genomics

tools and that comparability of data really requires comparability of the technical aspects of experiments as well as of the other aspects of experimental design. The meningococcal experiment compared a lack of iron with 100 μM of $FeNO_3$, whereas the gonococcal experiment used 10 μM of $FeNO_3$. The meningococcal experiment performed two biological replicate cultures and sampled at different time points, whereas the gonococcal study performed four biological replicates at one time point. The meningococcal experiment used 1.5 μg of RNA per channel and direct labeling; the gonococcal experiment used 16 μg of RNA per channel and indirect aminoallyl-dUTP labeling. The meningococcal experiment used a relatively unusual statistical test (Cyber-T) based on analysis of variance and Bayesian estimates, which is specifically designed to be sensitive when low numbers of replicates are used (although the paper is not explicit about which of the available tests within Cyber-T was used), whereas the gonococcal study used the less sensitive but more standard and widely understood Student's *t* test. The meningococcal study then used a statistical cutoff of a fold ratio of 2-fold and $P < 0.01$ ($n \geq 3$), whereas the gonococcal study used a fold ratio of 1.5-fold and $P < 0.02$ ($n = 4$). Actually, because of the differences in the statistical tests used, despite appearing to be less stringent, the latter is more likely to generate false-negative findings than the former. In addition to these differences, the meningococcal experiment used the Chiron *N. meningitidis* strain MC58 microarray; the gonococcal experiment used the pan-*Neisseria* microarray-v1 (supplied by Julian Davies). The studies therefore use different microarray PCR product probes, which for the former are not publicly available, so they cannot be compared with those of the latter. Although the gonococcal study includes a comparison of the findings of the two investigations, really these technical differences provide an obstacle to performing such a comparison, despite the fact that there is nothing technically better or inherently more correct or incorrect in either study. This is a general problem for functional genomics and systems biology in the context of many different species, and it is in no way a problem unique to the *Neisseria* spp. These issues must be considered whenever such studies are performed with differing protocols and tools.

To further characterize the iron response, microarray analysis was performed with a Fur mutant of *N. meningitidis* MC58. Comparisons of the mutant with the parent strain were used to identify the Fur regulon, and comparisons of the wild type and the *fur*-null mutant with and without iron were used to identify genes regulated by Fur, as well as those regulated by iron independently of Fur. This study identified 83 genes that respond to the absence of Fur. Three main categories of genes were repressed by Fur in the presence of iron, including those involved in iron uptake, secreted or cell envelope proteins including Frp proteins, and genes involved in energy metabolism. Genes induced by Fur and iron included those responding to oxidative stress and protein complexes involved in electron transfer. The chaperones and proteases of the heat shock response, which are altered in the Fur mutant, did not respond to iron limitation (27). This study did not identify the large-scale changes seen in response to iron limitation in the study by Grifantini et al. (46), but the regulation they do see as a result of iron limitation is almost entirely controlled by Fur, showing the importance of this regulator for the iron response.

The Role of PilT in Adhesion. PilT is involved in the retraction of the neisserial pilus and its action apparently prevents firm adhesion of the bacteria to host cells (117). Yasukawa et al. (164) therefore assessed the changes in transcriptional profiles between a *pilT* mutant and its parent strain, *N. meningitidis* 8013-2C4.3, by means of the Eurogentec glass slide microarray. This showed that *pilC*1, encoding a protein essential for pilus assembly (64) and adhesion to host cells (103, 122), was upregulated in the mutant. From this microarray-based information, the investigators pursued this observation and concluded that PilT represses the expression of *pilC*1 from transcriptional start site PC1.4.

Responses to Host Cells. Grifantini et al. (43) determined the transcriptional profiles of *N. meningitidis* and *N. lactamica* in contact with human epithelial cells. Changes were observed between adherent and nonadherent cells in 347 and 285 genes, respectively. Of these, only 167 were common to the two species. Because this was performed as a time course study with RNA samples being analyzed at several stages of infection, it was noticeable that the persistence induced by RNA on adhesion was longer in *N. lactamica* than in *N. meningitidis* (43). This may be as a result of RNA stability or transcriptional regulation; however, RNase expression has been shown to be important for meningococcal invasion, suggesting a need for rapid RNA turnover (148). The known adhesion molecules were found to be upregulated in both species, except for *mafA*-1 (NMB0652), which was not detected in *N. lactamica*, but which has been shown to be present in this species (134). One of the most striking differences was the upregulation of sulfur acquisition and metabolism genes, which were not required for *N. lactamica* adhesion (43).

Grifantini et al. (44) then went on to look for potential vaccine candidates by using the microarray data to identify the specific expression of outer membrane proteins on the surface of adherent and nonadherent bacterial cells. Fluorescence-activated cell sorter (FACS) analysis was performed with mouse sera raised against 12 proteins (NMB0207 *gapA*-1, NMB0214 *prlC*, NMB0315, NMB0652 *mafA*-1, NMB0655, NMB0741, NMB0787 ABC transporter homologue, NMB0995 MIP homologue, NMB1061, NMB1119, NMB1875, and NMB1876 *argA*) that were upregulated in the expression profiles during adhesion of the meningococcus in cell culture to detect the presence of the gene products on the surface of the bacteria during adhesion. Antisera against these 12 proteins were also used to test for complement-mediated killing; 5 of these showed bactericidal activity against the strain from which they were derived (44). These were to *mafA*-1 (NMB0652), MIP-related protein gene (NMB0995), *N*-acetylglutamate synthetase gene (*argA*; NMB1876), and two hypothetical genes (NMB0315 and NMB1119). The killing effect was also observed with meningococci grown in serum, suggesting that these proteins are potentially expressed in the bloodstream. Unlike the previous work of Pizza et al. (114), which was based on protein localization predictions, this microarray-based approach was also able to find vaccine candidates that were not computationally predicted to be membrane proteins (NMB1119 and NMB1876).

By means of a 40-mer oligonucleotide microarray designed by using the *N. meningitidis* MC58 genome sequence, Dietrich et al. (29) investigated the transcriptional changes in *N. meningitidis* during in vitro contact with human epithelial and endothelial cells. This study reported that 72 genes were differentially regulated on adherence to host epithelial cells and 48 in the response to binding to endothelial cells. In addition to many genes of unknown function, the genes identified as regulated on contact with these host cells included those encoding membrane proteins *porA* (NMB1429), *opa* (of which there are multiple alleles), and ABC transporters (NMB0787/NMB0788 and NMB0880/NMB0881), as well as a regulatory protein (*farR*; NMB1843) (29).

An interaction with host cells has also been investigated by using *N. gonorrhoeae* (34). Human epithelial cells derived from the endocervix were infected with *N. gonorrhoeae* MS11 and the expression profile studied by a gonococcal whole-gene probe microarray. These results indicate that adherence to epithelial cells induces the gonococcal general stress response, mediated by RpoH (34). This paper demonstrates a particular example of microarray data that cannot be readily compared with others. This microarray uses whole gene probes, which have a wide size range and are maximally cross-hybridizing. In addition, the paired spot replicates printed onto the slides are adja-

cent, which makes this functionally similar to an unreplicated microarray. The analysis of the data itself is based on the identification of genes that lie outside of a range determined by the standard deviation, which will find a similar number of genes whether a large number, a small number, or even no genes show real biological responses. Additionally, the raw data have not been made available, or submitted to a public repository, so it is not possible to address anything other than what the authors have chosen to describe.

The Regulon of Alternative Sigma Factor RpoH. RpoH and the general stress response of *N. gonorrhoeae* were studied by Gunesekere et al. (48) in *N. gonorrhoeae* strain FA1090 by using the pan-*Neisseria* microarray-v1. In *E. coli*, the RpoH sigma factor governs the expression of genes involved in protein homeostasis during stress responses. In *Bordetella* and *Burkholderia*, the expression of such genes is regulated by the HrcA repressor, which recognizes the sequence of the CIRCE repeats (controlling inverted repeat of chaperone expression). Searches of the neisserial genome sequences found no homologues of HrcA or the CIRCE repeats (111). The neisserial RpoH, however, is essential, is involved in the regulation of chaperones and other components of the heat shock response, and is increased in expression on heat shock (47, 75, 150). Through overexpression of RpoH, its regulon in *N. gonorrhoeae* was determined and compared with the genes regulated during a heat shock response (48), showing that these responses and their control are distinctly different from those described previously in other species.

Response to Heat Shock. In contrast, a study of the heat shock response of *N. meningitidis* MC58 suggested that very few genes are regulated in the meningococcus; only two genes were observed as being differentially regulated following exposure to 42 and 43.5°C, and 55 genes at a higher than usual test temperature of 45°C (47). The majority of genes that were upregulated were those involved in protein folding and the typical stress response (e.g., *groES*, *groEL*, and *dnaK*), as well as the heat shock response sigma factor *rpoH*. Downregulated genes included many involved in aerobic respiration and a cold shock regulator (NMB0838) (47).

The Regulon of Alternative Sigma Factor ECF. Only three functionally intact sigma factors have been identified within the genome sequences of the *Neisseria* spp.: *rpoD* (NMB1538), *rpoH* (NMB0712), and *ecf* (NMB2144). There is also a degenerate and nonfunctional *rpoN* (NMB0217), which lacks the helix-turn-helix motif required for DNA binding (74). The effects of the overexpression of the alternative sigma factor extracytoplasmic function (ECF) on the expression profile of *N. gonorrhoeae* strain FA1090 was assessed by the pan-*Neisseria* microarray-v1 (49). The ECF sigma factor was shown to control the expression of eight genes: *ecf* itself and five adjacent genes (NMB2140-5); a hypothetical gene (NGO0518); and *msrA/B* (NMB0044), which encodes methionine sulfoxide reductase, indicating that this regulon is primarily associated with a response to oxidative surface protein damage and repair.

The Response to Serum. Another attempt to characterize antigens as targets for vaccines by using expression data analysis was performed on *N. meningitidis* MC58 (73). Gene expression was measured after 30 min of exposure to serum compared with a similar time of incubation in phosphate-buffered saline (PBS) and a differential regulation of 279 genes was observed, including known virulence factors. These included upregulation of *pilC1/pilC2* (essential for cell interaction), *frpC* (a toxin of unknown function present in invasive strains), and *tonB* (involved in iron uptake). Two proteins that are being considered as vaccine candidates, NspA (88) and FetA/FrpB (3), were actually downregulated, showing the value of the use of complementary techniques in vaccine candidate identification and prioritization (73).

Comparison of the serum versus PBS microarray data (73) with that of cell contact on the same microarray system (29) showed that only three genes (*mdaB, tonB*, and NMB1575) were differentially regulated on serum exposure, epithelial cell adherence, and endothelial cell adherence; 36 were differentially regulated in serum and either endothelial or epithelial cell adherence (29). However, this study cannot be readily compared with the other studies of cell contact responses because it made use of a very different type of microarray that used relatively short (40 mer) oligonucleotide probes. Also, it may not be safe to assume that there is not a separate response related to incubation (and effectively starvation) in PBS for 30 min that may complicate the interpretation of this study.

The Genes Regulated through a Two-Component Regulatory System. Four two-component regulatory systems are present in the neisserial genome sequences on the basis of homology (NMB0114/NMB0115, NMB 0594/NMB0595, NMB1249/NMB1250, and NMB1606/NMB1607). One of these (NMB 0594/NMB0595) has been investigated by Newcombe et al. (105) by using the BμG@S neisserial microarray because it is the regulator with the greatest similarity to the PhoP/Q system of other bacterial species, and because their previous studies have found a mutant of this system to be attenuated. These authors concluded that this regulator is involved in the response to magnesium, which would be consistent with this being a functional PhoP/Q equivalent system in these species, and that its deletion resulted in altered expression of many genes encoding meningococcal surface structures, such as lipid A synthesis genes *kdsA* (NMB1283), *lpxC/envA* (NMB0017), and *lptA* (NMB1638) and surface proteins, consistent with its observed effects on virulence.

The Response to Hydrogen Peroxide. Grifantini et al. (46), in their study of iron starvation in the meningococcus, discovered that NMB1436-NMB1438 were upregulated. Two of these genes have homology to oxidoreductases, and the combination of this with the response to iron suggested these genes were involved in redox stress protection. It was shown that this is an iron- and Fur-responsive gene cluster involved in protection against hydrogen peroxide. Indeed, in an adult mouse model of meningococcal infection, it was seen that the deletion of these genes made the bacteria less virulent, possibly because of greater susceptibility to killing by neutrophils (45, 46). The transcriptional response of *N. meningitidis* strain MC58 to hydrogen peroxide revealed that only 17 genes were differentially regulated, including known oxidative stress response genes *kat*, *recN*, and *nifU* genes but not NMB1436-38. However, in the NMB1436-38 deletion mutant, another 143 genes were differentially regulated in response to hydrogen peroxide, suggesting an essential role for these genes in responding and limiting the effects of hydrogen peroxide (45). However, this effect, although clearly real, might well be reasonably considered to be primarily or also as importantly potentially adaptive to the physiological response to iron and in the response to free radicals produced in the presence of ferric iron by the Fenton reaction. This illustrates the care that must be taken in functionally interpreting experiments of this type.

These meningococcal results are in sharp contrast to the hydrogen peroxide response observed by Stohl et al. (146) in the gonococcus. More than 150 genes were differentially regulated in *N. gonorrhoeae* FA1090, suggesting that either the genes identified by Grifantini et al. (45) do not have the same effect as in *N. meningitidis* or that these differences are due to differences in the hydrogen peroxide concentrations used. In the gonococcal study, there was also upregulation of genes known to be involved in protection against oxidative stress as well as genes involved in iron acquisition (146). The latter were shown to be due to the upregulation of *fur* in the presence of hydrogen peroxide. Stohl et al. (146) characterized the protective responses of two hypothetical proteins: NGO1686 (NMB0315) is protective against peroxide damage, and the gonococcus-specific protein NGO554 is protective against

hydrogen peroxide. Again, there are differences in the experimental test conditions as well as between the microarrays, processing, and analysis, which make the direct comparison of these data sets problematic.

MICROARRAYS TO ASSESS GENE COMPLEMENT

Microarrays are most commonly thought of as a technology to assess differences in gene expression; however, they can also be used to assess gene presence. Through hybridization of chromosomal DNA, rather than cDNA, the presence of sequences within a strain of interest can be determined. This has made the genome sequences that are available more broadly applicable for the study of other strains for which there is no genome sequence currently available. It also facilitates the identification of things that are common, strain specific, and species specific, as well as potentially associated with differences in behavior.

Presence of a Filamentous Bacteriophage. Nylon membrane DNA arrays were used by Bille et al. (8) to investigate the presence of 1,950 of the 2,121 annotated genes in the *N. meningitidis* Z2491 genome sequence in a collection of 49 meningococcal strains. Each of these strains had been typed using multilocus sequence typing, and the DNA hybridization profiles to the membrane arrays were consistent within the same sequence type, demonstrating congruence between these two techniques in *N. meningitidis* (8). However, this clear association was not seen for gonococcal clinical isolates, perhaps reflecting the differing characters of these two species' population structures (135). From the meningococcal study, a group of genes (NMA1792 to NMA1799) homologous to those from filamentous bacteriophage were identified as being significantly overrepresented in the hypervirulent *N. meningitidis* isolates tested (8).

The Genes Necessary to Cause Disseminated Gonococcal Infections. The pan-*Neisseria* microarray was specifically designed with genomic microarray analyses in mind;

therefore, each of the gene probes was specifically chosen from within the most conserved regions of the genes, as assessed through comparative genomics of the available genome sequences (108, 152), *N. gonorrhoeae* FA1090 and *N. meningitidis* FAM18 (genome sequences unpublished but available). This has produced a microarray that is broadly applicable to other strains of *N. meningitidis* and *N. gonorrhoeae*, and that has also been successfully applied to more distantly related commensal *Neisseria* spp. *N. gonorrhoeae* is capable of causing the sexually transmitted infection gonorrhea (uncomplicated gonorrhea), but in some cases it can also cause a disseminated gonococcal infection (DGI). Of the four most widely used gonococcal strains in laboratory investigations, two of these come from cases of uncomplicated gonorrhea and two from DGI; therefore, Snyder et al. conducted a genomic microarray study of these strains (138). Although no association between the disease states and the presence of the genes assessed could be found, this study did conclude that the genes within the *N. gonorrhoeae* FA1090 genome sequence are also all present in *N. gonorrhoeae* F62; some of the genes in the genome sequence are absent from *N. gonorrhoeae* MS11 and FA19; and that regions where genes are absent in MS11 and FA19 are minimal mobile elements (MMEs).

The Presence of the GGI in *N. meningitidis*. Hybridizations of *N. meningitidis* strains to the pan–*Neisseria* microarray (139) revealed the presence of the GGI, a large genetic island encoding a type IV secretion system previously reported to be present only in *N. gonorrhoeae* (54). This demonstrates not only the utility of a multispecies microarray, but also the need for additional sequencing to be done beyond those done to date. It also suggests that genomic hybridization should be one basis for future strain selection for whole-genome sequencing. If these meningococcal strains had been hybridized to a meningococcal microarray or to a microarray that only contained probes to those genes from the published and unpublished genome sequences, the island

would have been missed. This 57-kb region is not present in *N. gonorrhoeae* FA1090 but is present in the majority of gonococcal strains that have been tested (54). The use and incorporation of the additional, highly complementary sequencing work on other strains and genes outside of that done by the genome-sequencing projects are highlighted by the utility of this additional sequence. Although the completed genome sequences provide an incomparable framework from which to conduct these studies, a genomics and population perspective must incorporate these additional contributions.

Gene Complement Differences between the Pathogenic and Commensal *Neisseria* spp. Although *N. meningitidis* and *N. gonorrhoeae* are the only human pathogens of the *Neisseria* spp., they are not the only ones to colonize humans as hosts. Studies of the genes present in the commensal *Neisseria* spp., such as *N. lactamica*, may provide insight into those genes required for colonization of the host and those that are involved in the species-specific virulence characteristics of the pathogens.

Perrin et al. (112) used comparative genome hybridization (to a nylon membrane array targeting 1-kb sequence regions) to identify genes that were meningococcus specific and virulence specific by comparing the genes present in several *N. meningitidis*, *N. gonorrhoeae*, and *N. lactamica* isolates. In addition to the known virulence genes of the capsule region (NMA0184 to NMA0186) and the *frp* genes (NMA1617 to NMA1626), this study also identified the FhaB filamentous hemagglutinin homologue (NMA0687 to NMA0696), an amino acid metabolism cluster (NMA0028 to NMA0033), and several hypothetical proteins as being specific to the meningococcus. Comparisons of *N. lactamica* identified many genes that were specific to the pathogenic species, including the IgA protease (NMA0905), hemoglobin receptor (*hmbR* NMA1925), *pilC* (NMA0609), and several putative siderophore receptors (NMA0575 to NMA0578). In all, 69 genes were identified as being pathogen specific in this study. No large pathogenicity islands were found, as observed in many other pathogenic bacterial species, that could be used to define the pathogenic properties of the *Neisseria* spp., which is consistent with the findings of the genome-sequencing projects themselves (108, 152).

By means of the BµG@S neisserial microarray, Stabler et al. (143) also sought to identify the pathogen-specific genes through hybridization of a collection of pathogen and commensal strains. Stabler et al. (143) were able to report a core set of genes from *N. meningitidis* serogroup B that are absent in the commensal *Neisseria* spp. However, only two *N. lactamica* strains of undefined diversity were used; the authors acknowledged that the use of their protocol resulted in poor hybridization of the commensal strains to their microarray; and there was incongruence with the results from other studies. These points raise some questions about the use of some of the findings of this study.

Snyder and Saunders (134) also investigated the commensal *Neisseria* spp., reporting that the majority of genes that others had previously defined as virulence genes (85 of 127) and of the common genes between the genome sequences (1,373 of 1,473) were present in *N. lactamica*. They concluded that the virulence of the pathogenic *Neisseria* spp. may be due to the combinations of genes they possess and perhaps differing regulation of common genes, rather than being a simple function of the presence or absence of a key set of virulence genes that are absent from the commensal species (134). This would be similar to the finding in *N. gonorrhoeae* isolates from DGI and uncomplicated gonorrhea (135). Although previously reported as being pathogen specific, hybridization to gene probes for the following were detected in this study with *N. lactamica*: *lipA* (NMB0082), *lipB* (NMB0083), NMB0084, *secY* (NMB0162), NMB0226, NMB0227, NMB0229, NMB0239, NMB0240, a TonB-dependent receptor (NMB0293), *dsbA* (NMB0294), *galM* (NMB0389), *pprC* (NMB0431), *ackA*-1

(NMB0435), NMB0473, *bioC* (NMB0474), NMB0486/NMB0970/NMB1741, *mafA* (NMB0652), NMB0654, *bioA* (NMB0732), an anticodon nuclease (NMB0832), an ABC transporter family protein (NMB1400), *mtrF* (NMB1719), and IS30.

Proteomics

The investigation of a bacterial proteome is greatly facilitated by the availability of complete genome sequences, providing predictable patterns of tryptic digestion mass fragments and the means to interpret short de novo N-terminal and internal sequences. Obviously, this approach is most powerful when working with the strains that have been sequenced, but this has not always been the approach adopted. Bernardini et al. (7) have performed the largest neisserial study of this type to date in which they investigated the expressed proteins of *N. meningitidis* 4970 by 2D gel electrophoresis. This resolved approximately 1,220 proteins when stained with colloidal Coomassie blue and 1,518 protein spots when stained with silver (7). A total of 287 protein spots were excised, and 273 of these were identified by matrix-assisted laser desorption ionization–time of flight mass spectroscopy. Mignogna et al. (99) also conducted a proteomic study with *N. meningitidis* MC58 that identified 238 protein species. These studies can confirm that annotated sequence features encode expressed proteins, can establish the identity of proteins that consistently feature in 2D gels, and form the foundation for comparative proteomic studies in the *Neisseria* spp. Indeed, comparative 2D gels of *N. meningitidis* were able to complement the results of a transcriptional microarray study (33) by showing that there are no global protein expression changes in response to AI-2 (126). However, it is not currently clear whether gel-based methodologies, which have a number of technical drawbacks, or other separation techniques will form the basis of whole cell proteomics in the future. This is particularly the case for bacteria, in which the size of the proteome is close to the existing resolving power of multidimensional column separation techniques.

Identification of Sequence Repeats

THE CORREIA REPEATS AND CORREIA REPEAT-ENCLOSED ELEMENTS

In the *Neisseria* spp., there are hundreds of copies of the Correia repeat-enclosed element (CREE) scattered around the genome (108, 152). It is made up of a 26-bp inverted repeat that flanks a characteristic core region, which may vary in length (19, 20, 80). With the availability of the genome sequences, several research groups have investigated the CREE, including Liu et al. (80), who developed the terminology to differentiate the 26-bp Correia repeats from the larger CREE and who defined the characteristics of the core regions which are flanked by inverted Correia repeats in a CREE. Buisine et al. (13) investigated the differences between CREE, their distribution in the genome sequences, and the lack of CREE in some regions that might suggest their recent horizontal acquisition. De Gregorio et al. (25) also determined the presence of the CREE in various locations on the chromosome and were then able to monitor their presence in other strains by PCR. Prompted by their observation of a CREE within the neisserial *dcw* cluster, Francis et al. (38) also investigated the distribution of these elements in the neisserial genomes.

Although these investigations have served to identify a different distribution of the CREE on each of the genome sequences, and thus to support the concept that they are or have been mobile, the studies that have addressed the function of these elements have not required the genome sequences. Functional studies of the CREE have instead focused on those present within a region of interest, and their functions in these locations would appear to be dependent on their chromosomal context (10, 14, 26, 120, 141).

Mazzone et al. (95) performed a genome-wide search for CREE (which they call "ne-mis") in the meningococcal strains MC58 and Z2491 and the gonococcal strain FA1090 genome sequences. Much higher copy numbers of the CREE were observed in the meningococcal genomes than within the gonococcal genome. Like Liu et al. (80), these researchers also subdivided the CREEs into different subtypes depending on the length and type of the repeats at each end. These subfamilies were also present in different ratios in the gonococcal as compared with the meningococcal genomes. To try and provide some answers as to the roles of these repeats, when present in single copies closely associated with genes, they looked at the conservation of the elements between different strains. It was found that approximately two-thirds of them were not conserved, suggesting that the actual sequence is not as important as the presence of such an element at a particular site. Although they did not exclude a role in transcriptional activation, they make a case for a role in affecting gene expression at the level of RNA turnover rather than transcription. This is based on the observation that at least some forms of this repeat can act as a site for ribonuclease cleavage, and that its stem-loop structure can stabilize mature or nascent mRNA (95).

PHASE-VARIABLE REPEATS

The presence and location of repeats and any differences in repeat tract lengths can be determined by using the complete genome sequences. Although there are many repeats present within the intergenic regions, which may or may not be functionally important, it is known that simple sequence repeats within coding regions and promoter regions can mediate phase variation in the *Neisseria* spp. These were therefore investigated, first by using the *N. meningitidis* MC58 genome sequence (123), and later by using a comparative genomics approach that evaluated potential phase variation mediating repeats in *N. meningitidis* MC58 and Z2491 and *N. gonorrhoeae* FA1090 (136). The available neisserial genome sequences were analyzed to determine the presence of simple sequence repeats within the right context to potentially mediate phase variation of the associated gene through a change in repeat copy number, either through affecting the promoter region or by generating a frameshift (136).

The repertoire of potential phase-variable genes has then been used to explore whether differences in phase-variable gene repertoires can account for the ability of some gonococcal strains to cause DGI (66), whether bacterial protein microarrays to the phase variably expressed proteins can be used diagnostically (145), and whether the predicted phase-variable genes could be demonstrated to be subject to phase-variable expression (90).

CODING TANDEM REPEATS

A different class of repeats, coding tandem repeats, was investigated by Jordan et al. (65). Changes in the repeat copy number of coding tandem repeats do not mediate phase variation or lead to a frameshift because these repeats are composed of nucleotide sequences that are multiples of 3. Hence, the variation in the length of the coding tandem repeats will not alter the expression of the gene, but rather the encoded amino acid sequence, which can potentially alter the function or antigenicity of the protein. From analyses of the genome sequences, 28 genes containing coding tandem repeats within their coding regions were identified, and changes in these repeat sequences were assessed in a collection of strains (65). It was concluded that 18 of these genes display variation in their coding tandem repeat tract lengths between strains of the same species, which may indicate that variation in repeat length is possible within a strain, leading to antigenic variation or alterations of protein function (65). Subsets of these repeats, in combination with simple sequence repeats, have subsequently been proposed by two groups as a basis for clinical study of epidemiology, population, and outbreak (79, 165, 166).

Reverse Vaccinology

It was anticipated that the genome sequences would reveal targets for an effective *N. meningitidis* vaccine, capable of preventing meningo-coccal meningitis and septicemia, regardless of the serogroup of the strain. Certainly, all of the proteins expressed by *N. meningitidis* MC58 must be encoded by genes on the chromosome (assuming there are no plasmids); therefore, all of the potential protein therapeutic and vaccine targets for this strain are potentially identifiable by using the genome sequences. The genome sequences are therefore being used in a process called reverse vaccinology, in which the genome sequence is the starting point that will ultimately lead to the elusive vaccine (114).

Pizza et al. (114) analyzed the *N. meningitidis* MC58 genome sequence to identify sequences encoding potential surface exposed or exported proteins. Of the 570 predicted proteins, 350 were cloned, expressed, and used to generate immune sera in mice. These serum samples were then subjected to enzyme-linked immunosorbent assay and FACS analysis to detect which of these proteins were present on the surface of a variety of invasive meningococcal strains. The serum samples were also used in bactericidal assays. Of the proteins tested, 85 were strongly positive in one of the assays and 7 were positive in all assays (GNA33/NMB0033, GNA992/NMB0992, GNA1162/NMB1162, GNA1220/NMB1220, GNA1946/NMB1946, GNA2001/NMB2001, and GNA2132/NMB2132). Of these seven proteins, all but NMB0992 and NMB2132 did not contain hypervariable regions, which have been shown to limit the effectiveness of vaccines because of differences between strains. Sera raised against the five proteins with low interstrain sequence variability were shown to be more effective at complement-mediated killing than that raised against outer membrane vesicles (a common method of generating vaccines), and were also effective against a general cross section of serogroup B meningococcal strains (114).

Thus far, this strategy has identified the autotransporters NadA (17), App (51), Hsf (110), and AutA (2) as potential vaccine targets; outer membrane lipoproteins NMB1870, NMB1946, and NMB2132 (also referred to as GNA1870, GNA1946, and GNA2132) are being pursued

(41, 62, 92, 159). In addition, vaccine candidates were identified by Grifantini et al. (44) on the basis of their microarray-based expression studies.

GNA33 (NMB0033), identified by Pizza et al. (114) as one of the first, best candidates by means of this approach and highly effective at generating an immune response, was subsequently shown to be a muramidase (63). Although anti-GNA33 monoclonal antibodies were able to bind all strains within a collection of meningococcal strains, they were able to cause complement-mediated lysis of strains producing the *porA* P1.2 epitope because of molecular mimicry of GNA33 with the PorA loop 4 structure (42). However, the GNA33 antibody was not as effective at causing complement-mediated lysis as the *porA* antibody.

Subsequent work with NMB1870 has shown this to be a more promising potential vaccine component, and it has been found that this lipoprotein is expressed by all strains tested (41), although at different levels, and was recognized by more than 50% of sera from convalescing patients, but only 20% of healthy patients. It has been shown to be immunogenic in mice, and even when expressed at low levels, antisera can cause complement-mediated lysis (41).

Horizontal Gene Transfer

By assessing sequence features such as percentage of G+C (%G+C) and dinucleotide signatures, investigators have been able to identify regions of the neisserial genomes that are likely to have been horizontally acquired, including the IHTs (152). In addition, through the comparison of the different genome sequences available, regions can be identified that might also have an origin outside of the *Neisseria* spp.

PROPHAGE GENOMES

Bacteriophage infection can result in the transfer of genes from one species to another. A study that looked at the genome sequence to identify phage infections in *N. meningitidis* MC58 showed that there is a region corresponding to a Mu phage genome insertion which has subsequently become degenerate (93). This region contains several genes with different %G+C when compared with the remainder of the region, suggesting that they have been horizontally acquired by the phage and transferred from other species. Subsequent analysis of three genes by FACS and bactericidal assays showed that these genes were surface expressed and could cause complement-mediated lysis, suggesting that these may be potential vaccine candidates against strains harboring this bacteriophage (93).

Kawai et al. (68) investigated the genome sequences of *N. meningitidis* MC58, Z2491, and FAM18, and of *N. gonorrhoeae* FA1090 to identify the filamentous prophages that are present in these genomes. This search identified 23 filamentous prophages in the four genome sequences, which they divided into four subtypes. The majority of these can be recognized to be truncated or incomplete, or to have internal deletions; only 11 were thought to be intact. In 17 of these 23 identified prophages, a *piv* homologue is associated with the right-hand side of the filamentous prophage sequence. Those regions that are not associated with a *piv* homologue are truncated on their right-hand side (68). The microarray-based genomics investigation of Bille et al. (8) has investigated one of the four subtypes of these filamentous prophages.

MINIMAL MOBILE ELEMENTS

Comparative analysis of the *N. meningitidis* strain MC58 genome sequence with that from *N. meningitidis* strain Z2491 identified regions in which strain-specific genes are located among clusters of conserved genes. This led to the development of a new model of horizontal gene transfer (124). MMEs are sites identified in the genome sequences in which strain-specific genes are located between flanking genes that have conserved sequence and chromosomal organization. These flanking genes are the substrates for homologous recombination following natural transformation. Whole gene transfer from one chromosome to another is therefore facilitated by the presence of flanking homologous sequences. This system does not rely on exogenous integrases, nor are any integrases contained within the MMEs that have led to their mobilization. The MME sites are identified through comparative analysis of the genomic region in different strains. A region is definitively identified as an MME only when different genes are present between the conserved flanking genes in different strains. A simple polymorphism in an intergenic region or divergent copies of the same gene do not constitute an MME. If the conserved flanking genes are contiguous in one strain, yet flank a gene or genes in another strain, the region can only be classified as a putative MME until such time as a strain that contains alternative genes between the flanks can be identified. Although there are instances of MMEs that contain IS elements, these MME sites have been first identified as a result of differences between strains which do not have the IS element in this location, and thus the IS element is not a consistent feature of this region and cannot be responsible for the mobilization of the genes. MMEs are not associated with tRNA sequences, as are pathogenicity islands, contain no repeats within the flanks, and are not limited to the *Neisseria* spp., having been subsequently identified in *E. coli* (132), *Helicobacter pylori* (16), *Streptococcus agalactiae* (56), and *Streptococcus suis* (130). Although some of the early identified MMEs predominantly contained genes for different restriction-modification systems (124, 132), these are not the only type of gene found within MMEs. Also identified are a penicillin acylase gene (132), meningococcal capsule biosynthesis genes (137), pilin modification genes (138) that play a key role in gonococcal dissemination (135), neisserial iron acquisition gene *hmbR* (67), and genes for the meningococcal adhesin Opc (167). The first of the MMEs to be specifically characterized, between the *pheS* and *pheT* genes of the *Neisseria* spp., contains nine different genes or combinations of genes between the conserved flanking sequences in the strain set used; therefore, each MME site may contain many different alternative gene cassettes (124).

Studies Using the Genome Sequences from Many Different Species

Several studies have been conducted that utilize the available genome sequences to answer questions addressable to bioinformatics. Largely, these are not directed at an understanding of the *Neisseria* spp., and they are therefore not discussed in this review. There is one publication, however, that investigated the uptake signal sequences of several species and was able to draw some conclusions that were specific to the presence and utility of these sequences in the *Neisseria* spp. Davidsen et al. (22) observed that genes involved in DNA repair and recombination tended to contain the neisserial uptake sequence within their coding regions, whereas those encoding surface proteins did not. There is therefore a bias toward the presence of uptake sequences being associated with genes that are important in DNA maintenance (22). This perhaps suggests that important genes can be preferentially repaired through the process of natural transformation and recombination, which would also suggest that these genes are subject to frequent recombination events.

KEY POSTGENOMIC FINDINGS

It has become apparent from both postgenomic studies and those done without the need for genome sequences that the *Neisseria* spp. do not behave in the same way as existing model organisms. For example, unlike *E. coli*, the model gram-negative bacterial species, the *Neisseria* spp. lack RpoN (74), regulate their genes differently with RpoH (48), have only four two-component regulatory systems (105), have only type I, IV, and V secretion systems (54, 156, 163), and do not respond to AI-2 (33, 126). A great deal has been learned by exploring what the *Neisseria* spp. do have and what they actually do with what they possess.

A constant and repeatedly evident characteristic of these species that follows from the ability to address them at the genome level is the immense diversity that exists between any two strains of the same species, as well as those between the pathogenic *Neisseria* spp. One might assume that this variation can account for the differences in behavior between different strains and species, and although this has been shown to be true in some cases, it is not universally so. It is, in fact, the combination of genes present in a strain that dictates the ability of *N. gonorrhoeae* to cause DGI (135). Although a few pathogen-specific genes have been identified, the diversity of both the pathogens and the commensal species such as *N. lactamica* complicates studies that try to define a core set of virulence genes on the basis of their absence from the commensals. Indeed, the genes encoding key meningococcal virulence factors such as the capsule (57) and Opc (129) have been found to be absent in *N. meningitidis* isolates from patients with invasive disease, which challenges some long-established models of meningococcal pathogenesis. The lack of a single gene, set of genes, or island has created a general perspective that the differences in behavior are dependent on polymorphisms within common genes and perhaps that differences between hosts are the primary determinant of who develops invasive disease. However, the alternative possibility—that there are several virulent gene combinations that can be formed and dispersed in the wider population—is a new model. This may form the basis of these differences in behavior in these species, and this has been appreciable only in the context of working from the complete genome sequences.

This issue of strain diversity is further complicated by the discovery of MMEs (124). The implication is that any of the genes within the MME can be transferred to any bacterial strain in which the same conserved flanking genes are present. Although the pathogenic *Neisseria* spp. would appear to be physically separated through their preferred colonizing niche, this does not preclude colonization of the nasopharynx by *N. gonorrhoeae* or the genital tract by *N. meningitidis*. In fact, some isolates from cases of gonorrhea are typed as *N. meningitidis*, and cases of gonococcal pharyngitis are not rare. In addition, the commensal *Neisseria* spp. may act as a reservoir for MME genes and

transfer from the commensal species to the pathogens has been observed in the case of the *penA* penicillin-resistance mutation (142).

In addition to the regulation of gene expression through its two alternative sigma factors, RpoH (48) and ECF (49), and the many regulatory proteins that have been identified, it appears that the *Neisseria* spp. are also subject to posttranscriptional regulation. The stability of RNA between *N. meningitidis* and *N. lactamica* has been noted, as has the activity of RNases on mRNA (43, 61, 148), particularly those containing CREE (95). This level of regulatory complexity has hardly been addressed to date and may ascribe functional importance to intergenic elements that have previously only represented curiosities and obstacles to sequence assembly.

In addition, the *Neisseria* spp. have the largest repertoire of phase-variable genes identified in any species to date, which may switch on or off independently at any time, generating a vast potential number of phenotypic combinations (123, 136). Although the on or off phenotypes may appear to be preferentially expressed under certain conditions, this is the result of a mixed effect of regulation and selective pressures being applied to the local population of bacteria, in which subpopulations with alternate phenotypes for as many as 60 genes are in a constant state of flux.

GENOME ANNOTATIONS

The annotation of the *N. meningitidis* MC58 genome sequence is available from the original publication (152), GenBank (NC_003112), and the TIGR Comprehensive Microbial Resource (CMR; http://cmr.tigr.org/tigrscripts/ CMR/CmrHomePage.cgi). All annotations from this genome-sequencing project are prefixed NMB. The annotation of the *N. meningitidis* strain Z2491 genome-sequence is available from the original publication (108), GenBank (NC_003116), the Wellcome Trust Sanger Institute (http://www.sanger.ac.uk/Projects/N_ meningitidis/seroA/seroA.shtml), and the TIGR CMR, where a TIGR annotation of this genome sequence is also available. All annotations from this genome-sequencing project are prefixed NMA. A preliminary annotation of the *N. meningitidis* strain FAM18 genome sequence is available from the Wellcome Trust Sanger Institute (http://www.sanger.ac.uk/Projects/N_meningitidis/seroC/NMC.embl). The annotation of the *N. gonorrhoeae* FA1090 genome sequence is available from GenBank (NC_002946).

All but the strain FAM18 genome sequence annotation, which will be added upon its publication, can be viewed comparatively within the GBrowse databases maintained by Nigel Saunders, Simon McGowan, and the Computational Biology Research Group (http://www.molbiol.ox.ac.uk/data.shtml).

Where possible, we have preferentially made use of the published annotation from *N. meningitidis* MC58 (NMB numbers) for consistency throughout the chapter. For genes that are not present in strain MC58, the annotation from *N. meningitidis* Z2491 (NMA numbers) was used or, failing that, the *N. gonorrhoeae* FA1090 annotation from GenBank (NGO numbers).

REFERENCES

1. Reference deleted.
2. **Ait-Tahar, K., K. G. Wooldridge, D. P. Turner, M. Atta, I. Todd, and D. A. Ala'Aldeen.** 2000. Auto-transporter A protein of *Neisseria meningitidis*: a potent CD4$^+$ T-cell and B-cell stimulating antigen detected by expression cloning. *Mol. Microbiol.* **37:**1094–1105.
3. **Ala'Aldeen, D. A., H. A. Davies, and S. P. Borriello.** 1994. Vaccine potential of meningococcal FrpB: studies on surface exposure and functional attributes of common epitopes. *Vaccine* **12:**535–541.
4. **Anjum, M. F., T. M. Stevanin, R. C. Read, and J. W. B. Moir.** 2002. Nitric oxide metabolism in *Neisseria meningitidis. J. Bacteriol.* **184:**2987–2993.
5. **Banerjee, A., R. Wang, S. L. Supernavage, S. K. Ghosh, J. Parker, N. F. Ganesh, P. G. Wang, S. Gulati, and P. A. Rice.** 2002. Implications of phase variation of a gene (*pgtA*) encoding a pilin galactosyl transferase in gonococcal pathogenesis. *J. Exp. Med.* **196:**147–162.
6. **Barbour, A. G.** 1981. Properties of penicillin-binding proteins in *Neisseria gonorrhoeae. Antimicrob. Agents Chemother.* **19:**316–322.
7. **Bernardini, G., G. Renzone, M. Comanducci, R. Mini, S. Arena, C. D'Ambrosio, S. Bambini, L. Trabalzini, G. Grandi, P. Martelli, M. Achtman, A. Scaloni, G. Ratti, and A. Santucci.** 2004. Proteome analysis of *Neisseria meningitidis* serogroup A. *Proteomics* **4:**2893–2926.
8. **Bille, E., J. R. Zahar, A. Perrin, S. Morelle, P. Kriz, K. A. Jolley, M. C. J. Maiden, C. Dervin, X. Nassif, and C. R. Tinsley.** 2005. A chromosomally integrated bacteriophage in invasive meningococci. *J. Exp. Med.* **201:**1905–1913.
9. **Binet, R., S. Letoffe, J. M. Ghigo, P. Delepelaire, and C. Wandersman.** 1997. Protein secretion by gram-negative bacterial ABC exporters—a review. *Gene* **192:**7–11.
10. **Black, C. G., J. A. Fyfe, and J. K. Davies.** 1995. A promoter associated with the neisserial repeat can be used to transcribe the *uvrB* gene from *Neisseria gonorrhoeae. J. Bacteriol.* **177:**1952–1958.
11. **Blondiau, C., P. Lagadec, P. Lejeune, N. Onier, J. M. Cavaillon, and J. F. Jeannin.** 1994. Correlation between the capacity to activate macrophages in vitro and the antitumor activity in vivo of lipopolysaccharides from different bacterial species. *Immunobiology* **190:**243–254.
12. **Brener, D., I. W. DeVoe, and B. E. Holbein.** 1981. Increased virulence of *Neisseria meningitidis* after in vitro iron-limited growth at low pH. *Infect. Immun.* **33:**59–66.
13. **Buisine, N., C. M. Tang, and R. Chalmers.** 2002. Transposon-like Correia elements: structure, distribution and genetic exchange between pathogenic *Neisseria* sp. *FEBS Lett.* **522:**52–58.
14. **Cantalupo, G., C. Bucci, P. Salvatore, C. Pagliarulo, V. Roberti, A. Lavitola, C. B. Bruni, and P. Alifano.** 2001. Evolution and function of the neisserial *dam*-replacing gene. *FEBS Lett.* **495:**178–183.
15. **Carson, S. D., P. E. Klebba, S. M. Newton, and P. F. Sparling.** 1999. Ferric enterobactin binding and utilization by *Neisseria gonorrhoeae. J. Bacteriol.* **181:**2895–2901.
16. **Chanto, G., A. Occhialini, N. Gras, R. A. Alm, F. Megraud, and A. Marais.** 2002. Identification of strain-specific genes located outside the plasticity zone in nine clinical isolates of *Helicobacter pylori. Microbiology* **148:**3671–3680.
17. **Comanducci, M., S. Bambini, B. Brunelli, J. Adu-Bobie, B. Arico, B. Capecchi, M. M. Giuliani, V. Masignani, L. Santini, S. Savino, D. M. Granoff, D. A. Caugant, M. Pizza, R. Rappuoli, and M. Mora.** 2002. NadA, a novel vaccine candidate of *Neisseria meningitidis. J. Exp. Med.* **195:**1445–1454.
18. **Cornelissen, C. N., G. D. Biswas, J. Tsai, D. K. Paruchuri, S. A. Thompson, and P. F.**

Sparling. 1992. Gonococcal transferrin-binding protein 1 is required for transferrin utilization and is homologous to TonB-dependent outer membrane receptors. *J. Bacteriol.* **174:**5788–5797.

19. **Correia, F. F., S. Inouye, and M. Inouye.** 1986. A 26-base-pair repetitive sequence specific for *Neisseria gonorrhoeae* and *Neisseria meningitidis* genomic DNA. *J. Bacteriol.* **167:**1009–1015.

20. **Correia, F. F., S. Inouye, and M. Inouye.** 1988. A family of small repeated elements with some transposon-like properties in the genome of *Neisseria gonorrhoeae*. *J. Biol. Chem.* **263:**12194–12198.

21. **Cox, A. D., J. C. Wright, J. J. Li, D. W. Hood, E. R. Moxon, and J. C. Richards.** 2003. Phosphorylation of the lipid a region of meningococcal lipopolysaccharide: identification of a family of transferases that add phosphoethanolamine to lipopolysaccharide. *J. Bacteriol.* **185:**3270–3277.

22. **Davidsen, T., E. A. Rodland, K. Lagesen, E. Seeberg, T. Rognes, and T. Tonjum.** 2004. Biased distribution of DNA uptake sequences towards genome maintenance genes. *Nucleic Acids Res.* **32:**1050–1058.

23. **Deghmane, A. E., D. Giorgini, M. Larribe, J. M. Alonso, and M. K. Taha.** 2002. Downregulation of pili and capsule of *Neisseria meningitidis* upon contact with epithelial cells is mediated by CrgA regulatory protein. *Mol. Microbiol.* **43:**1555–1564.

24. **Deghmane, A. E., M. Larribe, D. Giorgini, D. Sabino, and M. K. Taha.** 2003. Differential expression of genes that harbor a common regulatory element in *Neisseria meningitidis* upon contact with target cells. *Infect. Immun.* **71:**2897–2901.

25. **De Gregorio, E., C. Abrescia, M. S. Carlomagno, and P. P. Di Nocera.** 2003. Asymmetrical distribution of *Neisseria* miniature insertion sequence DNA repeats among pathogenic and nonpathogenic *Neisseria* strains. *Infect. Immun.* **71:**4217–4221.

26. **De Gregorio, E., C. Abrescia, M. S. Carlomagno, and P. P. Di Nocera.** 2003. Ribonuclease III–mediated processing of specific *Neisseria meningitidis* rnRNAs. *Biochem. J.* **374:**799–805.

27. **Delany, I., R. Grifantini, E. Bartolini, R. Rappuoli, and V. Scarlato.** 2006. Effect of *Neisseria meningitidis fur* mutations on global control of gene transcription. *J. Bacteriol.* **188:**2483–2492.

28. **Delany, I., R. Rappuoli, and V. Scarlato.** 2004. Fur functions as an activator and as a repressor of putative virulence genes in *Neisseria meningitidis.* *Mol. Microbiol.* **52:**1081–1090.

29. **Dietrich, G., S. Kurz, C. Hubner, C. Aepinus, S. Theiss, M. Guckenberger, U. Panzner, J. Weber, and M. Frosch.** 2003. Transcriptome analysis of *Neisseria meningitidis* during infection. *J. Bacteriol.* **185:**155–164.

30. **Dillard, J. P., and H. S. Seifert.** 2001. A variable genetic island specific for *Neisseria gonorrhoeae* is involved in providing DNA for natural transformation and is found more often in disseminated infection isolates. *Mol. Microbiol.* **41:**263–277.

31. **Doerrler, W. T., M. C. Reedy, and C. R. Raetz.** 2001. An *Escherichia coli* mutant defective in lipid export. *J. Biol. Chem.* **276:**11461–11464.

32. **Dorrell, N., J. A. Mangan, K. G. Laing, J. Hinds, D. Linton, H. Al-Ghusein, B. G. Barrell, J. Parkhill, N. G. Stoker, A. V. Karlyshev, P. D. Butcher, and B. W. Wren.** 2001. Whole genome comparison of *Campylobacter jejuni* human isolates using a low-cost microarray reveals extensive genetic diversity. *Genome Res.* **11:**1706–1715.

33. **Dove, J. E., K. Yasukawa, C. R. Tinsley, and X. Nassif.** 2003. Production of the signalling molecule, autoinducer-2, by *Neisseria meningitidis*: lack of evidence for a concerted transcriptional response. *Microbiology* **149:**1859–1869.

34. **Du, Y., J. Lenz, and C. G. Arvidson.** 2005. Global gene expression and the role of sigma factors in *Neisseria gonorrhoeae* in interactions with epithelial cells. *Infect. Immun.* **73:**4834–4845.

35. **Ducey, T. F., M. B. Carson, J. Orvis, A. P. Stintzi, and D. W. Dyer.** 2005. Identification of the iron-responsive genes of *Neisseria gonorrhoeae* by microarray analysis in defined medium. *J. Bacteriol.* **187:**4865–4874.

36. **Escolar, L., V. de Lorenzo, and J. Perez-Martin.** 1997. Metalloregulation in vitro of the aerobactin promoter of *Escherichia coli* by the Fur (ferric uptake regulation) protein. *Mol. Microbiol.* **26:**799–808.

37. **Forman, S., I. Linhartova, R. Osicka, X. Nassif, P. Sebo, and V. Pelicic.** 2003. *Neisseria meningitidis* RTX proteins are not required for virulence in infant rats. *Infect. Immun.* **71:**2253–2257.

38. **Francis, F., S. Ramirez-Arcos, H. Salimnia, C. Victor, and J. A. R. Dillon.** 2000. Organization and transcription of the division cell wall (*dcw*) cluster in *Neisseria gonorrhoeae*. *Gene* **251:**141–151.

39. **Gamian, A., M. Beurret, F. Michon, J. R. Brisson, and H. J. Jennings.** 1992. Structure of the L2 lipopolysaccharide core oligosaccharides of *Neisseria meningitidis*. *J. Biol. Chem.* **267:**922–925.

40. **Garcia-Horsman, J. A., B. Barquera, J. Rumbley, J. Ma, and R. B. Gennis.** 1994. The superfamily of heme-copper respiratory oxidases. *J. Bacteriol.* **176:**5587–5600.

41. Giuliani, M. M., L. Santini, B. Brunelli, A. Biolchi, B. Arico, F. Di Marcello, E. Cartocci, M. Comanducci, V. Masignani, L. Lozzi, S. Savino, M. Scarselli, R. Rappuoli, and M. Pizza. 2005. The region comprising amino acids 100 to 255 of *Neisseria meningitidis*, lipoprotein GNA 1870 elicits bactericidal antibodies. *Infect. Immun.* **73:**1151–1160.

42. Granoff, D. M., G. R. Moe, M. A. Giuliani, J. Adu-Bobie, L. Santini, B. Brunelli, F. Piccinetti, P. Zuno-Mitchell, S. S. Lee, P. Neri, L. Bracci, L. Lozzi, and R. Rappuoli. 2001. A novel mimetic antigen eliciting protective antibody to *Neisseria meningitidis*. *J. Immunol.* **167:**6487–6496.

43. Grifantini, R., E. Bartolini, A. Muzzi, M. Draghi, E. Frigimelica, J. Berger, F. Randazzo, and G. Grandi. 2002. Gene expression profile in *Neisseria meningitidis* and *Neisseria lactamica* upon host-cell contact: from basic research to vaccine development. *Ann. N.Y. Acad. Sci.* **975:**202–216.

44. Grifantini, R., E. Bartolini, A. Muzzi, M. Draghi, E. Frigimelica, J. Berger, G. Ratti, R. Petracca, G. Galli, M. Agnusdei, M. M. Giuliani, L. Santini, B. Brunelli, H. Tettelin, R. Rappuoli, F. Randazzo, and G. Grandi. 2002. Previously unrecognized vaccine candidates against group B meningococcus identified by DNA microarrays. *Nat. Biotechnol.* **20:**914–921.

45. Grifantini, R., E. Frigimelica, I. Delany, E. Bartolini, S. Giovinazzi, S. Balloni, S. Agarwal, G. Galli, C. Genco, and G. Grandi. 2004. Characterization of a novel *Neisseria meningitidis* Fur and iron-regulated operon required for protection from oxidative stress: utility of DNA microarray in the assignment of the biological role of hypothetical genes. *Mol. Microbiol.* **54:**962–979.

46. Grifantini, R., S. Sebastian, E. Frigimelica, M. Draghi, E. Bartolini, A. Muzzi, R. Rappuoli, G. Grandi, and C. A. Genco. 2003. Identification of iron-activated and -repressed Fur-dependent genes by transcriptome analysis of *Neisseria meningitidis* group B. *Proc. Natl. Acad. Sci. USA* **100:**9542–9547.

47. Guckenberger, M., S. Kurz, C. Aepinus, S. Theiss, S. Haller, T. Leimbach, U. Panzner, J. Weber, H. Paul, A. Unkmeir, M. Frosch, and G. Dietrich. 2002. Analysis of the heat shock response of *Neisseria meningitidis* with cDNA- and oligonucleotide-based DNA microarrays. *J. Bacteriol.* **184:**2546–2551.

48. Gunesekere, I. C., C. M. Kahler, D. R. Powell, L. A. S. Snyder, N. J. Saunders, J. I. Rood, and J. K. Davies. 2006. Comparison of the RpoH-dependent regulon and general stress response in *Neisseria gonorrhoeae*. *J. Bacteriol.* **188:**4769–4776.

49. Gunesekere, I. C., C. M. Kahler, C. S. Ryan, L. A. S. Snyder, N. J. Saunders, J. I. Rood, and J. K. Davies. 2006. Ecf, an alternative sigma factor from *Neisseria gonorrhoeae*, controls expression of *msrAB*, which encodes methionine sulfoxide reductase. *J. Bacteriol.* **188:**3463–3469.

50. Haas, R., S. Veit, and T. F. Meyer. 1992. Silent pilin genes of *Neisseria gonorrhoeae* MS11 and the occurrence of related hypervariant sequences among other gonococcal isolates. *Mol. Microbiol.* **6:**197–208.

51. Hadi, H. A., K. G. Wooldridge, K. Robinson, and D. A. A. Ala'Aldeen. 2001. Identification and characterization of App: an immunogenic autotransporter protein of *Neisseria meningitidis*. *Mol. Microbiol.* **41:**611–623.

52. Hagman, K. E., W. Pan, B. G. Spratt, J. T. Balthazar, R. C. Judd, and W. M. Shafer. 1995. Resistance of *Neisseria gonorrhoeae* to antimicrobial hydrophobic agents is modulated by the *mtrRCDE* efflux system. *Microbiology* **141(Pt 3):**611–622.

53. Hagman, K. E., and W. M. Shafer. 1995. Transcriptional control of the *mtr* efflux system of *Neisseria gonorrhoeae*. *J. Bacteriol.* **177:**4162–4165.

54. Hamilton, H. L., N. M. Dominguez, K. J. Schwartz, K. T. Hackett, and J. P. Dillard. 2005. *Neisseria gonorrhoeae* secretes chromosomal DNA via a novel type IV secretion system. *Mol. Microbiol.* **55:**1704–1721.

55. Henderson, I. R., F. Navarro-Garcia, M. Desvaux, R. C. Fernandez, and D. Ala'Aldeen. 2004. Type V protein secretion pathway: the autotransporter story. *Microbiol. Mol. Biol. Rev.* **68:**692–744.

56. Herbert, M. A., C. J. Beveridge, D. McCormick, E. Aten, N. Jones, L. A. Snyder, and N. J. Saunders. 2005. Genetic islands of *Streptococcus agalactiae* strains NEM316 and 2603VR and their presence in other group B streptococcal strains. *BMC Microbiol.* **5:**31.

57. Hoang, L. M., E. Thomas, S. Tyler, A. J. Pollard, G. Stephens, L. Gustafson, A. McNabb, I. Pocock, R. Tsang, and R. Tan. 2005. Rapid and fatal meningococcal disease due to a strain of *Neisseria meningitidis* containing the capsule null locus. *Clin. Infect. Dis.* **40:**e38–e42.

58. Holbein, B. E., K. W. Jericho, and G. C. Likes. 1979. *Neisseria meningitidis* infection in mice: influence of iron, variations in virulence among strains, and pathology. *Infect. Immun.* **24:** 545–551.

59. Howell-Adams, B., and H. S. Seifert. 1999. Insertion mutations in *pilE* differentially alter gonococcal pilin antigenic variation. *J. Bacteriol.* **181:**6133–6141.

60. **Howell-Adams, B., and H. S. Seifert.** 2000. Molecular models accounting for the gene conversion reactions mediating gonococcal pilin antigenic variation. *Mol. Microbiol.* **37:**1146–1158.

61. **Ieva, R., C. Alaimo, I. Delany, G. Spohn, R. Rappuoli, and V. Scarlato.** 2005. CrgA is an inducible LysR-type regulator of *Neisseria meningitidis*, acting both as a repressor and as an activator of gene transcription. *J. Bacteriol.* **187:**3421–3430.

62. **Jacobsson, S., S. Thulin, P. Molling, M. Unemo, M. Comanducci, R. Rappuoli, and P. Olcen.** 2006. Sequence constancies and variations in genes encoding three new meningococcal vaccine candidate antigens. *Vaccine* **24:**2161–2168.

63. **Jennings, G. T., S. Savino, E. Marchetti, B. Arico, T. Kast, L. Baldi, A. Ursinus, J. V. Holtje, R. A. Nicholas, R. Rappuoli, and G. Grandi.** 2002. GNA33 from *Neisseria meningitidis* serogroup B encodes a membrane-bound lytic transglycosylase (MltA). *Eur. J. Biochem.* **269:**3722–3731.

64. **Jonsson, A. B., G. Nyberg, and S. Normark.** 1991. Phase variation of gonococcal pili by frameshift mutation in *pilC*, a novel gene for pilus assembly. *EMBO J.* **10:**477–488.

65. **Jordan, P., L. A. S. Snyder, and N. J. Saunders.** 2003. Diversity in coding tandem repeats in related *Neisseria* spp. *BMC Microbiol.* **3:**23.

66. **Jordan, P. W., L. A. S. Snyder, and N. J. Saunders.** 2005. Strain-specific differences in *Neisseria gonorrhoeae* associated with the phase variable gene repertoire. *BMC Microbiol.* **5:**21.

67. **Kahler, C. M., E. Blum, Y. K. Miller, D. Ryan, T. Popovic, and D. S. Stephens.** 2001. *exl*, an exchangeable genetic island in *Neisseria meningitidis*. *Infect. Immun.* **69:**1687–1696.

68. **Kawai, M., I. Uchiyama, and I. Kobayashi.** 2006. Genome comparison in silico in *Neisseria* suggests integration of filamentous bacteriophages by their own transposase. *DNA Res.* **12:**389–401.

69. **Klee, S. R., X. Nassif, B. Kusecek, P. Merker, J. L. Beretti, M. Achtman, and C. R. Tinsley.** 2000. Molecular and biological analysis of eight genetic islands that distinguish *Neisseria meningitidis* from the closely related pathogen *Neisseria gonorrhoeae*. *Infect. Immun.* **68:**2082–2095.

70. **Kobayashi, N., K. Nishino, and A. Yamaguchi.** 2001. Novel macrolide-specific ABC-type efflux transporter in *Escherichia coli*. *J. Bacteriol.* **183:**5639–5644.

71. **Kogan, G., D. Uhrin, J. R. Brisson, and H. J. Jennings.** 1997. Structural basis of the *Neisseria meningitidis* immunotypes including the L4 and L7 immunotypes. *Carbohydr. Res.* **298:**191–199.

72. **Kulshin, V. A., U. Zahringer, B. Lindner, C. E. Frasch, C. M. Tsai, B. A. Dmitriev, and E. T. Rietschel.** 1992. Structural characterization of the lipid A component of pathogenic *Neisseria meningitidis*. *J. Bacteriol.* **174:**1793–1800.

73. **Kurz, S., C. Hubner, C. Aepinus, S. Theiss, M. Guckenberger, U. Panzner, J. Weber, M. Frosch, and G. Dietrich.** 2003. Transcriptome-based antigen identification for *Neisseria meningitidis*. *Vaccine* **21:**768–775.

74. **Laskos, L., J. P. Dillard, H. S. Seifert, J. A. Fyfe, and J. K. Davies.** 1998. The pathogenic neisseriae contain an inactive *rpoN* gene and do not utilize the *pilE* sigma54 promoter. *Gene* **208:**95–102.

75. **Laskos, L., C. S. Ryan, J. A. Fyfe, and J. K. Davies.** 2004. The RpoH-mediated stress response in *Neisseria gonorrhoeae* is regulated at the level of activity. *J. Bacteriol.* **186:**8443–8452.

76. **Lee, E. H., and W. M. Shafer.** 1999. The *farAB*-encoded efflux pump mediates resistance of gonococci to long-chained antibacterial fatty acids. *Mol. Microbiol.* **33:**839–845.

77. **Lee, V. T., and O. Schneewind.** 2001. Protein secretion and the pathogenesis of bacterial infections. *Genes Dev.* **15:**1725–1752.

78. **Lewis, L. A., E. Gray, Y. P. Wang, B. A. Roe, and D. W. Dyer.** 1997. Molecular characterization of *hpuAB*, the haemoglobin-haptoglobin-utilization operon of *Neisseria meningitidis*. *Mol. Microbiol.* **23:**737–749.

79. **Liao, J.-C., C.-C. Li, and C.-S. Chiou.** Use of a multilocus variable-number tandem repeat analysis method for molecular subtyping and phylogenetic analysis of *Neisseria meningitidis* isolates. *BMC Microbiol.* **6:**44.

80. **Liu, S. V., N. J. Saunders, A. Jeffries, and R. F. Rest.** 2002. Genome analysis and strain comparison of Correia repeats and Correia repeat-enclosed elements in pathogenic *Neisseria*. *J. Bacteriol.* **184:**6163–6173.

81. **Locht, C., P. Bertin, F. D. Menozzi, and G. Renauld.** 1993. The filamentous haemagglutinin, a multifaceted adhesion produced by virulent *Bordetella* spp. *Mol. Microbiol.* **9:**653–660.

82. **Lomovskaya, O., and K. Lewis.** 1992. Emr, an *Escherichia coli* locus for multidrug resistance. *Proc. Natl. Acad. Sci. USA* **89:**8938–8942.

83. **Lucas, C. E., J. T. Balthazar, K. E. Hagman, and W. M. Shafer.** 1997. The MtrR repressor binds the DNA sequence between the *mtrR* and *mtrC* genes of *Neisseria gonorrhoeae*. *J. Bacteriol.* **179:**4123–4128.

84. **Mackinnon, F. G., A. D. Cox, J. S. Plested, C. M. Tang, K. Makepeace, P. A. Coull, J. C. Wright, R. Chalmers, D. W. Hood, J. C. Richards, and E. R. Moxon.** 2002. Identification of a gene (lpt-3) required for the addition of phosphoethanolamine to the lipopolysaccharide inner core of *Neisseria meningitidis* and its role in mediating susceptibility to bactericidal killing and opsonophagocytosis. *Mol. Microbiol.* **43:**931–943.

85. **Mackman, N., J. M. Nicaud, L. Gray, and I. B. Holland.** 1985. Genetical and functional organisation of the *Escherichia coli* haemolysin determinant 2001. *Mol. Gen. Genet.* **201:**282–288.

86. **Maier, T. W., L. Zubrzycki, M. B. Coyle, M. Chila, and P. Warner.** 1975. Genetic analysis of drug resistance in *Neisseria gonorrhoeae*: production of increased resistance by the combination of two antibiotic resistance loci. *J. Bacteriol.* **124:**834–842.

87. **Marrs, C. F., F. W. Rozsa, M. Hackel, S. P. Stevens, and A. C. Glasgow.** 1990. Identification, cloning, and sequencing of *piv*, a new gene involved in inverting the pilin genes of *Moraxella lacunata*. *J. Bacteriol.* **172:**4370–4377.

88. **Martin, D., N. Cadieux, J. Hamel, and B. R. Brodeur.** 1997. Highly conserved *Neisseria meningitidis* surface protein confers protection against experimental infection. *J. Exp. Med.* **185:**1173–1183.

89. **Martin, P., K. Makepeace, S. A. Hill, D. W. Hood, and E. R. Moxon.** 2005. Microsatellite instability regulates transcription factor binding and gene expression. *Proc. Natl. Acad. Sci. USA* **102:**3800–3804.

90. **Martin, P., T. van de Ven, N. Mouchel, A. C. Jeffries, D. W. Hood, and E. R. Moxon.** 2003. Experimentally revised repertoire of putative contingency loci in *Neisseria meningitidis* strain MC58: evidence for a novel mechanism of phase variation. *Mol. Microbiol.* **50:**245–257.

91. **Masignani, V., E. Balducci, F. Di Marcello, S. Savino, D. Serruto, D. Veggi, S. Bambini, M. Scarselli, B. Arico, M. Comanducci, J. Adu-Bobie, M. M. Giuliani, R. Rappuoli, and M. Pizza.** 2003. NarE: a novel ADP-ribosyltransferase from *Neisseria meningitidis*. *Mol. Microbiol.* **50:**1055–1067.

92. **Masignani, V., M. Comanducci, M. M. Giuliani, S. Bambini, J. Adu-Bobie, B. Arico, B. Brunelli, A. Pieri, L. Santini, S. Savino, D. Serruto, D. Litt, S. Kroll, J. A. Welsch, D. M. Granoff, R. Rappuoli, and M. Pizza.** 2003. Vaccination against *Neisseria meningitidis* using three variants of the lipoprotein GNA1870. *J. Exp. Med.* **197:**789–799.

93. **Masignani, V., M. M. Giuliani, H. Tettelin, M. Comanducci, R. Rappuoli, and V. Scarlato.** 2001. Mu-like prophage in serogroup B *Neisseria meningitidis* coding for surface-exposed antigens. *Infect. Immun.* **69:**2580–2588.

94. **Masson, L., and B. E. Holbein.** 1985. Influence of nutrient limitation and low pH on serogroup B *Neisseria meningitidis* capsular polysaccharide levels: correlation with virulence for mice. *Infect. Immun.* **47:**465–471.

95. **Mazzone, M., E. De Gregorio, A. Lavitola, C. Pagliarulo, P. Alifano, and P. P. Di Nocera.** 2001. Whole-genome organization and functional properties of miniature DNA insertion sequences conserved in pathogenic *Neisseriae*. *Gene* **278:** 211–222.

96. **McFarland, L., T. A. Mietzner, J. S. Knapp, E. Sandstrom, K. K. Holmes, and S. A. Morse.** 1983. Gonococcal sensitivity to fecal lipids can be mediated by an Mtr-independent mechanism. *J. Clin. Microbiol.* **18:**121–127.

97. **Mekalanos, J. J., D. J. Swartz, G. D. Pearson, N. Harford, F. Groyne, and M. de Wilde.** 1983. Cholera toxin genes: nucleotide sequence, deletion analysis and vaccine development. *Nature* **306:**551–557.

98. **Meyer, T. F., E. Billyard, R. Haas, S. Storzbach, and M. So.** 1984. Pilus genes of *Neisseria gonorrheae*: chromosomal organization and DNA sequence. *Proc. Natl. Acad. Sci. USA* **81:**6110–6114.

99. **Mignogna, G., A. Giorgi, P. Stefanelli, A. Neri, G. Colotti, B. Maras, and M. E. Schinina.** 2005. Inventory of the proteins in *Neisseria meningitidis serogroup* B strain MC58. *J. Proteome Res.* **4:**1361–1370.

100. **Morelle, S., E. Carbonnelle, and X. Nassif.** 2003. The REP2 repeats of the genome of *Neisseria meningitidis* are associated with genes coordinately regulated during bacterial cell interaction. *J. Bacteriol.* **185:**2618–2627.

101. **Morita, Y., K. Kodama, S. Shiota, T. Mine, A. Kataoka, T. Mizushima, and T. Tsuchiya.** 1998. NorM, a putative multidrug efflux protein, of *Vibrio parahaemolyticus* and its homolog in *Escherichia coli*. *Antimicrob. Agents Chemother.* **42:** 1778–1782.

102. **Morse, S. A., P. G. Lysko, L. McFarland, J. S. Knapp, E. Sandstrom, C. Critchlow, and K. K. Holmes.** 1982. Gonococcal strains from homosexual men have outer membranes with reduced permeability to hydrophobic molecules. *Infect. Immun.* **37:**432–438.

103. **Nassif, X., J. L. Beretti, J. Lowy, P. Stenberg, P. O'Gaora, J. Pfeifer, S. Normark, and M. So.** 1994. Roles of pilin and PilC in adhesion of

Neisseria meningitidis to human epithelial and endothelial cells. *Proc. Natl. Acad. Sci. USA* **91:** 3769–3773.

104. **Newbury, S. F., N. H. Smith, E. C. Robinson, I. D. Hiles, and C. F. Higgins.** 1987. Stabilization of translationally active mRNA by prokaryotic REP sequences. *Cell* **48:**297–310.

105. **Newcombe, J., J. C. Jeynes, E. Mendoza, J. Hinds, G. L. Marsden, R. A. Stabler, M. Marti, and J. J. McFadden.** 2005. Phenotypic and transcriptional characterization of the meningococcal PhoPQ system, a magnesium-sensing two-component regulatory system that controls genes involved in remodeling the meningococcal cell surface. *J. Bacteriol.* **187:** 4967–4975.

106. **Osicka, R., J. Kalmusova, P. Krizova, and P. Sebo.** 2001. *Neisseria meningitidis* RTX protein FrpC induces high levels of serum antibodies during invasive disease: polymorphism of *frpC* alleles and purification of recombinant FrpC. *Infect. Immun.* **69:**5509–5519.

107. **Pan, W., and B. G. Spratt.** 1994. Regulation of the permeability of the gonococcal cell envelope by the *mtr* system. *Mol. Microbiol.* **11:**769–775.

108. **Parkhill, J., M. Achtman, K. D. James, S. D. Bentley, C. Churcher, S. R. Klee, G. Morelli, D. Basham, D. Brown, T. Chillingworth, R. M. Davies, P. Davis, K. Devlin, T. Feltwell, N. Hamlin, S. Holroyd, K. Jagels, S. Leather, S. Moule, K. Mungall, M. A. Quail, M. A. Rajandream, K. M. Rutherford, M. Simmonds, J. Skelton, S. Whitehead, B. G. Spratt, and B. G. Barrell.** 2000. Complete DNA sequence of a serogroup A strain of *Neisseria meningitidis* Z2491. *Nature* **404:**502–506.

109. **Peak, I. R., M. P. Jennings, D. W. Hood, and E. R. Moxon.** 1999. Tetranucleotide repeats identify novel virulence determinant homologues in *Neisseria meningitidis*. *Microb. Pathog.* **26:** 13–23.

110. **Peak, I. R., Y. Srikhanta, M. Dieckelmann, E. R. Moxon, and M. P. Jennings.** 2000. Identification and characterisation of a novel conserved outer membrane protein from *Neisseria meningitidis*. *FEMS Immunol. Med. Microbiol.* **28:**329–334.

111. **Permina, E. A., and M. S. Gelfand.** 2003. Heat shock (sigma32 and HrcA/CIRCE) regulons in beta-, gamma- and epsilon-proteobacteria. *J. Mol. Microbiol. Biotechnol.* **6:**174–181.

112. **Perrin, A., S. Bonacorsi, E. Carbonnelle, D. Talibi, P. Dessen, X. Nassif, and C. Tinsley.** 2002. Comparative genomics identifies the genetic islands that distinguish *Neisseria meningitidis*, the agent of cerebrospinal meningitis, from other *Neisseria* species. *Infect. Immun.* **70:**7063–7072.

113. **Pettersson, A., V. Klarenbeek, J. van Deurzen, J. T. Poolman, and J. Tommassen.** 1994. Molecular characterization of the structural gene for the lactoferrin receptor of the meningococcal strain H44/76. *Microb. Pathog.* **17:**395–408.

114. **Pizza, M., V. Scarlato, V. Masignani, M. M. Giuliani, B. Arico, M. Comanducci, G. T. Jennings, L. Baldi, E. Bartolini, B. Capecchi, C. L. Galeotti, E. Luzzi, R. Manetti, E. Marchetti, M. Mora, S. Nuti, G. Ratti, L. Santini, S. Savino, M. Scarselli, E. Storni, P. J. Zuo, M. Broeker, E. Hundt, B. Knapp, E. Blair, T. Mason, H. Tettelin, D. W. Hood, A. C. Jeffries, N. J. Saunders, D. M. Granoff, J. C. Venter, E. R. Moxon, G. Grandi, and R. Rappuoli.** 2000. Identification of vaccine candidates against serogroup B meningococcus by whole-genome sequencing. *Science* **287:**1816–1820.

115. **Pohlner, J., R. Halter, K. Beyreuther, and T. F. Meyer.** 1987. Gene structure and extracellular secretion of *Neisseria gonorrhoeae* IgA protease. *Nature* **325:**458–462.

116. **Powell, B. S., D. L. Court, T. Inada, Y. Nakamura, V. Michotey, X. Cui, A. Reizer, M. H. Saier, Jr., and J. Reizer.** 1995. Novel proteins of the phosphotransferase system encoded within the *rpoN* operon of *Escherichia coli*. Enzyme IIA^{Ntr} affects growth on organic nitrogen and the conditional lethality of an erats mutant. *J. Biol. Chem.* **270:**4822–4839.

117. **Pujol, C., E. Eugene, M. Marceau, and X. Nassif.** 1999. The meningococcal PilT protein is required for induction of intimate attachment to epithelial cells following pilus-mediated adhesion. *Proc. Natl. Acad. Sci. USA* **96:**4017–4022.

118. **Ren, J. S., S. Sainsbury, N. S. Berrow, D. Alderton, J. E. Nettleship, D. K. Stammers, N. J. Saunders, and R. J. Owens.** 2005. Crystal structure of nitrogen regulatory protein IIA(Ntr) from *Neisseria meningitidis*. *BMC Struct. Biol.* **5:**13.

119. **Rouquette-Loughlin, C., S. A. Dunham, M. Kuhn, J. T. Balthazar, and W. M. Shafer.** 2003. The NorM efflux pump of *Neisseria gonorrhoeae* and *Neisseria meningitidis* recognizes antimicrobial cationic compounds. *J. Bacteriol.* **185:** 1101–1106.

120. **Rouquette-Loughlin, C. E., J. T. Balthazar, S. A. Hill, and W. M. Shafer.** 2004. Modulation of the *mtrCDE*-encoded efflux pump gene complex of *Neisseria meningitidis* due to a Correia element insertion sequence. *Mol. Microbiol.* **54:**731–741.

121. **Rouquette-Loughlin, C. E., J. T. Balthazar, and W. M. Shafer.** 2005. Characterization of the MacA-MacB efflux system in *Neisseria gonorrhoeae. J. Antimicrob. Chemother.* **56:**856–860.

122. **Rudel, T., J. P. van Putten, C. P. Gibbs, R. Haas, and T. F. Meyer.** 1992. Interaction of two variable proteins (PilE and PilC) required for pilus-mediated adherence of *Neisseria gonorrhoeae* to human epithelial cells. *Mol. Microbiol.* **6:**3439–3450.

123. **Saunders, N. J., A. C. Jeffries, P. F. Peden, D. W. Hood, H. Tettelin, R. Rappuoli, and E. R. Moxon.** 2000. Repeat-associated phase variable genes in the complete genome sequence of *Neisseria meningitidis* strain MC58. *Mol. Microbiol.* **37:**207–215.

124. **Saunders, N. J., and L. A. S. Snyder.** 2002. The minimal mobile element. *Microbiology* **148:**3756-3760.

125. Reference deleted.

126. **Schauder, S., L. Penna, A. Ritton, C. Manin, F. Parker, and G. Renauld-Mongenie.** 2005. Proteomics analysis by two-dimensional differential gel electrophoresis reveals the lack of a broad response of *Neisseria meningitidis* to in vitro–produced AI-2. *J. Bacteriol.* **187:**392–395.

127. **Segal, E., E. Billyard, M. So, S. Storzbach, and T. F. Meyer.** 1985. Role of chromosomal rearrangement in *N. gonorrhoeae* pilus phase variation. *Cell* **40:**293–300.

128. **Seib, K. L., M. P. Jennings, and A. G. Mc-Ewan.** 2003. A Sco homologue plays a role in defence against oxidative stress in pathogenic *Neisseria. FEBS Lett.* **546:**411–415.

129. **Seiler, A., R. Reinhardt, J. Sarkari, D. A. Caugant, and M. Achtman.** 1996. Allelic polymorphism and site-specific recombination in the opc locus of *Neisseria meningitidis. Mol. Microbiol.* **19:**841–856.

130. **Sekizaki, T., D. Takamatsu, M. Osaki, and Y. Shimoji.** 2005. Different foreign genes incidentally integrated into the same locus of the *Streptococcus suis* genome. *J. Bacteriol.* **187:**872–883.

131. **Shafer, W. M., X. Qu, A. J. Waring, and R. I. Lehrer.** 1998. Modulation of *Neisseria gonorrhoeae* susceptibility to vertebrate antibacterial peptides due to a member of the resistance/nodulation/division efflux pump family. *Proc. Natl. Acad. Sci. USA* **95:**1829–1833.

132. **Sibley, M. H., and E. A. Raleigh.** 2004. Cassette-like variation of restriction enzyme genes in *Escherichia coli* C and relatives. *Nucleic Acids Res.* **32:**522–534.

133. **Skaar, E. P., B. LeCuyer, A. G. Lenich, M. P. Lazio, D. Perkins-Balding, H. S. Seifert, and A. C. Karls.** 2005. Analysis of the *piv* re-combinase-related gene family of *Neisseria gonorrhoeae. J. Bacteriol.* **187:**1276–1286.

134. **Snyder, L. A., and N. J. Saunders.** 2006. The majority of genes in the pathogenic *Neisseria* species are present in non-pathogenic *Neisseria lactamica,* including those designated as "virulence genes." *BMC Genomics* **7:**128.

135. Reference deleted.

136. **Snyder, L. A. S., S. A. Butcher, and N. J. Saunders.** 2001. Comparative whole-genome analyses reveal over 100 putative phase-variable genes in the pathogenic *Neisseria* spp. *Microbiology* **147:**2321–2332.

137. **Snyder, L. A. S., J. K. Davies, C. S. Ryan, and N. J. Saunders.** 2005. Comparative overview of the genomic and genetic differences between the pathogenic *Neisseria* strains and species. *Plasmid* **54:**191–218.

138. **Snyder, L. A. S., J. K. Davies, and N. J. Saunders.** 2004. Microarray genomotyping of key experimental strains of *Neisseria gonorrhoeae* reveals gene complement diversity and five new neisserial genes associated with minimal mobile elements. *BMC Genomics* **5:**23.

139. **Snyder, L. A. S., S. A. Jarvis, and N. J. Saunders.** 2005. Complete and variant forms of the "gonococcal genetic island" in *Neisseria meningitidis. Microbiology* **151:**4005–4013.

140. **Snyder, L. A. S., N. J. Saunders, and W. M. Shafer.** 2001. A putatively phase variable gene (*dca*) required for natural competence in *Neisseria gonorrhoeae* but not *Neisseria meningitidis* is located within the division cell wall (*dcw*) gene cluster. *J. Bacteriol.* **183:**1233–1241.

141. **Snyder, L. A. S., W. M. Shafer, and N. J. Saunders.** 2003. Divergence and transcriptional analysis of the division cell wall (*dcw*) gene cluster in *Neisseria* spp. *Mol. Microbiol.* **47:**431–441.

142. **Spratt, B. G., Q. Y. Zhang, D. M. Jones, A. Hutchison, J. A. Brannigan, and C. G. Dowson.** 1989. Recruitment of a penicillin-binding protein gene from *Neisseria flavescens* during the emergence of penicillin resistance in *Neisseria meningitidis. Proc. Natl. Acad. Sci. USA* **86:**8988–8992.

143. **Stabler, R. A., G. L. Marsden, A. A. Witney, Y. M. Li, S. D. Bentley, C. M. Tang, and J. Hinds.** 2005. Identification of pathogen-specific genes through microarray analysis of pathogenic and commensal *Neisseria* species. *Microbiology* **151:**2907–2922.

144. **Stefanova, M. E., J. Tomberg, C. Davies, R. A. Nicholas, and W. G. Gutheil.** 2004. Overexpression and enzymatic characterization of *Neisseria gonorrhoeae* penicillin-binding protein 4. *Eur. J. Biochem.* **271:**23–32.

145. Steller, S., P. Angenendt, D. J. Cahill, S. Heuberger, H. Lehrach, and J. Kreutzberger. 2005. Bacterial protein microarrays for identification of new potential diagnostic markers for *Neisseria meningitidis* infections. *Proteomics* 5:2048–2055.

146. Stohl, E. A., A. K. Criss, and H. S. Seifert. 2005. The transcriptome response of *Neisseria gonorrhoeae* to hydrogen peroxide reveals genes with previously uncharacterized roles in oxidative damage protection. *Mol. Microbiol.* 58:520–532.

147. Stojiljkovic, I., V. Hwa, L. de Saint Martin, P. O'Gaora, X. Nassif, F. Heffron, and M. So. 1995. The *Neisseria meningitidis* haemoglobin receptor: its role in iron utilization and virulence. *Mol. Microbiol.* 15:531–541.

148. Sun, Y. H., S. Bakshi, R. Chalmers, and C. M. Tang. 2000. Functional genomics of *Neisseria meningitidis* pathogenesis. *Nat. Med.* 6:1269–1273.

149. Takahashi, H., K. Hirose, and H. Watanabe. 2004. Necessity of meningococcal gamma-glutamyl aminopeptidase for *Neisseria meningitidis* growth in rat cerebrospinal fluid (CSF) and CSF-like medium. *J. Bacteriol.* 186:244–247.

150. Tauschek, M., C. W. Hamilton, L. A. Hall, C. Chomvarin, J. A. Fyfe, and J. K. Davies. 1997. Transcriptional analysis of the *groESL* operon of *Neisseria gonorrhoeae*. *Gene* 189:107–112.

151. Tefsen, B., M. P. Bos, F. Beckers, J. Tommassen, and H. de Cock. 2005. MsbA is not required for phospholipid transport in Neisseria meningitidis. *J. Biol. Chem.* 280:35961–35966.

152. Tettelin, H., N. J. Saunders, J. Heidelberg, A. C. Jeffries, K. E. Nelson, J. A. Eisen, K. A. Ketchum, D. W. Hood, J. F. Peden, R. J. Dodson, W. C. Nelson, M. L. Gwinn, R. DeBoy, J. D. Peterson, E. K. Hickey, D. H. Haft, S. L. Salzberg, O. White, R. D. Fleischmann, B. A. Dougherty, T. Mason, A. Ciecko, D. S. Parksey, E. Blair, H. Cittone, E. B. Clark, M. D. Cotton, T. R. Utterback, H. Khouri, H. Qin, J. Vamathevan, J. Gill, V. Scarlato, V. Masignani, M. Pizza, G. Grandi, L. Sun, H. O. Smith, C. M. Fraser, E. R. Moxon, R. Rappuoli, and J. C. Venter. 2000. Complete genome sequence of *Neisseria meningitidis* serogroup B strain MC58. *Science* 287:1809–1815.

153. Turner, D. P. J., K. G. Wooldridge, and D. A. A. Ala'Aldeen. 2002. Autotransported serine protease A of *Neisseria meningitidis*: an immunogenic, surface-exposed outer membrane, and secreted protein. *Infect. Immun.* 70:4447–4461.

154. Turner, P. C., C. E. Thomas, I. Stojiljkovic, C. Elkins, G. Kizel, D. A. A. Ala'Aldeen, and P. F. Sparling. 2001. Neisserial TonB-dependent outer-membrane proteins: detection, regulation and distribution of three putative candidates identified from the genome sequences. *Microbiology* 147:1277–1290.

155. Tzeng, Y. L., K. D. Ambrose, S. Zughaier, X. L. Zhou, Y. K. Miller, W. M. Shafer, and D. S. Stephens. 2005. Cationic antimicrobial peptide resistance in *Neisseria meningitidis*. *J. Bacteriol.* 187:5387–5396.

156. van Ulsen, P., L. van Alphen, C. T. P. Hopman, A. van der Ende, and J. Tommassen. 2001. In vivo expression of *Neisseria meningitidis* proteins homologous to the *Haemophilus influenzae* Hap and Hia autotransporters. *FEMS Immunol. Med. Microbiol.* 32:53–64.

157. van Ulsen, P., L. van Alphen, J. ten Hove, F. Fransen, P. van der Ley, and J. Tommassen. 2003. A neisserial autotransporter NalP modulating the processing of other autotransporters. *Mol. Microbiol.* 50:1017–1030.

158. Warren, M. J., and M. P. Jennings. 2003. Identification and characterization of *pptA*: a gene involved in the phase-variable expression of phosphorylcholine on pili of *Neisseria meningitidis*. *Infect. Immun.* 71:6892–6898.

159. Welsch, J. A., G. R. Moe, R. Rossi, J. Adu-Bobie, R. Rappuoli, and D. M. Granoff. 2003. Antibody to genome-derived neisserial antigen 2132, a *Neisseria meningitidis* candidate vaccine, confers protection against bacteremia in the absence of complement-mediated bactericidal activity. *J. Infect. Dis.* 188:1730–1740.

160. Winther-Larsen, H. C., F. T. Hegge, M. Wolfgang, S. F. Hayes, J. P. M. van Putten, and M. Koomey. 2001. *Neisseria gonorrhoeae* PilV, a type IV pilus-associated protein essential to human epithelial cell adherence. *Proc. Natl. Acad. Sci. USA* 98:15276–15281.

161. Winther-Larsen, H. C., M. Wolfgang, S. Dunham, J. P. M. van Putten, D. Dorward, C. Lovold, F. E. Aas, and M. Koomey. 2005. A conserved set of pilin-like molecules controls type IV pilus dynamics and organelle-associated functions in *Neisseria gonorrhoeae*. *Mol. Microbiol.* 56:903–917.

162. Wolfgang, M., J. P. van Putten, S. F. Hayes, and M. Koomey. 1999. The *comP* locus of *Neisseria gonorrhoeae* encodes a type IV prepilin that is dispensable for pilus biogenesis but essential for natural transformation. *Mol. Microbiol.* 31:1345–1357.

163. Wooldridge, K. G., M. Kizil, D. B. Wells, and D. A. A. Ala'Aldeen. 2005. Unusual genetic organization of a functional type I protein

secretion system in *Neisseria meningitidis. Infect. Immun.* **73:**5554–5567.

164. **Yasukawa, K., P. Martin, C. R. Tinsley, and X. Nassif.** 2006. Pilus-mediated adhesion of *Neisseria meningitidis* is negatively controlled by the pilus-retraction machinery. *Mol. Microbiol.* **59:**579–589.

165. **Yazdankhah, S. P., K. Kesanopoulos, G. Tzanakaki, J. Kremastinou, and D. A. Caugant.** 2005. Variable-number tandem repeat analysis of meningococcal isolates belonging to the sequence type 162 complex. *J. Clin. Microbiol.* **43:**4865–4867.

166. **Yazdankhah, S. P., B. A. Lindstedt, and D. A. Caugant.** 2005. Use of variable-number tandem repeats to examine genetic diversity of *Neisseria meningitidis. J. Clin. Microbiol.* **43:**1699–1705.

167. **Zhu, P. X., M. J. Klutch, J. P. Derrick, S. M. Prince, R. S. W. Tsng, and C. M. Tsai.** 2003. Identification of *opcA* gene in *Neisseria polysaccharea*: interspecies diversity of Opc protein family. *Gene* **307:**31–40.

THE STAPHYLOCOCCI: A POSTGENOMIC VIEW

Jodi A. Lindsay and Matthew T. G. Holden

5

The staphylococci are gram-positive cocci that occur as single cells, pairs, tetrads, or irregular grapelike clusters. Their name is derived from *staphyle*, Greek for "bunch of grapes" (64). They are nonmotile, non-spore-forming, facultative anaerobes and usually produce catalase. The genus *Staphylococcus* belongs to the *Bacillus-Lactobacillus-Streptococcus* cluster of gram-positive bacteria with a low G+C content and contains species that have been well studied because they frequently cause infections in humans and animals.

At the time of writing, the genus contained 38 recognized species (26) (http://www.bacterio.cict.fr/), which includes nonpathogens and pathogens with wide host ranges (Fig. 1). At least 16 species are associated with humans (17, 44) and can live on all parts of the human body surface, with certain species associated with particular niches. For example, *Staphylococcus aureus* is predominantly found in the nasal nares, as well as throat, perineum, and axillae; *S. epidermidis* prefer the skin; *S. capitis* the head; and *S. auricularis* the ears. Members of the genus can also be isolated from a wide range of other hosts including primates, pigs, cows, goats, sheep, dogs, cats, rodents, whales, otters, dolphins, and birds. The remaining members of the genus can be isolated from various source material, including insects, meat products, fermented fish, cheese, and soy sauce mash. The niche specificity of the species varies across the genus; some of the species, such as *S. pasteuri*, have been isolated from human, animal, and food specimens (17), whereas others, such as *S. delphini*, which was originally isolated from dolphins, appear to be more limited in their distribution (83).

Staphylococci that can cause disease in humans are *S. aureus*, *S. epidermidis*, *S. haemolyticus*, *S. saprophyticus*, *S. hominis*, *S. warneri*, *S. lugdunensis*, *S. schleiferi* subsp. *schleiferi*, *S. capitis* subsp. *ureolyticus*, and *S. simulans* (44). The most virulent species is *S. aureus*, which is identified by its ability to produce coagulase; the production of this enzyme has been used as a diagnostic marker and can subdivide members of the genus. The remaining human-associated species are predominantly coagulase-negative staphylococci (CoNS) and generally cause less severe diseases.

The genus was originally divided into two species, *Staphylococcus aureus* and *Staphylococcus albus*, on the basis of pigmented colony types

Jodi A. Lindsay Department of Cellular and Molecular Medicine, St. George's, University of London, Cranmer Terrace, London SW17 0RE, United Kingdom. *Matthew T. G. Holden* The Wellcome Trust Sanger Institute, Hinxton, Cambridge CB10 1SA, United Kingdom.

Bacterial Pathogenomics, Edited by M. J. Pallen et al.
© 2007 ASM Press, Washington, D.C.

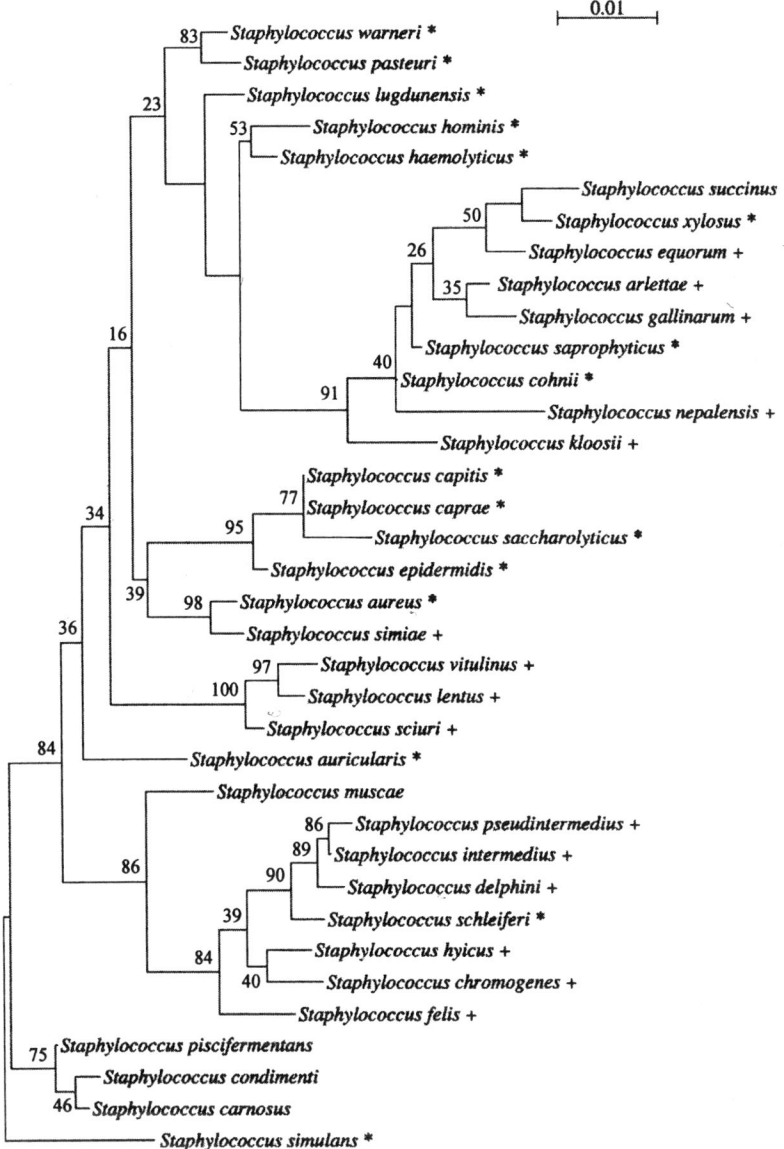

FIGURE 1 Phylogenetic relationships of staphylococcal species based on 16S rRNA sequences. Symbols after the names of the species indicate the host association: ★, human; +, animals and birds; no symbol, other. Maximum-likelihood tree built from 16S rRNA sequences downloaded from the Ribosomal Database Project-I (http://rdp.cme.msu.edu) (19) by ClustalX (81), Phylip (version 3.6) (29) and NJplot (70). The numbers at the tree branches are percentage bootstrap values indicating the confident levels at that node where congruent (28). The bar indicates the genetic distance between species (1 nucleotide substitution per 100) as displayed in the branch lengths. Sequences used to construct the tree are *S. aureus* (accession number X68417), *S. sciuri* (AJ421446), *S. condimenti* (Y15750), *S. nepalensis* (AJ517414), *S. muscae* (S83566), *S. arlettae* (AB009933), *S. carnosus* (AB009934), *S. caprae* (AB009935), *S. cohnii* (AB009936), *S. capitis* (AB009937), *S. delphini* (AB009938), *S. equorum* (AB009939), *S. kloosii* (AB009940), *S. lugdunensis* (AB009941), *S. piscifermentans* (AB009943), *S. schleiferi* (AB009945), *S. vitulinus* (AB009946), *S. succinus* (AF004220), *S. auricularis* (D83358), *S. chromogenes* (D83360), *S. epidermidis* (D83363), *S. felis* (D83364), *S. gallinarum* (D83366), *S. haemolyticus* (D83367), *S. hyicus* (D83368), *S. intermedius* (D83369), *S. lentus* (D83370), *S. saprophyticus* (D83371), *S. simulans* (D83373), *S. xylosus* (D83374), *S. hominis* (L37601), *S. saccharolyticus* (L37602), *S. warneri* (L37603), *S. pasteuri* (AF041361), *S. simiae* (AY727530), and *S. pseudintermedius* (AJ780976). Currently recognized *Staphylococcus* species missing from the tree are *Staphylococcus fleurettii* (84) and *Staphylococcus lutrae* (33); these species were not included in the tree because the 16S rRNA sequences available for these species are <600 bp.

(73). Over the last century, taxonomists have used changing microbiological and molecular techniques to define the genus. In recent decades, DNA relatedness, as defined by DNA-DNA hybridization experiments, has been central to the taxonomic definition of species within this genus. Members of the same species demonstrate relative DNA-binding values of greater than 70% under optimal DNA-DNA reassociation conditions (34). More recently, molecular genetic techniques such as ribotyping (86), pulsed-field gel electrophoresis (7), and PCR fingerprinting (40) have also been applied. The sequencing of several *Staphylococcus* species genomes in recent years has generated new insights into the genetics and biology of members of this genus. Nine *S. aureus* isolates, two *S. epidermidis* isolates, and one of each of *S. haemolyticus* and *S. saprophyticus* have been sequenced. Each species has different natural habitats, virulence, and resistance capabilities. In this chapter, we will briefly discuss the often complex biology of the organism before the genomic insights determined from the sequence. The knowledge gained from our initial genomic forays is helping to revisit the definitions of species, improve our understanding of the evolutionary mechanisms that have shaped the staphylococci, and identify how strains adapt to environmental niches and cause disease.

STAPHYLOCOCCUS AUREUS

S. aureus is a commensal of the human nose, with 20% of the population carrying *S. aureus* all of the time and another 60% carrying it intermittently (45). *S. aureus* is a common cause of superficial skin infections that rarely require treatment. In hospitals, *S. aureus* is the most common cause of hospital-acquired infections (42), particularly in elderly patients and those with wounds, catheters, or diabetes and those on ventilators. Infections in any tissue are described, including skin and soft-tissue infections, abscesses, bacteremia, endocarditis, pneumonia, osteomyelitis, infective arthritis, conjunctivitis, and even meningitis. Many infections are relatively minor, but

others lead to severe disease, occasionally with long-term sequelae or death. *S. aureus* is the most pathogenic species of the genus. It is capable of producing dozens of known toxins, host-tissue binding proteins, and immune evasion strategies. In addition, a few specific toxins are known to be associated with specific clinical syndromes, for example, toxic shock syndrome toxin-1 (*tst*) with toxic shock syndrome (74), enterotoxins with food poisoning (12), exfoliative toxins with scalded skin syndrome (60), and Panton-Valentine leukocidin (*PV-luk*) with hemolytic pneumonia (51). However, it is important to remember that healthy people rarely develop severe disease, and this is likely due to a rapid and effective neutrophil response. Patients with defective neutrophils, such as those with chronic granulomatous disease, are particularly susceptible to severe *S. aureus* disease (32).

S. aureus infections are treated with β-lactamase-resistant penicillins such as flucloxacillin; however, isolates that are resistant to these antibiotics (MRSA, methicillin-resistant *S. aureus*) are now widespread, with some hospitals reporting resistance rates of over 60% (82). Although MRSA were first reported in 1961, high levels of "endemic" MRSA in hospitals have become a problem only in the last 10 to 15 years (25). These MRSA can also be referred to as epidemic MRSA or hospital-acquired MRSA. MRSA infections are treated with glycopeptide antibiotics, such as vancomycin, but the first four cases of fully vancomycin-resistant *S. aureus* have now been described (2–4, 16). These isolates had acquired the *vanA* gene via horizontal transfer from vancomycin-resistant enterococci, commonly found in hospitals (87). In the last few years, increasing numbers of cases of MRSA have been reported outside of hospitals, with otherwise healthy patients developing severe skin and soft-tissue infections that require hospitalization (15). These community-acquired MRSA are also positive for PV-luk, which is implicated in the enhanced virulence. Thus, at least three new types of *S. aureus* have recently emerged (hospital- and community-acquired MRSA, and vancomycin-resistant MRSA), each with enhanced virulence and/or resistance and each causing novel clinical problems.

THE *STAPHYLOCOCCUS AUREUS* GENOME SEQUENCES

Nine *S. aureus* genomes have now been annotated and published, with more in the pipeline (for a list of the ongoing sequencing projects, see the Genomes OnLine Database: http://www.genomesonline.org/) (55). The comparison of genomes has enabled a greater understanding of *S. aureus* virulence, resistance, genome stability, variability, flexibility, and evolution.

N315 and Mu50

The first *S. aureus* sequences represented a milestone in our understanding of *S. aureus* metabolism and virulence potential. Two genomes were published in 2001 (49). The strains were N315 (accession number BA000018), an MRSA from a Japanese hospital, and Mu50 (accession number BA000017), a related hospital MRSA that showed intermediate level resistance to vancomycin. The genomes consisted of a circular chromosome of ~2.8 Mb and a plasmid of ~25 kb. The numbers of protein-coding sequences (CDSs) identified in the N315 and Mu50 genomes were 2,620 and 2,748, respectively. These first public sequences were valuable for identifying basic metabolic pathways in *S. aureus* and the diverse range of putative virulence factor genes.

The availability of a complete genomic inventory of *S. aureus* has allowed researchers to examine the metabolic networks of the cell. In silico metabolic reconstruction that used the N315 genome has been used to generate preliminary metabolic models of the cell (8, 38). Theoretical deletion studies using these models have been able to examine the effect of gene deletions on the metabolic fluxes of the cell and therefore predict putative drug targets.

The sequencing projects identified novel genes involved in cell division, cell wall synthesis, transport, osmoprotection, pigmentation, iron uptake and storage, capsule synthesis, and many others. Novel adhesins that bind to human host tissues were identified, particularly those with an LPXTG motif that indicated that they were anchored to the cell wall by sortase. Some very large surface proteins were also identified, and subsequent studies showed they bound host proteins. At up to 1 MDa in size, they were the largest bacterial protein described (18). Genes showing homology to major histocompatibility complex class II analogues were later shown to be phase variable (13). Novel exoenzymes that potentially encode lipases, proteases, and a thermonuclease were identified. A range of potential new toxins was also identified, including one with homology to a diarrheal toxin of *Bacillus cereus*, several hemolysins, leukocidins, and exotoxins. Dozens of novel genes involved in transcription, two-component response regulation, and a host of other gene regulators were identified, some with homology to known genes implicated in virulence. Thus, it was clear that regulation of *S. aureus* gene expression would be more complex than previously supposed.

The first two genomes also prompted the first genome comparisons. Despite the two strains being related and sharing an enormous amount of sequence homology, there was significant variation in the carriage of mobile genetic elements (MGEs) (Fig. 2).

MGEs in *S. aureus* include staphylococcal cassette chromosomes (SCC), bacteriophage, *S. aureus* pathogenicity islands (SaPI), plasmids, transposons, conjugative elements, and insertion sequences (IS) (52). Although some MGEs had already been described, as well as their ability to mobilize frequently (11, 54, 68, 72), it was now clear that individual strains carried many of these elements, and that they had the potential to vary greatly. The fact that two related strains could carry such a diverse range of MGEs indicated that horizontal transfer occurred frequently. And because many virulence and resistance genes are carried on these elements, the mobilization of MGEs has the potential to affect evolution and pathogenesis of *S. aureus* (52).

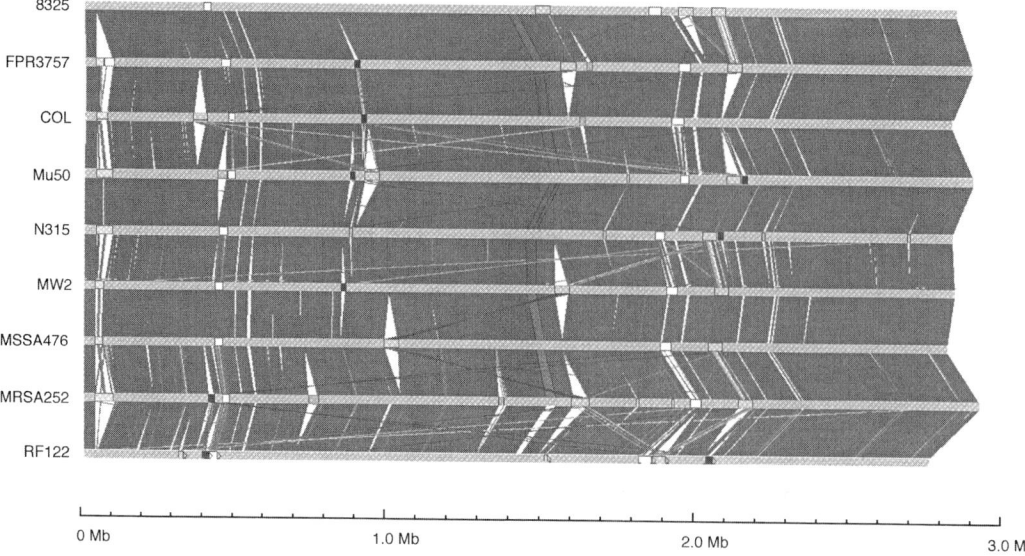

FIGURE 2   Pairwise comparisons of *S. aureus* NCTC8325, USA300, COL, Mu50, N315, MW2, MSSA476, MRSA252, and RF122 chromosomes displayed using the Artemis Comparison Tool (ACT) (14). The sequences have been aligned from the predicted replication origins (*oriC*; right), with the terminus of replication in the center. The dark gray bars separating each genome represent orthologous matches identified by reciprocal FASTA analysis (69), with an identity cutoff of 30% and a length-of-match cutoff of 80%. Variable regions of the chromosomes containing mobile genetic elements such as prophages, plasmids, transposons, SaPIs, and other genomic islands are marked as boxes within the genomes and vary significantly between each strain.

Interestingly, in the end, the intermediate-level vancomycin resistance phenotype was not caused by horizontal transfer of resistance genes, but by mutations in cell wall synthesis genes, leading to a thicker cell wall (22) that could absorb higher levels of vancomycin (21). Therefore, this represents an example where genome sequencing was unable to explain a phenotypic difference between two closely related strains. In *S. aureus*, it was due to the frequent movement of MGE, leading to significant genome variation in closely related isolates. This masked the subtle variation in relevant genes actually responsible for the resistance phenotype.

MW2

The next sequence was published by Baba et al. (5) for the isolate MW2 (accession number BA000033), a community-acquired MRSA from North Dakota that caused fatal septicemia in a 16-month-old child. This isolate is

regarded as typical of community-acquired MRSA that are becoming widespread in the United States and are causing severe skin and soft-tissue infections, with some invasive cases in healthy members of the community (1, 77). The MW2 sequence differed from the N315 and Mu50 sequences in a number of important virulence genes and generally shared less sequence homology. Even more striking was the variation in MGEs, with MW2 carrying a bacteriophage encoding the Panton-Valentine leukocidin gene (*PV-luk*), thought to be a major contributor to the virulence of this strain, and a type VI SCC*mec* element conferring methicillin resistance.

MRSA252 and MSSA476

The Wellcome Trust Sanger Institute published the next two sequences in 2004 (39). MRSA252 (accession number BX571856) is an epidemic hospital-acquired MRSA from a patient with bacteremia at the John Radcliffe

Hospital, Oxford, United Kingdom, and belongs to the epidemic MRSA-16 group, which is responsible for about half of all hospital MRSA infections in the United Kingdom (41) and is an important clonal MRSA type in the United States (known as USA200) (58). The sequence of this isolate was even more diverse in sequence homology than MW2, and again, this isolate carried a unique complement of MGEs (Fig. 2). The second isolate, MSSA476 (accession number BX571857), caused a community-acquired osteomyelitis in a healthy patient in the Oxford region in the United Kingdom. Its sequence was remarkably similar to MW2: 2,528 orthologous CDSs were identified, which contained only 285 single-nucleotide polymorphisms. Despite the high level of genetic relatedness, there was significant variation in some MGEs. For example, MSSA476 did not contain a prophage carrying *PV-luk* but did carry a unique SCC element with resistance to fusidic acid (SCC*far*) instead of methicillin. The similarity between MSSA476 and MW2 seemed surprising because they were isolated from the community in two different continents. Was it just coincidence?

COL and 8325

In 2005, The Institute for Genomic Research (TIGR) published the sixth *S. aureus* genome, that of COL (accession number CP000046), an early MRSA isolated in the United Kingdom in the 1960s (35). COL is a well-studied laboratory strain, but it is probably atypical of current epidemic hospital- or community-acquired MRSA. By this time, the sequence of NCTC8325 (RN1), an MSSA from the United Kingdom of around the 1940s, was complete and the unannotated sequence was available on the Internet (http://www.genome.ou.edu/staph. html). More recently, the annotated genome has been deposited in the public sequence databases (accession number CP000253) (36). Derivatives of NCTC8325 are genetically tractable, and these strains are the workhorse of *S. aureus* laboratory studies. Both COL and NCTC8325 were highly similar despite differing in their MGEs (Fig. 2); 2,435 or-

thologous CDS matched with an average identity at the protein level of 99.82%.

FPR3757

The sequence of a representative USA300 isolate, FPR3757, has just been published (accession number CP000255) (23). USA300 strains are community-acquired MRSA originally implicated in several outbreaks of severe skin and soft-tissue infection in healthy individuals in close proximity, such as being in prison (65) or on the same football team (43). USA300 isolates are positive for the toxin Panton-Valentine leukocidin (79). The first reports of this strain spreading to the hospital setting have now appeared (75), and they accounted for 34% of all bloodstream infections, with half arising from a skin and soft-tissue infection. Surprisingly, the mortality rate with this isolate was low (8% versus 29% with other MRSA types), and thus USA300 shows a unique type of MRSA infection. The sequence of this strain identified variations in MGE carriage compared to other strains, notably a prophage encoding the Panton-Valentine leukocidin gene (*PV-luk*; φSA2usa), a novel SaPI, and the presence of an island containing an arginine catabolism gene cluster (arginine catabolism mobile element, ACME) found next to the SCC*mec*. However, the role of these ACME genes in virulence is currently unknown. The genome of this community-acquired MRSA also contains three plasmids, two of which contain antibiotic resistance genes (pUSA02 and pUSA03) (23). The largest of the plasmids, pUSA03, is a conjugative plasmid that carries genes that encode constitutive resistance to macrolides–lincosamides–streptogramins B and high-level resistance to mupirocin. The plasmid shares much of its backbone with pLW1043, a conjugative plasmid carrying the vancomycin resistance transposon found in vancomycin-resistant *S. aureus* (87).

RF122

More recently, a bovine isolate of *S. aureus* that caused mastitis, RF122 (accession number AJ938182), has been deposited in the public domain. Bovine mastitis is a severe economic problem for dairy farmers because milking cows with mastitis must be removed from production and treated with antibiotics or destroyed. In addition, high white blood cell counts in milk are often used as a marker of low-quality milk and reduce its market price. RF122 is positive for toxic shock syndrome toxin (30), and the new sequence shows that this bovine strain has similarity with human isolates in many of its surface proteins and toxins.

A striking feature of all the sequencing projects is the number and range of known and putative virulence pathways. The amount of gene redundancy has become clear, particularly in enterotoxin and superantigen genes, cytotoxins, proteases, surface proteins, immune evasion pathways, and regulators. To date, the construction of isogenic mutant pairs for individual genes has identified few with a proven and substantial role in multiple infection models. It is difficult to genetically manipulate clinical isolates of *S. aureus*, and combined with the vast number of potential virulence genes and the problems of redundancy, understanding which factors are essential for virulence has been hampered. Furthermore, signature-tagged mutagenesis (20, 59), in vitro expression technology (56), and bursa aurealis (6) and Tn917 (10) transposon library screenings in *Caenorhabditis elegans* models have yielded a high proportion of basic metabolic and nutritional pathways, which have not clarified the picture. Therefore, it is still unclear exactly how *S. aureus* causes disease. Perhaps its ability to adapt to multiple environments, causing infection in potentially any tissue, is dependent on having so many potential virulence pathways. Many CDSs are still unannotated; they may play important roles in pathogenicity.

IN SILICO COMPARATIVE ANALYSIS OF *S. AUREUS*

All of the *S. aureus* genomes that have been sequenced thus far have a conserved chromosomal structure. Pairwise comparison of the nine *S. aureus* chromosomes shows that they are co-

linear with discrete regions of difference throughout their length (Fig. 2). The majority of the genome is composed of orthologous genes that have been vertically inherited from a common ancestor. These genes encode proteins that have equivalent functions. In addition to these core orthologs are variably distributed genes that are found in some or in individual strains. Many of these genes are found in regions that have been identified as MGEs that have been horizontally acquired and can be categorized as accessory genes (52) The crude partitioning of the genome into core and variable components provides an insight into the functional adaptation of *S. aureus* strains. A tripartite comparison of the MRSA252, N315, and MSSA476 genomes identified 2,274 CDSs orthologous gene matches that were shared between the three strains (Fig. 3). These genes encode central metabolic functions essential for growth, replication, and cell division. In addition, there are functions that may not be considered to be essential but that phenotypically define the species, such as virulence and survival traits.

Supplementing the core genome in each strain is a variable genome. The tripartite comparison reveals that 259, 167, and 148 CDSs were unique to MRSA252, N315, and MSSA476, respectively, with the remaining genes in the variable genomes differentially shared between the strains. This limited in silico analysis predicts that at least 17% of the genome of MRSA252 is composed of variable genes that encode some of the strain-specific functions and may promote success of this epidemic strain. Comparison of the functional categories of the genes in the shared and variable components of the genomes reveals that genes of unknown function and genes encoding functions associated with lateral gene transfer are overrepresented in the variable sets (Fig. 3). Genes associated with protective responses (including antibiotic resistance), transport, macromolecular metabolism, energy metabolism, small-molecule degradation, cell envelope, and regulation are also present. Many of these genes are located in MGEs or in vari-

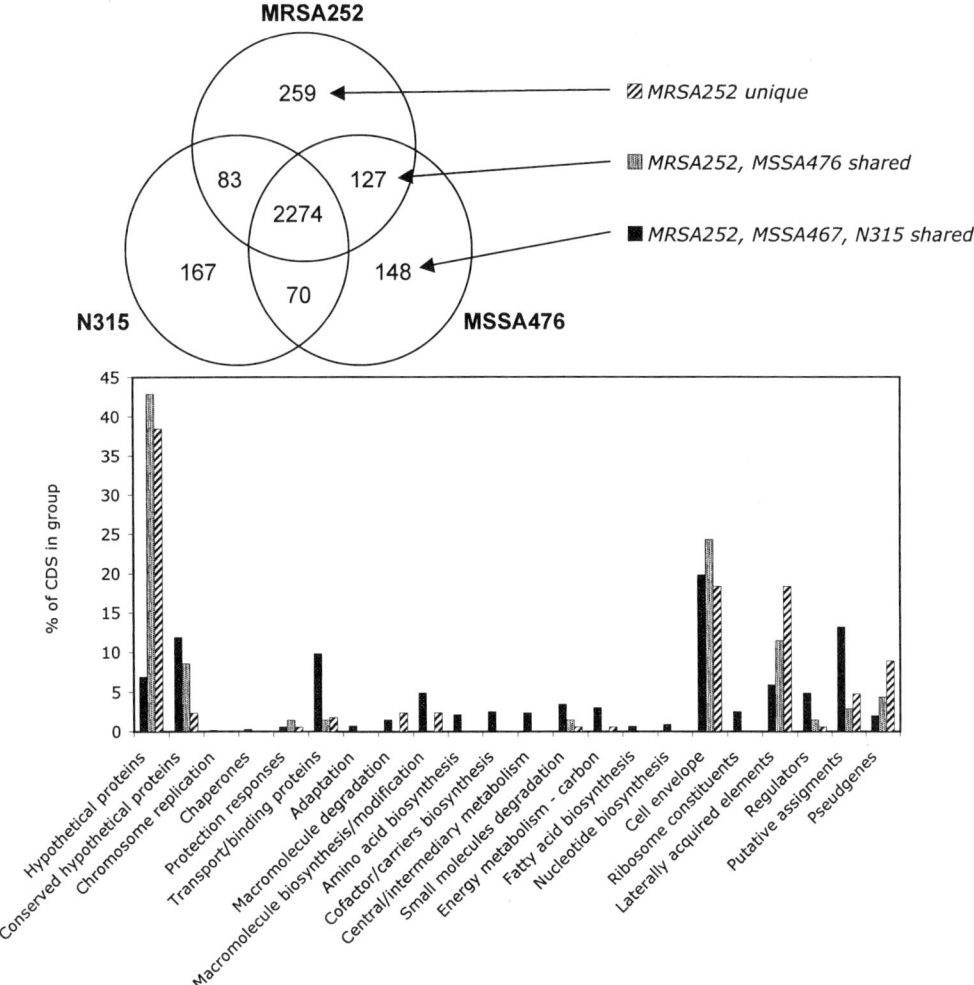

FIGURE 3 Distribution of orthologous CDSs in *S. aureus*. Venn diagram showing the number of genes unique or shared between three *S. aureus* strains (MRSA252, MSSA476, and N315). The associated pie charts show the breakdown of the functional groups assigned for CDSs in relevant sections of the Venn diagram. Orthologous matches were identified as previously described for Fig. 2. Grayscale code for the functional groups in the pie charts is displayed.

able regions of the chromosome such as genomic islands. Although comparisons such as these can help home in on the *S. aureus* genotype that defines the essence of the species, it is apparent that defining the full gene complement of *S. aureus*, or the *S. aureus* pangenome, is going to be a far more laborious task. With each new genome, the variable component of the *S. aureus* genome grows; the most recently published genome, FPR3757, yielded new

MGEs—for example, SaPI5 and ACME—not previously seen in the other sequenced strains (23). Recent investigation into the species genome of *Streptococcus agalactiae* identified that ~80% of the genome was core; the rest was composed of dispensable genome consisting of partially shared and strain-specific genes (80). By examining the redundancy of the genes in the variable genomes of 8 *S. agalactiae* strains, Tettelin et al. were able to predict that even with hundreds of additional genomes sequenced, the *S. agalactiae* pan genome would remain incomplete (80).

STAPHYLOCOCCUS AUREUS MULTILOCUS SEQUENCE TYPING

The comparison of the available *S. aureus* genomes has yielded much information regarding the genetic diversity of the species. However, it is not always apparent how much of the observed diversity is due to genetic unrelatedness of the compared strains or to diverse selection in response to niche differences. The context of these genome sequences in relation to each other, and the entire *S. aureus* population, is therefore important to understand. Interpreting the genomes in the context of a population framework can help elucidate the evolutionary mechanisms that have shaped the genome.

Pinpointing the sequenced *S. aureus* genomes within the population is best achieved by considering multilocus sequence typing (MLST) types. Enright et al., working in Brian Spratt's laboratory, developed MLST for *S. aureus*, and this has gone on to become the gold standard typing method for genotyping *S. aureus* (24). MLST involves sequencing seven housekeeping genes and identifying any variation in these genes. Each variant sequence, or allele, is given a corresponding number. Each isolate is therefore assigned a series of seven numbers. If two isolates share seven identical alleles, they are given the same sequence type (ST) number, and if two isolates share five or more alleles, they are given the same clonal complex (CC) number. Isolates with the same ST and/or CC numbers are presumed to be related. MLST is highly reproducible and allows isolates from all over the world to be compared. A large database containing the ST of 1,530 isolates (715 ST in total) is accessible via the Internet (http://saureus.mlst.net/). Studies of over 1,000 *S. aureus* isolates have revealed that *S. aureus* has a clonal population structure with about 10 dominant clonal complexes accounting for 85% of human isolates

TABLE 1 Clonal complex (CC) and sequence type (ST) of the sequenced isolates

CC	ST	Isolate
1	1	MW2
1	1	MSSA476
5	5	N315
5	5	Mu50
8	8	NCTC8325
8	8	FPR3757
8	250	COL
30	36	MRSA252
151	151	RF122

(27). They are CC1, CC5, CC8, CC12, CC15, CC22, CC25, CC30 (the most common), CC45, and CC51. More recently, *spa* typing, which involves the sequencing of a highly variant region of the gene encoding protein A, has also grouped strains into essentially the same major lineages (57).

The ST and CC numbers of the sequenced strains are listed in Table 1. As expected, this corresponded well with the similarity of the whole-genome "core" sequences. In other words, strains MW2 and MSSA476 were surprisingly similar in genome sequence because they belonged to the same lineage, as were N315 and Mu50. The suggestion now was that isolates of the same ST or CC group had highly similar core genome sequences, but on top of this, MGEs could move between isolates belonging to different clonal complexes. It also became clear that the sequencing projects did not cover all of the dominant human lineages.

STAPHYLOCOCCUS AUREUS COMPARATIVE GENOMICS USING MICROARRAYS

One of the most valuable tools that can arise from whole genome sequences are whole-genome microarrays. Microarrays are glass slides printed with high-density spots of DNA representing every gene in the genome. Several *S. aureus* microarrays have been constructed and used to investigate gene expression (9, 46, 48, 89). They are typically used to take a snapshot of gene expression in a bacterial strain under a given growth condition. This allows the bacterial gene expression responses to environmental growth conditions to be determined, or the role and interactions of gene regulators under different conditions to be defined.

A second application of microarrays is to compare genomes of different bacterial isolates and identify genes associated with particular phenotypic traits. The greater the number and diversity of genes on the microarray, the greater its power for identifying these genes or other patterns of gene carriage. *S. aureus* is a prime example of a bacterium that has been exploited by this approach because of the number of genome sequences and significant variation in MGEs. The use of microarrays for comparative genomic analysis has increased the breadth and depth of knowledge first gleaned from in silico comparisons. The increased number of strains that have been screened is generating a clearer picture of the genetic diversity of the species—so much so that we are able to refine our definitions of the core and variable genomes.

The pioneering study by Fitzgerald et al. was only able to exploit a single incomplete genome, COL, but still identified substantial variation in MGEs between 34 clinical isolates (31). More recently, Witney et al. have designed a seven-strain *S. aureus* microarray and used it to compare the genomes of a range of isolates (89). In a study of 161 community invasive (predominantly bacteremia) versus nasal carriage isolates from the Oxford region in the United Kingdom, enormous genomic variation between isolates was seen. This included variation in dozens of toxins, regulators, host-binding proteins, and other putative virulence factors. However, no genes were present in the invasive isolates that were consistently or largely absent in the carriage isolates (53). This strongly suggests that *S. aureus* is an opportunistic pathogen in healthy patients in the community setting and that typical *S. aureus* causes invasive disease due to host factors.

Core Variable Genes Define Lineages

This study also revealed new information about the *S. aureus* genome and its evolution. The sequencing projects had suggested that there was genetic variation between isolates of different CCs. By microarray, it was shown that there are hundreds of genes and gene variants that are not found in all strains (i.e., not core genes) and are not on MGEs. The distribution of these hundreds of genes is strongly linked with different CCs or lineages, and these genes are referred to as core variable (CV) genes. CV genes include clusters of genes, single genes, or variant regions within genes and are scattered randomly throughout the chromosome. A typical genome, such as MRSA252, consists of 71.2% core genes, 11.9% CV genes, and 16.8% MGE genes. A large number of CV genes encode surface proteins or structures or their regulators and are predicted to interact with the host. For example, CV genes include variants of genes for capsule, fibronectin-binding proteins, coagulase, protein A, *ebh*, accumulation-associated proteins, collagen-binding protein, hemagglutinin-like protein, a toxin related to a *B. cereus* diarrheal toxin, and variations in regulators *agr*, *sar* homologs, and TRAP. The genomic islands νSaα and νSaβ are CV and carry multiple variants of enterotoxins, proteases, lipoproteins, and leukocidins. Although each lineage has different combinations of CV genes, there are relatively few cases of lineages carrying unique CV genes, which suggests that recombination was common. The likely explanation is that a common *S. aureus* ancestor branched into multiple lineages, probably by recombination of lineages, and about 10 of these lineages successfully colonized the human nose. Because the nose is the normal ecological niche of *S. aureus* and each carrier has predominantly one type of *S. aureus* in his or her nose (63), selection has probably been driven by this environment.

The results of microarray analysis of the core genome reflect much of the in silico predictions. The microarray-defined core genome of *S. aureus* includes many characteristic virulence factors, such as hemolysins, proteases, clumping factors *clfA* and *clfB*, intercellular adhesin, autolysins, lipases, hyaluronate lyase, thermonuclease, siderophore production and iron uptake pathways, superoxide dismutase and catalase, urease, complement inhibitor, phenol soluble modulins, efflux pumps, various two-component signal transduction systems, and regulators *sarA*, *arl*, and *sae*.

MGE Distribution

The microarrays also enabled the most detailed picture of MGE distribution in *S. aureus* so far. It has been suggested that MGEs frequently move in and out of isolates, possibly during colonization or infection (37, 62). The arrays showed that MGEs exhibit enormous recombination potential, with most consisting of a series of mosaic fragments of other MGE. Some MGEs are clearly found in multiple lineages, which suggests frequent horizontal transfer. Others seem more stable because they are found in nearly all isolates of a particular lineage, and some seem to have moved into specific lineages only. The movement and distribution of MGEs are the keys to understanding *S. aureus* pathogenesis because many resistance and virulence genes are encoded on them. For example, SCC can encode methicillin resistance and/or resistance to erythromycin, kanamycin, bleomycin, or fusidic acid; transposons can carry many other resistance genes, including vancomycin resistance derived from the enterococci; bacteriophages carry *PV-luk*, exfoliative toxins, enterotoxin A, staphylokinase, and host evasion mechanisms; and SaPI encode toxic shock syndrome toxin and other superantigen enterotoxins. Therefore, the newly emerging epidemic MRSA in hospitals, community-acquired MRSA causing infections in healthy patients, and fully vancomycin-resistant MRSA have all evolved as a result of the horizontal transfer of MGEs.

We know that certain toxins are associated with certain *S. aureus* pathogenesis. For example, toxic shock syndrome is caused by *S. aureus* carrying the superantigen gene *tst* on a SaPI, although some surgical cases are associated with *seb* and *sec* carried on SaPIs. Approx-

imately 20% of clinical *S. aureus* are positive for one or more of these toxins. Scalded skin syndrome is caused by *S. aureus* carrying the exfoliative toxins *eta* or *etb* on phage or plasmids, and these toxins are carried by about 1% of *S. aureus* isolates. Acute hemolytic pneumonia is associated with isolates carrying the Panton-Valentine leukocidin (*PV-luk*) toxin on a prophage, and this toxin is typically found in only 1 to 2% of *S. aureus* isolates. These discoveries have been made by clinical observation and carefully collected clinical isolates combined with laboratory confirmation.

As the methods for classifying strains into dominant types (such as CC by MLST, or microarray, or *spa* typing) become more widespread, we are beginning to notice further epidemiological patterns and associations between certain lineages, toxins, and the infections they cause. This is promoting the collection of well-characterized clinical strains that can be compared by microarray, potentially identifying virulence factors associated with particular types of infection. However, these early studies are also suggesting a key role for host factors in *S. aureus* carriage and pathogenesis, suggesting that the interaction between host and pathogen will be a major area of research in the future.

STAPHYLOCOCCUS EPIDERMIDIS

S. epidermidis is a member of the CoNS and is by far the most common staphylococcus found on human skin. It is an opportunistic pathogen. Immunocompromised patients, especially those with indwelling medical devices such as catheters, shunts, and artificial valves and joints, can develop infections. The production of *S. epidermidis* "slime" on plastic devices is thought to be the major virulence factor (71), and these infections generally require removal of the device for successful therapy.

Two *S. epidermidis* isolates have been sequenced (35, 92). ATCC 12228 (accession number AE015929) is a nonpathogenic, non-biofilm-producing isolate that is used for the detection of residual antibiotics in food products (92). It has low virulence in animal mod-

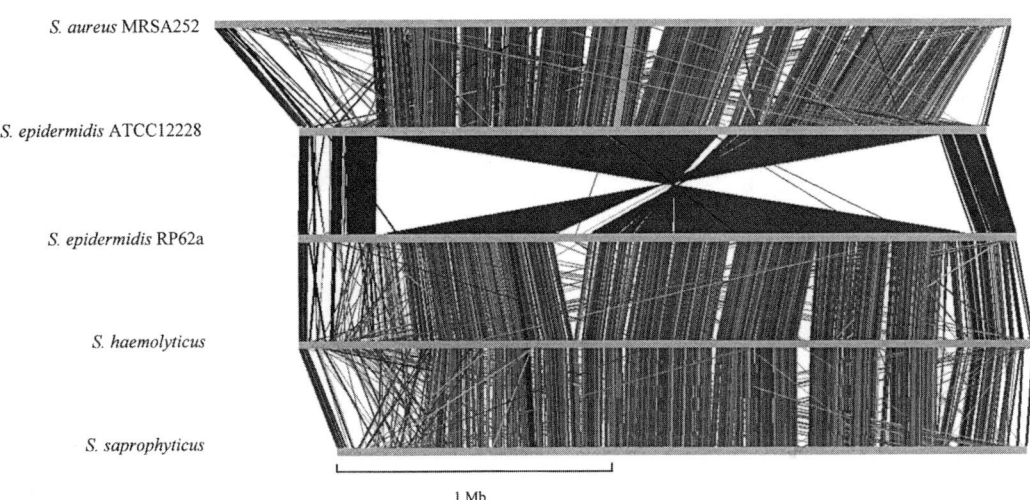

FIGURE 4   Pairwise comparisons of the *S. aureus* MRSA252, *S. epidermidis* ATCC 12228, *S. epidermidis* RP62A, *S. haemolyticus*, and *S. saprophyticus* chromosomes. The comparisons are displayed using the Artemis Comparison Tool (ACT) (14). The sequences have been aligned from the predicted replication origins (*oriC*; right). The gray bars separating each genome represent orthologous matches identified by reciprocal FASTA analysis as previously described for Fig. 2. The intensity of the gray matches indicates the relative identity of the orthologous match; the darker the gray, the higher the identity.

els of infection. This strain is negative for the intracellular adhesin (*ica*) operon that is implicated in the production of biofilm and/or slime. The second strain, RP62A (accession number CP000029) (ATCC 35984), is a well-characterized slime-producing, *ica*-positive isolate that is resistant to methicillin. It was isolated from a case of catheter-associated sepsis in Tennessee in 1979–1980 (35).

The core genomes of *S. epidermidis* and *S. aureus* are related and have a conserved gene order (Fig. 4). In general, the DNA-DNA similarity is higher within coding regions than intergenic regions. Homology between orthologous gene products varies widely from 100% identity down to borderline homology, >30% identity. The average identity is 76.3% at the protein level. *S. epidermidis* is missing genes for a number of *S. aureus*–specific virulence characteristics such as protein A, coagulase, alpha- and gamma-hemolysin, siderophore production, and *clfA*, and there are differences in many metabolic pathways. Most *S. aureus* CV genes are variants or not found in *S. epidermidis*, including many surface proteins, capsule, enterotoxin, exotoxin,

leukocidin, and regulatory genes. *S. epidermidis* has extra copies of small cytokine stimulating peptides, which could play a role in virulence, as well as two cell wall–associated biofilm proteins, *bap* and *bhp*. Notably, the two *S. epidermidis* strains sequenced thus far are missing the νSaα and νSaβ genomic islands that contain high concentrations of virulence determinant genes in *S. aureus*.

The MGEs in *S. epidermidis* ATCC 12228 include six free plasmids (pSE-12228-01 to pSE-12228-06), two integrated plasmids (νSe1 and νSe2), and a novel SCC*pbp4* element that lacks *mec* but carries penicillin-binding protein 4 (*pbp4*) (61). In addition to the intact SCC element, there also appeared to be the remnants of other SCCs. Next to the SCC*pbp4* element is a composite ACME island that contains genes encoding resistance to mercury and cadmium, as well as a variant of the ACME (ACME II) described in *S. aureus* USA300 (ACME I) (23). As in FPR3757, the *arc* gene cluster encoding arginine catabolism is flanked by IS elements and constitutes a putative catabolic composite transposon (23).

In RP62A, the MGEs include a novel prophage (φSPβ), and an integrated type I plasmid (νSe1) and a free plasmid (pSERP) carrying drug resistance genes. This strain also has an SCC*mec*II element that is very closely related to elements found in some strains of *S. aureus*, supporting the hypothesis that SCC*mec* regions originated in CoNS and moved to *S. aureus*. Aside from the high number of plasmids, the sequenced *S. epidermidis* generally carry fewer MGEs than *S. aureus*.

An MLST scheme for *S. epidermidis* has been developed (http://sepidermidis.mlst.net/), and it suggests that most strains belong to one large clonal cluster or lineage (85, 88) with many STs. Interestingly, ST27 contains the majority of clinical isolates. The presence of a key virulence operon (*ica*) is found in only some strains of *S. epidermidis*, including all ST27, but also some other STs (47). ST27 also contains the majority of methicillin-resistant *S. epidermidis*, and most have an SCC*mec* related to those found in *S. aureus* (88). This suggests that certain ST types may act as a repository for MGEs encoding virulence and resistance genes, and these cause the majority of infections. Furthermore, an *S. epidermidis* microarray based on the RP26A genome has been constructed (90). By comparative genomics, genes found at greater frequency in clinical isolates (*n* = 22) versus commensal isolates (*n* = 20) included *ica* and a hemagglutinin binding protein (91).

STAPHYLOCOCCUS HAEMOLYTICUS

S. haemolyticus is a CoNS that frequently colonizes human skin and mucous membranes. After *S. epidermidis*, *S. haemolyticus* is the second most common cause of infections due to CoNS, and it is often resistant to multiple antibiotics. Strain JCSC1435 was isolated from a patient in Japan in 2000 (accession number AP006716) and was sequenced by Takeuchi et al. (78).

The *S. haemolyticus* genome is missing most of the *S. aureus* virulence factor genes such as coagulase, DNase, and enterotoxins, as well as some genes found in both *S. aureus* and *S. epidermidis* such as urease, protease, and toxin regulators. Novel *S. haemolyticus* genes encoding potential virulence factors include several hemolysins.

A striking feature of the *S. haemolyticus* genome is the large number of MGEs. This strain carries three free plasmids (pSHaeA, pSHaeB, and pSHaeC), two integrated plasmids (πSh1 and πSh2), two prophages (φSh1 and φSh2), an intact SCC (SCC*h1435*) and remnants of several SCCs, and three genomic islands distinct from those in *S. aureus* (νSh1, νSh2, and νSh3). It also carries a large number of intact IS (60) and transposons (Tn*552* ×1; Tn*554* ×3; Tn*554*-like ×1). *S. haemolyticus* are generally more resistant to antibiotics than other staphylococci, and it could be that they accumulate more of these genes than other species.

STAPHYLOCOCCUS SAPROPHYTICUS

S. saprophyticus is unique among the staphylococci as a common cause of uncomplicated urinary tract infection in women of childbearing age. The strain sequenced, ATCC 15305 (accession number AP008934), was isolated from a human urine specimen before 1940 and sequenced by Kuroda et al. (50).

The *S. saprophyticus* genome is missing many *S. aureus* virulence genes, such as coagulase, hemolysins, many enterotoxins and exoenzymes, and most extracellular matrix-binding surface proteins. The virulence gene contents of the genome would appear to suggest that this organism has refined its virulence arsenal to its specific pathological niche. *S. saprophyticus* has extra ion transport genes and high urease activity that may contribute to survival in urine and the urinary tract. Most striking is that *S. saprophyticus* encodes only a single cell wall–anchored protein, compared with the dozens found in *S. aureus*. This large protein, uroadherence factor A, UafA, was subsequently shown to have hemagglutination properties and to be responsible for *S. saprophyticus* adherence to human bladder cells (50). Thus, the sequencing project has directly identified the key novel interaction between *S. saprophyticus* and the human urinary tract. This could represent a target for future therapy.

Like the other staphylococci, the genome contains a diverse collection of MGEs. These include two SCC elements, a novel SCC with genes encoding a new capsule type ($SCC_{15305cap}$), and an SCC carrying a restriction-modification system ($SCC_{15305RM}$); a prophage remnant; two free plasmids (pSSP1 and pSSP2); and an unusual genomic island (νSs_{15305}). Unlike the genomic islands in other staphylococci that are often associated with virulence, this island is associated with drug resistance. νSs_{15305} contains 26 genes, confers resistance to streptomycin and fosfomycin, and has an integrase related to the phage integrases.

THE *STAPHYLOCOCCUS* CORE GENOME

There is clearly variation in the staphylococcal genomes, but which genes are common to multiple species? The majority of the core *Staphylococcus* genome is composed of orthologous genes that have been inherited from a common ancestor and that encode proteins that have comparable functions. Unsurprisingly, the number of orthologs in related genomes mirrors the taxonomic relationships of the strains; intraspecies comparisons yield more orthologs matches than interspecies or intergenus comparisons. For example, 2,401 orthologs were identified in a comparison of *S. aureus* strains (MRSA252 versus MSSA476) (Fig. 3), as opposed to 1,850 for *S. aureus* versus *S. epidermidis* (MRSA252 versus RP62A), 1,862 for *S. aureus* versus *S. haemolyticus*, and 1,773 for *S. aureus* versus *S. saprophyticus* (Fig. 5A). A comparison of the functional distributions of these orthologous CDSs in the different species reveals some interesting differences (Fig. 5B). As a percentage of the total orthologous matches, *S. epidermidis* shares more hypothetical proteins with *S. aureus* than the other staphylococci. *S. saprophyticus*, by comparison, shares more transport proteins, proteins associated with small-molecule degradation, and amino acid biosynthesis proteins. Interestingly, there has been remarkable expansion of transport systems proteins in the variable genome of *S. saprophyticus*, which is thought to be an

adaptation to the urinary environment in which this organism grows (50). *S. haemolyticus*, on the other hand, contain a greater percentage of cell envelope proteins, laterally transferred element proteins, and regulators. Identification of the core genes common to all staphylococci will require additional species to be sequenced.

GENOME ORGANIZATION OF THE STAPHYLOCOCCI

Inversions

The gross architecture of the chromosome is not conserved across the genus. The chromosomes of *S. epidermidis* RP62A, *S. haemolyticus*, and *S. saprophyticus* (but not *S. epidermidis* ATCC 12228) have a conserved structure, but relative to the *S. aureus* chromosomes, they have large genomic inversions near the origin of replication (*oriC*). The *S. epidermidis* ATCC 12228 chromosome, in contrast, contains a large genomic inversion compared with *S. epidermidis* RP62A, resulting in a similar conserved structure to that of the *S. aureus* chromosomes (Fig. 4). A comparison of the two *S. epidermidis* genomes suggests that the rearrangement is due to reciprocal recombination between identical IS elements on opposite strands of the replication fork. The breakpoint of the large inversion in ATCC 12228 is flanked by two ISSep2 elements (SE_0282 and SE_2260). In RP62A, a single ISSep2 element is found at one edge of the inversion site (SERP0160), and the other edge contains the *ica* operon. Notably, the *ica* operon is absent in ATCC 12228. It is therefore tempting to speculate that this region may have been deleted by recombination between ISSep2 elements flanking this region, which subsequently recombined with the IS element on the opposite replicore to generate the large chromosomal inversion. The breakpoint of this IS element-mediated recombination is at a different site in the chromosome to the breakpoint associated with the *S. epidermidis* RP62A–*S. aureus* (and *S. haemolyticus*, *S. saprophyticus*) inversion. The region of the *S. epidermidis* ATCC 12228 chromosome between the IS-mediated

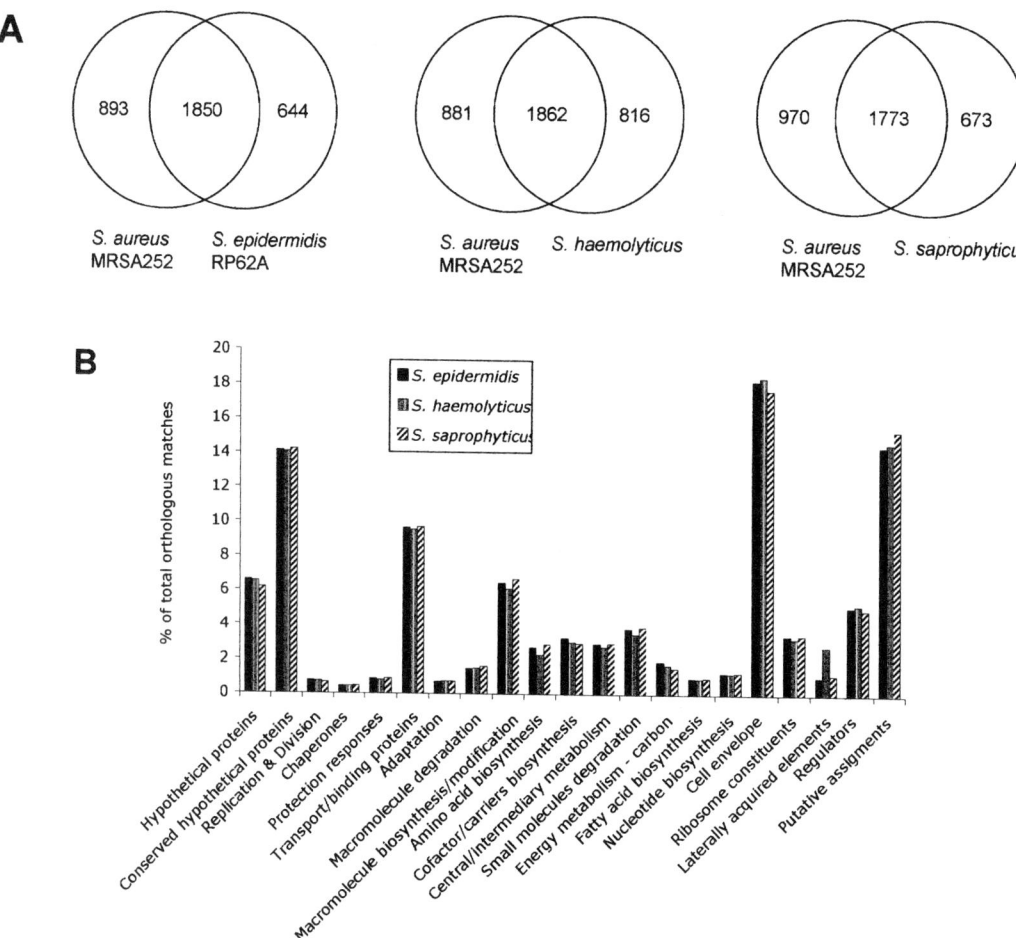

FIGURE 5 Distribution of orthologous CDSs in staphylococci. (A) Circular diagrams show the number of genes unique or shared between *S. aureus* MRSA252 and *S. epidermidis* RP62A; *S. aureus* MRSA252 and *S. haemolyticus*; and *S. aureus* MRSA252 and *S. saprophyticus*. (B) Functional distribution of orthologous CDSs. Orthologous matches were identified as previously described for Fig. 2.

inversion breakpoint and the origin of replication is in the same orientation as the other non-*S. aureus* species. This suggests that the IS-mediated inversion is a more recent event than the one that differentiates the *S. aureus* chromosomes from those of the other staphylococcal chromosomes.

oriC Environ

Although MGEs appear to insert throughout the staphylococcal chromosomes, a region near to the origin of replication appears to be particularly prone to the trafficking of MGEs and genetic rearrangement (Fig. 4). Comparisons of the chromosomes of the various species identified a region downstream of the *oriC* that contains the lowest density of orthologous genes in the chromosome. Takeuchi et al. found that many species-specific genes are contained within this region and designated it the "*oriC* environ" (78). Notably, this region contains the SCC insertion site and contains intact or partial copies of variants of this important mobile cassette. In *S. aureus*, this region includes many of the species-specific virulence genes (*spa*, coagulase, capsule), as well as the SCC element insertion site. In *S. epidermidis*, the region includes the *ica* genes (in *S. aureus*, orthologs of *ica* genes are to be found outside the *oriC* environ). In *S. haemolyticus*, the *oriC* region includes a novel hemagglutinin protein, capsule cluster, and several immunodominant antigens, whereas in *S. saprophyticus*, the *uafA* gene is in the *oriC* region. The density of MGEs carrying advantageous genes in the *oriC* environ may not be coincidental; there may be selective advantages to their chromosomal location. Studies looking at gene location within bacterial replicores have identified position-dependent changes in expression levels (76). Variation in expression may be linked to an increase in gene dosage associated with replication in regions close to *oriC*. It is therefore tempting to speculate that in the genomes of staphylococci, it would be advantageous for genes associated with strong selective pressures, such as drug resistance, to be located near the origin of replication to maximize their expression. The selective pressures associated with virulence may not be as stringent because of the redundancy of virulence factors and the complexity of the host environment; therefore, the distribution of MGE that carry virulence factors may be more randomly distributed in the replicores.

The genetic diversity of the *oriC* environ makes this an attractive target for diagnostic typing because it can be difficult to reliably discriminate among staphylococcal species by biochemical methods. In *S. haemolyticus*, the *oriC* environ is genetically unstable; spontaneous genomic rearrangements were identified by pulsed-field gel electrophoresis and PCR in this region and shown to be due to recombination between IS elements (78). Interestingly, *S. haemolyticus* contains a relatively high number of IS elements (a total of 82, with 60 of them intact), many of which are repetitive (78) and are distributed throughout the chromosome. Despite this, genetic rearrangements in the *S. haemolyticus* chromosome are confined to the *oriC* environ and have not me-

diated gross chromosomal rearrangement that can accompany IS expansion in some bacterial species, such as *Yersinia pestis* and *Bordetella pertusis* (66, 67). In these cases, the expansion in IS element number and subsequent recombination is thought to be due to the transition through an evolutionary bottleneck that results from a recent change in niche and the subsequent removal of selective pressure on a large number of genes.

CONCLUSIONS

Sequencing has revealed the enormous variability in staphylococcal genome sequences and has identified dozens of potential new virulence pathways. This is perhaps not surprising considering the wide variety of human niches that staphylococci thrive in, as well as the substantial variation in the types of infections caused and the severity of disease. At this time, only 4 of the 10 dominant human lineages of *S. aureus* have been sequenced, and only 4 of 38 species. Each of the sequenced staphylococcal genomes carries genes common to all other sequenced staphylococci, genes specific to particular species, genes specific to particular lineages, and MGEs. The MGEs carry many virulence and resistance genes, can move frequently between isolates, and can potentially move between species, contributing to the evolution of isolates presenting unique clinical challenges. The sequencing projects will lead to improved rapid and accurate diagnostic tools for speciation and lineage identification. As we accurately classify our isolates, we notice that certain staphylococci and/or virulence markers are strongly associated with specific clinical disease traits or outcomes. For some species, these markers have been identified—for example, the hemagglutinin in *S. saprophyticus,* toxic shock syndrome toxin in toxic shock syndrome, and exfoliative toxins in scalded skin syndrome. The identification of more of these markers will lead to improved diagnostic tools and future therapies, as well as a better understanding of staphylococcal pathogenesis and epidemiology. Perhaps most importantly, these studies have strongly implicated host response in the pathogenesis of staphylococcal disease, and the interaction between pathogen and host is where much future research will be directed.

REFERENCES

1. **Anonymous.** 1999. Four pediatric deaths from community-acquired methicillin-resistant *Staphylococcus aureus*: Minnesota and North Dakota, 1997–1999. *Morb. Mortal. Wkly. Rep.* **48:**707–710.
2. **Anonymous.** 2002. *Staphylococcus aureus* resistant to vancomycin—United States, 2002. *Morb. Mortal. Wkly. Rep.* **51:**565–567.
3. **Anonymous.** 2004. Vancomycin resistant *Staphylococcus aureus*—New York, 2004. *Morb. Mortal. Wkly. Rep.* **53:**322–323.
4. **Anonymous.** 2002. Vancomycin-resistant *Staphylococcus aureus*—Pennsylvania, 2002. *Morb. Mortal. Wkly. Rep.* **51:**902.
5. **Baba, T., F. Takeuchi, M. Kuroda, H. Yuzawa, K. Aoki, A. Oguchi, Y. Nagai, N. Iwama, K. Asano, T. Naimi, H. Kuroda, L. Cui, K. Yamamoto, and K. Hiramatsu.** 2002. Genome and virulence determinants of high virulence community-acquired MRSA. *Lancet* **359:**1819–1827.
6. **Bae, T., A. K. Banger, A. Wallace, E. M. Glass, F. Aslund, O. Schneewind, and D. M. Missiakas.** 2004. *Staphylococcus aureus* virulence genes identified by bursa aurealis mutagenesis and nematode killing. *Proc. Natl. Acad. Sci. USA* **101:**12312–12317.
7. **Bannerman, T. L., G. A. Hancock, F. C. Tenover, and J. M. Miller.** 1995. Pulsed-field gel electrophoresis as a replacement for bacteriophage typing of *Staphylococcus aureus*. *J. Clin. Microbiol.* **33:**551–555.
8. **Becker, S. A., and B. O. Palsson.** 2005. Genome-scale reconstruction of the metabolic network in *Staphylococcus aureus* N315: an initial draft to the two-dimensional annotation. *BMC Microbiol.* **5:**8.
9. **Beenken, K. E., P. M. Dunman, F. McAleese, D. Macapagal, E. Murphy, S. J. Projan, J. S. Blevins, and M. S. Smeltzer.** 2004. Global gene expression in *Staphylococcus aureus* biofilms. *J. Bacteriol.* **186:**4665–4684.
10. **Begun, J., C. D. Sifri, S. Goldman, S. B. Calderwood, and F. M. Ausubel.** 2005. *Staphylococcus aureus* virulence factors identified by using a high-throughput *Caenorhabditis elegans*–killing model. *Infect. Immun.* **73:**872–877.
11. **Betley, M. J., and J. J. Mekalanos.** 1985. Staphylococcal enterotoxin A is encoded by phage. *Science* **229:**185–187.
12. **Bohach, G. A., D. J. Fast, R. D. Nelson, and P. M. Schlievert.** 1990. Staphylococcal and streptococcal pyrogenic toxins involved in toxic shock syndrome and related illnesses. *Crit. Rev. Microbiol.* **17:**251–272.
13. **Buckling, A., J. Neilson, J. Lindsay, R. ffrench-Constant, M. Enright, N. Day, and R. C. Massey.** 2005. Clonal distribution and phase-variable expression of a major histocompatibility complex analogue protein in *Staphylococcus aureus*. *J. Bacteriol.* **187:**2917–2919.
14. **Carver, T. J., K. Rutherford, M. Berriman, M. A. Rajandream, B. Barrell, and J. Parkhill.** 2005. ACT: the Artemis comparison tool. *Bioinformatics* **21:**3422–3423.
15. **Chambers, H. F.** 2005. Community-associated MRSA—resistance and virulence converge. *N. Engl. J. Med.* **352:**1485–1487.
16. **Chang, S., D. M. Sievert, J. C. Hageman, M. L. Boulton, F. C. Tenover, F. P. Downes, S. Shah, J. T. Rudrik, G. R. Pupp, W. J. Brown, D. Cardo, and S. K. Fridkin.** 2003. Infection with vancomycin-resistant *Staphylococcus aureus* containing the *vanA* resistance gene. *N. Engl. J. Med.* **348:**1342–1347.
17. **Chesneau, O., A. Morvan, F. Grimont, H. Labischinski, and N. el Solh.** 1993. *Staphylococcus pasteuri* sp. nov., isolated from human, animal, and food specimens. *Int. J. Syst. Bacteriol.* **43:**237–244.
18. **Clarke, S. R., L. G. Harris, R. G. Richards, and S. J. Foster.** 2002. Analysis of Ebh, a 1.1-megadalton cell wall–associated fibronectin-binding protein of *Staphylococcus aureus*. *Infect. Immun.* **70:**6680–6687.
19. **Cole, J. R., B. Chai, R. J. Farris, Q. Wang, S. A. Kulam, D. M. McGarrell, G. M. Garrity, and J. M. Tiedje.** 2005. The Ribosomal Database Project (RDP-II): sequences and tools for high-throughput rRNA analysis. *Nucleic Acids Res.* **33:**D294–D296.
20. **Coulter, S. N., W. R. Schwan, E. Y. Ng, M. H. Langhorne, H. D. Ritchie, S. Westbrock-Wadman, W. O. Hufnagle, K. R. Folger, A. S. Bayer, and C. K. Stover.** 1998. *Staphylococcus aureus* genetic loci impacting growth and survival in multiple infection environments. *Mol. Microbiol.* **30:**393–404.
21. **Cui, L., A. Iwamoto, J. Q. Lian, H. M. Neoh, T. Maruyama, Y. Horikawa, and K. Hiramatsu.** 2006. Novel mechanism of antibiotic resistance originating in vancomycin-intermediate *Staphylococcus aureus*. *Antimicrob. Agents Chemother.* **50:**428–438.
22. **Cui, L., X. Ma, K. Sato, K. Okuma, F. C. Tenover, E. M. Mamizuka, C. G. Gemmell, M. N. Kim, M. C. Ploy, N. El-Solh, V. Ferraz,**

and K. Hiramatsu. 2003. Cell wall thickening is a common feature of vancomycin resistance in *Staphylococcus aureus. J. Clin. Microbiol.* **41:**5–14.

23. Diep, B. A., S. R. Gill, R. F. Chang, T. H. Phan, J. H. Chen, M. G. Davidson, F. Lin, J. Lin, H. A. Carleton, E. F. Mongodin, G. F. Sensabaugh, and F. Perdreau-Remington. 2006. Complete genome sequence of USA300, an epidemic clone of community-acquired methicillin-resistant *Staphylococcus aureus. Lancet* **367:** 731–739.

24. Enright, M. C., N. P. J. Day, C. E. Davies, S. J. Peacock, and B. G. Spratt. 2000. Multilocus sequence typing for characterization of methicillin-resistant and methicillin-susceptible clones of *Staphylococcus aureus. J. Clin. Microbiol.* **38:**1008–1015.

25. Enright, M. C., D. A. Robinson, G. Randle, E. J. Feil, H. Grundmann, and B. G. Spratt. 2002. The evolutionary history of methicillin-resistant *Staphylococcus aureus* (MRSA). *Proc. Natl. Acad. Sci. USA* **99:**7687–7692.

26. Euzeby, J. P. 1997. List of bacterial names with standing in nomenclature: a folder available on the Internet. *Int. J. Syst. Bacteriol.* **47:**590–592.

27. Feil, E. J., J. E. Cooper, H. Grundmann, D. A. Robinson, M. C. Enright, T. Berendt, S. J. Peacock, J. M. Smith, M. Murphy, B. G. Spratt, C. E. Moore, and N. P. J. Day. 2003. How clonal is *Staphylococcus aureus? J. Bacteriol.* **185:**3307–3316.

28. Felsenstein, J. 1985. Confidence limits on phylogenies: an approach using the bootstrap. *Evolution* **39:**783–791.

29. Felsenstein, J. 1989. PHYLIP—Phylogeny Inference Package (version 3.2). *Cladistics* **5**.

30. Fitzgerald, J. R., S. R. Monday, T. J. Foster, G. A. Bohach, P. J. Hartigan, W. J. Meaney, and C. J. Smyth. 2001. Characterization of a putative pathogenicity island from bovine *Staphylococcus aureus* encoding multiple superantigens. *J. Bacteriol.* **183:**63–70.

31. Fitzgerald, J. R., D. E. Sturdevant, S. M. Mackie, S. R. Gill, and J. M. Musser. 2001. Evolutionary genomics of *Staphylococcus aureus*: insights into the origin of methicillin-resistant strains and the toxic shock syndrome epidemic. *Proc. Natl. Acad. Sci. USA* **98:**8821–8826.

32. Forehand, J. R., and R. B. Johnston, Jr. 1994. Chronic granulomatous disease: newly defined molecular abnormalities explain disease variability and normal phagocyte physiology. *Curr. Opin. Pediatr.* **6:**668–675.

33. Foster, G., H. M. Ross, R. A. Hutson, and M. D. Collins. 1997. *Staphylococcus lutrae* sp. nov., a new coagulase–positive species isolated from otters. *Int. J. Syst. Bacteriol.* **47:**724–726.

34. Freney, J., W. E. Kloos, V. Hajek, J. A. Webster, M. Bes, Y. Brun, and C. Vernozy-Rozand. 1999. Recommended minimal standards for description of new staphylococcal species. Subcommittee on the taxonomy of staphylococci and streptococci of the International Committee on Systematic Bacteriology. *Int. J. Syst. Bacteriol.* **49** (Pt. 2):489–502.

35. Gill, S. R., D. E. Fouts, G. L. Archer, E. F. Mongodin, R. T. DeBoy, J. Ravel, I. T. Paulsen, J. F. Kolonay, L. Brinkac, M. Beanan, R. J. Dodson, S. C. Daugherty, R. Madupu, S. V. Angiuoli, A. S. Durkin, D. H. Haft, J. Vamathevan, H. Khouri, T. Utterback, C. Lee, G. Dimitrov, L. X. Jiang, H. Y. Qin, J. Weidman, K. Tran, K. Kang, I. R. Hance, K. E. Nelson, and C. M. Fraser. 2005. Insights on evolution of virulence and resistance from the complete genome analysis of an early methicillin-resistant *Staphylococcus aureus* strain and a biofilm-producing methicillin-resistant *Staphylococcus epidermidis* strain. *J. Bacteriol.* **187:**2426–2438.

36. Gillaspy, A. F., V. Worrell, J. Orvis, B. A. Roe, D. W. Dyer, and J. J. Iandolo. 2006. The *Staphylococcus aureus* NCTC8325 genome, p. 381–412. *In* V. Fischetti, R. Novick, J. Ferretti, D. Portnoy, and J. Rood (ed.), *Gram-Positive Pathogens*. ASM Press, Washington, D.C.

37. Goerke, C., S. M. Y. Papenberg, S. Dasbach, K. Dietz, R. Ziebach, B. C. Kahl, and C. Wolz. 2004. Increased frequency of genomic alterations in *Staphylococcus aureus* during chronic infection is in part due to phage mobilization. *J. Infect. Dis.* **189:**724–734.

38. Heinemann, M., A. Kummel, R. Ruinatscha, and S. Panke. 2005. In silico genome-scale reconstruction and validation of the *Staphylococcus aureus* metabolic network. *Biotechnol. Bioeng.* **92:**850–864.

39. Holden, M. T. G., E. J. Feil, J. A. Lindsay, S. J. Peacock, N. P. J. Day, M. C. Enright, T. J. Foster, C. E. Moore, L. Hurst, R. Atkin, A. Barron, N. Bason, S. D. Bentley, C. Chillingworth, T. Chillingworth, C. Churcher, L. Clark, C. Corton, A. Cronin, J. Doggett, L. Dowd, T. Feltwell, Z. Hance, B. Harris, H. Hauser, S. Holroyd, K. Jagels, K. D. James, N. Lennard, A. Line, R. Mayes, S. Moule, K. Mungall, D. Ormond, M. A. Quail, E. Rabbinowitsch, K. Rutherford, M. Sanders, S. Sharp, M. Simmonds, K. Stevens, S. Whitehead, B. G. Barrell, B. G. Spratt, and J. Parkhill. 2004. Complete genomes of two clinical *Staphylococcus aureus* strains: evidence for the rapid evolution of virulence and drug resistance. *Proc. Natl. Acad. Sci. USA* **101:**9786–9791.

40. **Jensen, M. A., J. A. Webster, and N. Straus.** 1993. Rapid identification of bacteria on the basis of polymerase chain reaction-amplified ribosomal DNA spacer polymorphisms. *Appl. Environ. Microbiol.* **59:**945–952.

41. **Johnson, A. P., H. M. Aucken, S. Cavendish, M. Ganner, M. C. J. Wale, M. Warner, D. M. Livermore, and B. D. Cookson.** 2001. Dominance of EMRSA-15 and -16 among MRSA causing nosocomial bacteraemia in the United Kingdom: analysis of isolates from the European Antimicrobial Resistance Surveillance System (EARSS). *J. Antimicrob. Chemother.* **48:** 143–144.

42. **Jones, R. N.** 2003. Global epidemiology of antimicrobial resistance among community-acquired and nosocomial pathogens: a five-year summary from the SENTRY Antimicrobial Surveillance Program (1997–2001). *Semin. Respir. Crit. Care Med.* **24:**121–134.

43. **Kazakova, S. V., J. C. Hageman, M. Matava, A. Srinivasan, L. Phelan, B. Garfinkel, T. Boo, S. McAllister, J. Anderson, B. Jensen, D. Dodson, D. Lonsway, L. K. McDougal, M. Arduino, V. J. Fraser, G. Killgore, F. C. Tenover, S. Cody, and D. B. Jernigan.** 2005. A clone of methicillin-resistant *Staphylococcus aureus* among professional football players. *N. Engl. J. Med.* **352:**468–475.

44. **Kloos, W. E., and T. L. Bannerman.** 1994. Update on clinical significance of coagulase-negative staphylococci. *Clin. Microbiol. Rev.* **7:**117–140.

45. **Kluytmans, J., A. vanBelkum, and H. Verbrugh.** 1997. Nasal carriage of *Staphylococcus aureus*: epidemiology, underlying mechanisms, and associated risks. *Clin. Microbiol. Rev.* **10:**505–520.

46. **Koessler, T., P. Francois, Y. Charbonnier, A. Huyghe, M. Bento, S. Dharan, G. Renzi, D. Lew, S. Harbarth, D. Pittet, and J. Schrenzel.** 2006. Use of oligoarrays for characterization of community-onset methicillin-resistant *Staphylococcus aureus*. *J. Clin. Microbiol.* **44:**1040–1048.

47. **Kozitskaya, S., M. E. Olson, P. D. Fey, W. Witte, K. Ohlsen, and W. Ziebuhr.** 2005. Clonal analysis of *Staphylococcus epidermidis* isolates carrying or lacking biofilm-mediating genes by multilocus sequence typing. *J. Clin. Microbiol.* **43:**4751–4757.

48. **Kuroda, M., H. Kuroda, T. Oshima, F. Takeuchi, H. Mori, and K. Hiramatsu.** 2003. Two-component system VraSR positively modulates the regulation of cell-wall biosynthesis pathway in *Staphylococcus aureus*. *Mol. Microbiol.* **49:** 807–821.

49. **Kuroda, M., T. Ohta, I. Uchiyama, T. Baba, H. Yuzawa, I. Kobayashi, L. Z. Cui, A. Oguchi, K. Aoki, Y. Nagai, J. Q. Lian, T. Ito,** M. Kanamori, H. Matsumaru, A. Maruyama, H. Murakami, A. Hosoyama, Y. Mizutani-Ui, N. K. Takahashi, T. Sawano, R. Inoue, C. Kaito, K. Sekimizu, H. Hirakawa, S. Kuhara, S. Goto, J. Yabuzaki, M. Kanehisa, A. Yamashita, K. Oshima, K. Furuya, C. Yoshino, T. Shiba, M. Hattori, N. Ogasawara, H. Hayashi, and K. Hiramatsu. 2001. Whole genome sequencing of meticillin-resistant *Staphylococcus aureus*. *Lancet* **357:**1225–1240.

50. **Kuroda, M., A. Yamashita, H. Hirakawa, M. Kumano, K. Morikawa, M. Higashide, A. Maruyama, Y. Inose, K. Matoba, H. Toh, S. Kuhara, M. Hattori, and T. Ohta.** 2005. Whole genome sequence of *Staphylococcus saprophyticus* reveals the pathogenesis of uncomplicated urinary tract infection. *Proc. Natl. Acad. Sci. USA* **102:**13272–13277.

51. **Lina, G., Y. Piemont, F. Godail-Gamot, M. Bes, M. O. Peter, V. Gauduchon, F. Vandenesch, and J. Etienne.** 1999. Involvement of Panton-Valentine leukocidin-producing *Staphylococcus aureus* in primary skin infections and pneumonia. *Clin. Infect. Dis.* **29:**1128–1132.

52. **Lindsay, J. A., and M. T. G. Holden.** 2004. *Staphylococcus aureus*: superbug, super genome? *Trends Microbiol.* **12:**378–385.

53. **Lindsay, J. A., C. E. Moore, N. P. Day, S. J. Peacock, A. A. Witney, R. A. Stabler, S. E. Husain, P. D. Butcher, and J. Hinds.** 2006. Microarrays reveal that each of the ten dominant lineages of *Staphylococcus aureus* has a unique combination of surface-associated and regulatory genes. *J. Bacteriol.* **188:**669–676.

54. **Lindsay, J. A., A. Ruzin, H. F. Ross, N. Kurepina, and R. P. Novick.** 1998. The gene for toxic shock toxin is carried by a family of mobile pathogenicity islands in *Staphylococcus aureus*. *Mol. Microbiol.* **29:**527–543.

55. **Liolios, K., N. Tavernarakis, P. Hugenholtz, and N. C. Kyrpides.** 2006. The Genomes On Line Database (GOLD) v.2: a monitor of genome projects worldwide. *Nucleic Acids Res.* **34:**D332–D334.

56. **Lowe, A. M., D. T. Beattie, and R. L. Deresiewicz.** 1998. Identification of novel staphylococcal virulence genes by in vivo expression technology. *Mol. Microbiol.* **27:**967–976.

57. **Malachowa, N., A. Sabat, M. Gniadkowski, J. Krzyszton-Russjan, J. Empel, J. Miedzobrodzki, K. Kosowska-Shick, P. C. Appelbaum, and W. Hryniewicz.** 2005. Comparison of multiple-locus variable-number tandem-repeat analysis with pulsed-field gel electrophoresis, *spa* typing, and multilocus sequence typing for clonal characterization of *Staphylococcus aureus* isolates. *J. Clin. Microbiol.* **43:**3095–3100.

58. **McDougal, L. K., C. D. Steward, G. E. Kill-gore, J. M. Chaitram, S. K. McAllister, and F. C. Tenover.** 2003. Pulsed-field gel electrophoresis typing of oxacillin-resistant *Staphylococcus aureus* isolates from the United States: establishing a national database. *J. Clin. Microbiol.* **41:**5113–5120.

59. **Mei, J. M., F. Nourbakhsh, C. W. Ford, and D. W. Holden.** 1997. Identification of *Staphylococcus aureus* virulence genes in a murine model of bacteraemia using signature-tagged mutagenesis. *Mol. Microbiol.* **26:**399–407.

60. **Melish, M. E., and L. A. Glasgow.** 1970. The staphylococcal scalded-skin syndrome. *N. Engl. J. Med.* **282:**1114–1119.

61. **Mongkolrattanothai, K., S. Boyle, T. V. Murphy, and R. S. Daum.** 2004. Novel non-*mecA*-containing staphylococcal chromosomal cassette composite island containing *pbp4* and *tagF* genes in a commensal staphylococcal species: a possible reservoir for antibiotic resistance islands in *Staphylococcus aureus*. *Antimicrob. Agents Chemother.* **48:**1823–1836.

62. **Moore, P. C. L., and J. A. Lindsay.** 2001. Genetic variation among hospital isolates of methicillin-sensitive *Staphylococcus aureus*: evidence for horizontal transfer of virulence genes. *J. Clin. Microbiol.* **39:**2760–2767.

63. **Nouwen, J., H. Boelens, A. van Belkum, and H. Verbrugh.** 2004. Human factor in *Staphylococcus aureus* nasal carriage. *Infect. Immun.* **72:**6685–6688.

64. **Ogston, A.** 1881. Report upon micro-organisms in surgical diseases. *Br. Med. J.* **1:**369–375.

65. **Pan, E. S., B. A. Diep, H. A. Carleton, E. D. Charlebois, G. F. Sensabaugh, B. L. Haller, and F. Perdreau-Remington.** 2003. Increasing prevalence of methicillin-resistant *Staphylococcus aureus* infection in California jails. *Clin. Infect. Dis.* **37:**1384–1388.

66. **Parkhill, J., M. Sebaihia, A. Preston, L. D. Murphy, N. Thomson, D. E. Harris, M. T. G. Holden, C. M. Churcher, S. D. Bentley, K. L. Mungall, A. M. Cerdeno-Tarraga, L. Temple, K. James, B. Harris, M. A. Quail, M. Achtman, R. Atkin, S. Baker, D. Basham, N. Bason, I. Cherevach, T. Chillingworth, M. Collins, A. Cronin, P. Davis, J. Doggett, T. Feltwell, A. Goble, N. Hamlin, H. Hauser, S. Holroyd, K. Jagels, S. Leather, S. Moule, H. Norberczak, S. O'Neil, D. Ormond, C. Price, E. Rabbinowitsch, S. Rutter, M. Sanders, D. Saunders, K. Seeger, S. Sharp, M. Simmonds, J. Skelton, R. Squares, S. Squares, K. Stevens, L. Unwin, S. Whitehead, B. G. Barrell, and D. J. Maskell.** 2003. Comparative analysis of the genome sequences of Bor-*detella pertussis, Bordetella parapertussis* and *Bordetella bronchiseptica. Nat. Genet.* **35:**32–40.

67. **Parkhill, J., B. W. Wren, N. R. Thomson, R. W. Titball, M. T. G. Holden, M. B. Prentice, M. Sebaihia, K. D. James, C. Churcher, K. L. Mungall, S. Baker, D. Basham, S. D. Bentley, K. Brooks, A. M. Cerdeno-Tarraga, T. Chillingworth, A. Cronin, R. M. Davies, P. Davis, G. Dougan, T. Feltwell, N. Hamlin, S. Holroyd, K. Jagels, A. V. Karlyshev, S. Leather, S. Moule, P. C. F. Oyston, M. Quail, K. Rutherford, M. Simmonds, J. Skelton, K. Stevens, S. Whitehead, and B. G. Barrell.** 2001. Genome sequence of *Yersinia pestis*, the causative agent of plague. *Nature* **413:**523–527.

68. **Paulsen, I. T., N. Firth, and R. A. Skurray.** 1997. Resistance to antimicrobial agents other than b-lactams, p. 175–212. *In* K. B. Crossley and G. L. Archer (ed.), *The Staphylococci in Human Disease.* Churchill Livingstone, New York, N.Y.

69. **Pearson, W. R., and D. J. Lipman.** 1988. Improved tools for biological sequence comparison. *Proc. Natl. Acad. Sci. USA* **85:**2444–2448.

70. **Perriere, G., and M. Gouy.** 1996. WWW-query: an on-line retrieval system for biological sequence banks. *Biochimie* **78:**364–369.

71. **Peters, G., R. Locci, and G. Pulverer.** 1982. Adherence and growth of coagulase-negative staphylococci on surfaces of intravenous catheters. *J. Infect. Dis.* **146:**479–482.

72. **Phillips, S., and R. P. Novick.** 1979. Tn*554*—site-specific repressor-controlled transposon in *Staphylococcus aureus. Nature* **278:**476–478.

73. **Rosenbach, F. J.** 1884. *Mikroorganismen bei Wundinfektionskrankheiten des Menschen.* Wiesbaden, Germany.

74. **Schlievert, P. M., K. N. Shands, B. B. Dan, G. P. Schmid, and R. D. Nishimura.** 1981. Identification and characterization of an exotoxin from *Staphylococcus aureus* associated with toxic-shock syndrome. *J. Infect. Dis.* **143:**509–516.

75. **Seybold, U., E. V. Kourbatova, J. G. Johnson, S. J. Halvosa, Y. F. Wang, M. D. King, S. M. Ray, and H. M. Blumberg.** 2006. Emergence of community-associated methicillin-resistant *Staphylococcus aureus* USA300 genotype as a major cause of health care-associated blood stream infections. *Clin. Infect. Dis.* **42:**647–656.

76. **Sousa, C., V. de Lorenzo, and A. Cebolla.** 1997. Modulation of gene expression through chromosomal positioning in *Escherichia coli. Microbiology* **143**(Pt. 6):2071–2078.

77. **Stemper, M. E., S. K. Shukla, and K. D. Reed.** 2004. Emergence and spread of community-associated methicillin-resistant *Staphylococcus aureus* in rural Wisconsin, 1989 to 1999. *J. Clin. Microbiol.* **42:**5673–5680.

78. Takeuchi, F., S. Watanabe, T. Baba, H. Yuzawa, T. Ito, Y. Morimoto, M. Kuroda, L. Cui, M. Takahashi, A. Ankai, S. Baba, S. Fukui, J. C. Lee, and K. Hiramatsu. 2005. Whole-genome sequencing of *Staphylococcus haemolyticus* uncovers the extreme plasticity of its genome and the evolution of human-colonizing staphylococcal species. *J. Bacteriol.* **187:**7292–7308.

79. Tenover, F. C., L. K. McDougal, R. V. Goering, G. Killgore, S. J. Projan, J. B. Patel, and P. M. Dunman. 2006. Characterization of a strain of community-associated methicillin-resistant *Staphylococcus aureus* widely disseminated in the United States. *J. Clin. Microbiol.* **44:**108–118.

80. Tettelin, H., V. Masignani, M. J. Cieslewicz, C. Donati, D. Medini, N. L. Ward, S. V. Angiuoli, J. Crabtree, A. L. Jones, A. S. Durkin, R. T. Deboy, T. M. Davidsen, M. Mora, M. Scarselli, I. Margarit y Ros, J. D. Peterson, C. R. Hauser, J. P. Sundaram, W. C. Nelson, R. Madupu, L. M. Brinkac, R. J. Dodson, M. J. Rosovitz, S. A. Sullivan, S. C. Daugherty, D. H. Haft, J. Selengut, M. L. Gwinn, L. Zhou, N. Zafar, H. Khouri, D. Radune, G. Dimitrov, K. Watkins, K. J. O'-Connor, S. Smith, T. R. Utterback, O. White, C. E. Rubens, G. Grandi, L. C. Madoff, D. L. Kasper, J. L. Telford, M. R. Wessels, R. Rappuoli, and C. M. Fraser. 2005. Genome analysis of multiple pathogenic isolates of *Streptococcus agalactiae*: implications for the microbial "pangenome." *Proc. Natl. Acad. Sci. USA* **102:**13950–13955.

81. Thompson, J. D., T. J. Gibson, F. Plewniak, F. Jeanmougin, and D. G. Higgins. 1997. The CLUSTAL_X windows interface: flexible strategies for multiple sequence alignment aided by quality analysis tools. *Nucleic Acids Res.* **25:**4876–4882.

82. Tiemersma, E. W., S. Bronzwaer, O. Lyytikainen, J. E. Degener, P. Schrijnemakers, N. Bruinsma, J. Monen, W. Witte, and H. Grundmann. 2004. Methicillin-resistant *Staphylococcus aureus* in Europe, 1999–2002. *Emerg. Infect. Dis.* **10:**1627–1634.

83. Varaldo, P. E., R. Kilpper-Bälz, F. Biavasco, G. Satta, and K. H. Schleifer. 1988. *Staphylococcus delphini* sp. nov., a coagulase-positive species isolated from dolphins. *Int. J. Syst. Bacteriol.* **38:**436–439.

84. Vernozy-Rozand, C., C. Mazuy, H. Meugnier, M. Bes, Y. Lasne, F. Fiedler, J. Etienne, and J. Freney. 2000. *Staphylococcus fleurettii* sp. nov., isolated from goat's milk cheeses. *Int. J. Syst. Evol. Microbiol.* **50**(Pt 4)**:**1521–1527.

85. Wang, X. M., L. Noble, B. N. Kreiswirth, W. Eisner, W. McClements, K. U. Jansen, and A. S. Anderson. 2003. Evaluation of a multilocus sequence typing system for *Staphylococcus epidermidis*. *J. Med. Microbiol.* **52:**989–998.

86. Webster, J. A., T. L. Bannerman, R. J. Hubner, D. N. Ballard, E. M. Cole, J. L. Bruce, F. Fiedler, K. Schubert, and W. E. Kloos. 1994. Identification of the *Staphylococcus sciuri* species group with *Eco*RI fragments containing rRNA sequences and description of *Staphylococcus vitulus* sp. nov. *Int. J. Syst. Bacteriol.* **44:**454–460.

87. Weigel, L. M., D. B. Clewell, S. R. Gill, N. C. Clark, L. K. McDougal, S. E. Flannagan, J. F. Kolonay, J. Shetty, G. E. Killgore, and F. C. Tenover. 2003. Genetic analysis of a high-level vancomycin-resistant isolate of *Staphylococcus aureus*. *Science* **302:**1569–1571.

88. Wisplinghoff, H., A. E. Rosato, M. C. Enright, M. Noto, W. Craig, and G. L. Archer. 2003. Related clones containing SCC*mec* type IV predominate among clinically significant *Staphylococcus epidermidis* isolates. *Antimicrob. Agents Chemother.* **47:**3574–3579.

89. Witney, A. A., M. Marsden, T. G. Holden, R. A. Stabler, S. E. Husain, J. K. Vass, P. D. Butcher, J. Hinds, and J. A. Lindsay. 2005. Design, validation, and application of a seven-strain *Staphylococcus aureus* PCR product microarray for comparative genomics. *Appl. Environ. Microbiol.* **71:**7504–7514.

90. Yao, Y., D. E. Sturdevant, and M. Otto. 2005. Genomewide analysis of gene expression in *Staphylococcus epidermidis* biofilms: insights into the pathophysiology of *S. epidermidis* biofilms and the role of phenol-soluble modulins in formation of biofilms. *J. Infect. Dis.* **191:**289–298.

91. Yao, Y. F., D. E. Sturdevant, A. Villaruz, L. Xu, Q. Gao, and M. Otto. 2005. Factors characterizing *Staphylococcus epidermidis* invasiveness determined by comparative genomics. *Infect. Immun.* **73:**1856–1860.

92. Zhang, Y. Q., S. X. Ren, H. L. Li, Y. X. Wang, G. Fu, J. Yang, Z. Q. Qin, Y. G. Miao, W. Y. Wang, R. S. Chen, Y. Shen, Z. Chen, Z. H. Yuan, G. P. Zhao, D. Qu, A. Danchin, and Y. M. Wen. 2003. Genome-based analysis of virulence genes in a non-biofilm-forming *Staphylococcus epidermidis* strain (ATCC 12228). *Mol. Microbiol.* **49:**1577–1593.

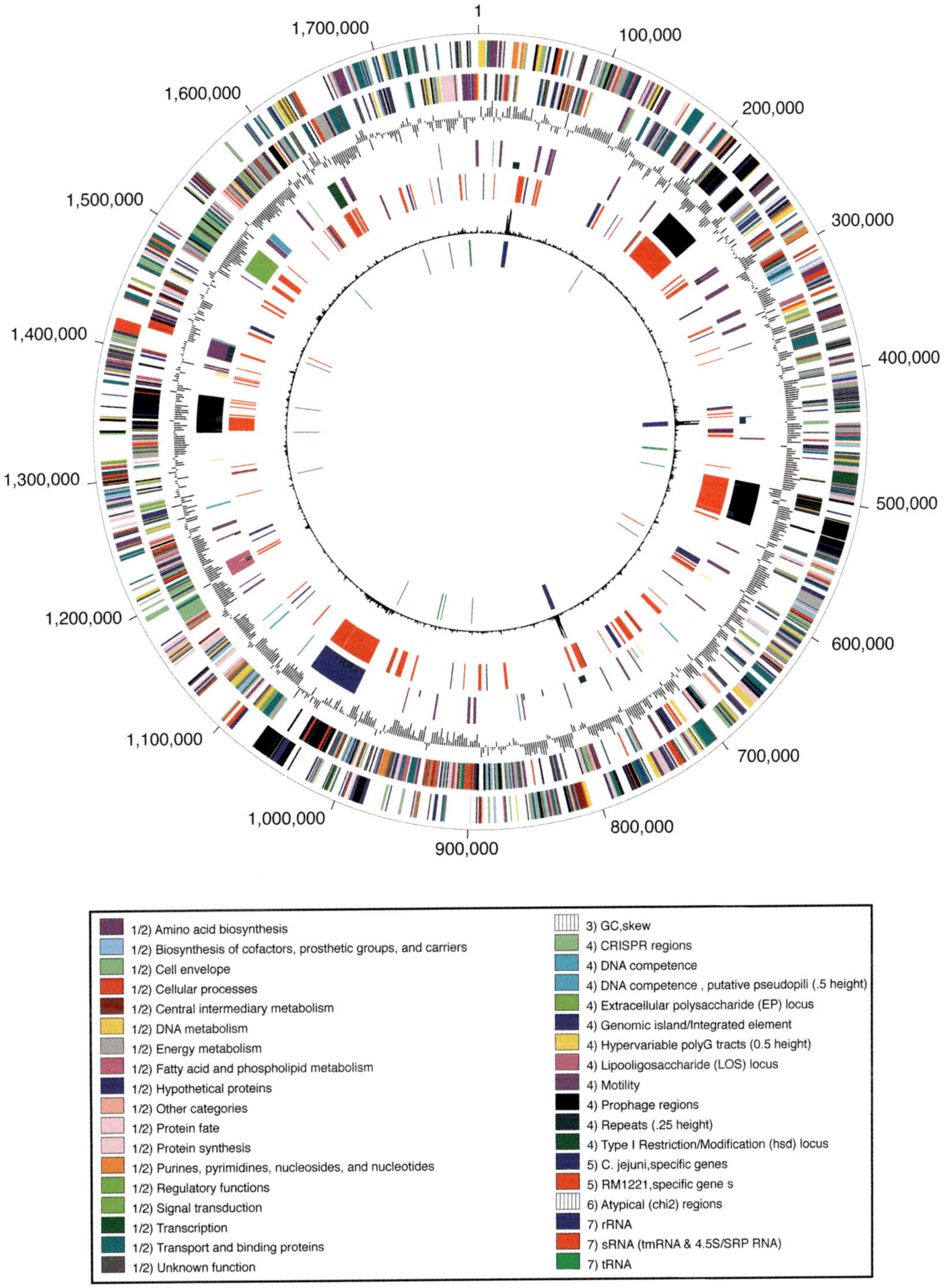

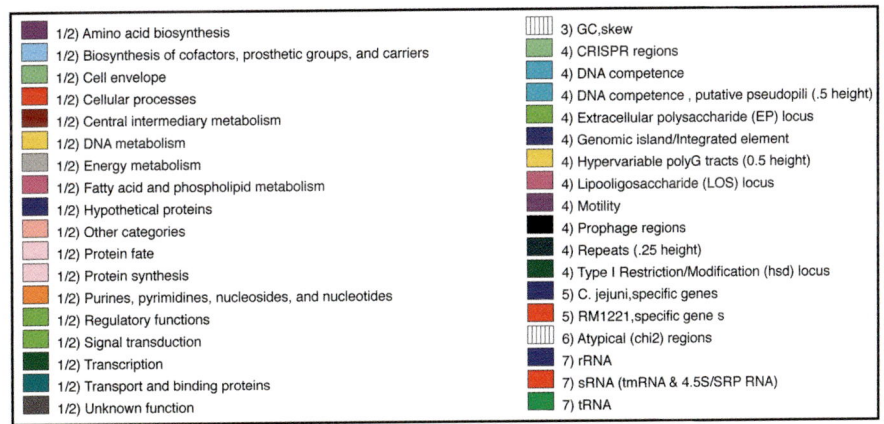

COLOR PLATE 1 Circular representation of closed *C. jejuni* RM1221 genome. Each concentric circle represents genomic data and is numbered from the outermost to the innermost circle. Refer to the key for details on color representations. The first and second circles represent predicted ORFs on the positive (+) and negative (−) strands, respectively. The third circle shows the GC skew. The fourth circle depicts genetic loci of interest: CRISPR (clustered regularly interspaced short palindromic repeats), DNA competence, EP, LOS, prophage and genomic island regions, motility, repeats, and type I restriction/modification regions. The fifth circle demarcates *C. jejuni*-specific and *C. jejuni* RM1221-specific ORFs. The sixth circle plots atypical regions (χ^2 value). The seventh circle denotes tRNA, rRNA, and sRNA (tmRNA and 4.5S RNA) loci.

A.

COLOR PLATE 2 Whole-genome comparison of five *Campylobacter* strains. Line figures were the result of PROmer analysis. (A) Colored lines denote percent identity of protein translations and were plotted according to the location in the reference (*C. jejuni* RM1221, *x* axis) and query (*C. jejuni* NCTC 11168 [upper *y* axis] and *C. coli* RM2228 [lower *y* axis]) genomes. (B and C) The Venn diagrams show the number of proteins shared (black) or unique (red) within a particular relationship for all five *Campylobacter* strains (B) and for members of the sequenced ε-*Proteobacteria* (C) compared in this study. Protein sequences binned as "unique" are unique within the context of the genomes plotted and the cut-offs used to parse the BLASTP data. The pie charts plot the number of protein sequences by main functional role categories for *C. jejuni* RM1221 ORFs. (D) A frequency distribution of protein percent identity was computed: specifically, the number of protein sequences within class intervals of 5% amino acid identity from 35 to 100% that match *C. jejuni* RM1221 reference sequences were plotted.

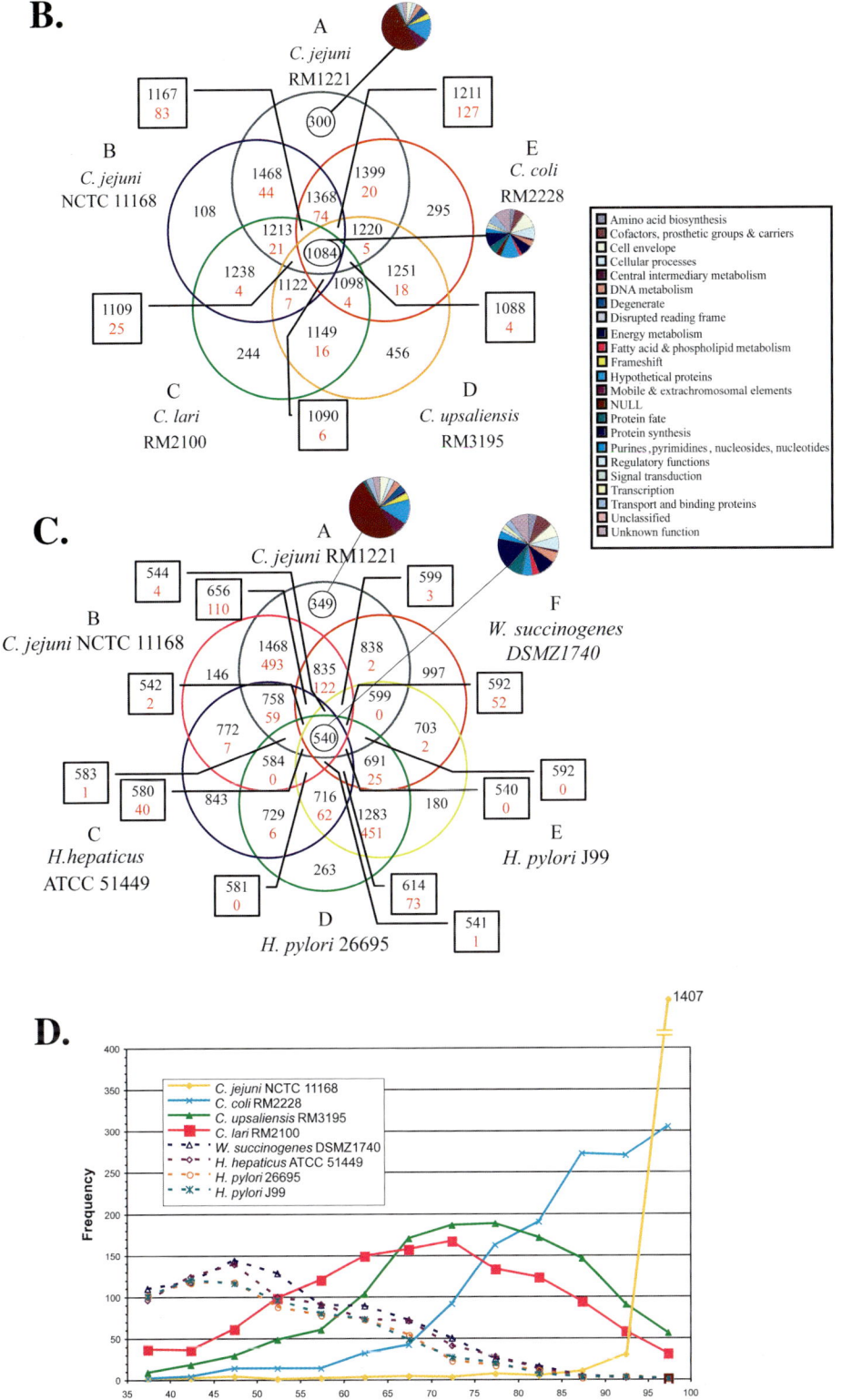

COLOR PLATE 2 *Continued*

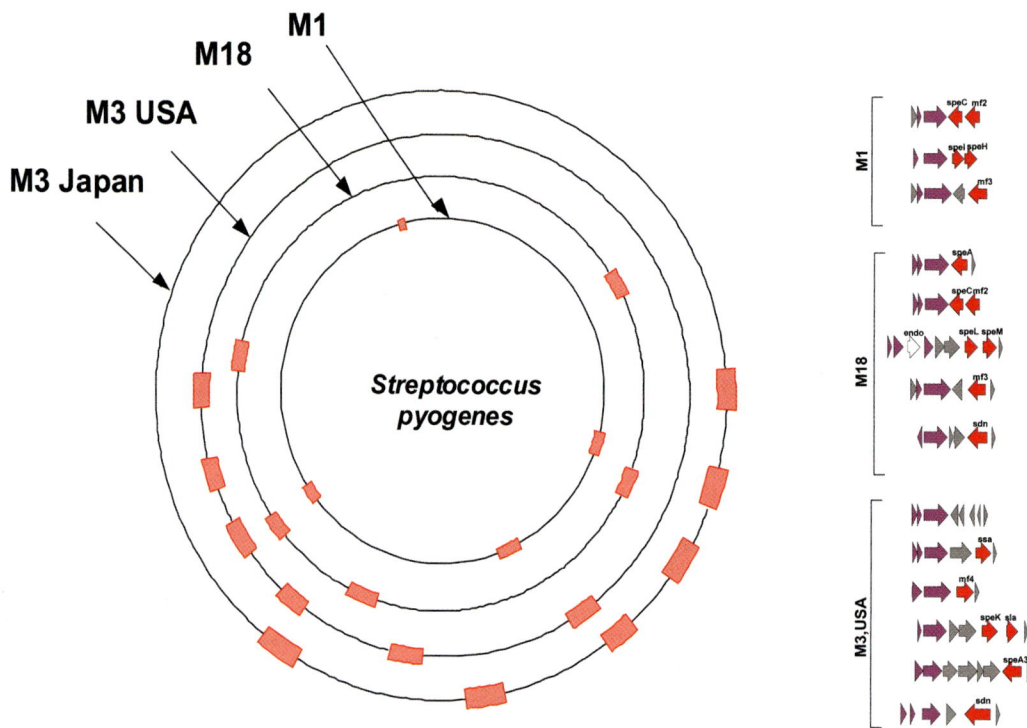

COLOR PLATE 3 Pathogenicity by the permutation principle. (Left) Genome maps of different *S. pyogenes* strains. Red rectangles are prophages (to scale). (Right) Downstream of the lysis genes (violet), virulence genes (red) are found in the vast majority of the prophages.

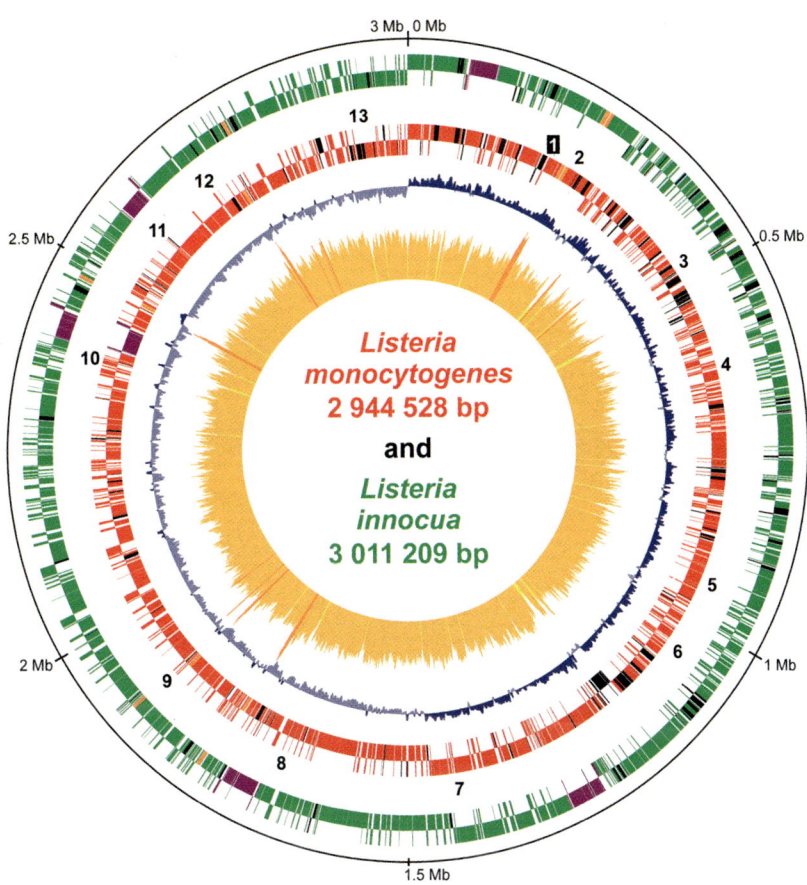

COLOR PLATE 4 Circular genome maps of *L. monocytogenes* EGDe and *L. innocua* CLIP11262 showing the position and orientation of genes. From the outside: circles 1 and 2, *L. innocua* and *L. monocytogenes* genes on the + and − strands, respectively; circle 3, G/C bias (G+C/G−C) of *L. monocytogenes*; circle 4, G+C content of *L. monocytogenes* with higher G+C content indicated by longer bars. The scale in megabases is indicated on the outside of the genomes with the origin of replication being at position 0 (adapted from Glaser et al., *Science* **294:**849–852, 2001).

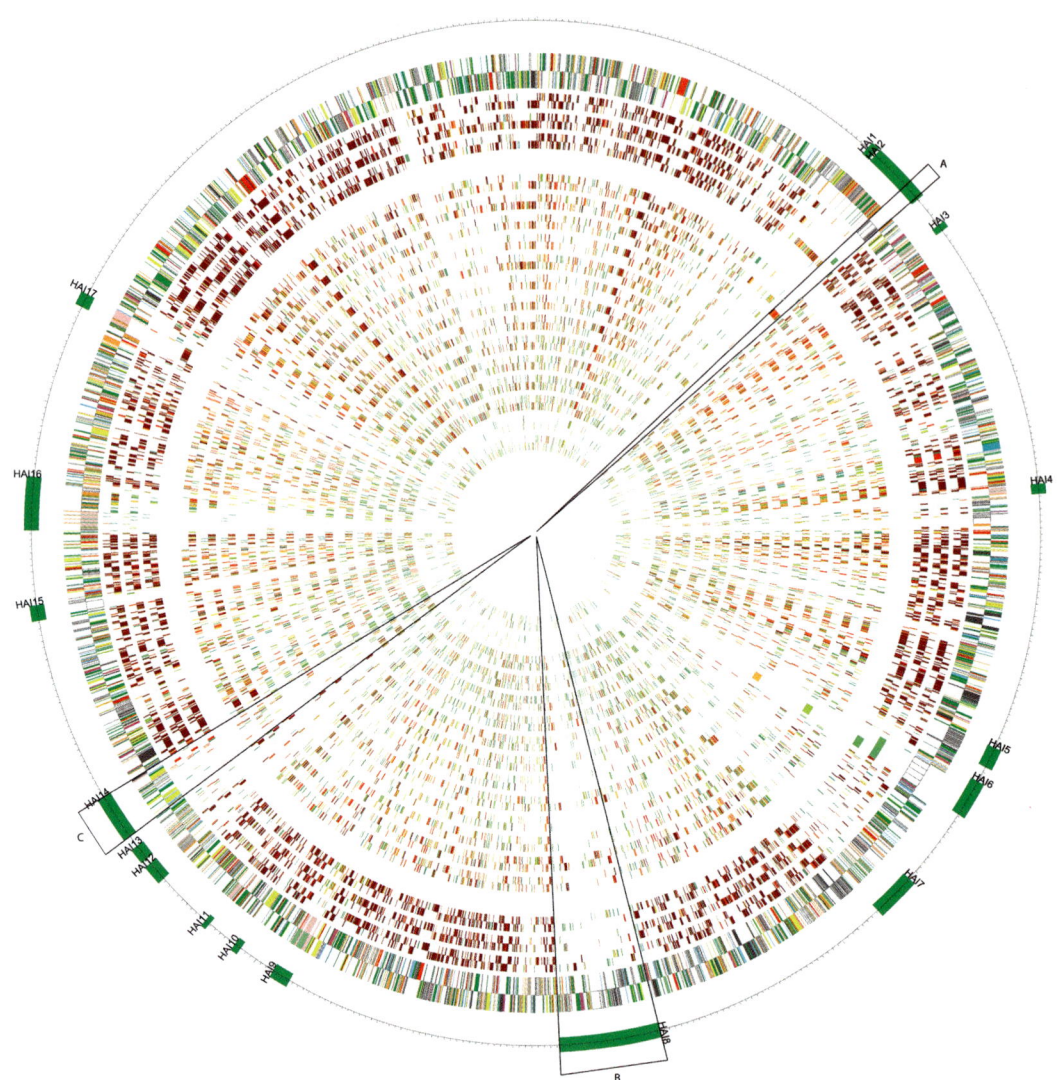

COLOR PLATE 5 Circular diagram of *Erwinia carotovora* subsp. *atroseptica* (strain SCRI1043) genome generated by GenomeDiagram software, showing from outer rings to inner rings: locations of 17 horizontally acquired islands; Eca1043 coding sequences in forward and reverse directions and colored by functional class; reciprocal best hits to coding sequence from plant- and animal-associated bacterial genomes, ordered by overall percentage identity between amino acid sequences: *Yersinia pestis* CO92, *Escherichia coli* CFT073, *Salmonella enterica* serovar Typhi CT18, *Pseudomonas putida* KT2440, *Pseudomonas syringae* pv. *tomato* DC3000, *Xanthomonas campestris* pv. *campestris* ATCC 33913, *Xylella fastidiosa* Temecula1, *Ralstonia solanacearum* GMI1000, *Xylella fastidiosa* 9a5c, *Xanthomonas oryzae* pv. *oryzae* KACC10331, *Xanthomonas axonopodis* pv. *citri* 306, *Bradyrhizobium japonicum* USDA 110, *Mesorhizobium loti* MAFF303099, *Agrobacterium tumefaciens* C58, *Sinorhizobium meliloti* 1021, onion yellows phytoplasma OY-M, *Leifsonia xyli* subsp. *xyli* CTCB07. Inner rings indicating reciprocal best hits to coding sequences from other genomes are colored individually on a scale from 30% amino acid identity (cyan) to 70% identity (brick red). Examples of horizontally acquired islands show an increase in the number of coding sequences with reciprocal best hits and/or a high amino acid sequence similarity to plant-associated bacteria (PAB) compared to animal-pathogenic enterobacteria (APE). (A) Coronafacic acid phytotoxin production and *Pseudomonas syringae* pv. *tomato* DC3000. (B) Type III secretion system. (C) Nitrogen fixation and the *Rhizobiaceae*.

COMPARATIVE PATHOGENOMICS OF SPIROCHETES

George M. Weinstock, David Šmajs, Petra Matějková, Timothy Palzkill, and Steven J. Norris

6

The spirochetes are a diverse group of bacteria that share many morphological characteristics, hence the name of this group. They are highly motile but contain the novel adaptation of periplasmic flagella. This group contains a number of pathogens and may be one of the oldest pathogenic groups of bacteria. In this chapter, we report on three pathogenic spirochetes: *Treponema pallidum*, causative agent of syphilis, and other treponemes; *Borrelia burgdorferi*, causative agent of Lyme disease, and related species; and *Leptospira* species, responsible for leptospirosis. The spirochetes are quite different from many of the other well-studied pathogens that are principally found in the proteobacteria and gram-positive groups. Indeed, *T. pallidum* and *B. burgdorferi* have many novel aspects to their pathogenic processes. These three genera of spirochetes show extreme differences in their genome structure,

from the extremely reduced and stable genome of treponemes, to the bizarre *Borrelia* genome with linear chromosomes and many plasmids, with *Leptospira* in the middle, carrying two chromosomes but undergoing significant rearrangement.

TREPONEMA

The genus *Treponema* comprises several pathogenic spirochete species and subspecies. The symptoms and epidemiological aspects of infections caused by pathogenic treponemes show decreasing extent from venereal syphilis to pinta (Table 1). The most invasive bacterium of this group is *Treponema pallidum* subsp. *pallidum*, the causative agent of syphilis (lues). Nonvenereal *T. pallidum* subsp. *pertenue* and *endemicum*, causing yaws and endemic syphilis, respectively, are moderately invasive. *Treponema carateum* is a noninvasive spirochete that causes local dermal lesions (pinta). *Treponema paraluiscuniculi* is not pathogenic for humans and causes veneral spirochetosis in rabbits.

The release of the complete genome sequence of *T. pallidum* subsp. *pallidum* strain Nichols in 1998 (24) increased our understanding of this unusual pathogen (91). The exclusive human pathogenicity and growth dependence were found to correlate with a relatively small chromosome (1.14 Mbp) with

George M. Weinstock Human Genome Sequencing Center, Baylor College of Medicine, 1 Baylor Plaza, Alkek N1619, Houston, Texas 77030. *David Šmajs and Petra Matějková* Department of Biology, Faculty of Medicine, Masaryk University, Kamenice 5, Building A6, 625 00 Brno, Czech Republic. *Timothy Palzkill* Department of Molecular Virology and Microbiology, Baylor College of Medicine, 1 Baylor Plaza, Houston, Texas 77030. *Steven J. Norris* Department of Pathology and Laboratory Medicine, University of Texas Medical School at Houston, MSB 2.120, P.O. Box 20708, Houston, Texas 77030.

Bacterial Pathogenomics, Edited by M. J. Pallen et al.
© 2007 ASM Press, Washington, D.C.

TABLE 1 Pathogenic spirochetes and diseases they cause

Treponema pallidum subspecies/species	Disease caused	Invasiveness
T. pallidum subsp. *pallidum*	Venereal syphilis (lues)	Highly invasive
T. pallidum subsp. *pertenue*	Yaws (framboesia)	Moderately invasive
T. pallidum subsp. *endemicum*	Endemic syphilis (bejel)	Moderately invasive
T. carateum	Pinta	Not invasive
T. paraluiscuniculi	Venereal spirochetosis in rabbits	Not pathogenic for humans

1,039 predicted open reading frames (ORFs) (including ORFs with frameshifts). At the same time, the genomic data provided new opportunities to study these organisms with the tools of comparative genomics.

T. pallidum subsp. *pallidum* SS14, *T. pallidum* subsp. *pertenue* Samoa D, and *T. paraluiscuniculi* Cuniculi A were compared with the genome of *T. pallidum* subsp. *pallidum* Nichols by means of whole genome fingerprints (WGF), DNA microarray hybridizations, and DNA sequencing of divergent chromosomal loci. In the WGF approach (92) (Šmajs and Weinstock, unpublished), the genomes were analyzed by using 97 overlapping XL-PCR amplicons covering the entire treponemal genomes with subsequent restriction mapping with several restriction endonucleases. DNA microarrays of *T. pallidum* subsp. *pallidum* Nichols containing PCR products corresponding to all 1,039 predicted ORFs (75) were used for hybridization of labeled chromosomal DNA of test and reference strains. By means of these approaches, four regions with insertions and deletions (indels) were found in the SS14 genome compared with the Nichols genome. Indels in all these regions were also found in Samoa D and Cuniculi A genomes, suggesting specificity of the Nichols genome structure in these regions (Šmajs and Weinstock, unpublished). After subtraction of regions containing indels present in the SS14 genome, seven additional chromosomal regions with indels were identified in the Samoa D genome, with one region specific to Samoa D and six heterologous regions that were also found in the Cuniculi A genome (Fig. 1). Seven additional regions with indels

were found to be specific for the Cuniculi A genome. Besides deletions identified by both WGF and DNA microarray approaches, the DNA microarray experiments identified 11 additional genes in the Cuniculi A genome with lower hybridization signal. These genes were sequenced and were found to contain prominent sequence changes (Šmajs and Weinstock, unpublished). Interestingly, most of these genes belonged to the paralogous families PGF2 and PGF14, which are colocalized on the treponemal chromosome (24) and which code for Tpr and hypothetical proteins, respectively. Both DNA microarray and WGF approaches represent complementary methods

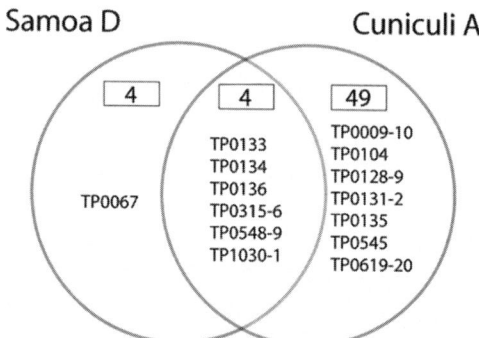

FIGURE 1 Comparison of genome differences found in *T. pallidum* subsp. *pertenue* Samoa D and in *T. paraluiscuniculi* Cuniculi A genomes when compared with the *T. pallidum* subsp. *pallidum* Nichols and SS14 genomes. Numbers in boxes represent the number of additional or missing restriction target sites in Samoa D and Cuniculi A genomes compared with the Nichols genome. Four chromosomal regions with indels present in SS14, Samoa D, and Cuniculi A genomes are not shown.

that allow selective detection of sequence diverse chromosomal regions by DNA microarray and selective identification of insertions within the genes and indels in intergenic regions by WGF. Although the DNA microarray represents a relatively quick method for mapping the presence and absence of genes in related genomes, its detection limit only allows genes to be identified with considerable deletions and/or prominent sequence changes. No insertions or indels in the intergenic regions can be identified by this approach. The WGF approach represents a relatively robust method for detection of indels in tested genomes (with the detection limit of indel length about 50 to 100 nucleotides). One of the possible limitations of this method is that larger sequentially divergent chromosomal regions do not allow primer annealing within the limited length of extra-long PCR product.

Closely related genomes with nucleotide identity higher than 99.5% can be mapped (sequenced) by using the comparative genome sequencing (CGS) strategy (2). In the first mapping stage (Fig. 2), maskless array synthesis was used to construct oligonucleotide arrays of 29 mers derived from the reference (Nichols) sequence, covering both strands, and spaced seven nucleotides apart throughout the genome. These arrays were hybridized separately to fluorescently labeled genomic DNA from *T. pallidum* Nichols and tested strain (Šmajs and Weinstock, unpublished). Oligonucleotides showing lower hybridization contained possible sequence variants and were selected for resequencing. In the resequencing second stage, a custom oligonucleotide array was synthesized representing all variant sequences of these regions from the tested genome (four alternative bases at each position for each strand). Fluorescently labeled genomic DNA was hybridized to this array, and the sequences were identified according to oligonucleotides showing the strongest hybridization.

By means of CGS (Šmajs and Weinstock, unpublished), more than 300 single nucleotide changes and small indels were found in the SS14 genome. When the Samoa D strain was

compared with the reference Nichols strain, more than 1,300 single nucleotide changes and small indels were found. On the basis of the preliminary estimates, about 10,000 single nucleotide changes and small indels differentiate the Cuniculi A genome from its Nichols counterpart. One of the most important limitations of the CGS approach is the inability to detect possible insertions in the tested genome and the low frequency in identification of short nucleotide indels (Matějková, Šmajs, and Weinstock, unpublished). These genome differences potentially affect gene function.

Another recent strategy of comparative genomics is the sequencing by synthesis using a pyrosequencing protocol, developed by 454 Life Sciences, optimized for solid support and picoliter-scale volumes (47). This strategy allows straightforward sequencing of bacterial genomes without bias in representation of genome regions due to vector cloning and replication in bacteria. The pyrosequencing protocol was used for sequencing of several treponemal strains including *T. pallidum* subsp. *pertenue* Samoa D, *T. pallidum* subsp. *pertenue* CDC-2, and *Treponema paraluiscuniculi* strain Cuniculi A (Šmajs and Weinstock, unpublished). In all cases, from 97.6 to 98.6% of the reference Nichols sequence was covered by the new sequences and the average sequence depth of coverage was from 17.8 to 36 times per base. Compared with the Sanger dideoxy-termination sequencing method, assembly of complete genome sequences based on pyrosequencing reads is more difficult in the regions containing repetitive sequences. This is a result of considerably shorter read lengths and no read pairs in the current pyrosequencing protocol.

Most of the genomic differences that were found among examined strains were localized in *tpr* loci and in the vicinity thereof. This suggests their possible role in the host range and pathogenicity in *T. pallidum* subsp. *pallidum*. In addition, the genome sequence of *T. pallidum* subsp. *pertenue* Samoa D appears to be an intermediate between *T. pallidum* subsp. *pallidum* and *T. paraluiscuniculi*. Unlike in several other bacterial species, the differences between pathogenic

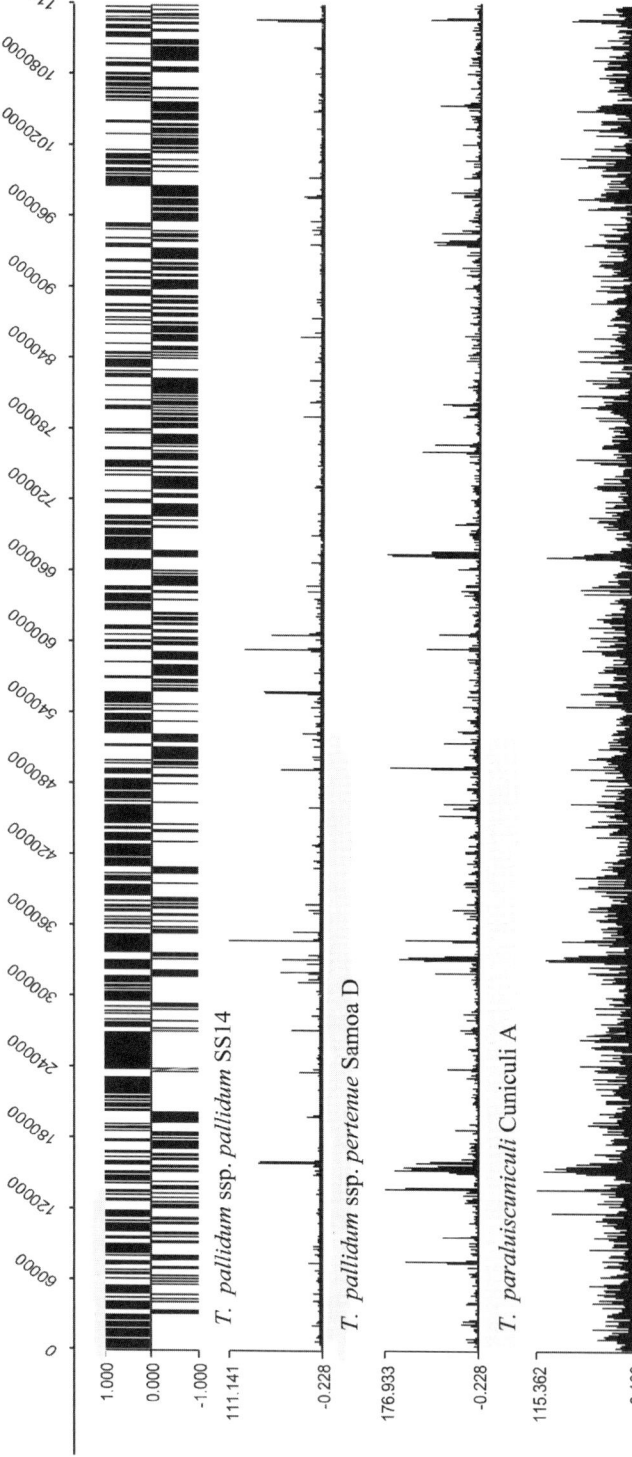

FIGURE 2 Mapping stage of CGS strategy. Results obtained for *T. pallidum* subsp. *pallidum* SS14, *T. pallidum* subsp. *pertenue* Samoa D, and *T. paraluiscuniculi* Cuniculi A are shown. Oligonucleotides showing significantly lower hybridization in the tested genome are shown as vertical bars and indicate sequence heterogeneity in these regions when compared with the Nichols reference genome. Note the increasing amount of sequence diversity in the order SS14, Samoa D, and Cuniculi A strains. ORFs of Nichols (both strands) are shown at the top.

and nonpathogenic strains are not related to the presence or absence of a set of genes encoding virulence factors. In contrast, differences in host range, disease symptoms, and course of disease appear to depend on relatively small differences in their genomes, including single nucleotide changes. Sequencing of additional treponemal strains will help to define genetic variability within and between subspecies and species. This will help define a set of genetic changes associated with different disease manifestations and could lead to identification of treponemal virulence determinants. Differences in the chromosome sequence can be used for molecular diagnostic testing of treponemal subspecies and strains and for epidemiological studies.

TREPONEMA PALLIDUM: A SET OF CLONED OPEN READING FRAMES

The genome sequences of many infectious microorganisms reveal many genes of unknown function. One of the goals of functional genomics studies is to determine the role of these genes with respect to host-pathogen interactions. The genome sequence information enables new approaches to be developed to determine the function of genes and their possible role in pathogenesis. A functional genomics approach has been used identify proteins important for the *Treponema pallidum* host-pathogen interaction (11, 48, 49). Several features of *T. pallidum* make it an excellent system on which to develop and test functional genomics technologies. First, with a size of 1 million base pairs, the genome is one of the smallest known (24). Second, there are a total of 1,031 open reading frames (ORFs), which makes it feasible to systematically construct libraries containing each ORF in a relatively short period of time. Finally, little is known of the biology or pathogenesis of this organism because a continuous culture system is not available. This severely limits the experimental options for study of the organism and puts an emphasis on developing alternate approaches to investigate *T. pallidum* biology.

Functional genomic studies are critically dependent on efficient cloning of the ORFs identified by genome sequences. The ORFs must be cloned into plasmid vectors that permit large-scale expression or facilitate functional analysis of the encoded proteins. Because all proteins are not expressed equally well from a single system, it is necessary to obtain constructs whereby the gene of interest is present in multiple protein expression systems. It was reasoned that a wide range of functional genomics experiments would be possible if the set of all *T. pallidum* ORFs were available in a number of vectors instead of a single plasmid. For example, it would be useful to have all of the ORFs cloned into an expression vector for production and purification of *T. pallidum* proteins in *Escherichia coli*, as well as having the complete set of genes cloned into the plasmids necessary for the yeast two-hybrid system for the ready evaluation of interactions between *T. pallidum* proteins.

One approach to establishing clones of *T. pallidum* genes in several different functional vectors would be to individually clone PCR products encoding ORFs into each desired vector. This approach, however, is time-consuming and subject to error. An alternative to the repeated cloning of PCR products is a recombination-based approach developed by Liu et al. (45) to permit the cloning of a PCR product into a plasmid and the rapid conversion of the plasmid to a number of different expression systems without the necessity of cloning the PCR product multiple, independent times. The method, termed the univector plasmid-fusion system, involves the insertion of the PCR product into a particular type of plasmid, called the univector, which can then be placed under the control of a variety of promoters or fused in frame to various tag sequences. The system is based on plasmid fusion using the Cre-*lox* site-specific recombination system of bacteriophage P1 (78). The Cre enzyme is a site-specific recombinase that catalyzes recombination between two 34 bp *loxP* sequences and is involved in the resolution of dimers formed during replication of the circular P1 chromosome (78). Cre can perform the recombination reaction both in vivo and in vitro (1, 45). The univector cloning system is illustrated in Fig. 3. The pUNI plasmid is used for the initial cloning of PCR products. The pHOST plasmid contains the appropriate promoter sequences or tag sequences for creating fusion proteins. The recombinant protein expression construct is made by fusion of the pUNI and pHOST plasmids mediated by Cre-*loxP* site-specific recombination (45).

One of the advantages of the univector system is that a number of different pHOST vectors can be fused to a pUNI plasmid containing the gene of interest. For example, the pHOST vector could allow the generation of protein fusions to glutathione-S-transferase (GST), a polyhistidine sequence, green fluorescent protein, or any tag protein of interest (Fig. 3). Importantly, the Cre-*loxP*-mediated fusion does not necessitate that the gene of interest be sequenced again because PCR has not been used to make the fusion. Finally, the pUNI vector contains a conditional origin of plasmid replication that is dependent on the *pir* gene, which must be present in the *E. coli* host (50).

Thus, the original PCR product is inserted in the pUNI vector and maintained in a *pir*⁺ strain of *E. coli*. However, after the fusion reaction, the plasmid is introduced into a strain of *E. coli* lacking the *pir* gene, and therefore only fusion plasmids will be propagated. This feature promotes the rapid generation of pUNI-pHOST fusion plasmids (45).

The pUNI vector and Cre-*loxP* recombination greatly facilitates the conversion of a set of clones to a set of functional vectors. However, before these steps can occur, it is necessary to insert the gene of interest into the pUNI vector. If only a few genes are being examined, conventional cloning methods that use restriction enzymes and DNA ligase are sufficiently efficient. For functional genomics, however, it is necessary to construct hundreds to thousands of plasmids, and more efficient methods are required. One such approach exploits the ability of the Vaccinia DNA topoisomerase I to both cleave and rejoin DNA strands with high sequence specificity (73). In the reaction, the enzyme recognizes the sequence 5′-CCCTT and cleaves at the final T

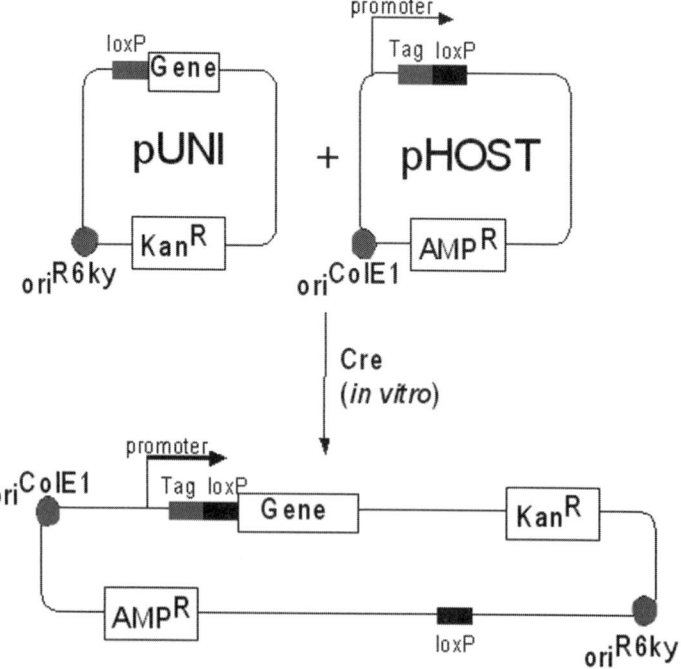

FIGURE 3 Univector plasmid fusion system. Cre-*loxP*–mediated site-specific recombination fuses the pUNI and pHOST plasmids at the *loxP* site. As a result, the gene of interest is placed under the control of the pHOST promoter and fused to any Tag sequences present in the pHOST plasmid. Figure adapted from references 45 and 49.

whereby a covalent adduct is formed between the 3′ phosphate of the cleaved strand and a tyrosine residue in the enzyme. The covalent complex can combine with a heterologous acceptor DNA that has a 5′ hydroxyl tail complementary to the sequence on the covalent adduct to create a recombinant molecule (73).

Topoisomerase I–based cloning exploits the reaction described above to join DNA fragments containing 5′ hydroxyl groups to acceptor plasmids to which the Vaccinia topoisomerase I enzyme is covalently attached. Because the joining reaction occurs only if the incoming DNA has a free 5′ hydroxyl, it is ideal for cloning PCR products because these do not possess 5′ phosphates as the oligonucleotide primers used for amplification do not have the 5′ phosphate group (73). The efficiency of this cloning method has been demonstrated by its use in the cloning of 6,035 ORFs from *Saccharomyces cerevisiae* into a plasmid vector for protein expression in yeast as well as a vector for expression in mammalian cells (31). The only requirement for PCR primer design for this system is to include the sequence 5′-CACC at the 5′ end of the PCR product. The 5′-CACC sequence is complementary to a sequence on the 5′ side of the plasmid–topoisomerase I adduct and as such controls the orientation of insertion of the PCR product. This topoisomerase I–based method was used to systematically clone *T. pallidum* ORFs into the pUNI vector (49).

IDENTIFICATION OF *T. PALLIDUM* ANTIGENS EXPRESSED DURING RABBIT INFECTION

The identification of *T. pallidum* antigens useful in diagnosis and vaccine development has been the object of intensive research efforts over the past two decades (8, 53, 61, 85). The goal of our experiments was to systematically identify the *T. pallidum* proteins important for eliciting a humoral immune response. This was accomplished by the use of the *T. pallidum* clone set constructed in the pUNI plasmid. For the purposes of monitoring the humoral response, the clone set was converted to vectors that expressed *T. pallidum* proteins as GST fusion proteins. The pUNI clone set was converted by using Cre recombinase, and the resultant plasmids were used to express approximately 900 of the *T. pallidum* ORFs as GST fusion proteins in *E. coli*. For this purpose, each *E. coli* strain containing a GST-*Treponema* ORF expression plasmid was inoculated into wells of 96-well 2-ml-deep well plates, and the cultures were grown with isopropyl-β-A-thiogalactopyranoside induction of GST–fusion gene expression. The cultures were then centrifuged and resuspended in protein extraction reagent to lyse the cells (49). Each fusion protein lysate was then added to microtiter wells that had been coated with glutathione in order to create an array of immobilized fusion proteins to be used in an enzyme-linked immunosorbent assay. The sera used in the assay were pooled from three rabbits that were each infected with 4×10^8 *Treponema pallidum* Nichols strain organisms by intratesticular inoculation (48, 49). Sera were collected 2 days before infection, and at 7, 14, 28, 56, and 84 days after infection. Sera from each of the three rabbits were pooled for use in the ELISA experiments to detect antigenic proteins. The sera were preabsorbed to *E. coli* protein lysates and then added to the microtiter wells coated with the GST fusion proteins (48). This array was used to systematically identify 106 *T. pallidum* proteins that serve as antigens during the humoral immune response of rabbits that had been infected for 84 days with *T. pallidum* (Fig. 4). Thus, approximately 1/10 of the *T. pallidum* proteome appears to be antigenic during an experimental rabbit infection.

Functional classifications are available for 56 of the 106 antigens identified (24). Proteins associated with the cell envelope are the most represented class (35%). Twenty-four (69%) of the 35 proteins exhibiting the highest reactivity were predicted to encode signal peptides or an N-terminal transmembrane helix (35); 10 of these were predicted or experimentally verified lipoproteins, consistent with prior observations that lipoproteins are among the most immunogenic proteins of *T. pallidum*. Therefore, there is

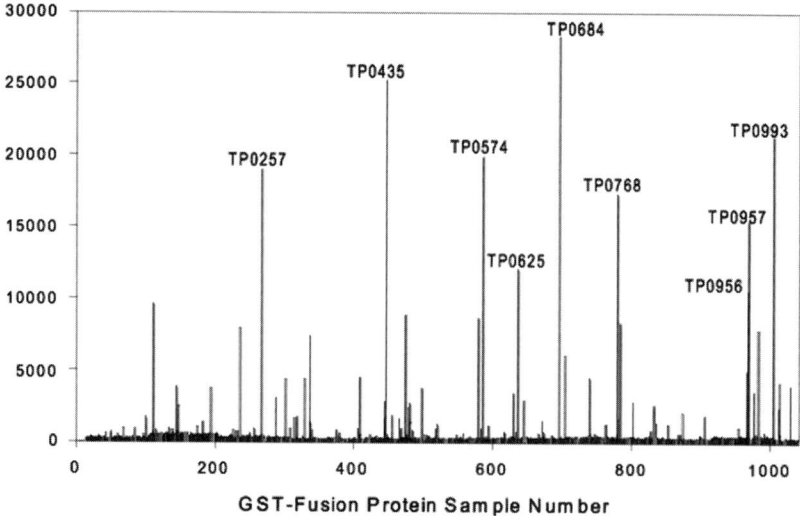

FIGURE 4 Identification of antigenic proteins in the *T. pallidum* proteome. The proteins arrayed for the immunoassay are presented numerically along the *x* axis according to their ORF number, TP0001 to TP1041. Chemiluminescence from a secondary anti-rabbit antibody conjugated to horseradish peroxidase was used to monitor rabbit antibody binding to GST fusion proteins and is measured in relative light units, shown along the *y* axis. *T. pallidum* fusion proteins that exhibited the highest levels of antibody binding are labeled. Figure adapted from reference 48.

a strong bias toward proteins that are secreted or membrane associated among the identified antigens, which is consistent with the idea that these proteins would have increased exposure to the immune system.

The GST fusion proteins were also used to monitor the timing of the humoral immune response in rabbits during *T. pallidum* infection (48). This was accomplished by using the rabbit sera that had been collected at 7, 14, 28, and 56 days after infection. These experiments were performed with 74 *T. pallidum* proteins that were representative of the 106 reactive proteins identified in the global screen at 84 days after infection (48). Little reactivity was observed between serum antibodies from uninfected rabbits and *T. pallidum* proteins. At 7 days after infection, however, antibody binding at levels at least twofold higher than binding to the GST protein alone was observed for a few *T. pallidum* proteins, including TP0319 membrane lipoprotein (TmpC), TP0574 47-kDa carboxypeptidase, TP0971 membrane antigen

TpD, and TP0684 methylgalactoside ABC transporter (MglB-2). These proteins were also found to bind high levels of antibody at the other postinfection time points (48). The induction of an early and continuous antibody response is a property that would make these proteins useful as diagnostics of *T. pallidum* infection.

The immune response expanded at 14 days after infection in that antibodies bound at levels twofold greater than GST alone to 19 of the 74 proteins arrayed. The most reactive proteins at this time point were the same as those seen at 7 days after infection. By 28 days after infection, the humoral response was extensive, with antibody binding at significant levels to 47 of the 74 proteins tested. The response continued to expand, however, and by 84 days after infection, 70 of the 74 proteins assayed exhibited twofold greater antibody binding compared to the GST-only control. Thus, the humoral response to an experimental *T. pallidum* infection in rabbits features a strong and early antibody response to a few proteins that continues to be strong during the course of the infection. The response broadens, however, to include a robust response late in infection to dozens of proteins and includes a statistically significant response to over 100 proteins (48).

IDENTIFICATION OF *T. PALLIDUM* ANTIGENS EXPRESSED DURING HUMAN INFECTION

To identify antigens important in the human immune response to syphilis, the serum antibody reactivity of patients with syphilis was examined with 908 *T. pallidum* proteins by using the same techniques as those described above for the rabbit sera (11). Thirty-four proteins exhibited significant antibody binding when assayed with human sera from patients in the early latent stage of syphilis. The early latent stage follows secondary syphilis, and patients in this stage are known to maintain a humoral response to *T. pallidum* proteins. The 34 proteins reactive with early latent human sera were also found to be significantly reactive in the rabbit antigen screen described above (48). The fact that the human antigens are a subset of the rabbit antigens validates the use of the rabbit model system for antigen identification. A total of 90 proteins, 32 of those reactive with early latent human sera, as well as additional proteins known to be antigenic in previous studies with rabbit sera, were selected for further analysis of reactivity with human sera from different stages of disease progression, including sera from patients with primary or secondary syphilis, or normal human sera.

Interestingly, the proteins exhibited a variety of patterns of reactivity with these sera. For example, some *T. pallidum* proteins were reactive with sera from patients at all stages of syphilis, while other proteins were reactive in a stage-specific manner (11). The biological significance of these findings is unclear and is a subject of future investigation. It is likely that some false-negative results were obtained in both the rabbit and human antigen screens as a result of a loss of antibodies during the preabsorption of sera with *E. coli* proteins, low expression levels,

solubility, or lability of some of the protein products in *E. coli*, or to mutations introduced into ORFs during the cloning process (49). It is also possible that the number of antigens detected in these experiments is limited by the number of patient sera used for detection and that further studies with additional sera would detect additional antigens. Therefore, the list of antigens is not inclusive of all *T. pallidum* antigens. Nevertheless, it is the first systematic search for *T. pallidum* antigens, and a number of novel antigens with high reactivity to antibodies in both rabbit and human sera were identified in these studies (11, 48, 49).

In summary, the genome scale-studies described above used a clone set consisting of ~90% of *T. pallidum* ORFs (49) to conduct the first systematic search for antigenic proteins in this organism. A large set of antigens was identified by using sera from experimental rabbit infections, which were further characterized with respect to the timing of the immune response. In addition, the clone set was used to identify 34 proteins that were bound by antibody from sera of humans with early latent syphilis (11). Proteins that elicit an early response that continues throughout experimental rabbit infection and are also human antigens may be useful in immunodiagnosis. Furthermore, novel vaccine candidates, including potential surface-exposed outer membrane proteins, may be present among the previously undescribed antigens identified in the human anti–*T. pallidum* immunoproteome. The use of a clone set is obviously not restricted to antigen discovery, and studies are underway that use the clone set to investigate genome-scale *T. pallidum* protein-protein interactions, as well as to investigate interactions between *T. pallidum* with host proteins.

BORRELIA

The genus *Borrelia* comprises spirochetes with loose wavelike coils that survive exclusively by transmission back and forth between vertebrate and arthropod hosts. *Borrelia* species can be divided roughly into four groups on the basis of genetic markers and patterns of pathogenesis

TABLE 2 *Borrelia* species

Lyme disease–associated species transmitted by *Ixodes* ticks (*B. burgdorferi* sensu lato)

   Borrelia afzelii

   Borrelia andersonii

   Borrelia bissettii

   Borrelia burgdorferi

   Borrelia garinii

   Borrelia japonica

   Borrelia lusitaniae

   Borrelia spielmanii

   Borrelia tanukii

   Borrelia turdi

   Borrelia valaisiana

Endemic relapsing fever–associated species transmitted by *Ornithodoros* ticks

   Borrelia crocidurae

   Borrelia duttonii

   Borrelia hermsii

   Borrelia hispanica

   Borrelia parkeri

   Borrelia persica

   Borrelia turicatae

Epidemic relapsing fever–associated species transmitted by the human body louse

   Borrelia recurrentis

Relapsing fever–related species transmitted by *Ixodes* or *Amblyomma* ticks

   Borrelia lonestari

   Borrelia miyamotoi

and transmission (Table 2). The first group is transmitted by hard-bodied ticks of the *Ixodes ricinus* group and includes species that cause Lyme disease in humans (*B. burgdorferi* in North America and *B. burgdorferi*, *B. afzelii*, *B. garinii*, and provisionally *B. spielmanii* in Eurasia). This group is referred to collectively as *B. burgdorferi* sensu lato. The second group is transmitted by soft-bodied ticks of the genus *Ornithodoros* and causes sporadic, endemic cases of relapsing fever in humans. The third group consists of *B. recurrentis*, the causative agent of epidemic relapsing fever, which is transmitted between humans by

body lice. Last, an "indeterminate" group containing *B. miyamotoi* and *B. lonestari* is genetically more closely aligned with the relapsing fever *Borrelia*, but it is harbored by hard-bodied ticks of the *Ixodes* or *Amblyomma* genera.

When an infected *Ixodes* tick bites a mammalian or avian host, *B. burgdorferi* sensu lato organisms migrate from the tick midgut to the salivary glands, followed by transmission in the saliva to the host (65, 77). During this process (which usually requires 24 to 72 h of feeding), Lyme disease *Borrelia* undergoes extensive changes in gene expression, exemplified by downregulation of the outer surface lipoprotein OspA and upregulation of OspC (Fig. 5). In humans, Lyme disease is a multistage infection initiated by a localized lesion called erythema migrans at the site of the tick bite (76). This erythematous rash with an expanding margin contains large numbers of organisms, which readily disseminate through the bloodstream to virtually all other tissues. Weeks to months after infection, disseminated manifestations such as neurologic symptoms and blockage of cardiac conductance (resulting in an irregular heartbeat) can occur. The chronic stage is characterized by migrating arthritis in the large joints, neurologic deficits, and long-term skin atrophy (called acroderma chronicum atrophicans). If untreated, symptoms of Lyme disease can persist for years.

In contrast, relapsing fever *Borrelia* is transmitted during the rapid feeding cycle of *Ornithodoros* ticks, a process typically taking less than 30 min. The infection in humans and other mammals is predominantly a bacteremia, with concentrations reaching as high as 10^8 *Borrelia* per ml of blood. High fever results and persists for several days until antibody responses against the expressed variable major surface lipoprotein (either a variable large protein [Vlp] or a variable small protein [Vsp]) result in elimination of most of the spirochetes. However, an elaborate antigenic variation mechanism (see below) results in outgrowth of a subpopulation of spirochetes expressing a different Vlp/Vsp type. In this

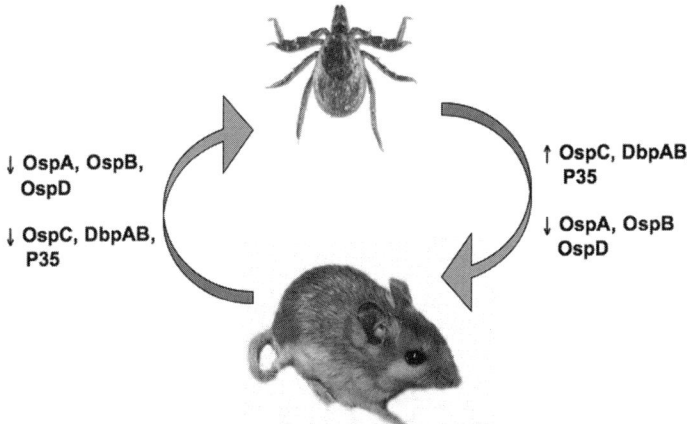

↓ OspA, OspB,
OspD

↓ OspC, DbpAB,
P35

↑ OspC, DbpAB,
P35

↓ OspA, OspB
OspD

FIGURE 5 Transmission cycle and protein expression changes of the Lyme disease spirochete, *Borrelia burgdorferi*. The expression of many proteins is upregulated or downregulated in the transition between arthropod and mammalian hosts, as exemplified by OspA, OspC, DbpA, and DbpB. From reference 54, used with permission.

manner, as many as seven relapses can occur, with each relapse consisting of a different Vlp/Vsp serotype (6).

Despite divergence in terms of pathogenesis and arthropod hosts, all *Borrelia* are morphologically indistinguishable and share a common genomic structure (14, 23, 25, 60). The *Borrelia* species characterized to date have a single linear chromosome of ~910 kb and multiple linear and circular plasmids (Fig. 6). The telomeres of the chromosome and linear plasmids consist of hairpin loops in which the 5′ and 3′ ends are linked by a phosphate bond (32). Replication of these linear moieties occurs by contiguous DNA synthesis in both directions at an origin of replication, resulting in formation of a dimeric circular intermediate. The duplicated telomere regions are subjected to single-strand cleavage, reformation of the hairpin loop, and ligation by the DNA resolvase ResT, encoded on the plasmid cp26 (5, 15, 37, 38, 87, 88, 93).

The chromosome of *B. burgdorferi* encodes most of the housekeeping genes required for in vitro survival and growth, as demonstrated by strain *B. burgdorferi* B313, which has lost most of the plasmids (66). The plasmids are populated primarily by hypothetical genes without homologs in other organisms, along with clusters of genes involved in plasmid replication (14, 23). Plasmid-encoded proteins include several prominently expressed lipopro-

teins, such as outer surface proteins OspA through OspG, decorin binding proteins DbpA and DbpB, and the VMP-like sequence (Vls) antigenic variation protein VlsE. The *B. burgdorferi* B31 genome contains 161 paralogous gene families; 107 of these have at least one plasmid-encoded member. The number of paralogs in each plasmid-associated family ranges from 2 to 41 (14), and most of these have no predicted function. In addition, the plasmids contain a large number of pseudogenes, and in comparisons of different species and strains, they have undergone extensive recombination and shuffling of DNA segments, consistent with a genome in flux that is undergoing rapid and extensive changes (14, 25, 60). The chromosome, in contrast, is relatively stable and exhibits extensive homology and synteny across different species and strains (14, 23, 25, 60).

Shortly after *B. burgdorferi* was first cultured in 1981, it was noted that the organism lost its ability to infect laboratory animals after 10 to 15 in vitro passages; this change was accompanied by loss of as yet unidentified plasmids. Xu et al. (93) determined that two plasmids approximately 24.7 and 27.5 kb in size were absent in *B. burgdorferi* B31 clones that had reduced infectivity in hamsters. Similarly, Zhang et al. (96) found by subtractive hybridization methods that a 28-kb plasmid (lp28-1) was absent from clones that had low infectivity in mice. Availability of the

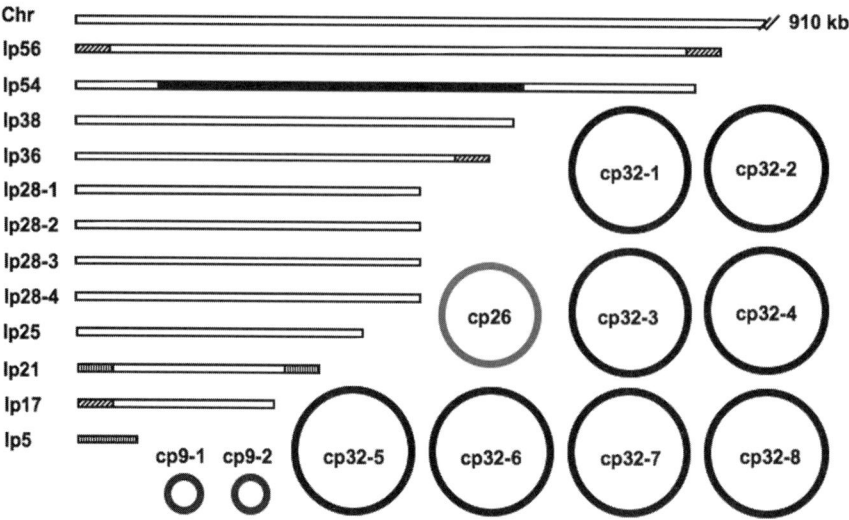

FIGURE 6 Complex genome structure of *Borrelia* species, as exemplified by *B. burgdorferi* B31 (14, 23). Extrachromosomal elements include both linear plasmids (lp) and circular plasmids; the approximate sizes in kilobase pairs are specified in the plasmid names. Closely related sequences are indicated by dark shading and patterns. The cp32 plasmids resemble prophages, and an additional copy of a cp32-related sequence is embedded in lp54. Similarly, lp21 contains a complete copy of lp5 bifurcated by an intervening sequence.

B31 genome sequence in 1997 permitted a more detailed analysis of the correlation between plasmid content and infectivity, and confirmed that the presence of lp25 and lp28-1 is required for full infectivity in the mouse model (28, 40, 41, 59). In contrast, loss of plasmids cp9, cp32-3, lp21, lp28-2, lp28-4, and lp56 had no apparent effect on infection of mice by needle inoculation, suggesting that these plasmids are not needed during infection of mammals. However, this finding does not preclude the involvement of these plasmids during the tick phase of the *Borrelia* life cycle or during transmission between ticks and vertebrate hosts.

Initial efforts to identify virulence factors of *B. burgdorferi* were hampered by an inability to transform and mutate low-passage infectious strains. This barrier was overcome in part by optimization of electroporation and chemical transformation conditions and identification of clones with a higher transformation rate (19, 67, 68); fusion of *B. burgdorferi* promoters with antibiotic resistance genes (9, 18); development of effective shuttle vectors (69, 83);

and inactivation of a putative restriction/modification gene, *BBE02,* and selection of clones lacking lp56, which carries another predicted restriction/modification gene, *BBQ67* (36, 42). Combination of these factors has led to successful transformation and gene activation of infectious *B. burgdorferi*, although at rates much lower than those obtained with other bacteria.

B. burgdorferi does not express any known toxins or secreted proteins. However, a number of potential virulence determinants likely to be involved in pathogenesis have been identified (Table 3). Transient expression of OspC is required for infection of mice; conflicting results have been obtained regarding its role in the migration of *B. burgdorferi* from the tick midgut to the salivary glands. The nicotinamidase PncA (encoded on lp25) apparently plays a role in the biosynthesis of NAD via a scavenging pathway, and it is essential for survival of *B. burgdorferi* in mice. Because it fulfills a metabolic function, PncA is considered a virulence determinant but not a "true" virulence factor. DbpA, DbpB, BBK32, and P66 are adhesins that bind

TABLE 3 Potential virulence determinants of *Borrelia burgdorferi* (partial listing)

Gene(s)[a]	Location	Evidence
Mammalian infection		
ospC (BBB19)	cp26	Increased expression under mammalian tissue conditions (37°C, pH 7.6) (71); inactivation results in loss of infectivity in mice (27, 84, 86); conflicting data exist on the effect of inactivation on migration of *B. burgdorferi* in ticks (27, 56, 84, 86)
OspEF-related proteins (*erpA-Q*) (17 paralogs)	cp32s, lp28-1	Increased expression in mammalian tissue (21, 90); differential expression in joint, heart, and spleen but not skin (20); binding to complement inhibitory factor H (3, 30, 39, 81)
crasp1 BBA68	lp54	Complement regulator–acquiring surface protein-1, binds to complement inhibitory factors H and FHL-1 and provides resistance to killing by human serum (12)
vlsE	lp28-1	Antigenic variation system (96)
dbpAB (BBA24,25)	lp54	Binding to extracellular matrix component decorin, GAGs (29)
bbk32	lp36	Binding to fibronectin (57); conflicting results exist on the effects of *bbk32* inactivation on pathogenesis in mice (44, 72)
p66 (BB0603)	Chromosome	Porin protein, involved in binding of *B. burgdorferi* to integrins (16, 17)
bmpAB, bmpCD (BB0382-BB0385)	Chromosome	Differential expression in mammalian tissue (4, 26, 62, 74)
rpoN (BB0450), *rpoS* (BB0771)	Chromosome	Regulation of temperature- and pH-dependent gene expression (34); interacts with *ospC* promoter (94); *rpoN* (*sigma54*) inactivation results in loss of mouse infectivity and mammal-arthropod transmission, but does not affect survival in ticks (22)
luxS (BB0377)	Chromosome	Potential involvement in quorum sensing (33, 79, 89); inactivation does not affect mouse infection by needle inoculation (33)
pncA (BBE22)	lp25	Nicotinamidase involved in NAD metabolism; inactivation results in loss of infectivity in mice (58)
guaAB (BBB17,18)	cp26	Purine biosynthesis (46)
Tick infection		
ospAB (BBA15,16)	lp54	Differential expression in tick midgut; involved in adherence to midgut epithelium (7, 70); required for tick colonization (95)
bptA (BBE16)	lp25	*Borrelia* persistence in ticks A (*bptA*). Function unknown; inactivation results in loss of persistence in *Ixodes* ticks, decreased infectivity in mice (64)
ospD (BBJ09)	lp38	Not required for mammalian infection; required during tick phase? (55, 80)

[a]Numbers in parentheses are gene designations according to Fraser et al. (23).

the mammalian extracellular matrix components decorin and fibronectin or host integrins, and hence are thought to be involved in the dissemination of *Borrelia*. Complement regulator-acquiring surface proteins (CRASPs), including CRASP-1 and OspE-related proteins (Erps), bind complement factor H and factor H–like protein-1 (FHL-1) and are thought to inhibit complement-dependent killing of *B. burgdorferi*

during infection. The sigma factors RpoN and RpoS form a cascade that regulates the expression of OspC, DbpA, DbpB, and other proteins during the tick-mammal cycle. In terms of the tick environment, OspA and BptA are required for survival of *B. burgdorferi* in the tick midgut. Although progress has been made during the past decade, our knowledge of the virulence factors involved in Lyme disease is at best fragmentary. Mutation of relapsing fever isolates has not been reported as yet.

Transposon mutagenesis using *Himar1*-based transposons has been described recently and provides another avenue to identifying *B. burgdorferi* factors involved in pathogenesis (51, 82). Combination of the *B. burgdorferi*–adapted transposon pMarGent with the transformable, infectious *B. burgdorferi* strain 5A18-NP1 resulted in the generation of random mutants in an infectious background (10). Expansion of this approach to provide saturation mutagenesis of the *B. burgdorferi* genome may provide a global view of the genes required for pathogenesis and hence may provide insight into the ability of invasive, nontoxigenic organisms to cause persistent infection and disease.

LEPTOSPIRA

The genus *Leptospira* is a large and diverse group of spirochetes, of which four genome sequences are available (13, 52, 63). *Leptospira* species cause leptospirosis, a major zoonotic disease worldwide (43). *Leptospira* can be highly pathogenic, causing acute human infections with mortality rates that can exceed 20%. *Leptospira* is found in many animal species, and this broad reservoir provides a continuous source for infection, making this one of the most widespread zoonotic infections.

Leptospira is classified into hundreds of serovars, relating mainly to the carbohydrate moiety of the lipopolysaccharide. Serovars can

be specific for a mammalian reservoir, where they colonize the renal tubules and are shed in urine. *L. interrogans* serovars are harbored by the brown Norway rat (*Rattus norvegicus*) and are a leading cause of disease in humans. The genomes of the Lai and Copenhageni serovars of *L. interrogans* have been sequenced (52, 63). *L. borgpetersenii* is the other major species of pathogenic *Leptospira*, with cattle as its primary maintenance host. *L. interrogans* and *L. borgpetersenii* account for most cases of leptospirosis. Although their diseases are similar, the mode of transmission differs, with *L. interrogans* being capable of transmission through contaminated water, while *L. borgpetersenii* may follow a more direct host-to-host route. The genomes of two strains of the Hardjo serovar of *L. borgpetersenii* have been sequenced (13).

The characteristics of the four sequenced *Leptospira* genomes are shown in Table 4. All of the sequenced *Leptospira* have two chromosomes, with the larger chromosome accounting for over 90% of the genome. The *L. borgpetersenii* genomes are found to be smaller and have fewer genes compared with *L. interrogans*. All four genomes contain numerous mobile genetic elements, accounting for over 7% of the sequences. These repeated sequences contribute to chromosome rearrangements and creation of pseudogenes, and they are an important mechanism in the expansion or reduction in genome size.

TABLE 4 *Leptospira genomes*

Leptospira strain	Genome size (Mb)			No. of CDS		
	Total	Large chromosome	Small chromosome	Total	Large chromosome	Small chromosome
L. interrogans serovar Lai	4,691,184	4,332,241	358,943	4,727	4,360	367
L. interrogans serovar Copenhageni	4,627,366	4,277,185	350,181	3,728	3,454	274
L. borgpetersenii serovar Hardjo						
Strain L550	3,931,782	3,614,446	317,336	3,211	2,949	262
Strain JB197	3,876,235	3,576,473	299,762	3,166	2,909	257

Bulach et al. (13) reannotated the *L. interrogans* genomes for consistency with the *L. borgpetersenii* analysis. They concluded that *L. borgpetersenii* strains were undergoing genome reduction, whereas *L. interrogans* strains were expanding their gene content by gene duplication as well as acquisition by lateral transfer—for example, of phage genes. Moreover, the functions lost in *L. borgpetersenii* were often required for adaptation to and survival in diverse environments, accounting for this species' more limited niche and inability to survive outside of its host. In contrast, *L. interrogans* has greater regulatory and metabolic capabilities, allowing it to survive in environments ranging from aquatic to mammalian.

CONCLUSION

The three genera of spirochetes described here illustrate the extreme diversity in genome structure found in this group. The continuing work in studying these genera also illustrates the value of having genome sequences of multiple representatives rather than relying on a single reference sequence. With the advent of new sequencing methodologies, such as those being applied to the treponemes, it will be possible to deeply sample these and other genera and develop a much more complete picture of their biology.

REFERENCES

1. **Abremski, K., R. Hoess, and N. Sternberg.** 1983. Studies on the properties of P1 site-specific recombination: evidence for topologically unlinked products following recombination. *Cell* **32:**1301–1311.
2. **Albert, T. J., D. Dailidiene, G. Dailide, J. E. Norton, A. Kalia, T. A. Richmond, M. Molla, J. Singh, R. D. Green, and D. E. Berg.** 2005. Mutation discovery in bacterial genomes: metronidazole resistance in *Helicobacter pylori. Nat. Methods* **2:**951–953.
3. **Alitalo, A., T. Meri, L. Ramo, T. S. Jokiranta, T. Heikkila, I. J. Seppala, J. Oksi, M. Viljanen, and S. Meri.** 2001. Complement evasion by *Borrelia burgdorferi*: serum-resistant strains promote C3b inactivation. *Infect. Immun.* **69:**3685–3691.
4. **Aron, L., M. Alekshun, L. Perlee, I. Schwartz, H. P. Godfrey, and F. C. Cabello.** 1994. Cloning and DNA sequence analysis of *bmpC*, a gene encoding a potential membrane lipoprotein of *Borrelia burgdorferi. FEMS Microbiol. Lett.* **123:**75–82.
5. **Bankhead, T., and G. Chaconas.** 2004. Mixing active-site components: a recipe for the unique enzymatic activity of a telomere resolvase. *Proc. Natl. Acad. Sci. USA* **101:**13768–13773.
6. **Barbour, A. G.** 2003. Antigenic variation in Borrelia: relapsing fever and Lyme borreliosis, p. 319–356. *In* A. Craig and A. Scherf (ed.), *Antigenic Variation.* Academic Press, Ltd., London, United Kingdom.
7. **Bergstrom, S., V. G. Bundoc, and A. G. Barbour.** 1989. Molecular analysis of linear plasmid-encoded major surface proteins, OspA and OspB, of the Lyme disease spirochaete *Borrelia burgdorferi. Mol. Microbiol.* **3:**479–486.
8. **Blanco, D. R., J. N. Miller, and M. A. Lovett.** 1997. Surface antigens of the syphilis spirochete and their potential as virulence determinants. *Emerg. Infect. Dis.* **3:**11–20.
9. **Bono, J. L., A. F. Elias, J. J. Kupko III, B. Stevenson, K. Tilly, and P. Rosa.** 2000. Efficient targeted mutagenesis in *Borrelia burgdorferi. J. Bacteriol.* **182:**2445–2452.
10. **Botkin, D. J., A. N. Abbott, P. E. Stewart, P. A. Rosa, H. Kawabata, H. Watanabe, and S. J. Norris.** 2006. Identification of potential virulence determinants by *Himar1* transposition of infectious *Borrelia burgdorferi* B31. *Infect. Immun.* **74:**6690–6699.
11. **Brinkman, M. B., M. McKevitt, M. McLoughlin, C. Perez, J. K. Howell, G. M. Weinstock, S. J. Norris, and T. Palzkill.** 2006. Reactivity of antibodies from syphilis patients to a protein array representing the *Treponema pallidum* proteome. *J. Clin. Microbiol.* **44:**888–891.
12. **Brooks, C. S., S. R. Vuppala, A. M. Jett, A. Alitalo, S. Meri, and D. R. Akins.** 2005. Complement regulator-acquiring surface protein 1 imparts resistance to human serum in *Borrelia burgdorferi. J. Immunol.* **175:**3299–3308.
13. **Bulach, D. M., R. L. Zuerner, P. Wilson, T. Seemann, A. McGrath, P. A. Cullen, J. Davis, M. Johnson, E. Kuczek, D. P. Alt, B. Peterson-Burch, R. L. Coppel, J. I. Rood, J. K. Davies, and B. Adler.** 2006. Genome reduction in *Leptospira borgpetersenii* reflects limited transmission potential. *Proc. Natl. Acad. Sci. USA* **103:**14560–14565.
14. **Casjens, S., N. Palmer, R. van Vugt, W. M. Huang, B. Stevenson, P. Rosa, R. Lathigra, G. Sutton, J. Peterson, R. J. Dodson, D. Haft, E. Hickey, M. Gwinn, O. White, and C. M. Fraser.** 2000. A bacterial genome in flux: the twelve linear and nine circular extrachromosomal DNAs in an infectious isolate of the Lyme disease

spirochete *Borrelia burgdorferi*. *Mol. Microbiol.* **35**:490–516.

15. **Chaconas, G., P. E. Stewart, K. Tilly, J. L. Bono, and P.** 2001. Telomere resolution in the Lyme disease spirochete. *EMBO J.* **20**:3229–3237.

16. **Coburn, J., W. Chege, L. Magoun, S. C. Bodary, and J. M. Leong.** 1999. Characterization of a candidate *Borrelia burgdorferi* beta3-chain integrin ligand identified using a phage display library. *Mol. Microbiol.* **34**:926–940.

17. **Coburn, J., and C. Cugini.** 2003. Targeted mutation of the outer membrane protein P66 disrupts attachment of the Lyme disease agent, *Borrelia burgdorferi*, to integrin alphavbeta3. *Proc. Natl. Acad. Sci. USA* **100**:7301–7306.

18. **Elias, A. F., J. L. Bono, J. J. Kupko III, P. E. Stewart, J. G. Krum, and P. A. Rosa.** 2003. New antibiotic resistance cassettes suitable for genetic studies in *Borrelia burgdorferi*. *J. Mol. Microbiol. Biotechnol.* **6**:29–40.

19. **Elias, A. F., P. E. Stewart, D. Grimm, M. J. Caimano, C. H. Eggers, K. Tilly, J. L. Bono, D. R. Akins, J. D. Radolf, T. G. Schwan, and P. Rosa.** 2002. Clonal polymorphism of *Borrelia burgdorferi* strain B31 MI: implications for mutagenesis in an infectious strain background. *Infect. Immun.* **70**:2139–2150.

20. **Fikrig, E., M. Chen, S. W. Barthold, J. Anguita, W. Feng, S. R. Telford III, and R. A. Flavell.** 1999. *Borrelia burgdorferi* erpT expression in the arthropod vector and murine host. *Mol. Microbiol.* **31**:281–290.

21. **Fikrig, E., W. Feng, J. Aversa, R. T. Schoen, and R. A. Flavell.** 1998. Differential expression of *Borrelia burgdorferi* genes during erythema migrans and Lyme arthritis. *J. Infect. Dis.* **178**:1198–1201.

22. **Fisher, M. A., D. Grimm, A. K. Henion, A. F. Elias, P. E. Stewart, P. A. Rosa, and F. C. Gherardini.** 2005. *Borrelia burgdorferi* sigma54 is required for mammalian infection and vector transmission but not for tick colonization. *Proc. Natl. Acad. Sci. USA* **102**:5162–5167.

23. **Fraser, C. M., S. Casjens, W. M. Huang, G. G. Sutton, R. Clayton, R. Lathigra, O. White, K. A. Ketchum, R. Dodson, E. K. Hickey, M. Gwinn, B. Dougherty, J. F. Tomb, R. D. Fleischmann, D. Richardson, J. Peterson, A. R. Kerlavage, J. Quackenbush, S. Salzberg, M. Hanson, R. van Vugt, N. Palmer, M. D. Adams, J. Gocayne, J. Weidman, T. Utterback, L. Watthey, L. McDonald, P. Artiach, C. Bowman, S. Garland, C. Fujii, M. D. Cotton, K. Horst, K. Roberts, B. Hatch, H. O. Smith, and J. C. Venter.** 1997. Genomic sequence of a Lyme disease spirochaete, *Borrelia burgdorferi*. *Nature* **390**:580–586.

24. **Fraser, C. M., S. J. Norris, G. M. Weinstock, O. White, G. G. Sutton, R. Dodson, M. Gwinn, E. K. Hickey, R. Clayton, K. A. Ketchum, E. Sodergren, J. M. Hardham, M. P. McLeod, S. Salzberg, J. Peterson, H. Khalak, D. Richardson, J. K. Howell, M. Chidambaram, T. Utterback, L. McDonald, P. Artiach, C. Bowman, M. D. Cotton, C. Fujii, S. Garland, B. Hatch, K. Horst, K. Roberts, M. Sandusky, J. Weidman, H. O. Smith, and J. C. Venter.** 1998. Complete genome sequence of *Treponema pallidum*, the syphilis spirochete. *Science* **281**:375–388.

25. **Glöckner, G., R. Lehmann, A. Romualdi, S. Pradella, U. Schulte-Spechtel, M. Schilhabel, B. Wilske, J. Sühnel, and M. Platzer.** 2004. Comparative analysis of the *Borrelia garinii* genome. *Nucleic Acids Res.* **32**:6038–6046.

26. **Gorbacheva, V. Y., H. P. Godfrey, and F. C. Cabello.** 2000. Analysis of the bmp gene family in *Borrelia burgdorferi* sensu lato. *J. Bacteriol.* **182**:2037–2042.

27. **Grimm, D., C. H. Eggers, M. J. Caimano, K. Tilly, P. E. Stewart, A. F. Elias, J. D. Radolf, and P. A. Rosa.** 2004. Experimental assessment of the roles of linear plasmids lp25 and lp28-1 of *Borrelia burgdorferi* throughout the infectious cycle. *Infect. Immun.* **72**:5938–5946.

28. **Grimm, D., K. Tilly, R. Byram, P. E. Stewart, J. G. Krum, D. M. Bueschel, T. G. Schwan, P. F. Policastro, A. F. Elias, and P. A. Rosa.** 2004. Outer-surface protein C of the Lyme disease spirochete: a protein induced in ticks for infection of mammals. *Proc. Natl. Acad. Sci. USA* **101**:3142–3147.

29. **Guo, B. P., E. L. Brown, D. W. Dorward, L. C. Rosenberg, and M. Hook.** 1998. Decorin-binding adhesins from *Borrelia burgdorferi*. *Mol. Microbiol.* **30**:711–723.

30. **Hellwage, J., T. Meri, T. Heikkila, A. Alitalo, J. Panelius, P. Lahdenne, I. J. Seppala, and S. Meri.** 2001. The complement regulator factor H binds to the surface protein OspE of *Borrelia burgdorferi*. *J. Biol. Chem.* **276**:8427–8435.

31. **Heyman, J. A., J. Cornthwaite, L. Foncerrada, J. R. Gilmore, E. Gontang, K. J. Hartman, C. L. Hernandez, R. Hood, H. M. Hull, W.-Y. Lee, R. Marcil, E. J. Marsh, K. M. Mudd, M. J. Patino, T. J. Purcell, J. J. Rowland, M. L. Sindici, and J. P. Hoeffler.** 1999. Genome-scale cloning and expression of individual open reading frames using topoisomerase I–mediated ligation. *Genome Res.* **9**:383–392.

32. **Hinnebusch, J., S. Bergstrom, and A. G. Barbour.** 1990. Cloning and sequence analysis of linear plasmid telomeres of the bacterium *Borrelia burgdorferi*. *Mol. Microbiol.* **4**:811–820.

33. **Hübner, A., A. T. Revel, D. M. Nolen, K. E. Hagman, and M. V. Norgard.** 2003. Expression of a luxS gene is not required for *Borrelia burgdorferi* infection of mice via needle inoculation. *Infect. Immun.* **71:**2892–2896.

34. **Hübner, A., X. Yang, D. M. Nolen, T. G. Popova, F. C. Cabello, and M. V. Norgard.** 2001. Expression of Borrelia burgdorferi OspC and DbpA is controlled by a RpoN-RpoS regulatory pathway. *Proc. Natl. Acad. Sci. USA* **98:**12724–12729.

35. **Juncker, A. S., H. Willenbrock, G. Von Heijne, S. Brunak, H. Nielsen, and A. Krogh.** 2003. Prediction of lipoprotein signal peptides in gram-negative bacteria. *Protein Sci.* **12:**1652–1662.

36. **Kawabata, H., S. J. Norris, and H. Watanabe.** 2004. BBE02 disruption mutants of *Borrelia burgdorferi* B31 have a highly transformable, infectious phenotype. *Infect. Immun.* **72:**7147–7154.

37. **Kobryn, K., and G. Chaconas.** 2002. ResT, a telomere resolvase encoded by the Lyme disease spirochete. *Mol. Cell* **9:**195–201.

38. **Kobryn, K., and G. Chaconas.** 2001. The circle is broken: telomere resolution in linear replicons. *Curr. Opin. Microbiol.* **4:**558–564.

39. **Kraiczy, P., C. Skerka, M. Kirschfink, V. Brade, and P. F. Zipfel.** 2001. Immune evasion of *Borrelia burgdorferi* by acquisition of human complement regulators FHL-1/reconectin and factor H. *Eur. J. Immunol.* **31:**1674–1684.

40. **Labandeira-Rey, M., E. Baker, and J. Skare.** 2001. Decreased infectivity in *Borrelia burgdorferi* strain B31 is associated with loss of linear plasmid 25 or 28-1. *Infect. Immun.* **69:**446–455.

41. **Labandeira-Rey, M., J. Seshu, and J. T. Skare.** 2003. The absence of linear plasmid 25 or 28-1 of *Borrelia burgdorferi* dramatically alters the kinetics of experimental infection via distinct mechanisms. *Infect. Immun.* **71:**4608–4613.

42. **Lawrenz, M. B., H. Kawabata, J. E. Purser, and S. J. Norris.** 2002. Decreased electroporation efficiency in *Borrelia burgdorferi* containing linear plasmids lp25 and lp56: impact on transformation of infectious *B. burgdorferi*. *Infect. Immun.* **70:**4798–4804.

43. **Levett, P. N.** 2001. Leptospirosis. *Clin. Rev. Microbiol.* **14:**296–326.

44. **Li, X., X. Liu, D. S. Beck, F. S. Kantor, and E. Fikrig.** 2006. *Borrelia burgdorferi* lacking BBK32, a fibronectin-binding protein, retains full pathogenicity. *Infect Immun.* **74:**3305–3313.

45. **Liu, Q., M. Z. Li, D. Leibham, D. Cortez, and S. J. Elledge.** 1998. The univector plasmid-fusion system, a method for rapid construction of recombinant DNA without restriction enzymes. *Curr. Biol.* **8:**1300–1309.

46. **Margolis, N., D. Hogan, K. Tilly, and P. A. Rosa.** 1994. Plasmid location of *Borrelia* purine biosynthesis gene homologs. *J. Bacteriol.* **176:**6427–6432.

47. **Margulies, M., M. Egholm, W. E. Altman, S. Attiya, J. S. Bader, L. A. Bemben, J. Berka, M. S. Braverman, Y. J. Chen, Z. Chen, S. B. Dewell, L. Du, J. M. Fierro, X. V. Gomes, B. C. Godwin, W. He, S. Helgesen, C. H. Ho, G. P. Irzyk, S. C. Jando, M. L. Alenquer, T. P. Jarvie, K. B. Jirage, J. B. Kim, J. R. Knight, J. R. Lanza, J. H. Leamon, S. M. Lefkowitz, M. Lei, J. Li, K. L. Lohman, H. Lu, V. B. Makhijani, K. E. McDade, M. P. McKenna, E. W. Myers, E. Nickerson, J. R. Nobile, R. Plant, B. P. Puc, M. T. Ronan, G. T. Roth, G. J. Sarkis, J. F. Simons, J. W. Simpson, M. Srinivasan, K. R. Tartaro, A. Tomasz, K. A. Vogt, G. A. Volkmer, S. H. Wang, Y. Wang, M. P. Weiner, P. Yu, R. F. Begley, and J. M. Rothberg.** 2005. Genome sequencing in microfabricated high-density picolitre reactors. *Nature* **437:**376–380.

48. **McKevitt, M., M. B. Brinkman, M. McLoughlin, C. Perez, J. K. Howell, G. M. Weinstock, S. J. Norris, and T. Palzkill.** 2005. Genome scale identification of *Treponema pallidum* antigens. *Infect. Immun.* **73:**4445-4450.

49. **McKevitt, M., K. Patel, D. Smajs, M. Marsh, M. McLoughlin, S. J. Norris, G. M. Weinstock, and T. Palzkill.** 2003. Systematic cloning of *Treponema pallidum* open reading frames for protein expression and antigen discovery. *Genome Res.* **13:**1665–1674.

50. **Metcalf, W. W., W. Jiang, and B. L. Wanner.** 1994. Use of the rep technique for allele replacement to construct new *Escherichia coli* hosts for maintenance of R6K gamma origin plasmids at different copy numbers. *Gene* **138:**1–7.

51. **Morozova, O. V., L. P. Dubytska, L. B. Ivanova, C. X. Moreno, A. V. Bryksin, M. L. Sartakova, E. Y. Dobrikova, H. P. Godfrey, and F. C. Cabello.** 2005. Genetic and physiological characterization of 23S rRNA and *ftsJ* mutants of *Borrelia burgdorferi* isolated by *mariner* transposition. *Gene* **357:**63–72.

52. **Nascimento, A. L., A. I. Ko, E. A. Martins, C. B. Monteiro-Vitorello, P. L. Ho, D. A. Haake, S. Verjovski-Almeida, R. A. Hartskeerl, M. V. Marques, M. C. Oliveira, C. F. Menck, L. C. Leite, H. Carrer, L. L. Coutinho, W. M. Degrave, O. A. Dellagostin, H. El-Dorry, E. S. Ferro, M. I. Ferro, L. R. Furlan, M. Gamberini, E. A. Giglioti, A. Goes-Neto, G. H. Goldman, M. H. Goldman, R. Harakava, S. M. Jeronimo, I. L. Junqueira-de-Azevedo, E. T. Kimura, E. E. Kuramae,**

E. G. Lemos, M. V. Lemos, C. L. Marino, L. R. Nunes, R. C. de Oliveira, G. G. Pereira, M. S. Reis, A. Schriefer, W. J. Siqueira, P. Sommer, S. M. Tsai, A. J. Simpson, J. A. Ferro, L. E. Camargo, J. P. Kitajima, J. C. Setubal, and M. A. Van Sluys. 2004. Comparative genomics of two Leptospira interrogans serovars reveals novel insights into physiology and pathogenesis. *J. Bacteriol.* **186:**2164–2172.

53. **Norris, S. J.** 1993. Polypeptides of *Treponema pallidum*: progress toward understanding their structural, functional, and immunologic roles. *Microbiol. Rev.* **57:**750–779.

54. **Norris, S. J.** 2006. The dynamic proteome of Lyme disease *Borrelia*. *Genome Biol.* **7:**209.

55. **Norris, S. J., C. J. Carter, J. K. Howell, and A. G. Barbour.** 1992. Low-passage-associated proteins of *Borrelia burgdorferi* B31: characterization and molecular cloning of OspD, a surface-exposed, plasmid-encoded lipoprotein. *Infect. Immun.* **60:**4662–4672.

56. **Pal, U., X. Yang, M. Chen, L. K. Bockenstedt, J. F. Anderson, R. A. Flavell, M. V. Norgard, and E. Fikrig.** 2004. OspC facilitates *Borrelia burgdorferi* invasion of Ixodes scapularis salivary glands. *J. Clin. Invest.* **113:**220–230.

57. **Probert, W. S., and B. J. Johnson.** 1998. Identification of a 47 kDa fibronectin-binding protein expressed by *Borrelia burgdorferi* isolate B31. *Mol. Microbiol.* **30:**1003–1015.

58. **Purser, J. E., M. B. Lawrenz, M. J. Caimano, J. K. Howell, J. D. Radolf, and S. J. Norris.** 2003. A plasmid-encoded nicotinamidase (PncA) is essential for infectivity of *Borrelia burgdorferi* in a mammalian host. *Mol. Microbiol.* **48:**753–764.

59. **Purser, J. E., and S. J. Norris.** 2000. Correlation between plasmid content and infectivity in *Borrelia burgdorferi*. *Proc. Natl. Acad. Sci. USA* **97:**13865–13870.

60. **Qiu, W. G., S. E. Schutzer, J. F. Bruno, O. Attie, Y. Xu, J. J. Dunn, C. M. Fraser, S. R. Casjens, and B. J. Luft.** 2004. Genetic exchange and plasmid transfers in *Borrelia burgdorferi* sensu stricto revealed by three-way genome comparisons and multilocus sequence typing. *Proc. Natl. Acad. Sci. USA* **101:**14150–14155.

61. **Radolf, J. D.** 1995. *Treponema pallidum* and the quest for outer membrane proteins. *Mol. Microbiol.* **16:**1067–1073.

62. **Ramamoorthy, R., and M. T. Philipp.** 1998. Differential expression of *Borrelia burgdorferi* proteins during growth in vitro. *Infect. Immun.* **66:**5119–5124.

63. **Ren, S. X., G. Fu, X. G. Jiang, R. Zeng, Y. G. Miao, H. Xu, Y. X. Zhang, H. Xiong, G. Lu, L. F. Lu, H. Q. Jiang, J. Jia, Y. F. Tu, J. X. Jiang, W. Y. Gu, Y. Q. Zhang, Z. Cai, H. H. Sheng, H. F. Yin, Y. Zhang, G. F. Zhu, M. Wan, H. L.** Huang, Z. Qian, S. Y. Wang, W. Ma, Z. J. Yao, Y. Shen, B. Q. Qiang, Q. C. Xia, X. K. Guo, A. Danchin, I. Saint Girons, R. L. Somerville, Y. M. Wen, M. H. Shi, Z. Chen, J. G. Xu, and G. P. Zhao. 2003. Unique physiological and pathogenic features of *Leptospira interrogans* revealed by whole-genome sequencing. *Nature* **422:**888–893.

64. **Revel, A. T., J. S. Blevins, C. Almazan, L. Neil, K. M. Kocan, J. de la Fuente, K. E. Hagman, and M. V. Norgard.** 2005. bptA (bbe16) is essential for the persistence of the Lyme disease spirochete, *Borrelia burgdorferi*, in its natural tick vector. *Proc. Natl. Acad. Sci. USA* **102:**6972–6977.

65. **Rosa, P. A., K. Tilly, and P. E. Stewart.** 2005. The burgeoning molecular genetics of the Lyme disease spirochaete. *Nat. Rev. Microbiol.* **3:**129–143.

66. **Sadziene, A., D. D. Thomas, and A. G. Barbour.** 1995. *Borrelia burgdorferi* mutant lacking Osp: biological and immunological characterization. *Infect. Immun.* **63:**1573–1580.

67. **Samuels, D. S.** 1995. Electrotransformation of the spirochete *Borrelia burgdorferi*, p. 253–259. *In* J. A. Nickoloff (ed.), *Methods in Molecular Biology*, vol. 47. Humana Press, Inc., Ottawa, Ontario, Canada.

68. **Samuels, D. S., K. E. Mach, and C. F. Garon.** 1994. Genetic transformation of the Lyme disease agent *Borrelia burgdorferi* with coumarin-resistant gyrB. *J. Bacteriol.* **176:**6045–6049.

69. **Sartakova, M. L., E. Y. Dobrikova, D. A. Terekhova, R. Devis, J. V. Bugrysheva, O. V. Morozova, H. P. Godfrey, and F. C. Cabello.** 2003. Novel antibiotic-resistance markers in pGK12-derived vectors for *Borrelia burgdorferi*. *Gene* **303:**131–137.

70. **Scharn, D., H. Wenschuh, U. Reineke, J. Schneider-Mergener, and L. Germeroth.** 2000. Spatially addressed synthesis of amino- and amino-oxy-substituted 1,3,5-triazine arrays on polymeric membranes. *J. Comb. Chem.* **2:**361–369.

71. **Schwan, T. G., and J. Piesman.** 2000. Temporal changes in outer surface proteins A and C of the Lyme disease–associated spirochete, *Borrelia burgdorferi*, during the chain of infection in ticks and mice. *J. Clin. Microbiol.* **38:**382–388.

72. **Seshu, J., M. D. Esteve-Gassent, M. Labandeira-Rey, J. H. Kim, J. P. Trzeciakowski, M. Hook, and J. T. Skare.** 2006. Inactivation of the fibronectin-binding adhesin gene bbk32 significantly attenuates the infectivity potential of *Borrelia burgdorferi*. *Mol. Microbiol.* **59:**1591–1601.

73. **Shuman, S.** 1994. Novel approach to molecular cloning and polynucleotide synthesis using vaccinai DNA topoisomerase. *J. Biol. Chem.* **269:**32678–32684.

74. **Simpson, W. J., W. Cieplak, M. E. Schrumpf, A. G. Barbour, and T. G. Schwan.** 1994. Nucleotide sequence and analysis of the gene in *Borrelia burgdorferi* encoding the immunogenic P39 antigen. *FEMS Microbiol. Lett.* **119:**381–387.

75. **Smajs, D., M. McKevitt, J. K. Howell, S. J. Norris, W. W. Cai, T. Palzkill, and G. M. Weinstock.** 2005. Transcriptome of *Treponema pallidum*: gene expression profile during experimental rabbit infection. *J. Bacteriol.* **187:**1866–1874.

76. **Steere, A. C.** 2001. Lyme disease. *N. Engl. J. Med.* **345:**115–125.

77. **Steere, A. C., J. Coburn, and L. Glickstein.** 2004. The emergence of Lyme disease. *J. Clin. Invest.* **113:**1093–1101.

78. **Sternberg, N., D. Hamilton, S. Austin, M. Yarmolinsky, and R. Hoess.** 1981. Site-specific recombination and its role in the life cycle of P1. *Cold Spring Harbor Symp. Quant. Biol.* **45:**297–309.

79. **Stevenson, B., and K. Babb.** 2002. LuxS-mediated quorum sensing in *Borrelia burgdorferi*, the Lyme disease spirochete. *Infect. Immun.* **70:** 4099–4105.

80. **Stevenson, B., L. K. Bockenstedt, and S. W. Barthold.** 1994. Expression and gene sequence of outer surface protein C of *Borrelia burgdorferi* reisolated from chronically infected mice. *Infect. Immun.* **62:**3568–3571.

81. **Stevenson, B., N. El-Hage, M. A. Hines, J. C. Miller, and K. Babb.** 2002. Differential binding of host complement inhibitor factor H by *Borrelia burgdorferi* Erp surface proteins: a possible mechanism underlying the expansive host range of Lyme disease spirochetes. *Infect. Immun.* **70:**491–497.

82. **Stewart, P. E., J. Hoff, E. Fischer, J. G. Krum, and P. A. Rosa.** 2004. Genome-wide transposon mutagenesis of *Borrelia burgdorferi* for identification of phenotypic mutants. *Appl. Environ. Microbiol.* **70:**5973–5979.

83. **Stewart, P. E., R. Thalken, J. L. Bono, and P. Rosa.** 2001. Isolation of a circular plasmid region sufficient for autonomous replication and transformation of infectious *Borrelia burgdorferi*. *Mol. Microbiol.* **39:**714–721.

84. **Stewart, P. E., X. Wang, D. M. Bueschel, D. R. Clifton, D. Grimm, K. Tilly, J. A. Carroll, J. J. Weis, and P. A. Rosa.** 2006. Delineating the requirement for the *Borrelia burgdorferi* virulence factor OspC in the mammalian host. *Infect. Immun.* **74:**3547–3553.

85. **Strugnell, R., A. Cockayne, and C. W. Penn.** 1990. Molecular and antigenic analysis of treponemes. *Crit. Rev. Microbiol.* **17:**231–250.

86. **Tilly, K., J. G. Krum, A. Bestor, M. W. Jewett, D. Grimm, D. Bueschel, R. Byram, D. Dorward, M. J. Vanraden, P. Stewart, and P.** Rosa. 2006. *Borrelia burgdorferi* OspC protein required exclusively in a crucial early stage of mammalian infection. *Infect. Immun.* **74:**3554–3564.

87. **Tourand, Y., T. Bankhead, S. L. Wilson, A. D. Putteet-Driver, A. G. Barbour, R. Byram, P. A. Rosa, and G. Chaconas.** 2006. Differential telomere processing by *Borrelia* telomere resolvases in vitro but not in vivo. *J Bacteriol.* **188:**7378–7386.

88. **Tourand, Y., K. Kobryn, and G. Chaconas.** 2003. Sequence-specific recognition but position-dependent cleavage of two distinct telomeres by the *Borrelia burgdorferi* telomere resolvase, ResT. *Mol. Microbiol.* **48:**901–911.

89. **von Lackum, K., K. Babb, S. P. Riley, R. L. Wattier, T. Bykowski, and B. Stevenson.** 2006. Functionality of *Borrelia burgdorferi* LuxS: the Lyme disease spirochete produces and responds to the pheromone autoinducer-2 and lacks a complete activated-methyl cycle. *Int. J. Med. Microbiol.* **296**(Suppl. 40)**:**92–102.

90. **Wallich, R., C. Brenner, M. D. Kramer, and M. M. Simon.** 1995. Molecular cloning and immunological characterization of a novel linear-plasmid-encoded gene, pG, of *Borrelia burgdorferi* expressed only in vivo. *Infect Immun.* **63:**3327–3335.

91. **Weinstock, G. M., J. M. Hardham, M. P. McLeod, E. J. Sodergren, and S. J. Norris.** 1998. The genome of *Treponema pallidum*: new light on the agent of syphilis. *FEMS Microbiol. Rev.* **22:**323–332.

92. **Weinstock, G. M., S. J. Norris, E. Sodergren, and D. Smajs.** 2000. Identification of virulence genes in silico: infectious disease genomics, p. 251–261. *In* K. Brogden (ed.), *Virulence Mechanisms of Bacterial Pathogens*, 3rd ed. ASM Press, Washington, DC.

93. **Xu, Y., C. Kodner, L. Coleman, and R. C. Johnson.** 1996. Correlation of plasmids with infectivity of *Borrelia burgdorferi* sensu stricto type strain B31. *Infect. Immun.* **64:**3870–3876.

94. **Yang, X. F., M. C. Lybecker, U. Pal, S. M. Alani, J. Blevins, A. T. Revel, D. S. Samuels, and M. V. Norgard.** 2005. Analysis of the ospC regulatory element controlled by the RpoN-RpoS regulatory pathway in *Borrelia burgdorferi*. *J. Bacteriol.* **187:**4822–4829.

95. **Yang, X. F., U. Pal, S. M. Alani, E. Fikrig, and M. V. Norgard.** 2004. Essential role for OspA/B in the life cycle of the Lyme disease spirochete. *J. Exp. Med.* **199:**641–648.

96. **Zhang, J. R., J. M. Hardham, A. G. Barbour, and S. J. Norris.** 1997. Antigenic variation in Lyme disease borreliae by promiscuous recombination of VMP-like sequence cassettes. *Cell* **89:** 275–285.

CAMPYLOBACTER PATHOGENOMICS: GENOMES AND BEYOND

Derrick E. Fouts, Emmanuel F. Mongodin, and Karen E. Nelson

7

Members of the genus *Campylobacter* are defined as fastidious, microaerophilic, oxidase positive, nonfermentative, gram-negative bacteria (210). They form spirally curved or corkscrew-shaped rods with a size ranging from 0.2 to 0.8 μm wide and 0.5 to 5 μm long (210); however, some species form straight rods and grow under anaerobic conditions (122). Currently the genus *Campylobacter* is composed of 16 official species (http://www.bacterio.cict.fr/c/campylobacter.html#r), which include *C. coli* (4, 44, 214), *C. concisus* (199), *C. curvus* (200, 212), *C. fetus* (182, 189), *C. gracilis* (199, 211), *C. helveticus* (190), *C. hominis* (122), *C. hyointestinalis* (67), *C. jejuni* (102, 214), *C. lanienae* (130), *C. lari* (14), *C. mucosalis* (123, 175), *C. rectus* (199, 212), *C. showae* (52), *C. sputorum* (172, 214), and *C. upsaliensis* (177). When the name of a newly discovered species, *C. insulaenigrae* (58), is finalized, there will be 17 species of *Campylobacter*.

Campylobacter fetus (*Vibrio fetus*) was the first member and type species of the *Campylobacteriaceae* family. It was first documented as a veterinary pathogen in 1909; it causes infertility and infectious abortion in cattle. In 1938, *C. jejuni* (*Vibrio jejuni*) was the first recognized member of the *Campylobacteriaceae* to cause human illness. Even though *C. jejuni* is now recognized as the most common cause of human bacterial gastroenteritis, responsible for at least 400 to 500 million cases per year (62), it too was first documented as an animal pathogen, causing dysentery in calves (102) and swine (43, 44). It was not until the invention of selective culturing techniques in the 1970s and 1980s that researchers were routinely able to isolate *C. jejuni* (and other *Campylobacter* spp.) from the stools of patients with gastroenteritis (23, 26, 75, 76, 105), which led to the observations that *C. jejuni* and *C. coli* play an important role in causing human disease. Certain strains of *C. jejuni* are commensal in cattle, swine, and birds (140, 192). *C. jejuni* is subdivided into two subspecies, *jejuni* and *doylei*, that can be biochemically differentiated by a nitrate reduction test: *jejuni* reduces nitrate, and *doylei* does not (191). The pathogenic role of *doylei* is unknown; however, it has been isolated from ulcerated gastric tissue, diarrhea, and the blood of humans (191).

Human illness is also associated, albeit infrequently, with other *Campylobacter* species, such as *C. upsaliensis, C. lari, C. fetus, C. concisus,* and *C. curvus*. New isolation techniques that do not rely on antibiotic resistance have revealed *C. upsalien-*

Derrick E. Fouts and Emmanuel F. Mongodin Department of Microbial Genomics, The Institute for Genomic Research, Rockville, Maryland 20850. *Karen E. Nelson* Department of Microbial Genomics, The Institute for Genomic Research, Rockville, Maryland 20850, and Department of Biology, Howard University, Washington DC 20059.

Bacterial Pathogenomics, Edited by M. J. Pallen et al.

sis to be more associated with human disease than previously known, suggesting the species to be an emerging pathogen (121). The only known nonpathogenic *Campylobacter* species is the human commensal *C. hominis* (122). The main route of *C. jejuni* infection is through improperly handled or undercooked poultry. *C. coli*, and to a lesser extent *C. lari* and *C. upsaliensis*, have also been isolated from poultry products. *C. coli*, however, is associated predominantly with swine (134, 135, 224), and *C. lari* has been isolated from shellfish (49, 213, 228); it is unclear what food, if any, is the main source of *C. upsaliensis* infection, but it has been isolated from domestic dogs and cats (50, 83, 149, 178, 186).

In 1947, *C. fetus* was first recognized as a human pathogen when it caused fever and abortion in humans (216). Today, an increased awareness of human illness from *C. fetus* has led this organism to be recognized as causing emerging infectious disease in humans, presumably as a result of the mishandling and undercooking of meat (19). Two closely related subspecies are recognized: *C. fetus* subsp. *venerealis* and *C. fetus* subsp. *fetus*. In cattle, *C. fetus* subsp. *venerealis* is transmitted from infected bulls to the heifer, resulting in infertility (16). In sheep, cattle, and other ungulates, *C. fetus* subsp. *fetus* is carried in the gallbladder; newly infected animals develop bacteremia, and if pregnant, abortion is common (205). In both the genital tract (*C. fetus* subsp. *venerealis*) and the gastrointestinal tract (*C. fetus* subsp. *fetus*), *C. fetus* can persist for years on mucosal surfaces (181), increasing the likelihood of its spreading to both humans and other animals.

PATHOGENESIS: DISEASE RAMIFICATIONS

Clinical symptoms of *C. jejuni* subsp. *jejuni* can range from asymptomatic to afebrile or febrile dysentery, but the most common symptoms are abdominal cramping and acute watery diarrhea that can last from 2 to 12 days. During the acute phase of infection, a subset of infected individuals experience fever, headache, nausea, vomiting, sweats and chills, and myal-

gia. Although most *C. jejuni* subsp. *jejuni* infections result in uncomplicated gastroenteritis, a small percentage of individuals develop the immune-mediated neurological disorders Guillain-Barré syndrome and Miller-Fisher syndrome (232), the rheumatic disorder reactive arthritis (106), or, recently, the tumorigenic disorder immunoproliferative small intestinal disease (125).

Guillain-Barré syndrome (GBS) is clinically defined as an acute paralytic neuropathy characterized by a symmetrical, ascending paralysis accompanied by loss of muscle tone, the absence of a reflex, and cerebrospinal fluid with increased protein levels but lacking cells (81). Two forms of GBS are associated with prior *C. jejuni* infection: acute inflammatory demyelinating polyneuropathy and acute motor axonal neuropathy (90). A component of Schwann cells is the target of autoimmune attack in the acute inflammatory demyelinating polyneuropathy form of GBS, while the attack is targeting the axolemma and nodes of Ranvier in the acute motor axonal neuropathis form (82). It is estimated that one in every 1,000 *C. jejuni* infections manifest to GBS in the United States (5). A greater number of GBS-associated *C. jejuni* strains have heat-stable (Penner) serotypes O:19 in Japan and O:41 in South Africa, but this relationship is not always present and is not as strong in other countries (152). GM_1 and GD_{1a}-like ganglioside-mimicking structures expressed within the lipooligosaccharide (LOS) are observed more frequently in neuropathy-associated C. *jejuni* stains than those strains causing only enteritis (153). More specifically, the class A LOS locus is associated with GBS and the presence of the GM_1-like structure, but not a related neuropathy Miller-Fisher syndrome (72). There is still no definitive diagnostic test that can differentiate GBS-associated *C. jejuni* from non-GBS-associated *C. jejuni* strains. Non-GBS-causing strains of *C. jejuni* can also produce ganglioside mimicks; therefore, there might be additional factors expressed by GBS-causing *C. jejuni* strains that remain to be characterized, or the difference is host based.

Miller-Fisher syndrome (MFS) is the most common clinical variant of GBS and is characterized as an acute onset of ataxia (the failure of muscle coordination), areflexia (absence of a reflex), and ophthalmoplegia (paralysis or weakness of the muscles that control the eye movement) (56). During the acute phase of the illness, most patients with MFS generate autoantibodies directed against some of the gangliosides expressed by the ocular motor nerves and the dorsal root ganglia, causing the onset of clinical symptoms (118). It is believed that certain *C. jejuni* strains can produce surface structures that mimic these ganglioside structures, stimulating the host antibody-mediated autoimmune response (233).

Reactive arthritis is typically an oligoarticular or polyarticular arthritis that is asymmetric and involves the knees, ankles, or wrists (106). Reactive arthritis is an autoimmune inflammatory arthritis that results from previous infection by certain gastrointestinal bacteria, including *C. jejuni, C. coli, Salmonella enterica* serovar Typhimurium, *Yersinia enterocolitica, Shigella flexneri, S. dysenteriae,* and the urogenital bacterium *Chlamydia trachomatis.* Recent PCR studies that used genus-specific primers from the synovial fluid of patients with a history of recent *Campylobacter* enteritis found the presence of *C. concisus* and *C. (Bacteroides) ureolyticus* 16S rRNA sequences (33). However, they were unable to detect the presence of *C. jejuni* or *C. coli* in these joints. Another form of inflammatory arthritis is septic arthritis where bacteria grow in the affected joint, causing inflammation. *C. jejuni* has been isolated from infected joints, but it is rare and is observed only in immunocompromised patients (165).

Immunoproliferative small intestinal disease, or alpha (α) chain disease, is a mucosa-associated lymphoid-tissue lymphoma that accounts for approximately one-third of gastrointestinal lymphomas in the Middle East and has been recently associated with *C. jejuni* infection (125). The disease has a number of clinical phenotypes, ranging from benign lymphoid infiltration to malignant diffuse large-B-cell lymphoma (125). The pathogenesis of immunoproliferative small intestinal disease by *C.*

jejuni is analogous to gastric mucosa–associated lymphoid-tissue lymphomas generated by *Helicobacter pylori* infection with some differences. During the course of *C. jejuni* infection, B and T cells are recruited to the intestinal mucosa, where an autoreactive B-cell clone grows, driven by *C. jejuni*–reactive T cells. This clone then secretes abnormal α heavy chain antibodies that lack a light chain, resulting in proliferative differentiation of plasma cells (162). This is in contrast to *H. pylori* mucosa-associated lymphoid-tissue lymphoma, where it is rare to have such robust differentiation of plasma cells and where secreted antibodies are normal (162). In both situations, application of antibiotics can cause the lymphoma to regress if given during the early stages of the disease. *C. jejuni* was identified by PCR from frozen intestinal tissue but has not been cultured.

In addition to causing acute diarrhea, the immune-mediated neurological disorders GBS and MFS, the rheumatic disorder reactive arthritis, and the tumorigenic disorder immunoproliferative small intestinal disease, *C. jejuni* has been isolated from human blood as the cause of both symptomatic (120) and asymptomatic (20) bacteremia, from cerebrospinal fluid as the cause of bacterial meningitis (20), from urine as the cause of a bladder infection (34), and from a gastric biopsy sample as the cause of gastritis (157).

PATHOGENESIS: MOLECULAR PATHOGENESIS

The molecular mechanisms responsible for acute intestinal infections by *C. jejuni,* although still poorly understood, are thought to involve adherence, cellular invasion, and toxin production (Fig. 1), but not all clinical isolates of *C. jejuni* are able to invade cultured human cells or produce defined toxins (109). However, a common feature of *Campylobacter* infectious enterocolitis is a localized acute inflammatory response that can lead to tissue damage and may be responsible for many of the clinical symptoms (109). Motility is the major factor that has been implicated directly in intestinal colonization (223). The two-component system RacR-RacS has been involved in a temperature-dependent signaling pathway and is required for the organism to colonize the chicken intestinal tract (24). Adherence of *C. jejuni* to epithelial cells (Fig. 1C) is thought to be mediated by multiple adhesins, including CadF, PEB1, JlpA, and a 43-kDa major outer membrane protein. General protein glycosylation has also been shown to be important for adherence and invasion in vitro, colonization of chicks, and colonization in a mouse model (104). Additionally, glycosylation of the *Campylobacter* flagellin is thought to provide protection from the host immune system (204).

There are a number of reports of a variety of enterotoxin and cytotoxin activities (70-kDa cytotoxin, Vero/HeLa cell cytotoxin, cytolethal distending toxin [CDT], Shiga-like toxin, hemolytic cytotoxins, and hepatotoxin) from different *Campylobacter* strains, but only CDT has been sequenced and reproducibly characterized (220). CDTs from enteropathogenic *Escherichia coli* have been shown to disrupt the barrier function of host intestinal epithelial tight junctions (150), causing cell cycle arrest in the G_2/M phase with progressive distension and death of Chinese hamster ovary (CHO) cells (101). The three CDTs, A, B, and C, are conserved across the five sequenced *Campylobacter* strains (60), and although they were shown to be important in causing gastroenteritis in an NF-κB-deficient mouse model (61), their precise role in the *Campylobacter* pathogenesis is still under investigation. LOS and capsular (extracellular) polysaccharide (EP) are important surface structures in *C. jejuni* that function in the interactions of the organism with the environment. Interesting aspects of *C. jejuni* LOS are their molecular mimicry of host gangliosides and their presumed roles in evading host immune responses and autoimmunity (147), decreased immunogenicity (79), and attachment and invasion (11).

GENOME SEQUENCE OF *C. JEJUNI* NCTC 11168

The genome sequence of the *C. jejuni* NCTC 11168, a human clinical isolate (serotype HS:2), was completed in 2000 (161). The purpose for sequencing this genome, as with many bacterial genome-sequencing projects,

was to understand the biology of *Campylobacter* from the complete genome sequence in the hopes that disease prevention and food-chain-control strategies can be developed. Unlike the more commonly studied food-borne pathogens *E. coli* and *Salmonella*, little was known about *Campylobacter* genetics, physiology, and virulence before the genome sequence project.

Genome Features

The genome consists of a single circular chromosome of 1,641,481 bp, 30.6% G+C, and coding for 1,654 proteins (161). Oddly enough, this genome was unique in that it was completely devoid of insertion sequence (IS) elements, transposons, retron elements, prophages, and plasmids and lacking repetitive DNA outside of the usual rRNA operon repeat elements. Another noteworthy genomic feature was the observation that there were very few operon structures—coordinately transcribed genes serving a similar biological pathway or function. Specifically, only certain amino acid biosynthetic genes, LOS, EP, flagellar, and ribosomal protein genes appeared in operons.

Hypervariable polyG:C Tracts

The analysis of regions of the genome that had sequence discrepancies revealed the presence of hypervariable stretches of homopolymeric G and C residues. This phenomenon had been previously observed in other pathogens such as *Neisseria meningitidis*, where it was presumed to enable rapid adaptation to the selective pressure of the host immune system by altering the coding frame or promoter spacing through slipped-strand mispairing. The authors suggested that the frequency of occurrence of the polyG:C tracts in the NCTC 11168 assemblies was greater than that observed in *Neisseria*. The genes that contain variable lengths of G or C tracts are called contingency genes. The presence of these hypervariable G:C tracts suggest a role in pathogenesis for these genes. Some contingency genes whose function could be determined include LOS biosynthesis, EP biosynthesis, and flagellar modification genes. The presence of such a high frequency

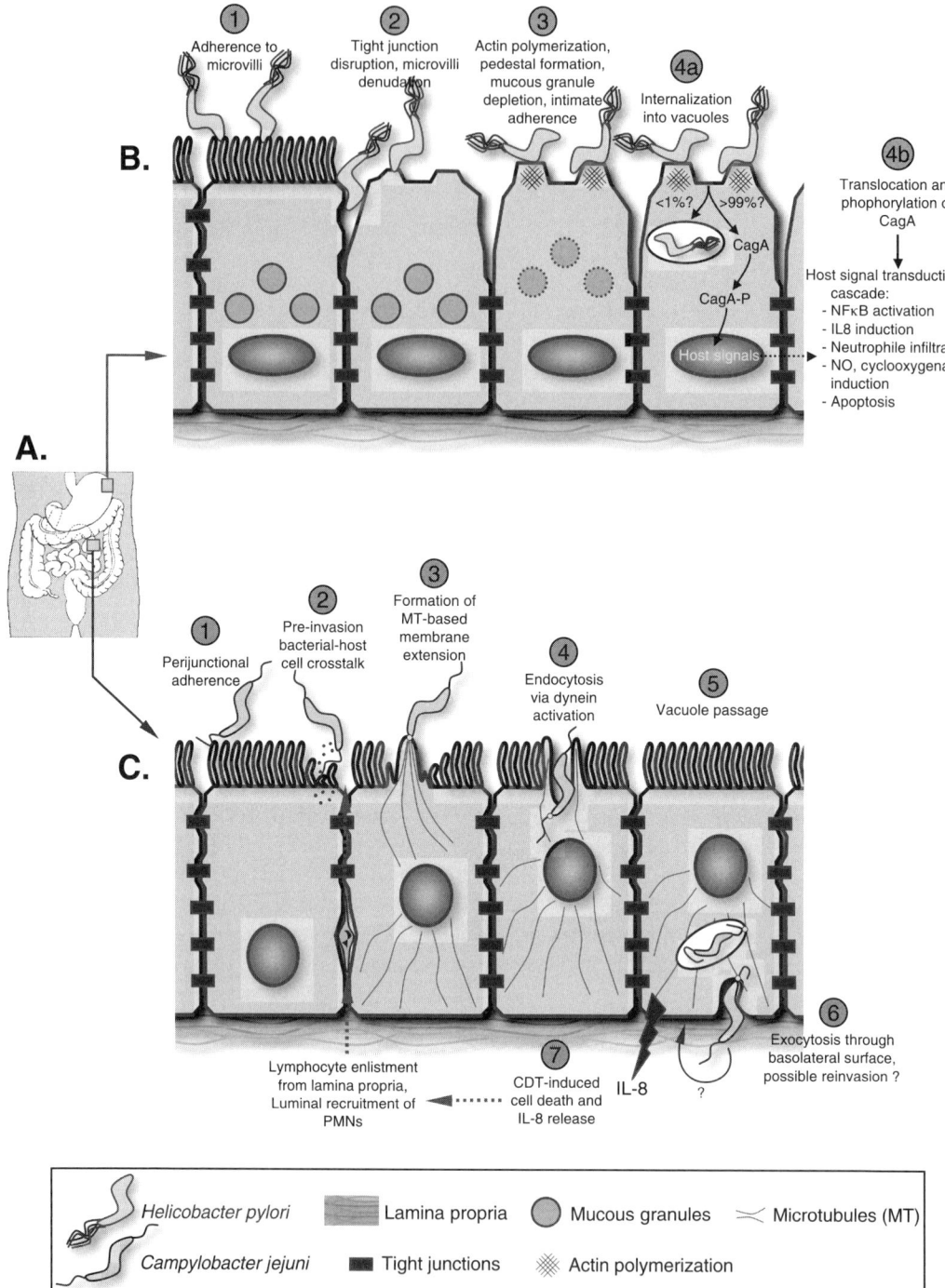

FIGURE 1 Schematic illustration comparing the steps of invasion and pathogenesis of the gastrointestinal pathogens *H. pylori* and *C. jejuni*. (A) Schematic representation of the human gastrointestinal tract and sites of colonization of *H. pylori*, largely confined to the antrum (part of the stomach that lacks acid-secreting parietal cells) and *C. jejuni*, which targets the ileal-colonic epithelial cells. (B) Invasion mechanisms of *H. pylori*. *H. pylori* probably adheres first to microvilli-containing regions of gastric epithelial cells (step 1). After adherence, microvilli are denuded and tight junctions disrupted under the action of urease; the VacA toxin is probably also involved in

of these variants suggested that *Campylobacter* lacks DNA repair capabilities. The authors noted the absence of *ada, phr, tag, alkA, mutM, nfo, vsr, mutH, mutL, sbcB, lexA, umuC,* and *umuD,* which are involved in DNA repair processes.

Virulence Genes

The only toxins found in the NCTC 11168 genome were the already-known CDT (subunits *cdtA-C*) and a putative hemolysin. The only other virulence factor mentioned in the genome paper was a gene encoding phospholipase A. There were no S-layer genes, which are important virulence factors for *C. fetus* (21) and *C. rectus* (155), or any homologs to the *H. pylori* virulence factors required for host cell invasion (e.g., the Hop porin, urease, vaculating toxin [*vacA*] or the cytotoxin-associated gene [*cag*] pathogenicity island [Fig. 1B]). Furthermore, a search for type III (other than the flagellar inner membrane apparatus) and type IV secretion systems was also futile, yielding negative results. If this strain of *Campylobacter* is secreting proteins outside of the cell, it uses type I systems, the TAT secretion system or the flagellar secretion system. If one considers iron acquisition/uptake systems virulence factors, then NCTC 11168 has five major systems: the enterochelin uptake operon, the siderophore receptor ortholog *cfrA*, a putative hemin uptake system, a periplasmic binding-protein-dependent system, and another predicted siderophore receptor.

Regulatory networks can be very important for pathogenesis, enabling the pathogen to tightly control its arsenal of virulence factors to be made at the appropriate time and location to enable a successful infection. It was shown that *Campylobacter* encodes a larger array of regulatory components than does the related pathogen *H. pylori*, the bulk of which are putative two-component regulatory proteins (sensor kinase and response regulator). The abundance of these sensory units undoubtedly enables *Campylobacter* cells to respond to a greater number of environmental signals than does *Helicobacter*, which is only known to grow in the stomach of animals. Interestingly, only three sigma factors were identified (*rpoD, rpoN,* and *fliA*), which means that any regulatory signal must pass through at least one of these sigma factors.

In contrast to other enteric pathogens, little biochemical characterization of *Campylobacter* metabolism was conducted before the genome-sequencing project. Analysis of the genome sequence revealed few genes for the degradation of carbohydrates and amino acids, an incomplete glycolytic pathway, and an intact TCA cycle. The glycolytic pathway is lacking the genes for glucokinase and 6-phosphofructokinase. Although the TCA cycle appears to have a complete set of genes, *Campylobacter* was shown experimentally (95) (and verified by the genome sequence [161]) to substitute 2-oxoglutarate ferredoxin oxidoreductase for 2-oxoglutarate dehydrogenase. It was noted

---

this process (step 2). Cuplike projections and pedestals form, accompanied by actin polymerization and cytoskeletal rearrangements (steps 2 and 3). Bacterial genes involved in the steps immediately after adherence are still poorly understood, and it is not clear at what stage rare *H. pylori* bacteria become internalized (step 4a). Genes from the *cag* (cytotoxin-associated gene) pathogenicity island are implicated in the postadherence mechanism of infection (step 4b) and are responsible for the induction of an intense inflammatory reaction. (C) *C. jejuni* pathogenesis. *C. jejuni* adheres to the apical cell surface of the perijunctional region of the intestinal epithelium (step 1). Among the multiple bacterial adhesions that are thought to be involved in *C. jejuni* adherence to intestinal epithelial cells are the 37k-Da adhesin CadF (which is believed to target the host fibronectin receptors), PEB1, JlpA, and a 43-kDa major outer membrane protein. Attached bacteria then secrete invasion effectors into the host cells (step 2), among which the 73-kDa secreted protein CiaB, resulting in the phosphorylation of host proteins and the release of Ca^{2+} from intracellular stores. Host signaling cascades trigger a localized disruption of cortical actin filaments and an extension of microtubules to form a membrane extension (step 3), resulting in the endocytosis of the *C. jejuni* cells (step 4). The vacuole-engulfed bacterium (step 5) then moves via dynein along the microtubules to the basolateral surface for exocytosis (step 6). Infected cells release interleukin-8 basolaterally, which enlists lymphocytes from the lamina propria. The two schematic illustrations in (B) and (C) were modified from (203) and (115).

that *Campylobacter* has the genes to synthesize purines, thiamine, and amino acids, whereas *H. pylori* does not.

COMPARATIVE GENOMICS OF *CAMPYLOBACTER* SPECIES

Knowledge of the genes and gene products of one strain is insufficient to provide a complete picture of other aspects of *C. jejuni* biology, including colonization of chickens and other animals (3), variation in LOS (80, 171) and capsule (12), and potential adaptations of *C. jejuni* in poultry production and processing environments (176).

To gain additional information about the genetic basis of *Campylobacter* attachment and infection, its evolution and persistence in the food chain, and the potential virulence of the "emerging" *Campylobacter* species, we fully sequenced and finished the genome of the *C. jejuni* poultry isolate RM1221 and compared it with the incomplete genomes (with eightfold coverage) of three additional *Campylobacter* species: *C. coli*, *C. lari*, and *C. upsaliensis* (60). *C. jejuni* RM1221 (HS:53) was sequenced because RM1221 was isolated from a market chicken carcass and passaged minimally (139); preliminary suppressive subtractive hybridization analysis revealed the presence of many putative RM1221 strain-specific genes; RM1221 colonized chicken ceca efficiently (e.g., 10^{12} CFU/g of tissue); RM1221 attached well to chicken skin and invaded Caco-2 cells (139); RM1221 had different LOS and capsule loci compared with NCTC 11168 and other characterized *C. jejuni* strains; and RM1221 responded to a variety of carbon substrates differently than NCTC 11168. *C. coli* RM2228 was isolated from a chicken carcass obtained from an inspected slaughter plant and was chosen for sequencing because it possesses resistance to multiple antibiotics, including aminoglycosides, cephalosporins, macrolides, sulfonamides, and tetracyclines. *C. lari* RM2100 was a human isolate obtained from the Centers for Disease Control and Prevention, Atlanta, GA (CDC strain D67, "case 6") (201). *C. upsaliensis* RM3195 was obtained from the feces of a 4-year-old GBS patient. The isolation procedure

involved a filtration method with selection of *Campylobacter* cells in diluted feces by their migration through a 0.6-μm membrane filter and subsequent growth on nonselective media (124).

All predicted proteins from the five *Campylobacter* genomes were compared with data from other published microbial genomes by BLASTP (7). For the identification of genus-specific, strain-specific, and species-specific open reading frames (ORFs), the protein sequences from five sequenced *Campylobacter* genomes (*C. jejuni* NCTC 11168 [161]), *C. jejuni* RM1221, *C. coli* RM2228, *C. lari* RM2100, and *C. upsaliensis* RM3195), three *Helicobacter* genomes (*H. pylori* 26695 [206], *H. pylori* J99 [6], and *H. hepaticus* ATCC51449 [196]), and the genome of *Wolinella succinogenes* DSMZ1740 (9) were compared, by WU-BLASTP (http://blast.wustl.edu) against a database composed of 195 prokaryotic, 8 eukaryotic, 175 phage, 63 virus, and 46 plasmid genomes. Specifically, bidirectional best matches were scored that met the following prerequisites: a P value of $\leq 1 \times 10^{-5}$ or less, >35% or more identity, and match lengths of at least 75% of the length of both query and subject. Novel ORFs were those proteins that had no WU-BLASTP match to anything in the database.

Hypervariable homopolymeric G or C tracts were identified by analyzing the underlying sequences for each nucleotide within a tract of six or more G or C nucleotides. A hypervariable tract was considered of high quality when its underlying sequence comprised at least three sequencing reads with an average Phred score greater than 30 (53, 54). By mapping the position of the variable nucleotides to the annotation, it was possible to determine the location of the polymorphism (intergenic vs. intragenic) and its effect on the deduced polypeptide (altered peptide sequence, truncation, no translation).

Comparative Genome Features

The genome of *C. jejuni* RM1221 is a single circular chromosome 1,777,831 bp in length with an average G+C content of 30.31%. There are a total of 1,884 predicted coding regions in the genome with an average ORF length of 885 bp. Ninety-four percent of the genome represents coding sequence. Putative functional assignments could be made for 1,124 of the ORFs (60%) (Color Plate 1). The bacterium was found to belong to multilocus sequence type 354 and FlaA short variable region 33, which belongs to clonal complex 354, whose members are associated with human disease, chickens, or chicken meat (39). The genome features for the unfinished *Campylobacter* genomes were based on automated analysis. The average coverage of the unfinished genomes was found to be 8.5-fold for *C. coli* RM2228, 16.5-fold for *C. lari* RM2100, and 9.0-fold for *C. upsaliensis* RM3195 for those contigs used to construct the pseudomolecules. The ambiguity rate (number of consensus-altering ambiguities per base pair) was determined to be between 1:54,000 and 1:93,000 for these unedited, unfinished genomes at eightfold depth of coverage. The genomic structure of *C. jejuni* RM1221 is syntenic with the genome of *C. jejuni* NCTC 11168 and is disrupted by inserted prophages/genomic islands in RM1221 (see below) and ORFs within the EP loci in NCTC 11168 (Color Plate 1). The *C. coli* RM2228 genomic structure also has a considerable amount of synteny with *C. jejuni* RM1221, sharing similar breakpoints as observed in the *C. jejuni* comparisons, but displaying evidence of rearrangements about the oriC as described for other bacterial genomes (46). In contrast, *C. lari* and *C. upsaliensis* possess little synteny with *C. jejuni* RM1221.

Comparison of *C. jejuni* RM1221 protein sequences with those of other fully sequenced members of the ε-*Proteobacteria* revealed 540 shared protein sequences, many of which are proposed to have housekeeping functions (Color Plate 2C). Of the 1,084 protein sequences shared by all the *Campylobacter* species in this study, 46 had no match to any other organism in the database (*P* value cutoff ≤ 10^{-5}) (Color Plate 2B). Eleven of these were assigned functions related to cell envelope biosynthesis, or fatty acid and phospholipid

metabolism. Further analysis revealed 44 proteins considered *C. jejuni* specific, of which 12 mostly hypothetical proteins were truly novel, with no match to other organisms in the database. Of the 300 *C. jejuni* RM1221–specific protein sequences, only 95 were not in phage or genomic island regions.

To quantify relatedness among the sequenced ε-*Proteobacteria*, the average percentage of protein identity was computed for all proteins matching the reference strain *C. jejuni* RM1221 with a *P* value of ≤10^{-5}, identity 35% or better, and match lengths of at least 75% of the length of both query and subject sequence. Not surprisingly, *C. jejuni* NCTC 11168 had the highest average percentage of protein identity (1,468 proteins averaging 98.41% identity) with *C. jejuni* RM1221 proteins. *C. coli* RM2228 was second, with 1,399 proteins averaging 85.81% identity. Surprisingly, *C. upsaliensis* RM3195 had the third highest average percentage of protein identity with *C. jejuni* RM1221 (1,261 proteins, 74.72%) followed by *C. lari* RM2100, with 1,251 proteins having 68.91% average identity. This was surprising because a 16S rRNA tree depicts *C. upsaliensis* to be more dissimilar to *C. jejuni*, *C. coli*, and *C. lari* (210). *Wollinela succinogenes* DSMZ1740 was next, with 838 proteins averaging 53.77% identity, *H. hepaticus* ATCC 51449 (770 proteins, 53.66%), *H. pylori* 26695 (675 proteins, 52.39%), and *H. pylori* J99 (682 proteins, 52.28%).

Phylogenetic Comparisons

To resolve the apparent discrepancy between the relatedness of the ε-*Proteobacteria* from the results of average percentage of protein identities from this study and the previously published 16S rRNA tree based on the percentage sequence similarity (210), a consensus bootstrapped maximum-likelihood tree was generated on the basis of trimmed alignments with gaps removed (Fig. 2A). One of the advantages of generating a whole-genome sequence is the magnitude of information available for resolving differences between closely related organisms. To better resolve the *Campylobacter* species, we took advantage of the wealth of sequence information to construct a maximum-likeli-hood concatenated protein tree by using a set of 12 conserved protein sequences that have been previously shown to be reliable markers for phylogenetic analysis (Fig. 2B) (25, 179). A frequency distribution of percentage of pro-tein identity was plotted with 5% class inter-vals to visualize the similarities of these genomes at the protein level (Color Plate 2D). The 16S rRNA tree of sequenced members of the ε-*Proteobacteria* suggests that *C. jejuni* RM1221 is more related to *C. coli* RM2228 than to the other *C. jejuni* strain NCTC 11168. However, the concatenated protein tree of these same organisms showed the two *C. je-juni* strains to be more related to each other than either is to *C. coli* RM2228, a finding that agrees with the distributions of the percentage of protein identities (Color Plate 2D). Both trees indicated that *W. succinogenes* is more re-lated to *Helicobacter* than to *Campylobacter*. Most likely, the protein tree is more accurate and the rRNA tree is incorrect because the 16S rRNA does not have enough variation to resolve these close relationships (179). Whole-genome sequencing of more members of the ε-*Pro-teobacteria* will enable a clearer resolution of the evolutionary relationships within this group of related organisms.

Phages/Genomic Islands

The major difference between the *C. jejuni* NCTC 11168 and *C. jejuni* RM1221 genomes is the presence within the strain RM1221 genome of four large, integrated elements (Color Plate 1). This characteristic has been observed in whole-genome intraspecies com-parisons of both gram-positive and gram-negative microorganisms (10, 15, 117, 154, 164). The first element, *Campylobacter* Mu-like phage (CMLP1) (30.5% G+C content), lo-cated upstream of *argC* (CJE0275), encodes several proteins with similarity to bacterio-phage Mu and other Mu-like prophage pro-teins (148), including putative MuA and MuB transposase homologs. Another feature consis-tent with the identification of CMLP1 as a novel Mu-like prophage is the presence of ter-

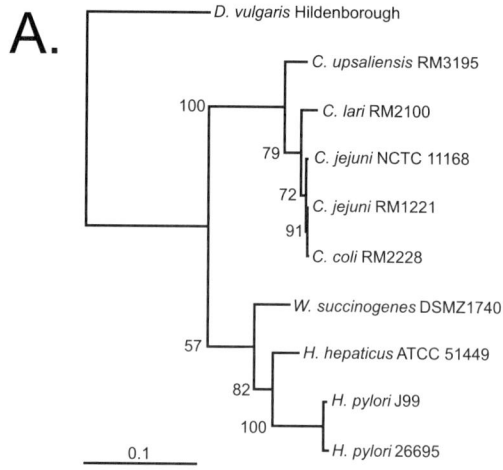

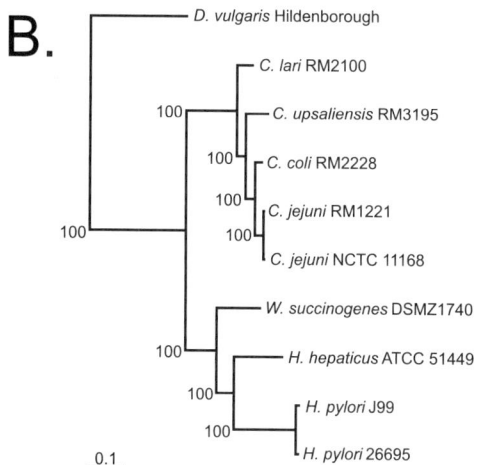

FIGURE 2 Phylogenetic analysis. Consensus maximum-likelihood trees are depicted using multi-ple alignments of 16S rRNA (A) or 12 concatenated protein data sets (B). The numbers along the branches denote percent occurrence of nodes among 100 bootstrap replicates. The scale bar represents the number of nucleotide (A) or amino acid (B) substi-tutions.

minal 5′-TG-3′ dinucleotides flanked by a 5-bp direct repeat (TATGC). Preliminary results suggest that this prophage is inducible with mitomycin C and that other *C. jejuni* strains harbor a related prophage (unpublished data). Genetic manipulation of this phage could yield useful molecular tools analogous to the

Mu derivatives for the construction of random gene fusions or mini-Mu elements for in vivo cloning. Although this Mu-like prophage contains no characterized virulence determinants, it could potentially alter pathogenicity or other phenotypes via insertional inactivation.

In contrast to CMLP1, *C. jejuni* RM1221 integrated elements 2 and 4 (CJIE2 and CJIE4) have integrated into the 3′ end of arginyl- and methionyl-tRNA genes, respectively. Several ORFs predicted to encode phage-related endonucleases, methylases, or repressors are present within these elements; however, unlike CMLP1, few ORFs encoding phage structural proteins were identified within CJIE4. CJIE4 is similar to a putative prophage contained within the *C. lari* RM2100 genome (CLIE1), 66% (35 of 53) of predicted proteins have BLASTP matches (P value of $\leq 10^{-5}$, identity 30% or better). CLIE1 is integrated into a leucinyl-tRNA. The inability to identify matches to major capsid, portal, and scaffold protease proteins within CJIE2 or *C. upsaliensis* RM3195 integrated element 1 (CUIE1) suggests that they represent either intact prophage with novel head morphogenesis proteins, satellite phages, nonfunctional prophages, or genomic islands.

The absence of any phage-related ORFs within CJIE3 (located within an arginyl-tRNA) suggests that CJIE3 is not a prophage but rather a genomic island or integrated plasmid. Seventy-three percent (45 of 62) of the CJIE3-predicted proteins are similar to predicted proteins encoded on the *C. coli* RM2228 megaplasmid (pCC178), suggesting that CJIE3 was plasmid derived. However, the observed lack of synteny between CJIE3 and the *C. coli* RM2228 megaplasmid suggests that CJIE3 was not derived from pCC178 but possibly from a related *Campylobacter* mega-plasmid. Although most of the ORFs contained within CJIE3 encode hypothetical proteins (23%, 14 of 62), many are similar to proteins found within the 71-kb genomic island of *H. hepaticus* ATCC 51449 (HHGI1), suggesting that this genomic island could also be plasmid derived (196). Furthermore, 33% (23 of 70) of

HHGI1 proteins match pCC178-encoded proteins.

Bacteriophages are vehicles for the lateral or horizontal movement of genes that can increase bacterial fitness (37, 85). Additionally, it has been demonstrated that bacteriophage-carried genes can play a role in many aspects of bacterial virulence (adhesion, invasion, host evasion, and toxin production) (218). Although only one of the *Campylobacter* prophages (CMLP1) has been shown to be inducible, we cannot predict whether the other putative prophages or plasmid-like element can be excised. Because the majority of ORFs that lie within prophage regions are hypothetical proteins, we are unable to deduce any putative functions from them; however, we cannot rule out possible functions that either directly affect virulence or increase fitness of the host to its environment.

Plasmids

Comparison of the plasmids of these *Campylobacter* species could provide some insight into their habitat and lifestyle. *C. coli* and *C. lari* each contain a single plasmid (~178 and ~46 kb, respectively), whereas *C. upsaliensis* contains two plasmids (~3.1 and ~110 kb). In the current study, neither *C. jejuni* isolate sequenced harbors a plasmid; however, previously a *C. jejuni* virulence plasmid, pVir, was sequenced and shown to play a role in pathogenesis (11). Interestingly, the coding regions of *C. jejuni* 81-178 pVir are entirely in one orientation except for a single coding region, which is uncharacteristic for a plasmid of this size. However, the *C. upsaliensis* plasmid pCU110 also demonstrates a similar bias, as does *C. lari* pCL46, although to a lesser extent. However, the orientation of the *C. coli* plasmid pCC178 coding regions does not exhibit the same bias. This may be the result of continued recombination by this plasmid. pCC178 is the largest plasmid and encodes a number of antibiotic resistance mechanisms flanked by possible mobile genetic elements, suggesting that the lack of coding region bias has been skewed to be more neutral over time by these insertions. Additionally, a large proportion of the plasmid proteins are unique or are similar to proteins of unknown function, and those with similarity are similar to unidentified proteins in the previously sequenced *C. jejuni* (161) or *H. pylori* (6, 206) genomes.

We have had difficulty in identifying the replication machinery of the *Campylobacter* plasmids; only the *C. upsaliensis* 3.1-kb plasmid (pCU3) has a clearly defined plasmid replication region. One plasmid-partitioning protein is conserved between *C. upsaliensis* pCU110 (CUPA0136) and *C. lari* pCL46 (CLAA0007) in association with a putative helicase; however, this does not allow these plasmids to be placed into the same group. Single-stranded binding proteins are conserved among all of the plasmids, indicating that the function of this protein is to bind *Campylobacter* DNA and suggesting that these plasmids are from similar origin; however, the nickase proteins on the plasmids are not conserved, indicating that some divergence has occurred over time so that each nickase is specific to the plasmid or organism.

One conserved feature of all of the large *Campylobacter* plasmids is the presence of a type IV secretion system. The plasmid-encoded T4SS on the non-*jejuni Campylobacter* species plasmids are most similar to each other on an amino acid and gene-order level. However, they share only gene order with the T4SS encoded by pVir of *C. jejuni* strain 81-176 or the *Agrobacterium tumefaciens* C58 Ti plasmid. The *C. jejuni* T4SS contains only a core complement of T4SS proteins, resembling systems that translocate protein effectors into host cells or are responsible for competence, whereas the T4SS of the other *Campylobacter* species plasmids are more complete and thus similar to the classical T4SS encoded by the *A. tumefaciens* Ti plasmid used for the movement of DNA (41). Additionally, *C. jejuni* pVir does not encode any pilin subunits, whereas each of the other *Campylobacter* plasmids contains at least a single pilin subunit, and in some cases, multiple distinct pilin proteins are present, which allows for the formation of a conjugative pilus. All of the *Campylobacter* plasmids contain additional proteins that do not appear to be directly involved in the T4SS but that are conserved between the plasmids to some degree, forming a gene cluster that ends with a topoisomerase, which in all cases is most similar to transfer proteins from the *E. coli* plasmid RP4.

pCC178 also encodes a number of antibiotic resistance mechanisms, including tetracycline (CCOA0206), kanamycin/neomycin (CCOA0067/68), and hygromycin B (CCOA0070)–containing antibiotics such as gentamicin. These antibiotic resistance proteins are often associated with transposable elements. These resistance mechanisms, combined with the potential for transfer of this plasmid, would allow for dissemination of these antimicrobial resistances among this group of organisms. In addition to the T4SS system, pCU110 also appears to contain a number of proteins that are similar to the conjugal transfer proteins of other plasmids, which may constitute an ancestral conjugative system that has diverged over time or a functioning system that in concert with the T4SS transfers plasmid DNA to donor cells.

pCL46 also encodes a single peptide that may be involved in virulence. CLAA0034 is similar to the invasin proteins encoded by *Yersinia* species that allow the bacterium to penetrate host cells (96). It is unclear whether this protein serves the same purpose in this species. Additionally, pCC178 and pCU110 both contain bacteriocin resistance proteins that are ~70% identical (amino acid) to chromosomally encoded proteins with similar function. However, it is unclear whether these proteins are functionally interchangeable.

Transposable Elements

Both strains of *C. jejuni* are notable for the nearly complete absence of IS elements. With the exception of one copy of a degenerate transposase resembling IS605, which is located between the locus containing the *tonB* gene and a copy of a 5S rRNA gene, their genomes are devoid of IS elements. In contrast, *C. coli* contains several copies of an IS element (IS-Cco1 of the IS605 family) at three positions in the chromosome and at least two positions in the megaplasmid, which hints at recent acquisition and transposition competence. In addition, the relatedness of *C. coli* and *C. jejuni* species is evident from the presence of the same degenerate tranposase within the *tonB*-5S rRNA locus in the three sequenced strains.

At the current level of sequencing coverage, both of the chromosomal pseudomolecules of *C. upsaliensis* and *C. lari* lack the *tonB*-5S rRNA locus. Therefore, we cannot currently compare the relatedness of these two strains to the *C. jejuni* strains on the basis of the presence or absence of the degenerate IS605 family transposase mentioned above. However, as best we can tell, IS elements appear to be absent from both the chromosomal pseudomolecules and the plasmids of these species. Thus, it is notable that these *Campylobacter* isolates are mostly devoid of IS elements.

CRISPR Analysis

Analysis of a particular class of DNA repeats, the clustered regularly interspaced short palindromic repeats (CRISPR elements), reveals relatively small genotypic differences that can be used to differentiate between the two sequenced strains of *C. jejuni* (180). The chromosomes of all five *Campylobacter* strains in this comparative study were examined for the presence or absence of CRISPR elements in intergenic regions. In this study, a strain is considered CRISPR positive when it contains two or more direct repeats of a 21-bp or larger DNA segment separated by unique spacer sequences of a similar size. This analysis identified CRISPR elements in only the two strains of *C. jejuni*. Perhaps it is not surprising that *C. upsaliensis* and *C. lari* lack the CRISPR element at the same chromosomal location as *C. jejuni*, because we have found that the two genes flanking the CRISPR locus in *C. jejuni* appear to have undergone shuffling and are widely separated in *C. upsaliensis* and *C. lari* (data not shown). However, a previous study found that CRISPR elements are sometimes detectable in *C. coli* (180). In findings that are also consistent with the previous study, the two

strains of *C. jejuni* examined here can be differentiated by both the unique sequence of the spacer sequences and the number of CRISPR repeats in the element (five in NCTC 11168 and four in RM1221).

Restriction-Modification Systems

The type I restriction-modification (RM) loci from 65 *C. jejuni* strains have been characterized previously (142). In contrast to the *C. jejuni*, *C. coli*, and *C. lari* strains sequenced in this study, the *C. upsaliensis* RM3195 genome is predicted to contain at least three type I RM loci. *C. upsaliensis* RM3195 also contains a putative fourth locus where the *hsdR* gene is absent. The sequenced genomes of *C. jejuni* NCTC 11168 and RM1221, *C. coli* RM2228, and *C. lari* RM2100 encode few type II or type III RM systems. *C. upsaliensis* RM3195 encodes one putative type II and two putative type III restriction enzymes. In addition, *C. upsaliensis* RM3195 encodes 15 putative adenine- or cytosine-specific DNA methyltransferases. It is noteworthy that the sequenced genome of *H. hepaticus* ATCC 51449, like *C. jejuni* RM1221, *C. coli* RM2228, and *C. lari* RM2100, has a paucity of RM loci (196) and would therefore be considered "*Campylobacter*-like," whereas RM3195 would be considered "*H. pylori*-like," with respect to RM systems. At least four of the *C. upsaliensis* RM3195 RM systems lie within regions of atypical nucleotide composition, suggesting recent horizontal transfer as selfish mobile elements (111).

Diversity within the *Campylobacter* RM systems has implications for *Campylobacter* biology, specifically DNA uptake and phage infection. *Campylobacter* are naturally competent (219), and horizontal gene transfer through natural transformation is thought to play an important role in the evolution of *C. jejuni* (197). Natural competence, as well as experimental introduction of DNA by electroporation, would presumably be influenced by host RM systems. Indeed, strain-specific differences in competence have been noted in *Campylobacter* (140, 222). RM system variation would also impact infection by both lytic and lysogenic bacteriophage. Future studies will be able to determine the functional status of the RM systems and their role in natural competence and phage restriction.

Campylobacter Metabolism

The campylobacters undoubtedly encounter different environments, and long-term survival on meats, in fecal matter, and in water has been documented (140). In the laboratory, *Campylobacter* spp. require a complex growth medium. There are few studies of the metabolic capabilities of *Campylobacter* species, but they are known to have a respiratory type of metabolism, with some species growing under both aerobic and anaerobic conditions (28, 188). Carbohydrates in general are not utilized. The genomes of *C. jejuni*, *C. coli*, *C. lari*, and *C. upsaliensis* have essentially identical metabolic profiles with intact TCA cycles as well as respiratory chains. The TCA cycle most likely serves a dual role of generating biosynthetic compounds and providing intermediates that feed into electron transport. All species have pathways for the metabolism and biosynthesis of a number of amino acids (Fig. 3), and acetate, formate, and lactate appear to be the main end products from carbon metabolism. *C. coli* RM2228, *C. upsaliensis* RM3195, and *C. lari* RM2100 apparently lack a succinate dehydrogenase gene, and none of the strains appear to encode SucAB (oxogutarate dehydrogenase).

Comparative analysis of the genomes does, however, reveal many genes that are strain and or species specific, a number of which are related to cellular processes, and may contribute to observed differences in physiology of the members of this genus. Preliminary Biolog (http://www.biolog.com) unpublished data from the Mandrell laboratory (U.S. Food and Drug Administration, Albany, CA) also demonstrated differences in substrate utilization patterns across the campylobacters (unpublished data). For example, from Biolog data, *C. jejuni* RM1221, *C. coli* RM2228, and *C. lari* RM2100 all respire in the presence of arabinose, fucose, and formic and lactic acid. In ad-

dition, only *C. jejuni* RM1221 respires in the presence of fructose, mannose, hydroxybutyric acid, asparagine, and aspartic acid, in contrast to the other species. These observed phenotypic differences from the preliminary Biolog data either may be a reflection of the conditions under which the substrates were tested or may be a result of *C. jejuni* having pathways or genes that are absent in the other strains. Because of the lack of complete genomes from the other strains, we cannot say with confidence what the reason is for the observed differences, but variable patterns in substrate utilization by *Campylobacter* species have also previously been described (144).

Some of these substrate utilization differences can likely be explained through an investigation of the strain- and species-specific genes present in these isolates, or in simple gene mutations that cannot be detected at the genome level. In *C. jejuni*, for example, the inability to grow on sugars that are added to the growth media is thought to be a reflection of the missing phosphofructokinase that is necessary for glycolysis (161). Interestingly, for all the ε-*Proteobacteria* included in the whole-genome comparisons described above, no phosphofructokinase could be identified except for *W. succinogenes*, enabling *Wolinella* to metabolize a wider range of carbohydrates than *Campylobacter*. For the strain-specific genes, *C. jejuni* 11168 has three strain-specific ORFs that are likely involved in the metabolism of neuramic acid, a constituent of the host cell surfaces. These genes (Cj1141 to Cj1143 and NTORF1083 to NTORF1085) may assist in the pathogenicity of this strain through the breakdown of host cell-wall polymers where neuramic acid may be a component, and may also result in substrate availability in the form of glucosamine to the bacterium. Similarly, *C. coli* RM2228 encodes a dihydroxy-acid dehydratase gene (CCO0046), an Na^+/H^+ antiporter (CCO00159-160), as well as genes for the methylcitric acid cycle (ORFA00358 to ORFA00362) analogous to that described for *Salmonella enterica* serovar

Typhimurium, where it enables growth on proprionate and fatty acid intermediates (93).

Chromosomally Encoded Protein Secretion Systems

The five *Campylobacter* strains analyzed in this study have the Sec-dependent and Sec-independent (twin-arginine translocation "TAT") protein export pathways for the secretion of proteins across the inner or periplasmic membrane. In addition, *Campylobacter* has the signal recognition particle (SRP) pathway. We have found no evidence for chromosomally encoded *lol*, type III, or type IV secretion systems other than the flagellar export apparatus (114). In all five strains, there are putative proteins that comprise components of a transformation system with similarity to type II secretion systems (227). A putative pre-pilin peptidase, as well as several putative pseudopilins, has been identified based on BLASTP similarity or the presence of an N-terminal pre-pilin peptidase cleavage signal. The two-partner secretion/single accessory pathway is used by gram-negative bacteria to secrete adhesins and cytolysins (97). There are undisrupted copies of putative pore-forming single-accessory factors (generically termed *TpsB homologs*) in *C. coli* RM2228 (CCO0190), *C. lari* RM2100 (CLA0150), and *C. jejuni* NCTC 11168 (Cj0975); however, CCO1305 in *C. coli*, and CJE0841/842/843 and CJE1056 in *C. jejuni* RM1221 are disrupted. It is unclear whether these disruptions are real in the unfinished genomes or whether there would be any consequence for the disruption in *C. jejuni* RM1221.

Virulence

A very important set of steps in the cycle of *C. jejuni* passage through the environment involves steps in the poultry production process. These steps could involve entry, survival, and growth in poultry during breeding, vertical or horizontal transmission in layers, in and on preharvest birds (skin, feathers, gastrointestinal tract), in soil and feed, on processing equipment, in hot and cold water (stress response), in biofilms, in and on processed birds and package liquid, and many other steps. This may be part of the explanation why poultry isolates of *C. jejuni* are rarely, if ever, clonal.

The pathogenic mechanisms responsible for acute intestinal infections by *Campylobacter*, although still poorly understood, are thought to involve adherence, cellular invasion, and toxin production, but not all clinical isolates of *C. jejuni* are able to invade cultured human cells or produce defined toxins (109). However, a common feature of *Campylobacter* infectious enterocolitis is a localized acute inflammatory response (Fig. 1C) that can lead to tissue damage and may be responsible for many of the clinical symptoms (109).

Motility is the major factor that has been implicated directly in intestinal colonization (223). Of the 580 ORFs conserved between the *Campylobacter* and *Helicobacter* species included in this study (Color Plate 2C), 27 ORFs involved in flagellar biosynthesis and function were conserved between *Campylobacter* and *Helicobacter*. Another set of 18 ORFs involved in chemotaxis and motility was found to be conserved across the *Campylobacter* strains, but with no bidirectional match in *Helicobacter* (P value of $\leq 10^{-5}$, identity 35% or better, match lengths of at least 75% of the length of both query and subject sequence), emphasizing the importance of bacterial motility and adhesion for virulence (231).

Two-component regulatory (TCR) systems are commonly used by bacteria to respond to specific environmental signals. We identified five TCR systems (pairs of adjacent histidine kinase and response regulator genes) that appear to be conserved across the *Campylobacter* strains: CJE0968–CJE0969, CJE1357–CJE1358, CJE1361–CJE1362, *racR–racS* (CJE1397–CJE1398), and CJE1664–CJE1665. In addition, another four putative response regulator genes (CJE0746, CJE0404, CJE1168, and CJE1780) and one putative histidine kinase gene (CJE0884) could be found in the finished *C. jejuni* genomes. Brás et al. (24) showed that the RacR-RacS system is involved in a temperature-dependent signaling pathway and is required for the organism to colonize the

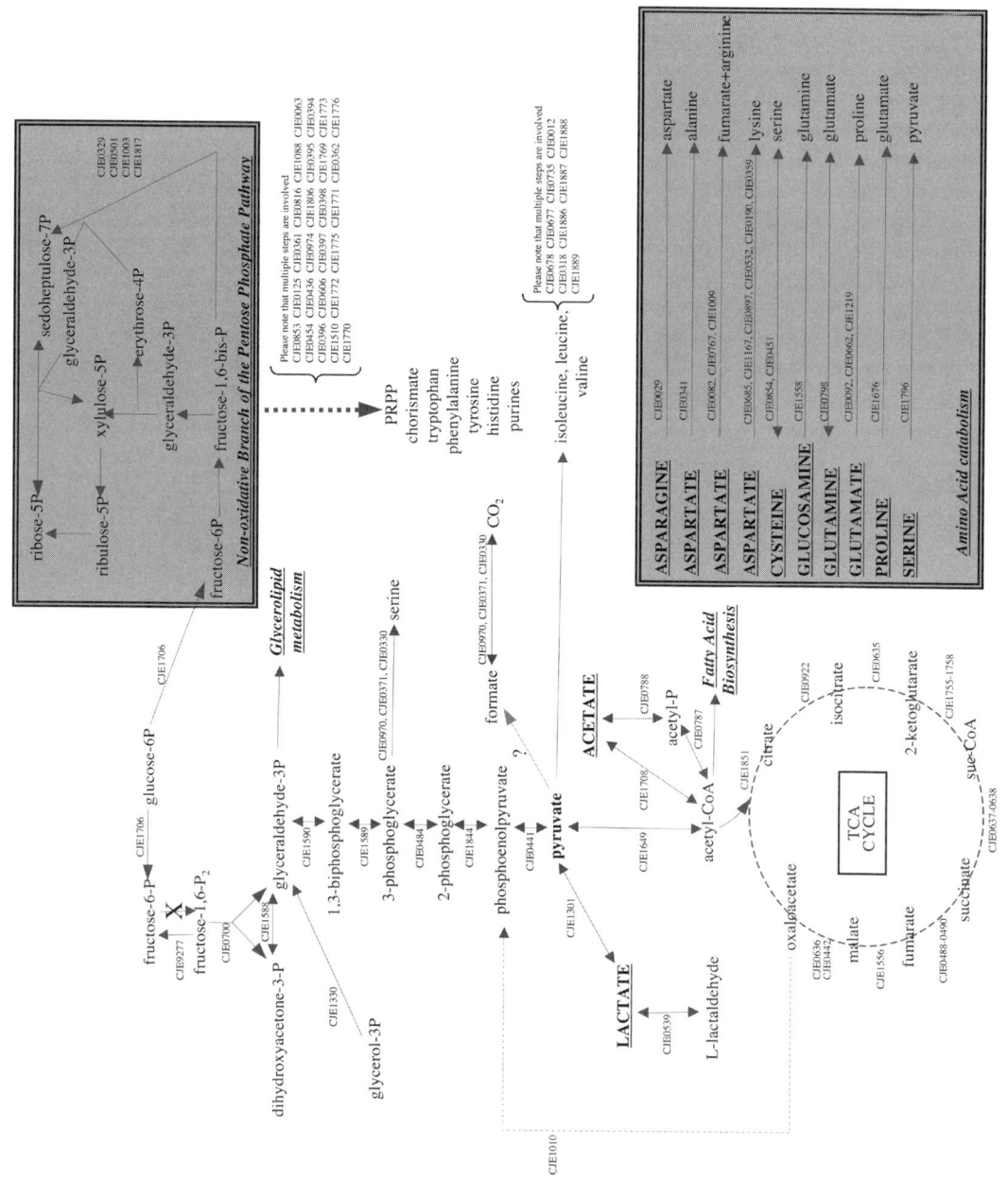

chicken intestinal tract. The high degree of conservation of these ORFs suggests an importance in *Campylobacter* pathogenicity, given the likely exposure of the bacteria to temperature stress during the infectious process.

Adherence of *C. jejuni* to epithelial cells is mediated by multiple adhesins, including CadF (CJE1651) and PEB1 (CJE0997-CJE1000), JlpA (CJE1065), and a 43-kDa major outer membrane protein (CJE1395) (Fig. 1C). Fibronectin (FN) has been implicated in *C. jejuni* adherence to epithelial cells via the protein CadF (113). In addition to CadF, we found two putative FN-binding proteins (CJE1415 and CJE1538) that are conserved across the five *Campylobacter* strains. The FN host cell–surface receptor is the α5β1 integrin. In intact epithelia, these integrins are restricted to the basolateral membrane and thus are not available for interaction with luminally positioned microbial pathogens (136). However, experimental evidence suggests that adherence and internalization of *C. jejuni* are greatly increased by exposure of cellular basolateral surfaces and that FN is the receptor (145). This suggests that *C. jejuni* invasion preferentially occurs via a paracellular route, rather than via an intracellular route. Additionally, inspection of loci adjacent to putative TpsB proteins revealed two intact filamentous hemagglutinin (FHA)-like adhesions: *C. lari* RM2100 (CLA0151) and *C. coli* RM2228 (CCO1312). The regions upstream of the remaining TpsB-like proteins have fragmented adhesion-like ORFs. Only *C. lari* RM2100 has both an undisrupted TpsB-like transporter (CLA0150) and an adjacent putative FHA-like adhesion (CLA0151), which, if functional, could enable *C. lari* RM2100 to attach to cell surfaces.

CDTs from enteropathogenic *E. coli* have been shown to disrupt the barrier function of host intestinal epithelial tight junctions (150). It was recently shown that the disruption of tight junctions by *C. jejuni* represents a potential mechanism for the formation of diarrhea (132). The three CDTs, A, B, and C (CJE0075, CJE0074, and CJE0073), were conserved across the five *Campylobacter* strains. CdtA and CdtC were recently shown to bind specifically to HeLa cells (126). In addition, *C. lari* RM2100 encodes a single peptide (CLAA0034) in pCL46 that is similar to the *Yersinia* invasin proteins that enable *Yersinia* to penetrate host cells (96), suggesting that this *C. lari* strain might also have the ability to penetrate host cells.

Identification of a Novel *Campylobacter* Putative Virulence Locus

Examination of the *C. upsaliensis* RM3195 sequence revealed a putative *licABCD* (CUP0277–CUP0274) locus with varying but significant identity to genes present in *Haemophilus influenzae* (225), commensal *Neisseria* species (184), and *Streptococcus pneumoniae* (236). *licABCD* genes in these microorganisms encode proteins involved in the acquisition of choline (*licB*, CUP0276), synthesis of phosphorylcholine (PCho) (*licA*, CUP0277; *licC*, CUP0275), and transfer of PCho (*licD*, CUP0274) to LOS or teichoic/lipoteichoic acids to facilitate attachment to host cells (184). Preliminary studies indicate that other strains of *C. upsaliensis* from South Africa also contain *licA* (unpublished results). It is noteworthy that *licA* expression in *H. influenzae* is regulated by variation in the number of intragenic tandem tetranucleotide repeats (CAAT) at the 5′ end, resulting in translational on/off synthesis of PCho and expression on LOS (226). A poly G tract within the *licA* gene (bp 132 to 146) of *C. upsaliensis* RM3195 probably regulates synthesis of PCho and decoration of LOS by a similar mechanism.

Hypervariable Homopolymeric Tracts

The presence of the homopolymeric repeat sequences in the genome of *C. jejuni* NCTC 11168 has been described (161). However, by

---

FIGURE 3 Main pathways for metabolism derived from an analysis of five *Campylobacter* genomes. The tricarboxylic (TCA) cycle has major variations based on comparative analysis across the strains (see text). Differences in substrate respiration are based on an analysis of Biolog data, and species-specific pathways are presented in the text.

comparing these five *Campylobacter* strains, a number of other phenomena related to these repetitive regions were observed. First, when a homopolymeric repeat region was associated with a potential coding region, the most included base in the repeated region was a G on the coding strand, resulting in polyglycine, not polyproline, in the peptide. Second, the *C. upsaliensis* RM3195 genome contains nearly 3 times as many variable homopolymeric repeats (22) when compared with *C. jejuni* RM1221 (8), 7 times as many as *C. lari* RM2100 (3), and 22 times as many as *C. coli* RM2228 (1). These variable *C. upsaliensis* RM3195 poly G:C tracts come from a pool of almost 5 times as many total poly G:C tracts as *C. jejuni* RM1221 and *C. coli* RM2228 and nearly 10 times as many total poly G:C tracts as *C. lari* RM2100. Of these 22 varied poly G:C tracts, 11 (50%) are strain specific. It appears that the excess variable poly G:C tracts are due to the presence of unique ORFs; however, it is unclear why *C. upsaliensis* RM3195 contains so many more total homopolymeric repeated regions, because only 61 of the 209 regions are within unique ORFs. These variable regions encode a combination of hypothetical, cell envelope, and virulence-associated ORFs, which in other pathogenic bacteria has been shown to be the molecular basis of lipopolysaccharide phase variation (98), has been used to identify novel virulence genes in *H. influenzae* (92), and has been speculated to play a similar role in *C. jejuni* (161). However, these observed differences could be the result of different culturing conditions before library construction.

LOS and Capsular Polysaccharide Biosynthesis

LOS and EP are important surface structures in *C. jejuni* that function in the interactions of the organism with the environment. Interesting aspects of *C. jejuni* LOS are their molecular mimicry of host gangliosides and their presumed roles in evading host immune responses and autoimmunity (147), decreased immunogenicity (79), and attachment and invasion (11). The capsule of *C. jejuni* 81-176 has been

reported to have a role in increasing serum resistance, invasion of cell lines, and surface hydrophilicity (12).

The LOS biosynthesis loci of all sequenced campylobacters are organized as previously observed in other *C. jejuni* strains (71). At either end of the loci are the heptosyltransferase genes, *waaC* and *waaF*, that surround regions exhibiting significant variation in ORF content. Thus, these organisms likely synthesize novel LOS structures (71). In particular, the LOS of *C. jejuni* RM1221 is distinct from the LOS of NCTC 11168 as seen on polyacrylamide gels, possessing three LOS bands as compared to one LOS band for NCTC 11168 (unpublished data). In fact, Parker et al. went on to determine that there are at least eight LOS classes from 123 strains studied and that RM1221 has a class F structure while NCTC 11168 has a class C organization (160). Two LOS genes from *C. jejuni* RM1221 possess homopolymeric G:C tracts, which may explain the additional bands. Comparison of the LOS genes from the sequenced campylobacters with the LOS genes from *C. jejuni* strains that produce ganglioside mimics (70) demonstrates that these four strains do not possess the genes involved in the synthesis of *N*-acetylneuramic (sialic) acid or the associated sialic acid transferase and are not likely to produce ganglioside mimics. Within the LOS loci of *C. lari* RM2100 and *C. upsaliensis* RM3195, there are ORF clusters that have homologs in NCTC 11168, which are unrelated to LOS biosynthesis. It is unclear what role this genomic reorganization plays in the biosynthesis of LOS.

Like the EP locus of *C. jejuni* NCTC 11168, *C. jejuni* RM1221, *C. coli* RM2228, and *C. lari* RM2100 possess *kps* orthologs involved in polysaccharide export; however, many putative EP biosynthesis genes from *C. jejuni* RM1221 and *C. coli* RM2228 are unique compared with *C. jejuni* NCTC 11168. The *kps* orthologs are present in *C. upsaliensis* RM3195, but they are not clustered with other polysaccharide biosynthetic genes, as observed in the other strains. Specifically, there are three clusters of EP genes: CUP0615–CUP0619, CUP1248–CUP1270, and CUP1328–CUP1329. The second cluster contains many ORFs that are unique to *C. upsaliensis*, including two of the three copies of a putative GDP-fucose synthetase (CUP1255, CUP1257, and CUP1258). Only *C. jejuni* strains (Cj1428c, CJE1612) and *C. upsaliensis* RM3195 encode this enzyme. Of these, only CUP1257 was shown to contain variable poly G tracts.

Antibiotic Resistance

The sequenced *Campylobacter* strains have adapted or acquired many mechanisms of antibiotic resistance. All strains are resistant to cloxacillin, nafcillin, oxacillin, trimethoprim-sulfamethoxazole, trimethoprim, and vancomycin, and this resistance is likely inherent to all *Campylobacter* strains. Every strain but *C. upsaliensis* RM3195 is resistant to most β-lactam antibiotics. This general lack of resistance to β-lactam antibiotics for RM3195 is likely due to the disruption of a class D β-lactamase matching GenBank accession AAT01092 (CUP0345), but was found to be an intact single copy in NCTC 11168 (Cj0299), RM1221 (CJE0344), and RM2100 (CLA0304). The corresponding sequence in *C. coli* RM2228 may reside in unsequenced regions. Only *C. lari* RM2100 was resistant to a broad range of quinolone and fluoroquinolone antibiotics. This broad quinolone and fluoroquinolone resistance is most likely the result of adaptation by mutation of DNA gyrase (*gyrA*) that changed codon 86 from Thr to Val (167). The macrolide antibiotics azithromycin, clindamycin, erythromycin, and tilmicosin were effective against all but *C. coli* RM2228. This is likely due to a mutation in all three copies of the 23S rRNA (A2122G), corresponding to position 2143 of *H. pylori* (202). *C. coli* RM2228 has acquired resistance to the aminoglycosides kanamycin and neomycin, tetracycline, oxytetracycline, minocycline, and presumably hygromycin B (but not gentamicin) from the megaplasmid pCC178. It is possible that *C. coli* has acquired resistance to macrolides and tetracyclines as a result of the application of these drugs during poultry production. The resistance of *C. upsaliensis* RM3195 to oxytetracycline and intermediate resistance to tetracycline may be due to the action of multidrug efflux pumps or a novel mechanism because there is no evidence for tetracycline resistance genes (31) or known mutations in the 16S rRNA (207). Similarly, no known mutations in *gyrA* or *gyrB* were found in *C. upsaliensis* RM3195 to explain the resistance to nalidixic acid (167) and novobiocin (78). There were no obvious known mutations of dihydropteroate synthase (*folP*) (69) to explain the observed variable resistance to sulfonamide-class drugs. Rifampin resistance was observed in all strains but *C. lari* RM2100 but was not due to the classic mutations in the β subunit of RNA polymerase (84).

Conclusions Drawn from Comparison of Five *Campylobacter* Genomes

The comparison of five sequenced *Campylobacter* genomes has provided the core genetic blueprint of the species. Although the blueprint presented obvious differences in genome structure and content, additional epidemiological data are needed to correlate these findings or other elusive differences (regulation, point mutations) with differences in virulence. Some obvious differences were the presence of drug resistance genes that may have been the result of adaptation in the animal production environment, where antibiotics are frequently used to eliminate bacterial infections. It is anticipated that the analysis of the *Campylobacter* genomes presented here will lay the foundation for the development of systems for fingerprinting strains for phylogenetics, epidemiology, and source tracking, as well as the development of alternative treatments for controlling *Campylobacter* in the food chain and in human infection.

POSTGENOMIC STUDIES

Data from the genome sequence have provided an excellent starting point for studying genes encoding proteins involved in outer surface structures and glycosylation (70, 129), and for identifying hypervariable homopolymeric tracts (usually polyG:C) involved in the contingency expression of gene products such as glycosyl transferases and restriction enzymes (161). Additionally, many genes have been mutated, and their effect on transcription has been documented by microarrays. The genome sequence has also stimulated research into the development of diagnostic PCR- and microarray-based technologies. Comparative genome hybridization and informatic studies have been conducted to better understand *Campylobacter* diversity and evolution. The postgenomic progress that has been made since the release of the first *Campylobacter* genome sequence has been grouped into six areas of research and discussed below. This list is far from comprehensive because the number of publications that deal with *Campylobacter* biology has exploded since the release of the first genome sequence. Many aspects of *Campylobacter* biology, like LOS and EP structure determination (195), the glycome (104), quorum sensing (47, 99), GBS (166), and virulence mechanisms (*cdt* [2, 126] and invasion [146]), will not be covered in detail.

Genome-Based Detection

One of the expectations from the genome sequence efforts was the creation of robust tools for species and strain diagnostics and classification. This is particularly important to the U.S. Department of Agriculture for the purpose of source tracking of *Campylobacter* outbreaks to individual farms or processing plants. The medical community would also like to be able to accurately fingerprint *Campylobacter* strains for epidemiological purposes. Previous attempts to detect, differentiate (type), and classify *Campylobacter* isolates were based on biochemical, bacteriological, and serological properties (214); DNA hybridization (159); 16S rRNA hybridization (174); pulsed-field gel electrophoresis (229); PCR of the *flaA* gene (17); FlaA short variable region (38, 39, 137); multilocus enzyme electrophoresis (138); amplified fragment length polymorphism (157, 180); multilocus sequence type (38, 40); and randomly amplified polymorphic DNA (197).

Because of the extreme genetic diversity between the *Campylobacter* strains, no single method accurately classifies the strains or captures the diversity. Since the genome sequences were made available, many researchers have rushed to the bench to develop novel detection procedures using the genomic data. Randomly amplified polymorphic DNA and serotyping data were used in conjunction with a new *C. jejuni* microarray to demonstrate the high level of specificity of the DNA microarray for epidemiological purposes (127). Another study used a microarray-based and PCR-based approach on five genes to discriminate among *C. jejuni*, *C. coli*, *C. lari*, and *C. upsaliensis* (217). Some studies focused on differentiating the closely related *C. jejuni* and *C. coli* species from chicken feces (107, 108). A multipathogen oligonucleotide microarray has been constructed, including *Campylobacter* spp., for the detection of biodefense agents (183). Having the sequence of every gene in *Campylobacter* enabled new PCR-based diagnostic methods to be tested. One new PCR-based approach proposed combining a hippurate test with PCR amplification of the *omp50* gene to differentiate *Campylobacter* species (35). It is clear that the genome sequences of *Campylobacter* spp. are being used for the generation of very precise detection tools.

Differential Expression Studies

One of the benefits of having complete genome sequence data is the ability to design microarrays for the purpose of studying transcriptional regulation. Looking at the final level in the central dogma—translation—proteomics makes use of the genome sequence to identify which proteins are produced. The research topics for differential expression in *Campylobacter* can be divided into three areas: specific regulatory mutant studies, different growth conditions, and analysis of archived versus multiply passaged strains. A discussion of the proteomic work that has been done is also presented.

Microarray transcriptional analysis has been a very useful tool for mapping regulatory networks (regulons) within bacteria. *C. jejuni* genes that are involved in the heat-shock response were identified through DNA microarray transcript profiling of an *hspR* mutant, a known negative regulator (8). Another stress response pathway, the stringent response, is a global stress response used by bacteria to survive under a wide variety of stressful environmental conditions. Gene expression differences between wild-type and a *spoT* mutant, a gene involved in the stringent response in other organisms, identified genes that were likely involved in rifampicin resistance and intracellular survival (66). Two-component signal transduction systems enable bacteria to sense environmental cues and respond to them through altered gene expression. Mutants of the two-component system genes *dccR* and *dccS* (diminished capacity to colonize) in *C. jejuni* were used to identify genes involved in colonization (133). Finally, an iron acquisition regulon (91), the flagellar apparatus regulon (29), and a nitrosative stress-responsive regulon that neutralizes the toxic effects of nitric oxide (48) all used wild-type versus mutant conditions to map out regulatory networks using microarrays. Other microarray studies monitored mRNA transcript levels of wild-type strains under varying growth conditions. Some of these studies looked at temperature variation (143, 158, 193, 194), different atmospheric conditions (65, 66, 143, 158, 194), and nutrient starvation (91, 194).

For reasons that are unclear, *C. jejuni* NCTC 11168 was chosen as the representative *Campylobacter* strain to be sequenced. It was originally isolated in 1977 from a patient with diarrhea, but the original frozen culture of NCTC 11168 was not the one used as the source of genomic DNA in the 2000 Parkhill et al. genome paper (161). The sequenced version of strain NCTC 11168 (dubbed 11168-GS by Gaynor et al. [65]) was repeatedly passaged and may no longer be pathogenic for humans. Indeed, it has been shown, after sequencing, to have dramatic differences in the expression of flagellar genes (29), and genes involved in respiration and metabolism (65). Sequencing of sigma factors revealed major dif-

ferences between the original archived strain and the sequenced strain (65). The sequenced strain NCTC 11168 has also been shown to have lost the ability to colonize chickens (11, 65), to invade intestinal epithelial cells, and to cause disease in a ferret diarrhea model (11). However, the sequenced strain was observed to adapt to high-level colonization of chickens after 5 weeks (103).

Comparative Genome Hybridization Studies

A number of studies have been conducted using the NCTC 11168 genomic data that monitor DNA-DNA hybridizations on a microarray slide. These studies can be divided into three types: gene discovery, genotyping, and genetic diversity studies. By using comparative genome hybridization, genes specific to *C. jejuni* 81-176 (3, 168) and *C. jejuni* ATCC 43431 (169) were discovered. Genes specific for GBS were looked for by comparative genome hybridization of GBS-positive and GBS-negative strains, but were not found because of a high degree of heterogeneity between the two genomes (128). Comparative genome hybridization has been used as yet another tool for genotyping (110) and as a tool for quantitating genome diversity (42, 198).

Transposon Mutagenesis Studies

An extremely important genetic tool that has been used long before DNA sequencing technologies is transposon mutagenesis. IS/transposable elements, or transposons, are genetic entities that move from one piece of genetic space to another (either chromosomal or extrachromosomal). Their mobility has been useful and manipulated in the design of genetic screens and selections aimed at the identification of specific classes of insertion mutations. The gene(s) of interest may have been the target of insertion or so-called downstream polar mutations. Whole-genome sequence data have made life much easier for the geneticist by enabling easy mapping of insertion events from short stretches of sequences obtained from transposon-specific primers (59). David Acheson's group was the first to use a *mariner*-based transposon called *Himar1* (119) to mutagenize *C. jejuni* strain 480 in vivo (74). The location of 12 transposon mutants was mapped using the sequence of *C. jejuni* NCTC 11168. They later used this technique to identify genes involved in motility and autoagglutination (73). Victor DiRita's group later used an in vitro *mariner*-based transposition system to obtain a higher frequency of transposition and applied the technique to identify genes involved in motility in *C. jejuni* 81-176 (86), a strain that we recently sequenced under the National Institutes of Health National Institute of Allergy and Infectious Diseases (NIAID)-sponsored microbial sequencing contract. They followed up this work to show that flagellar genes were regulated by σ^{54}, but not σ^{28} (88). Two years later, the *Himar1* construct from DiRita's group (88) was adapted for signature-tagged transposon mutagenesis (89) to identify genes involved in colonization of the chicken gastrointestinal tract (87) and ceca (77).

Promoter and Transcriptional Start Mapping

Knowing the complete genome sequence of a bacterium makes discovery of promoter sequences much easier. Promoters are short sequences ~25 bp in length that are ligands of bacterial sigma factors, which are part of the cell's transcription machinery. The binding of a sigma factor to a promoter sequence is a critical initial step in transcription. By having the genome sequence, one can use computational means to predict promoter sequences. Alternatively, results of more traditional runoff or 5′ race experiments to determine the transcriptional start site are greatly strengthened by prior knowledge of the genomic sequence by giving information about genes and operon structure. Jeon et al. recently determined that the genes that constitute the cytolethal distending toxin (*cdtA*, *B*, and *C*) are regulated by the quorum-sensing protein LuxS, are polycistronic, and begin transcription 81 bp upstream of the *cdtA* start codon (100). Christine Szymanski's group used informatics to identify

putative σ^{28} and σ^{54} promoters and studied the expression profiles of a *fliA* (σ^{28}) and *flhA* (homolog of type III secretion core genes *hrcV*/*invA*/*lcrD*) mutants (29).

Informatics Studies

Perhaps the field of study that has benefited most from genomic sequence data—or better yet, would not exist otherwise—is bioinformatics. This field of study uses computers to make predictions about classification, coding capacity, expression, and biological function from genomic data. Some laboratories have done whole-genome computational comparisons of the *Campylobacterales* (51, 60), while other groups have focused on codon usage (63, 156), dinucleotide frequencies (32), and evolutionary relationships (13) of *Campylobacter*. Finally, an online database resource has been developed to facilitate genome-wide comparisons among the *Campylobacterales* (30).

FUTURE DIRECTIONS

Microbial Sequencing Contract

NIAID has awarded exclusive contracts to The Institute for Genomic Research (TIGR) (http://www.tigr.org) and the Massachusetts Institute of Technology Broad Institute (http://www.broad.mit.edu/seq/msc/) to sequence microorganisms that are possible biodefense threats. The TIGR contract was $65 million for 5 years to sequence bacterial, viral, parasite, and invertebrate vector genomes that cause or transmit infectious disease in humans. The organisms on the NIAID priority pathogens list are broken down into three classes, which represent different safety levels (http://www3.niaid.nih.gov/biodefense/band c_priority.htm). Category A infectious agents include organisms like *Bacillus anthracis* and the Ebola virus and represent agents that are highly infectious at low doses or that have a high mortality rate, or those organisms that can be made into bioterror agents. *Campylobacter jejuni*, *Shigella*, and hepatitis A are some examples of category B food- and waterborne pathogens. Category C agents represent the lowest level bioterror threat and include organisms that are emerging infectious disease threats like influenza, severe acute respiratory syndrome, and multidrug resistant tuberculosis.

Despite the great deal of information obtained from the analysis of the genome sequences of the two *C. jejuni* strains, RM1221 and NCTC 11168, fundamental questions concerning what makes *C. jejuni* an emerging human pathogen remain to be answered. For instance, what makes some *C. jejuni* isolates more virulent or cause different clinical symptoms than other isolates? Why do some isolates cause GBS or MFS but not others? Where are the mysterious Shiga-like toxins that have been reported in the literature? Perhaps these questions would be answered with better choice of strains. In contrast to NCTC 11168, *C. jejuni* RM1221 was isolated from the skin of a market chicken carcass, with its potential as a human pathogen unknown (139). Attempting to correlate the difference in source between the NCTC 11168 and RM1221 with differences in their gene content has been unrewarding. In the future, if we hope to use genome sequencing to identify determinants crucial to human infection, it will be useful to target minimally passaged isolates that are clearly associated with human disease and compare them with strains that are known or suspected to be less pathogenic.

In immunocompetent human hosts, *C. fetus* subsp. *fetus* commonly causes an acute diarrheal illness, similar to that caused by *C. jejuni*, or septic abortion in pregnant women. However, the ratio of bloodstream infection to diarrheal infection for *C. fetus* is nearly 400-fold higher than for *C. jejuni*, indicating a greater propensity for invasive disease in humans than *C. jejuni* (19). While in immunocompromised human hosts (infancy, alcoholism, liver disease, malignancy, infection with human immunodeficiency virus), *C. fetus* subsp. *fetus* usually causes systemic illnesses including bacteremia, meningitis, peritonitis, abscesses, and endocarditis (27). In contrast to *C. jejuni*, *C. fetus* has only one well-defined virulence factor, a paracrystalline S layer (64), which confers serum resistance by inhibiting binding of complement (C3b) to the cell surface, thus preventing the opsonization of the bacteria (22). This S layer covers the underlying immunogenic LPS layer, rendering it inaccessible or undetectable by the host (57). The only other *Campylobacter* species to have an S layer is *C. rectus* (155).

The original aim of the *Campylobacter* microbial sequencing contract proposal was to obtain eightfold (draft) sequence coverage from eight *C. jejuni* subsp. *jejuni* genomes representing the diversity of clinical symptoms associated with human *Campylobacter* infection and eightfold average sequence coverage with full gap closure (complete) of 2 *C. jejuni* subsp. *doylei* strains representing 2 different clinical symptoms and 10 *Campylobacter* species that cause human illness. The details of this goal included:

- Prepare random genomic shotgun libraries of *Campylobacter* chromosomal DNA as previously described (60). One small insert (2 to 3 kb) and one medium insert (10 to 12 kb) library will be constructed.
- Sequence both ends of randomly selected clones from both libraries with the ratio fivefold coverage with the small insert library (~7,580 good reads for 1.8-Mbp genome) and threefold coverage with the medium insert library (~12,620 good reads for 1.8-Mbp genome). This will total eightfold average sequence coverage. In our experience, we were able to achieve at least 713 bp average edited read length for *Campylobacter* genomes. The medium insert library is necessary to provide linking information and to help span small repetitive regions. *C jejuni* RM1221 had two repeat classes greater than 1 kb (one was 2,061 bp and the other ~6,200 bp [rRNA operons]).
- Assemble with the Celera Assembler (151), and order scaffolds with NUCMER (36) and BAMBUS (170).
- Close sequencing gaps for *C. jejuni* subsp. *jejuni* (draft) genomes that are unique (for *C. jejuni* RM1221, there were four unique regions that were bacteriophage or plasmid-like regions [60]).

- For the non–*C. jejuni* subsp. *jejuni* genomes, close the sequencing gaps, as done previously for *C. jejuni* RM1221 (60). At TIGR, there is a new cost-saving, automated sequencing, gap-closing pipeline that is available that is being used.
- Use automated annotation tools to identify all ORFs and transfer annotation from the finished, manually annotated *C. jejuni* RM1221. For regions of interest, some manual curation will be performed.

ISOLATES TO BE SEQUENCED IN ORDER OF PRIORITY

Information on these isolates can be found at http://msc.tigr.org/campy/index.shtml.

C. jejuni subsp. *jejuni* 81-176 (to Eightfold Draft)

This is the best-characterized *C. jejuni* strain and one that has caused disease in two human volunteer studies. It was originally isolated from the feces of an 9-year-old girl with diarrhea as the result of an outbreak of campylobacteriosis associated with the consumption of raw milk on a dairy farm in Minnesota during a third-grade-class field trip in 1981 (116). It was passaged through human volunteer trials by Black et al. in 1987 (18) and passaged again by Patricia Guerry's group, where it is proven to still cause human gastroenteritis. It is virulent in rhesus macaques, ferrets, and colostrum-deprived piglets, and it is virulent when delivered intranasally in mice. The strain invades intestinal epithelial cells at levels that are as much as 3 logs higher than other invasive strains.

C. jejuni subsp. *jejuni* 260.94 (to Eightfold Draft)

This is a representative of the clonal population of Penner serotype O:41 GBS-associated *Campylobacter* strains from the Red Cross Children's Hospital in Cape Town, South Africa (221). It has been associated with the acute inflammatory demyelinating polyneuropathy form of GBS, which is the most common form of GBS. This strain has been compared with HB93-13 (below) in a previous study. It has been shown to express the GM_1

ganglioside-like structure. This strain was chosen because it is part of the second known overrepresented serotype that is highly associated with GBS.

C. fetus subsp. *fetus* 82-40 (to Completion)

This is a serotype A *C. fetus* subsp. *fetus* isolated from the blood of a human patient who was undergoing renal transplantation (163, 208). The ratio of bloodstream infection to diarrheal illnesses for *C. fetus* is nearly 400-fold higher than for *C. jejuni*, indicating its marked propensity for invasive disease compared with *C. jejuni*. This strain was chosen because it is the best-characterized human *C. fetus* isolate.

C. jejuni subsp. *jejuni* HB93-13 (to Eightfold Draft)

This is a well-characterized strain that was isolated from the feces of an 8-year-old boy in China who was diagnosed with the acute motor axonal neuropathy form of GBS (185, 208). It is of Penner serotype O:19, which is overrepresented among GBS-associated serotypes in certain parts of the world. It has been shown to express the GM_1 and GD_{1a} ganglioside-like structures. Many different laboratories study this strain.

C. jejuni subsp. *jejuni* CF93-6 (to Eightfold Draft)

This is a representative strain isolated in Japan from a patient diagnosed with MFS (234). It has been recently shown to produce a GT_{1a}-like ganglioside mimic that is cross-reactive to anti-GQ_{1b} antibodies (112).

C. jejuni subsp. *jejuni* 84-25 (to Eightfold Draft)

This strain was isolated in the United States from the cerebrospinal fluid of a child with meningitis (20). This strain is of interest because it represents a *C. jejuni* strain capable of systemic infection. It is also auxotrophic (Met$^-$ Ser$^-$ and requires cysteine and cystine). The sequence of this isolate might reveal previously unknown mechanisms for nongastrointestinal tropism.

C. jejuni subsp. *doylei* 269.97 (to Completion)

This strain is a representative of the *C. jejuni* subsp. *doylei* that causes bacteremia in South Africa. At the Red Cross Children's Hospital in Cape Town, South Africa, *C. jejuni* subsp. *doylei* has been isolated from the blood more frequently than *C. jejuni* subsp. *jejuni*. The complete genomic sequence for *C. jejuni* subsp. *doylei* will reveal the differences between *C. jejuni* subsp. *doylei* and *C. jejuni* subsp. *jejuni* and may help determine why these emerging *C. jejuni* subsp. *doylei* isolates cause more blood-borne disease than *C. jejuni* subsp. *jejuni*.

C. concisus 13826 (to Completion)

Strains of this species were first isolated from gingival crevices of persons with gingivitis, periodontitis, and periodontosis, but they have subsequently been isolated from the feces of patients with gastroenteritis. This strain was isolated in Denmark from the feces of a patient with bloody diarrhea (1). No sequence is available for this *Campylobacter* species. The complete genome sequence of this isolate may enable identification of possible mechanisms for the ability of *Campylobacter* to cause diarrhea and any horizontal transfer events from *C. jejuni* or *C. coli*, which are more common diarrhea-causing *Campylobacter* species. Perhaps we can determine from the complete genomic sequence how this species is emerging as a gastrointestinal infectious agent.

C. curvus 525.92 (to Completion)

Like *C. concisus*, this *Campylobacter* species has been historically associated with periodontal disease, but it has been recently isolated from human patients with gastroenteritis. This strain was isolated in South Africa from the feces of a patient with diarrhea. It will be interesting to compare the genomic sequence of this strain to the genome of *C. concisus*. Genes or pathways shared between *C. curvus* and *C. concisus* might reveal mechanisms for causing gastroenteritis or periodontal disease. As with *C. concisus*, we might be able to explain common mechanisms for their emergence as human gastrointestinal infectious agents.

C. hominis ATCC BAA-381 (to Completion)

This is the only known species of *Campylobacter* that is not pathogenic to humans. It is the type strain that was isolated in 2001 from the feces of a healthy human (122). The complete sequence of this strain will be useful for subtracting all the genes involved in survival in the human gut and attachment to intestinal epithelial cells, which will facilitate identification of novel genes in the pathogenic species.

Timeline for *Campylobacter* Sequencing Projects

On the basis of the current capacity of the Joint Technology Center, we estimate that it will take ~11 months to sequence one strain of *Campylobacter* by random shotgun with full closure. To eliminate all risk of cross-contamination, each of the 10 projects will be started 4 weeks apart, depending on the availability of the genomic DNA for each strain. Library construction and validation will take approximately 4 weeks. Template preparation and high-throughput sequencing to eightfold coverage will require 4 weeks for each strain. The closure of five strains may require ~15 months, but it may be less if the existing *Campylobacter* genome sequences can be used effectively in the closure process. All 10 genomes will be analyzed by using comparative genomic tools developed at TIGR that extract detailed information on the polymorphic regions when each genome is compared with a reference genome. Assemblies and autoannotation will be deposited into GenBank 1 or 2 days after the assembly has been completed and quality tested.

PROTEOMICS

Although some proteomic studies have been conducted for *C. jejuni*, these were limited to differential analysis of specific mutations (8, 91), planktonic versus biofilm growth (45), and on NCTC 11168 stocks with different amounts of passaging (29). It is important to know this information, but to fully understand *Campylobacter* biology and virulence, more complete proteomic analysis needs to be conducted. For example, it would be useful to know the localization of the protein components and which proteins were produced during initial and late stages of infection. It is a straightforward experiment to fractionate the cells grown under laboratory or pseudopathogenic conditions into three parts (secreted, membrane, and cytocolic). This should be standard operating procedure for every sequenced microorganism. This not only tells us which proteins are made and their size, but also their location. These fractions can also be studied further by probing Western blots with antibodies to detect posttranslational modifications. It is much more difficult to do proteomics on bacteria that have been in contact with eukaryotic cells (cell culture, or ileum loop models), but procedures could be created to separate the bacterial cells from the host either before or after chromatographic steps.

YEAST TWO-HYBRID STUDIES

The yeast two-hybrid system has been around since 1989 (55) and has been a very useful tool to identify protein-protein (173, 209) and protein-peptide (230) interactions in vivo (131). Though there are cases of false-positive and false-negative results, many of the interactions identified are real and have been supported by follow-up biochemical experiments. Many new twists to the original yeast two-hybrid have been developed that enable interaction mapping of RNA-protein interactions (235), that allow interaction mapping of membrane proteins (94), that enable positive selection (reverse two-hybrids) of protein mutants that fail to interact (187, 215), and that increase the throughput of the assay (68). There has been a protein-protein interaction map published for the related bacterium *Helicobacter pylori* (173), which has been used to infer interacting protein partners in *Campylobacter*, but at the time of this writing, we could not find any published *Campylobacter* yeast two-hybrid data. It is true that many core proteins are conserved between *Campylobacter* and *Helicobacter*, but there are just as many genes that are only in *Campylobacter*. For example, *C. jejuni* RM1221 and NCTC 11168 have ~500 genes that are shared between them, but no other members of the ε-*Proteobacteria* and *H. pylori* strains share 451 genes not seen in the other ε-*Proteobacteria* (Color Plate 2C) (60).

To get a full understanding of the function of *Campylobacter* proteins, it is critical to complete a protein-protein interaction map with a supporting public database framework. Many so-called hypothetical proteins could become proteins with generally defined functions (i.e., localized to a specific location inside or outside of the cell or localized to a specific pathway) through the analysis of protein-protein interaction networks. For example, if a particular hypothetical protein was shown in the yeast two-hybrid system and in biochemical studies to interact with proteins involved in LOS production, then perhaps this hypothetical protein also serves some role in LOS catabolism. We could also obtain clues about human protein targets, such as receptors or toxin targets, by completing a yeast two-hybrid interaction map with human proteins as bait.

It is also entirely possible that some interesting *Campylobacter* proteins might not be expressed in yeast for good reason. They could be toxins whose targets are conserved in humans and yeast. This phenomenon was observed by D.E.F. with the *Pseudomonas syringae* effector protein HrmA but was never followed up. If such toxicity in yeast is real and not due to overexpression, then mutations in yeast could help define the target and specific amino acid residues necessary for the interaction of toxin with target.

REFERENCES

1. **Aabenhus, R., H. Permin, S. L. On, and L. P. Andersen.** 2002. Prevalence of *Campylobacter concisus* in diarrhoea of immunocompromised patients. *Scand. J. Infect. Dis.* **34:**248–252.
2. **Abuoun, M., G. Manning, S. A. Cawthraw, A. Ridley, I. H. Ahmed, T. M. Wassenaar, and D. G. Newell.** 2005. Cytolethal distending toxin

(CDT)-negative *Campylobacter jejuni* strains and anti-CDT neutralizing antibodies are induced during human infection but not during colonization in chickens. *Infect. Immun.* **73:**3053–3062.

3. **Ahmed, I. H., G. Manning, T. M. Wassenaar, S. Cawthraw, and D. G. Newell.** 2002. Identification of genetic differences between two *Campylobacter jejuni* strains with different colonization potentials. *Microbiology* **148:**1203–1212.

4. **Alderton, M. R., V. Korolik, P. J. Coloe, F. E. Dewhirst, and B. J. Paster.** 1995. *Campylobacter hyoilei* sp. nov., associated with porcine proliferative enteritis. *Int. J. Syst. Bacteriol.* **45:**61–66.

5. **Allos, B. M.** 1997. Association between *Campylobacter* infection and Guillain-Barré syndrome. *J. Infect. Dis.* **176:**S125–S128.

6. **Alm, R. A., L. S. Ling, D. T. Moir, B. L. King, E. D. Brown, P. C. Doig, D. R. Smith, B. Noonan, B. C. Guild, B. L. deJonge, G. Carmel, P. J. Tummino, A. Caruso, M. Uria-Nickelsen, D. M. Mills, C. Ives, R. Gibson, D. Merberg, S. D. Mills, Q. Jiang, D. E. Taylor, G. F. Vovis, and T. J. Trust.** 1999. Genomic-sequence comparison of two unrelated isolates of the human gastric pathogen *Helicobacter pylori*. *Nature* **397:**176–180.

7. **Altschul, S. F., W. Gish, W. Miller, E. W. Myers, and D. J. Lipman.** 1990. Basic local alignment search tool. *J. Mol. Biol.* **215:**403–410.

8. **Andersen, M. T., L. Brondsted, B. M. Pearson, F. Mulholland, M. Parker, C. Pin, J. M. Wells, and H. Ingmer.** 2005. Diverse roles for HspR in *Campylobacter jejuni* revealed by the proteome, transcriptome and phenotypic characterization of an *hspR* mutant. *Microbiology* **151:**905–915.

9. **Baar, C., M. Eppinger, G. Raddatz, J. Simon, C. Lanz, O. Klimmek, R. Nandakumar, R. Gross, A. Rosinus, H. Keller, P. Jagtap, B. Linke, F. Meyer, H. Lederer, and S. C. Schuster.** 2003. Complete genome sequence and analysis of *Wolinella succinogenes*. *Proc. Natl. Acad. Sci. USA* **100:**11690–11695.

10. **Baba, T., F. Takeuchi, M. Kuroda, H. Yuzawa, K. Aoki, A. Oguchi, Y. Nagai, N. Iwama, K. Asano, T. Naimi, H. Kuroda, L. Cui, K. Yamamoto, and K. Hiramatsu.** 2002. Genome and virulence determinants of high virulence community-acquired MRSA. *Lancet* **359:**1819–1827.

11. **Bacon, D. J., R. A. Alm, L. Hu, T. E. Hickey, C. P. Ewing, R. A. Batchelor, T. J. Trust, and P. Guerry.** 2002. DNA sequence and mutational analyses of the pVir plasmid of *Campylobacter jejuni* 81-176. *Infect. Immun.* **70:**6242–6250.

12. **Bacon, D. J., C. M. Szymanski, D. H. Burr, R. P. Silver, R. A. Alm, and P. Guerry.** 2001. A phase-variable capsule is involved in virulence of *Campylobacter jejuni* 81–176. *Mol. Microbiol.* **40:**769–777.

13. **Bansal, A. K., and T. E. Meyer.** 2002. Evolutionary analysis by whole-genome comparisons. *J. Bacteriol.* **184:**2260–2272.

14. **Benjamin, J., S. Leaper, R. J. Owen, and M. B. Skirrow.** 1983. Description of *Campylobacter laridis*, a new species comprising the nalidixic acid resistant thermophilic *Camplybacter* (NARTC) group. *Curr. Microbiol.* **8:**231–238.

15. **Beres, S. B., G. L. Sylva, K. D. Barbian, B. Lei, J. S. Hoff, N. D. Mammarella, M. Y. Liu, J. C. Smoot, S. F. Porcella, L. D. Parkins, D. S. Campbell, T. M. Smith, J. K. McCormick, D. Y. Leung, P. M. Schlievert, and J. M. Musser.** 2002. Genome sequence of a serotype M3 strain of group A *Streptococcus*: phage-encoded toxins, the high-virulence phenotype, and clone emergence. *Proc. Natl. Acad. Sci. USA* **99:**10078–10083.

16. **Berg, R. L., J. W. Jutila, and B. D. Firehammer.** 1971. A revised classification of *Vibrio fetus*. *Am. J. Vet. Res.* **32:**11–22.

17. **Birkenhead, D., P. M. Hawkey, J. Heritage, D. M. Gascoyne-Binzi, and P. Kite.** 1993. PCR for the detection and typing of campylobacters. *Lett. Appl. Microbiol.* **17:**235–237.

18. **Black, R. E., M. M. Levine, M. L. Clements, T. P. Hughes, and M. J. Blaser.** 1988. Experimental *Campylobacter jejuni* infection in humans. *J. Infect. Dis.* **157:**472–479.

19. **Blaser, M. J.** 1998. *Campylobacter fetus*—emerging infection and model system for bacterial pathogenesis at mucosal surfaces. *Clin. Infect. Dis.* **27:**256–258.

20. **Blaser, M. J., G. P. Perez, P. F. Smith, C. Patton, F. C. Tenover, A. J. Lastovica, and W. I. Wang.** 1986. Extraintestinal *Campylobacter jejuni* and *Campylobacter coli* infections: host factors and strain characteristics. *J. Infect. Dis.* **153:**552–559.

21. **Blaser, M. J., P. F. Smith, J. A. Hopkins, I. Heinzer, J. H. Bryner, and W. L. Wang.** 1987. Pathogenesis of *Campylobacter fetus* infections: serum resistance associated with high-molecular-weight surface proteins. *J. Infect. Dis.* **155:**696–706.

22. **Blaser, M. J., P. F. Smith, J. E. Repine, and K. A. Joiner.** 1988. Pathogenesis of *Campylobacter fetus* infections. Failure of encapsulated *Campylobacter fetus* to bind C3b explains serum and phagocytosis resistance. *J. Clin. Invest.* **81:**1434–1444.

23. **Bolton, F. J., D. N. Hutchinson, and D. Coates.** 1984. Blood-free selective medium for isolation of *Campylobacter jejuni* from feces. *J. Clin. Microbiol.* **19:**169–171.

24. Brás, A. M., S. Chatterjee, B. W. Wren, D. G. Newell, and J. M. Ketley. 1999. A novel *Campylobacter jejuni* two-component regulatory system important for temperature-dependent growth and colonization. *J. Bacteriol.* **181:**3298–3302.

25. Brown, J. R., C. J. Douady, M. J. Italia, W. E. Marshall, and M. J. Stanhope. 2001. Universal trees based on large combined protein sequence data sets. *Nat. Genet.* **28:**281–285.

26. Butzler, J. P., M. De Boeck, and H. Goossens. 1983. New selective medium for isolation of *Campylobacter jejuni* from faecal specimens. *Lancet* **i:**818.

27. Carbone, K. M., M. C. Heinrich, and T. C. Quinn. 1985. Thrombophlebitis and cellulitis due to *Campylobacter fetus* ssp. *fetus*. Report of four cases and a review of the literature. *Medicine (Baltimore)* **64:**244–250.

28. Carlone, G. M., and J. Lascelles. 1982. Aerobic and anaerobic respiratory systems in *Campylobacter fetus* subsp. *jejuni* grown in atmospheres containing hydrogen. *J. Bacteriol.* **152:**306–314.

29. Carrillo, C. D., E. Taboada, J. H. Nash, P. Lanthier, J. Kelly, P. C. Lau, R. Verhulp, O. Mykytczuk, J. Sy, W. A. Findlay, K. Amoako, S. Gomis, P. Willson, J. W. Austin, A. Potter, L. Babiuk, B. Allan, and C. M. Szymanski. 2004. Genome-wide expression analyses of *Campylobacter jejuni* NCTC11168 reveals coordinate regulation of motility and virulence by *flhA*. *J. Biol. Chem.* **279:**20327–20338.

30. Chaudhuri, R. R., and M. J. Pallen. 2006. xBASE, a collection of online databases for bacterial comparative genomics. *Nucleic Acids Res.* **34:** D335–D337.

31. Chopra, I., and M. Roberts. 2001. Tetracycline antibiotics: mode of action, applications, molecular biology, and epidemiology of bacterial resistance. *Microbiol. Mol. Biol. Rev.* **65:**232–260.

32. Coenye, T., and P. Vandamme. 2004. Use of the genomic signature in bacterial classification and identification. *Syst. Appl. Microbiol.* **27:**175–185.

33. Cox, C. J., K. E. Kempsell, and J. S. Gaston. 2003. Investigation of infectious agents associated with arthritis by reverse transcription PCR of bacterial rRNA. *Arthritis Res. Ther.* **5:**R1–R8.

34. Davies, J. S., and J. B. Penfold. 1979. *Campylobacter* urinary infection. *Lancet* **i:**1091–1092.

35. Dedieu, L., J. M. Pages, and J. M. Bolla. 2004. Use of the *omp50* gene for identification of *Campylobacter* species by PCR. *J. Clin. Microbiol.* **42:**2301–2305.

36. Delcher, A. L., A. Phillippy, J. Carlton, and S. L. Salzberg. 2002. Fast algorithms for large-scale genome alignment and comparison. *Nucleic Acids Res.* **30:**2478–2483.

37. Desiere, F., W. M. McShan, D. van Sinderen, J. J. Ferretti, and H. Brüssow. 2001. Comparative genomics reveals close genetic relationships between phages from dairy bacteria and pathogenic *Streptococci*: evolutionary implications for prophage-host interactions. *Virology* **288:**325–341.

38. Dingle, K. E., F. M. Colles, D. Falush, and M. C. Maiden. 2005. Sequence typing and comparison of population biology of *Campylobacter coli* and *Campylobacter jejuni*. *J. Clin. Microbiol.* **43:**340–347.

39. Dingle, K. E., F. M. Colles, R. Ure, J. A. Wagenaar, B. Duim, F. J. Bolton, A. J. Fox, D. R. Wareing, and M. C. Maiden. 2002. Molecular characterization of *Campylobacter jejuni* clones: a basis for epidemiologic investigation. *Emerg. Infect. Dis.* **8:**949–955.

40. Dingle, K. E., F. M. Colles, D. R. Wareing, R. Ure, A. J. Fox, F. E. Bolton, H. J. Bootsma, R. J. Willems, R. Urwin, and M. C. Maiden. 2001. Multilocus sequence typing system for *Campylobacter jejuni*. *J. Clin. Microbiol.* **39:**14–23.

41. Ding, Z., K. Atmakuri, and P. J. Christie. 2003. The outs and ins of bacterial type IV secretion substrates. *Trends Microbiol.* **11:**527–535.

42. Dorrell, N., J. A. Mangan, K. G. Laing, J. Hinds, H. Linton, H. Al-Ghusein, B. G. Barrell, J. Parkhill, N. G. Stoker, A. V. Karlyshev, P. D. Butcher, and B. W. Wren. 2001. Whole genome comparison of *Campylobacter jejuni* human isolates using a low-cost microarray reveals extensive genetic diversity. *Genome Res.* **11:**1706–1715.

43. Doyle, L. 1944. A *Vibrio* associated with swine dysentery. *Am. J. Vet. Res.* **5:**3–5.

44. Doyle, L. 1948. The etiology of swine dysentery. *Am. J. Vet. Res.* **9:**50–51.

45. Dykes, G. A., B. Sampathkumar, and D. R. Korber. 2003. Planktonic or biofilm growth affects survival, hydrophobicity and protein expression patterns of a pathogenic *Campylobacter jejuni* strain. *Int. J. Food Microbiol.* **89:**1–10.

46. Eisen, J. A., J. F. Heidelberg, O. White, and S. L. Salzberg. 2000. Evidence for symmetric chromosomal inversions around the replication origin in bacteria. *Genome Biol.* **1:** RESEARCH0011.

47. Elvers, K. T., and S. F. Park. 2002. Quorum sensing in *Campylobacter jejuni*: detection of a *luxS* encoded signalling molecule. *Microbiology* **148:**1475–1481.

48. Elvers, K. T., S. M. Turner, L. M. Wainwright, G. Marsden, J. Hinds, J. A. Cole, R. K. Poole, C. W. Penn, and S. F. Park. 2005. NssR, a member of the Crp-Fnr superfamily from *Campylobacter jejuni*, regulates a nitrosative stress-

responsive regulon that includes both a single-domain and a truncated haemoglobin. *Mol. Microbiol.* **57:**735–750.

49. **Endtz, H. P., J. S. Vliegenthart, P. Vandamme, H. W. Weverink, N. P. van den Braak, H. A. Verbrugh, and A. van Belkum.** 1997. Genotypic diversity of *Campylobacter lari* isolated from mussels and oysters in The Netherlands. *Int. J. Food Microbiol.* **34:**79–88.

50. **Engvall, E. O., B. Brandstrom, L. Andersson, V. Baverud, G. Trowald-Wigh, and L. Englund.** 2003. Isolation and identification of thermophilic *Campylobacter* species in faecal samples from Swedish dogs. *Scand. J. Infect. Dis.* **35:**713–718.

51. **Eppinger, M., C. Baar, G. Raddatz, D. H. Huson, and S. C. Schuster.** 2004. Comparative analysis of four *Campylobacterales*. *Nat. Rev. Microbiol.* **2:**872–885.

52. **Etoh, Y., F. E. Dewhirst, B. J. Paster, A. Yamamoto, and N. Goto.** 1993. *Campylobacter showae* sp. nov., isolated from the human oral cavity. *Int. J. Syst. Bacteriol.* **43:**631–639.

53. **Ewing, B., and P. Green.** 1998. Base-calling of automated sequencer traces using phred. II. Error probabilities. *Genome Res.* **8:**186–194.

54. **Ewing, B., L. Hillier, M. C. Wendl, and P. Green.** 1998. Base-calling of automated sequencer traces using phred. I. Accuracy assessment. *Genome Res.* **8:**175–185.

55. **Fields, S., and O. Song.** 1989. A novel genetic system to detect protein-protein interactions. *Nature* **340:**245–246.

56. **Fisher, M.** 1956. An unusual variant of acute idiopathic polyneuritis (syndrome of ophthalmoplegia, ataxia and areflexia). *N. Engl. J. Med.* **255:**57–65.

57. **Fogg, G. C., L. Y. Yang, E. Wang, and M. J. Blaser.** 1990. Surface array proteins of *Campylobacter fetus* block lectin-mediated binding to type A lipopolysaccharide. *Infect. Immun.* **58:**2738–2744.

58. **Foster, G., B. Holmes, A. G. Steigerwalt, P. A. Lawson, P. Thorne, D. E. Byrer, H. M. Ross, J. Xerry, P. M. Thompson, and M. D. Collins.** 2004. *Campylobacter insulaenigrae* sp. nov., isolated from marine mammals. *Int. J. Syst. Evol. Microbiol.* **54:**2369–2373.

59. **Fouts, D. E., R. B. Abramovitch, J. R. Alfano, A. M. Baldo, C. R. Buell, S. Cartinhour, A. K. Chatterjee, M. D'Ascenzo, M. L. Gwinn, S. G. Lazarowitz, N. C. Lin, G. B. Martin, A. H. Rehm, D. J. Schneider, K. van Dijk, X. Tang, and A. Collmer.** 2002. Genomewide identification of *Pseudomonas syringae* pv. *tomato* DC3000 promoters controlled by the HrpL alternative sigma factor. *Proc. Natl. Acad. Sci. USA* **99:**2275–2280.

60. **Fouts, D. E., E. F. Mongodin, R. E. Mandrell, W. G. Miller, D. A. Rasko, J. Ravel, L. M. Brinkac, R. T. Deboy, C. T. Parker, S. C. Daugherty, R. J. Dodson, A. S. Durkin, R. Madupu, S. A. Sullivan, J. U. Shetty, M. A. Ayodeji, A. Shvartsbeyn, M. C. Schatz, J. H. Badger, C. M. Fraser, and K. E. Nelson.** 2005. Major structural differences and novel potential virulence mechanisms from the genomes of multiple *Campylobacter* species. *PLoS Biol.* **3:** e15.

61. **Fox, J. G., A. B. Rogers, M. T. Whary, Z. Ge, N. S. Taylor, S. Xu, B. H. Horwitz, and S. E. Erdman.** 2004. Gastroenteritis in NF-kappaB-deficient mice is produced with wild-type *Campylobacter jejuni* but not with *C. jejuni* lacking cytolethal distending toxin despite persistent colonization with both strains. *Infect. Immun.* **72:** 1116–1125.

62. **Friedman, C. R., J. Neimann, H. C. Wegener, and R. V. Tauxe.** 2000. Epidemiology of *Campylobacter jejuni* infections in the United States and other industrialized nations, p. 121–138. *In* I. Nachamkin and M. J. Blaser (ed.), Campylobacter. ASM Press, Washington, DC.

63. **Fuglsang, A.** 2003. The genome of *Campylobacter jejuni*: codon and amino acid usage. *APMIS* **111:**605–618.

64. **Fujimoto, S., A. Takade, K. Amako, and M. J. Blaser.** 1991. Correlation between molecular size of the surface array protein and morphology and antigenicity of the *Campylobacter fetus* S layer. *Infect. Immun.* **59:**2017–2022.

65. **Gaynor, E. C., S. Cawthraw, G. Manning, J. K. MacKichan, S. Falkow, and D. G. Newell.** 2004. The genome-sequenced variant of *Campylobacter jejuni* NCTC 11168 and the original clonal clinical isolate differ markedly in colonization, gene expression, and virulence-associated phenotypes. *J. Bacteriol.* **186:**503–517.

66. **Gaynor, E. C., D. H. Wells, J. K. MacKichan, and S. Falkow.** 2005. The *Campylobacter jejuni* stringent response controls specific stress survival and virulence-associated phenotypes. *Mol. Microbiol.* **56:**8–27.

67. **Gebhart, C. J., P. Edmonds, G. E. Ward, H. J. Kurtz, and D. J. Brenner.** 1985. "*Campylobacter hyointestinalis*" sp. nov.: a new species of *Campylobacter* found in the intestines of pigs and other animals. *J. Clin. Microbiol.* **21:**715–720.

68. **Gera, J. F., T. R. Hazbun, and S. Fields.** 2002. Array-based methods for identifying protein-protein and protein-nucleic acid interactions. *Methods Enzymol.* **350:**499–512.

69. **Gibreel, A., and O. Skold.** 1999. Sulfonamide resistance in clinical isolates of *Campylobacter jejuni*: mutational changes in the chromosomal

dihydropteroate synthase. *Antimicrob. Agents Chemother.* **43**:2156–2160.

70. **Gilbert, M., J. R. Brisson, M. F. Karwaski, J. Michniewicz, A. M. Cunningham, Y. Wu, N. M. Young, and W. W. Wakarchuk.** 2000. Biosynthesis of ganglioside mimics in *Campylobacter jejuni* OH4384. Identification of the glycosyltransferase genes, enzymatic synthesis of model compounds, and characterization of nanomole amounts by 600-mhz (1)h and (13)c NMR analysis. *J. Biol. Chem.* **275**:3896–3906.

71. **Gilbert, M., P. C. R. Godschalk, C. T. Parker, H. P. Endtz, and W. W. Wakarchuk.** Genetic basis for the variation in the lipooligosaccharide outer core of *Campylobacter jejuni* and possible association of glycosyltransferase genes with postinfectious neurophathies, p. 219–248. *In* J. Ketley and M. E. Konkel (ed.), Campylobacter jejuni: *New Perspectives in Molecular and Cellular Biology*, in press. Horizon Scientific Press, Norfolk, United Kingdom.

72. **Godschalk, P. C., A. P. Heikema, M. Gilbert, T. Komagamine, C. W. Ang, J. Glerum, D. Brochu, J. Li, N. Yuki, B. C. Jacobs, A. van Belkum, and H. P. Endtz.** 2004. The crucial role of *Campylobacter jejuni* genes in anti-ganglioside antibody induction in Guillain-Barré syndrome. *J. Clin. Invest.* **114**:1659–1665.

73. **Golden, N. J., and D. W. Acheson.** 2002. Identification of motility and autoagglutination *Campylobacter jejuni* mutants by random transposon mutagenesis. *Infect. Immun.* **70**:1761–1771.

74. **Golden, N. J., A. Camilli, and D. W. Acheson.** 2000. Random transposon mutagenesis of *Campylobacter jejuni*. *Infect. Immun.* **68**:5450–5453.

75. **Goossens, H., M. De Boeck, H. Coignau, L. Vlaes, C. Van den Borre, and J. P. Butzler.** 1986. Modified selective medium for isolation of *Campylobacter* spp. from feces: comparison with Preston medium, a blood-free medium, and a filtration system. *J. Clin. Microbiol.* **24**:840–843.

76. **Goossens, H., L. Vlaes, I. Galand, C. Van den Borre, and J. P. Butzler.** 1989. Semisolid blood-free selective-motility medium for the isolation of campylobacters from stool specimens. *J. Clin. Microbiol.* **27**:1077–1080.

77. **Grant, A. J., C. Coward, M. A. Jones, C. A. Woodall, P. A. Barrow, and D. J. Maskell.** 2005. Signature-tagged transposon mutagenesis studies demonstrate the dynamic nature of cecal colonization of 2-week-old chickens by *Campylobacter jejuni*. *Appl. Environ. Microbiol.* **71**:8031–8041.

78. **Gross, C. H., J. D. Parsons, T. H. Grossman, P. S. Charifson, S. Bellon, J. Jernee, M. Dwyer, S. P. Chambers, W. Markland, M. Botfield, and S. A. Raybuck.** 2003. Active-site residues of *Escherichia coli* DNA gyrase required in coupling ATP hydrolysis to DNA supercoiling and amino acid substitutions leading to novobiocin resistance. *Antimicrob. Agents Chemother.* **47**:1037–1046.

79. **Guerry, P., C. P. Ewing, T. E. Hickey, M. M. Prendergast, and A. P. Moran.** 2000. Sialylation of lipooligosaccharide cores affects immunogenicity and serum resistance of *Campylobacter jejuni*. *Infect. Immun.* **68**:6656–6662.

80. **Guerry, P., C. M. Szymanski, M. M. Prendergast, T. E. Hickey, C. P. Ewing, D. L. Pattarini, and A. P. Moran.** 2002. Phase variation of *Campylobacter jejuni* 81-176 lipooligosaccharide affects ganglioside mimicry and invasiveness in vitro. *Infect. Immun.* **70**:787–793.

81. **Guillain, G., J. A. Barré, and A. Strohl.** 1916. Sur un syndrome de radiculonévrite avec hyperalbuminose du liquide céphalo-rachidien sans réaction cellulaire. Remarques sur les caractères cliniques et graphiques des réflexes tendineux. *Bull. Soc. Med. Hop. Paris* **40**:1462–1470.

82. **Hafer-Macko, C., S. T. Hsieh, C. Y. Li, T. W. Ho, K. Sheikh, D. R. Cornblath, G. M. McKhann, A. K. Asbury, and J. W. Griffin.** 1996. Acute motor axonal neuropathy: an antibody-mediated attack on axolemma. *Ann. Neurol.* **40**:635–644.

83. **Hald, B., and M. Madsen.** 1997. Healthy puppies and kittens as carriers of *Campylobacter* spp., with special reference to *Campylobacter upsaliensis*. *J. Clin. Microbiol.* **35**:3351–3352.

84. **Heep, M., D. Beck, E. Bayerdorffer, and N. Lehn.** 1999. Rifampin and rifabutin resistance mechanism in *Helicobacter pylori*. *Antimicrob. Agents Chemother.* **43**:1497–1499.

85. **Hendrix, R. W., J. G. Lawrence, G. F. Hatfull, and S. Casjens.** 2000. The origins and ongoing evolution of viruses. *Trends Microbiol.* **8**:504–508.

86. **Hendrixson, D. R., B. J. Akerley, and V. J. DiRita.** 2001. Transposon mutagenesis of *Campylobacter jejuni* identifies a bipartite energy taxis system required for motility. *Mol. Microbiol.* **40**:214–224.

87. **Hendrixson, D. R., and V. J. DiRita.** 2004. Identification of *Campylobacter jejuni* genes involved in commensal colonization of the chick gastrointestinal tract. *Mol. Microbiol.* **52**:471–484.

88. **Hendrixson, D. R., and V. J. DiRita.** 2003. Transcription of σ^{54}-dependent but not σ^{28}-dependent flagellar genes in *Campylobacter jejuni* is associated with formation of the flagellar secretory apparatus. *Mol. Microbiol.* **50**:687–702.

89. **Hensel, M., J. E. Shea, C. Gleeson, M. D. Jones, E. Dalton, and D. W. Holden.** 1995. Simultaneous identification of bacterial virulence genes by negative selection. *Science* **269**:400–403.

90. **Ho, T. W., B. Mishu, C. Y. Li, C. Y. Gao, D. R. Cornblath, J. W. Griffin, A. K. Asbury, M. J. Blaser, and G. M. McKhann.** 1995. Guillain-Barré syndrome in northern China. Relationship to *Campylobacter jejuni* infection and antiglycolipid antibodies. *Brain* 118(Pt. 3):597–605.

91. **Holmes, K., F. Mulholland, B. M. Pearson, C. Pin, J. McNicholl-Kennedy, J. M. Ketley, and J. M. Wells.** 2005. *Campylobacter jejuni* gene expression in response to iron limitation and the role of Fur. *Microbiology* 151:243–257.

92. **Hood, D. W., M. E. Deadman, M. P. Jennings, M. Bisercic, R. D. Fleischmann, J. C. Venter, and E. R. Moxon.** 1996. DNA repeats identify novel virulence genes in *Haemophilus influenzae. Proc. Natl. Acad. Sci. USA* 93:11121–11125.

93. **Horswill, A. R., and J. C. Escalante-Semerena.** 1999. *Salmonella typhimurium* LT2 catabolizes propionate via the 2-methylcitric acid cycle. *J. Bacteriol.* 181:5615–5623.

94. **Hubsman, M., G. Yudkovsky, and A. Aronheim.** 2001. A novel approach for the identification of protein-protein interaction with integral membrane proteins. *Nucleic Acids Res.* 29:E18.

95. **Hughes, N. J., C. L. Clayton, P. A. Chalk, and D. J. Kelly.** 1998. *Helicobacter pylori porCDAB* and *oorDABC* genes encode distinct pyruvate:flavodoxin and 2-oxoglutarate:acceptor oxidoreductases which mediate electron transport to NADP. *J. Bacteriol.* 180:1119–1128.

96. **Isberg, R. R., D. L. Voorhis, and S. Falkow.** 1987. Identification of invasin: a protein that allows enteric bacteria to penetrate cultured mammalian cells. *Cell* 50:769–778.

97. **Jacob-Dubuisson, F., C. Locht, and R. Antoine.** 2001. Two-partner secretion in gram-negative bacteria: a thrifty, specific pathway for large virulence proteins. *Mol. Microbiol.* 40:306–313.

98. **Jennings, M. P., Y. N. Srikhanta, E. R. Moxon, M. Kramer, J. T. Poolman, B. Kuipers, and P. van der Ley.** 1999. The genetic basis of the phase variation repertoire of lipopolysaccharide immunotypes in *Neisseria meningitidis. Microbiology* 145(Pt. 11):3013–3021.

99. **Jeon, B., K. Itoh, N. Misawa, and S. Ryu.** 2003. Effects of quorum sensing on *flaA* transcription and autoagglutination in *Campylobacter jejuni. Microbiol. Immunol.* 47:833–839.

100. **Jeon, B., K. Itoh, and S. Ryu.** 2005. Promoter analysis of cytolethal distending toxin genes (*cdtA, B,* and *C*) and effect of a *luxS* mutation on CDT production in *Campylobacter jejuni. Microbiol. Immunol.* 49:599–603.

101. **Johnson, W. M., and L. H.** 1987. Response of Chinese hamster ovary cells to a cytolethal distending toxin (CDT) of *Escherichia coli* and possible misinterpretation as heat labile (LB) enterotoxin. *FEMS Microbiol. Lett.* 3:19–23.

102. **Jones, F. S., M. Orcutt, and R. B. Little.** 1931. Vibrios (*Vibrio jejuni* n. sp.) associated with intestinal disorders of cows and calves. *J. Exp. Med.* 53:853–864.

103. **Jones, M. A., K. L. Marston, C. A. Woodall, D. J. Maskell, D. Linton, A. V. Karlyshev, N. Dorrell, B. W. Wren, and P. A. Barrow.** 2004. Adaptation of *Campylobacter jejuni* NCTC11168 to high-level colonization of the avian gastrointestinal tract. *Infect. Immun.* 72:3769–3776.

104. **Karlyshev, A. V., J. M. Ketley, and B. W. Wren.** 2005. The *Campylobacter jejuni* glycome. *FEMS Microbiol. Rev.* 29:377–390.

105. **Karmali, M. A., A. E. Simor, M. Roscoe, P. C. Fleming, S. S. Smith, and J. Lane.** 1986. Evaluation of a blood-free, charcoal-based, selective medium for the isolation of *Campylobacter* organisms from feces. *J. Clin. Microbiol.* 23:456–459.

106. **Keat, A.** 1983. Reiter's syndrome and reactive arthritis in perspective. *N. Engl. J. Med.* 309:1606–1615.

107. **Keramas, G., D. D. Bang, M. Lund, M. Madsen, H. Bunkenborg, P. Telleman, and C. B. Christensen.** 2004. Use of culture, PCR analysis, and DNA microarrays for detection of *Campylobacter jejuni* and *Campylobacter coli* from chicken feces. *J. Clin. Microbiol.* 42:3985–3991.

108. **Keramas, G., D. D. Bang, M. Lund, M. Madsen, S. E. Rasmussen, H. Bunkenborg, P. Telleman, and C. B. Christensen.** 2003. Development of a sensitive DNA microarray suitable for rapid detection of *Campylobacter* spp. *Mol. Cell. Probes* 17:187–196.

109. **Ketley, J. M.** 1997. Pathogenesis of enteric infection by *Campylobacter. Microbiology* 143(Pt. 1):5–21.

110. **Kim, C. C., E. A. Joyce, K. Chan, and S. Falkow.** 2002. Improved analytical methods for microarray-based genome-composition analysis. *Genome Biol.* 3:RESEARCH0065.

111. **Kobayashi, I.** 2001. Behavior of restriction-modification systems as selfish mobile elements and their impact on genome evolution. *Nucleic Acids Res.* 29:3742–3756.

112. **Koga, M., M. Gilbert, J. Li, S. Koike, M. Takahashi, K. Furukawa, K. Hirata, and N. Yuki.** 2005. Antecedent infections in Fisher syndrome: a common pathogenesis of molecular mimicry. *Neurology* 64:1605–1611.

113. **Konkel, M. E., S. G. Garvis, S. L. Tipton, D. E. Anderson, Jr., and W. Cieplak, Jr.** 1997. Identification and molecular cloning of a gene encoding a fibronectin-binding protein (CadF) from *Campylobacter jejuni. Mol. Microbiol.* 24:953–963.

114. Konkel, M. E., J. D. Klena, V. Rivera-Amill, M. R. Monteville, D. Biswas, B. Raphael, and J. Mickelson. 2004. Secretion of virulence proteins from *Campylobacter jejuni* is dependent on a functional flagellar export apparatus. *J. Bacteriol.* **186:**3296–3303.

115. Kopecko, D. J., L. Hu, and K. J. Zaal. 2001. *Campylobacter jejuni*—microtubule-dependent invasion. *Trends Microbiol.* **9:**389–396.

116. Korlath, J. A., M. T. Osterholm, L. A. Judy, J. C. Forfang, and R. A. Robinson. 1985. A point-source outbreak of campylobacteriosis associated with consumption of raw milk. *J. Infect. Dis.* **152:**592–596.

117. Kuroda, M., T. Ohta, I. Uchiyama, T. Baba, H. Yuzawa, I. Kobayashi, L. Cui, A. Oguchi, K. Aoki, Y. Nagai, J. Lian, T. Ito, M. Kanamori, H. Matsumaru, A. Maruyama, H. Murakami, A. Hosoyama, Y. Mizutani-Ui, N. K. Takahashi, T. Sawano, R. Inoue, C. Kaito, K. Sekimizu, H. Hirakawa, S. Kuhara, S. Goto, J. Yabuzaki, M. Kanehisa, A. Yamashita, K. Oshima, K. Furuya, C. Yoshino, T. Shiba, M. Hattori, N. Ogasawara, H. Hayashi, and K. Hiramatsu. 2001. Whole genome sequencing of meticillin-resistant *Staphylococcus aureus*. *Lancet* **357:**1225–1240.

118. Kusunoki, S., A. Chiba, and I. Kanazawa. 1999. Anti-GQ1b IgG antibody is associated with ataxia as well as ophthalmoplegia. *Muscle Nerve* **22:**1071–1074.

119. Lampe, D. J., B. J. Akerley, E. J. Rubin, J. J. Mekalanos, and H. M. Robertson. 1999. Hyperactive transposase mutants of the *Himar1* mariner transposon. *Proc. Natl. Acad. Sci. USA* **96:**11428–11433.

120. Lastovica, A. J., and J. L. Penner. 1983. Serotypes of *Campylobacter jejuni* and *Campylobacter coli* in bacteremic, hospitalized children. *J. Infect. Dis.* **147:**592.

121. Lastovica, A. J., and M. B. Skirrow. 2000. Clinical significance of *Campylobacter* and related species other than *Campylobacter jejuni* and *C. coli*, p. 89–120. *In* I. Nachamkin and M. J. Blaser (ed.), Campylobacter. ASM Press, Washington, DC.

122. Lawson, A. J., S. L. On, J. M. Logan, and J. Stanley. 2001. *Campylobacter hominis* sp. nov., from the human gastrointestinal tract. *Int. J. Syst. Evol. Microbiol.* **51:**651–660.

123. Lawson, G. H. K., J. L. Leaver, G. W. Pettigrew, and A. C. Rowland. 1981. Some features of *Campylobacter sputorum* subsp. *mucosalis* subsp. nov., norm. rev., and their taxonomic significance. *Int. J. Syst. Bacteriol.* **31:**385–391.

124. le Roux, E., and A. J. Lastovica. 1998. The Cape Town Protocol: how to isolate the most campylobacters for your dollar, pound, franc, yen, etc., p. 30–33. *In* A. J. Lastovica, D. G. Newell, and E. E. Lastovica (ed.), Campylobacter, Helicobacter *and Related Organisms*. Institute of Child Health, Cape Town, South Africa.

125. Lecuit, M., E. Abachin, A. Martin, C. Poyart, P. Pochart, F. Suarez, D. Bengoufa, J. Feuillard, A. Lavergne, J. I. Gordon, P. Berche, L. Guillevin, and O. Lortholary. 2004. Immunoproliferative small intestinal disease associated with *Campylobacter jejuni*. *N. Engl. J. Med.* **350:**239–248.

126. Lee, R. B., D. C. Hassane, D. L. Cottle, and C. L. Pickett. 2003. Interactions of *Campylobacter jejuni* cytolethal distending toxin subunits CdtA and CdtC with HeLa cells. *Infect. Immun.* **71:**4883–4890.

127. Leonard, E. E., II, T. Takata, M. J. Blaser, S. Falkow, L. S. Tompkins, and E. C. Gaynor. 2003. Use of an open-reading frame-specific *Campylobacter jejuni* DNA microarray as a new genotyping tool for studying epidemiologically related isolates. *J. Infect. Dis.* **187:**691–694.

128. Leonard, E. E., II, L. S. Tompkins, S. Falkow, and I. Nachamkin. 2004. Comparison of *Campylobacter jejuni* isolates implicated in Guillain-Barré syndrome and strains that cause enteritis by a DNA microarray. *Infect. Immun.* **72:**1199–1203.

129. Linton, D., A. V. Karlyshev, P. G. Hitchen, H. R. Morris, A. Dell, N. A. Gregson, and B. W. Wren. 2000. Multiple N-acetyl neuraminic acid synthetase (*neuB*) genes in *Campylobacter jejuni*: identification and characterization of the gene involved in sialylation of lipo-oligosaccharide. *Mol. Microbiol.* **35:**1120–1134.

130. Logan, J. M. J., A. Burnens, D. Linton, A. J. Laswon, and J. Stanley. 2000. *Campylobacter lanienae* sp. nov., a new species isolated from workers in an abattoir. *Int. J. Syst. Evol. Microbiol.* **50:**865–872.

131. Luban, J., and S. P. Goff. 1995. The yeast two-hybrid system for studying protein-protein interactions. *Curr. Opin. Biotechnol.* **6:**59–64.

132. MacCallum, A., S. P. Hardy, and P. H. Everest. 2005. *Campylobacter jejuni* inhibits the absorptive transport functions of Caco-2 cells and disrupts cellular tight junctions. *Microbiology* **151:**2451–2458.

133. MacKichan, J. K., E. C. Gaynor, C. Chang, S. Cawthraw, D. G. Newell, J. F. Miller, and S. Falkow. 2004. The *Campylobacter jejuni dccR S* two-component system is required for optimal *in vivo* colonization but is dispensable for in vitro growth. *Mol. Microbiol.* **54:**1269–1286.

134. Madden, R. H., L. Moran, and P. Scates. 1996. Sub-typing of animal and human *Campy-*

lobacter spp. using RAPD. *Lett. Appl. Microbiol.* **23:**167–170.

135. **Manser, P. A., and R. W. Dalziel.** 1985. A survey of *Campylobacter* in animals. *J. Hyg. (Lond.)* **95:**15–21.

136. **McCormick, B. A., A. Nusrat, C. A. Parkos, L. D'Andrea, P. M. Hofman, D. Carnes, T. W. Liang, and J. L. Madara.** 1997. Unmasking of intestinal epithelial lateral membrane beta1 integrin consequent to transepithelial neutrophil migration in vitro facilitates *inv*-mediated invasion by *Yersinia pseudotuberculosis. Infect. Immun.* **65:**1414–1421.

137. **Meinersmann, R. J., L. O. Helsel, P. I. Fields, and K. L. Hiett.** 1997. Discrimination of *Campylobacter jejuni* isolates by fla gene sequencing. *J. Clin. Microbiol.* **35:**2810–2814.

138. **Meinersmann, R. J., C. M. Patton, G. M. Evins, I. K. Wachsmuth, and P. I. Fields.** 2002. Genetic diversity and relationships of *Campylobacter* species and subspecies. *Int. J. Syst. Evol. Microbiol.* **52:**1789–1797.

139. **Miller, W. G., A. H. Bates, S. T. Horn, M. T. Brandl, M. R. Wachtel, and R. E. Mandrell.** 2000. Detection on surfaces and in Caco-2 cells of *Campylobacter jejuni* cells transformed with new *gfp, yfp,* and *cfp* marker plasmids. *Appl. Environ. Microbiol.* **66:**5426–5436.

140. **Miller, W. G., and R. E. Mandrell.** 2005. Prevalence of *Campylobacter* in the food and water supply: incidence, outbreaks, isolation, and detection, p. 101–163. *In* M. E. Konkel and J. M. Ketley (ed.), Campylobacter: *Molecular and Cellular Biology.* Horizon Scientific Press, Norwich, United Kingdom.

141. Reference deleted.

142. **Miller, W. G., B. M. Pearson, J. M. Wells, C. T. Parker, V. V. Kapitonov, and R. E. Mandrell.** 2005. Diversity within the *Campylobacter jejuni* type I restriction-modification loci. *Microbiology* **151:**337–351.

143. **Moen, B., A. Oust, O. Langsrud, N. Dorrell, G. L. Marsden, J. Hinds, A. Kohler, B. W. Wren, and K. Rudi.** 2005. Explorative multifactor approach for investigating global survival mechanisms of *Campylobacter jejuni* under environmental conditions. *Appl. Environ. Microbiol.* **71:**2086–2094.

144. **Mohammed, K. A., R. J. Miles, and M. A. Halablab.** 2004. The pattern and kinetics of substrate metabolism of *Campylobacter jejuni* and *Campylobacter coli. Lett. Appl. Microbiol.* **39:**261–266.

145. **Monteville, M. R., and M. E. Konkel.** 2002. Fibronectin-facilitated invasion of T84 eukaryotic cells by *Campylobacter jejuni* occurs preferentially at the basolateral cell surface. *Infect. Immun.* **70:**6665–6671.

146. **Monteville, M. R., J. E. Yoon, and M. E. Konkel.** 2003. Maximal adherence and invasion of INT 407 cells by *Campylobacter jejuni* requires the CadF outer-membrane protein and microfilament reorganization. *Microbiology* **149:**153–165.

147. **Moran, A. P., B. J. Appelmelk, and G. O. Aspinall.** 1996. Molecular mimicry of host structures by lipopolysaccharides of *Campylobacter* and *Helicobacter* spp: implications in pathogenesis. *J. Endotoxin Res.* **3:**521–531.

148. **Morgan, G. J., G. F. Hatfull, S. Casjens, and R. W. Hendrix.** 2002. Bacteriophage Mu genome sequence: analysis and comparison with Mu-like prophages in *Haemophilus, Neisseria* and *Deinococcus. J. Mol. Biol.* **317:**337–359.

149. **Moser, I., B. Rieksneuwohner, P. Lentzsch, P. Schwerk, and L. H. Wieler.** 2001. Genomic heterogeneity and O-antigenic diversity of *Campylobacter upsaliensis* and *Campylobacter helveticus* strains isolated from dogs and cats in Germany. *J. Clin. Microbiol.* **39:**2548–2557.

150. **Muza-Moons, M. M., A. Koutsouris, and G. Hecht.** 2003. Disruption of cell polarity by enteropathogenic *Escherichia coli* enables basolateral membrane proteins to migrate apically and to potentiate physiological consequences. *Infect. Immun.* **71:**7069–7078.

151. **Myers, E. W., G. G. Sutton, A. L. Delcher, I. M. Dew, D. P. Fasulo, M. J. Flanigan, S. A. Kravitz, C. M. Mobarry, K. H. Reinert, K. A. Remington, E. L. Anson, R. A. Bolanos, H. H. Chou, C. M. Jordan, A. L. Halpern, S. Lonardi, E. M. Beasley, R. C. Brandon, L. Chen, P. J. Dunn, Z. Lai, Y. Liang, D. R. Nusskern, M. Zhan, Q. Zhang, X. Zheng, G. M. Rubin, M. D. Adams, and J. C. Venter.** 2000. A whole-genome assembly of *Drosophila. Science* **287:**2196–2204.

152. **Nachamkin, I.** 2002. Chronic effects of *Campylobacter* infection. *Microbes Infect.* **4:**399–403.

153. **Nachamkin, I., J. Liu, M. Li, H. Ung, A. P. Moran, M. M. Prendergast, and K. Sheikh.** 2002. *Campylobacter jejuni* from patients with Guillain-Barré syndrome preferentially expresses a GD(1a)-like epitope. *Infect. Immun.* **70:**5299–5303.

154. **Nelson, K. E., D. E. Fouts, E. F. Mongodin, J. Ravel, R. T. DeBoy, J. F. Kolonay, D. A. Rasko, S. V. Angiuoli, S. R. Gill, I. T. Paulsen, J. Peterson, O. White, W. C. Nelson, W. Nierman, M. J. Beanan, L. M. Brinkac, S. C. Daugherty, R. J. Dodson, A. S. Durkin, R. Madupu, D. H. Haft, J. Selengut, S. Van Aken, H. Khouri, N. Fedorova, H. Forberger, B. Tran, S. Kathariou, L. D. Wonderling, G. A. Uhlich, D. O.**

Bayles, J. B. Luchansky, and C. M. Fraser. 2004. Whole genome comparisons of serotype 4b and 1/2a strains of the food-borne pathogen *Listeria monocytogenes* reveal new insights into the core genome components of this species. *Nucleic Acids Res.* **32:**2386–2395.

155. **Nitta, H., S. C. Holt, and J. L. Ebersole.** 1997. Purification and characterization of *Campylobacter rectus* surface layer proteins. *Infect. Immun.* **65:**478–483.

156. **Ohno, H., H. Sakai, T. Washio, and M. Tomita.** 2001. Preferential usage of some minor codons in bacteria. *Gene* **276:**107–115.

157. **On, S. L., and C. S. Harrington.** 2000. Identification of taxonomic and epidemiological relationships among *Campylobacter* species by numerical analysis of AFLP profiles. *FEMS Microbiol. Lett.* **193:**161–169.

158. **Oust, A., B. Moen, H. Martens, K. Rudi, T. Naes, C. Kirschner, and A. Kohler.** 2006. Analysis of covariance patterns in gene expression data and FT-IR spectra. *J. Microbiol. Methods* **65:**573–584.

159. **Owen, R. J.** 1983. Nucleic acids in the classification of campylobacters. *Eur J. Clin. Microbiol.* **2:**367–377.

160. **Parker, C. T., S. T. Horn, M. Gilbert, W. G. Miller, D. L. Woodward, and R. E. Mandrell.** 2005. Comparison of *Campylobacter jejuni* lipooligosaccharide biosynthesis loci from a variety of sources. *J. Clin. Microbiol.* **43:**2771–2781.

161. **Parkhill, J., B. W. Wren, K. Mungall, J. M. Ketley, C. Churcher, D. Basham, T. Chillingworth, R. M. Davies, T. Feltwell, S. Holroyd, K. Jagels, A. V. Karlyshev, S. Moule, M. J. Pallen, C. W. Penn, M. A. Quail, M. A. Rajandream, K. M. Rutherford, A. H. van Vliet, S. Whitehead, and B. G. Barrell.** 2000. The genome sequence of the food-borne pathogen *Campylobacter jejuni* reveals hypervariable sequences. *Nature* **403:**665–668.

162. **Parsonnet, J., and P. G. Isaacson.** 2004. Bacterial infection and MALT lymphoma. *N. Engl. J. Med.* **350:**213–215.

163. **Perez, G. I., J. A. Hopkins, and M. J. Blaser.** 1985. Antigenic heterogeneity of lipopolysaccharides from *Campylobacter jejuni* and *Campylobacter fetus*. *Infect. Immun.* **48:**528–533.

164. **Perna, N. T., G. Plunkett, 3rd, V. Burland, B. Mau, J. D. Glasner, D. J. Rose, G. F. Mayhew, P. S. Evans, J. Gregor, H. A. Kirkpatrick, G. Posfai, J. Hackett, S. Klink, A. Boutin, Y. Shao, L. Miller, E. J. Grotbeck, N. W. Davis, A. Lim, E. T. Dimalanta, K. D. Potamousis, J. Apodaca, T. S. Anantharaman, J. Lin, G. Yen, D. C. Schwartz, R. A. Welch, and F. R. Blattner.** 2001. Genome sequence of entero-

haemorrhagic *Escherichia coli* O157:H7. *Nature* **409:**529–533.

165. **Peterson, M. C.** 1994. Clinical aspects of *Campylobacter jejuni* infections in adults. *West. J. Med.* **161:**148–152.

166. **Phongsisay, V., V. N. Perera, and B. N. Fry.** 2006. Exchange of lipooligosaccharide synthesis genes creates potential Guillain-Barré syndrome—inducible strains of *Campylobacter jejuni*. *Infect. Immun.* **74:**1368–1372.

167. **Piddock, L. J., V. Ricci, L. Pumbwe, M. J. Everett, and D. J. Griggs.** 2003. Fluoroquinolone resistance in *Campylobacter* species from man and animals: detection of mutations in topoisomerase genes. *J. Antimicrob. Chemother.* **51:**19–26.

168. **Poly, F., D. Threadgill, and A. Stintzi.** 2005. Genomic diversity in *Campylobacter jejuni*: identification of *C. jejuni* 81-176-specific genes. *J. Clin. Microbiol.* **43:**2330–2338.

169. **Poly, F., D. Threadgill, and A. Stintzi.** 2004. Identification of *Campylobacter jejuni* ATCC 43431–specific genes by whole microbial genome comparisons. *J. Bacteriol.* **186:**4781–4795.

170. **Pop, M., D. S. Kosack, and S. L. Salzberg.** 2004. Hierarchical scaffolding with Bambus. *Genome Res.* **14:**149–159.

171. **Prendergast, M. M., D. R. Tribble, S. Baqar, D. A. Scott, J. A. Ferris, R. I. Walker, and A. P. Moran.** 2004. In vivo phase variation and serologic response to lipooligosaccharide of *Campylobacter jejuni* in experimental human infection. *Infect. Immun.* **72:**916–922.

172. **Prévot, A. R.** 1940. Etude de la systématique bactérienne. V—Essai de classification des vibrions anaérobies. *Ann. Inst. Pasteur* **64:**117–125.

173. **Rain, J. C., L. Selig, H. De Reuse, V. Battaglia, C. Reverdy, S. Simon, G. Lenzen, F. Petel, J. Wojcik, V. Schachter, Y. Chemama, A. Labigne, and P. Legrain.** 2001. The protein-protein interaction map of *Helicobacter pylori*. *Nature* **409:**211–215.

174. **Romaniuk, P. J., and T. J. Trust.** 1989. Rapid identification of *Campylobacter* species using oligonucleotide probes to 16S ribosomal RNA. *Mol. Cell. Probes* **3:**133–142.

175. **Roop, R. M. I., R. M. Smibert, J. L. Johnson, and N. R. Krieg.** 1985. *Campylobacter mucosalis* (Lawson, Leaver, Pettigrew, and Rowland 1981) comb. nov.: emended description. *Int. J. Syst. Bacteriol.* **35:**189–192.

176. **Sanchez, M. X., W. M. Fluckey, M. M. Brashears, and S. R. McKee.** 2002. Microbial profile and antibiotic susceptibility of *Campylobacter* spp. and *Salmonella* spp. in broilers processed in air-chilled and immersion-chilled environments. *J. Food Prot.* **65:**948–956.

177. **Sandstedt, K., and J. Ursing.** 1991. Description of *Campylobacter upsaliensis* sp. nov. previously known as the CNW group. *Syst. Appl. Microbiol.* **14:**39–45.

178. **Sandstedt, K., J. Ursing, and M. Walder.** 1983. Thermotolerant *Campylobacter* with no or weak catalase activity isolated from dogs. *Curr. Microbiol.* **8:**209–213.

179. **Santos, S. R., and H. Ochman.** 2004. Identification and phylogenetic sorting of bacterial lineages with universally conserved genes and proteins. *Environ. Microbiol.* **6:**754–759.

180. **Schouls, L. M., S. Reulen, B. Duim, J. A. Wagenaar, R. J. Willems, K. E. Dingle, F. M. Colles, and J. D. Van Embden.** 2003. Comparative genotyping of *Campylobacter jejuni* by amplified fragment length polymorphism, multilocus sequence typing, and short repeat sequencing: strain diversity, host range, and recombination. *J. Clin. Microbiol.* **41:**15–26.

181. **Schurig, G. D., C. E. Hall, K. Burda, L. B. Corbeil, J. R. Duncan, and A. J. Winter.** 1973. Persistent genital tract infection with *Vibrio fetus* intestinalis associated with serotypic alteration of the infecting strain. *Am. J. Vet. Res.* **34:**1399–1403.

182. **Sebald, M., and M. Véron.** 1963. Teneur en bases de l'ADN et classification des vibrions. *Ann. Inst. Pasteur* **105:**897–910.

183. **Sergeev, N., M. Distler, S. Courtney, S. F. Al-Khaldi, D. Volokhov, V. Chizhikov, and A. Rasooly.** 2004. Multipathogen oligonucleotide microarray for environmental and biodefense applications. *Biosens. Bioelectron.* **20:** 684–698.

184. **Serino, L., and M. Virji.** 2002. Genetic and functional analysis of the phosphorylcholine moiety of commensal *Neisseria* lipopolysaccharide. *Mol. Microbiol.* **43:**437–448.

185. **Sheikh, K. A., I. Nachamkin, T. W. Ho, H. J. Willison, J. Veitch, H. Ung, M. Nicholson, C. Y. Li, H. S. Wu, B. Q. Shen, D. R. Cornblath, A. K. Asbury, G. M. McKhann, and J. W. Griffin.** 1998. *Campylobacter jejuni* lipopolysaccharides in Guillain-Barré syndrome: molecular mimicry and host susceptibility. *Neurology* **51:**371–378.

186. **Shen, Z., Y. Feng, F. E. Dewhirst, and J. G. Fox.** 2001. Coinfection of enteric *Helicobacter* spp. and *Campylobacter* spp. in cats. *J. Clin. Microbiol.* **39:**2166–2172.

187. **Shih, H. M., P. S. Goldman, A. J. DeMaggio, S. M. Hollenberg, R. H. Goodman, and M. F. Hoekstra.** 1996. A positive genetic selection for disrupting protein-protein interactions: identification of CREB mutations that prevent association with the coactivator CBP. *Proc. Natl. Acad. Sci. USA* **93:**13896–13901.

188. **Smibert, R. M.** 1984. *Campylobacter*, p. 111–118. *In* J. G. Holt and N. R. Krieg (ed.), *Bergey's Manual of Systematic Bacteriology*, vol. 1. Williams & Wilkins, Baltimore, MD.

189. **Smith, T., and M. S. Taylor.** 1919. Some morphological and biochemical characters of the spirilla (*Vibrio fetus*, n. sp.) associated with disease of the fetal membranes in cattle. *J. Exp. Med.* **30:**299–311.

190. **Stanley, J., A. P. Burnens, D. Linton, S. L. On, M. Costas, and R. J. Owen.** 1992. *Campylobacter helveticus* sp. nov., a new thermophilic species from domestic animals: characterization, and cloning of a species-specific DNA probe. *J. Gen. Microbiol.* **138:**2293–2303.

191. **Steele, T. W., and R. J. Owen.** 1988. *Campylobacter jejuni* subsp. *doylei* subsp. nov., a subspecies of nitrate-negative campylobacters isolated from human clinical specimens. *Int. J. Syst. Bacteriol.* **38:**316–318.

192. **Stern, N. J., and J. E. Line.** 2000. Campylobacter, p. 1040–1056. *In* B. M. Lund, A. C. Baird-Parker, and G. W. Gould (ed.), *The Microbiological Safety and Quality of Food*, vol. 2. Aspen Publishers, Gaithersburg, MD.

193. **Stintzi, A.** 2003. Gene expression profile of *Campylobacter jejuni* in response to growth temperature variation. *J. Bacteriol.* **185:**2009–2016.

194. **Stintzi, A., D. Marlow, K. Palyada, H. Naikare, R. Panciera, L. Whitworth, and C. Clarke.** 2005. Use of genome-wide expression profiling and mutagenesis to study the intestinal lifestyle of *Campylobacter jejuni*. *Infect. Immun.* **73:**1797–1810.

195. **St. Michael, F., C. M. Szymanski, J. Li, K. H. Chan, N. H. Khieu, S. Larocque, W. W. Wakarchuk, J. R. Brisson, and M. A. Monteiro.** 2002. The structures of the lipooligosaccharide and capsule polysaccharide of *Campylobacter jejuni* genome sequenced strain NCTC 11168. *Eur. J. Biochem.* **269:**5119–5136.

196. **Suerbaum, S., C. Josenhans, T. Sterzenbach, B. Drescher, P. Brandt, M. Bell, M. Droge, B. Fartmann, H. P. Fischer, Z. Ge, A. Horster, R. Holland, K. Klein, J. Konig, L. Macko, G. L. Mendz, G. Nyakatura, D. B. Schauer, Z. Shen, J. Weber, M. Frosch, and J. G. Fox.** 2003. The complete genome sequence of the carcinogenic bacterium *Helicobacter hepaticus*. *Proc. Natl. Acad. Sci. USA* **100:**7901–7906.

197. **Suerbaum, S., M. Lohrengel, A. Sonnevend, F. Ruberg, and M. Kist.** 2001. Allelic diversity and recombination in *Campylobacter jejuni*. *J. Bacteriol.* **183:**2553–2559.

198. **Taboada, E. N., R. R. Acedillo, C. D. Carrillo, W. A. Findlay, D. T. Medeiros, O. L.**

Mykytczuk, M. J. Roberts, C. A. Valencia, J. M. Farber, and J. H. Nash. 2004. Large-scale comparative genomics meta-analysis of *Campylobacter jejuni* isolates reveals low level of genome plasticity. *J. Clin. Microbiol.* **42:**4566–4576.

199. Tanner, A. C. R., S. Badger, C. H. Lai, M. A. Listgarten, R. A. Visconti, and S. S. Socransky. 1981. *Wolinella* gen. nov., *Wolinella succinogenes* (*Vibrio succinogenes* Wolin et al.) comb. nov., and description of *Bacteroides gracilis* sp. nov., *Wolinella recta* sp. nov., *Campylobacter concisus* sp. nov., and *Eikenella corrodens* from humans with periodontal disease. *Int. J. Syst. Bacteriol.* **31:**432–445.

200. Tanner, A. C. R., M. A. Listgarten, and J. L. Ebersole. 1984. *Wolinella curva* sp. nov.: "*Vibrio succinogenes*" of human origin. *Int. J. Syst. Bacteriol.* **34:**275–282.

201. Tauxe, R. V., C. M. Patton, P. Edmonds, T. J. Barrett, D. J. Brenner, and P. A. Blake. 1985. Illness associated with *Campylobacter laridis*, a newly recognized *Campylobacter* species. *J. Clin. Microbiol.* **21:**222–225.

202. Taylor, D. E., Z. Ge, D. Purych, T. Lo, and K. Hiratsuka. 1997. Cloning and sequence analysis of two copies of a 23S rRNA gene from *Helicobacter pylori* and association of clarithromycin resistance with 23S rRNA mutations. *Antimicrob. Agents Chemother.* **41:**2621–2628.

203. Testerman, T. L., D. J. McGee, and H. L. T. Mobley. 2001. Adherence and colonization, p. 381–405. *In* H. L. T. Mobley, G. L. Mendz, and S. L. Hazell (ed.), Helicobacter pylori: *Physiology and Genetics.* ASM Press, Washington, DC.

204. Thibault, P., S. M. Logan, J. F. Kelly, J. R. Brisson, C. P. Ewing, T. J. Trust, and P. Guerry. 2001. Identification of the carbohydrate moieties and glycosylation motifs in *Campylobacter jejuni* flagellin. *J. Biol. Chem.* **276:**34862–34870.

205. Thompson, S. A., and M. J. Blaser. 2000. Pathogenesis of *Campylobacter fetus* infections, p. 321–347. *In* I. Nachamkin and M. J. Blaser (ed.), Campylobacter. ASM Press, Washington, DC.

206. Tomb, J. F., O. White, A. R. Kerlavage, R. A. Clayton, G. G. Sutton, R. D. Fleischmann, K. A. Ketchum, H. P. Klenk, S. Gill, B. A. Dougherty, K. Nelson, J. Quackenbush, L. Zhou, E. F. Kirkness, S. Peterson, B. Loftus, D. Richardson, R. Dodson, H. G. Khalak, A. Glodek, K. McKenney, L. M. Fitzegerald, N. Lee, M. D. Adams, J. C. Venter, et al. 1997. The complete genome sequence of the gastric pathogen *Helicobacter pylori*. *Nature* **388:**539–547.

207. Trieber, C. A., and D. E. Taylor. 2002. Mutations in the 16S rRNA genes of *Helicobacter pylori* mediate resistance to tetracycline. *J. Bacteriol.* **184:**2131–2140.

208. Tu, Z. C., F. E. Dewhirst, and M. J. Blaser. 2001. Evidence that the *Campylobacter fetus sap* locus is an ancient genomic constituent with origins before mammals and reptiles diverged. *Infect. Immun.* **69:**2237–2244.

209. Uetz, P., L. Giot, G. Cagney, T. A. Mansfield, R. S. Judson, J. R. Knight, D. Lockshon, V. Narayan, M. Srinivasan, P. Pochart, A. Qureshi-Emili, Y. Li, B. Godwin, D. Conover, T. Kalbfleisch, G. Vijayadamodar, M. Yang, M. Johnston, S. Fields, and J. M. Rothberg. 2000. A comprehensive analysis of protein-protein interactions in *Saccharomyces cerevisiae*. *Nature* **403:**623–627.

210. Vandamme, P. 2000. Taxonomy of the family *Campylobacteraceae*, p. 3–26. *In* I. Nachamkin and M. J. Blaser (ed.), Campylobacter. ASM Press, Washington, DC.

211. Vandamme, P., M. I. Daneshvar, F. E. Dewhirst, B. J. Paster, K. Kersters, H. Goossens, and C. W. Moss. 1995. Chemotaxonomic analyses of *Bacteroides gracilis* and *Bacteroides ureolyticus* and reclassification of *B. gracilis* as *Campylobacter gracilis* comb. nov. *Int. J. Syst. Bacteriol.* **45:**145–152.

212. Vandamme, P., E. Falsen, R. Rossau, B. Hoste, P. Segers, R. Tytgat, and J. De Ley. 1991. Revision of *Campylobacter*, *Helicobacter*, and *Wolinella* taxonomy: emendation of generic descriptions and proposal of *Arcobacter* gen. nov. *Int. J. Syst. Bacteriol.*. **41:**88–103.

213. Van Doorn, L. J., A. Verschuuren-Van Haperen, A. Van Belkum, H. P. Endtz, J. S. Vliegenthart, P. Vandamme, and W. G. Quint. 1998. Rapid identification of diverse *Campylobacter lari* strains isolated from mussels and oysters using a reverse hybridization line probe assay. *J. Appl. Microbiol.* **84:**545–550.

214. Véron, M., and R. Chatelain. 1973. Taxonomic study of the genus *Campylobacter* Sebald and Véron and designation of the neotype strain for the type species *Camplybacter fetus* (Smith and Taylor) Sebald and Véron. *Int. J. Syst. Bacteriol.* **23:**122–134.

215. Vidal, M., R. K. Brachmann, A. Fattaey, E. Harlow, and J. D. Boeke. 1996. Reverse two-hybrid and one-hybrid systems to detect dissociation of protein-protein and DNA-protein interactions. *Proc. Natl. Acad. Sci. USA* **93:**10315–10320.

216. Vinzent, R., J. Dumas, and N. Picard. 1947. Septicémie grave au cours de la grossesse due à

un Vibrion. Avortement consécutif. *Bull. Acad. Nat. Med. Paris* **131**:90–92.

217. **Volokhov, D., V. Chizhikov, K. Chumakov, and A. Rasooly.** 2003. Microarray-based identification of thermophilic *Campylobacter jejuni, C. coli, C. lari,* and *C. upsaliensis. J. Clin. Microbiol.* **41**:4071-4080.

218. **Wagner, P. L., and M. K. Waldor.** 2002. Bacteriophage control of bacterial virulence. *Infect. Immun.* **70**:3985–3993.

219. **Wang, Y., and D. E. Taylor.** 1990. Natural transformation in *Campylobacter* species. *J. Bacteriol.* **172**:949–955.

220. **Wassenaar, T. M.** 1997. Toxin production by *Campylobacter* spp. *Clin. Microbiol. Rev.* **10**:466–476.

221. **Wassenaar, T. M., B. N. Fry, A. J. Lastovica, J. A. Wagenaar, P. J. Coloe, and B. Duim.** 2000. Genetic characterization of *Campylobacter jejuni* O:41 isolates in relation with Guillain-Barré syndrome. *J. Clin. Microbiol.* **38**:874–876.

222. **Wassenaar, T. M., B. N. Fry, and B. A. van der Zeijst.** 1993. Genetic manipulation of *Campylobacter:* evaluation of natural transformation and electro-transformation. *Gene* **132**:131–135.

223. **Wassenaar, T. M., B. A. van der Zeijst, R. Ayling, and D. G. Newell.** 1993. Colonization of chicks by motility mutants of *Campylobacter jejuni* demonstrates the importance of flagellin A expression. *J. Gen. Microbiol.* **139**(Pt. 6):1171–1175.

224. **Weijtens, M. J., P. G. Bijker, J. Van der Plas, H. A. Urlings, and M. H. Biesheuvel.** 1993. Prevalence of *Campylobacter* in pigs during fattening; an epidemiological study. *Vet. Q.* **15**:138–143.

225. **Weiser, J. N., A. A. Lindberg, E. J. Manning, E. J. Hansen, and E. R. Moxon.** 1989. Identification of a chromosomal locus for expression of lipopolysaccharide epitopes in *Haemophilus influenzae. Infect. Immun.* **57**:3045–3052.

226. **Weiser, J. N., M. Shchepetov, and S. T. Chong.** 1997. Decoration of lipopolysaccharide with phosphorylcholine: a phase-variable characteristic of *Haemophilus influenzae. Infect. Immun.* **65**:943–950.

227. **Wiesner, R. S., D. R. Hendrixson, and V. J. DiRita.** 2003. Natural transformation of *Campylobacter jejuni* requires components of a type II secretion system. *J. Bacteriol.* **185**:5408–5418.

228. **Wilson, I. G., and J. E. Moore.** 1996. Presence of *Salmonella* spp. and *Campylobacter* spp. in shellfish. *Epidemiol. Infect.* **116**:147–153.

229. **Yan, W., N. Chang, and D. E. Taylor.** 1991. Pulsed-field gel electrophoresis of *Campylobacter jejuni* and *Campylobacter coli* genomic DNA and its epidemiologic application. *J. Infect. Dis.* **163**: 1068–1072.

230. **Yang, M., Z. Wu, and S. Fields.** 1995. Protein-peptide interactions analyzed with the yeast two-hybrid system. *Nucleic Acids Res.* **23**:1152–1156.

231. **Yao, R., D. H. Burr, P. Doig, T. J. Trust, H. Niu, and P. Guerry.** 1994. Isolation of motile and non-motile insertional mutants of *Campylobacter jejuni:* the role of motility in adherence and invasion of eukaryotic cells. *Mol. Microbiol.* **14**:883–893.

232. **Yuki, N.** 2001. Infectious origins of, and molecular mimicry in, Guillain-Barré and Fisher syndromes. *Lancet Infect. Dis.* **1**:29–37.

233. **Yuki, N., K. Susuki, M. Koga, Y. Nishimoto, M. Odaka, K. Hirata, K. Taguchi, T. Miyatake, K. Furukawa, T. Kobata, and M. Yamada.** 2004. Carbohydrate mimicry between human ganglioside GM1 and *Campylobacter jejuni* lipooligosaccharide causes Guillain-Barré syndrome. *Proc. Natl. Acad. Sci. USA* **101**:11404–11409.

234. **Yuki, N., T. Taki, M. Takahashi, K. Saito, H. Yoshino, T. Tai, S. Handa, and T. Miyatake.** 1994. Molecular mimicry between GQ1b ganglioside and lipopolysaccharides of *Campylobacter jejuni* isolated from patients with Fisher's syndrome. *Ann. Neurol.* **36**:791–793.

235. **Zhang, B., B. Kraemer, D. SenGupta, S. Fields, and M. Wickens.** 2000. Yeast three-hybrid system to detect and analyze RNA-protein interactions. *Methods Enzymol.* **318**:399–419.

236. **Zhang, J. R., I. Idanpaan-Heikkila, W. Fischer, and E. I. Tuomanen.** 1999. Pneumococcal *licD2* gene is involved in phosphorylcholine metabolism. *Mol. Microbiol.* **31**:1477–1488.

GENOMIC SIGNATURES OF INTRACELLULARITY: EVOLUTIONARY PATTERNS AND PACES IN BACTERIAL MUTUALISTS AND PARASITES

Jennifer J. Wernegreen

8

Bacteria that live solely within host cells, or intracellular endosymbionts, include a diverse group of pathogenic and mutualistic species. Full genome sequence data have helped to clarify the metabolic capabilities these bacteria possess, thereby offering clues about their functional significance in host biology and evolutionary processes that shape their intimate associations. Genome comparisons have revealed similar evolutionary trajectories for many intracellular parasites and mutualists, characterized by convergent patterns of genome reduction, accelerated rates of DNA sequence evolution, and strong base compositional biases. However, comparative approaches also reveal important differences in the gene contents of harmful and beneficial associates that may underlie their distinct effects on host fitness. Moreover, studies of genome dynamics within intracellular groups are detecting links between bacterial lifestyle and levels of genome fluidity. These trends, identified from recent genome data, guide predictions about forces that influence the pace and mode of molecular evolution within a host cell, including roles of natural selection, mutation, and genetic drift.

The rapid expansion of bacterial genome datasets promises numerous opportunities to test such predictions, to identify notable exceptions, and to continually refine our understanding of the genomic consequences of cellular associations.

Bacteria represent an astonishing range of physiological, ecological, and genetic diversity. Much of this variation can be linked to the wide array of lifestyles that bacteria have adopted, ranging from free-living species that pervade all terrestrial, aquatic, and marine environments, to symbiotic species that occupy a similar span of niches but live in close partnerships with hosts. Broadly speaking, symbionts include both obligate and facultative associates whose status as parasitic, commensal, or mutualistic depends on the fitness consequences for both partners involved. Among endosymbionts, which can live within host tissues or cells, certain highly specialized lineages can replicate solely within the domain of a host cell, having lost any free-living phase to their life cycle (Fig. 1). These obligately intracellular bacteria have acted as major evolutionary catalysts throughout the history of life, e.g., giving rise to organelles that helped to fuel the diversification of eukaryotic lineages (49).

Today, intracellular bacteria range from parasites that impose devastating effects on host

Jennifer J. Wernegreen   Josephine Bay Paul Center for Comparative Molecular Biology and Evolution, Marine Biological Laboratory, 7 MBL Street, Woods Hole, Massachusetts 02543.

Bacterial Pathogenomics, Edited by M. J. Pallen et al.

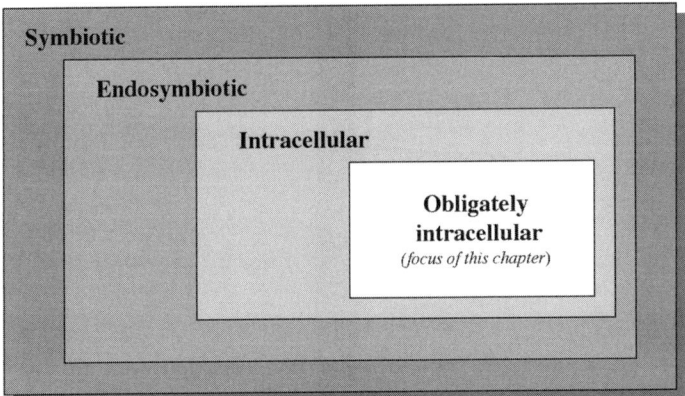

FIGURE 1 Schematic representation of associates that bacteria form with hosts, nested by increasing intimacy of the association. Symbiotic bacteria, the broadest category, associate with hosts for some or all of the bacterium's life cycle. Of these symbionts, some are endosymbiotic and can live within the tissues or cells. Endosymbionts that replicate within host cells are termed intracellular. Of these intracellular associates, certain highly specialized lineages have lost the ability to replicate outside of host cells and are obligately intracellular. The final category is the focus of this chapter.

fitness to essential mutualists that play key, beneficial roles in the ecology, physiology, and long-term divergence of their hosts. The medical and ecological significance of species, in addition to the small size of their typically streamlined genomes, has brought many into the spotlight of genome sequence studies. In recent years, numerous intracellular bacteria have been the targets of genome-sequencing projects that seek to identify specific genes that catalyze host-microbe associations and to examine the genetic consequences of this specialized lifestyle (Table 1).

This chapter highlights some intriguing patterns of genome change identified from genome-sequencing projects and molecular evolution studies. Trends include striking similarities between intracellular mutualists and parasites that may reflect common pressures of an intracellular lifestyle, as well as key differences that highlight distinct genomic consequences of their host associations. Given the rapid expansion of bacterial genome data sets, this overview is necessarily incomplete and will be rapidly outdated. However, I hope to offer a general framework to interpret the

wealth of recent data, to identify notable exceptions to observed trends, and to generate hypotheses about specific ties between bacterial lifestyle and genome evolution.

LIFESTYLE DIVERSITY OF INTRACELLULAR BACTERIA

Bacteria form a variety of associations with host cells. At one end of the spectrum, the replication and spread of intracellular parasites impose a fitness cost to hosts. Fully sequenced intracellular parasites include *Mycobacterium leprae* and *Coxiella burnetii* that represent relatively recent transitions to an intracellular lifestyle, as well as older intracellular groups such as *Phytoplasma*, a plant parasite and close relative of the epicellular *Mycoplasmas,* and the exclusively intracellular *Rickettsiaceae* and *Anaplasmataceae,* two families that apparently have been intracellular since they diverged ~400 mya (75). *Anaplasmataceae* includes vertebrate pathogens and the invertebrate endosymbiont, *Wolbachia*. This endosymbiont can spread by hijacking the reproduction of arthropod hosts (65), yet it also acts as a mutualist involved in the development and oogenesis in nearly all filarial nematodes

TABLE 1 Obligately intracellular bacteria for which full genome sequence data were published as of July 2005[a]

Organism	Genome size (Mbp)	Genome released	Host	Host effects
γ-Proteobacteria				
Enterobacteriales				
Buchnera aphidicola APS	0.66	2000	Aphid, Acyrthosiphon pisum	Nutritional mutualist
Buchnera aphidicola Sg	0.64	2002	Aphid, Schizaphis graminum	Nutritional mutualist
Buchnera aphidicola Bp	0.62	2003	Aphid, Baizongia pistaciae	Nutritional mutualist
Wigglesworthia glossinidia	0.7	2002	Tsetse fly, Glossina brevipalpis	Nutritional mutualist
Blochmannia floridanus	0.71	2003	Ant, Camponotus floridanus	Nutritional mutualist
Blochmannia pennsylvanicus	0.79	2005	Ant, Camponotus pennsylvanicus	Nutritional mutualist
Legionellales				
Coxiella burnetii	2.03	2003	Reptiles, birds, and mammals	Q fever
α-Proteobacteria				
Rickettsiales				
Rickettsiaceae				
Rickettsia conorii	1.27	2000	Mammals, via insect vectors	Rocky Mountain spotted fever
Rickettsia prowazekii	1.11	1998	Mammals, via insect vectors	Typhus
Rickettsia typhi	1.11	2003	Mammals, via insect vectors	Murine typhus
Rickettsia felis	1.46	2005	Mammals, via insect vectors	Spotted fever
Anaplasmataceae				
Anaplasma marginale	1.2	2004	Mammals, via insect vectors	Bovine anaplasmosis, human granulocytic ehrlichiosis

Ehrlichia ruminantium (2 strains)	1.5–1.52	2005	Wild ruminants, via tick host	Heartworm disease
Wolbachia wMel	1.27	2004	Fruit fly, *Drosophila melanogaster*	Cytomplasmic incompatibility
Wolbachia wBm[b]	1.08	2005	Filarial nematode, *Brugia malayi*	Worm development and fertility
Wolbachia wAna		2005 (95% of genome recovered from Trace Archive)	Fruit fly, *Drosophila ananassae*	Cytomplasmic incompatibility
Wolbachia wSim		2005 (75–80% of genome recovered from Trace Archive)	Fruit fly, *Drosophila simulans*	Cytomplasmic incompatibility
Mollicutes				
Acholeplasmatales				
Phytoplasma asteris	0.86	2003	Plants, via insect vector	Stunted plant growth and other symptoms
Actinobacteria				
Actinomycetales				
Mycobacterium leprae	3.27	2001	Humans and other vertebrates	Leprosy
Chlamydiae				
Chlamydiales				
Parachlamydia sp.[b]	2.41	2004	Free-living amoebae	Mutualist of amoebae, occasional parasite of humans
Chlamydiaceae				
Chlamydia muridarum	1.08	2000	Rodents	Mouse lung or genital tract infections
Chlamydia trachomatis	1.04	1998	Human and other mammals	Chronic genital and ocular infections
Chlamydophila abortus	1.14	2005	Ruminants and swine	Ruminant abortion
Chlamydophila caviae	1.18	2003	Guinea pig	Guinea pig inclusion conjunctivitis
Chlamydophila pneumoniae (4 strains)	1.23	1999–2003	Human and other mammals	Pneumonia, bronchitis, and pharyngitis

[a]Reprinted in part, with permission from Wernegreen, 2005. © (2005) Elsevier Science.
[b]Mutualistic association.

(43). Even more ancient, the ~700-my-old order *Chlamydiales* (44) includes parasitic *Chlamydiaceae* that cause respiratory, genital, and ocular infections, as well as *Parachlamydia* sp., a
mutualistic symbiont of free-living amoebae
and occasional opportunistic human pathogen.

At the other end of the symbiotic spectrum, certain intracellular bacterial groups
form exclusively mutualistic associations with
hosts. In addition to the nematode and amoebae hosts above, insects as a group frequently
form long-term, stable associations with beneficial bacteria, called primary endosymbionts
because of their essential role in host survival
and fecundity (8, 13, 58). When functions are
known, these insect endosymbionts often play
key roles in supplementing their hosts' nutri-
tionally unbalanced diets with essential nutrients. For example, the bacterial mutualist *Buchnera* provides essential amino acids that are
lacking in the plant sap diet of its aphid host
(83), and the related mutualist *Wigglesworthia*
supplies its tsetse fly host with B-complex vitamins that are missing from the vertebrate
blood on which the fly feeds (64). The function of *Blochmannia*, an endosymbiont of *Camponotus* and related ant genera, is less certain
because the diet of the ant host varies considerably among species and can be relatively
complex. In addition, removing the symbiont
from laboratory-reared hosts has no apparent
side effects (84). However, endosymbiont
functions may be important during times of
food scarcity in the life cycle of ant individuals
or colonies in the field. By allowing insects to
exploit otherwise inadequate food sources and
habitats, the acquisition of these mutualists can
be viewed as a key innovation in the evolution
of their hosts (55).

The relationship between primary endosymbionts and their insect hosts is reciprocally
beneficial because the bacteria have lost the
ability to grow outside the host cell. Rather,
they live within specialized host cells called
bacteriocytes, either directly in the host cytoplasm or within host-derived vacuoles. The
endosymbionts also infect female reproductive
structures through which they are maternally

transmitted to host offspring (13). Agreement
of host and symbiont phylogenies shows that
primary endosymbionts have cospeciated with
their particular insect host group for tens to
hundreds of millions of years. For example,
the *Blochmannia*-ant and *Wigglesworthia*-tsetse
associations originated at least 30 mya (16, 23),
and the even older *Buchnera*-aphid symbiosis
was established on the order of 200 mya (60).
Primary endosymbionts often coexist with
Wolbachia and secondary endosymbionts that
form transient associations with insects (3, 32,
62).

With the acquisition of full genome sequence data for several lineages of primary endosymbionts, these mutualists have become
models in which to study the changes in rates
and patterns of bacterial evolution that are
coupled with an intracellular lifestyle. Molecular phylogenetic analysis has shown that many
insect endosymbionts, including the aphid,
tsetse fly, and ant mutualists described above,
group with gamma-3 subdivision of *Proteobacteria*. Their close relationship to *Escherichia coli*
and other well-characterized *Enterobacteria* has
facilitated genome annotation, reconstruction
of metabolic pathways, and tracking of
genetic changes that occur upon a transition
to intracellularity. Moreover, comparative approaches have helped to clarify the similarities
and differences between intracellular mutualists
and pathogens.

INTRACELLULARITY MAY ALTER
BASIC EVOLUTIONARY FORCES

Regardless of whether a bacterial associate exploits or assists its host, the transition to obligate intracellularity may produce shifts in three
fundamental evolutionary processes: (i) the
strength of natural selection, (ii) mutational
processes such as point mutation and recombination, and (iii) random genetic drift, or
changes in the frequencies of alleles or genotypes due to chance alone (6, 56). The latter
stochastic effect plays a particularly important
role in small populations.

First, living within a host cell may lead to
relaxed purifying selection on metabolic func-

tions that are dispensable in a resource-rich intracellular niche. That is, intracellular replication and survival may require genes for transcription, translation, and other basic cellular processes, often viewed as the minimal gene set required for life (e.g., 46). However, abilities to import and metabolize many substrates are expected to be superfluous within host cells.

Second, mutation may differ in bacteria that are sequestered within host cells. For instance, such species might have limited opportunities for DNA exchange with genetically distinct species (30). Severe limitation on recombination may lead to the extinction of mobile DNA elements that rely on horizontal transfer for their maintenance, as well as the loss of repeated DNA and many recombination functions. Because many *rec* genes are also involved in DNA repair, their deletion may contribute to high rates of nucleotide mutation.

Third, intracellular bacteria may undergo a reduction in effective population sizes (N_e) compared to free-living bacteria. Not only do the initial infections at the origin of host associations likely purge genetic diversity of these bacteria, but ongoing transmission of both mutualists and parasites frequently entails population bottlenecks that will reduce N_e compared with free-living relatives (6). This effect may be most severe in vertically transmitted, intracellular mutualists of insects thought to experience severe population bottlenecks upon each inoculation of developing host eggs or embryos (53). That is, like mitochondria, the N_e of endosymbionts will depend on the size, dynamics, and genetic structure of (female) host populations (33, 80). Because population sizes of insects ($N_e \approx 10^3$ to 10^5) are typically several orders of magnitude smaller than those of free-living bacteria (e.g., $N_e \approx 10^9$ for the enterics; 86), the N_e of endosymbionts are much smaller than those of related, free-living bacterial species.

This reduction in N_e is expected to increase the rates of fixation of slightly deleterious mutations by random genetic drift (56, 71). Specifically, under the nearly neutral model of evolution (71, 72), many mutations are expected to experience mild selection pressures, with selection coefficients (s) near the reciprocal of N_e. The fate of mutations in this category is determined by a balance between weak selection and genetic drift. Particularly in small populations, genetic drift may result in the fixation of these mutations despite their slightly deleterious effects on fitness. Reduced gene exchange would exacerbate this effect by preventing the recovery of beneficial alleles or of entire gene regions that are lost (56, 63).

Considering the effects of these three evolutionary forces—selection, mutation, and genetic drift—offers a valuable framework to explain commonalities among intracellular bacteria, to identify exceptions, and to develop predictions that can be tested with new genome data. The abundance of superb reviews in this area illustrates the utility of this conceptual framework in integrating a wealth of new genome information (e.g., 7, 35, 46, 61, 79, 90, 100). As detailed below, features shared by many intracellular species include a drastic reduction in genome size; accelerated rates of DNA sequence evolution; strong nucleotide compositional biases; and, in many cases, reduced rates of recombination, inversions, translocations, and other forms of genome fluidity. The overview below highlights mechanisms that may explain these similarities, as well as notable exceptions to these trends and evidence for varied genomic signatures of parasitic and mutualistic lifestyles.

GENOME REDUCTION: A COMMON THREAD IN INTRACELLULAR LIFE

From the first bacterial genomes sequenced in the mid- to late 1990s, to the more than 300 published today (as of April 2006), data point to a substantial impact of lifestyle on genome size and content. Most strikingly, obligately intracellular associates typically have extremely reduced genomes that fall in the range of 1 Mb or much less, including the smallest known bacterial genomes (Fig. 2). As expected from their small genome sizes, metabolisms of intracellular bacteria are simpler than free-living or facultatively intracellular species with larger genomes.

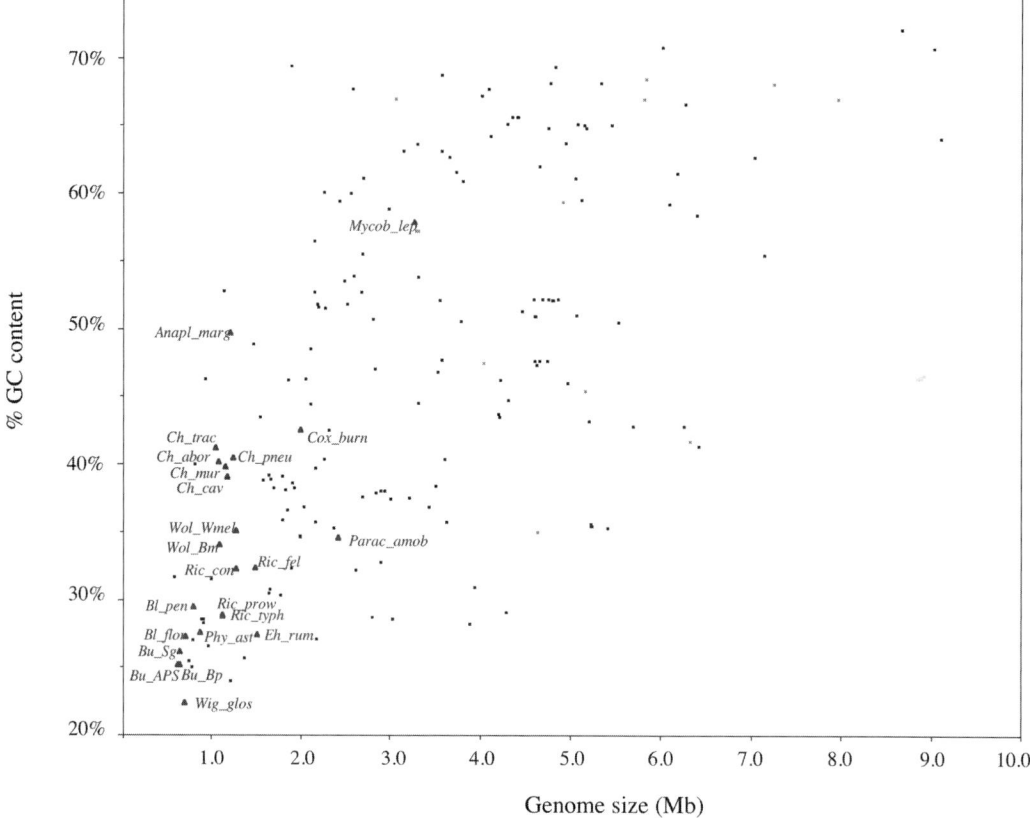

FIGURE 2 Genome size and %GC content of bacterial chromosome sequences, illustrating the small size and AT richness of obligate intracellular associates (labeled triangles). Chart is similar to published figures (e.g., reference 57) and includes genomes that were publicly available as of July 2005, with the exception of multiple, closely related strains. Gray x, species possessed two or more chromosomes (mark reflects values for single chromosome). Reprinted with permission from reference 94. © (2005) Elsevier Science.

Severe gene loss in intracellular species may reflect relaxed selection on many genes for metabolic diversity, combined with a mutational bias toward deletions (over insertions) that occurs across bacterial taxa and that will eliminate dispensable DNA regions (4, 5, 54). Reconstruction of specific deletion events suggests massive gene loss soon after the transition to intracellularity. For example, early evolution in *Buchnera* involved large deletions often encompassing 20 or more open reading frames (ORFs), as well as very high levels of gene rearrangements compared with other γ-*Proteobacteria* (9, 59). Genome size variation within endosymbiont groups indicates that

streamlining has continued in the context of intracellular associations, but more gradually. That is, in contrast to large early deletions, subsequent gene loss in *Buchnera* has tended to occur through gene disruption and gradual erosion, with inactivated genes requiring ~40 to 60 my to erode completely (38).

Sequence data from facultatively or younger intracellular bacteria provide a view into the early stages of lifestyle transitions, suggesting that the proliferation of insertion sequences and other mobile DNA may catalyze gene disruption and overall genome instability (61, 66). For example, the larger genomes, numerous pseudogenes, and/or dispersed repeats in relatively young intracellular associates (e.g., *Coxiella burnetii, M. leprae,* and the *Sitophilus oryzae* [weevil] primary endosymbiont; 18, 22, 87) and in *Sodalis glossinidius* (92), an endosymbiont of tsetse flies that can be cultured outside of the host (50), point to initial genetic turbulence on the road to obligate intracellularity.

Metabolic Implications of Severe Genome Reduction

Consistent with the prediction that many deleted genes were dispensable in a host cell, small intracellular genomes tend to lose many loci encoding metabolic diversity, yet retain those encoding transcription, translation, and other basic processes that are important regardless of ecological niche (6). The preferential retention of informational genes also holds true within endosymbiont groups. Analysis of partial genome regions indicates that, compared with its *Buchnera* relatives, the exceptionally small genome of *Buchnera* associated with the aphid *Cinara cedri* shows more extensive loss of metabolic than informational functions (76). The extensive loss of metabolic pathways apparently is a general feature of intracellular parasites and mutualists alike.

Holding onto Biosynthetic Functions, for Hosts' Sake

Although all intracellular associates rely on their host for certain nutrients, the gene contents of parasites and mutualists reflect the differing nature of the level of their dependency. As expected from their important roles in host nutrition, primary endosymbionts of insects retain a wide spectrum of genes to biosynthesize specific nutrients, often those that are conspicuously absent from the diet of their insect host group. These endosymbionts devote a higher fraction of their genomes to biosynthetic functions compared to free-living bacteria or pathogens (reviewed in reference 100) (Fig. 3). For example, the genomes of γ-3 proteobacterial insect mutualists show clear evidence of host-level selection on nutritional functions. *Wigglesworthia* retains numerous genes for the synthesis of vitamin B cofactors,

which are required by the host (2). Similarly, *Buchnera* genomes available to date retain the genetic potential to synthesize nearly all essential amino acids, as expected from its role in supplementing the plant sap of aphids (89). Furthermore, two available genomes of *Blochmannia* retain a wide array of metabolic functions that may benefit their ant hosts, including the biosynthesis of many amino acids, cofactors, and fatty acids, as well as sulfate reduction and nitrogen recycling (24, 36). On the other hand, related γ-*Proteobacteria* parasites, such as *Coxiella burnetii*, devote a much smaller fraction of their genomes to biosynthetic functions and apparently rely on their eukaryotic host cell for nearly all amino acids, cofactors, nucleotides, and other compounds (87).

Likewise, within the *Rickettsiales* and *Chlamydiales,* mutualists also encode a wider array of biosynthetic functions than parasites, but the difference is more subtle in these orders. The moderately sized (2.41 Mbp) genome of *Parachlamydia* complicates direct comparison of genome proportions with the ~1 to 1.2 Mbp genomes of parasitic chlamydia, but the amoebae mutualist's retention of twice as many amino acid and cofactor biosynthetic genes suggests it imposes fewer metabolic demands on its host cell. Within *Rickettsiales,* the mutualistic *Wolbachia* of the filarial nematode (*Brugia malayi,* or *Wolbachia* wBm) retains the ability to synthesize riboflavin and other coenzymes that *Rickettsia* spp. cannot make (28). In this bacterial-nematode mutualism, biosynthesis of riboflavin and heme may be key symbiont functions because neither pathway has been detected in the *Brugia malayi* genome that is in progress (34). Because *Wolbachia* wBm also retains the ability to biosynthesize purines and pyrimidines, the mutualist may also supplement the nematode's nucleotide pools during oogenesis and embryogenesis (28). Interestingly, *Wolbachia* that acts as a reproductive parasite in *Drosophila melanogaster* (*Wolbachia* wMel) also retains many of these same functions (99), but mutualistic *Wolbachia* devotes a higher proportion of its smaller genome to these potentially host-beneficial traits.

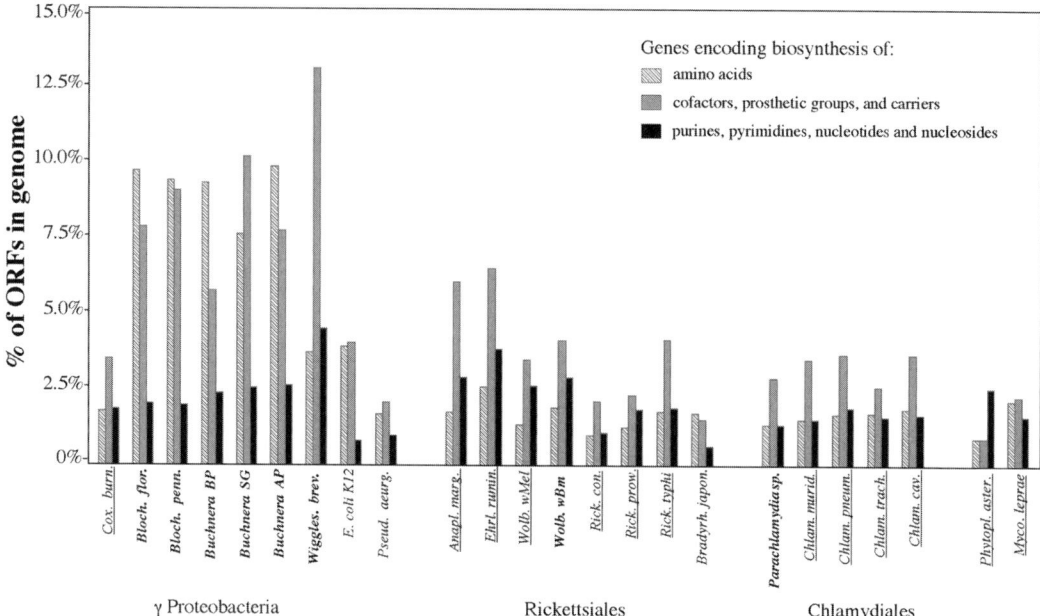

FIGURE 3 Percentage of genes encoding particular biosynthetic functions. *Y* axis indicates proportion of ORFs involved in biosynthesis of (i) amino acids, (ii) cofactors, prosthetic groups, and carriers, and (iii) purines, pyrimidines, nucleotides and nucleosides. Boldface, obligately intracellular mutualists; underline, obligately intracellular pathogen. *E. coli, P. aeuriginosa,* and *B. japonicum* retain a free-living phase and were included for comparisons. Values for *E. coli* and the gamma-*Proteobacterial* nutritional mutualists were based on reanalysis of genome sequences (24). Data for other species were downloaded from the Comprehensive Microbial Resource at TIGR for a more consistent comparison across genomes (77). However, as a result of differences in annotation and functional categorization, these counts may differ from original genome papers. Readers interested in particular taxa are encouraged to refer to original genome publications. In the rare cases where TIGR counted two largely overlapping, putative ORFs as two separate genes, these were counted as a single ORF for the purposes of this figure. Reprinted with permission from (94). © (2005) Elsevier Science.

Infection Strategies Shared by Diverse Intracellular Associates

Consistent with their infectious lifestyles, obligately intracellular parasites typically encode numerous mechanisms to invade host tissue and cells and to escape the host immune system. Such mechanisms include type III secretion (e.g., in *Chlamydiales*), an assemblage of ~20 proteins that spans the cell membrane, transports proteins out of the cell, and mediates the delivery of specific proteins that suppress defenses or otherwise facilitate cell invasion. In addition, type IV secretion (e.g., found in *Coxiella* and *Rickettsiales*) has been derived repeatedly from conjugation systems (29) and exports distinct DNA or protein substrates that

cause various physiological changes in host cells during infection.

Perhaps more surprisingly, these and other pathways associated with pathogenesis have also been discovered in mutualists (37, 40). For instance, the presence of type III secretion in the amoebae symbiont *Parachlamydia* implies the ancestor of the ancient order *Chlamydiales* could infect cells ~700 mya (20, 26, 44), before the appearance of most animal phyla. Like certain insect endosymbionts that possess type III secretion (21), *Parachlamydia* may deliver specific proteins into host cells in order to facilitate cell invasion. In addition, *Parachlamydia* apparently acquired type IV secretion through horizontal gene transfer, and it is the only chlamydia known to possess this pathway (44). Like its pathogenic relatives, *Parachlamydia* also imports ATP from the host cytosol by using an ATP/ADP translocase, an "energy-parasite" transport system unique to rickettsiae, chlamydiae, and plant plastids (85). Infection mechanisms are also shared between mutualistic and parasitic *Wolbachia*. In addition to sharing type IV secretion, the presence of ankyrin repeats in both *Wolbachia* genomes (although fewer in the mutualistic *Wolbachia* wBm) may mediate attachment of these bacteria to the cytoskeleton, alteration of host gene expression, or other functions involved in an intracellular lifestyle. This arsenal of infection mechanisms in mutualists and parasites alike suggests that "pathogenicity" functions play generally important roles for host-associated bacteria and might have evolved in the context of beneficial interactions.

Long-term insect mutualists also possess genes that are typically considered characteristic of pathogens. For example, *Blochmannia* retains the urease gene cluster, a significant virulence factor in some bacterial and fungal pathogens (14). In the ant mutualist, however, urease genes may benefit hosts by the hydrolysis of insect waste (urea) to ammonia, which the mutualist can then use to synthesize amino acids (36). In addition, the highly expressed GroEL protein is a major antigen and candidate for vaccine development in pathogens (10, 41, 78). The constitutive overexpression of this chaperonin in certain insect mutualists may help the folding of proteins that have experienced deleterious amino acid changes (27) or may counter oxidative modifications during stationary phase (31) where endosymbionts probably spend much of their life cycle. Other parallels with pathogens include the retention of certain outer membrane proteins and flagellar genes, the latter of which may be involved in secretion (2).

Role of Deleterious Changes

The above examples highlight metabolic adaptation to an intracellular lifestyle. Namely, relaxed selection in an intracellular niche may explain the convergent loss of many metabolic capabilities, yet selection is sufficient to maintain key genes for basic cellular processes, cell invasion mechanisms, and, especially in the case of mutualists, biosynthetic functions. A striking match between observed gene contents and simulated reconstructions of minimal metabolic networks also supports a central role of selection in shaping gene content (74).

However, in addition to adaptive changes, deleterious deletions may influence the metabolism of intracellular associates. That is, certain instances of gene loss may reflect the fixation of a deletion by random genetic drift in small populations, despite that deletion having negative fitness consequences. Although the fitness effects of historical gene losses are difficult to infer, candidates for harmful deletions include the loss of several DNA repair functions. In *Buchnera,* many DNA repair loci were lost in the large, early deletions that included numerous genes of varied functions (59), a pattern that is difficult to reconcile with adaptive fine-tuning of the gene content. In addition, the erosion of certain amino acid biosynthetic functions in various *Buchnera* lineages (47, 91, 96) may reflect deleterious changes that become fixed by genetic drift. As detailed below, the lack of gene transfer implies that gene silencing is irreversible and may impose evolutionary constraints on both the host and the symbiont (91).

RAPID, AT-BIASED SEQUENCE EVOLUTION IS COUPLED TO INTRACELLULARITY

It is increasingly clear that transitions to obligate host associations can drastically affect not only genome size and content, but also the tempo of DNA sequence evolution. A striking feature of most intracellular bacteria is their accelerated rates of sequence evolution. Relative-rate tests have demonstrated significant rate acceleration in host-restricted bacterial species compared to free-living ones (15, 56, 98). Supporting those results, divergences among two or more members of a given parasitic or mutualist group often indicate rapid sequence change within the context of a host association (17, 24, 67).

This observed rate acceleration may reflect a combination of evolutionary forces, including relaxed functional constraint on encoded proteins, so they are free to evolve more quickly, faster mutation rates in species that lack many DNA repair genes, and/or accelerated fixation of deleterious mutations by genetic drift. The predicted signatures of these forces on protein evolution are different (Table 2). First, because genetic drift and mutational pressure act genome-wide, they are expected to lead to rate acceleration across all genes. By contrast, relaxed selection is expected to affect particular loci with reduced functional significance. In addition, both relaxed selection and reduced efficacy of that selection due to genetic drift will disproportionately affect changes with fitness consequences—notably, nonsynonymous mutations that alter the amino acid encoded, as opposed to synonymous changes that experience far weaker, if any, selection. Thus, both forces are expected to lead to a greater acceleration of nonsynonymous changes and thereby elevate the ratio of nonsynonymous to synonymous divergences (dN/dS). By contrast, if mutation alone explained the rate elevation, dN/dS is not expected to increase.

To date, genome-wide rate comparisons strongly suggest a consistent, genome-wide rate increase in intracellular species, as expected under the influences of increased mutation and/or genetic drift. Although selection constrains the extent of this rate elevation at functionally important loci, existing data support a consistent increase across genes (15, 17, 24). Second, comparisons of dN/dS in intracellular versus free-living species are complicated by differences in base compositions that can affect sequence divergence estimates. In addition, varying effects of selection on codon usage bias, known to constrain synonymous divergences in many bacteria (88), can influence dN/dS and confound comparisons across lineages. Despite these potential complications, several studies show consistent increase in dN/dS (or reduced dS/dN) in the best-studied obligate mutualist, *Buchnera* (12, 17, 56, 97). Although mutational explanations have been put forward (e.g., 45) the combination of genome-wide acceleration and elevated dN/dS is most consistent with effects of genetic drift in small populations. Moreover, population genetic analyses also support the hypothesis that genetic drift plays a major role in the molecular evolution of this aphid mutualist (1, 33, 42). Tracing genetic changes within other mutualistic and parasitic groups is necessary to assess whether these signatures of genetic drift characterize intracellular species generally.

Although current evidence supports a strong contribution of genetic drift to the observed rate acceleration, particularly well documented in *Buchnera*, it is also clear that underlying GC→AT bias drives many of the nucleotide changes that occur. The exception-

TABLE 2 Signatures of distinct mechanisms that may contribute to rapid protein evolution in intracellular bacteria[a]

Mechanism	Rate acceleration	Increased dN/dS
Relaxed selection	At specific loci	At specific loci
Genetic drift	Genome-wide★	Genome-wide★
Mutational pressure	Genome-wide★	No increase

[a]This framework distinguishes the effects of three processes: (i) a change in selective regime, including relaxed selection on dispensable functions, (ii) a strong effect of genetic drift, expected to reduce efficiency of selection across all loci, or (iii) increased mutation rates and biases, also expected to act genome-wide. The ratio of nonsynonymous to synonymous divergence (dN/dS) within an intracellular group is expected to increase as the strength or efficacy of selection weakens. By contrast, elevated mutation alone is expected to affect all sites proportionally with no expected increase in dN/dS. ★, Patterns observed in intracellular bacteria (see text).

ally low %GC content (percentage of guanine and cytosine in the genome sequences) of endosymbiont sequences (less than ~35% in most cases) indicates a strong AT bias across phylogenetically independent species (Fig. 2). Likewise, evolutionary studies show a strong directionality to base changes within intracellular groups (4, 17, 95). This convergent trend toward AT bias may reflect greater exposure of an underlying GC→AT mutational pressure in small genomes that lack many DNA repair functions (but see reference 81 for a metabolic hypothesis for AT bias). Many common categories of mutation in bacteria produce a bias from GC to AT base pairs, such as the deamination of cytosine to uracil (coded as thymine). If left uncorrected, these mutations would reduce the overall genomic %GC content.

As expected, mutational bias has the strongest impact on synonymous positions, where selection is expected to be very weak. However, the bias also influences changes at nonsynonymous sites and thus affects the amino acid compositions of proteins encoded by intracellular species (e.g., 56). For example, compared with bacteria with moderate GC contents, proteins of AT-rich intracellular species are strongly biased toward the use of amino acids encoded by relatively AT-rich codons, such as Ile (AT{TCA}) and Lys (AA{AG}). Shifts toward these AT-biased amino acids often entail physiochemical changes and may contribute to the lower protein stability inferred for diverse intracellular species (93).

GENOME DYNAMICS: FROM STATIC TO PLASTIC

Among bacteria, various mechanisms of genome fluidity—e.g., horizontal gene acquisition from foreign donors, recombination among related bacterial strains, and inversions, translocations, and other intramolecular changes—can play important roles in the evolution of new host associations (68). However, rather than fluidity, genome stasis characterizes the evolution of several intracellular groups. The availability of two or more genomes for a particular group offers a window into the evolution of genome architecture, or changes in the order and strand orientation of shared loci. In the most extreme cases, two groups of insect mutualists—*Buchnera* of aphids and *Blochmannia* of ants—show complete genome stability. Genome data revealed a surprising lack of gene acquisitions, inversions, or translocations throughout 50 to 70 million years of evolution within aphids (91) and near-perfect conservation of gene order since the establishment of this association 150 to 200 mya (93). Within the ant mutualist, *Blochmannia,* strains that diverged ~20 mya also exhibit chromosomal stasis, with no changes in the order or strand orientation of shared genes (24). Among parasites, genome stability has also been discovered in certain long-term intracellular groups, in which very rare examples of horizontal transfer (51) appear to be the exceptions, synteny between species implies few intragenomic rearrangements (52, 91), and repeated DNA is scarce (30).

However, recent data suggest greater fluidity in other intracellular pathogens. The genome of *Rickettsia felis*, the agent of flea-borne spotted fever rickettsiosis, possesses the first conjugative plasmid discovered in intracellular bacteria (69). Unlike other *Rickettsia* spp*., R. felis* also has abundant transposases, a >sixfold higher density of repeats, paralogous gene families encoding surface proteins, a substantial number of horizontally acquired loci, and evidence for frequent inversions and/or translocations that disrupt synteny. During transmission through flea vectors, frequent coinfection with *Bartonella henselae, B. quintana,* and *Wolbachia* may provide ample opportunities for gene transfer (69). Another possible exception to genome stasis, *Parachlamydia* sp. possesses a genomic island encoding genes for F-like conjugative DNA transfer, which might occur among the bacteria tightly packed in vacuoles of the amoebae host (39, 44). Although compositional analysis does not support recent lateral gene acquisition, transfer of the F-like conjugative DNA transfer and type IV secretion (noted above) may have played important roles in the evolution of host interactions (44).

In addition to conjugation machinery, certain intracellular bacteria—particularly parasites—possess more mobile DNA than previously suspected (11, 25). Although elements such as bacteriophage, transposable elements, and mobile plasmids are missing from nutritional mutualists, one or more of these elements persist in *Parachlamydia* sp., *R. felis*, *Phytoplasma asteris*, *Wolbachia* wMel, and to a lesser extent *Wolbachia* wBm (28, 48, 69, 73, 99). These intracellular bacteria also have surprisingly abundant repeated DNA sequences, devoting a relatively large fraction of their chromosomes to global direct and inverted repeats that can mediate large-scale intragenomic rearrangements and may explain frequent inversions and translocations that disrupt synteny in certain groups (28, 44, 82, 99).

At a more local genomic scale, recombination among paralogous gene copies or tandem repeats allows certain parasites to evade the adaptive immune response of vertebrate hosts. Intracellular parasites devote a high percentage of tiny genomes to paralogous families of polymorphic surface molecules, indicating that host immunity is among the most important challenges they face (75). For example, *Anaplasma* achieves a persistent infection by generating sequential diversity of membrane proteins through recombination of various pseudogenes into a single expression site. In *Ehrlichia*, duplications among numerous tandemly repeated genes create new loci that are important in immune evasion (19).

In short, despite the exceptional genome stasis documented in some intracellular associates, it is increasingly clear that life with host cells does not necessarily constrain genome flux. In some lineages, genome fluidity, ranging from broad transfer of pathways across deep phylogenetic distances, to local shuffling among tandem repeats, may be important in mediating host interactions.

CONCLUSIONS

The rapid growth of genomics has elucidated processes that shape wide variation in genome size, gene content, patterns of DNA sequence evolution, and levels of genome fluidity in the bacterial world. The central importance of bacterial lifestyle in shaping these genome features is strikingly clear, particularly among species that engage in intimate associations with host cells. Across distinct phylogenetic lineages, intracellular bacteria have experienced severe gene loss and now encode minimal metabolisms that reflect their specialized lifestyles. A corresponding reduction in genomic %GC content may reflect greater exposure of underlying GC→AT mutations, and genetic drift in small populations apparently contributes to consistent acceleration in evolutionary rates in these species.

Many intracellular genomes are exceptionally stable, showing little evidence of gene acquisition by lateral transfer or intramolecular changes that disrupt synteny. However, recent genome data teach us that intracellular associates can experience various forms of genome fluidity, ranging from gene acquisition from phylogenetically distant donors, to specific recombination mechanisms for generating antigenic diversity. Surprising discoveries of conjugation machinery in the parasite *Rickettsia felis* and the amoebae symbiont *Parachlamydia* sp. suggest a key role for DNA transfer. Such discoveries have prompted new research into mechanisms of lateral gene transfer and recombination in these species, including roles of mobile elements that persist within certain long-term associates. The rapid pace of bacterial genomics, coupled with new experimental approaches and bioinformatic tools, promises an even richer context to explore mechanisms of host-microbe interactions and the evolutionary consequences of an intracellular existence.

REFERENCES

1. **Abbot, P., and N. A. Moran.** 2002. Extremely low levels of genetic polymorphism in endosymbionts (*Buchnera*) of aphids (*Pemphigus*). *Mol. Ecol.* **11:**2649–2660.
2. **Akman, L., A. Yamashita, H. Watanabe, K. Oshima, T. Shiba, M. Hattori, and S. Aksoy.** 2002. Genome sequence of the endocellular obligate symbiont of tsetse flies, *Wigglesworthia glossinidia. Nat. Genet.* **32:**402–407.

3. **Aksoy, S., and R. V. Rio.** 2005. Interactions among multiple genomes: tsetse, its symbionts and trypanosomes. *Insect Biochem. Mol. Biol.* **35:**691–698.

4. **Andersson, J. O., and S. G. Andersson.** 1999. Genome degradation is an ongoing process in *Rickettsia. Mol. Biol. Evol.* **16:**1178–1191.

5. **Andersson, J. O., and S. G. Andersson.** 2001. Pseudogenes, junk DNA, and the dynamics of *Rickettsia* genomes. *Mol. Biol. Evol.* **18:**829–839.

6. **Andersson, S. G., and C. G. Kurland.** 1998. Reductive evolution of resident genomes. *Trends Microbiol.* **6:**263–268.

7. **Batut, J., S. G. Andersson, and D. O'Callaghan.** 2004. The evolution of chronic infection strategies in the alpha-Proteobacteria. *Nat. Rev. Microbiol.* **2:**933–945.

8. **Baumann, P., N. Moran, and L. Baumann.** 2000. Bacteriocyte-associated endosymbionts of insects. *In* M. Dworkin (ed.), *The Prokaryotes: A Handbook on the Biology of Bacteria; Ecophysiology, Isolation, Identification, Applications.* Springer-Verlag, New York, NY.

9. **Belda, E., A. Moya, and F. J. Silva.** 2005. Genome rearrangement distances and gene order phylogeny in gamma-Proteobacteria. *Mol. Biol. Evol.* **22:**1456–1467.

10. **Bencina, D., B. Slavec, and M. Narat.** 2005. Antibody response to GroEL varies in patients with acute *Mycoplasma pneumoniae* infection. *FEMS Immunol. Med. Microbiol.* **43:**399–406.

11. **Bordenstein, S. R., and W. S. Reznikoff.** 2005. Mobile DNA in obligate intracellular bacteria. *Nat. Rev. Microbiol.* **3:**688–699.

12. **Brynnel, E. U., C. G. Kurland, N. A. Moran, and S. G. Andersson.** 1998. Evolutionary rates for *tuf* genes in endosymbionts of aphids. *Mol. Biol. Evol.* **15:**574–582.

13. **Buchner, P.** 1965. *Endosymbiosis of Animals with Plant Microorganisms.* Interscience Publishers, New York, NY.

14. **Burne, R. A., and Y. Y. Chen.** 2000. Bacterial ureases in infectious diseases. *Microbes Infect.* **2:**533–542.

15. **Canback, B., I. Tamas, and S. G. Andersson.** 2004. A phylogenomic study of endosymbiotic bacteria. *Mol. Biol. Evol.* **21:**1110–1122.

16. **Chen, X., S. Li, and S. Aksoy.** 1999. Concordant evolution of a symbiont with its host insect species: molecular phylogeny of genus *Glossina* and its bacteriome-associated endosymbiont, *Wigglesworthia glossinidia. J. Mol. Evol.* **48:**49–58.

17. **Clark, M. A., N. A. Moran, and P. Baumann.** 1999. Sequence evolution in bacterial endosymbionts having extreme base compositions. *Mol. Biol. Evol.* **16:**1586–1598.

18. **Cole, S. T., K. Eiglmeier, J. Parkhill, K. D. James, N. R. Thomson, P. R. Wheeler, N. Honore, T. Garnier, C. Churcher, D. Harris, K. Mungall, D. Basham, D. Brown, T. Chillingworth, R. Connor, R. M. Davies, K. Devlin, S. Duthoy, T. Feltwell, A. Fraser, N. Hamlin, S. Holroyd, T. Hornsby, K. Jagels, C. Lacroix, J. Maclean, S. Moule, L. Murphy, K. Oliver, M. A. Quail, M. A. Rajandream, K. M. Rutherford, S. Rutter, K. Seeger, S. Simon, M. Simmonds, J. Skelton, R. Squares, S. Squares, K. Stevens, K. Taylor, S. Whitehead, J. R. Woodward, and B. G. Barrell.** 2001. Massive gene decay in the leprosy bacillus. *Nature* **409:**1007–1011.

19. **Collins, N. E., J. Liebenberg, E. P. de Villiers, K. A. Brayton, E. Louw, A. Pretorius, F. E. Faber, H. van Heerden, A. Josemans, M. van Kleef, H. C. Steyn, M. F. van Strijp, E. Zweygarth, F. Jongejan, J. C. Maillard, D. Berthier, M. Botha, F. Joubert, C. H. Corton, N. R. Thomson, M. T. Allsopp, and B. A. Allsopp.** 2005. The genome of the heartwater agent *Ehrlichia ruminantium* contains multiple tandem repeats of actively variable copy number. *Proc. Natl. Acad. Sci. USA* **102:**838–843.

20. **Corsaro, D., and D. Venditti.** 2004. Emerging chlamydial infections. *Crit. Rev. Microbiol.* **30:**75–106.

21. **Dale, C., G. R. Plague, B. Wang, H. Ochman, and N. A. Moran.** 2002. Type III secretion systems and the evolution of mutualistic endosymbiosis. *Proc. Natl. Acad. Sci. USA* **99:** 12397–12402.

22. **Dale, C., B. Wang, N. Moran, and H. Ochman.** 2003. Loss of DNA recombinational repair enzymes in the initial stages of genome degeneration. *Mol. Biol. Evol.* **20:**1188–1194.

23. **Degnan, P., A. Lazarus, C. Brock, and J. Wernegreen.** 2004. Host-symbiont stability and fast evolutionary rates in an ant-bacterium association: cospeciation of *Camponotus* species and their endosymbionts, *Candidatus* Blochmannia. *Syst. Biol.* **53:**95–110.

24. **Degnan, P. H., A. B. Lazarus, and J. J. Wernegreen.** 2005. Genome sequence of *Blochmannia pennsylvanicus* indicates parallel evolutionary trends among bacterial mutualists of insects. *Genome Res.* **15:**1023–1033.

25. **Duron, O., J. Lagnel, M. Raymond, K. Bourtzis, P. Fort, and M. Weill.** 2005. Transposable element polymorphism of *Wolbachia* in the mosquito *Culex pipiens*: evidence of genetic diversity, superinfection and recombination. *Mol. Ecol.* **14:**1561–1573.

26. **Everett, K., M Thao, M. Horn, G. Dyszynski, and P. Baumann.** 2005. Novel Chlamydiae in whiteflies and scale insects: endosymbionts

"*Candidatus* Fritschea bemisiae" strain Falk and "*Candidatus* Fritschea eriococci" strain Elm. *Int. J. Syst. Evol. Microbiol.* **55:**1581–1587.

27. **Fares, M. A., A. Moya, and E. Barrio.** 2004. GroEL and the maintenance of bacterial endosymbiosis. *Trends Genet.* **20:**413–416.

28. **Foster, J., M. Ganatra, I. Kamal, J. Ware, K. Makarova, N. Ivanova, A. Bhattacharyya, V. Kapatral, S. Kumar, J. Posfai, T. Vincze, J. Ingram, L. Moran, A. Lapidus, M. Omelchenko, N. Kyrpides, E. Ghedin, S. Wang, E. Goltsman, V. Joukov, O. Ostrovskaya, K. Tsukerman, M. Mazur, D. Comb, E. Koonin, and B. Slatko.** 2005. The *Wolbachia* genome of *Brugia malayi*: endosymbiont evolution within a human pathogenic nematode. *PLoS Biol.* **3:**e121.

29. **Frank, A. C., C. M. Alsmark, M. Thollesson, and S. G. Andersson.** 2005. Functional divergence and horizontal transfer of type IV secretion systems. *Mol. Biol. Evol.* **22:**1325–1336.

30. **Frank, A. C., H. Amiri, and S. G. Andersson.** 2002. Genome deterioration: loss of repeated sequences and accumulation of junk DNA. *Genetica* **115:**1–12.

31. **Fredriksson, A., M. Ballesteros, S. Dukan, and T. Nystrom.** 2005. Defense against protein carbonylation by DnaK/DnaJ and proteases of the heat shock regulon. *J. Bacteriol.* **187:**4207–4213.

32. **Fukatsu, T., N. Nikoh, R. Kawai, and R. Koga.** 2000. The secondary endosymbiotic bacterium of the pea aphid *Acyrthosiphon pisum* (Insecta: Homoptera). *Appl. Environ. Microbiol.* **66:**2748–2758.

33. **Funk, D. J., J. J. Wernegreen, and N. A. Moran.** 2001. Intraspecific variation in symbiont genomes: bottlenecks and the aphid-*Buchnera* association. *Genetics* **157:**477–489.

34. **Ghedin, E., S. Wang, J. M. Foster, and B. E. Slatko.** 2004. First sequenced genome of a parasitic nematode. *Trends Parasitol.* **20:**151–153.

35. **Gil, R., A. Latorre, and A. Moya.** 2004. Bacterial endosymbionts of insects: insights from comparative genomics. *Environ. Microbiol.* **6:**1109–1122.

36. **Gil, R., F. J. Silva, E. Zientz, F. Delmotte, F. Gonzalez-Candelas, A. Latorre, C. Rausell, J. Kamerbeek, J. Gadau, B. Holldobler, R. C. van Ham, R. Gross, and A. Moya.** 2003. The genome sequence of *Blochmannia floridanus*: comparative analysis of reduced genomes. *Proc. Natl. Acad. Sci. USA* **100:**9388–9393.

37. **Goebel, W., and R. Gross.** 2001. Intracellular survival strategies of mutualistic and parasitic prokaryotes. *Trends Microbiol.* **9:**267–273.

38. **Gomez-Valero, L., A. Latorre, and F. J. Silva.** 2004. The evolutionary fate of nonfunctional DNA in the bacterial endosymbiont *Buchnera aphidicola*. *Mol. Biol. Evol.* **21:**2172–2181.

39. **Greub, G., F. Collyn, L. Guy, and C. A. Roten.** 2004. A genomic island present along the bacterial chromosome of the *Parachlamydiaceae* UWE25, an obligate amoebal endosymbiont, encodes a potentially functional F-like conjugative DNA transfer system. *BMC Microbiol.* **4:**48.

40. **Hacker, J., U. Hentschel, and U. Dobrindt.** 2003. Prokaryotic chromosomes and disease. *Science* **301:**790–793.

41. **Hechard, C., O. Grepinet, and A. Rodolakis.** 2004. Molecular cloning of the *Chlamydophila abortus groEL* gene and evaluation of its protective efficacy in a murine model by genetic vaccination. *J. Med. Microbiol.* **53:**861–868.

42. **Herbeck, J. T., D. J. Funk, P. H. Degnan, and J. J. Wernegreen.** 2003. A conservative test of genetic drift in the endosymbiotic bacterium *Buchnera*: slightly deleterious mutations in the chaperonin *groEL*. *Genetics* **165:**1651–1660.

43. **Hoerauf, A., K. Nissen-Pahle, C. Schmetz, K. Henkle-Duhrsen, M. L. Blaxter, D. W. Buttner, M. Y. Gallin, K. M. Al-Qaoud, R. Lucius, and B. Fleischer.** 1999. Tetracycline therapy targets intracellular bacteria in the filarial nematode *Litomosoides sigmodontis* and results in filarial infertility. *J. Clin. Invest.* **103:**11–18.

44. **Horn, M., A. Collingro, S. Schmitz-Esser, C. L. Beier, U. Purkhold, B. Fartmann, P. Brandt, G. J. Nyakatura, M. Droege, D. Frishman, T. Rattei, H. W. Mewes, and M. Wagner.** 2004. Illuminating the evolutionary history of Chlamydiae. *Science* **304:**728–730.

45. **Itoh, T., W. Martin, and M. Nei.** 2002. Acceleration of genomic evolution caused by enhanced mutation rate in endocellular symbionts. *Proc. Natl. Acad. Sci. USA* **99:**12944–12948.

46. **Klasson, L., and S. G. Andersson.** 2004. Evolution of minimal-gene-sets in host-dependent bacteria. *Trends Microbiol.* **12:**37–43.

47. **Lai, C. Y., P. Baumann, and N. Moran.** 1996. The endosymbiont (*Buchnera* sp.) of the aphid *Diuraphis noxia* contains plasmids consisting of *trpEG* and tandem repeats of *trpEG* pseudogenes. *Appl. Environ. Microbiol.* **62:**332–339.

48. **Lee, I. M., Y. Zhao, and K. D. Bottner.** 2005. Novel insertion sequence–like elements in *Phytoplasma* strains of the aster yellows group are putative new members of the IS3 family. *FEMS Microbiol. Lett.* **242:**353–360.

49. **Margulis, L.** 1975. Symbiotic theory of the origin of eukaryotic organelles; criteria for proof. *Symp. Soc. Exp. Biol.* **1975**(2a):21–38.

50. **Matthew, C. Z., A. C. Darby, S. A. Young, L. H. Hume, and S. C. Welburn.** 2005. The rapid isolation and growth dynamics of the tsetse sym-

biont *Sodalis glossinidius*. *FEMS Microbiol. Lett.* **248:**69–74.

51. **McLeod, M. P., X. Qin, S. E. Karpathy, J. Gioia, S. K. Highlander, G. E. Fox, T. Z. Mc-Neill, H. Jiang, D. Muzny, L. S. Jacob, A. C. Hawes, E. Sodergren, R. Gill, J. Hume, M. Morgan, G. Fan, A. G. Amin, R. A. Gibbs, C. Hong, X. J. Yu, D. H. Walker, and G. M. Weinstock.** 2004. Complete genome sequence of *Rickettsia typhi* and comparison with sequences of other rickettsiae. *J. Bacteriol.* **186:**5842–5855.

52. **Mira, A., L. Klasson, and S. G. Andersson.** 2002. Microbial genome evolution: sources of variability. *Curr. Opin. Microbiol.* **5:**506–512.

53. **Mira, A., and N. A. Moran.** 2002. Estimating population size and transmission bottlenecks in maternally transmitted endosymbiotic bacteria. *Microb. Ecol.* **44:**137–143.

54. **Mira, A., H. Ochman, and N. A. Moran.** 2001. Deletional bias and the evolution of bacterial genomes. *Trends Genet.* **17:**589–596.

55. **Moran, N., and A. Telang.** 1998. Bacteriocyte-associated symbionts of insects. *Bioscience* **48:**295–304.

56. **Moran, N. A.** 1996. Accelerated evolution and Muller's rachet in endosymbiotic bacteria. *Proc. Natl. Acad. Sci. USA* **93:**2873–2878.

57. **Moran, N. A.** 2002. Microbial minimalism: genome reduction in bacterial pathogens. *Cell* **108:**583–586.

58. **Moran, N. A.** 2002. The ubiquitous and varied role of infection in the lives of animals and plants. *Am. Nat.* **60:**S1–S8.

59. **Moran, N. A., and A. Mira.** 2001. The process of genome shrinkage in the obligate symbiont *Buchnera aphidicola*. *Genome Biol.* **2:**RESEARCH 0054.

60. **Moran, N. A., M. A. Munson, P. Baumann, and H. Ishikawa.** 1993. A molecular clock in endosymbiotic bacteria is calibrated using the insect hosts. *Proc. R. Soc. Lond. B* **253:**167–171.

61. **Moran, N. A., and G. R. Plague.** 2004. Genomic changes following host restriction in bacteria. *Curr. Opin. Genet. Dev.* **14:**627–633.

62. **Moran, N. A., J. A. Russell, R. Koga, and T. Fukatsu.** 2005. Evolutionary relationships of three new species of Enterobacteriaceae living as symbionts of aphids and other insects. *Appl. Environ. Microbiol.* **71:**3302–3310.

63. **Muller, J.** 1964. The relation of recombination to mutational advance. *Mutat. Res.* **1:**2–9.

64. **Nogge, G.** 1981. Significance of symbionts for the maintenance of an optimal nutritional state for successful reproduction in hematophagous arthropods. *Parasitology* **82:**299–304.

65. **O'Neill, S., A. Hoffman, and J. Werren.** 1998. *Influential Passengers: Inherited Microorganisms and Arthropod Reproduction*. Oxford University Press, New York, NY.

66. **Ochman, H., and L. M. Davalos.** 2006. The nature and dynamics of bacterial genomes. *Science* **311:**1730–1733.

67. **Ochman, H., S. Elwyn, and N. A. Moran.** 1999. Calibrating bacterial evolution. *Proc. Natl. Acad. Sci. USA* **96:**12638–12643.

68. **Ochman, H., and N. A. Moran.** 2001. Genes lost and genes found: evolution of bacterial pathogenesis and symbiosis. *Science* **292:**1096–1099.

69. **Ogata, H., P. Renesto, S. Audic, C. Robert, G. Blanc, P. E. Fournier, H. Parinello, J. M. Claverie, and D. Raoult.** 2005. The genome sequence of *Rickettsia felis* identifies the first putative conjugative plasmid in an obligate intracellular parasite. *PLoS Biol.* **3:**e248.

70. Reference deleted.

71. **Ohta, T.** 1973. Slightly deleterious mutant substitutions in evolution. *Nature* **246:**96–98.

72. **Ohta, T., and M. Kimura.** 1971. On the constancy of the evolutionary rate of cistrons. *J. Mol. Evol.* **1:**18–25.

73. **Oshima, K., S. Kakizawa, H. Nishigawa, H. Y. Jung, W. Wei, S. Suzuki, R. Arashida, D. Nakata, S. Miyata, M. Ugaki, and S. Namba.** 2004. Reductive evolution suggested from the complete genome sequence of a plant-pathogenic phytoplasma. *Nat. Genet.* **36:**27–29.

74. **Pal, C., B. Papp, M. J. Lercher, P. Csermely, S. G. Oliver, and L. D. Hurst.** 2006. Chance and necessity in the evolution of minimal metabolic networks. *Nature* **440:**667–670.

75. **Palmer, G. H.** 2002. The highest priority: what microbial genomes are telling us about immunity. *Vet. Immunol. Immunopathol.* **85:**1–8.

76. **Perez-Brocal, V., A. Latorre, R. Gil, and A. Moya.** 2005. Comparative analysis of two genomic regions among four strains of *Buchnera aphidicola*, primary endosymbiont of aphids. *Gene* **345:**73–80.

77. **Peterson, J. D., L. A. Umayam, T. Dickinson, E. K. Hickey, and O. White.** 2001. The comprehensive microbial resource. *Nucleic Acids Res.* **29:**123–125.

78. **Renesto, P., S. Azza, A. Dolla, P. Fourquet, G. Vestris, J. P. Gorvel, and D. Raoult.** 2005. Proteome analysis of *Rickettsia conorii* by two-dimensional gel electrophoresis coupled with mass spectrometry. *FEMS Microbiol. Lett.* **245:**231–238.

79. **Renesto, P., H. Ogata, S. Audic, J. M. Claverie, and D. Raoult.** 2005. Some lessons from *Rickettsia* genomics. *FEMS Microbiol. Rev.* **29:**99–117.

80. **Rispe, C., and N. A. Moran.** 2000. Accumulation of deleterious mutations in endosymbionts:

Muller's ratchet with two levels of selection. *Am. Nat.* **156:**425–441.

81. **Rocha, E. P., and A. Danchin.** 2002. Base composition bias might result from competition for metabolic resources. *Trends Genet.* **18:**291–294.

82. **Salzberg, S. L., J. C. Hotopp, A. L. Delcher, M. Pop, D. R. Smith, M. B. Eisen, and W. C. Nelson.** 2005. Serendipitous discovery of *Wolbachia* genomes in multiple *Drosophila* species. *Genome Biol.* **6:**R23.

83. **Sandström, J., A. Telang, and N. A. Moran.** 2000. Nutritional enhancement of host plants by aphids—a comparison of three aphid species on grasses. *J. Insect Physiol.* **46:**33–40.

84. **Sauer, C., D. Dudaczek, B. Holldobler, and R. Gross.** 2002. Tissue localization of the endosymbiotic bacterium "*Candidatus* Blochmannia floridanus" in adults and larvae of the carpenter ant *Camponotus floridanus. Appl. Environ. Microbiol.* **68:**4187–4193.

85. **Schmitz-Esser, S., N. Linka, A. Collingro, C. L. Beier, H. E. Neuhaus, M. Wagner, and M. Horn.** 2004. ATP/ADP translocases: a common feature of obligate intracellular amoebal symbionts related to chlamydiae and rickettsiae. *J. Bacteriol.* **186:**683–691.

86. **Selander, R. K., D. A. Caugant, and T. S. Whittam.** 1987. Genetic structure and variation in natural populations of *Escherichia coli*, p. 1625–1648. *In* F. Neidhardt (ed.), Escherichia coli *and* Salmonella typhimurium: *Cellular and Molecular Biology.* ASM Press, Washington, DC.

87. **Seshadri, R., I. T. Paulsen, J. A. Eisen, T. D. Read, K. E. Nelson, W. C. Nelson, N. L. Ward, H. Tettelin, T. M. Davidsen, M. J. Beanan, R. T. Deboy, S. C. Daugherty, L. M. Brinkac, R. Madupu, R. J. Dodson, H. M. Khouri, K. H. Lee, H. A. Carty, D. Scanlan, R. A. Heinzen, H. A. Thompson, J. E. Samuel, C. M. Fraser, and J. F. Heidelberg.** 2003. Complete genome sequence of the Q-fever pathogen *Coxiella burnetii. Proc. Natl. Acad. Sci. USA* **100:**5455–5460.

88. **Sharp, P. M.** 1991. Determinants of DNA sequence divergence between *Escherichia coli* and *Salmonella typhimurium*: codon usage, map position, and concerted evolution. *J. Mol. Evol.* **33:**23–33.

89. **Shigenobu, S., H. Watanabe, M. Hattori, Y. Sakaki, and H. Ishikawa.** 2000. Genome sequence of the endocellular bacterial symbiont of aphids *Buchnera* sp. APS. *Nature* **407:**81–86.

90. **Subtil, A., and A. Dautry-Varsat.** 2004. *Chlamydia*: five years A.G. (after genome). *Curr. Opin. Microbiol.* **7:**85–92.

91. **Tamas, I., L. Klasson, B. Canback, A. K. Naslund, A. S. Eriksson, J. J. Wernegreen, J. P. Sandstrom, N. A. Moran, and S. G. Andersson.** 2002. 50 million years of genomic stasis in endosymbiotic bacteria. *Science* **296:** 2376–2379.

92. **Toh, H., B. L. Weiss, S. A. Perkin, A. Yamashita, K. Oshima, M. Hattori, and S. Aksoy.** 2006. Massive genome erosion and functional adaptations provide insights into the symbiotic lifestyle of *Sodalis glossinidius* in the tsetse host. *Genome Res.* **16:**149–156.

93. **Van Ham, R. C., J. Kamerbeek, C. Palacios, C. Rausell, F. Abascal, U. Bastolla, J. M. Fernandez, L. Jimenez, M. Postigo, F. J. Silva, J. Tamames, E. Viguera, A. Latorre, A. Valencia, F. Moran, and A. Moya.** 2003. Reductive genome evolution in *Buchnera aphidicola. Proc. Natl. Acad. Sci. USA* **100:**581–586.

94. **Wernegreen, J. J.** 2005. For better or worse: Genomic consequences of intracellular mutualism and parasitism. *Curr. Opin. Genet. Dev.* **15:** 572–583.

95. **Wernegreen, J. J., and D. J. Funk.** 2004. Mutation exposed: a neutral explanation for extreme base composition of an endosymbiont genome. *J. Mol. Evol.* **59:**849–858.

96. **Wernegreen, J. J., and N. A. Moran.** 2000. Decay of mutualistic potential in aphid endosymbionts through silencing of biosynthetic loci: *Buchnera* of *Diuraphis. Proc. R. Soc. Lond. B Biol. Sci.* **267:**1423–1431.

97. **Wernegreen, J. J., and N. A. Moran.** 1999. Evidence for genetic drift in endosymbionts (Buchnera): analyses of protein-coding genes. *Mol. Biol. Evol.* **16:**83–97.

98. **Woolfit, M., and L. Bromham.** 2003. Increased rates of sequence evolution in endosymbiotic bacteria and fungi with small effective population sizes. *Mol. Biol. Evol.* **20:**1545–1555.

99. **Wu, M., L. V. Sun, J. Vamathevan, M. Riegler, R. Deboy, J. C. Brownlie, E. A. McGraw, W. Martin, C. Esser, N. Ahmadinejad, C. Wiegand, R. Madupu, M. J. Beanan, L. M. Brinkac, S. C. Daugherty, A. S. Durkin, J. F. Kolonay, W. C. Nelson, Y. Mohamoud, P. Lee, K. Berry, M. B. Young, T. Utterback, J. Weidman, W. C. Nierman, I. T. Paulsen, K. E. Nelson, H. Tettelin, S. L. O'Neill, and J. A. Eisen.** 2004. Phylogenomics of the reproductive parasite *Wolbachia pipientis* wMel: a streamlined genome overrun by mobile genetic elements. *PLoS Biol.* **2:**E69.

100. **Zientz, E., T. Dandekar, and R. Gross.** 2004. Metabolic interdependence of obligate intracellular bacteria and their insect hosts. *Microbiol. Mol. Biol. Rev.* **68:**745–770.

MODELING MICROBIAL VIRULENCE IN A GENOMIC ERA: IMPACT OF SHARED GENOMIC TOOLS AND DATA SETS

Daniel G. Lee, Nicole T. Liberati, Jonathan M. Urbach, Gang Wu, and Frederick M. Ausubel

9

Attempts to model the complex problem of human pathogenesis in the laboratory have typically taken one of two (non–mutually exclusive) approaches. One approach simplifies the problem by reducing the host (or sometimes both the host and the pathogen) to a particular activity that can be studied in vitro or in the context of a mammalian cell culture system. Rather than studying a pathogen's ability to attach to a host tissue, invade the cells, propagate, and then spread systemically, the process by which a bacterium attaches to an inert plastic surface or becomes internalized in a human cell-culture assay is analyzed. In this manner, the environment becomes more controlled and complicating variables can be minimized, allowing for cheaper, higher-throughput experiments with a sophisticated level of detail. However, by the very act of reduction, the focus may be placed on processes that are not relevant enough in the context of a true human infection to be critical for virulence or viable therapeutic

targets. In contrast, the alternative experimental extreme replaces the human with a surrogate host such as a mouse or other small mammal, allowing for a closer approximation of a complex infection in which the pathogen has to contend with a variety of host cell types and defense responses. Unfortunately, whole vertebrate (mammalian) animal experiments have traditionally placed financial and ethical limitations on the size and scope of experiments that can reasonably be performed. Certainly, much work has focused on being able to straddle both extremes, or design experiments that use both in vitro experiments and mammalian hosts in a complementary fashion. In recent years, tremendous advances in tools available for studying both the host and the pathogen have allowed for the merging of these two extremes to a much greater degree.

In this chapter, we will discuss the relevance of model host pathogenesis as a third general approach to studying virulence, in which infections are studied in the context of nonvertebrate whole animal hosts, thereby reducing the financial and ethical barriers to high-throughput experimentation while maintaining a more complex set of host defense responses than can be mimicked in vitro. During the past few years,

Daniel G. Lee, Nicole T. Liberati, Jonathan M. Urbach, Gang Wu, and Frederick M. Ausubel   Department of Genetics, Harvard Medical School, and Department of Molecular Biology, Massachusetts General Hospital, Boston, Massachusetts 02114.

Bacterial Pathogenomics, Edited by M. J. Pallen et al.
© 2007 ASM Press, Washington, D.C.

a variety of pathogenesis systems have been developed that capitalize on the unexpected finding that a variety of human bacterial and fungal pathogens can infect and cause disease in well-studied model genetic hosts, including the nematode *Caenhorhabidits elegans*, the insect *Drosophila melanogaster*, and the plant *Arabidopsis thaliana*. This has allowed the systematic generation and characterization of either avirulent pathogen mutants or host innate immune response mutants at a genome-wide scale. In many cases, important features of mammalian pathogenesis are mimicked by the interactions of human pathogens with these relatively simple alternative hosts, suggesting that the underlying mechanism of pathogenesis has been evolutionarily conserved. In particular, we will focus on the manner in which advances in genomics are affecting the nature and the scope of experiments performed by using model-host pathogenesis systems to characterize microbial and host genes involved in pathogenesis and host defense.

Our discussion is presented from the perspective of a model studied in our laboratory, the pathogenic interaction between *C. elegans* and the opportunistic human pathogen *Pseudomonas aeruginosa*, but the general approach can be applied to any pathosystem for which genome sequences are known and for which genetic tools are available for both the pathogen and the host. For the combination of model host analysis and genomics to reach its fullest potential, we will discuss four key steps that we view as essential: (i) the development of the model host-pathogen system, (ii) the development of genomic tools in both the pathogen and host, (iii) the distribution and use of these tools by the greater research community beyond the laboratories involved in their initial development, and (iv) the collection and ultimate integration of experimental data from a wide variety of research groups made possible by the widespread use of a common resource and its accompanying Web-accessible public database.

STAGE 1: DEVELOPMENT OF A HOST-PATHOGEN SYSTEM, AND *C. ELEGANS* AS AN APPROPRIATE MODEL HOST FOR MICROBIAL PATHOGENS

C. elegans as a Model for Studying Bacterial Pathogenesis

In 1995, our laboratory reported that *P. aeruginosa* PA14, a primary human isolate, causes disease in both *Arabidopsis* and mice by using a shared subset of virulence factors (65). Because *P. aeruginosa* is a ubiquitous gram-negative bacterium commonly found in soil, we reasoned that *C. elegans* and *P. aeruginosa* might have an antagonistic relationship in their natural environment. *C. elegans* feeds on bacteria, and when *C. elegans* is fed *P. aeruginosa* PA14 as its sole food source, it lives approximately 2 days instead of the standard 2 weeks when fed their usual laboratory food, *Escherichia coli* OP50 (80). We consider the shortened longevity to be indicative of an active pathogenic process, and we refer to the shortened longevity as worm killing due to a lethal intestinal infection (see below). Because of its broad host range, *P. aeruginosa* PA14 is also well suited to study pathogenesis and host response in a variety of nonmammalian hosts in addition to *Arabidopsis* and *C. elegans*, including the insect *Drosophila melanogaster* (44) and the slime mold *Dictyostelium discoideum* (62).

In addition to *P. aeruginosa*, a remarkably large number of human microbial pathogens have been shown to "kill" *C. elegans*, including the gram-negative pathogens *Serratia marcescens* (39, 49, 61), *Salmonella enterica* (3, 42), and *Yersinia pestis* (13), and the gram-positive pathogens *Enterococcus faecalis* (19) and *Staphylococcus aureus* (19, 74). *C. elegans* also has a shortened life span when fed the pathogenic yeast *Cryptococcus neoformans* but has a normal life span and produces normal broods when fed nonpathogenic yeast such as *Cryptococcus laurentii* (56).

For most of the pathogens studied in the *C. elegans* system, including *P. aeruginosa*, *S. enterica*, *S. aureus*, *E. faecalis*, and *C. neoformans*, shortened longevity has been shown to involve active killing of the worms and is not a trivial consequence of the failure of the worms to eat the pathogens. *C. elegans* exhibits normal longevity when fed heat- or antibiotic-killed pathogenic bacteria or yeast. Moreover, most pathogen mutations that adversely affect pathogenesis in mammalian hosts also result in diminished killing of *C. elegans*. These virulence factors include two-component regulators (*gacA/gacS* of *P. aeruginosa* [81], *phoP/ phoQ* of *S. enterica* serovar Typhimurium [3]), quorum-sensing systems (*lasR* of *P. aeruginosa*, *agr* of *S. aureus*, *fsr* of *E. faecalis* [19, 74, 75, 81]), and alternative sigma factors (*rpoN* of *P. aeruginosa*, *rpoS* of *S. enterica* serovar Typhimurium, and s^B of *S. aureus* [23, 42, 74]). These results validate *C. elegans* as a surrogate host in which to identify novel pathogen virulence factors required for mammalian pathogenicity. Indeed, novel virulence factors have been identified in *P. aeruginosa* (48, 81), *E. faecalis* (19), *S. enterica* serovar Typhimurium (82), *S. aureus* (6), *S. marcescens* (38), and *C. neoformans* (57) by screening insertion mutant libraries for clones that exhibited diminished levels of killing.

A second reason for concluding that *C. elegans* killing by pathogenic microorganisms is an infectious process is that killing typically correlates with the accumulation of intact organisms in the *C. elegans* intestine (3, 19, 56, 74, 80). When *C. elegans* is feeding on a relatively benign food source, such as *E. coli* OP50 or *C. laurentii*, essentially no intact bacteria or yeast, respectively, can be seen in the intestine. In contrast, when feeding on all of the microbial pathogens that kill worms, large quantities of intact pathogenic bacteria or yeast can accumulate in the intestine, which can become grossly distended, as shown in Fig. 1 for the case of *C. elegans* feeding on *E. faecalis*. On the other hand, accumulation in the intestinal tract is not sufficient for killing. *Enterococcus faecium* accumulates to high levels but does not kill (19). Some pathogenic bacteria, like *P. aeruginosa* and *S. aureus*, transiently colonize

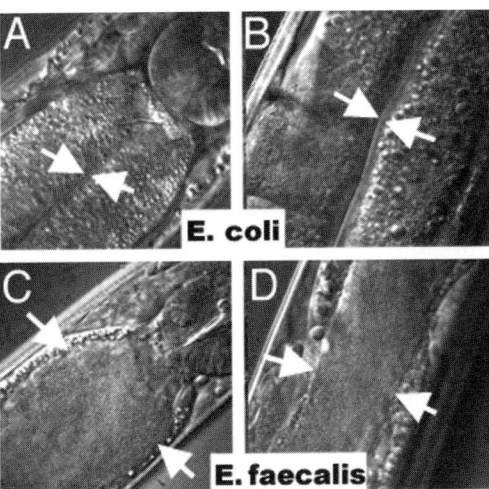

FIGURE 1 Accumulation of *E. faecalis* in the *C. elegans* intestine. Micrographs were taken of worms fed on lawns of bacteria on brain heart infusion agar for 3 days. Arrows point to the borders of the intestinal lumen. When *C. elegans* feed on their normal laboratory food, *E. coli* strain OP50, very few, if any, intact bacterial cells accumulate in the intestinal lumen, which appears as a narrow channel (A and B). In contrast, there is dramatic distension of the intestine when *C. elegans* are fed on *E. faecalis*, and numerous densely packed *E. faecalis* cells are visible. In the upper right-hand corner of (A) and (C), the round structure depicted is the pharyngeal grinder organ that physically disrupts ingested bacteria. (A) and (C) show the proximal portion of the intestine immediately following the pharyngeal grinder, and (B) and (D) show a middle portion of the intestinal tract.

the alimentary tract and can be "washed out" by the normal defecation process if the worms are transferred to a benign food source (74, 80). Others, like *S. enterica* serovar Typhimurium and *E. faecalis* (3, 19), persistently colonize and proliferate within the intestinal tract.

A particular advantage of the *C. elegans* pathogenicity model and of similar models that use model genetic hosts is the ability to carry out so-called interactive genetic analysis, involving the identification of *C. elegans* mutants that "suppress" the phenotypes of pathogen mutants and vice versa. This potentially permits the identification of pathogen virulence factors and host functions that interact with each other in vivo. For example, our laboratory has shown that a *C. elegans* mutant defective in efflux pumping of small molecules is more susceptible to *P. aeruginosa* PA14-mediated killing that involves the production of low-molecular-weight PA14 toxins synthesized when PA14 is grown on high-osmolarity media (48). Because one class of *P. aeruginosa* mutants defective in this type of killing (referred to as "fast killing") secretes reduced levels of phenazines (48), we hypothesized that phenazines are toxic to *C. elegans* but are normally effectively pumped out of nematode cells by efflux pumps.

Phenazines are low-molecular-mass tricyclic redox active toxins that have been implicated in human disease (reviewed in reference 60). In support of this hypothesis, when a *C. elegans pgp-1 pgp-3* mutant defective in two particular efflux pumps was tested with two classes of *P. aeruginosa* mutants that are defective in fast killing (mutants that synthesize reduced levels of phenazines and mutants that synthesize normal levels of phenazines), attenuated killing by pyocyanin-positive *P. aeruginosa* mutants (but not pyocyanin-negative mutants) was suppressed (48). Moreover, worm mutants that are more resistant to oxidative stress, such as *age-1*, are more resistant to *P. aeruginosa* PA14 fast killing. Conversely, the *C. elegans* mutants *mev-1* and *rad-8*, which are more sensitive to oxidative stress, are more sensitive to fast killing (48).

Although pyocyanin, a well-characterized phenazine, was initially implicated as the primary phenazine compound involved in *C. elegans* killing, subsequent unpublished work in our laboratory has shown that *P. aeruginosa* PA14 mutants defective in pyocyanin production are still competent for fast killing, implicating an as yet unidentified phenazine toxin. In any case, when combined with the data on the *C. elegans pgp* mutants, the phenazine data nicely illustrate how interactive genetic analysis can be carried out in the *C. elegans*–*P. aeruginosa* pathogenicity model.

C. elegans as a Model for Studying Host Immunity

Recent work summarized in three reviews (33, 39, 54) demonstrates that *C. elegans* possesses a sophisticated immune system. Because of their unique experimental versatility, *C. elegans*–based host-pathogen systems are attractive models in which to elucidate evolutionarily conserved innate immune mechanisms and in which to investigate the evolutionary origins of metazoan immune response pathways. Innate immunity is characterized by receptors that are activated by signals indicative of pathogens, often referred to as pathogen-associated molecular patterns (PAMPs). PAMPs include conserved microbial cell wall molecules such as lipopolysaccharide and peptidoglycan. In insects and mammals, a family of conserved Toll-like receptors are involved in PAMP recognition (28, 50–53). The Toll-like receptors are coupled to signaling adapters, some of which contain protein-protein interaction TIR (Toll interleukin-1 receptor) domains that activate downstream mitogen-activated protein kinase (MAPK)-signaling cascades and the nuclear translocation of Rel/NF-κB transcription factors. Ultimately, a set of immune effectors is activated, including signaling molecules such as cytokines. In addition to the conserved Toll pathway, insects and mammals share an additional immune pathway, referred to as the IMD (immunodeficiency) pathway in flies (21, 25, 45) and the TNF (tumor necrosis factor) pathway in mammals (reviewed in reference 25). In flies, the IMD pathway responds to gram-negative bacteria via a peptidoglycan recognition protein (10, 22, 66), whereas the Toll pathway responds to fungi and gram-positive bacteria (21, 25, 45). Like the Toll and Toll-like receptor (TLR) pathways, the IMD and TNF pathways also lead to the activation of Rel-like transcription factors. In vertebrates, the innate immune system functions as a first line of defense against microbial invaders, as well as activating the adaptive immune system.

Compared to mammals and *Drosophila*, however, relatively little is known about the innate immune response in *C. elegans*. Bioinformatic analysis of the fully sequenced *C. elegans* genome identifies a single TLR but no Rel/NF-κB transcription factors or homologues of the IMD/TNF pathways (except for a homologue of the TAK-1/MEKK kinase). Interestingly, mutation of the single *C. elegans* TLR does not confer an immunocompromised phenotype (61). Nevertheless, transcriptional profiling analysis from our laboratory and others has shown that *C. elegans* responds to bacterial pathogens by the activation of a variety of genes, homologues of which are known to be involved in antimicrobial responses in insects and mammals (12, 49; D. Kim, E. Troemel, and F. Ausubel, unpublished data).

Both forward and reverse genetic approaches have been used to study the *C. elegans* innate immune response. For example, our laboratory has isolated and characterized *C. elegans* "Esp" (*e*nhanced *s*usceptibility to *p*athogens) mutants that die more quickly than wild-type worms when feeding on *P. aeruginosa* PA14 but that have the same longevity as wild type when feeding on a nonpathogenic food source (34). Because *P. aeruginosa* PA14 does not kill *C. elegans* eggs and because *C. elegans* are self-fertilizing hermaphrodites, putative Esp mutants can be recovered by transferring the eggs from dead carcasses of putative Esp mutants to *E. coli* OP50.

The isolation and characterization of Esp mutants in our laboratory led to the identification of a conserved p38 MAPK signaling pathway that plays a key role in the *C. elegans* immune response. Specifically, the *esp-8* gene encodes the *C. elegans* orthologue of the mammalian MAP kinase kinase kinase (MAPKKK) gene, *ASK-1* (34), previously identified in *C. elegans* as NSY-1. Similarly, the *esp-2* gene encodes a *C. elegans* homologue of the mammalian MKK3/MKK6-type MAP kinase kinase (MAPKK), previously identified in *C. elegans* as SEK-1 (34). Cell biological and biochemical studies have shown that the mammalian orthologues of NSY-1 and SEK-1 (ASK1 and MKK3/6, respectively) activate

p38 and JNK MAPKs in biochemical assays and cell culture systems (41). The p38 MAPK, in particular, plays a key a role in innate immunity (41). Importantly, mutations in the *C. elegans pmk-1* gene, which encodes a p38 MAPK homologue, result in an Esp phenotype (35), as does knockdown of the *pmk-1* gene expression by using RNA interference (RNAi) (34).

As an example of a reverse genetic approach to the study of *C. elegans* immunity, a gene that functions upstream of the PMK-1 p38 pathway was identified by RNAi technology (46). In *Drosophila* and mammals, transmembrane Toll and TLRs as well as their corresponding intracellular "adapter" proteins, all of which contain TIR domains, function upstream of the conserved p38 MAPK cascade (for review, see references 5 and 89). TIR domains allow these receptors and adapter proteins to homo- and heterooligomerize, forming multiprotein complexes that initiate signaling. There are only two proteins encoded in the *C. elegans* genome that contain a recognizable TIR domain. One is the TOL-1 protein, a homologue of mammalian and insect TLRs, which, as described above, does not appear to be involved in the immune response (61). The other TIR domain containing protein is referred to as TIR-1, which is homologous to the mammalian gene known as *SARM*. The absence of transmembrane motifs in the primary amino acid sequence of both TIR-1 and SARM suggests that these proteins may function as so-called immune adapters, similar to a subgroup of mammalian TIR domain proteins that include MyD88, Mal (also known as TIRAP), TRIF (also known as TICAM-1), and TRAM (also known as TIRP and TICAM-2) that transduce signals from transmembrane TIR domain receptors to the downstream components of the immune pathway (for reviews, see references 5, 7, 15, 64, and 79). Whether SARM mediates TLR signaling like other members of this protein subgroup, however, has not yet been established. Importantly, nematodes fed *tir-1* RNAi died significantly faster than control worms feeding on a variety of bacterial and fungal pathogens, similar to *pmk-1* RNAi-inactivated worms (12, 46).

Moreover, levels of activated PMK-1 (p38) in *tir-1* RNAi-targeted worms is significantly reduced, consistent with the conclusion that TIR-1 functions upstream of PMK-1 (p38). Because among the family of mammalian TIR-domain adapter proteins, only SARM is conserved in *C. elegans*, it is possible that TIR-1 may be an ancient member of this class of signaling proteins and that the evolution of innate immunity involving TLR signaling through NF-κB in insects and mammals required the further evolution of the TIR domain adapter family. This line of reasoning offers an explanation why SARM does not appear to function in TLR-dependent signaling in mammalian cells.

In summary, as illustrated in Fig. 2, *C. elegans* possesses an immune system that resembles, at least in part, the immune system of flies and the innate immune system of mammals. A variety of work has shown that an ancient, highly conserved p38 MAPK signaling cassette plays a key role in *C. elegans* immunity (33, 34). A TIR-domain protein has been identified that functions upstream of p38 MAPK (12, 46) in *C. elegans*. In work not summarized here, it has

been shown that programmed cell death (14) functioning downstream of the p38 MAPK pathway (1, 2) is important for *C. elegans* immunity. It has also been shown that the transforming growth factor-β signaling pathway (40) and an insulin-like signaling pathway (20) are also involved in mounting the *C. elegans* innate immune response.

STAGE 2: DEVELOPMENT OF GENOMIC TOOLS THAT FACILITATE MODEL HOST-PATHOGEN ANALYSES

Genomic Tools for Pathogens

A wide variety of genomic tools has been developed in both potential hosts and pathogens during the past several years, and many have been directly applied to the study of virulence. Although a genome sequence is not an absolute prerequisite, a complete or draft sequence generally precedes the development of genomic resources, and it is certainly important for the refinement of any such tools. The greatest contribution has come from the use of microarrays for transcriptional profiling studies to identify both host and pathogen genes that are potentially important during infection (re-

FIGURE 2 Immune signaling pathways in mammals and nematodes. In mammals, a family of TLRs mediates the recognition of PAMPs. Immune receptors have not yet been identified in *C. elegans*. Mammals and *C. elegans* share a conserved p38 MAPK signaling module, but *C. elegans* does not encode a transcription factor homologous to mammalian NF-κB. Interestingly, *C. elegans* has a TIR domain-containing protein, TIR-1, that functions upstream of the p38 MAPK that is homologous to the mammalian SARM protein, but the role of the SARM protein in mammalian immunity is not known.

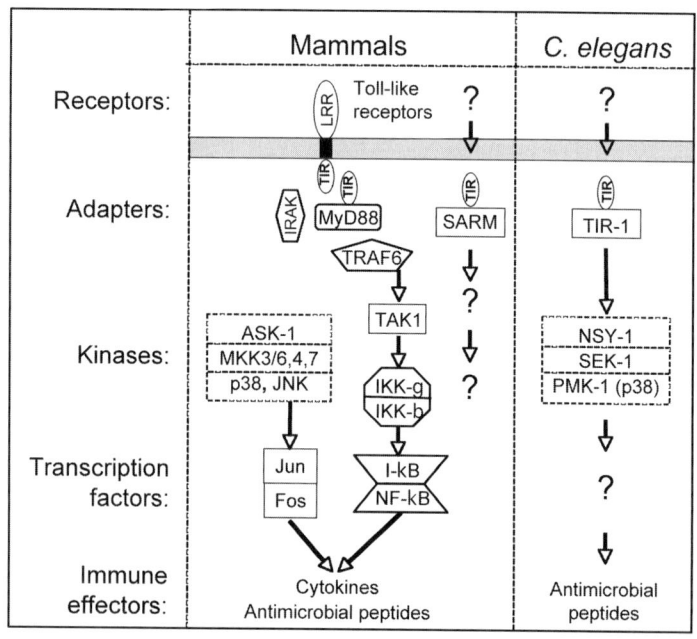

viewed in reference 67; D. Kim, E. Troemel, and F. M. Ausubel, unpublished data). These same microarrays have also been invaluable tools for determining the genome composition (or "genomotyping") of related pathogenic strains by hybridizing labeled genomic DNA to identify loci that are absent in a given isolate (16, 32, 72, 76, 87). Finally, proteomic analysis has started to be applied to the analysis of pathogenicity, identifying putative secreted virulence factors in *Vibrio cholerae* and *Mycobacterium tuberculosis* (17, 63).

In addition to the above-mentioned tools, which can be used to examine the wild-type pathogen, genome-wide collections of mutants are becoming more common within the pathogenesis community. In one scenario, large collections of transposon insertion mutants have been generated and the site of the lesion has been identified by sequencing of genomic regions adjacent to the transposon, as has been performed in the well-characterized laboratory strain of *P. aeruginosa*, PAO1 (26). Mutant libraries such as these allow researchers to rapidly obtain lesions in genes of interest for directed experiments; in more ambitious settings, the entire collection has been subjected to a phenotypic screen (in the case of the PAO1, for defects in twitching motility and prototrophic growth phenotypes). As described below, to facilitate genome-wide screens, the mutants can be further refined to generate a nonredundant library, in which a single transposon insertion in each disrupted gene is selected from the starting collection of mutants (47). In this manner, the larger initial library is streamlined into a smaller subset that is amenable to a wider range of phenotypic screens. Finally, an alternative refinement of mutant libraries allows mutant isolates to be analyzed in pools rather than individually. Similar to signature-tagged mutagenesis (24) (reviewed in reference 70), in which individual transposons are marked with a unique tag to allow distinct insertion strains to be identified from within a mixed population of mutants, transposon site hybridization (TraSH) analysis (4, 71, 88) utilizes the unique genomic DNA

adjacent to each transposon as a means to distinguish mutants from each other (see below). In this system, mutants are subjected to various test conditions as pools, and those mutants that survive the condition (and, by inference, those that fail to survive) can be identified from the complex mixture at the end of the experiment. This ability to group mutants into large pools greatly increases the ability to screen large numbers of genes in situations where analyzing individual mutants would be prohibitive, such as whole-animal models of infection.

In our model host infection system, we benefit from the use of small, quickly reproducing organisms that make high-throughput analysis feasible. A major goal of pathogenesis research is to identify the entire suite of genes in a particular pathogen that are specifically involved in the pathogenic process. An important advantage of the *C. elegans* pathogenesis model, as well as analogous models that use a variety of nonvertebrate genetic hosts, is that each putative mutant in a transposon library can be tested in an individual host to determine whether virulence has been affected. Combining this model of disease with a nonredundant library of mutants would therefore allow every (nonessential) gene in the pathogen to be readily assayed to identify novel virulence factors. Therefore, of all the bacterial genomic tools outlined above, a defined mutant library most obviously integrates into the model host system of studying pathogenesis in a way that does not alter the basic experimental design except to expand its scope to survey the entire genome. We have therefore constructed a nonredundant library of transposon insertions in *P. aeruginosa* strain PA14, a library that has been designed to allow for analysis of individual mutants as well as pools of mutants using TraSH (47), and we will describe it in greater detail below.

Genomic Tools for Hosts

In parallel with nonredundant transposon mutation libraries of pathogens, it is possible to take advantage of analogous tools in model hosts that are genetically tractable to identify host genes important in the infectious process or host genes that encode products that interact with pathogen virulence factors. Although our laboratory has successfully performed forward genetic screens in *C. elegans* for Esp mutants, generating the host mutants and ultimately cloning them by traditional positional mapping techniques are tedious because the Esp phenotype is scored as a population assay and requires phenotyping in the F3 or F4 generation. However, RNAi is particularly robust and straightforward in *C. elegans* because RNAi can be efficiently delivered to the worms simply by feeding the worms an *E. coli* strain that expresses a dsRNA corresponding to a particular *C. elegans* gene (31, 84, 85). Full-genome *C. elegans* RNAi libraries in *E. coli* are publicly available and have been initially used for the study of development in the nematode (18, 29, 30). However, these RNAi libraries have proven to be invaluable tools to complement the forward genetic analyses of the host response to pathogen attack, allowing candidate genes for involvement in the innate immune response to be quickly tested (12, 34, 35, 46), and genome-wide RNAi screens are being carried out by a green fluorescent protein (GFP) reporter construct that is activated in response to pathogen attack (D. Kim and F. Ausubel, unpublished).

A Nonredundant Transposon Mutation Library for *P. aeruginosa* PA14

Our laboratory has chosen *P. aeruginosa* PA14 for the construction of a nonredundant mutant library because it is remarkably virulent in the greatest number of model hosts tested, as well as in a number of murine systems of infection (11, 27, 37, 44, 55, 62, 65, 80). There are four major advantages to this approach as opposed to the traditional approach of screening a random collection of strains for avirulent or attenuated mutants. First, a nonredundant library significantly decreases the workload for each attempted screen. A mutant library large enough to saturate a typical open reading frame-rich bacterial genome (that is, large enough such that there is a 95% probability

that each nonessential gene has been hit at least once) needs to contain approximately five times as many mutants as the number of nonessential genes in the genome. *P. aeruginosa* has a relatively large genome; for example, *P. aeruginosa* PA01 and PA14 encode approximately 5,570 and 5,962 genes, respectively (47), and corresponding saturated transposon libraries would need to contain 27,900 and 29,800 different mutants. However, the nonredundant library currently reduces the number of mutants that need to be screened to approximately 5,400 (see below). Second, because the insertion site in each member of the library is known, screening the nonredundant library identifies genes that are not required for pathogenesis as well as genes that are required for pathogenesis in a particular host. Third, screens for avirulent mutants can be repeated in each of the model hosts (nematodes, insects, and plants). Knowing the identity of each mutant before screening allows us to easily integrate all of the data (both positive and negative) from various experiments, which would not be possible if each screen were initiated with a random (uncharacterized) library. Fourth, the pool of more than 30,000 mutants used to construct the nonredundant library will contain multiple insertions in most genes (approximately four on average; see below), which can be used to verify the mutant phenotype of the insertion mutation in the nonredundant library, thereby definitively correlating the mutant phenotype with a mutation in a particular gene.

To construct a nonredundant transposon library for *P. aeruginosa* PA14, our laboratory generated a library of random insertion mutations by using *MAR2xT7*, a derivative of the *D. melanogaster mariner* family transposon *Himar1*, which transposes in both prokaryotic and eukaryotic genomes and is thought to exhibit minimal insertion site specificity (43, 69). Because the *Himar1* transposase is located on a suicide vector outside the *MAR2xT7* coding sequence, subsequent *MAR2xT7* transposition to secondary sites is not possible. (The DNA sequence of *MAR2xT7* can be downloaded at http://ausubellab.mgh.harvard.edu/cgi-bin/pa14/downloads.cgi.) To date, a PCR protocol utilizing a transposon-specific and an "arbitrary" primer (ARB PCR) (8) has been used to map the insertion sites of approximately 20,530 unique *MAR2xT7* insertions onto the PA14 genome sequence, which has been determined in our laboratory (http://ausubellab.mgh.harvard.edu/pa14sequencing/). *MAR2xT7* insertions have been isolated in 4,469 or 75% of the predicted PA14 genes and, on average, 4.3 transposon insertions were mapped to each gene. Although *MAR2xT7* insertions appear to be relatively random across the genome, near saturation of the PA14 genome was achieved with the current set of *MAR2xT7* ~25,000 insertions, with approximately 1,400 genes not yet targeted, indicating some amount of insertion site bias of *MAR2xT7* transposition. Future efforts to expand the PA14 library will include the use of a different transposon.

An important feature of the *MAR2xT7* transposon is that it can also be used for genetic footprinting of the mutants by TraSH analysis (as described above). This makes it possible to determine the fitness of particular mutants in a pool in different environments, such as in live-animal infection models, in biofilms, and in media containing antibiotics. T7 promoters placed at both ends of *MAR2xT7* allow each *MAR2xT7* mutant to be uniquely identified via PCR-amplified genomic sequence adjacent to the transposon. The T7 promoters are then used to direct transcription from these PCR products, and the resulting RNA is used to create probes for hybridization to a *P. aeruginosa* DNA microarray to identify the amplified genomic fragments (and therefore the individual mutants) present in the pool. We have shown that T7-directed RNA can be successfully generated from PCR-amplified genomic DNA extracted from *MAR2xT7* mutants, confirming that *MAR2xT7*-generated mutants can be used for TraSH analysis (47).

A subset of the parental PA14 *MAR2xT7* transposon insertion library, called the PA14 nonredundant (NR) set, was created to facilitate genome-wide screening (Fig. 3). The PA14NR set consists of a selected group of mutants from the library in which a single insertion mutant (or in some cases two mutants) represents each PA14 gene that was targeted by *MAR2xT7*. As discussed above, the PA14NR set does not represent all of the PA14 nonessential genes because at least 1,400 PA14 genes were not hit in the library. A comparison of the *MAR2xT7* library and a similar library constructed in *P. aeruginosa* strain PAO1 using a derivative of transposon Tn5 (26) indicates that there are only about 335 *P. aeruginosa* essential genes, suggesting that about 1,000 nonessential genes are missing from the PA14NR set. Currently, a total of 5,459 mutants are included in the PA14NR set, which corresponds to 4,596 predicted PA14 genes (the library contains additional mutants using the TnPhoA transposon to supplement the 4,469 genes disrupted by *MAR2xT7* insertions). In choosing particular mutants to include in the PA14NR set, a prioritization scheme was used in which most mutants in the NR set correspond to the insertion that was closest to the 5′ end. The PA14NR set is arrayed in 96-well microtiter plates, and a great amount of attention was paid to developing procedures to minimize the amount of cross-contamination between the wells of the library, including colony purification of all members of the PA14NR set. PCR-based analysis of the PA14NR set indicated that at least 97% of the wells contain the expected mutation. Details concerning the construction and features of the PA14NR set can be found at http://ausubellab.mgh.harvard.edu/cgi-bin/pa14/mutantrequest.cgi.

To assess the utility of the PA14NR set for high-throughput functional screens, our laboratory screened the entire PA14NR set (5,459 mutants) for attachment to polyvinyl chloride (PVC) plastic (47), a phenotype known to require several *P. aeruginosa* genes (58, 59). Because attachment to PVC plastic is correlated with the ability to form biofilm (58, 59), mutants with altered PVC attachment profiles may form biofilm improperly. A total of 416 PA14NR set mutants with a PVC attachment phenotype were identified in the primary screen, including insertions in *pilC, rpoN, algR, clpP, crc, fleR, fliP, sadB, sadA,* and *sadR,* which had previously been shown to be required for PVC attachment and biofilm formation (9, 36, 58, 59, 83, 86). In addition to these previously known genes, the PA14NR set screening for attachment to PVC also identified many genes without previously reported attachment phenotypes. These findings validate the PA14NR set as an extremely efficient tool to scan the *P. aeruginosa* genome to identify genes critical for a specific phenotype.

STAGE 3: DISTRIBUTION AND ADOPTION OF GENOMIC TOOLS

The PA14NR set is a public resource, currently available to all members of the research community, even though handling and ship-

ping the library present some practical challenges because of its relatively large physical size. The nonredundant library currently exists as stocks frozen in a set of 59 96-well microtiter plates. Attempts to reduce the physical size of the library by rearraying the strains in a 384-well format have proven unsuccessful because cross-contamination between wells could not be completely eliminated during the robotic handling of the liquid cultures. To offset the tremendous labor required to reproduce and disseminate the library, our plan is to produce approximately 40 copies for initial distribution to *P. aeruginosa* researchers, with the expectation that other laboratories will duplicate the library, thereby further dissemi-

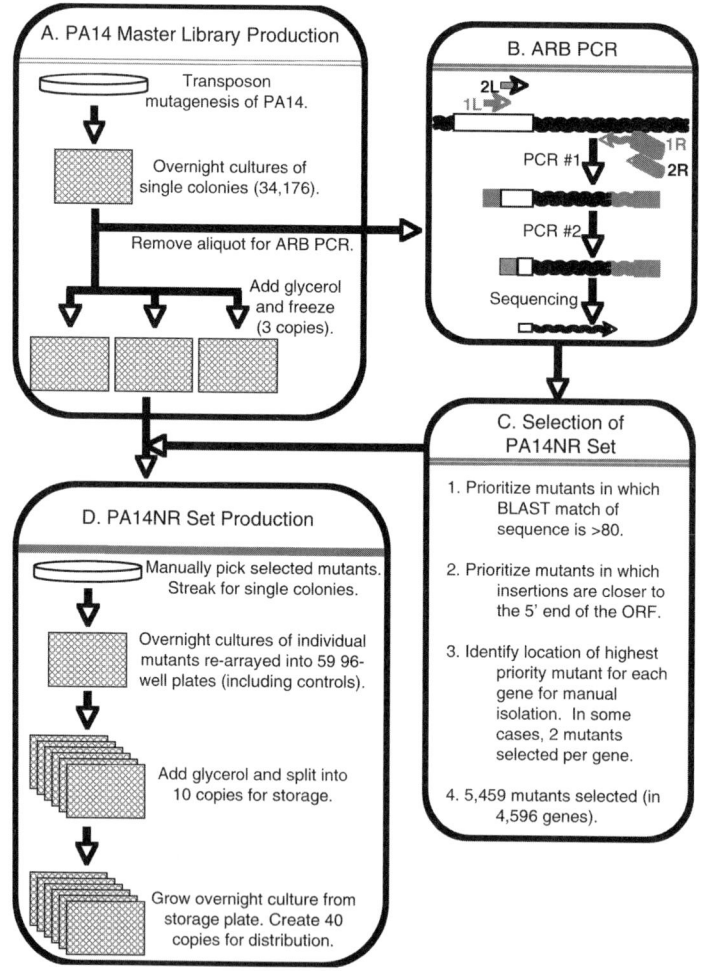

nating it widely in the *P. aeruginosa* research community. We hope to formalize the process of having several distributed sources for this shared genomic tool, a process that invariably happens in an informal (and less documented) manner at each institution possessing a copy.

Genomic tools that rely on a particular experimental platform such as microarrays may potentially be prohibitive to research groups that lack the expertise, the infrastructure, or the funds to invest in such large-scale experiments. In contrast, the individual unit of the PA14NR set (a mutant strain) is a familiar tool already used in most laboratory settings. Therefore, once the PA14NR set is acquired,

our hope is that researchers find that integrating the PA14NR set into their existing work flows should not be an overly disruptive or expensive transition, and that this relative ease of use will encourage its wide adoption. Certainly, laboratories interested in and able to perform whole-genome screens with the entire nonredundant set have that option, whether it is in the context of studying mutants one by one, or in pools of mutants by TraSH analysis. In this manner, by distributing the library to as many laboratories as possible for use in as diverse a range of experiments as possible, the community can collectively compile a body of data that is extremely rich and broad, affecting our understanding of both

---

FIGURE 3 Schematic of PA14 library and PA14NR construction. (A) PA14 master library production. Transposon mutagenesis of PA14 was performed and individual colonies were picked and grown overnight in 96-well plates. An aliquot of the overnight culture was set aside for arbitrary (ARB) PCR analysis to map each transposon insertion site. Glycerol was added to the remaining culture and mixed, and the sample was split into three aliquots for storage at −80°C. (B) ARB PCR. ARB PCR was used to map the site of insertion for each mutant. In the schematic diagram, the transposon is indicated by a white rectangle within the PA14 genome (wavy lines). The positions of four PCR primers are indicated in this schematic. An aliquot of the original overnight culture was used as a template for the first PCR reaction. The left primer for the first reaction (1L) anneals to the transposon. The right primer (1R) contains a variable sequence in the 3′ end that (randomly) anneals to the PA14 genome at some distance from the transposon; the 5′ end of the 1R primer has a constant region. A subset of PCR products from this reaction will contain the terminal sequence of the transposon and the flanking genomic sequence (varying lengths of genomic sequence for each mutant strain); other PCR products will be the result of pairs of 1R primers annealing to other regions of the PA14 genome. An aliquot of the first completed PCR reaction was used as a template for a second PCR reaction. The left primer for the second reaction (2L) is a nested primer annealing to the transposon sequence, and the right primer (2R) is identical to the constant 5′ end of the 1R primer. In this manner, only fragments that contain the transposon will be amplified as the dominant product in the second completed PCR reaction. The PCR reaction was processed to remove primers and free nucleotides, and the remaining double-stranded product was subjected to sequencing with a transposon-specific sequencing primer. The resulting sequences were batch processed to (i) trim off low-quality sequence, (ii) identify the transposon sequence, and (iii) check the remaining sequence against the PA14 genome via BLAST to map the site of insertion. (C) Selection of PA14NR set. The insertion sites for each mutant identified in step 3B were mapped onto the predicted PA14 ORFs. An automated script was designed to identify the best candidates for inclusion in the PA14 nonredundant set (PA14NR set). For predicted ORFs in which only one insertion was available, that mutant was selected. In cases where more than one insertion was available, all sequences were filtered to retain only those in which the BLAST match of the obtained high-quality sequence to the PA14 genome had a score of 80 or greater. Of the remaining cases, the insertion that was the most 5′ in the ORF was generally chosen. In some cases, two insertions were chosen, such that 5,459 mutants with insertions in 4,596 genes comprise the current version of the PA14NR set. Finally, the script identified the original position of each chosen mutant in the original master library and remapped the location onto a new set of 96-well plates for the NR set. (D) PA14NR set production. By using the list of mutants identified in step 3C, individual strains were manually picked from one of the three frozen copies of the PA14 master library. Each mutant was streaked onto selective media to isolate a pure colony, which was then used to inoculate overnight cultures in 96-well plates (using the remapped NR set locations in each plate). The overnight cultures were grown in deep-well plates to allow for a sufficient volume to freeze 10 copies of the PA14NR set. One of these copies will subsequently be used to create copies that will be distributed to other laboratories.

pathogenesis and other biological processes in *P. aeruginosa.*

STAGE 4: COLLECTION AND INTEGRATION OF HOST AND PATHOGEN DATA FROM THE RESEARCH COMMUNITY

Collecting Genome-Wide Data with the PA14NR Set

How will genomic tools most advance the field of model host pathogenesis? The initial and simplest manifestation will be to allow experiments that are conceptually the same as what has been performed previously, but on a larger, whole-genome scale. Screens for PA14 mutants with attenuated virulence have been performed in our laboratory with *C. elegans, Arabidopsis,* and the wax moth, *Galleria mellonella* (48, 55, 81). However, none of these screens was exhaustive, because the requirement to examine roughly 30,000 random transposon mutants for full-genome coverage was prohibitive. However, the small size of the PA14NR set makes a genome-wide screen in all three hosts possible for the first time. Furthermore, features added to the *MAR2xT7* transposon that enable TraSH analysis now allow for new experimental designs that use mixed populations of mutants. However, we believe that the true impact of the library will not be realized until a wide variety of data is collected by the larger *P. aeruginosa* community sharing a common data repository. Our previous model host screens were each initiated with distinct insertion libraries, making direct comparisons between screens impossible. Furthermore, only those isolates with a positive phenotype were sequenced to identify the disrupted gene; therefore, no negative data were collected. In contrast, the PA14NR set can be reused in multiple screens in which both positive and negative results can be recorded for each gene tested, thereby compiling an extensive list of testable phenotypes for the mutants.

Certainly, the range of phenotypes that can be documented will go beyond direct measures of virulence used in model host screens or assays of processes thought to be involved in virulence, such as the PVC attachment screen already described for the PA14NR set (47). Bacterial virulence is greatly impacted by many other aspects of basic biology. For example, the ability to acquire sufficient iron while inside a host is a key factor in determining whether or not a pathogen can be successful (reviewed in reference 68). Therefore, the ability to index and access data pertaining to a wide variety of processes can ultimately reveal important factors regulating virulence that have been studied in other settings, but whose links to pathogenicity may not be immediately apparent.

A Public Pathogenomics Database Based on the PA14NR Set

An essential feature of the PA14 transposon mutation library in general and the PA14NR set in particular was the concomitant development of a relational database, the PA14 Transposon Insertion Mutant Database (PATIMDB). During the construction phase of the library, the database served two primary functions. It was initially created to organize location and status information for each sample as it moved from transposon integration, through sequencing, and on to its eventual relocation from the initial library to its new position in the PA14NR set. The second function was to perform automated sequence analysis of each mutant to aid in identifying the locus disrupted by the insertion. Raw genomic sequences adjacent to the transposon were stored, filtered to remove poor-quality sequences, and aligned by BLAST (Basic Local Alignment Search Tool) against the genomes of *P. aeruginosa* PA14 and PAO1 (http://ausubellab.mgh.harvard.edu/pa14sequencing/; http://www.pseudomonas.com) (78).

As we are preparing to make the library available to the general community, two new features have been added to the PATIMDB to facilitate accessibility to this genomic tool. Information describing the construction, contents, and use of the library has been posted at http://ausubellab.mgh.harvard.edu/cgi-bin/pa14/downloads.cgi in a downloadable detailed users' manual. The contents of the library can further be searched (http://ausubellab

.mgh.harvard.edu/cgi-bin/pa14/search.cgi) or browsed using a transposon insertion map of all PA14 transposon insertions (http://ausubellab.mgh.harvard.edu/cgi-bin/pa14/tnmap.cgi). This map allows users to both search for specific mutant sequence and BLAST alignments and visually scan predicted PA14 genes for insertion locations. Second, the website allows users to request copies of the NR set (http://ausubellab.mgh.harvard.edu/cgi-bin/pa14/mutantrequest.cgi), and the database will be used to process and track all mutant requests.

The next version of the database will incorporate the ability to store, search, and download experimental data collected using the mutants in the library that have been deposited by members of the *P. aeruginosa* research community. The flexibility of the library will allow it to be used to collect data in a variety of experimental settings, adding to the potential value of the resource. However, this flexibility will also present logistical problems in how the collected data can be standardized for inclusion in the database and subsequent searches and analyses. A similar effort to create centralized public repositories for transcriptional profiling data has required the coordination of a consortium of researchers known as the Microarray Gene Expression Data Society (http://www.mged.org/) to establish standards for microarray data annotation and exchange. This effort has resulted in the creation of a microarray gene expression markup language (MAGE-ML), an extensible, self-descriptive, XML-based language that ensures that reported expression data are in a defined format containing sufficient information to allow disparate experiments to be compared (77). These technical hurdles were substantial, even though all microarray experiments are conceptually similar with respect to the type of data collected (intensities for a given spot on an array) and the range of possible manipulations for the raw data. The nature of the data collected using the PA14 transposon mutation library will necessarily be more diverse, making a complete and inclusive set of standards for the raw data formats unlikely. An alternative approach is to define a controlled vocabulary for specifying experiment types and phenotypes. For example, the WormBase repository for *C. elegans* data (http://www.wormbase.org) includes a list of observed mutant phenotypes for each annotated gene, as well as serving as a repository for RNAi data that is organized and searchable based on predefined observable phenotypes. Although this approach functions well for the most common uses of this library (many of the phenotypic classes are developmental and morphological in nature, reflecting the predominantly developmental focus of the *C. elegans* community during its early days), many phenotypes that are being assayed in less conventional screens cannot currently be included in the search.

We are taking the pragmatic approach of anticipating a wide variety of types of user-submitted data relating to the PA14NR set, defining preset categories' experimental data templates where appropriate. Users will be able to define new categories and data fields when submitting their data, and those that are frequently used will be incorporated as presets in new versions of PATIMDB. New versions will also allow for open-ended entry of text information that will still be searchable by the Web-accessible interface to the database. For example, basic organizational information will describe the experimenter and his or her affiliation, and the more straightforward aspects of the experiment (a unique identifier for each experiment, date of execution, date of publication, duration). More detailed descriptions of the experimental design will need to incorporate a combination of general categories to subdivide the submissions (i.e., general experimental categories such as genomic, genetic, biochemical, physiological, morphological, or other) and open-ended text fields for detailed experimental descriptions. The observed phenotype for each gene tested will include a general category of the process affected as defined by one of the 27 functional classes used in the annotation of ORFs in PAO1, which may or may not correspond to the functional class of the annotated gene (for example, a gene annotated as and previously demonstrated to belong to the "central intermediary metabolism" category may be identified in a screen examining a phenotype that belongs to the "chemotaxis" functional category). We will also record whenever a gene shows a wild-type phenotype in any given assay. However, more complex aspects of the phenotype, such as a detailed description or a listing of genetic interactions with other bacterial or host genes, will require text field entries.

At this time, it is neither feasible nor appropriate for our group to define a finished list of standards governing the formats for submission of experimental data because we cannot anticipate all of the current (and future) uses for the library that researchers might exploit. While we will develop and promote the adoption of standard operating procedures for frequently used experimental protocols associated with the library, there will be numerous specialized or creative uses for the library that are not easily standardized but that will nevertheless generate data valuable to other researchers. Therefore, ongoing dialogue with the community will be critical in helping us to expand and refine the list of criteria defining the nature and the format of submitted data as needed, with the current implementation remaining as flexible and all-inclusive as possible. We will continue to rely on the use of open standards for our database and will make our source code available to other developers to facilitate future interoperability with other databases and the creation of additional data modules, applications, and visualization tools.

CONCLUDING REMARKS

The approach of using simple model hosts to study human pathogens was developed as a high-throughput complement to existing mammalian models. The incorporation of genomic tools into this approach will further increase the scope and breadth of these studies and allow us to integrate data from a variety of sources through the availability of common re-

sources to the greater research community. As described in this chapter, important tools for this experimental approach have been the addition of defined genome-wide mutant libraries in both the pathogen and the host (Fig. 4). Rather than being limited to a small number of random mutants, libraries in PA14 and in *C. elegans* have made possible saturated genome-wide screens for virulence factors or host response genes, or greater numbers of combinatorial genetic experiments in which possible interactions between host and pathogen genes can be examined. In addition to performing traditional infection assays, the model-host pathogen interaction can also be probed using additional genomic tools. Transcriptional microarray and proteomic studies can be used to monitor changes in either PA14 or *C. elegans* to monitor global changes induced as a result of the interaction between the two organisms. These assays can be taken one step further to combine the use of microarrays with mutant libraries (again, in either the pathogen or the host). Indeed, iterative expression microarray methods have been used to compare the transcriptome of wild-type and mutant strains defective in putative virulence-related genes to identify candidate regulatory networks relevant for pathogenesis (73). A microarray experiment in a *C. elegans* mutant defective in a newly defined host defense response gene identified candidate downstream genes important in the response to pathogens (D. Kim, E. Troemel, and F. M. Ausubel, unpublished data). Clearly, the ability to combine various genomic tools will allow for a large and rich body of data that will help us define, in more precise terms, the interactions that occur between pathogens and their hosts.

With the PA14 transposon insertion library, we are aiming to provide not only a useful resource for other laboratories that can affect the scope and magnitude of individual experiments, but also a platform for integrating diverse sets of experimental data through a shared common resource. The bioinformatics tools required to store, index, and compare such dis-

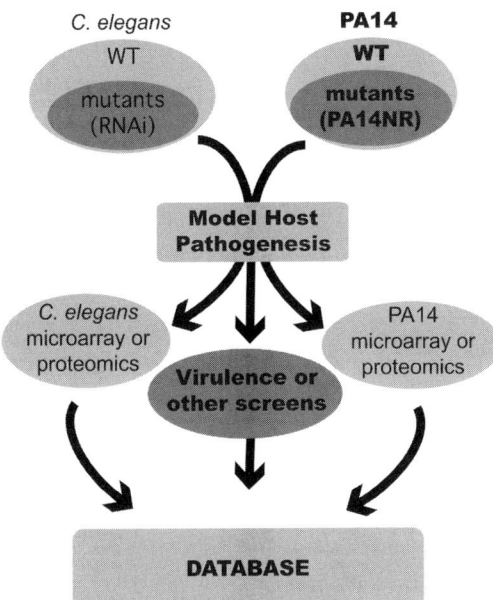

FIGURE 4 Interface between model host pathogenesis and genomics. Elements that are the major focus of this chapter are indicated in bold. The basic premise of the model host pathogenesis system is that the bacterial pathogen (in this case, *P. aeruginosa* PA14) is used to infect a simple, genetically tractable host (in this case, *C. elegans*) under conditions in which many aspects of the host-pathogen interaction are thought to reflect those that occur during a human infection. Infections are traditionally performed using wild-type or mutant isolates of the host or pathogen. Screens have been carried out in our laboratory for pathogen mutants with reduced virulence and host mutants with enhanced susceptibility and enhanced resistance to pathogens (Esp and Erp phenotypes, respectively). However, the availability of genome-wide mutant libraries in both organisms (RNAi libraries in *C. elegans*, and the PA14 nonredundant library of transposon insertions described in this chapter) allows these screens for novel pathogenesis-related genes to be performed to saturation. In the case of the PA14 nonredundant set (PA14NR set), the library is designed such that mutants can be assayed individually or in pools by TraSH techniques. The interaction of host and pathogen can also be examined by using microarrays or proteomic techniques to analyze changes in the transcriptome or proteome of either organism as a consequence of infection. The use of these additional genomic tools can be coupled with the available mutant libraries; transcriptional or proteomic profiling of candidate virulence mutants compared with their wild-type parent has been used to identify putative downstream targets. The collected experimental data will be deposited in a publicly accessible database. Initially, the database will focus on processing phenotypic data for mutants in the PA14NR set library, but it can ultimately be expanded to incorporate additional types of data gathered by other genomic tools.

parate types of information are still in an early stage. Our goal is to provide a minimal framework for the initial archiving of raw data generated in our laboratory through the use of the mutant library. However, the true value of such a shared repository will be expanded as more and more researchers contribute not only to the contents, but also to the design and implementation of such a database. Thinking ahead and encouraging other researchers to contribute to

this discussion as the mutant library is disseminated (rather than after it is in wide use) will be essential to this communal effort.

ACKNOWLEDGMENTS
This work was funded by National Institutes of Health grants R01 AI064332 and U01 HL66678, Department of Energy grant DE-FG02-ER63445, and Cystic Fibrosis Foundation grant AUSUBE04V0.

REFERENCES

1. **Aballay, A., and F. M. Ausubel.** 2001. Programmed cell death mediated by ced-3 and ced-4 protects *Caenorhabditis elegans* from *Salmonella typhimurium*–mediated killing. *Proc. Natl. Acad. Sci. USA* **98:**2735–2739.

2. **Aballay, A., E. Drenkard, L. R. Hilbun, and F. M. Ausubel.** 2003. *Caenorhabditis elegans* innate immune response triggered by *Salmonella enterica* requires intact LPS and is mediated by a MAPK signaling pathway. *Curr. Biol.* **13:**47–52.

3. **Aballay, A., P. Yorgey, and F. M. Ausubel.** 2000. *Salmonella typhimurium* proliferates and establishes a persistent infection in the intestine of *Caenorhabditis elegans. Curr. Biol.* **10:**1539–1542.

4. **Badarinarayana, V., P. W. Estep III, J. Shendure, J. Edwards, S. Tavazoie, F. Lam, and G. M. Church.** 2001. Selection analyses of insertional mutants using subgenic-resolution arrays. *Nat. Biotechnol.* **19:**1060–1065.

5. **Barton, G. M., and R. Medzhitov.** 2003. Toll-like receptor signaling pathways. *Science* **300:**1524–1525.

6. **Begun, J., C. D. Sifri, S. Goldman, S. B. Calderwood, and F. M. Ausubel.** 2005. *Staphylococcus aureus* virulence factors identified by using a high-throughput *Caenorhabditis elegans*–killing model. *Infect. Immun.* **73:**872–877.

7. **Beutler, B., and M. Rehli.** 2002. Evolution of the TIR, tolls and TLRs: functional inferences from computational biology. *Curr. Top. Microbiol. Immunol.* **270:**1–21.

8. **Caetano-Anolles, G., and B. J. Bassam.** 1993. DNA amplification fingerprinting using arbitrary oligonucleotide primers. *Appl. Biochem. Biotechnol.* **42:**189–200.

9. **Caiazza, N. C., and G. A. O'Toole.** 2004. SadB is required for the transition from reversible to irreversible attachment during biofilm formation by *Pseudomonas aeruginosa* PA14. *J. Bacteriol.* **186:**4476–4485.

10. **Choe, K. M., T. Werner, S. Stoven, D. Hultmark, and K. V. Anderson.** 2002. Requirement for a peptidoglycan recognition protein (PGRP) in Relish activation and antibacterial immune responses in *Drosophila. Science* **296:**359–362.

11. **Coleman, F. T., S. Mueschenborn, G. Meluleni, C. Ray, V. J. Carey, S. O. Vargas, C. L. Cannon, F. M. Ausubel, and G. B. Pier.** 2003. Hypersusceptibility of cystic fibrosis mice to chronic *Pseudomonas aeruginosa* oropharyngeal colonization and lung infection. *Proc. Natl. Acad. Sci. USA* **100:**1949–1954.

12. **Couillault, C., N. Pujol, J. Reboul, L. Sabatier, J. F. Guichou, Y. Kohara, and J. J. Ewbank.** 2004. TLR-independent control of innate immunity in *Caenorhabditis elegans* by the TIR domain adaptor protein TIR-1, an ortholog of human SARM. *Nat. Immunol.* **5:**488–494.

13. **Darby, C., J. W. Hsu, N. Ghori, and S. Falkow.** 2002. *Caenorhabditis elegans*: plague bacteria biofilm blocks food intake. *Nature* **417:**243–244.

14. **Dong, C., R. J. Davis, and R. A. Flavell.** 2002. MAP kinases in the immune response. *Annu. Rev. Immunol.* **20:**55–72.

15. **Dunne, A., and L. A. O'Neill.** 2003. The interleukin-1 receptor/Toll-like receptor superfamily: signal transduction during inflammation and host defense. *Sci. STKE* **2003:**re3.

16. **Ernst, R. K., D. A. D'Argenio, J. K. Ichikawa, M. G. Bangera, S. Selgrade, J. L. Burns, P. Hiatt, K. McCoy, M. Brittnacher, A. Kas, D. H. Spencer, M. V. Olson, B. W. Ramsey, S. Lory, and S. I. Miller.** 2003. Genome mosaicism is conserved but not unique in *Pseudomonas aeruginosa* isolates from the airways of young children with cystic fibrosis. *Environ. Microbiol.* **5:**1341–1349.

17. **Fortune, S. M., A. Jaeger, D. A. Sarracino, M. R. Chase, C. M. Sassetti, D. R. Sherman, B. R. Bloom, and E. J. Rubin.** 2005. Mutually dependent secretion of proteins required for mycobacterial virulence. *Proc. Natl. Acad. Sci. USA* **102:**10676–10681.

18. **Fraser, A. G., R. S. Kamath, P. Zipperlen, M. Martinez-Campos, M. Sohrmann, and J. Ahringer.** 2000. Functional genomic analysis of *C. elegans* chromosome I by systematic RNA interference. *Nature* **408:**325–330.

19. **Garsin, D. A., C. D. Sifri, E. Mylonakis, X. Qin, K. V. Singh, B. E. Murray, S. B. Calderwood, and F. M. Ausubel.** 2001. A simple model host for identifying gram-positive virulence factors. *Proc. Natl. Acad. Sci. USA* **98:**10892–10897.

20. **Garsin, D. A., J. M. Villanueva, J. Begun, D. H. Kim, C. D. Sifri, S. B. Calderwood, G. Ruvkun, and F. M. Ausubel.** 2003. Long-lived *C. elegans* daf-2 mutants are resistant to bacterial pathogens. *Science* **300:**1921.

21. **Georgel, P., S. Naitza, C. Kappler, D. Ferrandon, D. Zachary, C. Swimmer, C. Kopczynski, G. Duyk, J. M. Reichhart, and J. A. Hoffmann.** 2001. *Drosophila* immune deficiency (IMD) is a death domain protein that activates antibacterial defense and can promote apoptosis. *Dev. Cell.* **1:**503–514.

22. **Gottar, M., V. Gobert, T. Michel, M. Belvin, G. Duyk, J. A. Hoffmann, D. Ferrandon, and J. Royet.** 2002. The *Drosophila* immune response against gram-negative bacteria is mediated by a peptidoglycan recognition protein. *Nature* **416:**640–644.

23. Hendrickson, E. L., J. Plotnikova, S. Maha-jan-Miklos, L. G. Rahme, and F. M. Ausubel. 2001. Differential roles of the *Pseudomonas aeruginosa* PA14 rpoN gene in pathogenicity in plants, nematodes, insects, and mice. *J. Bacteriol.* **183:** 7126–7134.

24. Hensel, M., J. E. Shea, C. Gleeson, M. D. Jones, E. Dalton, and D. W. Holden. 1995. Simultaneous identification of bacterial virulence genes by negative selection. *Science* **269:**400–403.

25. Hoffmann, J. A., and J. M. Reichhart. 2002. Drosophila innate immunity: an evolutionary perspective. *Nat. Immunol.* **3:**121–126.

26. Jacobs, M. A., A. Alwood, I. Thaipisuttikul, D. Spencer, E. Haugen, S. Ernst, O. Will, R. Kaul, C. Raymond, R. Levy, L. Chun-Rong, D. Guenthner, D. Bovee, M. V. Olson, and C. Manoil. 2003. Comprehensive transposon mutant library of *Pseudomonas aeruginosa. Proc. Natl. Acad. Sci. USA* **100:**14339–14344.

27. Jander, G., L. G. Rahme, and F. M. Ausubel. 2000. Positive correlation between virulence of *Pseudomonas aeruginosa* mutants in mice and insects. *J. Bacteriol.* **182:**3843–3845.

28. Janeway, C. A., Jr., and R. Medzhitov. 2002. Innate immune recognition. *Annu. Rev. Immunol.* **20:**197–216.

29. Kamath, R. S., and J. Ahringer. 2003. Genome-wide RNAi screening in *Caenorhabditis elegans. Methods* **30:**313–321.

30. Kamath, R. S., A. G. Fraser, Y. Dong, G. Poulin, R. Durbin, M. Gotta, A. Kanapin, N. Le Bot, S. Moreno, M. Sohrmann, D. P. Welchman, P. Zipperlen, and J. Ahringer. 2003. Systematic functional analysis of the *Caenorhabditis elegans* genome using RNAi. *Nature* **421:**231–237.

31. Kamath, R. S., M. Martinez-Campos, P. Zipperlen, A. G. Fraser, and J. Ahringer. 2001. Effectiveness of specific RNA-mediated interference through ingested double-stranded RNA in *Caenorhabditis elegans. Genome Biol.* **2:** RESEARCH0002.

32. Kim, C. C., E. A. Joyce, K. Chan, and S. Falkow. 2002. Improved analytical methods for microarray-based genome-composition analysis. *Genome Biol.* **3:**RESEARCH0065.

33. Kim, D. H., and F. M. Ausubel. 2005. Evolutionary perspectives on innate immunity from the study of *Caenorhabditis elegans. Curr. Opin. Immunol.* **17:**4–10.

34. Kim, D. H., R. Feinbaum, G. Alloing, F. E. Emerson, D. A. Garsin, H. Inoue, M. Tanaka-Hino, N. Hisamoto, K. Matsumoto, M. W. Tan, and F. M. Ausubel. 2002. A conserved p38 MAP kinase pathway in *Caenorhabditis elegans* innate immunity. *Science* **297:**623–626.

35. Kim, D. H., N. T. Liberati, T. Mizuno, H. Inoue, N. Hisamoto, K. Matsumoto, and F. M. Ausubel. 2004. Integration of *Caenorhabditis elegans* MAPK pathways mediating immunity and stress resistance by MEK-1 MAPK kinase and VHP-1 MAPK phosphatase. *Proc. Natl. Acad. Sci. USA* **101:**10990–10994.

36. Kuchma, S. L., J. P. Connolly, and G. A. O'Toole. 2005. A three-component regulatory system regulates biofilm maturation and type III secretion in *Pseudomonas aeruginosa. J. Bacteriol.* **187:**1441–1454.

37. Kulesekara, H., V. Lee, A. Brencic, N. Liberati, J. Urbach, S. Miyata, D. G. Lee, A. N. Neely, M. Hyodo, Y. Hayakawa, F. M. Ausubel, and S. Lory. 2006. Analysis of *Pseudomonas aeruginosa* diguanylate cyclases and phosphodiesterases reveals a role for bis-(3′-5′)-cyclic-GMP in virulence. *Proc. Natl. Acad. Sci. USA* **21:**2839–2844.

38. Kurz, C. L., S. Chauvet, E. Andres, M. Aurouze, I. Vallet, G. P. Michel, M. Uh, J. Celli, A. Filloux, S. De Bentzmann, I. Steinmetz, J. A. Hoffmann, B. B. Finlay, J. P. Gorvel, D. Ferrandon, and J. J. Ewbank. 2003. Virulence factors of the human opportunistic pathogen *Serratia marcescens* identified by in vivo screening. *EMBO J.* **22:**1451–1460.

39. Kurz, C. L., and J. J. Ewbank. 2003. *Caenorhabditis elegans*: an emerging genetic model for the study of innate immunity. *Nat. Rev. Genet.* **4:**380–390.

40. Kurz, C. L., and M. W. Tan. 2004. Regulation of aging and innate immunity in *C. elegans. Aging Cell* **3:**185–193.

41. Kyriakis, J. M., and J. Avruch. 2001. Mammalian mitogen-activated protein kinase signal transduction pathways activated by stress and inflammation. *Physiol. Rev.* **81:**807–869.

42. Labrousse, A., S. Chauvet, C. Couillault, C. L. Kurz, and J. J. Ewbank. 2000. *Caenorhabditis elegans* is a model host for *Salmonella typhimurium. Curr. Biol.* **10:**1543–1545.

43. Lampe, D. J., T. E. Grant, and H. M. Robertson. 1998. Factors affecting transposition of the *Himar1 mariner* transposon in vitro. *Genetics* **149:** 179–187.

44. Lau, G. W., B. C. Goumnerov, C. L. Walendziewicz, J. Hewitson, W. Xiao, S. Maha-jan-Miklos, R. G. Tompkins, L. A. Perkins, and L. G. Rahme. 2003. The *Drosophila melanogaster* toll pathway participates in resistance to infection by the gram-negative human pathogen *Pseudomonas aeruginosa. Infect. Immun.* **71:** 4059–4066.

45. Lemaitre, B., E. Kromer-Metzger, L. Michaut, E. Nicolas, M. Meister, P. Georgel,

J. M. Reichhart, and J. A. Hoffmann. 1995. A recessive mutation, immune deficiency (imd), defines two distinct control pathways in the *Drosophila* host defense. *Proc. Natl. Acad. Sci. USA* **92:**9465–9469.

46. Liberati, N. T., K. A. Fitzgerald, D. H. Kim, R. Feinbaum, D. T. Golenbock, and F. M. Ausubel. 2004. Requirement for a conserved Toll/interleukin-1 resistance domain protein in the *Caenorhabditis elegans* immune response. *Proc. Natl. Acad. Sci. USA* **101:**6593–6598.

47. Liberati, N. T., J. M. Urbach, S. Miyata, D. G. Lee, E. Drenkard, G. Wu, J. Villanueva, T. Wei, and F. M. Ausubel. 2006. An ordered, nonredundant library of *Pseudomonas aeruginosa* strain PA14 transposon insertion mutants. *Proc. Natl. Acad. Sci. USA.* **103:**2833–2838.

48. Mahajan-Miklos, S., M. W. Tan, L. G. Rahme, and F. M. Ausubel. 1999. Molecular mechanisms of bacterial virulence elucidated using a *Pseudomonas aeruginosa–Caenorhabditis elegans* pathogenesis model. *Cell* **96:**47–56.

49. Mallo, G. V., C. L. Kurz, C. Couillault, N. Pujol, S. Granjeaud, Y. Kohara, and J. J. Ewbank. 2002. Inducible antibacterial defense system in *C. elegans. Curr. Biol.* **12:**1209–1214.

50. Medzhitov, R., and C. Janeway, Jr. 2000. Innate immune recognition: mechanisms and pathways. *Immunol. Rev.* **173:**89–97.

51. Medzhitov, R., and C. Janeway, Jr. 2000. The Toll receptor family and microbial recognition. *Trends Microbiol.* **8:**452–456.

52. Medzhitov, R., and C. A. Janeway, Jr. 2002. Decoding the patterns of self and nonself by the innate immune system. *Science* **296:**298–300.

53. Medzhitov, R., and C. A. Janeway, Jr. 2000. How does the immune system distinguish self from nonself? *Semin. Immunol.* **12:**185–188; discussion 257–344.

54. Millet, A. C., and J. J. Ewbank. 2004. Immunity in *Caenorhabditis elegans. Curr. Opin. Immunol.* **16:**4–9.

55. Miyata, S., M. Casey, D. W. Frank, F. M. Ausubel, and E. Drenkard. 2003. Use of the *Galleria mellonella* caterpillar as a model host to study the role of the type III secretion system in *Pseudomonas aeruginosa* pathogenesis. *Infect. Immun.* **71:**2404–2413.

56. Mylonakis, E., F. M. Ausubel, J. R. Perfect, J. Heitman, and S. B. Calderwood. 2002. Killing of *Caenorhabditis elegans* by *Cryptococcus neoformans* as a model of yeast pathogenesis. *Proc. Natl. Acad. Sci. USA* **99:**15675–15680.

57. Mylonakis, E., A. Idnurm, R. Moreno, J. El Khoury, J. B. Rottman, F. M. Ausubel, J. Heitman, and S. B. Calderwood. 2004. *Cryptococcus neoformans* Kin1 protein kinase homolog,

identified through a *Caenorhabditis elegans* screen, promotes virulence in mammals. *Mol. Microbiol.* **54:**407–419.

58. O'Toole, G. A., and R. Kolter. 1998. Flagellar and twitching motility are necessary for *Pseudomonas aeruginosa* biofilm development. *Mol. Microbiol.* **30:**295–304.

59. O'Toole, G. A., and R. Kolter. 1998. Initiation of biofilm formation in *Pseudomonas fluorescens* WCS365 proceeds via multiple, convergent signalling pathways: a genetic analysis. *Mol. Microbiol.* **28:**449–461.

60. Price-Whelan, A., L. E. Dietrich, and D. K. Newman. 2006. Rethinking "secondary" metabolism: physiological roles for phenazine antibiotics. *Nat. Chem. Biol.* **2:**71–78.

61. Pujol, N., E. M. Link, L. X. Liu, C. L. Kurz, G. Alloing, M. W. Tan, K. P. Ray, R. Solari, C. D. Johnson, and J. J. Ewbank. 2001. A reverse genetic analysis of components of the Toll signaling pathway in *Caenorhabditis elegans. Curr. Biol.* **11:**809–821.

62. Pukatzki, S., R. H. Kessin, and J. J. Mekalanos. 2002. The human pathogen *Pseudomonas aeruginosa* utilizes conserved virulence pathways to infect the social amoeba *Dictyostelium discoideum. Proc. Natl. Acad. Sci. USA* **99:**3159–3164.

63. Pukatzki, S., A. T. Ma, D. Sturtevant, B. Krastins, D. Sarracino, W. C. Nelson, J. F. Heidelberg, and J. J. Mekalanos. 2006. Identification of a conserved bacterial protein secretion system in *Vibrio cholerae* using the *Dictyostelium* host model system. *Proc. Natl. Acad. Sci. USA* **103:**1528–1533.

64. Qureshi, S. T., and R. Medzhitov. 2003. Toll-like receptors and their role in experimental models of microbial infection. *Genes Immun.* **4:**87–94.

65. Rahme, L. G., E. J. Stevens, S. F. Wolfort, J. Shao, R. G. Tompkins, and F. M. Ausubel. 1995. Common virulence factors for bacterial pathogenicity in plants and animals. *Science* **268:**1899–1902.

66. Ramet, M., P. Manfruelli, A. Pearson, B. Mathey-Prevot, and R. A. Ezekowitz. 2002. Functional genomic analysis of phagocytosis and identification of a *Drosophila* receptor for *E. coli. Nature* **416:**644–648.

67. Raskin, D. M., R. Seshadri, S. U. Pukatzki, and J. J. Mekalanos. 2006. Bacterial genomics and pathogen evolution. *Cell* **124:**703–714.

68. Ratledge, C., and L. G. Dover. 2000. Iron metabolism in pathogenic bacteria. *Annu. Rev. Microbiol.* **54:**881–941.

69. Rubin, E. J., B. J. Akerley, V. N. Novik, D. J. Lampe, R. N. Husson, and J. J. Mekalanos.

1999. In vivo transposition of *mariner*-based elements in enteric bacteria and mycobacteria. *Proc. Natl. Acad. Sci. USA* **96:**1645–1650.

70. **Saenz, H. L., and C. Dehio.** 2005. Signature-tagged mutagenesis: technical advances in a negative selection method for virulence gene identification. *Curr. Opin. Microbiol.* **8:**612–619.

71. **Sassetti, C. M., D. H. Boyd, and E. J. Rubin.** 2001. Comprehensive identification of conditionally essential genes in mycobacteria. *Proc. Natl. Acad. Sci. USA* **98:**12712–12717.

72. **Shah, P. H., R. C. MacFarlane, D. Bhattacharya, J. C. Matese, J. Demeter, S. E. Stroup, and U. Singh.** 2005. Comparative genomic hybridizations of *Entamoeba* strains reveal unique genetic fingerprints that correlate with virulence. *Eukaryot. Cell* **4:**504–515.

73. **Shelburne, S. A., III, P. Sumby, I. Sitkiewicz, C. Granville, F. R. DeLeo, and J. M. Musser.** 2005. Central role of a bacterial two-component gene regulatory system of previously unknown function in pathogen persistence in human saliva. *Proc. Natl. Acad. Sci. USA* **102:**16037–16042.

74. **Sifri, C. D., J. Begun, F. M. Ausubel, and S. B. Calderwood.** 2003. *Caenorhabditis elegans* as a model host for *Staphylococcus aureus* pathogenesis. *Infect. Immun.* **71:**2208–2217.

75. **Sifri, C. D., E. Mylonakis, K. V. Singh, X. Qin, D. A. Garsin, B. E. Murray, F. M. Ausubel, and S. B. Calderwood.** 2002. Virulence effect of *Enterococcus faecalis* protease genes and the quorum-sensing locus fsr in *Caenorhabditis elegans* and mice. *Infect. Immun.* **70:**5647–5650.

76. **Singh, U., P. H. Shah, and R. C. MacFarlane.** 2004. DNA content analysis on microarrays. *Methods Mol. Biol.* **270:**237–248.

77. **Spellman, P. T., M. Miller, J. Stewart, C. Troup, U. Sarkans, S. Chervitz, D. Bernhart, G. Sherlock, C. Ball, M. Lepage, M. Swiatek, W. L. Marks, J. Goncalves, S. Markel, D. Iordan, M. Shojatalab, A. Pizarro, J. White, R. Hubley, E. Deutsch, M. Senger, B. J. Aronow, A. Robinson, D. Bassett, C. J. Stoeckert, Jr., and A. Brazma.** 2002. Design and implementation of microarray gene expression markup language (MAGE-ML). *Genome Biol.* **3:**RESEARCH0046.

78. **Stover, C. K., X. Q. Pham, A. L. Erwin, S. D. Mizoguchi, P. Warrener, M. J. Hickey, F. S. Brinkman, W. O. Hufnagle, D. J. Kowalik, M. Lagrou, R. L. Garber, L. Goltry, E. Tolentino, S. Westbrock-Wadman, Y. Yuan, L. L. Brody, S. N. Coulter, K. R. Folger, A. Kas, K. Larbig, R. Lim, K. Smith, D. Spencer,** G. K. Wong, Z. Wu, I. T. Paulsen, J. Reizer, M. H. Saier, R. E. Hancock, S. Lory, and M. V. Olson. 2000. Complete genome sequence of *Pseudomonas aeruginosa* PA01, an opportunistic pathogen. *Nature* **406:**959–964.

79. **Takeda, K., and S. Akira.** 2003. Toll receptors and pathogen resistance. *Cell. Microbiol.* **5:**143–153.

80. **Tan, M. W., S. Mahajan-Miklos, and F. M. Ausubel.** 1999. Killing of *Caenorhabditis elegans* by *Pseudomonas aeruginosa* used to model mammalian bacterial pathogenesis. *Proc. Natl. Acad. Sci. USA* **96:**715–720.

81. **Tan, M. W., L. G. Rahme, J. A. Sternberg, R. G. Tompkins, and F. M. Ausubel.** 1999. *Pseudomonas aeruginosa* killing of *Caenorhabditis elegans* used to identify *P. aeruginosa* virulence factors. *Proc. Natl. Acad. Sci. USA* **96:**2408–2413.

82. **Tenor, J., B. A. McCormick, F. M. Ausubel, and A. Aballay.** 2004. *Caenorhabditis elegans*–based screen identifies *Salmonella* virulence factors required for conserved host-pathogen interactions. *Curr. Biol.* **14:**1018–1024.

83. **Thompson, L. S., J. S. Webb, S. A. Rice, and S. Kjelleberg.** 2003. The alternative sigma factor RpoN regulates the quorum sensing gene rhlI in *Pseudomonas aeruginosa. FEMS Microbiol. Lett.* **220:**187–195.

84. **Timmons, L., D. L. Court, and A. Fire.** 2001. Ingestion of bacterially expressed dsRNAs can produce specific and potent genetic interference in *Caenorhabditis elegans. Gene* **263:**103–112.

85. **Timmons, L., and A. Fire.** 1998. Specific interference by ingested dsRNA. *Nature* **395:**854.

86. **Whitchurch, C. B., T. E. Erova, J. A. Emery, J. L. Sargent, J. M. Harris, A. B. Semmler, M. D. Young, J. S. Mattick, and D. J. Wozniak.** 2002. Phosphorylation of the *Pseudomonas aeruginosa* response regulator AlgR is essential for type IV fimbria-mediated twitching motility. *J. Bacteriol.* **184:**4544–4554.

87. **Wolfgang, M. C., B. R. Kulasekara, X. Liang, D. Boyd, K. Wu, Q. Yang, C. G. Miyada, and S. Lory.** 2003. Conservation of genome content and virulence determinants among clinical and environmental isolates of *Pseudomonas aeruginosa. Proc. Natl. Acad. Sci. USA* **100:**8484–8489.

88. **Wong, S. M., and J. J. Mekalanos.** 2000. Genetic footprinting with mariner-based transposition in *Pseudomonas aeruginosa. Proc. Natl. Acad. Sci. USA* **97:**10191–10196.

89. **Zhang, G., and S. Ghosh.** 2001. Toll-like receptor–mediated NF-kappaB activation: a phylogenetically conserved paradigm in innate immunity. *J. Clin. Invest.* **107:**13–19.

PATHOGENOMICS OF BACTERIAL BIOTHREAT AGENTS

Timothy D. Read and Brendan Thomason

10

INTRODUCTION

Biodefense and Microbial Genomics Collide

The 2001 anthrax bioterror attacks served to remind both decision makers and the general public of the dangers of infectious disease agents deliberately targeted against humans. In the years after this act of terrorism there has been a massive increase in biodefense research funding in the United States (201, 202). Coincidental with this increased focus on biothreat agents, there has been a revolution in microbial genomics. Because the event that is conventionally considered the start of the bacterial genomic era, the publication of the *Haemophilus influenzae* sequence in 1995 (67), genomics has spread from the sequencing of model prokaryotes to the acquisition of almost all major bacterial and viral pathogens, their near neighbors, and bacteria important in industry and the environment. Close on the heels of genome sequencing have come functional genomics approaches, such as microarray expression analysis and genome-based proteomics.

The idea of "pathogenomics" ultimately encompasses every aspect of microbial pathogenesis. Before genome sequences for biothreat pathogens became available, the types of whole-genome approaches applied today were unimaginable to investigators. In a few years, it will be similarly unimaginable not to perform biology without reference to multiple whole-genome sequences. Researchers will doubtless take for granted existing genomic and functional genomic information and the ease of acquiring new information. Our current craze for designating an "-omics" suffix on an experiment will be redundant. This chapter therefore revolves around a period of transition in the study of pathogen biology in general and of biothreat agents in particular, much of it stemming from the recent biodefense funding impetus.

What Are Biothreat Agents?

There is no easy definition of what makes a biothreat agent. Lists of potential biothreat organisms created by federal agencies such as the Centers for Disease Control and Prevention (CDC), National Institutes of Allergy and Infectious Disease, and the United States Department of Agriculture include a large number of bacteria, fungi, and viruses, reflecting the fact that many common infectious disease

Timothy D. Read and Brendan Thomason   Genomics Group, Biological Defense Research Directorate, Naval Medical Research Center, Rockville, Maryland, 20852.

organisms are capable of causing loss of life, economic damage, and terrorism of the population if deliberately introduced in an effective manner. One of the best-documented episodes of bioterrorism in recent times involved deliberate spiking of the enteric pathogen *Salmonella enterica* serovar Typhimurium in a salad bar (223). (Whether accidental or not, the more virulent S. *enterica* serovar Typhi species was not used.) However, we have chosen to focus this chapter on a smaller group of bacteria causing human infections chosen by military offensive programs between World War I and the end of the cold war. These include *Bacillus anthracis, Yersinia pestis, Burkholderia mallei* (and *B. pseudomallei*), *Brucella melitensis, Coxiella burnetii*, and *Francisella tularensis* (Table 1). This group (particularly the first two listed) has been the principal beneficiary of the postgenomic funding increases since 2001. In order to focus the chapter, other important groups of biothreat agents cannot be discussed at length. These include highly infectious viruses with few countermeasures such as the hemorrhagic fever–causing Ebola and Marburg viruses, and smallpox, important agents that primarily are targeted at livestock (e.g., foot and mouth virus) or crops (e.g., wheat rust), or toxins of biological derivation, such as botulinum toxin and ricin.

Even though the small group of bacterial species that is the focus of the chapter is phylogenetically and phenotypically diverse, with few common features, there are commonalities. The classic bacterial biothreat agents tend to cause lethal or disabling systemic disease at low infectious doses. They are aerosolizable, allowing infection through inhalation (which for many is not the common route of infection in nature). A preponderance of these organisms consists primarily of zoonotic pathogens not specifically adapted to humans, as witnessed by their high levels of virulence typical of "unadapted" pathogenicity on a nonadapted host predicted in many models. Also, the fact that these organisms are not common causes of human disease in the Western world means that specific countermeasures have not yet been developed as a result of lack of both political interest and commercial demand. Another important distinction between classical agents mentioned above is one of regulation. Legal and institutional restrictions on movement of select agents such as *Y. pestis, B. anthracis*, and smallpox and the confinement of infection studies to biosafety level (BSL) 3 laboratories reduce the number of scientists that can work on the organisms directly. In several ways, this increases the value of genome sequences for researchers, particularly those without access to the live organism. Specific gene targets can be screened and selected in silico, cloned from PCR amplification of genomic DNA, and expressed in a permissive host to be analyzed.

The State of the Art of Functional Genomics Technologies for Biothreat Agents

By the time of writing (May 2006), all the primary bacterial biothreat agents have at least one closed, published genome sequence project (Table 2). Whole-genome shotgun sequencing using Sanger chain termination chemistry (198) has been the primary technology for sequence acquisition used in these studies. The ever-decreasing cost of sequencing at genome centers and the anticipated emergence of new ultra-low-cost sequencing technologies (207) will lead to many more strains of these agents and their near neighbors being sequenced in the near future. These genome sequences are, and will be, the starting point for further exploration using "postgenomic technologies." DNA microarrays (24) have been used to leverage the sequence information for RNA transcript level studies (82, 139) and for comparative genome hybridization (CGH) to detect genes missing from closely related strains (22, 87, 168, 186). High-density oligonucleotide microarrays, currently allowing as many as 2.4 million features on a single chip, can greatly add to the precision of transcript mapping when used in a whole-genome tiling design (194). The high-density microarrays also permit acquisition of sequences of closely related strains through resequencing

TABLE 1 NIAID biothreat agents, classes A to C[a]

Organism	Gram stain	Taxonomy	Disease caused	NIAID priority code	Validated or potential biological weapon	Live organism known to infect mammals via aerosol
*Francisella tularensis**	Negative	*γ-Proteobacteria*	Tularemia	A	Validated	Yes
*Yersinia pestis**	Negative	*γ-Proteobacteria*	Plague	A	Validated	Yes
*Bacillus anthracis**	Positive	*Firmicutes*	Anthrax	A	Validated	Yes
Clostridium botulinum	Positive	*Firmicutes*	Botulism	A		
*Brucella melitensis**	Negative	*α-Proteobacteria*	Brucellosis	B	Validated	Yes
Brucella melitensis biovar Abortus*	Negative	*α-Proteobacteria*	Brucellosis	B	Validated	Yes
Brucella melitensis biovar Canis	Negative	*α-Proteobacteria*	Brucellosis	B		
Brucella melitensis biovar Suis*	Negative	*α-Proteobacteria*	Brucellosis	B	Validated	Yes
Rickettsia prowazekii	Negative	*α-Proteobacteria*	Typhus fever	B	Validated	Yes
*Burkholderia mallei**	Negative	*β-Proteobacteria*	Glanders	B	Validated	Yes
*Burkholderia pseudomallei**	Negative	*β-Proteobacteria*	Melioidosis	B	Validated	Yes
Campylobacter jejuni	Negative	*ε-Proteobacteria*		B		
*Coxiella burnetii**	Negative	*γ-Proteobacteria*	Q fever	B	Validated	Yes
Escherichia coli	Negative	*γ-Proteobacteria*		B		
Escherichia coli O157:H7	Negative	*γ-Proteobacteria*	EHEC	B		
Salmonella enterica serovar Typhi	Negative	*γ-Proteobacteria*	Typhoid fever	B	Potential	
Shigella boydii	Negative	*γ-Proteobacteria*	Shigellosis	B		
Shigella dysenteriae	Negative	*γ-Proteobacteria*	Shigellosis	B	Validated	
Shigella sonnei	Negative	*γ-Proteobacteria*	Shigellosis	B		
Vibrio cholerae	Negative	*γ-Proteobacteria*	Cholera	B	Validated	
Vibrio cholerae O139	Negative	*γ-Proteobacteria*	Cholera	B	Validated	
Vibrio mimicus	Negative	*γ-Proteobacteria*		B		
Vibrio parahaemolyticus	Negative	*γ-Proteobacteria*		B		
Vibrio vulnificans	Negative	*γ-Proteobacteria*		B		
Vibrio enterocolitica	Negative	*γ-Proteobacteria*		B		
Listeria monocytogenes	Positive	*Firmicutes*	Listeriosis	B		
Rickettsia conorii	Negative	*α-Proteobacteria*		C		Yes
Rickettsia rickettsii	Negative	*α-Proteobacteria*	Rocky Mountain spotted fever	C	Potential	Yes
Mycobacterium tuberculosis	Positive	Actinobacteria	Tuberculosis	C		Yes

[a]Table also includes whether the biological weapon is validated or potential (58). Information is based on supplemental tables from Ecker et al. (58). Species marked with an asterisk are those that are described in detail in this chapter. NIAID, National Institutes of Allergy and Infectious Disease.

TABLE 2 Complete genomes of biological weapons agents and selected close relatives mentioned in this chapter[a]

Organism (strain)	Chromosome size in Mbp (RefSeq accession no.)	Extrachromosomal element sizes in Mbp (RefSeq accession no.)	G+C%[b]	Reference
B. anthracis*				
Ames (pXO1+, pXO2+)*	5.23 (NC_003997)		35	194
Ames 0581 (Ames ancestor)*	5.23 (NC_007530)	pXO1: 0.18 (NC_007322) pXO2: 0.09 (NC_007323)	35	Unpublished
Sterne (pXO2-)*	5.23 (NC_005945)	pXO1	35	Unpublished
B. cereus group*				
B. cereus G9241*	5.94 (draft) (NZ_AAEK00000000)	pBC218: 0.22 pBCXO1: 0.19	35	89
B. cereus ATCC 14579*	5.41 (NC_004722)	pClin15: 0.0015 (NC_004721)	35	99
B. cereus ATCC 10987*	5.24 (NC_003909)	pBC10987: 0.21 (NC_005707)	35	184
B. cereus E33L*	5.30 (NC_006274)	pZK467: 0.47 (NC_007103) pZK5: 0.005 (NC_007104) pZK54: 0.0053 (NC_007105) pZK8: 0.008 (NC_007106) pZK9: 0.009 (NC_007107)	35	Unpublished
B. thuringiensis Konkukian 97-27*	5.24 (NC_005957)	pBT9727: 0.08 (NC_006578)	35	Unpublished
Brucella melitensis				
Biovar Abortus strain 9-941	Chr I: 2.14 (NC_006932) Chr II: 1.16 (NC_006933)		57	80
Biovar Abortus strain 2308	Chr I: 2.12 (NC_007618) Chr II: 1.16 (NC_007624)		57	34
Biovar Melitensis strain 16M	Chr I: 2.12 (NC_003317) Chr II: 1.18 (NC_003318)		57	46
Biovar Suis strain 1330	Chr I: 2.11 (NC_004310) Chr II: 1.21 (NC_004311)		57	172
Burkholderia mallei/B. pseudomallei*				
B. mallei ATCC 23344*	Chr I: 3.51 (NC_006348) Chr II: 2.33 (NC_006349)		68	164
B. pseudomallei 1710b*	Chr I: 4.13 (NC_007434) Chr II: 3.18 (NC_007435)		68	Unpublished
B. pseudomallei K92643*	Chr I: 4.07 (NC_006350) Chr II: 3.17 (NC_006351)		68	90
Burkholderia thailandensis*				
B. thailandensis E264*	Chr I: 3.81 (NC_007651) Chr II: 3.81 (NC_007650)		68	114
Coxiella burnetii				
C. burnetii RSA 493	2.00 (NC_002971)	pQpH1: 0.027 (NC_004704)	42	205
Francisella tularensis*				
Subspecies holarctica*	1.90 (NC_007880)		32	Unpublished
Subspecies tularensis SchuS 4*	1.90 (NC_006570)		32	121

(Continued)

TABLE 2 *Continued*

Organism (strain)	Chromosome size in Mbp (RefSeq accession no.)	Extrachromosomal element sizes in Mbp (RefSeq accession no.)	G+C%[b]	Reference
Yersinia pestis				
Biovar Orientalis strain CO92	4.65 (NC_003143)	pCD1: 0.07 (NC_003131) pPCP1: 0.009 (NC_003132) pMT1: 0.096 (NC_003134)	47	171
Biovar Mediaevalis strain 91001	4.59 (NC_005810)	pCD1: 0.07 (NC_005813) pPCP1: 0.009 (NC_005816) pMT1: 0.106 (NC_005815) pCRY: 0.021 (NC_005814)	47	212
Biovar Mediaevalis strain KIM	4.60 (NC_4088)	pMT1: 0.101 (NC_004838)	47	48
Yersinia pseudotuberculosis				
IP32953	4.74 (NC_006155)	pYV: 0.068 (NC_006153) pYptb32953: 0.027 (NC_006154)	47	33

[a]Except for G9241, only completed genome sequences are listed (G9241 chromosome is a draft shotgun sequence; the plasmids are complete). Data from NCBI genomes page, April 2006. Species marked with an asterisk are those that are described in detail in this chapter.
[b]Average for all molecules.

(241) and the detection of single nucleotide polymorphisms (SNPs) and other small differences between strains that might have distinct phenotypes, such as drug resistance (4). Whole-genome sequences also play a vital role in proteomics, facilitating identification of proteins visible in two-dimensional gel electrophoresis and making possible mass spectroscopy–based shotgun approaches to proteome characterization such as multidimension protein identification technology (139, 150). Expression of large numbers of proteins from candidate genomes (54) and arraying them on protein microarrays are now possible (82). Traditional genetic approaches to bacterial gene functional characterization, such as mutagenesis, have also been upgraded in the light of genome sequence availability. As mentioned earlier, it is now possible to trace the location of single mutations in genomes (4). Methods developed for identification of genes necessary for bacterial survival and colonization in vivo, such as in vivo expression technology (143) and signature-tagged mutagenesis (86), can be performed much more efficiently when a whole sequence is available. In addition to experimen-

tal postgenomics approaches, there is increasing use of genomic data sets for in silico bioinformatic prediction, with applications varying from flux-based metabolic reconstructions (59) to prediction of highly expressed genes (106), to prioritizing targets from antimicrobial drug development (178) and vaccines (9). Although the specifics of these methodologies are beyond the scope of this chapter, the important point is that these models can generate hypotheses about pathogenesis that the researcher can test experimentally.

PATHOGENOMIC THEMES IN BIODEFENSE

In discussing the genome biology of classical biothreat pathogens, several common themes emerge. In this section, we will provide an overview of these ideas before launching into the specific pathogenomics of each agent. Almost all of these ideas originate in fields of study outside of biodefense. The study of genomics emphasizes the similarities of the major themes in biology and allows advances that occur in one field to be quickly transferred to others.

Pathogenicity Can Be Enhanced by Both Gains and Losses from the Genome

The enhanced pathogenicity of biothreat agents ultimately stems from changes in their genome sequences that distinguish them from close, less pathogenic neighbors. Patterns emerging from a general study of the field of bacterial genomics can be applied to learn more about the pathogenicity of these bacteria (234). Fundamentally, it is being recognized that bacterial genomes are in dynamic equilibrium between diversifying their gene content to perform better in their current niche and to invade new niches and reducing their genome size for the purposes of streamlining and, in certain cases, enhancing their survival by ditching genes (124); both gains and losses in the genome sequence have been shown to affect the virulence of biothreat agents.

The enormity of the impact of horizontal gene transfer (HGT) on the genome content of bacterial species is gradually coming to light (132, 170). The presence of key virulence systems on extrachromosomal elements is a feature of both *Y. pestis* (e.g., type III secretion genes on plasmid pCD1) and *B. anthracis* (e.g., lethal toxin on plasmid pXO1 and capsule genes on pXO2 [221]). *Y. pestis*, *B. pseudomallei*, *B. melitensis* biovar Suis, and *F. tularensis* have genomic islands (78, 79) that contain potential virulence genes in their chromosomes (90, 121, 123, 171, 172). In the case of the *Y. pestis* high-pathogenicity island region, virulence genes are on a cassette that can be transferred to other enterobacteria by site-specific recombination (200). Often, these genomic islands turn out to be prophages that are either functional and can make phage particles or defective due to gene decay. Easily identifiable prophages may make up more than 20% of pathogen genomes (18), although the true number of genes imported into genomes by phages may actually be higher (132). One instance of this is exemplified by the genes encoding botulinum toxin of *Clostridium botulinum*, which were introduced to the bacterium on a phage (26).

Although HGT seems to be the dominant mechanism of acquiring new functions in bacteria, gene duplication can also be important. Duplicated genes can serve either to raise the copy number of genes particularly important to pathogenesis or, through subsequent mutation, to allow the generation of new substrate specificities. For instance, there is a duplicated region in the *F. tularensis* SchuS4 genome that reduces the growth rate of the bacterium within macrophages only when both copies are deleted (121, 161).

Given that the acquisition of virulence genes by biothreat agents has been so visible by comparative genome analysis, the importance of genome reduction for development of pathogenesis is sometimes overlooked. Reduction of genome sizes of endosymbionts or intracellular pathogens relative to their free-living relatives has been consistently found in bacteria (195, 226). The process can broadly be seen occurring in two steps. First there is a rapid gene loss associated with a "facultative intracellular" pathogen niche (18). This involves accumulation of mobile DNA, pseudogenes, and deletions over a relatively short time (in the evolutionary sense). The most extreme example of this process yet seen is the leprosy-causing bacterium, *Mycobacterium leprae* (40), where more than half the genes in the 3.27 Mb genome have acquired loss-of-function mutations. Second, after the rapid gene loss, the organism achieves a much reduced genome size with little mobile or repetitive DNA and can survive only as an obligate intracellular bacterium, with little genomic change, for long periods of time (195, 226) (although we are finding now that there may be DNA exchange even between very reduced genomes [147, 166]). In the context of these findings, it is significant that *Y. pestis*, *F. tularensis*, *B. mallei*, *B. pseudomallei*, *C. burnetii*, *Brucella* spp., and to an extent *B. anthracis* all show distinctive symptoms of the first stages of genome reduction.

A question of interest regards the selective pressures that exist that drive genome reduction for a facultative intracellular pathogen. To a great extent, the changes in the genome may reflect transition from survival in the external environment to an exclusive parasitic relationship with the host, leading to an ability to scavenge host-derived metabolic intermediates and the requirement to respond to a more fixed and smaller range of stimuli. For example, all the pathogenic brucellae have lost completely the metabolic pathways for glycogen, biotin, NAD, and choline (34). This "metabolic streamlining" is a general feature of small genomes; larger genomes are relatively richer in genes for signal transduction, transcription factors, and motility (118).

Another type of selection pressure favoring loss of function mutations is "pathoadaption" (45): enhancements to virulence through loss of gene function. These changes may be highly significant for pathogenesis but not obvious at all from initial examination of the genomes. One example that has been discovered recently is the arabinose assimilation operon in *B. pseudomallei* and *B. mallei* (159). Here, comparative genomics revealed the deletion of a nine-gene operon for utilization of L-arabinose as the sole carbon source in multiple strains of *B. pseudomallei* that was present in a nonpathogenic relative, *B. thailandensis.* Introduction of the entire arabinose operon *in B. pseudomallei* 406e decreased the pathogenicity for hamsters.

As our understanding grows regarding genome modification effects on biothreat-pathogen evolution, more parallels in the mechanisms of pathogenesis may become apparent. One significant similarity of many of these biothreat pathogens of diverse origin is the absence of functional flagellar machinery. *Rickettsia, Coxiella,* and *Francisella* lack flagellum and motility genes; *B. anthracis, Y. pestis, B. pseudomallei, B. mallei,* and the pathogenic brucellae are all nonmotile, yet still code for flagellum biosynthesis genes. Because motility is vital for survival for many bacteria, and for some (e.g., *Helicobacter*) is an important virulence attribute (157), the loss of this function must be significant. At the present time, it is still uncertain whether this is a case of parallel evolution of metabolic streamlining (the flagellum is a structure that requires a lot of energy to maintain and may not be vital for intracellular pathogenesis) or a pathoadaptive mutation (the flagellum may be a primary target for the host immune system). Possibly, there is selection pressure for deletion for both reasons.

Comparative Genomics: the Importance of Biothreat Pathogen Near Neighbors

As outlined above, comparative genomic analysis between pathogens can lead to generalizations about virulence that could be translated into broad-range countermeasures. Equally important is comparative analysis of each pathogen with genomes of its phylogenetic near neighbors. Sequencing of organisms very similar to biothreat pathogens but lacking the enhanced pathogenicity can achieve several important goals. The most obvious is to identify genes that have been acquired by highly pathogenic strains by recent HGT events. In addition, the importance of putative virulence genes in biothreat pathogens can be assessed by comparative analysis. When the *B. anthracis* genome was sequenced (186), several novel chromosomal genes were identified as encoding proteins potentially involved in virulence such as anthrolysin-O, a cholesterol-dependent cytolysin and an enhancin-like protease. However, sequencing of several near neighbors in the *B. cereus* group (99, 183, 184) and microarray-based CGH (186) revealed that these genes were present in genomes not associated with anthrax-like pathogenesis. Anthrolysin-O and enhancin may play a role in anthrax pathogenesis, but they are not unique to *B. anthracis*. In fact, comparative analysis suggests that, aside from prophages, there are only a very few chromosomal genes not found in the other five genomes sequenced to date (99, 183, 184); most unique regions of *B. anthracis* turn out to be prophages (186).

Close-relative genome sequences also enhance the detection of pseudogenes, particularly those cases where the disrupted gene product lacks a known function. Several bioinformatic approaches for pseudogene detection based on comparative genomics have been published recently (133, 140, 177), and the number of pseudogenes found in genomes has been increasing as new close-neighbor genome projects are completed. (For instance, the *Y. pseudotuberculosis* genome resulted in a doubling of the number of pseudogenes found in *Y. pestis* [33].) In the search for pathoadaptive mutations, these could be valuable new leads.

The genome sequence of near neighbors can reveal surprising parallel evolution of pathogenesis. In one case, *B. cereus* G9241, a strain associated with an unusual fulminate case of pneumonia in humans (155), was found to contain sequences similar to the *B. anthracis* protective antigen genes (*pag*) (89). In addition, genome sequencing revealed the presence of a plasmid almost identical to the *B. anthracis* virulence plasmid pXO1 with an intact tripartite lethal toxin gene cluster (89). Although no *B. anthracis*–like pXO2 plasmid was detected, the G9241 strain contained another large plasmid with a polysaccharide capsule cluster and, completely unexpectedly, another cluster of three genes encoding proteins with distant similarity to the lethal toxin subunits. The finding that G9241 lethality for A/J mice matched the enhanced level of *B. anthracis* suggests a similar type of infectious etiology. However, the analysis of G9241 raised key questions about *B. anthracis* pathogenicity, such as if G9241 can cause a disease similar to anthrax, why has it only just now been discovered? The molecular phylogeny of pXO1 shows that the G9241 plasmid was not recently acquired from *B. anthracis*. Also, if G9241 can cause an infection of the same severity as *B. anthracis*, what does this say about the role of the pXO2-encoded poly D-glutamic acid capsule? Perhaps only the lethal toxin is key for anthrax toxicity and the presence of antiphagocytic capsules of poly–amino acid or polysaccharide capsule will suffice as a secondary factor.

Microarray-based CGH studies have been used to rapidly examine a large number of closely related strains than could otherwise have been achieved through comparative sequencing (22, 87, 168, 182, 186). In most cases, the same microarray design used for expression analysis can also be used for genomic DNA hybridizations. CGH data have been used to gauge the degree of distance of the nonpathogens based on the number of shared genes and also a sense of the genes that may be unique to the pathogen. In a study of 22 *Y. pestis* and 10 *Y. pseudotuberculosis* strains hybridized against a glass slide array with 4,221 coding sequence of *Y. pestis* CO92 (171), only 11 loci were found to be absent in all the *Y. pseudotuberculosis*.

Near-neighbor genomes serve as the starting point for functional analysis of these organisms. For biodefense studies, where the virulence of the pathogens involved and the regulatory and ethical issues that surround experimentation serve to make much functional genomics work difficult and expensive, less virulent near neighbors can be invaluable as model organisms.

Genome Sequences Have Revolutionized DNA-Based Genotyping and Detection Methods

The genome sequences of biothreat pathogens have led to the large-scale improvement of DNA-based genotyping and detection methods. Before sequencing, many biothreat pathogens had very few sequences in GenBank for development of assays and non-sequence-based methods such as amplified fragment length polymorphism (AFLP), restriction fragment length polymorphism, and insertion sequence typing prevailed. The availability of the genome sequence of *B. anthracis* before publication led to the bioinformatic identification of tandem repeat regions used to design a multilocus variable number of tandem repeat regions assay for subtyping *B. anthracis* strains (110). This method works well for very closely related strains and was extensively used in forensic attribution work after the 2001 anthrax attack. Multiple locus variable number of tandem repeats analysis (MLVA) schemes have been developed for a number of biothreat pathogens (64, 127). Another genotyping method largely developed using whole-genome data, multiple-locus sequence typing (144), uses Sanger sequencing of (typically) seven housekeeping genes to build phylogenies at the species or genus level (73). Microarray-based resequencing of portions of genomes essentially generates data similar to multiple-locus sequence typing but holds the promise of being able to sequence large portions of multiple genomes (241). Whole-genome Sanger shotgun sequencing has been used to subtype *B. anthracis* and to identify SNPs between recently derived isolates for forensic purposes (44, 173, 187).

Biothreat pathogen genomes have yielded new targets for environmental and clinical detection assays for real-time PCR (17, 53). Microarray-based detection schemes are also being developed to allow detection of multiple biothreat agents simultaneously (138, 203). In this regard, generation of multiple near-neighbor sequences is going to be vital for discrimination of true pathogen signatures from false-positive signatures. Genome sequencing of reference pathogens and nonneighbors will also be essential in development of community sequencing ("metagenomics") for pathogen detection. Some recent studies illustrate how these non-culture-based approaches can be used to supplement information on difficult to study organisms. In the first metagenomic analysis of uncultured viral community, Breitbart et al. (21) used partial Sanger shotgun sequencing to identify an estimated 1,200 virus genotypes. Most of the sequences were unrelated to anything previously reported. More effective and sensitive methods could play a major role in understanding biothreat pathogenicity—for instance, in tracking the spread of agents through tissue of infected human patients or model animals.

Importance of Host Genomics

One commonality in the biothreat pathogens discussed at length in this article is that the human species is the target for deliberate infection. Most pathogens also have mammalian models for experimental infection. In many cases, the response of the host cell is at least as important as the action of the infecting patho-

gen in predicting the course of disease. It is also, in many cases, easier to monitor the host cell proteome and mRNA variation than the invading virus or bacterium as a result of the low multiplicity of infection that occurs and the extra difficulty of isolating RNA and protein samples in the BSL3 and BSL4 environments. Thus, many of the future advances in biothreat pathogenomics could take advantage of fast developing postgenomic tools for human and model eukaryote genomics. In one of the earliest studies after the *B. anthracis* genome sequence, Bergman et al., using a glass-slide microarray of 28,000 mouse gene fragments, characterized the transcriptional changes within macrophages infected with *B. anthracis*, including the induction of apoptosis and suppression of immune system protein expression by lethal toxin (16). The expression patterns underlying *B. anthracis*–induced apoptosis led the authors to test further the importance of one very highly induced macrophage gene, that for ornithine decarboxylase, which was shown to play an important role in suppressing apoptosis in *B. anthracis*–infected cells. Another *B. anthracis* macrophage study used an Affymetrix GeneChip microarray to show that the expression of several mitogen-activated protein kinase kinase regulatory genes was affected within 1.5 h after exposure to lethal toxin (42). By 3 h, the expression of 103 genes was altered, including those involved in intracellular signaling, energy production, and protein metabolism (Fig. 1). It is also possible to differentiate human proteomic responses. Chromy et al. used two-dimensional gel electrophoresis gels and mass spectroscopy to define protein biosignatures that differentiate human monocyte-like cells exposed to *Y. pestis*, *Y. pseudotuberculosis*, and *Y. enterocolitica* (37).

Biodefense and Pathogen Ecology

As pathogenesis is viewed increasingly from an evolutionary viewpoint (234), the importance of understanding pathogen ecology grows. A number of biothreat agents are facultative intracellular pathogens having (or having had until recently) the ability to survive outside the

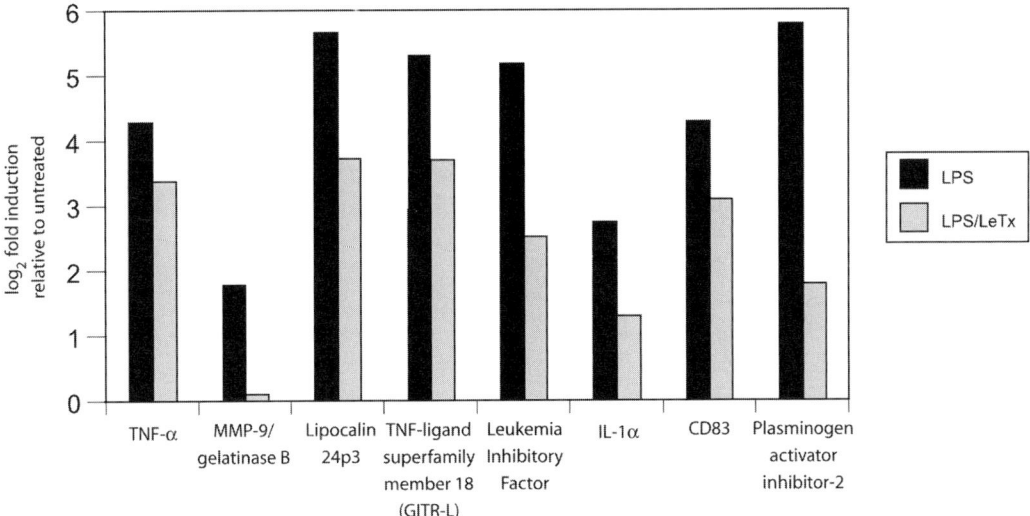

FIGURE 1 Dampening of macrophage immune response genes after treatment with anthrax lethal toxin measured by DNA microarray. Bars indicate the level of induction relative to expression in untreated cells, as measured by DNA microarray. For each gene indicated, induction levels are shown for cells treated with lipopolysaccharide (LPS) and cells treated with LPS and lethal toxin (LeTx). TNF, tumor necrosis factor; MMP-9, matrix metalloproteinase 9; IL, interleukin. From reference 16.

host. Genomics opens the door for dissection of the genetic basis for pathogen ecology. It is vital to disentangle the part of cellular metabolism needed to survive in the external environment from that needed to persist and cause infection in the host. In some cases, the "environmental" metabolome may even interfere with the "pathogenic" metabolome (e.g., the arabinose assimilation operon in *Burkholderia thailandensis* [159]). Also, a number of virulence genes may be specifically adapted to another host (e.g., the pMT plasmid of *Y. pestis* encodes murine toxin required for survival within the flea [240]). Proteins specifically active against mammalian targets may thus be viewed as accidentally maladapted from their true host. One potential way to investigate this is through the use of nonmammalian hosts, which may provide novel, low-cost animal models that accurately reflect natural pathogenic mechanisms (108). The roundworm *Caenorhabditis elegans,* an intensively studied model eukaryote (29), has been shown to be a convenient host for infection studies on *Y. pestis* (218) and *B. pseudomallei* (169). Although understanding how to

use postgenomic studies to determine the genes affecting environmental durability is in its infancy, data such as these may be important in predicting the potential for development of an emerging disease bacterium or virus as a biothreat pathogen; a recent study demonstrated a positive correlation between high levels of environmental durability and virulence across a phylogenetically diverse range of respiratory pathogens (230).

Several biothreat pathogens seem to share an affinity for insects as vectors, in common with several other large groups of pathogenic bacteria (231). *Y. pestis* is perhaps the most obvious example. *Y. pestis* exhibits a distinct infection phenotype in its flea vector, and a transmissible infection depends on genes that are specifically required in the flea, but not the mammal (88). Transmission factors identified to date suggest that the rapid evolutionary transition of *Y. pestis* to flea-borne transmission within the last 1,500 to 20,000 years involved at least three steps: acquisition of the two *Y. pestis*–specific plasmids by HGT and recruitment of endogenous chromosomal genes for new functions. Perhaps reflective of the recent adaptation, transmission of *Y. pestis* by fleas is inefficient, and this likely imposed selective pressure favoring the evolution of increased virulence in this pathogen.

B. anthracis has been proposed to have evolved from an ancestor within the *B. cereus* group that had some form of symbiotic or parasitic relationship with invertebrates (103). All *B. cereus* can grow saprophytically under nutrient-rich conditions, and analyses of the genomes sequenced confirm the metabolic flexibility of the group (183). Because members of the group have recently been discovered as common inhabitants of the invertebrate gut, it has been suggested that they are "sextons," surviving outside their host and waiting for it to die, where they are advantageously placed to degrade the carcass (103).

Single-celled protozoa have also been speculated to be a training ground for facultative intracellular bacteria to develop mechanisms advantageous for both macrophage invasion and survival (84). Additionally, this group of organisms has been implicated as a potential reservoir that maintains pathogenic *Burkholderia* spp. in tropical soils (95–97). Another alternative training ground for human biothreat pathogens was proposed when the genome sequencing of *Brucella melitensis* biovar Suis revealed striking functional conservation with plant pathogens like *Rhizobium* and *Agrobacterium* (176), suggesting commonalities in strategies to parasitize animal or plant hosts.

PATHOGENOMICS OF MAJOR BACTERIAL BIOTHREAT AGENTS

After discussing some of the overarching themes in pathogenomics, the next section will outline some of the specific findings emerging from work on the bacteria listed in Table 1. A brief description of the mode of virulence most related to the probable manifestation as a bioweapon will be outlined, as well as a summary of the major findings from genome sequencing. Each section will then attempt to describe some of the postgenomic analysis for each pathogen and some new ideas

about pathogenesis that have followed from the sequencing projects.

Burkholderia mallei and *pseudomallei*

The gram-negative bacterium *Burkholderia mallei* is the etiologic agent of glanders, a highly infectious disease spread by aerosol that results in a painful incapacitation of humans that is often fatal and usually hard to diagnose (39, 141). The bacterium is an obligate parasite of horses, donkeys, and mules, with no other known animal (or environmental) reservoir (11). *B. mallei* has been developed and deployed as a bioweapon on several occasions, from as early as the Civil War to probably the USSR's invasion of Afghanistan in 1980 (164). Recent genotyping studies on the *Burkholderia* group (73) have established that *B. mallei* is a clonal strain recently emerged from within the larger *B. pseudomallei* group. *B. pseudomallei* is the etiological agent of melioidosis, a disease that is endemic in humans and animals primarily in Southeast Asia and Northern Australia (97) and is itself on the Centers for Disease Control and Prevention biothreat list as a category B agent (http://www.cdc.gov/). Prevalent in these tropical environments as a soil saprophyte, *B. pseudomallei* accounts for 25% of community-acquired sepsis (35). Melioidosis is variable in its pathology and can present as an inhalational pneumonia or as a cutaneous infection leading to sepsis, depending on the portal of entry. The disease states of melioidosis range from an acute infection lasting less than 2 months, a chronic infection, and a latent infection in which the organism remains dormant for up to 60 years (163). A pneumonic mouse model relevant to biodefense studies has recently been developed (102). Complicating treatment is the resistance of *B. pseudomallei* to many frontline antibiotics. At the time of writing, there is no licensed human melioidosis vaccine, although several candidates are at an advanced stage of development.

The genome sequences of the type strain of *B. mallei* and a human clinical isolate of *B. pseudomallei* recently appeared in the same issue of *Proceedings of the National Academy of Sciences of the USA* (90, 164). From these reports, it is clear that *B. pseudomallei* consists of two chromosomes of 4.07 and 3.17 Mb, whereas the *B. mallei* chromosomes are 3.5 and 2.3 Mb; the difference in size is a result of large-scale deletion in the recently diverged *B. mallei* clone. Additionally, the *B. mallei* genome contains numerous insertion elements (IS) that have mediated extensive changes in the genome relative to *B. pseudomallei*. The accumulated deletions in *B. mallei* have occurred in accessory metabolic pathways as well as a flagellar biosynthesis operon (164). A microarray CGH study that used 24 strains illustrated that sequence deletion appears to be a common mechanism for niche adaptation in *Burkholderia*, and on the basis of these data, *B. pseudomallei* can be broken into three major subtypes (168). Both published *Burkholderia* genomes contain a large number of simple sequence repeats in genes that would be expected given their size and nucleotide composition (90, 164). Variation in simple sequence repeats in key genes can provide a mechanism for generating antigenic variation, enabling the organism to possibly evade the immune system during long-term infection.

Just as it has done for other biothreat agents, genome sequencing has revolutionized *Burkholderia pseudomallei* and *B. mallei* research. Until 2002, fewer than 200 genes had been sequenced and characterized. Even given the many deletions, the burkholderia have a large and complex genetic repertoire, almost matching that of the model eukaryote *Saccharomyces cerevisiae* in size. *B. pseudomallei* contains a large fraction of genes (many apparently acquired by HGT) that could be important for pathogenesis, functions that include antibiotic resistance, multiple secretion systems, adhesions, and lipopolysaccharide (90). A polysaccharide capsule not found in the nonpathogenic near neighbor *B. thailandensis* is known to be important for virulence (188). As discussed earlier, deletion of an L-arabinose assimilation operon may also be an adaptation of *B. pseudomallei* and *B. mallei* to the host environment

(159). Microarray analysis showed that L-arabinose seems to play a regulatory role in the cell, resulting in the downregulation of several putative virulence genes. One of the challenges facing researchers now is how to dissect the genetic basis for the complex set of disease outcomes these *Burkholderia* strains cause. An early result coming indirectly from genome sequencing was the identification of the BimA protein that is required for *B. pseudomallei* formation of actin tails during intracellular invasion (216). BimA contains proline-rich motifs, and it shares limited homology at the C terminus with the *Yersinia* YadA adhesin. BimA stimulates actin polymerization in vitro, and mutation of *bimA* abolished actin-based motility of the pathogen in J774.2 cells.

So far there have been few postgenomic microarray experiments that analyze virulence genes. Tuanyok et al. (224) showed that *B. mallei* and *B. pseudomallei* gene expression profiles were similar at low iron concentrations, regardless of growth phase. In low iron, genes coding for most respiratory metabolic systems were decreased (NADH-DH, cytochrome oxidases, and ATP synthases), whereas siderophore Fe transport genes, heme-hemin receptors, and alternative metabolism genes were increased.

Francisella tularensis

Although there is no clearly established historic use of *Francisella tularensis*, the gram-negative, facultatively intracellular bacterium and the causative agent of tularemia, as a biological weapon, its low infectious dose (~50 organisms) and ability to be aerosolized make the threat of weaponization high (49, 60, 209). Two clinically significant strains of *F. tularensis* have been characterized: type A (subspecies *tularensis*) is found primarily in North America and is much more virulent than the Europe- and Asia-localized type B (subspecies *holarctica*) (219). Primarily a zoonotic disease, tularemia is transmitted by arthropod bite, contact with blood or tissues of infected animals, ingestion of contaminated water or meat, or inhalation of aerosols. The inhalation form is an acute, nonspecific illness typically beginning 3 to 5 days after exposure, although in some cases, pleuropneumonia may develop after several days or weeks (49, 60, 209).

The fundamental mechanisms involved in virulence and pathogenesis of *F. tularensis* remain incompletely understood. The cell wall of the bacterium is unusually high in fatty acids. Loss of the capsule may lead to attenuation but not loss of viability; however, the capsule is neither toxic nor immunogenic (209). Infection with *F. tularensis* involves the reticuloendothelial system and results in bacterial replication in the lungs, liver, and spleen. After respiratory exposure, *F. tularensis* infects phagocytic cells, including pulmonary macrophages, where it prevents phagosomal acidification and lysosomal fusion, escapes from the vacuole, and replicates within the cytoplasm (199). These processes all require the elements present on the *Francisella* pathogenicity island, an ~30-kb region consisting of 17 open reading frames (ORFs) (161). In the liver, *F. tularensis* has been shown to invade and replicate in hepatocytes. Destruction of infected hepatocytes results in the release of bacteria and subsequent uptake by phagocytes. When lysis of hepatocytes was prevented via treatment with a monoclonal antibody, bacteria continued to replicate intracellularly, leading to rapid lethality (49).

The first whole-genome sequence of *F. tularensis* was published in 2005 when Larsson et al. completed the variant tularensis SchuS4 strain (121). The analysis uncovered previously uncharacterized genes encoding a type IV pilus, a surface polysaccharide, and iron-acquisition systems. Several virulence-associated genes were located in a putative pathogenicity island, which was found to be duplicated on the chromosome. The genome is rich in IS elements, including 50 copies of ISFtu1, a member of the IS630 Tc-1 *mariner* family of transposons. Ironically, Tc-1 elements are typically found in eukaryotes, including insects, a reported type of vector for *F. tularensis* transmission. The group also found an unexpectedly large proportion of disrupted coding sequence of predicted metabolic pathways, which helps

to explain its fastidious nature and host dependence (121).

A growing theme in the evolution of pathogenicity is that the inactivation of genes may play a significant role (33). Although analysis of *F. tularensis* has shown clear acquisition of DNA, the organism also appears to be undergoing this reductive evolution (121). In fact, more than 10% of the putative coding sequences contain insertions/deletions or substitution mutations. Additionally, analyses of unidirectional genomic deletion events and SNPs suggest that the four subspecies of *F. tularensis* have evolved by vertical descent, generally with the highly virulent type A appearing before the more attenuated type B (22, 219), a pattern that contrasts sharply with that seen in *Y. pestis*, which has presumably become more virulent through gene loss in its transition from its ancestor *Y. pseudotuberculosis* (33).

Coxiella burnetii

Coxiella burnetii is an obligate intracellular bacterial pathogen and the causative agent of human Q fever, a zoonosis that commonly presents as an acute influenza-like illness but at times results in a chronic infection manifesting as endocarditis or, less commonly, hepatitis (109). Although classically grouped with the *Rickettsiaceae*, 16S rRNA sequence analysis actually places *C. burnetii* in the gamma subdivision of the *Proteobacteria*, with *Legionella pneumophila*, a facultative intracellular human pathogen, and *Ricketsiella grylli*, an obligate intracellular parasite of insects, as its closest neighbors (43). Given its ease of aerosol dissemination; resistance to extreme heat, pressure, and chemicals; and high infectivity (50% infective dose, 1 to 10 bacteria), *C. burnetii* is a likely candidate for bioweaponization (85, 109, 148, 196).

In vivo, the initial host-cell target for *C. burnetii* is the alveolar macrophage, although the bacterium is capable of dissemination and subsequent replication within a wide variety of cell types. In the case of macrophages, newly infected bacteria-containing ("parasitophorous") vacuoles undergo typical maturation into a phagolysosome, including the acidification to a pH of about 4.8, a process essential to *C. burnetii* growth. A reduced oxidative burst accompanying this phagocytic process has been reported (13). Transmission electron microscopy by McCaul and Williams has allowed a model of *C. burnetii* infection to be formed (149): metabolically dormant small cell variants are phagocytosed by a eukaryotic host cell and subsequently sequestered in a phagolysosome. Exposure to low intraphagolysosomal pH and, perhaps, nutrient sources present in the vacuole triggers the bacterial differentiation into the more metabolically active large cell variant (85, 109, 196).

In 2003, the complete genome of the *C. burnetii* Nine Mile phase I strain was sequenced (205). The genome consists of a 1,995,275-bp chromosome and a 37,393-bp QpH1 plasmid. The genome is predicted to code for 2,134 coding sequences, of which 33.7% are hypothetical—a large portion considering the wide availability of other γ-proteobacterial sequences. A putative excisable intein is located in the C-terminal region of the replicative DNA helicase, and the sole 23S rRNA gene is interrupted by both an intervening sequence and a group I self-splicing intron, a phenomenon implicated in the slow replication rates of *C. burnetii* (205).

Although *C. burnetii* is an obligate intracellular pathogen and thus shares many characteristics with other bacteria with similar niches, the genome differs considerably, particularly in terms of the presence of mobile genetic elements and the extent of genome reduction (205). Genomes of other obligate intracellular pathogens contain few to no IS elements, resolvases, transposases, and prophages, likely because of the lack of opportunities for gene transfer. *C. burnetii*, however, has 29 IS elements dispersed around the chromosome. Genome analysis identified 83 pseudogenes and reduced metabolic and transport capacities, illustrating an organism undergoing reductive evolution as it becomes increasingly reliant on the host for nutritional requirements. The fact that many of the pseudogenes

are disrupted by a single frameshift implies that these mutations are recent. Additionally, a much higher percentage of the *C. burnetii* genome is coding (89.1%) than what has been observed for species known to have undergone extensive reductive evolution (205). Genome analysis has also demonstrated the presence of types I, II, and IV secretion systems that may be used to secrete host-cell effectors during its residence in the phagolysosome. Newly identified factors proposed to mediate survival in the parasitophorous vacuole include iron-regulation systems and enzymes to detoxify the compartment (catalase and superoxide dismutase) and downregulate the oxidative burst (acid phosphatase and a eukaryotic-like protein kinase) (205).

Protection against the low pH may be conferred by basic proteins that could serve as a buffer against the excess protons that may enter the bacterial cell. *C. burnetii* encodes an unusually high number of basic proteins; the average pI value for all predicted proteins is 8.25, higher than other sequenced genomes (except for *Helicobacter pylori*, which resides in the stomach mucosal layer). Furthermore, ~45% of proteins were found to have a pI of ≥9, in striking contrast to other γ-proteobacteria (204, 205).

Genomes from *C. burnetii* K and G isolates have also been analyzed. Although both are currently in the closing stages, it is clear that there exists a high degree of synteny with the Nine Mile phase 1 genome (15). There are, however, regions containing novel genes, rearrangements, or indels, with the regions of deletions, frequently flanked by transposon-like sequences, likely reflecting the ongoing reductive evolution (15).

These and ongoing efforts (http://msc.tigr .org/c_burnetii/index.shtml) to sequence genomes of heterogenous *C. burnetii* strains as the well the obligate insect parasite, *R. grylli*, may help to delineate the bacterial factors responsible for differences in virulence properties and host range and to identify a core set of proteins for the *Coxiella* spp. that may have facilitated this pathogen's transmission to—or virulence in—a vertebrate host.

Brucella

Species in the genus *Brucella* are the etiologic agent of brucellosis, the most frequent human zoonosis and one that causes abortions and infertility in animals and a systemic, undulating febrile illness in humans. Human infection most commonly is initiated on contact with tissues and fluids from infected animals, but brucellosis can also be contracted through consumption of contaminated foods or by inhalation of the bacteria (52, 105, 192).

Brucella spp. are facultative intracellular pathogens that enter the host through the mucosal surfaces and preferentially infect and survive within macrophages by, most of all, inhibition of phagosome-lysosome fusion (105). Although there are six recognized species, DNA hybridization and genomic sequencing studies reiterate that brucellae are part of a monospecific genus, and it is recommended that the organisms be grouped as biovars of the single species *B. melitensis*, with each biovar selectively infecting specific hosts (47).

The 3.3-Mb genome of *Brucella* spp., although somewhat variable, typically consists of two circular chromosomes, with the larger chromosome (Ch1) roughly twice the size of the smaller (Ch2) (47, 172). The larger chromosome resembles a classic bacterial circular chromosome with the likely origin of replication adjacent to a gene cluster consisting of *dnaA*, *dnaN*, and *recF*, whereas the smaller chromosome possesses a cluster of plasmid-like replication genes, including a replication initiation protein RepC and partitioning proteins RepA and RepB, similar to the plasmid replication genes from *Agrobacterium* Ti plasmids (47).

There is also an underscored asymmetry of genes in different functional categories between the two chromosomes. Ch1 codes for the majority of the core metabolic machinery for processes, such as transcription, translation, and protein synthesis. Ch2 is overrepresented in genes involved in processes such as membrane transport, central intermediary and energy metabolism, and regulation. Genes coding for products involved in cellular processes are also concentrated on Ch2, mostly because of the presence of three clusters of flagellar biosynthesis and secretion genes. Additionally, Ch2 houses many auxiliary pathways (e.g., homoprotocatechuate and beta-ketoadipate) relevant to the utilization of aromatic compounds encountered in the soil and also includes a number of essential ones, such as the single tRNA-Cys and three tRNA synthetases (172).

Comparison of the genomes of *B. melitensis* biovars Suis and Melitensis has revealed extensive similarity and gene synteny, with a total of 7,307 SNPs along a shared backbone of 3.2 Mb (172). More than 90% of the *B. melitensis* biovars Suis and Melitensis genes share 98 to 100% identity at the DNA level, with the more variable genes consisting first of hypothetical genes but also a UreE urease component, various outer membrane proteins, membrane transporters, a putative invasin, and an adhesin, which may explain the differences in species specificity. In fact, only 42 *B. melitensis* biovar Suis genes and 32 *B. melitensis* biovar Melitensis genes were identified that are completely absent in the other genome, many of which are hypothetical in nature and localized to genetic islands flanked by phage integrase orthologues. Approximately 50% of these hypothetical proteins are predicted to be surface genes and are implicated in determining host range and virulence potential (172). Two additional *B. melitensis* biovar Abortus genomes (strains 9-941 and 2308) have been added to the catalog of available genomes for the genus with further discoveries of polymorphisms, deletions, and insertions, as well as insights into the evolution of the group (34).

Rajashekara et al. constructed a microarray by using *B. melitensis* biovar Melitensis (highly pathogenic to humans), which revealed 217 ORFs altered in five species analyzed (182). The ORFs were often found in islands in the *B. melitensis* biovar Melitensis genome, many of which were acquired horizontally. Deletions of genetic content identified in *Brucella* are conserved in multiple strains of the same biovar, and genomic islands missing in a given biovar are often restricted to that particular biovar. Although missing hypothetical proteins may account for some of the varying levels of pathogenicity, natural disruptions of known virulence factors localized to genome islands must also play a role. For instance, mutations that inactivate the lipopolysaccharide biosynthesis gene *wbdA* in *B. melitensis* biovar Suis likely attenuate its capacity to replicate inside of macrophages (117). Also, *B. melitensis* biovar Ovis lacks two genome islands present in *B. melitensis* biovar Melitensis (182), including one that codes for the Opp system of ABC transporters, which may lead to increased uptake of peptides and thus explain its relative attenuation in nonprofessional phagocytes (69). This suggests that whereas the loss or gain of genetic material may be related to the host range and virulence restriction of certain *Brucella* biovars for humans, independent mechanisms involving gene inactivation or altered expression of determinants may also contribute to these differences.

Yersinia pestis

Currently, natural plague cases worldwide are estimated to be on the order of about 2,000 annually, historically, an unusually low rate (191). In normal cases, *Y. pestis* infections are initiated by flea or rodent bite and result in either bubonic or septicemic plague (191). Pneumonic plague is a secondary infection that may follow bubonic plague but that is rarely seen since the advent of effective antibiotic prophylaxis (175). However, in the context of biodefense, primary pneumonic plague caused by inhalation of even small numbers of aerosolized *Y. pestis* is a major concern (93). If untreated, the disease is highly infectious and has a mortality rate that approaches 100%. Recently, modern subunit vaccines based on the F1 and V antigens have been shown to have promise for protection against pneumonic infection (61, 235) but there is still much to be learned about how the bacterium actually causes disease.

As is the case with several biothreat pathogens, the gram-negative bacterium *Y. pestis* comes from a genus that generally shows very limited virulence for mammals. *Yersinia* is classified within the family *Enterobacteriaceae* and is composed of 11 species (119). Of these 11 species, only *Y. pseudotuberculosis*, *Y. enterocolitica*, and *Y. pestis* are pathogenic for humans and animals. Many virulence pathway determinants had been characterized before the genomic era (175, 189, 239). These include those encoding the Pla adhesin, lipopolysaccharide, F1 capsule antigen, and iron acquisition functions. Although all three pathogenic species require the type III secretion system (T3SS)-encoding plasmid pYV (also known as pCD) for virulence, *Y. pestis* contains additional plasmids of 110 kb (pMT1 or pFRa) and 9 kb (pPCP1 or pPla) that are essential for its unique virulence processes. The pMT1 encodes the murine toxin, Ymt, that is essential for *Y. pestis* maintenance in fleas (88), a unique feature of this species in the *Yersineae*. All three pathogenic species carry a 102-kb unstable *pgm* (pigmentation locus) in the chromosome containing several virulence genes, including the yersinia-bactin iron-uptake gene on an excisable high-pathogenicity island (27). The island can be transferred horizontally as an element, making it a problematic locus for *Y. pestis*-specific detection assays (134, 135, 200).

The genome sequence of *Y. pestis* CO92 (an Orientalis biovar strain) was published in 2001 (171), making it the first major biothreat pathogen to be fully sequenced. The striking features of the *Y. pestis* genome included a very high number of IS sequences (3.7% of the DNA sequence) and anomalies in GC base-composition bias, indicating frequent intragenomic recombination. Genomes of a *Y. pestis* biovar Mediavalis isolate, KIM (48), and a microtus biovar strain, 91001 (212), that followed soon after showed the species to be very recently formed, with only minimal changes in gene content and order. The *Y. pestis* CO92 genome was found to contain 149 pseudogenes (the number was upped to 298 after the *Y. pseudotuberculosis* genome sequence was completed [33]). Despite the very close sequence identity of orthologous genes, as many as 13% of the genes in *Y. pseudotuberculosis* appear to be missing from *Y. pestis* (33), a pattern typical of recent reductive evolution in pathogens. On the basis of aberrant GC base composition and genetic structures (e.g., flanking IS and/or tRNA sequences [79]), a number of putative laterally transferred islands were found in the genome. These include functions that may contribute to pathogenesis such as adhesins, type II and III secretion systems, and insecticidal toxins (171, 237). However, almost all of these islands are common to the *Y. pseudotuberculosis* genome (33). Other than the pMT1 and pPCP1 plasmids, only 112 chromosomal genes turned out to be unique to *Y. pestis*, and CGH experiments with different strains showed that only 32 genes, spread out over six regions, were conserved across the species. Many of these were hypothetical or phage-related functions, potentially important for pathogenesis but just as likely to be non-functional remnants of horizontal transfer (220).

The evolution of the pathogenic *Yersiniae* is turning out to be a fascinating and complex story (Fig. 2) (237). *Y. pseudotuberculosis* and *Y. pestis* are closely related at the genetic level, while *Y. enterocolitica* represents an earlier evolutionary divergence. *Y. pestis* appears to have evolved from *Y. pseudotuberculosis* by a process combining plasmid uptake and gene inactivation. The comparative genome approach has raised several questions about *Y. pestis* pathogenicity that may be followed up with functional studies. What genes in the genome of *Y. pestis* are important for pathogenicity that we have not yet characterized? What makes *Y. pestis* hypervirulent compared with *Y. pseudotuberculosis*? Are there key mutations that are responsible for enhanced pathogenicity aside from general "streamlining" (the loss of nonessential housekeeping genes such as those that code for a flagellar secretion system, making the bacterium nonmotile but saving energy)? Finally, because it is the most likely route for invasion of hosts in an attack, can we identify the specific set of genetic components necessary for the pneumonic infection specifically?

Several postgenomic studies have concentrated on global changes in *Y. pestis* gene expression upon change in temperature from 26 to 37°C (mimicking the transition from ambient temperature to that of the mammal) and with varying Ca^{2+} levels (Yops and other virulence genes are known to be regulated through the so-called low-calcium response [175]). Forde et al. (68) used a green fluorescent protein reporter system to show thermal induction of type III secretion component genes (*yopE*, *sycE*, *yopK*, *yopT*, *yscN*, and *lcrE/yopN*) that was repressed by calcium. Whole-ORF DNA microarrays of *Y. pestis* were used to characterize global low-calcium response and temperature regulatory changes after shifts from steady-state growth (160). Ca^{2+} levels were found to have relatively little effect on chromosomal genes, whereas the majority of genes carried by pCD were downregulated. In contrast, 235 chromosomal genes were thermally upregulated, and 274 were thermally downregulated. A number of genes on the plasmids were thermoinduced (including some that were not known to be regulated in this manner). The advantage of the whole-genome array is that global changes can be tracked. The transition from growth in 26 to 37°C was responsible for many such gross metabolic shifts, including turning off glycolysis, repressing nitrogen assimilation and amino acid biosynthesis, and switching to terminal oxidation of carbohydrates and other substrates (160). This confirms what was thought about the slow-growing lifestyle of *Y. pestis* in the flea gut compared with its explosive growth and acquisition of nutrients in its mammalian host. Other laboratory culture–based microarray studies have also been recently performed, including examining the effects of heat shock, cold shock, streptomycin, and salinity (81, 82, 181, 238). A protein microarray, consisting of 149 *Y. pestis* genes cloned and expressed in *Escherichia coli* and printed on glass microarray slides, has also been used to find 11 novel proteins that predominate in the antibody response of plague-immunized rabbits (in addition to the well-known F1 and V antigens) (136).

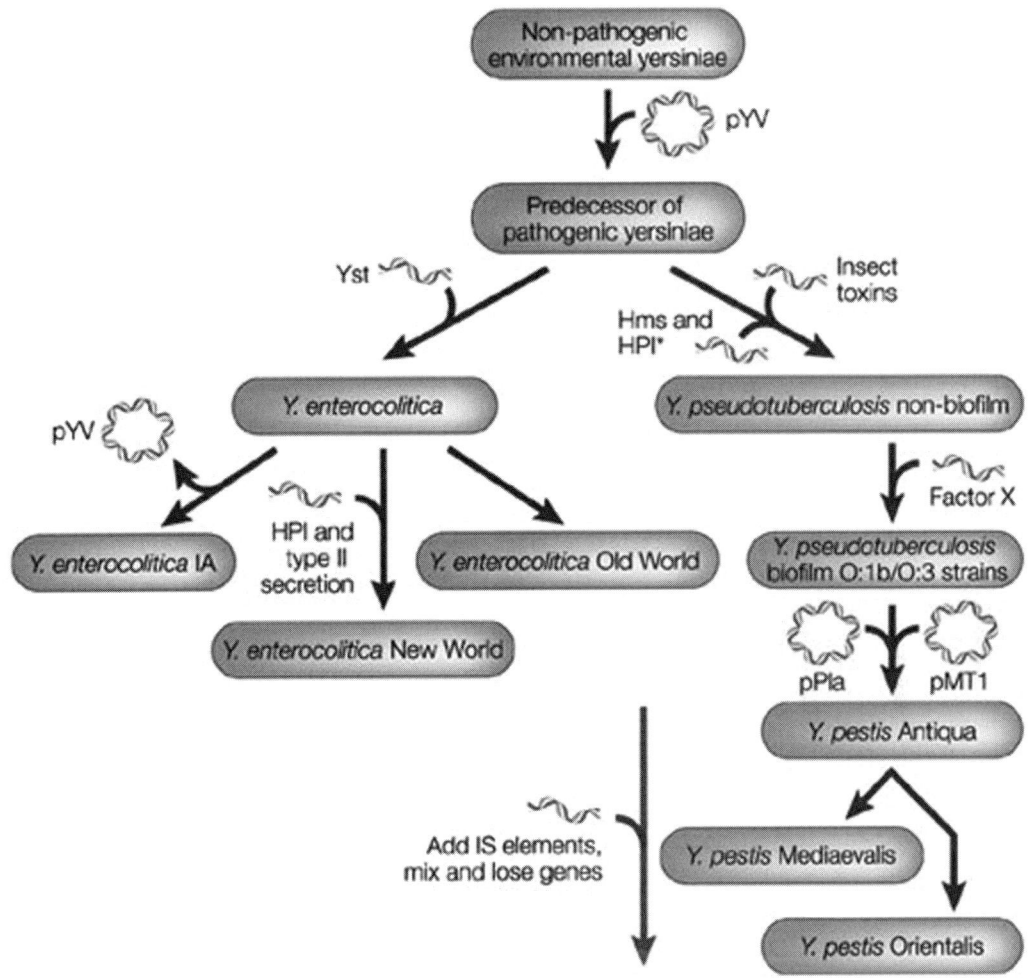

FIGURE 2 The nonpathogenic yersiniae gain the virulence plasmid pYV to form the predecessor of pathogenic yersinae. *Y. enterocolitica* diverges from *Y. pseudotuberculosis* and forms three lineages: 1A, Old World, and New World. *Y. pseudotuberculosis* gains the ability to parasitize insects and to form biofilms in hosts before evolving into *Y. pestis* through the acquisition of plasmids pPla and pMT1, genome mixing, and decay. Hms, hemin storage; HPI and HPI★, high-pathogenicity islands; IS, insertion sequence. Reprinted by permission from Macmillan Publishers Ltd.: *Nature Reviews Microbiology* (231), copyright 2004.

In addition to allowing for the dissection of expression patterns, postgenomic studies are identifying new factors that contribute to plague pathogenesis. In an effort to find novel attenuated strains, Flashner et al. (66) devised a signature-tagged mutagenesis strategy to find nonpathogenic mutants of *Y. pestis* Kimberley58 by means of an outbred mice model of infection. Out of 300 mutants screened, 16 were missing from the animal after infection. Several of these mutants were known to be important in mammalian growth in *Yersinia* or other enterobacteria (for instance, genes involved purine biosynthesis). However, there were several completely novel virulence genes isolated, including *pcm*, a stress-response gene. Another novel virulence locus identified postgenomically is *rip*, located on the *pgm* locus

(180). The three Rip products, which are all annotated as putative metabolic enzymes—acetyl CoA transferase, monamine oxygenase, and citrate lyase—are required for intracellular replication in interferon-activated macrophages. *Y. pestis pgm* mutants replicated in inducible nitric oxide synthase–deficient macrophages that were postactivated with interferon gamma, suggesting that killing of *Y. pestis* Δ*pgm* is NO dependent. Interestingly, the *rip* genes were similar to those found in a genomic island of another intracellular pathogen, *Salmonella enterica* serovar Enteritidis.

Finally, some progress has also been made in examining the pathogenomics of pneumonic plague. Lathem et al. (122) developed an intranasal mouse model and were able to follow *Y. pestis* gene expression by comparison of 48-h lung-lavage samples with bacteria grown in laboratory culture media at 37°C. About 10% of the genome was differentially regulated in the study, with the most conspicuous change being the upregulation of genes on the pCD plasmid. The result indicates that there may be additional in vivo signals that upregulate the genes expressing the T3SS on pCD, a process vital for its anti-inflammatory and antiphagocytotic roles in infection.

In summary, the *Y. pestis* genome carries a large number of virulence factors (175, 189, 240) that allow it a bewildering range of activities in mammals: cellular invasion and intracellular survival (probably involving macrophage invasion [179]); resistance to phagocytosis, complement, and antibodies; probable killing of immune cells (146); interference with innate immunity (111); and scavenging of iron. Virulence involves specialized reactions to a number of environmental stimuli and manipulation of the host environment through protein secreted by T3SS. *Y. pestis* also has the ability (unique to an enterobacterial species) of being transmitted by insects. All this is tied in with an ongoing process of genome reduction that seems to be crucial in making the bacterium far more dangerous than its close cousins in the *Yersinia* genus.

Bacillus anthracis

Anthrax is a potentially fatal infection caused by the gram-positive, toxigenic, spore-forming *B. anthracis*, a bacterium to which all warm-blooded animals are susceptible (51). Anthrax manifests in three basic forms: cutaneous (for which transmission has been reported to occur through insect bites) (20a, 224a), gastrointestinal, and inhalation. Although the inhalation route is the least commonly encountered nonintentional form of infection, it is the most likely scenario for use of *B. anthracis* as a biological weapon (94). By extrapolating data from nonhuman primates, the estimated 50% lethal dose for humans is 8,000 to 10,000 endospores (151); however, it is clear that unknown factors are significant in determining the establishment of disease, given the low incidence among workers in animal hide mills who are routinely exposed to high titers of anthrax endospores (5, 20). At the same time, aerosolized endospores present a major health hazard (even to those who are miles downwind from the point of release), as illustrated by the events on Gruinard Island in the United Kingdom, Sverdlovsk in the former Soviet Union, and the 2001 anthrax attacks in the United States (1, 100, 104, 145, 215).

Anthrax spores are roughly 1 to 2 μm in diameter, an ideal size for inhalation and deposition in the alveolar spaces (51). Although the lung is the initial site of contact, inhalation anthrax is not considered a true pneumonia. Lung infection is rare and, in those instances, only during the late stages of disease. Instead, alveolar phagocytes engulf the endospores and transport them to the mediastinal and peribronchial lymph nodes, providing the necessary environment for endospore germination (50, 75–77, 98, 125, 193). After a brief intracellular stage, anthrax bacilli are released from the phagocyte and freely replicate in the lymph nodes, causing hemorrhagic mediastinitis, and spread through the body extracellularly, where they can grow to titers as high as 10^9 bacteria/ml of blood by the time of death (51). Illness begins subtly, with onset of symptoms occurring 3 to 4 days after exposure and resembling those of an upper respiratory infection. This initial phase, typically lasting 2 to 3 days and often followed by a period of recovery, abruptly transitions into fulminant anthrax, characterized by hemorrhagic mediastinitis, dyspnea, a strident cough, and chills, ultimately leading to death (5, 19, 70, 91, 176).

It was recognized early in studies of *B. anthracis* that anthrax is an exotoxin-mediated disease. And, although the identification of anthrax toxin by Smith and Keppie (211) first established this principle, it was the subsequent work, illustrating that challenges with either endospores or anthrax toxin alone led essentially to identical pathophysiological changes in vivo, that fully solidified this tenet (115, 165, 210). The anthrax toxins, members of the AB-type toxin family, are composed of three polypeptides, edema factor (EF), lethal factor (LF), and protective antigen (PA), which act in binary combinations (131, 213) and are all encoded by the pXO1 plasmid (154, 167). AB toxins are composed of two functionally distinct moieties; the enzymatically active A component gains entry into sensitive cells via the B moiety's ability to bind to its extracellular target. The anthrax toxin is unique as it utilizes a single B subunit (PA) to mediate the entry into cells of two distinct A subunits (EF and LF); this leads to the production of two independent, binary toxins: edema toxin (EdTx), a calmodulin-dependent adenylate cyclase that acts in the cytoplasm of eukaryotic cells converting intracellular ATP to cyclic AMP (129, 130), and lethal toxin (LeTx), a zinc metalloprotease (116) that cleaves members of the mitogen-activated protein kinase kinase (MEK) family (57, 174, 197, 228). The structure and function of each component of the two *B. anthracis* toxins have been characterized and reviewed extensively (12, 23, 41, 51, 83, 158).

Although the roles of pXO1 and pXO2 plasmid genes in virulence had been well established, the contributions of factors coded for by the chromosome were less clearly understood. The chromosome codes for 5,508 predicted proteins with pronounced bias for genes on the replication-leading strand, a trait common in other low G+C% gram-positive replicons (186). A feature shared with the chromosomes of other endospore-forming bacteria of the *Bacillus* and *Clostridium* genera is the concentration of the rRNA, tRNA, and ribosomal protein genes around the origin of replication, a trait that may facilitate protein synthesis during early rounds of DNA replication after germination from the dormant endospore. The chromosome also contains four prophages unique to *B. anthracis* as well as two type I introns.

There were several features of the *B. anthracis* chromosome that were surprising. For instance, *B. anthracis* is well known for its lack of extracellular protease activity (153), in part, at least, because of the nonsense mutation in the *plcR*-positive regulator gene. In *B. thuringiensis* and *B. cereus*, PlcR is known to upregulate the production of numerous extracellular enzymes (2). Although the *B. anthracis plcR* orthologue is truncated, there are 56 putative PlcR-binding motifs in the chromosome and 2 on pX02. Initial reports supported the idea of an incompatibility between PlcR and AtxA, the master regulator of *B. anthracis* gene expression including the toxins and capsule (152); however, the discovery of *B. cereus* G9241, a strain that contains full-length PlcR and AtxA, has weakened that hypothesis.

One putative virulence gene standing out on the chromosome was a cholesterol-dependent cytolysin, dubbed anthrolysin-O (ALO) (206). Similar genes in another gram-positive pathogen, *Listeria monocytogenes*, mediate escape of the pathogen from the phagocytotic vacuole. ALO cloned into *L. monocytogenes* could functionally replace a mutant *Listeria* protein for intracellular growth and spread (232). ALO may also be apoptotic to macrophages (185), suggesting a multifunctional role in virulence for the protein.

Another surprise on the chromosome was an orthologue of the metalloprotease, enhancin, a virulence factor in insect-infecting viruses (128). *Y. pestis*, with an insect transmission stage, also carries this protein (although the amino acid sequences share only distant sequence similarity [171]). Although the presence of the enhancin gene is suggestive of a recent insect-infecting ancestor, there is no direct functional evidence yet. Bioassays suggested that the bacterial enzymes are cytotoxic but do not enhance infection of insect larvae in the same manner as the viral protein (71).

B. anthracis was found to have an expanded array of iron-acquisition genes, including two siderophore systems, one of which has been shown to be essential for maximal virulence in a mouse model (32). Also of interest, the *B. anthracis* genome codes for several proteins that mitigate damage by free oxygen radicals and are putatively involved in survival within the bacteriocidal phagolysosome, including five catalases and three superoxide dismutases (14).

The complete proteome of the *B. anthracis* infectious endospore was solved and found to be of surprising composition (139). A large portion (approximately 36%) of the *B. anthracis* genome was found to be regulated in a growth phase–dependent manner, and this regulation was marked by five distinct waves of gene expression as cells proceed from exponential growth through sporulation. The identities of more than 750 proteins present in the spore were determined by multidimensional chromatography and tandem mass spectrometry. Comparison of data sets revealed that although the genes responsible for assembly and maturation of the spore are tightly regulated in discrete stages, many of the components ultimately found in the spore are expressed throughout and even before sporulation, suggesting that gene expression during sporulation may be mainly related to the physical construction of the spore, rather than synthesis of eventual content. The spore also contains an assortment of specialized, but not obviously related, metabolic and protective proteins.

In another postgenomic study, comparative proteomics of the secretomes performed with two-dimensional gel electrophoresis between fully virulent *B. anthracis*, pXO1$^+$/pXO2$^+$, and pXO1$^-$/pXO2$^-$ was performed (120). In all, 57 protein spots, representing 26 different proteins coded for on the chromosome or pXO1, were identified. S-layer-derived proteins, such as Sap and EA1 (62, 63), were most frequently observed. Many sporulation-associated enzymes were overexpressed in strains containing pXO1. They also found evidence that pXO2 is necessary for the maximal expression of the pXO1-associated toxins LF, EF, and PA. Several newly identified putative virulence factors were observed, including a high-affinity zinc-uptake transporter and the peroxide stress-related alkyl hydroperoxide reductase.

The pathobiology of *B. anthracis*, stimulated by the genome sequence (186), is a highly active field. Recently, a novel FtsZ-like replication protein of pXO1 was characterized (222)—a significant advancement because the replication mechanism of this plasmid had been something of a mystery previously. The protein, highly conserved with orthologues on *B. cereus* megaplasmids, may offer a target for therapeutics that detoxify *B. anthracis* by plasmid curing. Many other genes discovered by sequencing have been analyzed as knowledge of the basic biology of the organism rapidly increases (10, 14, 25, 28, 55, 56, 107, 137, 142, 156, 177, 208, 214, 227, 229). Although lethal toxin proteins and capsule, characterized before the genomic era, are still primarily used as antigens for vaccine development, efforts have been made to discover additional targets by using vaccinogenomics methods (8, 9).

OUTPUTS OF BIOTHREAT PATHOGENOMIC STUDIES

Although each of the potential agents can be considered separately, what are the potential advantages for combining knowledge of the pathogenomics of multiple organisms? Two areas stand out where comparative pathogenomics can be used for tangible benefit for biodefense preparedness: development of multiagent countermeasures, and detection and treatment of genetically engineered pathogens.

Countermeasures that can simultaneously provide protection against multiple agents are to a great extent the "holy grail" of biodefense studies. Extensive comparative genomics within pathogen species will allow selection of targets that are common to all members. It may also be possible to use genomic approaches that find genes conserved between taxonomically divergent species with common virulence pathways. A recent study used a comparative genomic approach to identify common hypothetical virulence-associated genes in the bacterial pathogens *Y. pestis*, *Neisseria gonorrhoeae*, *Helicobacter pylori*, *Borrelia burgdorferi*, *Streptococcus pneumoniae*, and *Treponema pallidum* (72). Five common genes were found to be essential in vivo during *Y. pseudotuberculosis* infection of mice, offering the potential as targets for multiagent therapies. Although such a strategy may work across some organisms, the diversity of biothreat agents and the tendency for many virulence genes to be accessory rather than conserved cast doubt on its universal effectiveness. An alternative strategy used in vaccine development is to simultaneously protect by using epitopes from multiple organisms (113, 190).

As knowledge of biothreat pathogenesis grows, so does the possibility of identifying common pathways in the host. For instance, many of the bacteria described in this chapter invade macrophages at some point when causing disease (Table 3). With increased knowledge of the host-pathogen interaction, it may be possible to find common receptors or enzymes within human macrophages that can be targeted for downregulation (perhaps at the site of infection in the body). An alternative approach is find common mechanisms to stimulate human innate immunity (6).

Even with stockpiled multiagent therapies on hand, biodefense scientists have to contend with the possibility that an attack could be instigated by use of an organism not previously encountered. Agents that exist naturally—as an animal pathogen, for instance—could be introduced into humans deliberately. The recent experiences of severe acute respiratory syndrome (36, 236) and avian influenza serve as reminders of how such strains may originate. Alternatively, genetically modified, hypervirulent strains might be engineered by humans accidentally (101), through movement of virulence genes into novel genetic backgrounds or "improvement" of already known pathogens, or possibly even whole-genome synthesis in vitro (31). No matter how a genetically modified biothreat is formed, it is likely that

TABLE 3 Pathogen interactions with macrophage[a]

Organism	Early and late endosome markers	Phago-lysosome fusion	Escape from vacuole	Early release from macrophage	References
F. tularensis	Yes	No	Yes	No	38, 74, 199
C. burnetii	Yes	Yes[b]	Yes	No	3, 13, 65
B. pseudomallei	Yes	Yes	Yes[c]	No	112, 225
B. anthracis	Yes	Yes	Yes	Yes	50, 75, 193, 233
Y. pestis	Yes	Yes	Yes	Yes	179, 180, 217
B. melitensis	Yes	No[d]	No	No	1, 7, 30, 92, 162

[a]One common trait of the bacterial pathogens most likely to be bioweaponized is their ability to traffic through mammalian macrophage. All six bacteria have been shown to be taken up within endosomes on the basis of host-specific markers. Although there is variation in the outcome of infection from vacuolar growth (*B. melitensis*) to escape from the phagosome (*F. tularensis*, *B. pseudomallei*), all six can survive and replicate within the hostile intra-macrophage environment. *B. anthracis* and *Y. pestis* are released from macrophage within hours of initial infection, whereas *C. burnetii*, *B. pseudomallei*, *F. tularensis*, and *B. melitensis* are more specialized intracellular pathogens that can persist for much longer time periods.
[b]Requires acidic compartment.
[c]Induces host cell-to-cell fusion; nucleates host actin for motility.
[d]Fuses with endoplasmic reticulum.

genome sequencing can now be performed very early in analysis of the organism, serving as a starting block for the type of pathogenomic studies described for the known pathogens earlier in this chapter. Genomics will be used for comparison to known models of pathogenicity of biothreat agents possibly to find similar activities (e.g., macrophage invasion or toxin expression) where there may be vulnerabilities. It may be possible to identify therapeutics already existing that can be used to treat the genetically modified biothreat organism.

The era of genome sequencing has generated a massive increase in knowledge about biothreat organisms. This information is not without its dangers: it could potentially be exploited by sophisticated bioterrorists to engineer new pathogens; however, as Lederberg noted (in regard to the imminent onset of recombinant DNA studies in 1975), genomics offers "uncertain peril and certain benefit" (126). The DNA sequence of a pathogen's genome is such a fundamental biological property that it has the power to unite disparate fields of study and sharing of such information is a prerequisite of an open society. We remain confident that pathogenomics studies, openly published and openly discussed, will lead to better countermeasures against known biological weapons and will allow us to react rapidly to whatever the future may hold.

ACKNOWLEDGMENTS
We acknowledge the help of Jacques Ravel, Steve Cendrowski, Stan Goldman, Tim Inglis, Bart Legutki, Rehka Seshadri, and Richard Titball for reading the chapter in manuscript and making valuable suggestions.

Writing this chapter was made possible through funding from the Defense Threat Reduction Agency. The views expressed in this article are those of the authors and do not necessarily reflect the official policy or position of the Department of the Navy, the Department of Defense, or the U.S. government.

REFERENCES
1. **Abramova, F. A., L. M. Grinberg, O. V. Yampolskaya, and D. H. Walker.** 1993. Pathology of inhalational anthrax in 42 cases from the Sverdlovks outbreak of 1979. *Proc. Natl. Acad. Sci. USA* **90:**2291–2294.
2. **Agaisse, H., M. Gominet, O. A. Okstad, A. B. Kolsto, and D. Lereclus.** 1999. PlcR is a pleiotropic regulator of extracellular virulence factor gene expression in *Bacillus thuringiensis*. *Mol. Microbiol.* **32:**1043–1053.
3. **Akporiaye, E. T., J. D. Rowatt, A. A. Aragon, and O. G. Baca.** 1983. Lysosomal response of a murine macrophage-like cell line persistently infected with *Coxiella burnetii*. *Infect. Immun.* **3:** 1155–1162.
4. **Albert, T. J., D. Dailidiene, G. Dailide, J. E. Norton, A. Kalia, T. A. Richmond, M. Molla, J. Singh, R. D. Green, and D. E. Berg.** 2005. Mutation discovery in bacterial genomes: metronidazole resistance in *Helicobacter pylori*. *Nat. Methods* **2:**951–953.
5. **Albrink, W. S., S. M. Brooks, R. E. Biron, and M. Kopel.** 1960. Human inhalation anthrax: a report of three fatal cases. *Am. J. Pathol.* **36:**457–471.
6. **Amlie-Lefond, C., D. A. Paz, M. P. Connelly, G. B. Huffnagle, K. S. Dunn, N. T. Whelan, and H. T. Whelan.** 2005. Innate immunity for biodefense: a strategy whose time has come. *J. Allergy Clin. Immunol.* **116:**1334–1342.
7. **Arenas, G. N., A. S. Staskevich, A. Aballay, and L. S. Mayorga.** 2000. Intracellular trafficking of *Brucella abortus* in J774 macrophages. *Infect. Immun.* **68:**4255–4263.
8. **Ariel, N., A. Zvi, H. Grosfeld, O. Gat, Y. Inbar, B. Velan, S. Cohen, and A. Shafferman.** 2002. Search for potential vaccine candidate open reading frames in the *Bacillus anthracis* virulence plasmid pXO1: in silico and in vitro screening. *Infect. Immun.* **70:**6817–6827.
9. **Ariel, N., A. Zvi, K. S. Makarova, T. Chitlaru, E. Elhanany, B. Velan, S. Cohen, A. M. Friedlander, and A. Shafferman.** 2003. Genome-based bioinformatic selection of chromosomal *Bacillus anthracis* putative vaccine candidates coupled with proteomic identification of surface-associated antigens. *Infect. Immun.* **71:**4563–4579.
10. **Aronson, A. I., C. Bell, and B. Fulroth.** 2005. Plasmid-encoded regulator of extracellular proteases in *Bacillus anthracis*. *J. Bacteriol.* **187:**3133–3138.
11. **Arun, S., H. Neubauer, A. Gurel, G. Ayyildiz, B. Kuscu, T. Yesildere, H. Meyer, and W. Hermanns.** 1999. Equine glanders in Turkey. *Vet. Rec.* **144:**255–258.
12. **Ascenzi, P., P. Visca, G. Ippolito, A. Spallarossa, M. Bolognesi, and C. Montecucco.** 2002. Anthrax toxin: a tripartite lethal combination. *FEBS Lett.* **531:**384–388.

13. **Baca, O. G., Y.-P. Li, and H. Kumar.** 1994. Survival of the Q fever agent *Coxiella burnetii* in the phagolysosome. *Trends. Microbiol.* **2:**476–480.

14. **Baillie, L., S. Hibbs, P. Tsai, G. L. Cao, and G. M. Rosen.** 2005. Role of superoxide in the germination of *Bacillus anthracis* endospores. *FEMS Microbiol. Lett.* **245:**33–38.

15. **Beare, P. A., S. F. Porcella, R. Seshadri, J. E. Samuel, and R. A. Heinzen.** 2005. Preliminary assessment of genome differences between the reference Nine Mile Isolate and two human endocarditis isolates of *Coxiella burnetii. Ann. N.Y. Acad. Sci.* **1063:**64–67.

16. **Bergman, N. H., K. D. Passalacqua, R. Gaspard, L. M. Shetron-Rama, J. Quackenbush, and P. C. Hanna.** 2005. Murine macrophage transcriptional responses to *Bacillus anthracis* infection and intoxication. *Infect. Immun.* **73:**1069–1080.

17. **Bode, E., W. Hurtle, and D. Norwood.** 2004. Real-time PCR assay for a unique chromosomal sequence of *Bacillus anthracis. J. Clin. Microbiol.* **42:**5825–5831.

18. **Bordenstein, S. R., and W. S. Reznikoff.** 2005. Mobile DNA in obligate intracellular bacteria. *Nat. Rev. Microbiol.* **3:**688–699.

19. **Borio, L., T. Inglesby, C. J. Peters, A. L. Schmaljohn, J. M. Hughes, P. B. Jahrling, T. Ksiazek, K. M. Johnson, A. Meyerhoff, T. O'Toole, M. S. Ascher, J. Bartlett, J. G. Breman, E. M. Eitzen, Jr., M. Hamburg, J. Hauer, D. A. Henderson, R. T. Johnson, G. Kwik, M. Layton, S. Lillibridge, G. J. Nabel, M. T. Osterholm, T. M. Perl, P. Russell, and K. Tonat.** 2002. Hemorrhagic fever viruses as biological weapons: medical and public health management. *JAMA* **287:**2391–2405.

20. **Brachman, P. S., A. F. Kaufman, and F. G. Dalldorf.** 1966. Industrial inhalation anthrax. *Bacteriol. Rev.* **30:**646–659.

20a.**Bradaric, N., and V. Punda-Polic.** 1992. Cutaneous anthrax due to penicillin-resistant *Bacillus anthracis* transmitted by an insect bite. *Lancet* **340:**306–307.

21. **Breitbart, M., I. Hewson, B. Felts, J. M. Mahaffy, J. Nulton, P. Salamon, and F. Rohwer.** 2003. Metagenomic analyses of an uncultured viral community from human feces. *J. Bacteriol.* **185:**6220–6223.

22. **Broekhuijsen, M., P. Larsson, A. Johansson, M. Bystrom, U. Eriksson, E. Larsson, R. G. Prior, A. Sjostedt, R. W. Titball, and M. Forsman.** 2003. Genome-wide DNA microarray analysis of *Francisella tularensis* strains demonstrates extensive genetic conservation within the species but identifies regions that are unique to the highly virulent *F. tularensis* subsp. *tularensis. J. Clin. Microbiol.* **41:**2924–2931.

23. **Brossier, F., and M. Mock.** 2001. Toxins of *Bacillus anthracis. Toxicon* **39:**1747–1755.

24. **Brown, P. O., and D. Botstein.** 1999. Exploring the new world of the genome with DNA microarrays. *Nat. Genet.* **21:**33–37.

25. **Brunsing, R. L., C. La Clair, S. Tang, C. Chiang, L. E. Hancock, M. Perego, and J. A. Hoch.** 2005. Characterization of sporulation histidine kinases of *Bacillus anthracis. J. Bacteriol.* **187:**6972–6981.

26. **Brussow, H., C. Canchaya, and W. D. Hardt.** 2004. Phages and the evolution of bacterial pathogens: from genomic rearrangements to lysogenic conversion. *Microbiol. Mol. Biol. Rev.* **68:**560–602.

27. **Buchrieser, C., C. Rusniok, L. Frangeul, E. Couve, A. Billault, F. Kunst, E. Carniel, and P. Glaser.** 1999. The 102-kilobase pgm locus of *Yersinia pestis*: sequence analysis and comparison of selected regions among different *Yersinia pestis* and *Yersinia pseudotuberculosis* strains. *Infect. Immun.* **67:**4851–4861.

28. **Candela, T., and A. Fouet.** 2005. *Bacillus anthracis* CapD, belonging to the gamma-glutamyl-transpeptidase family, is required for the covalent anchoring of capsule to peptidoglycan. *Mol. Microbiol.* **57:**717–726.

29. *C. elegans* **Sequencing Consortium.** 1998. Genome sequence of the nematode *C. elegans*: a platform for investigating biology. *Science* **282:**2012–2018.

30. **Celli, J., C. de Chastellier, D.-M. Franchini, J. Cerda-Pizarro, E. Moreno, and J.-P. Gorvel.** 2006. *Brucella* evades macrophage killing via VirB-dependent sustained interactions with the endoplasmic reticulum. *J. Exp. Med.* **198:**545–556.

31. **Cello, J., A. V. Paul, and E. Wimmer.** 2002. Chemical synthesis of poliovirus cDNA: generation of infectious virus in the absence of natural template. *Science* **297:**1016–1018.

32. **Cendrowski, S., W. MacArthur, and P. Hanna.** 2004. *Bacillus anthracis* requires siderophore biosynthesis for growth in macrophages and mouse virulence. *Mol. Microbiol.* **51:**407–417.

33. **Chain, P. S., E. Carniel, F. W. Larimer, J. Lamerdin, P. O. Stoutland, W. M. Regala, A. M. Georgescu, L. M. Vergez, M. L. Land, V. L. Motin, R. R. Brubaker, J. Fowler, J. Hinnebusch, M. Marceau, C. Medigue, M. Simonet, V. Chenal-Francisque, B. Souza, D. Dacheux, J. M. Elliott, A. Derbise, L. J. Hauser, and E. Garcia.** 2004. Insights into the evolution of *Yersinia pestis* through whole-genome comparison with *Yersinia pseudotuberculosis. Proc. Natl. Acad. Sci. USA* **101:**13826–13831.

34. Chain, P. S., D. J. Comerci, M. E. Tolmasky, F. W. Larimer, S. A. Malfatti, L. M. Vergez, F. Aguero, M. L. Land, R. A. Ugalde, and E. Garcia. 2005. Whole-genome analyses of speciation events in pathogenic *Brucellae. Infect. Immun.* **73:**8353–8361.

35. Chaowagul, W., N. J. White, D. A. Dance, Y. Wattanagoon, P. Naigowit, T. M. Davis, S. Looareesuwan, and N. Pitakwatchara. 1989. Melioidosis: a major cause of community-acquired septicemia in northeastern Thailand. *J. Infect. Dis.* **159:**890–899.

36. Chow, K. Y., C. C. Hon, R. K. Hui, R. T. Wong, C. W. Yip, F. Zeng, and F. C. Leung. 2003. Molecular advances in severe acute respiratory syndrome-associated coronavirus (SARS-CoV). *Genomics Proteomics Bioinformatics* **1:**247–262.

37. Chromy, B. A., J. Perkins, J. L. Heidbrink, A. D. Gonzales, G. A. Murphy, J. P. Fitch, and S. L. McCutchen-Maloney. 2004. Proteomic characterization of host response to *Yersinia pestis* and near neighbors. *Biochem. Biophys. Res. Commun.* **320:**474–479.

38. Clemens, D. L., B. Y. Lee, and M. A. Horwitz. 2004. Virulent and avirulent strains of *Francisella tularensis* prevent acidification and maturation of their phagosomes and escape into the cytoplasm of human macrophages. *Infect. Immun.* **74:**3204–3217.

39. Coenye, T., and J. J. LiPuma. 2003. Molecular epidemiology of Burkholderia species. *Front. Biosci.* **8:**e55–e67.

40. Cole, S. T., K. Eiglmeier, J. Parkhill, K. D. James, N. R. Thomson, P. R. Wheeler, N. Honore, T. Garnier, C. Churcher, D. Harris, K. Mungall, D. Basham, D. Brown, T. Chillingworth, R. Connor, R. M. Davies, K. Devlin, S. Duthoy, T. Feltwell, A. Fraser, N. Hamlin, S. Holroyd, T. Hornsby, K. Jagels, C. Lacroix, J. Maclean, S. Moule, L. Murphy, K. Oliver, M. A. Quail, M. A. Rajandream, K. M. Rutherford, S. Rutter, K. Seeger, S. Simon, M. Simmonds, J. Skelton, R. Squares, S. Squares, K. Stevens, K. Taylor, S. Whitehead, J. R. Woodward, and B. G. Barrell. 2001. Massive gene decay in the leprosy bacillus. *Nature* **409:**1007–1011.

41. Collier, R. J., and J. A. T. Young. 2003. Anthrax toxin. *Ann. Rev. Cell. Dev. Biol.* **19:**45–70.

42. Comer, J. E., C. L. Galindo, A. K. Chopra, and J. W. Peterson. 2005. GeneChip analyses of global transcriptional responses of murine macrophages to the lethal toxin of *Bacillus anthracis. Infect. Immun.* **73:**1879–1885.

43. Crouzet, J., L. Naudin, C. Orsini, E. Vigne, L. Ferrero, A. Le Roux, P. Benoit, M. Latta,

C. Torrent, D. Branellec, P. Denefle, J. F. Mayaux, M. Perricaudet, and P. Yeh. 1997. Recombinational construction in *Escherichia coli* of infectious adenoviral genomes. *Proc. Natl. Acad. Sci. USA* **94:**1414–1419.

44. Cummings, C. A., and D. A. Relman. 2002. Genomics and microbiology. Microbial forensics—"cross-examining pathogens." *Science* **296:** 1976–1979.

45. Day, W. A., Jr., R. E. Fernandez, and A. T. Maurelli. 2001. Pathoadaptive mutations that enhance virulence: genetic organization of the cadA regions of *Shigella* spp. *Infect. Immun.* **69:**7471–7480.

46. DelVecchio, V. G., V. Kapatral, P. Elzer, G. Patra, and C. V. Mujer. 2002. The genome of *Brucella melitensis. Vet. Microbiol.* **90:**587–592.

47. DelVecchio, V. G., V. Kapatral, R. J. Redkar, G. Patra, C. Mujer, T. Los, N. Ivanova, I. Anders, A. Bhattacharyya, W. Lykidis, G. Reznik, L. Jablonski, N. Larsen, M. D'Souza, A. Bernal, M. Mazur, E. Goltsman, E. Selkov, P. H. Elzer, S. Hagius, D. O'Callaghan, J.-J. Letesson, R. Haselkorn, N. Kyrpides, and R. Overbeek. 2002. The genome sequence of the facultative intracellular pathogen *Brucella melitensis. Proc. Natl. Acad. Sci. USA* **99:**443–448.

48. Deng, W., V. Burland, G. Plunkett III, A. Boutin, G. F. Mayhew, P. Liss, N. T. Perna, D. J. Rose, B. Mau, S. Zhou, D. C. Schwartz, J. D. Fetherston, L. E. Lindler, R. R. Brubaker, G. V. Plano, S. C. Straley, K. A. McDonough, M. L. Nilles, J. S. Matson, F. R. Blattner, and R. D. Perry. 2002. Genome sequence of *Yersinia pestis* KIM. *J. Bacteriol.* **184:**4601–4611.

49. Dennis, D. T., T. V. Inglesby, D. A. Henderson, J. G. Bartlett, M. S. Ascher, E. Eitzen, A. D. Fine, A. M. Friedlander, J. Hauer, M. Layton, S. R. Lillibridge, J. E. McDade, M. T. Osterholm, T. O'Toole, G. Parker, T. M. Perl, P. K. Russell, and K. Tonat. 2001. Tularemia as a biological weapon. *JAMA* **285:**2763–2773.

50. Dixon, T. C., A. A. Fadl, T. M. Koehler, J. A. Swanson, and P. C. Hanna. 2000. Early *Bacillus anthracis*–macrophage interactions: intracellular survival and escape. *Cell. Microbiol.* **2:**453–463.

51. Dixon, T. C., M. Meselson, J. Guilemin, and P. C. Hanna. 1999. Anthrax. *N. Engl. J. Med.* **341:**815–826.

52. Dornand, J., Gross, A., Lafont, V., Liautard, J., Oliaro, J., and Liautard, J.-P. 2002. The innate immune response against *Brucella* in humans. *Vet. Microbiol.* **90:**383–394.

53. Draghici, S., P. Khatri, Y. Liu, K. J. Chase, E. A. Bode, D. A. Kulesh, L. P. Wasieloski, D. A. Norwood, and J. Reifman. 2005. Identification of genomic signatures for the design of assays

for the detection and monitoring of anthrax threats. *Pac. Symp. Biocomput.* **2005:**248–259.

54. **Dricot, A., J.-F. Rual, P. Lamesch, N. Bertin, D. Dupuy, T. Hao, C. Lambert, R. Hallez, J.-M. Delroisse, J. Vandenhaute, I. Lopez-Goñi, I. Moriyon, J. M. Garcia-Lobo, F. J. Sangari, A. P. MacMillan, S. J. Cutler, A. M. Whatmore, S. Bozak, R. Sequerra, L. Doucette-Stamm, M. Vidal, D. E. Hill, J.-J. Letesson, and X. De Bolle.** 2006. Generation of the *Brucella melatensis* ORFeome version 1.1. *Genome Res.* **10B:**2201–2206.

55. **Drysdale, M., A. Bourgogne, and T. M. Koehler.** 2005. Transcriptional analysis of the *Bacillus anthracis* capsule regulators. *J. Bacteriol.* **187:**5108–5114.

56. **Drysdale, M., S. Heninger, J. Hutt, Y. Chen, C. R. Lyons, and T. M. Koehler.** 2005. Capsule synthesis by *Bacillus anthracis* is required for dissemination in murine inhalation anthrax. *EMBO J.* **24:**221–227.

57. **Duesbery, N. S., C. P. Webb, S. H. Leppla, V. M. Gordon, K. R. Klimpel, T. D. Copeland, N. G. Ahn, M. K. Oskarsson, K. Fikasawa, K. D. Paull, and G. P. Vande Woude.** 1998. Proteolytic inactivation of MAP-kinase-kinase by anthrax lethal factor. *Science* **280:**734–737.

58. **Ecker, D. J., R. Sampath, P. Willett, J. R. Wyatt, V. Samant, C. Massire, T. A. Hall, K. Hari, J. A. McNeil, C. Buchen-Osmond, and B. Budowle.** 2005. The Microbial Rosetta Stone Database: a compilation of global and emerging infectious microorganisms and bioterrorist threat agents. *BMC Microbiol.* **5:**19.

59. **Edwards, J. S., M. Covert, and B. Palsson.** 2002. Metabolic modelling of microbes: the flux-balance approach. *Environ. Microbiol.* **4:**133–140.

60. **Ellis, J., P. C. F. Oyston, M. Green, and R. Titball.** 2002. Tularemia. *Clin. Microbiol. Rev.* **15:**631–646.

61. **Elvin, S. J., J. E. Eyles, K. A. Howard, E. Ravichandran, S. Somavarappu, H. O. Alpar, and E. D. Williamson.** 2006. Protection against bubonic and pneumonic plague with a single dose microencapsulated sub-unit vaccine. *Vaccine* **15:**4433–4439.

62. **Etienne-Toumelin, I., J. C. Sirard, E. Duflot, M. Mock, and A. Fouet.** 1995. Characterization of the *Bacillus anthracis* S-layer: clongin and sequence of the structural gene. *J. Bacteriol.* **177:**614–620.

63. **Ezzell, J. W. J., and T. G. Abshire.** 1988. Immunological analysis of cell-associated antigens of *Bacillus anthracis.* *Infect. Immun.* **56:**349–356.

64. **Farlow, J., K. L. Smith, J. Wong, M. Abrams, M. Lytle, and P. Keim.** 2001. *Francisella tularensis* strain typing using multiple-locus, variable-number tandem repeat analysis. *J. Clin. Microbiol.* **39:**3186–3192.

65. **Fernandez-Moreia, E., and M. S. Swanson.** 2002. A microbial strategy to multiply in macrophages: the pregnant pause. *Traffic* **3:**170–177.

66. **Flashner, Y., E. Mamroud, A. Tidhar, R. Ber, M. Aftalion, D. Gur, S. Lazar, A. Zvi, T. Bino, N. Ariel, B. Velan, A. Shafferman, and S. Cohen.** 2004. Generation of *Yersinia pestis* attenuated strains by signature-tagged mutagenesis in search of novel vaccine candidates. *Infect. Immun.* **72:**908–915.

67. **Fleischmann, R. D., M. D. Adams, O. White, R. A. Clayton, E. F. Kirkness, A. R. Kerlavage, C. J. Bult, J.-F. Tomb, B. A. Dougherty, J. M. Merrick, K. McKenny, G. Sutton, W. Fitzhugh, C. Fields, J. D. Gocayne, J. Scott, R. Shirley, L.-I. Liu, A. Glodek, J. M. Kelley, J. F. Wiedman, C. A. Phillips, T. Spriggs, E. Hedblom, M. D. Cotton, T. R. Utterback, M. C. Hanna, D. T. Nquyen, D. M. Saudek, R. C. Brandon, L. D. Fine, J. L. Fritchman, J. L. Fuhrman, N. S. M. Geoghagen, C. L. Gnehm, L. A. McDonald, K. V. Small, C. M. Fraser, H. O. Smith, and J. C. Venter.** 1995. Whole-genome random sequencing and assembly of *Haemophilus influenzae* Rd. *Science* **269:**496–512.

68. **Forde, C. E., J. M. Rocco, J. P. Fitch, and S. L. McCutchen-Maloney.** 2004. Real-time characterization of virulence factor expression in *Yersinia pestis* using a GFP reporter system. *Biochem. Biophys. Res. Commun.* **324:**795–800.

69. **Freer, E., J. Pizarro-Cerda, A. Weintraub, J. A. Bengoechea, I. Moriyon, K. Hultenby, J. P. Grovel, and E. Moreno.** 1999. The outer membrane of *Brucella ovis* shows increased permeability to hydrophobic probes and is more susceptible to cationic peptides than are the outer membranes of mutant rough *Brucella abortus* strains. *Infect. Immun.* **67:**6181–6186.

70. **Friedlander, A. M., S. L. Welkos, M. L Pitt, J. W. Ezzell, P. L. Worsham, K. J. Rose, B. E. Ivins, J. R. Lowe, G. B Howe, P. Mikesell, and W. B. Lawrence.** 1993. Postexposure prophylaxis against experimental inhalation antharx. *J. Infect. Dis.* **167:**1239–1243.

71. **Galloway, C. S., P. Wang, D. Winstanley, and I. M. Jones.** 2005. Comparison of the bacterial *Enhancin*-like proteins from *Yersinia* and *Bacillus* spp. with a baculovirus *Enhancin. J. Invertebr. Pathol.* **90:**134–137.

72. **Garbom, S., A. Forsberg, H. Wolf-Watz, and B. M. Kihlberg.** 2004. Identification of novel virulence-associated genes via genome analysis of hypothetical genes. *Infect. Immun.* **72:**1333–1340.

73. Godoy, D., G. Randle, A. J. Simpson, D. M. Aanensen, T. L. Pitt, R. Kinoshita, and B. G. Spratt. 2003. Multilocus sequence typing and evolutionary relationships among the causative agents of melioidosis and glanders, *Burkholderia pseudomallei* and *Burkholderia mallei*. *J. Clin. Microbiol.* **41:**2068–2079.

74. Golovliov, I., V. Baranov, Z. Krocova, H. Kovarova, and A. Sjostedt. 2006. An attenuated strain of the facultative intracellular bacterium *Francisella tularensis* can escape the phagosome of monocytic cells. *Infect. Immun.* **71:**5940–5950.

75. Guidi-Rontani, C. 2002. The alveolar macrophage: the Trojan horse of *Bacillus anthracis*. *Trends Microbiol.* **10:**405–409.

76. Guidi-Rontani, C., M. Levy, H. Ohayon, and M. Mock. 2001. Fate of germinated *Bacillus anthracis* spores in primary murine macrophages. *Mol. Microbiol.* **42:**931–938.

77. Guidi-Rontani, C., Y. Pereira, S. Ruffie, J. C. Sirard, M. Weber-Levy, and M. Mock. 1999. Identification and characterization of a germination operon on the virulence plasmid pXO1 of *Bacillus anthracis*. *Mol. Microbiol.* **33:**407–414.

78. Hacker, J., G. Blum-Oehler, I. Muhldorfer, and H. Tschape. 1997. Pathogenicity islands of virulent bacteria: structure, function and impact on microbial evolution. *Mol. Microbiol.* **23:**1089–1097.

79. Hacker, J., B. Hochhut, B. Middendorf, G. Schneider, C. Buchrieser, G. Gottschalk, and U. Dobrindt. 2004. Pathogenomics of mobile genetic elements of toxigenic bacteria. *Int. J. Med. Microbiol.* **293:**453–461.

80. Halling, S. M., B. D. Peterson-Burch, B. J. Bricker, R. L. Zuerner, Z. Qing, L. L. Li, V. Kapur, D. P. Alt, and S. C. Olsen. 2005. Completion of the genome sequence of *Brucella abortus* and comparison to the highly similar genomes of *Brucella melitensis* and *Brucella suis*. *J. Bacteriol.* **187:**2715–2726.

81. Han, Y., D. Zhou, X. Pang, L. Zhang, Y. Song, Z. Tong, J. Bao, E. Dai, J. Wang, Z. Guo, J. Zhai, Z. Du, X. Wang, J. Wang, P. Huang, and R. Yang. 2005. Comparative transcriptome analysis of *Yersinia pestis* in response to hyperosmotic and high-salinity stress. *Res. Microbiol.* **156:**403–415.

82. Han, Y., D. Zhou, X. Pang, L. Zhang, Y. Song, Z. Tong, J. Bao, E. Dai, J. Wang, Z. Guo, J. Zhai, Z. Du, X. Wang, J. Wang, P. Huang, and R. Yang. 2005. DNA microarray analysis of the heat- and cold-shock stimulons in *Yersinia pestis*. *Microbes Infect.* **7:**335–348.

83. Hanna, P. 1998. Anthrax pathogenesis and host response. *Curr. Top. Microbiol. Immunol.* **225:**13–35.

84. Harb, O. S., L. Y. Gao, and Y. Abu Kwaik. 2000. From protozoa to mammalian cells: a new paradigm in the life cycle of intracellular bacterial pathogens. *Environ. Microbiol.* **2:**251–265.

85. Heinzen, R. A., Hackstadt, T., and Samuel, J.E. 1999. Developmental biology of *Coxiella burnetii*. *Trends Microbiol.* **7:**149–154.

86. Hensel, M., J. E. Shea, C. Gleeson, M. D. Jones, E. Dalton, and D. W. Holden. 1995. Simultaneous identification of bacterial virulence genes by negative selection. *Science* **269:**400–403.

87. Hinchliffe, S. J., K. E. Isherwood, R. A. Stabler, M. B. Prentice, A. Rakin, R. A. Nichols, P. C. Oyston, J. Hinds, R. W. Titball, and B. W. Wren. 2003. Application of DNA microarrays to study the evolutionary genomics of *Yersinia pestis* and *Yersinia pseudotuberculosis*. *Genome Res.* **13:**2018–2029.

88. Hinnebusch, B. J. 2005. The evolution of flea-borne transmission in *Yersinia pestis*. *Curr. Issues Mol. Biol.* **7:**197–212.

89. Hoffmaster, A. R., J. Ravel, D. A. Rasko, G. D. Chapman, M. D. Chute, C. K. Marston, B. K. De, C. T. Sacchi, C. Fitzgerald, L. W. Mayer, M. C. Maiden, F. G. Priest, M. Barker, L. Jiang, R. Z. Cer, J. Rilstone, S. N. Peterson, R. S. Weyant, D. R. Galloway, T. D. Read, T. Popovic, and C. M. Fraser. 2004. Identification of anthrax toxin genes in a *Bacillus cereus* associated with an illness resembling inhalation anthrax. *Proc. Natl. Acad. Sci. USA* **101:**8449–8454.

90. Holden, M. T., R. W. Titball, S. J. Peacock, A. M. Cerdeno-Tarraga, T. Atkins, L. C. Crossman, T. Pitt, C. Churcher, K. Mungall, S. D. Bentley, M. Sebaihia, N. R. Thomson, N. Bason, I. R. Beacham, K. Brooks, K. A. Brown, N. F. Brown, G. L. Challis, I. Cherevach, T. Chillingworth, A. Cronin, B. Crossett, P. Davis, D. DeShazer, T. Feltwell, A. Fraser, Z. Hance, H. Hauser, S. Holroyd, K. Jagels, K. E. Keith, M. Maddison, S. Moule, C. Price, M. A. Quail, E. Rabbinowitsch, K. Rutherford, M. Sanders, M. Simmonds, S. Songsivilai, K. Stevens, S. Tumapa, M. Vesaratchavest, S. Whitehead, C. Yeats, B. G. Barrell, P. C. Oyston, and J. Parkhill. 2004. Genomic plasticity of the causative agent of melioidosis, *Burkholderia pseudomallei*. *Proc. Natl. Acad. Sci. USA* **101:**14240–14245.

91. Holty, J.-E. C., D. M. Bravata, H. Liu, R. A. Olshen, K. M. McDonald, and D. K. Owens. 2006. Systematic review: a century of inhalational anthrax cases from 1900 to 2005. *Ann. Intern. Med.* **144:**270–280.

92. Hong, P. C., R. M. Tsolis, and T. A. Ficht. 2000. Identification of genes required for chronic persistence of *Brucella abortus* in mice. *Infect. Immun.* **90:**281–297.

93. **Inglesby, T. V., D. T. Dennis, D. A. Henderson, J. G. Bartlett, M. S. Ascher, E. Eitzen, A. D. Fine, A. M. Friedlander, J. Hauer, J. F. Koerner, M. Layton, J. McDade, M. T. Osterholm, T. O'Toole, G. Parker, T. M. Perl, P. K. Russell, M. Schoch-Spana, and K. Tonat.** 2000. Plague as a biological weapon: medical and public health management. Working Group on Civilian Biodefense. *JAMA* **283:** 2281–2290.

94. **Inglesby, T. V., D. A. Henderson, J. G. Bartlett, M. S. Ascher, E. Eitzen, A. M. Friedlander, J. Hauer, J. McDade, M. T. Osterholm, T. O'Toole, G. Parker, T. M. Perl, P. K. Russell, and K. Tonat.** 1999. Anthrax as a biological weapon: medical and public health management. Working Group on Civilian Biodefense. *JAMA* **281:**1735–1745.

95. **Inglis, T. J., P. Rigby, T. A. Robertson, N. S. Dutton, M. Henderson, and B. J. Chang.** 2000. Interaction between *Burkholderia pseudomallei* and *Acanthamoeba* species results in coiling phagocytosis, endamebic bacterial survival, and escape. *Infect. Immun.* **68:**1681–1686.

96. **Inglis, T. J., T. Robertson, D. E. Woods, N. Dutton, and B. J. Chang.** 2003. Flagellum-mediated adhesion by *Burkholderia pseudomallei* precedes invasion of *Acanthamoeba astronyxis*. *Infect. Immun.* **71:**2280–2282.

97. **Inglis, T. J. J.** 2004. Mellioidosis in man and other animals: epidemiology, ecology and pathogenesis. *Vet. Bull.* **74:**39N–48N.

98. **Ireland, J. A., and P. C. Hanna.** 2002. Macrophage-enhanced germination of *Bacillus anthracis* endospores requires gerS. *Infect. Immun.* **70:**5870–5872.

99. **Ivanova, N., A. Sorokin, I. Anderson, N. Galleron, B. Candelon, V. Kapatral, A. Bhattacharyya, G. Reznik, N. Mikhailova, A. Lapidus, L. Chu, M. Mazur, E. Goltsman, N. Larsen, M. D'Souza, T. Walunas, Y. Grechkin, G. Pusch, R. Haselkorn, M. Fonstein, S. D. Ehrlich, R. Overbeek, and N. Kyrpides.** 2003. Genome sequence of *Bacillus cereus* and comparative analysis with *Bacillus anthracis*. *Nature* **423:**87–91.

100. **Jackson, P. J., M. E. Hugh-Jones, D. M. Adair, G. Green, K. K. Hill, C. R. Kuske, L. M. Grinberg, F. A. Abramova, and P. Keim.** 1998. PCR analysis of tissue samples from the 1979 Sverdlovsk anthrax victims: the presence of multiple *Bacillus anthracis* strains in different victims. *Proc. Natl. Acad. Sci. USA* **95:**1224–1229.

101. **Jackson, R. J., A. J. Ramsay, C. D. Christensen, S. Beaton, D. F. Hall, and I. A. Ramshaw.** 2001. Expression of mouse interleukin-4 by a recombinant ectromelia virus suppresses cytolytic lymphocyte responses and overcomes genetic resistance to mousepox. *J. Virol.* **75:**1205–1210.

102. **Jeddeloh, J. A., D. L. Fritz, D. M. Waag, J. M. Hartings, and G. P. Andrews.** 2003. Biodefense-driven murine model of pneumonic melioidosis. *Infect. Immun.* **71:**584–587.

103. **Jensen, G. B., B. M. Hansen, J. Eilenberg, and J. Mahillon.** 2003. The hidden lifestyles of *Bacillus cereus* and relatives. *Environ. Microbiol.* **5:**631–640.

104. **Jernigan, D. B., P. L. Raghunathan, B. P. Bell, R. Brechner, E. A. Bresnitz, J. C. Butler, M. Cetron, M. Choen, T. Doyel, M. Fisher, C. Greene, K. S. Griffith, J. Guarner, J. L. Hadler, J. A. Hayslett, R. Meyer, L. R. Petersen, M. Phillips, R. Pinner, T. Popovic, C. P. Quinn, J. Reefhuis, D. Reissman, N. Rosenstein, A. Schuchat, W. J. Shieh, L. Siegal, D. L. Swerdlow, F. C. Tenover, M. Traeger, J. W. Ward, I. Weisfurse, S. Wiersma, K. Yeskey, S. Zaki, D. A. Ashford, B. A. Perkins, S. Ostroff, J. Hughes, D. Fleming, J. P. Koplan, J. L. Gerberding, and the National Anthrax Epidemiologic Investigation Team.** 2002. Investigation of bioterrorism-related anthrax, United States, 2001: epidemiologic findings. *Emerg. Infect. Dis.* **8:**1019–1028.

105. **Jiménez de Bagüés, M. P., S. Dudal, J. Dornand, and A. Gross.** 2005. Cellular bioterrorism: how *Brucella* corrupts macrophage physiology to promote invasion and proliferation. *Clin. Immunol.* **114:**227–238.

106. **Karlin, S., M. J. Barnett, A. M. Campbell, R. F. Fisher, and J. Mrazek.** 2003. Predicting gene expression levels from codon biases in alpha-proteobacterial genomes. *Proc. Natl. Acad. Sci. USA* **100:**7313–7318.

107. **Kau, J. H., D. S. Sun, W. J. Tsai, H. F. Shyu, H. H. Huang, H. C. Lin, and H. H. Chang.** 2005. Antiplatelet activities of anthrax lethal toxin are associated with suppressed p42/44 and p38 mitogen-activated protein kinase pathways in the platelets. *J. Infect. Dis.* **192:**1465–1474.

108. **Kavanagh, K., and E. P. Reeves.** 2004. Exploiting the potential of insects for in vivo pathogenicity testing of microbial pathogens. *FEMS Microbiol. Rev.* **28:**101–112.

109. **Kazar, J.** 2005. *Coxiella burnetii* infection. *Ann. N.Y. Acad. Sci.* **1063:**105–114.

110. **Keim, P., A. M. Klevytska, L. B. Price, J. M. Schupp, G. Zinser, K. L. Smith, M. E. Hugh-Jones, R. Okinaka, K. K. Hill, and P. J. Jackson.** 1999. Molecular diversity in *Bacillus anthracis*. *J. Appl. Microbiol.* **87:**215–217.

111. **Kerschen, E. J., D. A. Cohen, A. M. Kaplan, and S. C. Straley.** 2004. The plague virulence

protein YopM targets the innate immune response by causing a global depletion of NK cells. *Infect. Immun.* **72:**4589–4602.

112. **Kespichayawattana, W., S. Rattanachetkul, T. Wanun, P. Utaisincharoen, and S. Sirisinha.** 2000. *Burkholderia pseudomallei* induces cell fusion and actin-associated membrane protrusion: a possible mechanism for cell-to-cell spread. *Infect. Immun.* **68:**5377–5384.

113. **Khan, A. S., C. V. Mujer, T. G. Alefantis, J. P. Connolly, U. B. Mayr, P. Walcher, W. Lubitz, and V. G. Delvecchio.** 2006. Proteomics and bioinformatics strategies to design countermeasures against infectious threat agents. *J. Chem. Inf. Model* **46:**111–115.

114. **Kim, H. S., M. A. Schell, Y. Yu, R. L. Ulrich, S. H. Sarria, W. C. Nierman, and D. DeShazer.** 2005. Bacterial genome adaptation to niches: divergence of the potential virulence genes in three *Burkholderia* species of different survival strategies. *BMC Genomics* **6:**174.

115. **Klein, F., J. S. Walker, D. F. Fitzpatrick, R. E. Lincoln, B. G. Mahlandt, W. I. Jones, Jr., J. P. Dobbs, and K. J. Hendrix.** 1966. Pathophysiology of anthrax. *J. Infect. Dis.* **116:**123–138.

116. **Klimpel, K. R., N. Arora, and S. H. Leppla.** 1994. Anthrax toxin lethal factor contains a zinc metalloprotease consensus sequence which is required for lethal toxin activity. *Mol. Microbiol.* **13:**1093–1100.

117. **Kohler, S., V. Foulongne, S. Ouahrani-Bettache, G. Bourg, J. Teyssier, M. Ramuz, and J. P. Liautard.** 2002. The analysis of the intramacrophage virulome of *Brucella suis* deciphers the environment encountered by the pathogen inside the macrophage host cell. *Proc. Natl. Acad. Sci. USA* **99:**15711–15716.

118. **Konstantinidis, K. T., and J. M. Tiedje.** 2004. Trends between gene content and genome size in prokaryotic species with larger genomes. *Proc. Natl. Acad. Sci. USA* **101:**3160–3165.

119. **Kotetishvili, M., A. Kreger, G. Wauters, J. G. Morris, Jr., A. Sulakvelidze, and O. C. Stine.** 2005. Multilocus sequence typing for studying genetic relationships among *Yersinia* species. *J. Clin. Microbiol.* **43:**2674–2684.

120. **Lamonica, J. M., M. Wagner, M. Eschenbrenner, L. E. Williams, T. L. Miller, G. Patra, and V. G. DelVecchio.** 2005. Comparative secretome analyses of three *Bacillus anthracis* strains with variant plasmid contents. *Infect. Immun.* **73:**3646–3658.

121. **Larsson, P., P. C. Oyston, P. Chain, M. C. Chu, M. Duffield, H. H. Fuxelius, E. Garcia, G. Halltorp, D. Johansson, K. E. Isherwood, P. D. Karp, E. Larsson, Y. Liu, S. Michell, J. Prior, R. Prior, S. Malfatti, A.** Sjostedt, K. Svensson, N. Thompson, L. Vergez, J. K. Wagg, B. W. Wren, L. E. Lindler, S. G. Andersson, M. Forsman, and R. W. Titball. 2005. The complete genome sequence of *Francisella tularensis*, the causative agent of tularemia. *Nat. Genet.* **37:**153–159.

122. **Lathem, W. W., S. D. Crosby, V. L. Miller, and W. E. Goldman.** 2005. Progression of primary pneumonic plague: a mouse model of infection, pathology, and bacterial transcriptional activity. *Proc. Natl. Acad. Sci. USA* **102:**17786–17791.

123. **Lavigne, J. P., A. C. Vergunst, G. Bourg, and D. O'Callaghan.** 2005. The IncP island in the genome of *Brucella suis* 1330 was acquired by site-specific integration. *Infect. Immun.* **73:**7779–7783.

124. **Lawrence, J. G.** 2005. Common themes in the genome strategies of pathogens. *Curr. Opin. Genet. Dev.* **15:**584–588.

125. **Leamon, J. H., W. L. Lee, K. R. Tartaro, J. R. Lanza, G. J. Sarkis, A. D. deWinter, J. Berka, M. Weiner, J. M. Rothberg, and K. L. Lohman.** 2003. A massively parallel PicoTiter-Plate based platform for discrete picoliter-scale polymerase chain reactions. *Electrophoresis* **24:**3769–3777.

126. **Lederberg, J.** 1975. DNA splicing: will fear rob us of its benefits? *Prism* **3:**33–37.

127. **Le Fleche, P., I. Jacques, M. Grayon, S. Al Dahouk, P. Bouchon, F. Denoeud, K. Nockler, H. Neubauer, L. A. Guilloteau, and G. Vergnaud.** 2006. Evaluation and selection of tandem repeat loci for a *Brucella* MLVA typing assay. *BMC Microbiol.* **6:**9.

128. **Lepore, L. S., P. R. Roelvink, and R. R. Granados.** 1996. Enhancin, the granulosis virus protein that facilitates nucleopolyhedrovirus (NPV) infections, is a metalloprotease. *J. Invertebr. Pathol.* **68:**131–140.

129. **Leppla, S. H.** 1982. Anthrax toxin edema factor: a bacterial adenylate cyclase that increases cyclic AMP concentrations in eukaryotic cells. *Proc. Natl. Acad. Sci. USA* **79:**3162–3166.

130. **Leppla, S. H.** 1984. *Bacillus anthracis* calmodulin-dependent adenylate cyclase: chemical and enzymatic properties and interactions with eucaryotic cells. *Adv. Cyclic Nucleotide Protein Phosphorylation Res.* **17:**189–198.

131. **Leppla, S. H.** 1991. *The Anthrax Toxin Complex.* Academic Press, London.

132. **Lerat, E., V. Daubin, H. Ochman, and N. A. Moran.** 2005. Evolutionary origins of genomic repertoires in bacteria. *PLoS Biol.* **3:**e130.

133. **Lerat, E., and H. Ochman.** 2004. Psi-Phi: exploring the outer limits of bacterial pseudogenes. *Genome Res.* **14:**2273–2278.

134. **Lesic, B., S. Bach, J. M. Ghigo, U. Dobrindt, J. Hacker, and E. Carniel.** 2004. Excision of

the high-pathogenicity island of *Yersinia pseudotuberculosis* requires the combined actions of its cognate integrase and Hef, a new recombination directionality factor. *Mol. Microbiol.* **52**:1337–1348.

135. **Lesic, B., and E. Carniel.** 2005. Horizontal transfer of the high-pathogenicity island of *Yersinia pseudotuberculosis. J. Bacteriol.* **187**:3352–3358.

136. **Li, B., L. Jiang, Q. Song, J. Yang, Z. Chen, Z. Guo, D. Zhou, Z. Du, Y. Song, J. Wang, H. Wang, S. Yu, J. Wang, and R. Yang.** 2005. Protein microarray for profiling antibody responses to *Yersinia pestis* live vaccine. *Infect. Immun.* **73**:3734–3739.

137. **Li, Z. N., S. N. Mueller, L. Ye, Z. Bu, C. Yang, R. Ahmed, and D. A. Steinhauer.** 2005. Chimeric influenza virus hemagglutinin proteins containing large domains of the *Bacillus anthracis* protective antigen: protein characterization, incorporation into infectious influenza viruses, and antigenicity. *J. Virol.* **79**:10003–10012.

138. **Lin, B., Z. Wang, G. J. Vora, J. A. Thornton, J. M. Schnur, D. C. Thach, K. M. Blaney, A. G. Ligler, A. P. Malanoski, J. Santiago, E. A. Walter, B. K. Agan, D. Metzgar, D. Seto, L. T. Daum, R. Kruzelock, R. K. Rowley, E. H. Hanson, C. Tibbetts, and D. A. Stenger.** 2006. Broad-spectrum respiratory tract pathogen identification using resequencing DNA microarrays. *Genome Res.* **16**:527–535.

139. **Liu, H., N. H. Bergman, B. Thomason, S. Shallom, A. Hazen, J. Crossno, D. A. Rasko, J. Ravel, T. D. Read, S. N. Peterson, J. Yates III, and P. C. Hanna.** 2004. Formation and composition of the *Bacillus anthracis* endospore. *J. Bacteriol.* **186**:164–178.

140. **Liu, Y., P. M. Harrison, V. Kunin, and M. Gerstein.** 2004. Comprehensive analysis of pseudogenes in prokaryotes: widespread gene decay and failure of putative horizontally transferred genes. *Genome Biol.* **5**:R64.

141. **Lopez, J., J. Copps, C. Wilhelmsen, R. Moore, J. Kubay, M. St.-Jacques, S. Halayko, C. Kranendonk, S. Toback, D. DeShazer, D. L. Fritz, M. Tom, and D. E. Woods.** 2003. Characterization of experimental equine glanders. *Microbes Infect.* **5**:1125–1131.

142. **Lu, Q., W. Wei, P. E. Kowalski, A. C. Chang, and S. N. Cohen.** 2004. EST-based genome-wide gene inactivation identifies ARAP3 as a host protein affecting cellular susceptibility to anthrax toxin. *Proc. Natl. Acad. Sci. USA* **101**: 17246–17251.

143. **Mahan, M. J., J. M. Slauch, and J. J. Mekalanos.** 1993. Selection of bacterial virulence genes that are specifically induced in host tissues. *Science* **259**:686–688.

144. **Maiden, M. C., J. A. Bygraves, E. Feil, G. Morelli, J. E. Russell, R. Urwin, Q. Zhang, J. Zhou, K. Zurth, D. A. Caugant, I. M. Feavers, M. Achtman, and B. G. Spratt.** 1998. Multilocus sequence typing: a portable approach to the identification of clones within populations of pathogenic microorganisms. *Proc. Natl. Acad. Sci. USA* **95**:3140–3145.

145. **Manchee, R. J., M. G. Broster, R. M. Henstridge, A. J. Stagg, and J. Melling.** 1982. Anthrax island. *Nature* **296**:598.

146. **Marketon, M. M., R. W. DePaolo, K. L. DeBord, B. Jabri, and O. Schneewind.** 2005. Plague bacteria target immune cells during infection. *Science* **309**:1739–1741.

147. **Masui, S., S. Kamoda, T. Sasaki, and H. Ishikawa.** 1999. The first detection of the insertion sequence ISW1 in the intracellular reproductive parasite *Wolbachia. Plasmid* **42**:13–19.

148. **Maurin, M., and D. Raoult.** 1999. Q fever. *Clin. Microbiol. Rev.* **12**:518–553.

149. **McCaul, T. F., and J. C. Williams.** 1981. Developmental cycle of *Coxiella burnetii*: structure and morphogenesis of vegetative and sporogenic differentiations. *J. Bacteriol.* **147**:1063–1076.

150. **McDonald, W. H., and J. R. Yates III.** 2002. Shotgun proteomics and biomarker discovery. *Dis. Markers* **18**:99–105.

151. **Meselson, M., J. Guillemin, M. Hugh-Jones, A. Langmuir, I. Popova, A. Shelokov, and O. Yampolskaya.** 1994. The Svedlovsk anthrax outbreak of 1979. *Science* **266**:1202–1208.

152. **Mignot, T., B. Denis, E. Couture-Tosi, A. B. Kolsto, M. Mock, and A. Fouet.** 2001. Distribution of S-layers on the surface of *Bacillus cereus* strains: phylogenetic origin and ecological pressure. *Environ. Microbiol.* **3**:493–501.

153. **Mignot, T., M. Mock, D. Robichon, A. Landier, D. Lereclus, and A. Fouet.** 2001. The incompatibility between the PlcR- and AtxA-controlled regulons may have selected a nonsense mutation in *Bacillus anthracis. Mol. Microbiol.* **42**:1189–1198.

154. **Mikesell, P., B. E. Ivins, J. D. Ristroph, and T. M. Dreier.** 1983. Evidence for plasmid-mediated toxin production in *Bacillus anthracis. Infect. Immun.* **39**:371–376.

155. **Miller, J. M., J. G. Hair, M. Hebert, L. Hebert, F. J. Roberts, Jr., and R. S. Weyant.** 1997. Fulminating bacteremia and pneumonia due to *Bacillus cereus. J. Clin. Microbiol.* **35**:504–507.

156. **Minakhin, L., E. Semenova, J. Liu, A. Vasilov, E. Severinova, T. Gabisonia, R. Inman, A.**

Mushegian, and K. Severinov. 2005. Genome sequence and gene expression of *Bacillus anthracis* bacteriophage Fah. *J. Mol. Biol.* **354:**1–15.

157. Moens, S., and J. Vanderleyden. 1996. Functions of bacterial flagella. *Crit. Rev. Microbiol.* **22:**67–100.

158. Mogridge, J., K. Cunningham, D. B. Lacy, M. Mourez, and R. J. Collier. 2002. The lethal and edema factors of anthrax toxin bind only to oligomeric forms of the protective antigen. *Proc. Natl. Acad. Sci. USA* **99:**7045–7048.

159. Moore, R. A., S. Reckseidler-Zenteno, H. Kim, W. Nierman, Y. Yu, A. Tuanyok, J. Warawa, D. DeShazer, and D. E. Woods. 2004. Contribution of gene loss to the pathogenic evolution of *Burkholderia pseudomallei* and *Burkholderia mallei. Infect. Immun.* **72:**4172–4187.

160. Motin, V. L., A. M. Georgescu, J. P. Fitch, P. P. Gu, D. O. Nelson, S. L. Mabery, J. B. Garnham, B. A. Sokhansanj, L. L. Ott, M. A. Coleman, J. M. Elliott, L. M. Kegelmeyer, A. J. Wyrobek, T. R. Slezak, R. R. Brubaker, and E. Garcia. 2004. Temporal global changes in gene expression during temperature transition in *Yersinia pestis. J. Bacteriol.* **186:**6298–6305.

161. Nano, F. E., N. Zhang, S. C. Cowley, K. E. Klose, K. K. Cheung, M. J. Roberts, J. S. Ludu, G. W. Letendre, A. I. Meierovics, G. Stephens, and K. L. Elkins. 2004. A *Francisella tularensis* pathogenicity island required for intramacrophage growth. *J. Bacteriol.* **186:**6430–6436.

162. Naroeni, A., N. Jouy, S. Ouahrani-Bettache, J. P. Liautard, and F. Porte. 2001. *Brucella suis*–impaired specific recognition of phagosomes by lysosomes due to phagosomal membrane modifications. *Infect. Immun.* **69:**486–493.

163. Ngauy, V., Y. Lemeshev, L. Sadkowski, and G. Crawford. 2005. Cutaneous melioidosis in a man who was taken as a prisoner of war by the Japanese during World War II. *J. Clin. Microbiol.* **43:**970–972.

164. Nierman, W. C., D. DeShazer, H. S. Kim, H. Tettelin, K. E. Nelson, T. Feldblyum, R. L. Ulrich, C. M. Ronning, L. M. Brinkac, S. C. Daugherty, T. D. Davidsen, R. T. Deboy, G. Dimitrov, R. J. Dodson, A. S. Durkin, M. L. Gwinn, D. H. Haft, H. Khouri, J. F. Kolonay, R. Madupu, Y. Mohammoud, W. C. Nelson, D. Radune, C. M. Romero, S. Sarria, J. Selengut, C. Shamblin, S. A. Sullivan, O. White, Y. Yu, N. Zafar, L. Zhou, and C. M. Fraser. 2004. Structural flexibility in the *Burkholderia mallei* genome. *Proc. Natl. Acad. Sci. USA* **101:**14246–14251.

165. Nordberg, B. K., C. G. Schmiterlow, and H. J. Hansen. 1961. Pathophysiological investi-

gations into the terminal course of experimental anthrax in the rabbit. *Acta Pathol. Microbiol. Scand.* **53:**295–318.

166. Ogata, H., P. Renesto, S. Audic, C. Robert, G. Blanc, P.-E. Fournier, H. Parinello, J.-M. Claverie, and D. Raoult. 2005. The genome sequence of *Rickettsia felis* identifies the first putative conjugative plasmid in an obligate intracellular parasite. *PLoS Biol.* **3:**e248.

167. Okinaka, R. T., K. Cloud, O. Hampton, A. R. Hoffmaster, K. K. Hill, P. Keim, T. M. Koehler, G. Lamke, S. Kuman, J. Mahillon, D. Manter, Y. Martinez, D. Ricke, R. Svensson, and P. J. Jackson. 1999. Sequence and organization of pXO1, the large *Bacillus anthracis* plasmid harboring the anthrax toxin genes. *J. Bacteriol.* **181:**6509–6515.

168. Ong, C., C. H. Ooi, D. Wang, H. Chong, K. C. Ng, F. Rodrigues, M. A. Lee, and P. Tan. 2004. Patterns of large-scale genomic variation in virulent and avirulent *Burkholderia* species. *Genome Res.* **14:**2295–2307.

169. Ou, K., C. Ong, S. Y. Koh, F. Rodrigues, S. H. Sim, D. Wong, C. H. Ooi, K. C. Ng, H. Jikuya, C. C. Yau, S. Y. Soon, D. Kesuma, M. A. Lee, and P. Tan. 2005. Integrative genomic, transcriptional, and proteomic diversity in natural isolates of the human pathogen *Burkholderia pseudomallei. J. Bacteriol.* **187:**4276–4285.

170. Pal, C., B. Papp, and M. J. Lercher. 2005. Adaptive evolution of bacterial metabolic networks by horizontal gene transfer. *Nat. Genet.* **37:**1372–1375.

171. Parkhill, J., B. W. Wren, N. R. Thomson, R. W. Titball, M. T. Holden, M. B. Prentice, M. Sebaihia, K. D. James, C. Churcher, K. L. Mungall, S. Baker, D. Basham, S. D. Bentley, K. Brooks, A. M. Cerdeno-Tarraga, T. Chillingworth, A. Cronin, R. M. Davies, P. Davis, G. Dougan, T. Feltwell, N. Hamlin, S. Holroyd, K. Jagels, A. V. Karlyshev, S. Leather, S. Moule, P. C. Oyston, M. Quail, K. Rutherford, M. Simmonds, J. Skelton, K. Stevens, S. Whitehead, and B. G. Barrell. 2001. Genome sequence of *Yersinia pestis*, the causative agent of plague. *Nature* **413:**523–527.

172. Paulsen, I. T., R. Seshadri, K. E. Nelson, J. A. Eisen, J. F. Heidelberg, T. D. Read, R. J. Dodson, L. Umayam, L. M. Brinkac, M. J. Beanan, S. C. Daugherty, R. T. Deboy, A. S. Durkin, J. F. Kolonay, R. Madupu, W. C. Nelson, B. Ayodeji, M. Kraul, J. Shetty, J. Malek, S. E. Van Aken, S. Riedmuller, H. Tettelin, S. R. Gill, O. White, S. L. Salzberg, D. L. Hoover, L. E. Lindler, S. M. Halling, S. M. Boyle, and C. M. Fraser. 2002. The *Bru-*

cella suis genome reveals fundamental similarities between animal and plant pathogens and symbionts. *Proc. Natl. Acad. Sci. USA* **99:**13148–13153.

173. **Pearson, T., J. D. Busch, J. Ravel, T. D. Read, S. D. Rhoton, J. M. U'Ren, T. S. Simonson, S. M. Kachur, R. R. Leadem, M. L. Cardon, M. N. Van Ert, L. Y. Huynh, C. M. Fraser, and P. Keim.** 2004. Phylogenetic discovery bias in *Bacillus anthracis* using single-nucleotide polymorphisms from whole-genome sequencing. *Proc. Natl. Acad. Sci. USA* **101:**13536–13541.

174. **Pellizzari, R., C. Guidi-Rontani, G. Vitale, M. Mock, and C. Montecucco.** 1999. Anthrax lethal factor cleaves MKK3 in macrophages and inhibits the LPS/IFN gamma-induced release of NO and TNF alpha. *FEBS Lett.* **462:**199–204.

175. **Perry, R. D., and J. D. Fetherston.** 1997. *Yersinia pestis*—etiologic agent of plague. *Clin. Microbiol. Rev.* **10:**35–66.

176. **Plotkin, S. A., P. S. Brachman, M. Utell, F. H. Bumford, and M. M. Stechinson.** 2002. An epidemic of inhalation anthrax, the first in the twentieth century: I. Clinical features. *Am. J. Med.* **112:**4–12.

177. **Price, M. N., K. H. Huang, E. J. Alm, and A. P. Arkin.** 2005. A novel method for accurate operon predictions in all sequenced prokaryotes. *Nucleic Acids Res.* **33:**880–892.

178. **Pucci, M. J.** 2006. Use of genomics to select antibacterial targets. *Biochem. Pharmacol.* **71:**1066–1072.

179. **Pujol, C., and J. B. Bliska.** 2005. Turning *Yersinia* pathogenesis outside in: subversion of macrophage function by intracellular yersiniae. *Clin. Immunol.* **114:**216–226.

180. **Pujol, C., J. P. Grabenstein, R. D. Perry, and J. B. Bliska.** 2005. Replication of *Yersinia pestis* in interferon gamma-activated macrophages requires *ripA*, a gene encoded in the pigmentation locus. *Proc. Natl. Acad. Sci. USA* **102:**12909–12914.

181. **Qiu, J., D. Zhou, Y. Han, L. Zhang, Z. Tong, Y. Song, E. Dai, B. Li, J. Wang, Z. Guo, J. Zhai, Z. Du, X. Wang, and R. Yang.** 2005. Global gene expression profile of *Yersinia pestis* induced by streptomycin. *FEMS Microbiol. Lett.* **243:**489–496.

182. **Rajashekara, G., J. D. Glasner, D. A. Glover, and G. A. Splitter.** 2004. Comparative whole-genome hybridization reveals genomic islands in *Brucella* species. *J. Bacteriol.* **186:**5040–5051.

183. **Rasko, D. A., M. R. Altherr, C. S. Han, and J. Ravel.** 2005. Genomics of the *Bacillus cereus* group of organisms. *FEMS Microbiol. Rev.* **29:**303–329.

184. **Rasko, D. A., J. Ravel, O. A. Okstad, E. Helgason, R. Z. Cer, L. Jiang, K. A. Shores, D. E. Fouts, N. J. Tourasse, S. V. Angiuoli, J. Kolonay, W. C. Nelson, A. B. Kolsto, C. M. Fraser, and T. D. Read.** 2004. The genome sequence of *Bacillus cereus* ATCC 10987 reveals metabolic adaptations and a large plasmid related to *Bacillus anthracis* pXO1. *Nucleic Acids Res.* **32:**977–988.

185. **Ratner, A. J., K. R. Hippe, J. L. Aguilar, M. H. Bender, A. L. Nelson, and J. N. Weiser.** 2006. Epithelial cells are sensitive detectors of bacterial pore-forming toxins. *J. Biol. Chem.* **281:**12994–12998.

186. **Read, T. D., S. N. Peterson, N. Tourasse, L. W. Baillie, I. T. Paulsen, K. E. Nelson, H. Tettelin, D. E. Fouts, J. A. Eisen, S. R. Gill, E. K. Holtzapple, O. A. Okstad, E. Helgason, J. Rilstone, M. Wu, J. F. Kolonay, M. J. Beanan, R. J. Dodson, L. M. Brinkac, M. Gwinn, R. T. DeBoy, R. Madpu, S. C. Daugherty, A. S. Durkin, D. H. Haft, W. C. Nelson, J. D. Peterson, M. Pop, H. M. Khouri, D. Radune, J. L. Benton, Y. Mahamoud, L. Jiang, I. R. Hance, J. F. Weidman, K. J. Berry, R. D. Plaut, A. M. Wolf, K. L. Watkins, W. C. Nierman, A. Hazen, R. Cline, C. Redmond, J. E. Thwaite, O. White, S. L. Salzberg, B. Thomason, A. M. Friedlander, T. M. Koehler, P. C. Hanna, A. B. Kolsto, and C. M. Fraser.** 2003. The genome sequence of *Bacillus anthracis* Ames and comparison to closely related bacteria. *Nature* **423:**81–86.

187. **Read, T. D., S. L. Salzberg, M. Pop, M. Shumway, L. Umayam, L. Jiang, E. Holtzapple, J. D. Busch, K. L. Smith, J. M. Schupp, D. Solomon, P. Keim, and C. M. Fraser.** 2002. Comparative genome sequencing for discovery of novel polymorphisms in *Bacillus anthracis*. *Science* **296:**2028–2033.

188. **Reckseidler, S. L., D. DeShazer, P. A. Sokol, and D. E. Woods.** 2001. Detection of bacterial virulence genes by subtractive hybridization: identification of capsular polysaccharide of *Burkholderia pseudomallei* as a major virulence determinant. *Infect. Immun.* **69:**34–44.

189. **Revell, P. A., and V. L. Miller.** 2001. Yersinia virulence: more than a plasmid. *FEMS Microbiol. Lett.* **205:**159–164.

190. **Riemenschneider, J., A. Garrison, J. Geisbert, P. Jahrling, M. Hevey, D. Negley, A. Schmaljohn, J. Lee, M. K. Hart, L. Vanderzanden, D. Custer, M. Bray, A. Ruff, B. Ivins, A. Bassett, C. Rossi, and C. Schmaljohn.** 2003. Comparison of individual and combination DNA vaccines for B. anthracis, Ebola virus,

Marburg virus and Venezuelan equine encephalitis virus. *Vaccine* **21:**4071–4080.

191. **Rollins, S. E., S. M. Rollins, and E. T. Ryan.** 2003. *Yersinia pestis* and the plague. *Am. J. Clin. Pathol.* **119** (Suppl.)**:**S78–S85.

192. **Roop, R. M. I., J. M. Gee, G. T. Robertson, J. M. Richardson, W.-L. Ng, and M. E. Winkler.** 2003. *Brucella* stationary-phase gene expression and virulence. *Annu. Rev. Microbiol.* **57:**57–76.

193. **Ross, J. M.** 1955. On the histopathology of experimental anthrax in the guinea pig. *Br. J. Exp. Pathol.* **36:**336–339.

194. **Royce, T. E., J. S. Rozowsky, P. Bertone, M. Samanta, V. Stolc, S. Weissman, M. Snyder, and M. Gerstein.** 2005. Issues in the analysis of oligonucleotide tiling microarrays for transcript mapping. *Trends Genet.* **21:**466–475.

195. **Sallstrom, B., and S. G. Andersson.** 2005. Genome reduction in the alpha-*Proteobacteria*. *Curr. Opin. Microbiol.* **8:**579–585.

196. **Samuel, J. E., Kiss, K., and Varghess, S.** 2003. Molecular pathogenesis of *Coxiella burnetii* in a genomics era. *Ann. N.Y. Acad. Sci.* **990:**653–663.

197. **Sanchez, A. M., and K. A. Bradley.** 2004. Anthrax toxin: can a little be a good thing? *Trends Microbiol.* **12:**143–145.

198. **Sanger, F., S. Nicklen, and A. R. Coulson.** 1977. DNA sequencing with chain terminating inhibitors. *Proc. Natl. Acad. Sci. USA* **74:**5463–5467.

199. **Santic, M., M. Molmeret, K. E. Klose, and Y. A. Kwaik.** 2006. *Francisella tularensis* travels a novel, twisted road within macrophages. *Trends Microbiol.* **14:**37–44.

200. **Schubert, S., A. Rakin, H. Karch, E. Carniel, and J. Heesemann.** 1998. Prevalence of the "high-pathogenicity island" of *Yersinia* species among *Escherichia coli* strains that are pathogenic to humans. *Infect. Immun.* **66:**480–485.

201. **Schuler, A.** 2005. Billions for biodefense: federal agency biodefense budgeting, FY2005–FY2006. *Biosecur. Bioterror.* **3:**94–101.

202. **Schuler, A.** 2004. Billions for biodefense: federal agency biodefense funding, FY2001–FY2005. *Biosecur. Bioterror.* **2:**86–96.

203. **Sergeev, N., M. Distler, S. Courtney, S. F. Al-Khaldi, D. Volokhov, V. Chizhikov, and A. Rasooly.** 2004. Multipathogen oligonucleotide microarray for environmental and biodefense applications. *Biosens. Bioelectron.* **20:**684–698.

204. **Seshadri, R., and J. E. Samuel.** 2001. Characterization of a stress-induced alternate sigma factor, RpoS, of *Coxiella burnetii* and its expres-sion during the developmental cycle. *Infect. Immun.* **69:**4874–4883.

205. **Seshadri, R., I. T. Paulsen, J. A. Eisen, T. D. Read, K. E. Nelson, W. C. Nelson, N. L. Ward, H. Tettelin, T. M. Davidsen, M. J. Beanan, R. T. Deboy, S. C. Daugherty, L. M. Brinkac, R. Madupu, R. J. Dodson, H. M. Khouri, K. H. Lee, H. A. Carty, D. Scanlan, R. A. Heinzen, H. A. Thompson, J. E. Samuel, C. M. Fraser, and J. F. Heidelberg.** 2003. Complete genome sequence of the Q-fever pathogen *Coxiella burnetii*. *Proc. Natl. Acad. Sci. USA* **100:**5455–5460.

206. **Shannon, J. G., C. L. Ross, T. M. Koehler, and R. F. Rest.** 2003. Characterization of anthrolysin O, the *Bacillus anthracis* cholesterol-dependent cytolysin. *Infect. Immun.* **71:**3183–3189.

207. **Shendure, J., R. D. Mitra, C. Varma, and G. M. Church.** 2004. Advanced sequencing technologies: methods and goals. *Nat. Rev. Genet.* **5:**335–344.

208. **Shukla, H. D., and S. K. Sharma.** 2005. *Clostridium botulinum*: a bug with beauty and weapon. *Crit. Rev. Microbiol.* **31:**11–18.

209. **Sjöstedt, A.** 2003. Virulence determinants and protective antigens of *Francisella tularensis*. *Curr. Opin. Microbiol.* **6:**66–71.

210. **Slein, M. W., and G. F. Logan, Jr.** 1960. Mechanism of action of the toxin of *Bacillus anthracis*. I. Effect in vivo on some blood serum components. *J. Bacteriol.* **80:**77–85.

211. **Smith, H., and J. Keppie.** 1954. Observations on experimental anthrax; demonstration of a specific lethal factor produced *in vivo* by *Bacillus anthracis*. *Nature* **173:**869–870.

212. **Song, Y., Z. Tong, J. Wang, L. Wang, Z. Guo, Y. Han, J. Zhang, D. Pei, D. Zhou, H. Qin, X. Pang, Y. Han, J. Zhai, M. Li, B. Cui, Z. Qi, L. Jin, R. Dai, F. Chen, S. Li, C. Ye, Z. Du, W. Lin, J. Wang, J. Yu, H. Yang, J. Wang, P. Huang, and R. Yang.** 2004. Complete genome sequence of *Yersinia pestis* strain 91001, an isolate avirulent to humans. *DNA Res.* **11:**179–197.

213. **Stanley, J. L., and H. Smith.** 1961. Purification of factor I and recognition of a third factor of the anthrax toxin. *J. Gen. Microbiol.* **26:**49–66.

214. **Steele, A. D., J. M. Warfel, and F. D'Agnillo.** 2005. Anthrax lethal toxin enhances cytokine-induced VCAM-1 expression on human endothelial cells. *Biochem. Biophys. Res. Commun.* **337:**1249–1256.

215. **Sterne, M.** 1982. Anthrax island. *Nature* **295:**362.

216. **Stevens, M. P., J. M. Stevens, R. L. Jeng, L. A. Taylor, M. W. Wood, P. Hawes, P.**

Monaghan, M. D. Welch, and E. E. Galyov. 2005. Identification of a bacterial factor required for actin-based motility of *Burkholderia pseudomallei*. *Mol. Microbiol.* **56**:40–53.

217. **Strayley, S. C., and P. A. Harmon.** 1984. *Yersinia pestis* growns within phagolysosomes in mouse peritoneal macrophages. *Infect. Immun.* **45**:655–659.

218. **Styer, K. L., G. W. Hopkins, S. S. Bartra, G. V. Plano, R. Frothingham, and A. Aballay.** 2005. *Yersinia pestis* kills *Caenorhabditis elegans* by a biofilm-independent process that involves novel virulence factors. *EMBO Rep.* **6**:992–997.

219. **Svensson, K., P. Larsson, D. Johansson, M. Byström, M. Forsman, and A. Johansson.** 2005. Evolution of subspecies of *Francisella tularensis*. *J. Bacteriol.* **187**:3903–3908.

220. **Taoka, M., Y. Yamauchi, T. Shinkawa, H. Kaji, W. Motohashi, H. Nakayama, N. Takahashi, and T. Isobe.** 2004. Only a small subset of the horizontally transferred chromosomal genes in *Escherichia coli* are translated into proteins. *Mol. Cell. Proteomics* **3**:780–787.

221. **Thorne, C. B.** 1993. *Bacillus anthracis*, p. 113–124. *In* Bacillus subtilis *and Other Gram-Positive Bacteria*. American Society for Microbiology, Washington, DC.

222. **Tinsley, E., and S. A. Khan.** 2006. A novel FtsZ-like protein is involved in replication of the anthrax toxin–encoding pXO1 plasmid in *Bacillus anthracis*. *J. Bacteriol.* **188**:2829–2835.

223. **Torok, T. J., R. V. Tauxe, R. P. Wise, J. R. Livengood, R. Sokolow, S. Mauvais, K. A. Birkness, M. R. Skeels, J. M. Horan, and L. R. Foster.** 1997. A large community outbreak of salmonellosis caused by intentional contamination of restaurant salad bars. *JAMA* **278**:389–395.

224. **Tuanyok, A., H. S. Kim, W. C. Nierman, Y. Yu, J. Dunbar, R. A. Moore, P. Baker, M. Tom, J. M. Ling, and D. E. Woods.** 2005. Genome-wide expression analysis of iron regulation in *Burkholderia pseudomallei* and *Burkholderia mallei* using DNA microarrays. *FEMS Microbiol. Lett.* **252**:327–335.

224a. **Turell, M. J., and G. B. Knudsen.** 1987. Mechanical transmission of *Bacillus anthracis* by stable flies (*Stomoxys calcitrans*) and mosquitos (*Aedes aegypti* and *Aedes taeniorhynchus*). *Infect. Immun.* **55**:1859–1861.

225. **Valvano, M. A., K. E. Keith, and S. T. Cardona.** 2005. Survival and persistence of opportunistic *Burkholdeia* species in host cells. *Curr. Opin. Microbiol.* **8**:99–105.

226. **van Ham, R. C., J. Kamerbeek, C. Palacios, C. Rausell, F. Abascal, U. Bastolla, J. M. Fernandez, L. Jimenez, M. Postigo, F. J. Silva, J. Tamames, E. Viguera, A. Latorre, A. Valencia, F. Moran, and A. Moya.** 2003. Reductive genome evolution in *Buchnera aphidicola*. *Proc. Natl. Acad. Sci. USA* **100**:581–586.

227. **Vetter, S. M., and P. M. Schlievert.** 2005. Glycerol monolaurate inhibits virulence factor production in Bacillus anthracis. *Antimicrob. Agents Chemother.* **49**:1302–1305.

228. **Vitale, G., R. Pellizzari, C. Recchi, G. Napolitani, M. Mock, and C. Montecucco.** 1998. Anthrax lethal factor cleaves the N-terminus of MAPKKs and induces tyrosine/threonine phosphorylation of MAPKs in cultured macrophages. *Biochem. Biophys. Res. Commun.* **248**:706–711.

229. **von Delwig, A., J. A. Musson, H. K. Shim, J. J. Lee, N. Walker, C. V. Harding, E. D. Williamson, and J. H. Robinson.** 2005. Distribution of productive antigen-processing activity for MHC class II presentation in macrophages. *Scand. J. Immunol.* **62**:243–250.

230. **Walther, B. A., and P. W. Ewald.** 2004. Pathogen survival in the external environment and the evolution of virulence. *Biol. Rev. Camb. Philos. Soc.* **79**:849–869.

231. **Waterfield, N. R., B. W. Wren, and R. H. Ffrench-Constant.** 2004. Invertebrates as a source of emerging human pathogens. *Nat. Rev. Microbiol.* **2**:833–841.

232. **Wei, Z., P. Schnupf, M. A. Poussin, L. A. Zenewicz, H. Shen, and H. Goldfine.** 2005. Characterization of *Listeria monocytogenes* expressing anthrolysin O and phosphatidylinositol-specific phospholipase C from *Bacillus anthracis*. *Infect. Immun.* **73**:6639–6646.

233. **Weiner, M. A., and P. C. Hanna.** 2003. Macrophage-mediated germination of *Bacillus anthracis* endospores requires the *gerH* operon. *Infect. Immun.* **71**:3954–3959.

234. **Whittam, T. S., and A. C. Bumbaugh.** 2002. Inferences from whole-genome sequences of bacterial pathogens. *Curr. Opin. Genet. Dev.* **12**:719–725.

235. **Williamson, E. D., S. M. Eley, A. J. Stagg, M. Green, P. Russell, and R. W. Titball.** 2000. A single dose sub-unit vaccine protects against pneumonic plague. *Vaccine* **19**:566–571.

236. **Wong, S. S. Y., and K. Y. Yuen.** 2005. The severe acute respiratory syndrome (SARS). *J. Neurovirol.* **11**:455–468.

237. **Wren, B. W.** 2003. The yersiniae—a model genus to study the rapid evolution of bacterial pathogens. *Nat. Rev. Microbiol.* **1**:55–64.

238. **Zhou, D., Y. Han, L. Qin, Z. Chen, J. Qiu, Y. Song, B. Li, J. Wang, Z. Guo, Z. Du, X. Wang, and R. Yang.** 2005. Transcriptome analysis of the Mg^{2+}-responsive PhoP regulator in *Yersinia pestis*. *FEMS Microbiol. Lett.* **250:**85–95.

239. **Zhou, D., Y. Han, Y. Song, P. Huang, and R. Yang.** 2004. Comparative and evolutionary genomics of *Yersinia pestis*. *Microbes Infect.* **6:**1226–1234.

240. **Zhou, D., Y. Han, and R. Yang.** 2006. Molecular and physiological insights into plague transmission, virulence and etiology. *Microbes Infect.* **8:**273–284.

241. **Zwick, M. E., F. McAfee, D. J. Cutler, T. D. Read, J. Ravel, G. R. Bowman, D. R. Galloway, and A. Mateczun.** 2005. Microarray-based resequencing of multiple *Bacillus anthracis* isolates. *Genome Biol.* **6:**R10.

IMPACT OF PHAGES ON EVOLUTION
OF BACTERIAL PATHOGENICITY

Harald Brüssow

11

INTRODUCTION

Phage-encoded toxins are not a novelty, but the overall role that phages have played in the evolution of bacterial pathogenicity has become apparent to microbiologists only recently. A number of laboratories have argued over the last years that phage-associated virulence factors are a logical evolutionary consequence of lysogeny, the integration of the phage DNA into the bacterial chromosome. I will summarize the major theoretical arguments for this connection without going into a detailed analysis presented in previous reviews (6, 19, 26, 30, 32, 57, 70, 125, 126). Instead, I will review recent progress in the analysis of prophages from two intensively investigated bacterial pathogens, *Streptococcus pyogenes* and *Staphylococcus aureus*, which were strongly shaped by prophage acquisitions. An interesting facet of these two medically important pathogens is their double role as commensals and pathogens. In this context, certain aspects of bacterial pathogenicity can be interpreted as a recall of antipredation strategies that bacteria evolved against phagocytosis by protozoan grazers. From an evolutionary perspective, bacterial pathogenesis can be seen as a special case of the "eat and be eaten" motif that pervades all interactions in biology when applied to bacteria, which live on us. It is indeed ironic that bacteria recruit genes from their viral predators to fight their eukaryotic hosts.

Extent and Types of Prophages

Many bacteria contain multiple genomes of bacterial viruses in their chromosome. Prophage DNA can constitute a sizable part of the total bacterial DNA. The most extreme case is the gram-negative food pathogen *Escherichia coli* O157:H7, strain Sakai. It contains 18 prophage genome elements, which amount to 16% of its total genome content. This is not an exceptional case; the gram-positive *Streptococcus pyogenes* contains up to eight prophages, which can represent 12% of the bacterial DNA content. These prophages do not represent exotic phage types: the *E. coli* O157 prophages resemble the well-known temperate *E. coli* phages λ, P2, its satellite phage P4, and phage Mu (89). The *S. pyogenes* prophages belong to phage types that are well known to the dairy microbiologist (25, 42).

Virulence and Variability

As mobile DNA elements, phage DNA is a vector for lateral gene transfer between bacteria (28). Not surprisingly, numerous virulence

Harald Brüssow Nestlé Research Centre, Nutrition and Health Department / Food and Health Microbiology, CH-1000 Lausanne 26, Vers-chez-les-Blanc, Switzerland.

Bacterial Pathogenomics, Edited by M. J. Pallen et al.
© 2007 ASM Press, Washington, D.C.

factors from bacterial pathogens are therefore phage encoded (19, 125, 126). Furthermore, prophages account in several bacterial species for a substantial amount of interstrain genetic variability (e.g., *Staphylococcus aureus* [4] and *Streptococcus pyogenes* [108]) (Fig. 1). When genomes from closely related bacteria were compared in a dot plot analysis, prophage sequences frequently accounted for a substantial, if not major, part of the differences between the genomes (76, 95). Microarray (96) and PCR (90) analyses confirmed these conclusions, irrespective of whether pathogenic or nonpathogenic bacteria were investigated.

Expression

When mRNA expression patterns were studied with microarrays in lysogenic bacteria that underwent physiologically relevant changes in growth conditions, prophage genes figured prominently among the mRNA species changing their expression pattern (110, 129). These data demonstrate that prophages are not a passive genetic cargo of the bacterial chromosome but likely an active player in the cell physiology. Subtractive mRNA hybridization analysis demonstrated that prophage genes are also a prominent share of the *E. coli* genes upregulated when the bacteria invaded the lungs of infected birds (45).

Forms

Prophages can be present in many different forms, ranging from inducible prophages via prophages showing deletions, insertion, and rearrangements, to prophage remnants that lost most of the phage genome. This makes it difficult to distinguish prophage remnants from conjugative transposon-like elements, from remnants of integrative plasmids, and from integrons and pathogenicity islands. The most prominent class of prophages are represented by lambda-like *Siphoviridae*, followed by *Inoviridae* (112, 127) and then P2- and Mu-like *Myoviridae* (82, 85). However, lysogeny need not necessarily lead to the integration of phage DNA into the bacterial chromosome. A prominent case for a toxin-encoding prophage

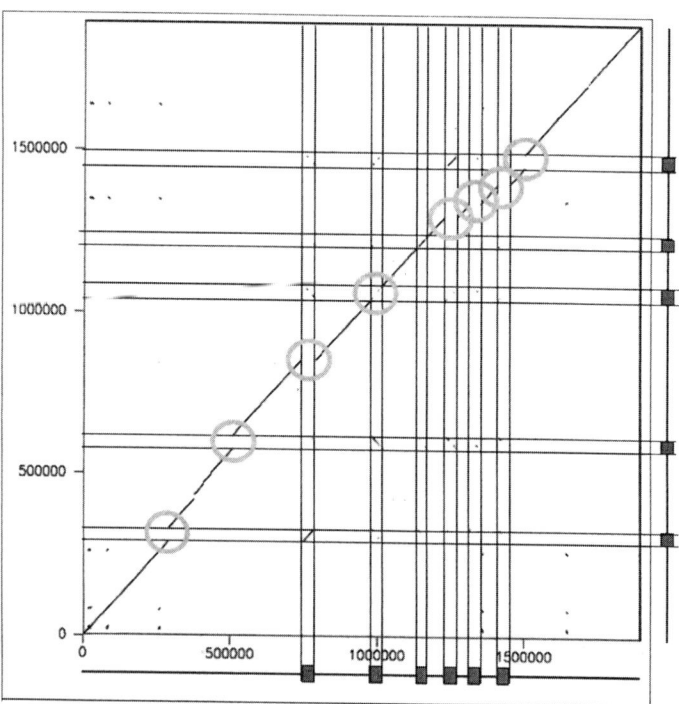

FIGURE 1 Genome alignment of *S. pyogenes* serotype M18 (vertical) against serotype M3 (horizontal) shows that practically all diversity is prophage induced. Regions of nonalignment are circled, and prophages are marked with rectangles.

that is not integrated into the bacterial chromosome is provided by *Clostridium botulinum* (105).

PHAGE-BACTERIUM COEVOLUTION

The Cooperating Phage

The viral DNA is subject to different selective pressures when replicating during lytic infection cycles as compared with prophage DNA maintained in the bacterial genome during lysogeny. Darwinian considerations along with the "selfish gene" concept lead to interesting conjectures (26, 30, 42, 70). One could anticipate that the prophage decreases the fitness of its lysogenic host by at least two processes: first by the metabolic burden to replicate extra DNA, and second by the lysis of the host after prophage induction. To compensate for these disadvantages, one has to assume that temperate phages encode functions that increase the fitness of the lysogen. According to the selective value of these postulated phage genes, the lysogenic cell will be maintained or even be overrepresented in the bacterial population. An obvious selective advantage for the lysogenic host is the immunity (phage repressor) and superinfection exclusion genes of the prophage that protect the lysogen against extra phage infection. These genes are also of direct advantage for the prophage because they exclude superinfecting phage DNA from competing with the resident prophage DNA for the same host. When phages from the environment are not a sufficiently strong selection pressure, other phage genes have to increase the fitness of the lysogenic host, frequently in unanticipated ways (lysogenic conversion genes). Classic examples of such phage-encoded genes that increase host fitness include the nonessential phage λ genes *bor* and *lom*, which confer serum resistance and better survival in macrophages, respectively, to the *Escherichia coli* lysogen (10). In these cases, the reproductive success of the lysogenic bacterium endowed with these new genes translates directly into an evolutionary success for the resident prophage. In this theoretical framework, it becomes understandable why

phages that integrate into bacterial pathogens became important carriers of virulence genes.

Arms Race

However, host-parasite relationships are also in an arms race and therefore represent a highly dynamic genetic equilibrium. Gains from prophages carrying genes that increase host fitness are short lived from a bacterial standpoint if the resident prophage ultimately destroys the bacterial lineage. In this way, prophages can be considered to be dangerous molecular time bombs that can kill the lysogenic cell upon their eventual induction (70). One would therefore expect evolution to select lysogenic bacteria with mutations in the prophage DNA. Mutations that inactivate the prophage induction process avoid the loss of the lysogenic clone from the bacterial population (Fig. 2). One would further expect that selection leads to large-scale deletion of prophage DNA in order to decrease the metabolic burden of extra DNA synthesis and a littering of the bacterial genomes by selfish DNA elements. One predicts, furthermore, that useful prophage genes (e.g., lysogenic conversion genes) are preferentially spared from this deletion process because their loss would actually decrease the fitness of the cell. It was proposed that a high genomic deletion rate is instrumental in removing dangerous genetic parasites from the bacterial genome (70). These deletion processes could explain why the bacterial genomes (in general) did not increase in size despite a constant bombardment with parasitic DNA over evolutionary time periods. The streamlined bacterial chromosome containing few pseudogenes might be the consequence of this deletion process of parasitic DNA.

THE PHAGE SIDE OF THE COIN. . .

Antiquity and Sheer Numbers

Phages and bacteria are linked by a long history of coevolution. The time dimension of this coevolution cannot at present be defined. Some phages may have originated from assemblages of host genes that split billions of years

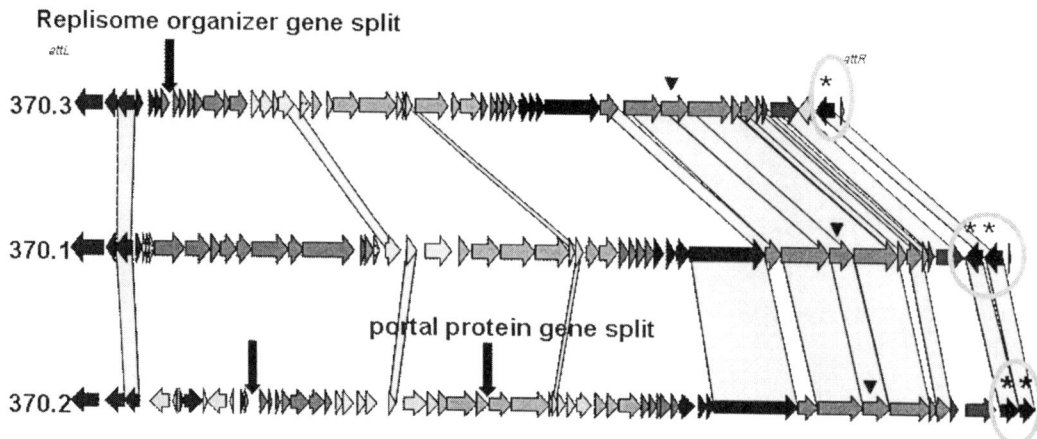

FIGURE 2 *S. pyogenes* prophage inactivation by point mutations in DNA replication or DNA assembly proteins (vertical black arrow). Virulence genes are circled.

ago from bacterial genomes, escaped from cellular control, and now lead a selfish life. Other phages might have originated more recently. Most importantly, there is ample evidence for continued exchange of genetic elements between phages, bacterial genomes, and various other mobile genetic elements. Microbial ecologists tell us that phages outnumber bacteria in many environments by a factor of 10 and represent, with approximately 10^{31} tailed phage particles, numerically the largest share of biological material on earth (27). Up to 10^7 particles/ml were found in ocean water and sediment (132). It has been estimated that 10^{25} phage infections are initiated every second worldwide (28), and this has probably occurred for the last 3 billion years. Phage-bacterium interaction has thus probably shaped the biosphere. Furthermore, random sequencing of viral DNA in different environments ("metagenome" analysis) argues in favor of 2 billion undiscovered phage open reading frames (ORFs), thus representing a large fraction of the total DNA sequence space (102).

The Modular Theory of Phage Evolution

When looking at a dot plot alignment for two prophages from, for example, two different *Staphylococcus aureus* strains, we typically see a long, straight line interrupted by small regions of nonalignment. In prophage alignments, the gaps are mainly DNA replacements. A DNA segment found in one phage is replaced in another phage by a sequence-unrelated DNA segment that frequently fulfills the same or a related function. These modular exchanges in phages were already described by heteroduplex analysis in phages from enterobacteria in the pregenomics era (18, 116, 128). According to that theory, the genomes from lambdoid coliphages can be subdivided into 11 modules, each representing an independent genetic functional unit. Each functional unit is represented by several alleles (modules) (33). The order of modules on the phage genome map is fairly conserved while different alleles of the modules can be freely assorted. This gives phage genomes a substantial genomic variability. Actually, one could describe the prophages from a given bacterial species as "swarms" of a few phage species that share variable amounts of genome segments on a relatively random basis (Fig. 3). Phage genomics has largely confirmed the modular theory of phage evolution (63).

Mode of Recombination

There is an ongoing discussion on the mechanism of phage module exchange, with much of the dissent centered on the relative role of

S. pyogenes Prophages

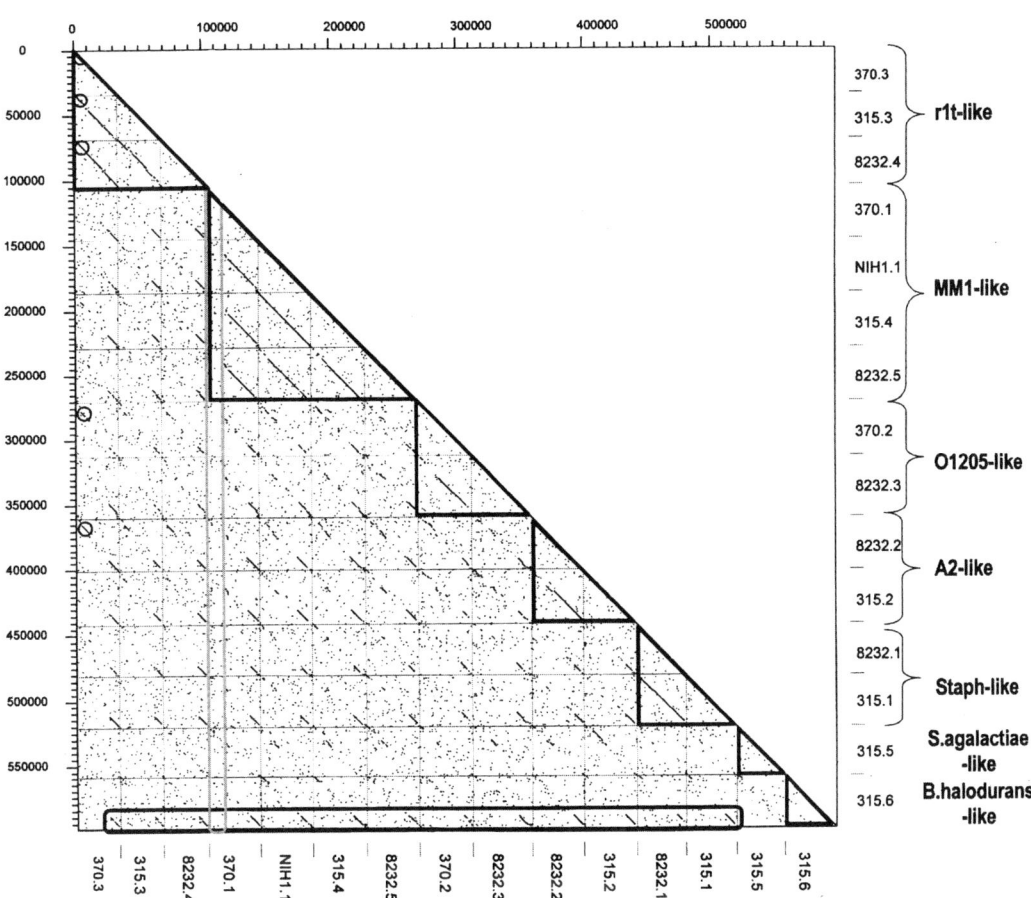

FIGURE 3 Dot plot matrix of *S. pyogenes* prophages identified in *S. pyogenes* sequencing projects.

illegitimate (63, 93) versus homologous recombination (25, 37, 80, 81, 88, 94) in this process. Many arguments seem to favor illegitimate recombination. Let us refer again to prophages from a well-known pathogen. In an alignment of *S. aureus* prophages Mu50B and ETA, the apparent units of DNA exchange between phages are small and might cover a few adjacent genes, only a single gene, or even a gene fragment encoding a single protein domain. The great variability seen in many phage genome alignments seems to exclude the use of a few predetermined conserved linking sites for recombination. Illegitimate recombination occurs with much lower frequency than ho-

mologous recombination, but the estimated size of the global phage population makes even unlikely recombination events observable, provided that they are conferring a sufficiently strong selective advantage to the recombinant phage.

Lysogenic Conversion

For the purpose of this review, another process of gene acquisition is of even greater importance: lysogenic conversion. The acquisition of prophages would be an irrelevant process for the evolution of pathogenic bacteria if phages did not transfer useful genes to the lysogen. Some phage genes are known to increase the survival fitness of lysogens. The selection value of prophage was illustrated in a recent study of different *Salmonella enterica* serovar Typhimurium strains harboring different sets of prophages. Lysogens like these often release low titers of phage (10^2 to 10^4 CFU/ml). These initially low titers of phage were sufficient to kick off an efficient decimation of a competing nonlysogenic strain (17).

Even more exciting was the discovery that phages can directly contribute to the virulence of pathogens (Fig. 4). This was recognized relatively early for toxins of *Corynebacterium diphtheriae*, *Clostridium botulinum* (botulism), *Streptococcus pyogenes* (scarlet fever), *Staphylococcus aureus* (food poisoning), and *E. coli* (Shiga toxin), which are all phage encoded. As far as we know, these genes do not play a role in the life cycle of the phages. The list of phage-encoded fitness factors is rapidly growing and now comprises a wide range of different genes, including ADP-ribosyl transferase toxins, superantigens, lipopolysaccharide-modifying enzymes, type III effector proteins, detoxifying enzymes, hydrolytic enzymes, or proteins conferring serum resistance. In exceptional cases, phage tail genes seem to have developed dual functions and also serve as ad-hesion proteins for bacterial host attachment (*Streptococcus mitis pblA, B* genes) (11).

Location of "Extra" Genes at Prophage Ends

An important lead was provided by the early observation that the toxin genes in *C. diphtheriae* and *S. pyogenes* were encoded next to the right phage attachment site *attR* (Fig. 5). This specific location led to the hypothesis that these prophage genes actually represent bacterial genes that were acquired by a faulty excision process from a previous bacterial host. Sometimes these toxin genes still showed a

Protein	Gene
Extracellular toxins	
Diphtheria toxin	*tox*
Neurotoxin	*C1*
Shiga toxins	*stx1, -2*
Enterohemolysin	*hly2*
Cytotoxin	*ctx*
Enterotoxin	*see, sel*
Enterotoxin P	*sep*
Enterotoxin A	*entA*
Enterotoxin A	*sea*
Exfoliative toxin A	*eta*
Leukocidin	*lukS, -F, -M*
Toxin A	*speA*
Toxin C	*speC*
Toxin A1, A3, C, I, H, M, L, K	*speA1, -A3, -C, -I, -H, -M, -L, -K*
Superantigens	*ssa*
Cholera toxin	*ctxAB*
Proteins that alter antigenicity	
Membrane proteins	Mu-like
Glucosylation	*rfb*
Glucosylation	*gtr*
O-antigen acetylase	*oac*
Glucosyl transferase	*gtrII*
Effector proteins involved in invasion	
Type III effector	*sopE*
Type III effector	*gogB*
Type III effector	*ssel (gtgB)*
Type III effector	*sspH1*

Protein	Gene
Enzymes required for intracellular survival	
Superoxide dismutase	*sodC*
Superoxide dismutase	*sodC-I*
Superoxide dismutase	*sodC-III*
Neuraminidase	*nanH*
Staphylokinase	*sak*
Hyaluronidase	*hylP*
Serum resistance	
OMP	*lom*
OMP	*bor*
OMP	*eib*
Adhesions for bacterial host attachment	
Vir	*vir*
Coat proteins	*pblA, pblB*
TCP pilus	*tcp*
Others	
IS-like	*gipA*
Antivirulence gene	*grvA*
G-protein-like	*glo*
Mitogenic factors	*mf2, -3, -4*
Streptodornases	*sdn, sda*
Phospholipase	*sla*

FIGURE 4 Selective list of virulence genes encoded on prophages.

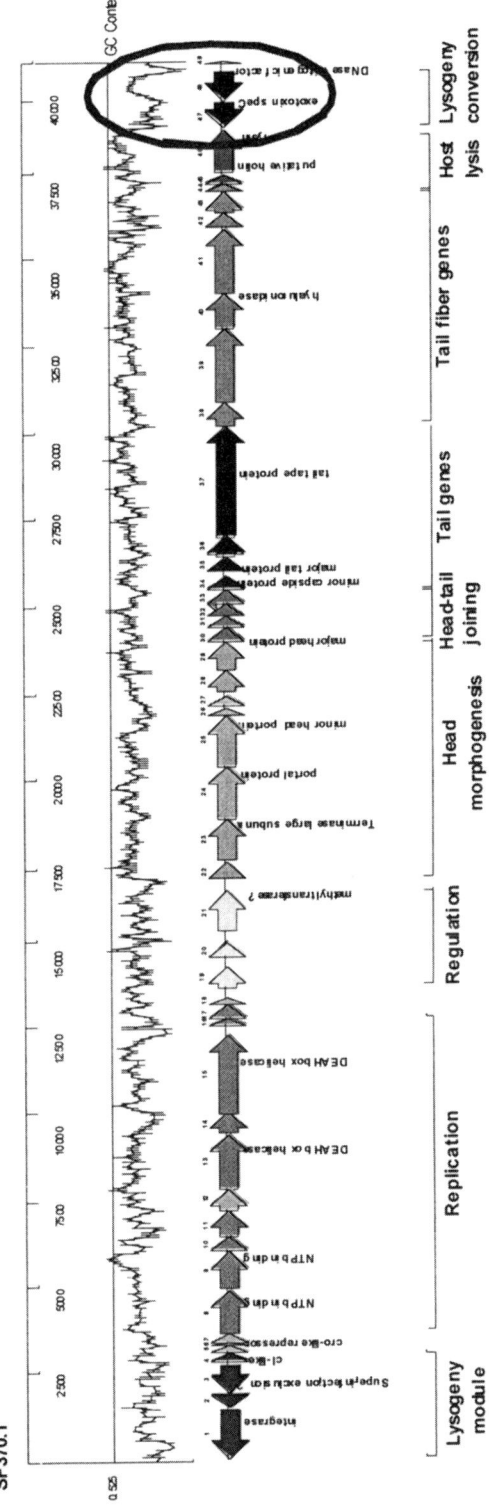

FIGURE 5 Genome organization of a typical *S. pyogenes* prophage encoding two virulence factors (circled).

clearly distinct GC content, pointing to an unusual bacterial host as source for this DNA (46).

In prophages from gram-negative bacteria, "extra" genes were identified near both prophage DNA ends. Examples are O antigen-modifying enzymes inserted between the phage integrase gene and the left attachment site *attL* in bacteriophage P22 or genes inserted downstream of the phage tail genes (19). The location at both prophage ends has an intrinsic logic: an extension of these prophage transcripts would run into the bacterial DNA or as an antimessenger into the prophage DNA, in both cases preventing an accidental induction of the prophage via transcription of prophage DNA.

Morons

Another frequent location for extra genes in lamboid prophages from gram-negative bacteria was next to the N- or Q-like antiterminator genes in the middle of the prophage genome (19). In these locations, an accidental induction of the prophage is a real risk. Therefore, it was postulated that these extra genes are flanked by an independent promoter and terminator structure. For this independently transcribed gene cassette, the Pittsburgh phage group coined the expression "moron" (more DNA; 63).

Fitness Factor: The USA300 Case

It was postulated that prophages from pathogens are only a special case for the general role that prophages play for bacterial evolution in general. If the bacterial host is a nonpathogenic species, the lysogenic conversion genes should take the form of fitness factors that further the progress of the lysogen by other means. I will illustrate the thin line between a fitness and a virulence factor with the arginine catabolic mobile element (ACME) from *S. aureus*. *S. aureus* poses increasing treatment problems as a result of the rapid rise of antibiotic-resistant strains. Methicillin-resistant *S. aureus* (MRSA) was initially observed in nosocomial infections and attributed to the high selection pressure of antibiotic use in the hospital. However, over the past years, MRSA was also ob-

served in community-acquired infections, and microbiologists identified a rapid clonal expansion of one *S. aureus* strain behind this phenomenon, named USA300 (92). The core genome of this strain did not differ from that of a much less virulent *S. aureus* strain isolated in the 1960s in Britain, but the strains differed substantially for their virulence in a mouse infection model (124). The only significant differences between these strains were associated with mobile DNA elements (43). These were:

- plasmids, two of which encode resistance to a variety of antibiotics,
- two prophages, which encode the Panton-Valentine leukocidin (PVL) toxin, staphylokinase, and a chemotaxis inhibitory protein,
- a pathogenicity island that encodes two enterotoxins of the pyrogenic toxin superantigen class,
- SCC*mec* IV, the type IV staphylococcal chromosomal cassette, which contains the *mecA* gene conferring resistance to β-lactam antibiotic methicillin,
- ACME, the arginine catabolic mobile element.

This latter nonphage element contains a complete arginine deiminase pathway that converts L-arginine to carbon dioxide, ATP, and ammonia. At first glance, this suite of catabolic enzymes is a classical fitness factor involved in peptide nutrition and energy provision under anoxic growth conditions. This diagnosis fits with the likely origin of this element from the skin commensal *S. epidermidis*. At the same time, it must be an important virulence factor, which can be easily rationalized: *S. aureus* encounters anoxic growth conditions during anaerobic growth in wounds. Furthermore, L-arginine is a substrate for nitric oxide production. Depletion of L-arginine will thus interfere with NO production, an important part of the innate and adaptive immune responses against microbial infection. Similar expansions were seen for clones of *S. pyogenes*. It is notable that these emerging clones did not only become predominant clinical isolates, but also represented major commensal isolates from healthy carriers (59).

Antibiotic Resistance and Prophages

USA300 can illustrate another aspect of mobile elements. We will see in this review that many different virulence/fitness factors are associated with prophages. This makes sense for the pathogen because it provides easy access to an assortment of different genes that constitute a type of toolbox for "how to make a successful bacterial pathogen." Notably, the different classes of mobile elements overtook different tool functions. For example, plasmids, transposons, and pathogenicity islands are privileged carriers of antibiotic resistance genes. Prophages, in contrast, easily carry toxin genes and genes interfering with the immune defense of the animal host, but not antibiotic resistance genes. There are only a few exceptions, like an element in the erythromycin-resistant serotype M6 clone of *Streptococcus pyogenes*, which was associated with a large outbreak of pharyngitis in schoolchildren from Pittsburgh. The clone contained the *mefA* gene encoding an antibiotic efflux pump. Mitomycin C–induced M6 cells showed the *mefA* gene in the cell supernatant in a DNase-resistant form (9). The supernatant demonstrated likewise many siphovirus particles, and the antibiotic resistance could be transferred to other strains by transduction. These observations suggested the incorporation of the *mefA* gene into a phage particle. Indeed, sequence analysis revealed a 59-kb chimeric genetic element composed of a transposon inserted into a prophage. It is currently not clear what evolutionary or ecological factors determine this unequal distribution of fitness/virulence factors with the various mobile elements.

....AND THE BACTERIAL SIDE OF THE COIN

Lateral Gene Transfer as a Source of Genetic Novelty

Comparative bacterial genomics led to the distinction of a conserved core genome sequence, which is shaped by the mechanisms of vertical evolution, and a variable part of the genome, which is dominated by processes of horizontal evolution. Phage transduction and prophage integration are major mechanisms of lateral DNA transfer in prokaryotes. Bacteria are therefore confronted with a dilemma: phages are a threat to their survival, and bacteria must mount defensive countermeasures against them (surface changes, restriction modification, and a variety of abortive phage infection mechanisms). At the same time, phages are an important tool for the acquisition of genes that could help defend their ecological place or gain new ones.

A Long Lineage

Bacterial evolution started long before the emergence of animals. In fact, bacteria and archaea were the first, and for some time the only, inhabitants of earth. So early evolution involved only competition, genetic exchange, and selection between bacteria. One can assume that phages also took part in this early phase of evolution. The genetic conflict between bacteria and their viral predators is thus a very ancient determinant of bacterial evolution. Eukaryotic protozoa that made a living by grazing bacteria appeared only much later on the scene and opened a new front line for bacteria. It is likely that in the meanwhile, some bacteria also discovered that attacking and eating other bacteria was a food option. The modern bdellovibrios can serve as an example for this lifestyle. By extension, it is also plausible that some bacteria specialized on eukaryotic protists as prey. When multicellular eukaryotes evolved about 1 billion years ago and mammals and birds proliferated massively over the past 65 million years, bacteria that used them as a food source would have followed them as pathogens. A new twist in this "eat or be eaten" strategy was introduced with the evolution of the adaptive immune system of birds and mammals, which presented a formidable defense barrier against unfriendly takeovers. Bacteria had to find answers to this new defense if they wanted to exploit this food source. Not surprisingly, numerous virulence factors from human and veterinary pathogens target this evolutionary recent resistance system.

Human-restricted pathogens like *Streptococcus pyogenes*, *Shigella* spp., and the human-adapted *Salmonella* strains must have become adapted to their hosts in 10^6 years or less, i.e., the time frame of human evolution. This is approximately the timescale for the separation of strains within the species *E. coli*. There are good reasons to anticipate that some human pathogens evolved only after human populations reached a critical size after the Neolithic revolution, which took place about 10,000 years ago. Therefore, the time for adaptation to this new niche has been relatively short and will be enriched for quick adaptation processes.

Two Modes of Evolution

Bacterial evolution requires modifying old functions and developing new ones. Nucleotide exchange, insertion, or deletions are the most frequent events. Mutation rates in bacteria are generally in the range of 10^{-6} to 10^{-9} per nucleotide per generation. In addition, module exchanges between different genes, gene disruptions, and deletions are occurring at appreciable frequency. These mechanisms are common to all living organisms and allow modification of existing functions in order to optimize fitness in an existing niche or to adapt to a new niche. In contrast to many higher eukaryotes, bacteria have no sexual life cycles to facilitate exchange of alleles within a population. In bacteria, this function is fulfilled by horizontal gene transfer; in this way, entire functional units can be imported from other sources, and these sources are principally not restricted by species barriers. The transferred DNA can range in size from less than 1 kb to more than 100 kb. It can encode entire metabolic pathways or complex surface structures. These genes can be taken up as naked DNA or transferred in the form of plasmids, conjugative transposons, or phages.

Bacteria apparently evolve with two gears. The slow mode (often in the 10^6 years range) is based on the usual mechanisms of vertical evolution mediating a step-by-step genetic adaptation to their approximate environment. Horizontal gene transfer can be regarded as the fast mode of evolution (time range of years to decades). New sets of genes are acquired by transduction, transposition, transformation, and lysogenization with phages. Most of these gains might be ephemeral and are as easily lost as gained (70). What counts is a momentary selective advantage over competing bacteria, especially in environments that are quickly changing. This changing environment can be the body surfaces of new host species that are rapidly proliferating in the ecosphere (e.g., humankind) or settings with unusual host densities (animal farming, human urbanization). New industrial food preparation and health care techniques, international travel and transportation, and wars create enormous possibilities for microbes that can invade these new niches. The impact of lateral gene transfer is obvious in some of these settings (e.g., antibiotic resistance gene acquisition via plasmids and transposons in the hospital environment), so it would be surprising if it were restricted to these examples. The acquisition of mobile DNA can provide the necessary genetic material on a timescale that allows rapid exploitation of these ecological "opportunities." Phage DNA fulfills a number of criteria for being an ideal vehicle for lateral gene transfer. Actually, some bacteria seem to use phages as gene transfer particles to shuttle pathogenicity islets or random samples of chromosomal DNA (*Bacillus subtilis* prophage PBSX) (91).

In this model, different combinations of mobile DNA can be explored, and suitable combinations are maintained and further developed, leading to genotypes that suddenly fill old and newly created niches. In the field of pathogenic bacteria, we see these events as emerging new infectious diseases or the outgrowth of specific clones of pathogens in the human population. As discussed below, phages can play a key role in these short-term adaptation processes.

The effect of this quick process on the long-term evolution of bacteria is less certain. Apparently only very small amounts of prophage DNA are fixed into the bacterial chromosome.

Time Line for Prophage-Bacterium Association

Prophages seem to be only transient passengers on the bacterial chromosomes, at least when seen on an evolutionary timescale. Theoretical arguments suggested a series of events leading from the accumulation of mutations to massive loss of prophage DNA and the ultimate disappearance of the prophage (31, 70). Inactivating point mutations were identified by bioinformatic analysis, e.g., introduction of stop codons into the replisome organizer gene and the portal protein-encoding gene in *S. pyogenes* prophages (42) or inactivation of the N antitermination genes in the lambdoid coliphages from O157 (70). Prophage remnants are the result of an ongoing prophage decay process. If the prophage decay process is slow with respect to bacterial speciation, one would expect the conservation of closely related prophage remnants in different bacterial isolates from a given species. A few closely related prophage remnants were detected in *S. pyogenes* (29) and *E. coli* (32). However, these are the exceptions and are not even widely distributed in the investigated species. In fact, even closely related bacteria generally do not share prophage remnants: the remnants in *E. coli* K-12 and *Shigella flexneri* (despite the name, a strain within the *E. coli* species) are distinct. This observation suggests that the average time of acquisition and subsequent loss of prophages is shorter than the timescale of strain differentiation within a bacterial species. Even such closely related bacterial strains as the two sequenced *E. coli* O157 strains already differ in prophage content and identity, underlining the rapid nature of phage evolution in their bacterial hosts.

THE ECOLOGICAL PERSPECTIVE

For a better understanding of bacterial pathogenesis, one should also complement the evolutionary reasoning by ecological arguments. When seen from an ecological perspective, bacterial mortality has three major causes: starvation, phage lysis, and protozoa grazing. Bacteria therefore had to respond to food shortage and to predation by viruses and heterotrophic protozoa. The prey-predator interaction can be studied in the laboratory when putting food bacteria and grazing protozoa in a vessel, which leads after some cycling to the selection of grazing-resistant bacteria.

The latter have developed some antipredator traits and resume growth by replacing the initial bacterial population (75). A formally very similar scenario occurs when *E. coli* and its phages are co-cultivated in a vessel, as was already recognized by the founders of molecular biology in a classic experiment (72). In the test tube, phage-resistant bacteria developed and variant phages tried to circumvent these defenses. In this war, as bacteria fought at two fronts against phages and protozoa, they "invented" a number of tricks. Pathogenic bacteria apparently reuse some of these lessons and apply them to their special problem: how to make a living from a human host.

One possible antipredation strategy of bacteria against grazing protozoa is surface masking. Amoebae recognize surface structures on the bacterial prey to set in the phagocytosis process. *Salmonella enterica* elaborates about 70 O types, chemically distinct forms of lipopolysaccharides, which decorate the surface of the bacterium. The classical interpretation is that this is a virulence factor important for escape from immune surveillance by the vertebrate host. However, evolutionary biologists argued that *Salmonella* is not much exposed to the immune system. It is either in the gut or inside a cell. Recent experiments demonstrated selective feeding of gut amoeba on *Salmonella* belonging to different O serotypes. The authors argue that the "ecological reason" for variability of the *Salmonella* O serotypes is to avoid protozoa feeding (130).

Once inside the food vacuole of protozoa, bacteria have to opt for other strategies. They must change from predigestion to postdigestion antipredation mechanisms—the most obvious being resistance to digestion. For example, a few minutes after ingestion of the cyanobacterium *Synechococcus*, flagellates reject the prey. They are apparently unpalatable, probably as a result of the proteinaceous S layer, with which the cyanobacterium surrounds itself (75). Under other conditions, attack might be the best defense. Some bacteria create toxins that have potent antiprotozoal activity (e.g., the pigment violacein produced by *Chromobacterium*). The predator flagellate *Ochromonas* does not distinguish between pigmented and nonpigmented bacteria. However, after digestion of as few as two pigmented bacteria, the predator lyses and releases its cell contents. The ingested bacterium is dead (the toxin is not released from the intact cell), but the cytoplasm of the lysed protists now feeds the surviving bacterial population (74).

Actually, one might hypothesize that measures deployed against protozoan predators were later reused by bacterial pathogens that target animals as a food source. The common denominator would be that animals deploy cells against bacterial invaders that formally closely resemble grazing amoeba that use phagocytosis as an eating mode. In fact, a surprising fraction of bacterial virulence factors address the problem of survival within the vacuole of phagocytes.

In this interpretation, one might see the evolution of bacterial pathogens and the immune system as the result of the fight of bacterial commensals-pathogens to exploit the animal body as a food source. Pathogenic bacteria apparently exploit their viral predators to co-opt phage-carried genes to fight the immune system of birds and mammals. I will illustrate these principles with two gram-positive pathogens that were over the last years intensively investigated along the phage–bacterial pathogen–human host axis. If confirmed, these considerations would place the evolution of bacterial pathogenicity just as a special case of the general quest for food motif in nature (24a).

STREPTOCOCCUS PYOGENES

A Paradigm for a Bacterial Pathogen Shaped by Lateral Gene Transfer

In 1927, long before lysogeny was described, it was demonstrated that a filterable agent from scarlet fever isolates could convert nonscarlatinal S. pyogenes to toxigenic strains. We know now that this conversion is mediated by bacteriophages and that 90% of S. pyogenes isolates are lysogenic. Transformation and conjugation play only minor roles for lateral DNA transfer

in this species, giving phages a special role (46, 108). S. pyogenes belongs to the lactic acid bacteria branch of low-GC-content gram-positive bacteria, and its phages resemble strongly dairy phages, which were intensively investigated for their economic impact in industrial milk fermentation (25).

The only known habitat of S. pyogenes is humans, where it is normally found on the skin and in the oral cavity. S. pyogenes comes in many M serotypes and causes an astonishing range of diseases, including pharyngitis (streptococcal sore throat), scarlet fever, pyodermitis, necrotizing fasciitis (flesh-eating disease), septicemia, erysipelas, cellulitis, rheumatic fever, and toxic shock syndrome. Careful prospective, population-based surveillance studies of invasive group A streptococcal diseases were conducted, which revealed an annual incidence of 1.5 cases per 100,000 inhabitants (39). Case rates were highest in people <10 and >60 years old; half of the patients had an underlying chronic disease. The clinical manifestations were mainly soft tissue infections, followed by bacteremia, toxic shock, pneumonia, and fasciitis. Overall mortality was 15% but reached 80% for toxic shock. Invasive infections were linked with a substantial risk of transmission in households and health care institutions. The most common serotypes were, in decreasing order, M1, M12, M4, M28, and M3. M1 strains have shown a remarkable increase over recent years. In addition, a strong association between M1 serotype and possession of the phage-encoded speA gene was found.

Five conditions make S. pyogenes a model case for a pathogen evolving under the influence of lateral gene transfer.

- First, different serotypes show significant association with disease types. For example, M1 strains are associated with wound infections, M3 strains were identified in patients with severe invasive infections, M18 strains caused rheumatic fever outbreaks, and M28 strains were associated with puerperal sepsis (6).
- Second, despite the protean character of this pathogen, the sequenced S. pyogenes

isolates are genetically closely related. For example, the M18 strain shared 1,532 of 1,696 ORFs identified in the M1 strain and sequence identity ranged from 83 to 100% at the base pair level. In fact, a dot plot analysis revealed essentially a straight line between both serotypes with 1.7 Mb of the 1.9 Mb chromosomal DNA shared. There were only five larger regions of difference; all were prophage sequences (108).

- Third, *S. pyogenes* has been the target for intensive sequencing efforts. In the meanwhile, complete genome sequences are available for five M serotypes that represent the major pathotypes: M1, M3, M6 (8), M18 (108), and M28 (54). Some are represented with multiple sequences. In the case of M1 isolates, one strain from the pre-1988 period was sequenced (46) and compared with the highly successful M1 clone emerging after 1988 (114). M3 isolates from the United States (13) and Japan (84) were sequenced, giving *S. pyogenes* genomics a temporal and geographical breadth.

- Fourth, large and systematic strain collections are available for this pathogen, which allowed a combination of modern technologies, like microarray analysis with classical medical epidemiology approaches (e.g., 12).

- Fifth, the possession of animal models ranging from mice to macaques permitted genetic approaches to associate genotype with phenotype by using knockout mutants (113) and microarray analysis for the temporal expression pattern of the *S. pyogenes* genome during the disease process (121).

Prophage Variability: Comparative Genomics

Microarray analysis of 36 M18 strains demonstrated that prophages are not only a significant source of genetic diversity between strains of M1, M3, and M18 serotypes, but also the predominant source of difference between M18 strains. The M18 strains differed by a maximum of 3%, and prophages were responsible for virtually all variation in gene content. Variation in the prophages ranged from presence or absence of the prophages to small differences in the gene content of the prophages (108). This observation leads to an interesting question: is the specific pathogenic potential of a given *S. pyogenes* strain influenced by its prophage content? Intensive and concerted efforts, especially those of the James Musser laboratory, point to an affirmative answer for this question.

When the DNA sequences from *S. pyogenes* prophages of the sequenced M serotypes were compared against each other in a dot plot matrix, six clusters of phages consisting of two to five members could be distinguished; two further phages shared only limited DNA sequence similarity with the rest of the phages (54). Recently, it was proposed to base the taxonomy of *Siphoviridae* on the genetic organization of the structural gene cluster (100). Overall, lactic acid bacteria prophages showed a conserved overall gene order: left attachment site *attL*—lysogeny—DNA replication—transcriptional regulation—DNA packaging—head-joining-tail—tail fiber—lysis modules—right attachment site *attR*. Particularly conserved is the order of the structural genes allowing the distinction of three major forms of head gene clusters exemplified by the *cos*-site *Streptococcus* phage Sfi21, the *pac*-site *Streptococcus* phage Sfi11, and the *cos*-site *Lactococcus* phage r1t as prototypes. On the basis of sequence similarities, subtypes could be distinguished within the Sfi11-like (MM1-like or O1205-like phages) and Sfi21-like phages (A2-like or staphylococcal-like phages) (100). Most of the observed *S. pyogenes* prophages fit into this scheme.

Except for the shared tail fiber and lysis genes and part of the lysogeny genes, M3 prophage 315.6 and a closely related M6 prophage lack DNA sequence similarity with the other *S. pyogenes* prophages. M3 prophage 315.5 even differs over the tail fiber module from the other *S. pyogenes* prophages. Over these genes, 315.5 shared about 40% aa sequence identity with *S. agalactiae* prophage SA1, suggesting a possible cross-species infection. In fact, these two phages shared some DNA sequence similarity across their genomes. Another peculiar case is the M6 prophage 10394.4, the above-mentioned transposon-prophage chimera carrying an antibiotic resistance gene (9).

Polylysogeny

One would predict competition and exclusion between prophages during the establishment of a polylysogenic cell. Exclusion could be mediated by three proteins encoded in the lysogeny module, namely, the phage integrase, the superinfection exclusion gene *sie* (77), and the phage repressor (immunity function). It is therefore not surprising that a substantial diversification was observed over this region between the *S. pyogenes* prophages. In the latest account, 16 distinct prophage integration sites were identified in the sequenced *S. pyogenes* strains (54). Five sites were occupied in more than one strain, and in these cases, the corresponding phage integrases shared at least 92% aa sequence identity. Other prophages used unique integration sites; however, not all possessed unique integrases. Apparently, competition for integration sites and repressor and superinfection immunity functions did not prevent the integration of up to eight prophages in a single M6 strain (8).

Fine Structure Analysis

The similarity of *S. pyogenes* prophages to well-characterized dairy phages allowed an in-depth comparison of their genomes. These alignments pointed to the accumulation of genetic changes in *S. pyogenes* prophages that concur with theoretical predictions about prophage inactivation and decay. For example, one of the r1t-like prophages showed a point mutation that interrupts and truncates the tail tape gene (prophage 8232.4), and several r1t-like prophages showed a disrupted replisome organizer gene (31).

Within the O1205-like prophages, disrupting point mutations were seen in the tail tape and a tail fiber gene (prophage 8232.3); prophage 370.2 contained a stop codon within the portal gene.

Within the MM1-like *S. pyogenes* prophages, the holin gene was shifted to the opposite strand and was thereby probably inactivated; a point mutation led to an inactivating frameshift in a tail fiber gene.

A clear trend for prophage DNA loss was seen in the M1, M6, and M28 *S. pyogenes* strain presenting 8- to 14-kb-long prophage remnants. One remnant (370.4) encoding only lysogeny and DNA replication genes was seen in three different M serotypes. The prophages shared both flanking *att* sites but differed by internal insertion/deletions and gene replacements (29). In addition, three prophage remnants only 2 kb in length were identified; they consisted of the phage integrase accompanied by the phage repressor and a potential lysogenic conversion gene in the R-1092 remnant (29).

Virulence Factors

When 21 of the *S. pyogenes* prophages with a standard phage genome organization were investigated for potential virulence factors, 20 prophages encoded at least one and 9 encoded even two virulence genes, characteristically located between the lysin gene and the *attR* site (6, 19, 42, 46) (Color Plate 3). The horizontal spread of these genes is also suggested by the presence of sequence-identical genes in the horse pathogen *Streptococcus equi* (29). Notably, there is a short stretch of sequence conservation adjacent to the right attachment site between different *S. pyogenes* prophages (31). This conserved segment and the highly conserved region around the hyaluronidase gene in the tail fiber module could allow an exchange of lysogenic conversion genes between different *S. pyogenes* prophages by homologous recombination. This hypothesis was confirmed when the Malak Kotb laboratory compared the prophages from the globally disseminated clonal M1T1 strain with the older M1 strain SF370. The scientists identified three prophages representing 5% M1T1-specific DNA (1). Phages commonly carry unimaginative names like Sfi21 or SF370. Not so in the Kotb laboratory, where the new prophages were called SphinX, MemPhiS, and PhiRamid. MemPhiS is 99% bp identical with an M3 prophage, suggesting the sharing of prophages between different M serotypes. SphinX and PhiRamid are highly mosaic prophages. The authors called the conserved ORF that follows the "toxin gene paratox gene" and postulated that it provides with the conserved genes at the other side of the toxin genes (lysin, hyaluronidase) target sites for homologous recombination between phages and thus a mechanism for toxin gene transfer in the streptococcal phage pool.

The virulence factors from *S. pyogenes* prophages comprise pyrogenic toxin superantigens (*speA, C, H, I, K, L, M,* and *ssa*), DNases (*spd1,3,4,sdn,* and *sda*), and a phospholipase A2 (*slaA*). All of these proteins are encoded near the *attR* site of the prophage. We will now investigate these phage-encoded virulence genes in greater detail.

Tail Fibers and Hyaluronidase

Older publications had the phage hyaluronidase on the list of virulence factors. The prophage-encoded hyaluronidase was suspected of promoting bacterial spread through host tissue by its ability to hydrolyze glucosamine bonds in hyaluronic acid, a major component of the extracellular matrix in the connective tissue. Notably, antibodies against the phage hyaluronidase and DNase were found in some patients (56).

Phage tail fibers and lysins have to interact with the cell surface and the cell wall of their bacterial host cell and should therefore be submitted to a strong adaptive selection pressure. Not surprisingly, all but one of the sequenced *S. pyogenes* prophages shared a highly related tail fiber module. Central to this module is the phage hyaluronidase, whose physiological function is to split the hyaluronic acid–containing capsule surrounding the bacterium. The hyaluronan layer can be thicker than the cell diameter; the phage then has the problem of injecting its DNA. Biochemical studies demonstrated that the *S. pyogenes* hyaluronidase is a lyase catalyzing a β elimination reaction,

thereby cutting the polysaccharide strand. Its structure of a triple-stranded β helix identifies it as a phage tail protein resembling the bacteriophage T4 gp5 and gp12 proteins (107). This lytic enzyme frees the way for the phage to the cell wall of the target bacterial cell. The unrelated hyaluronate lyases from group B streptococci function in the degradation of connective tissue from the host organism and thereby facilitate pathogen invasion. In contrast to the group B streptococci enzyme, the group A streptococci (GAS) hyaluronidase is inactive on other connective tissue glycosaminoglycans.

DNase

Early reports associated the phage-encoded DNase with the liquefaction of pus when degrading the DNA from decaying lymphocytes. Then mitogen and bacterial superantigen activities were attributed to streptococcal phage DNase (111). However, Spd1, which belongs to a family of DNases found in a variety of *S. pyogenes* strains, showed DNase activity, but no superantigen activity (24). It has a leader sequence and is secreted via the Sec pathway. The authors suggested that in vivo streptococcal phage induction occurs in the pharynx, where other strains of *S. pyogenes* are likely to be present. Spd1 is secreted just before lysis of the induced lysogenic bacterial cell and could thus digest the bacterial DNA, which is spilled out after lysis. This would reduce the DNA-mediated viscosity of the tissue fluid and facilitate the spread of the released phage to the next target bacterium. Lysogenic conversion of Tox⁻ pharyngeal *S. pyogenes* by phages released from Tox⁺ lysogens or free phage was demonstrated in mice (22). Another function of the phage DNase is to liquefy the pus, which contains substantial amounts of DNA from dying leukocytes. DNase would thus also increase the spread of the bacteria on the pharyngeal mucosa. An older hypothesis stated that the prophage DNase provides a growth advantage by increasing the nucleotide offer via DNA hydrolysis. The Kotb laboratory characterized the streptodornase Sda1 biochemically as DNase that degrades both streptococcal and mammalian DNA (2). Recent experiments of the Musser laboratory went a step further when suggesting an interesting function for the prophage DNase: according to this hypothesis, it assists pathogenesis by enhancing evasion of the innate immune response (113). A major difficulty in the analysis of its function was the multicopy presence of prophage- and chromosomally encoded DNases, necessitating the construction of sequential knockout mutants. Extracellular DNase activity was only eliminated in a triple mutant where all three DNase genes were deleted. The triple mutant had a characteristic phenotype: it formed large aggregates of GAS chains, which were dissipated by the addition of exogenous DNase. The triple mutant was only of low pathogenicity in a mouse infection model. Wild-type cells formed abscesses with a central necrosis containing viable bacteria, while in mice infected with the triple mutant, immune cells dominated the scene. The wild-type cell was less susceptible to killing by neutrophils than the triple mutant.

Likewise, the triple knockout mutant failed to induce pharyngitis in a macaque model. Apparently, extracellular DNase is required for the development of GAS pharyngitis. Although the connection is not yet proven, the Musser laboratory proposed that the protective activity of GAS DNases is mediated via degradation of NETs. This abbreviation stands for neutrophil extracellular traps; "net" is, however, also an adequate morphological description. NETs are fragile fibers consisting of DNA strands decorated with histones and antibacterial proteases like neutrophil elastase (21). Neutrophils release NETs as early as 10 min after activation by interleukin-8 and lipopolysaccharide. It is conceivable that NET formation is an early event of programmed cell death of activated neutrophils. The neutrophil elastase degrades virulence factors of gram-negative bacteria. The fibrous structure of NETs is necessary for efficient killing of bacteria. Notably, killing of bacteria became negligible when the NETs were destroyed by the addition of DNase.

Phospholipase

S. pyogenes prophages encode another excreted enzyme, the prophage 315.4-encoded Sla protein. The recombinant Sla protein has phospholipase A2 activity and structural similarity to a snake venom toxin (14). It is the target for antibody response in infected patients and may account for bleeding disorders, which accompany some invasive S. pyogenes infections. In fact, inflammation and coagulopathy are seen as prominent clinical manifestations in the streptococcal toxic shock syndrome (13). M3 strains lacked this gene until the mid-1980s, but afterward it was detected in 10% of the isolates. Only one sla allele was identified. The extracellular enzyme activity was quickly detected in the pathogen upon co-cultivation with human pharyngeal epithelial cells (83).

Superantigens

Many S. pyogenes prophages encoded streptococcal pyrogenic exotoxins (Spe) in the lysogenic conversion region (6). However, the specific combination of toxins differed between the sequenced S. pyogenes strains: the M1 strain showed speC, speH, and speI genes; the M3 strain demonstrated ssa, speK, and speA3 genes; and the sequenced M18 contained genes speA1, speC, and speL/M (6). These are all distinct members of a large family of superantigens, and they include the scarlet fever toxin. These proteins bind the T-cell receptor and the major histocompatibility complex (MHC) protein outside of the usual peptide binding site and lead to a pathological activation of the immune system, possibly allowing the escape of S. pyogenes from immune surveillance (99). It is conceivable that the variable combination of superantigens provided by multiple prophages influences the pathogenic potential of the polylysogenic host. This is a theoretically interesting hypothesis capable of explaining the strikingly distinct symptomatology associated with pathogens that are so similar with respect to the bacterial genome sequence. For example, prophage NIH1.1 was identified in an M3 S. pyogenes strain from a patient with toxic shock syndrome. It resembled SF370.1

prophage over the entire structural gene cluster, but encoded a distinct superantigen (SpeL instead of SpeC) (59). Notably, possession of prophage NIH1.1 was a genetic marker for newly emerging M3 *S. pyogenes* strains in Japan (60) that had replaced an otherwise genetically identical strain, which just lacked the prophage NIH1.1 (59). Prophage acquisition might thus be a major mechanism of short-term evolution in this epidemiologically highly dynamic bacterial species (6, 13).

In an appealing model, the emergence of new, unusually virulent subclones of M3 strains is explained by the sequential acquisition of prophages (13), suggesting bacterial pathogenicity evolution in the fast lane. More specifically, the authors proposed a model where the contemporary highly virulent M3 strains are clonal and the result of the sequential acquisition of three prophages (phi315.5 acquired circa 1920, phi 315.2 acquired circa 1940, and phi 315.4 acquired circa 1980), leading to the constellation of three particular superantigens (SpeA3, SSA, SpeK) combined with Sla.

To provide evidence for a causal relationship between the prophage lysogenic conversion genes and the high virulence of this strain, the Musser laboratory investigated 255 invasive M3 *S. pyogenes* isolates recovered during a 10-year survey in Ontario, Canada (12). Prophage screening revealed that all 255 strains showed between four and seven prophages. One prophage profile, characterized by the six prophages found in the sequenced M3 strain, was predominant in both infection peaks during the observation period. The core genome content was identical to the sequenced 315 M3 serotype. All differences in gene content were located in regions of the genome that contained prophages. The authors concluded that the acquisition and loss of prophages were the major generator of distinct genotypes providing novel combinations of virulence factors. The distinct genotypes can undergo very rapid population expansion and can cause infections that differ significantly in character. Strains with some prophage combinations are

fitter than others, and strains lacking certain prophages can undergo clonal extinction. Next to prophages, only a duplication of amino acids in the M protein was observed, which conferred an altered immune recognition to M proteins. The Emm3.2 variant rose to prominence as a consequence of host selective pressure. Subsequently, the Musser laboratory extended the analysis to a large collection of M1 serotypes isolated in six countries (114). Isolates recovered before the mid-1980s were genetically heterogeneous, but all recent isolates had virtually the same genotype, regardless of their place of origin. The recent isolates differed from the former isolates mainly in three prophages, encoding the superantigen SpeA2 and the DNases Spd3 and SdaD2. In addition, a horizontal transfer of a 36-kb fragment, apparently derived from an M12 strain, was observed in the recent M1 clonal expansion. In the view of the authors, these horizontal DNA transfer events, combined with minor transcriptional alterations, are the genetic basis for the global dissemination of this clone with increased invasive disease capacity.

Streptococcal superantigens are members of a growing family of proteins that simultaneously bind MHC class II molecules and T-cell receptors. In contrast to normal antigens presented by MHC II, which activate 0.001 to 0.0001% of all T cells, superantigens activate up to 20% of all T cells. This results in massive proliferation and subsequent release of inflammatory cytokines. These factors are thought to cause the high fever and shock or autoimmune sequelae in some patients with streptococcal infections who develop acute rheumatic fever. Epidemiologically, acute rheumatic fever is associated with M18 *S. pyogenes* serotypes. In the sequenced M18 strain, one prophage encodes the superantigens SpeL and SpeM, which are sequence related to SpeC and SpeK supcrantigens (108). SpeL and SpeM cause proliferation of blood cells, are pyrogenic, and in combination with endotoxin are lethal in rabbits. They stimulated T cells with three different β chains. Their transcription was enhanced in exponentially growing cells, and all patients wth acute rheumatic fever showed markedly increased antibody titers to both proteins (210).

Prophage 315.4 from the sequenced M3 strain showed *speK* and *sla* genes in the lysogenic conversion region.

Prophage Gene Expression

Lytic induction of *S. pyogenes* prophages has been studied in considerable detail. In bacterial growth media, phages were either spontaneously released from the lysogenic *S. pyogenes* strain MGAS315 (two prophages) or induced by more or less physiological stimuli. Hydrogen peroxide in vivo, produced by attacking phagocytes, induced three prophages, whereas mitomycin C, which causes damage to the bacterial DNA, induced all five prophages, although with variable efficiency (7). Mitomycin C might not be a natural inducer, but this type of induction of prophages is interesting for the understanding of the paradoxical effects of antibiotics on bacterial infections. Fluoroquinolone antibiotics, which poison DNA topoisomerase II, do not only induce Shiga toxin–encoding bacteriophages in *E. coli* O157, but also a mitogen-encoding bacteriophage in *Streptococcus canis* (61). *S. canis* is a commensal bacterium in dog and may cause various opportunistic infections. Notably, after the introduction of fluoroquinolone into canine veterinary medicine, streptococcal toxic shock syndrome and necrotizing fasciitis increased steadily (98).

The possession of phage-encoded toxin genes does not automatically lead to the expression of these genes. Clinical isolates containing toxin genes showed a variable pattern of toxin expression when grown in broth culture (66). However, growth of these strains in the mouse or co-culture with human pharyngeal cells led to the production of the phage-encoded streptococcal pyrogenic exotoxin SpeC, Spd1, and a number of other bacterial proteins. Spd1 and SpeC are encoded next to each other on the same prophage. A small, heat-stable factor released from the pharyngeal cells was identified as an inducer of the prophages (23). This is a fascinating observation because it means that streptococcal prophages respond via bacterial regulation systems to signals emitted from the eukaryotic host. Mobile DNA and prophages were also the most prominent group of genes that showed expression changes when mRNA from *S. pyogenes* cells grown at 29 or 37°C were assayed on an M1 strain–based microarray (110).

SpeA versus SpeB

A basic problem with all studies of *S. pyogenes* virulence is that most of the time, these bacteria behave as commensals and do not cause disease. For this reason, much effort was spent to correlate virulence factor (i.e., superantigen) secretion with certain disease phenotypes. Recently, Kotb and colleagues observed that the expression of the phage-encoded SpeA was either very low or undetectable in about half of the clinical M1 *S. pyogenes* isolates (34). When these isolates were introduced into Teflon tissue chambers within mice for 5 days, the expression of SpeA was turned on and, concomitantly, the expression of SpeB was downregulated (66). SpeB is a chromosomally encoded secreted cysteine protease.

Interestingly, the same authors had before observed an inverse relationship between SpeB expression and disease severity in clinical M1 *S. pyogenes* infections (65). Isolates recovered from the chambers continued to produce SpeA for extended passages in vitro, suggesting a stable genetic switch for SpeA expression. In those cases where in vitro SpeA expression was finally downregulated, SpeB expression was again turned on. Electrophoretic two-dimensional gel analysis of the secreted M1 *S. pyogenes* proteome, coupled with matrix-assisted laser desorption ionization–time of flight mass spectroscopy, revealed that expression of active SpeB caused the degradation of most of the secreted bacterial proteins, including several known virulence factors (3). Deletion of the *speB* gene or addition of a cysteine protease inhibitor inactivating SpeB yielded cells that revealed more than 150 spots in the secreted proteome, including the prophage-encoded

Sda (streptodornase). A complex secreted proteome was also reported for the strain recovered from the mouse tissue chamber. The proteome included SpeA representing a clear case of in vivo upregulation of a phage-encoded virulence protein.

Prophage Gene Regulation

The regulation of gene expression in *S. pyogenes* grown under different conditions was the focus of a series of studies from the James Musser laboratory. At the extremes of its human habitats, *S. pyogenes* has to adapt to a range of temperatures extending from about 25°C on the surface of the skin to 40°C or more in deep tissue infections. In microarray analysis, globally, 9% of the genes from the M1 strain SF370 were differentially transcribed by organisms grown at 29°C compared with 37°C (110). Genes from mobile DNA (mainly phages) belonged to the most prominently upregulated genes at 29°C, followed by transport and binding proteins.

During *S. pyogenes*–phagocyte interaction, about 16% of the bacterial genes were differentially transcribed. The largest fraction (69 genes) represented hypothetical upregulated genes. The next prominent group comprised prophages (23 genes). Because most of the prophage genes showed increased transcription, the authors suggested that these phage genes play a role in the host-pathogen interaction (123). The expression pattern of putative virulence and regulatory genes was investigated in *S. pyogenes* recovered from children with pharyngitis. Notably, the most prominently upregulated gene (60-fold increase in transcript level) was the prophage *sda* gene, encoding a DNase (120).

Two regulators were identified that controlled the prophage lysogenic conversion gene expression. One was the two-component regulator CovR/S. When it was inactivated, a mucoid colony phenotype was observed that was associated with overexpression of the hyaluronic acid capsule. The phage-encoded DNase was more abundant in the strain (52) because it was relieved from repressor-binding

to the *sda* promoter (78). The second identified transcriptional regulator was Rgg, which controlled the expression of the prophage-encoded genes *sda* and *mf3* (35).

However, the exact role of the upregulated phage genes in *S. pyogenes* cells making contact with the mammalian cells is not yet clear: do they directly benefit the propagation of phage, or do they benefit the lysogen through a bacteriocin-like effect or the expression of phage-encoded fitness factors? In summary, it is fair to say that perhaps more than in any other system, the recent evolution of *S. pyogenes* has been guided by bacteriophage. Even when making contact with the mammalian host, *S. pyogenes* strains alter not only their gene expression pattern, but, by lysogenization of bystander cells, also the genomes of commensal *S. pyogenes* (22). In addition, prophages also change the bacterial genome while residing silently in the chromosome by serving as anchor points for homologous recombination, leading to bacterial chromosome rearrangements independent of lysogenization (30).

In Vivo Growth

S. pyogenes is well adapted to growth in the human oropharynx. Saliva is normally a major component of the innate immune defense at this anatomical site and contributes a number of antimicrobial components active against many bacteria. M1 strains from *S. pyogenes* did not only grow in saliva, but also maintained increased bacterial colony counts in the stationary phase and proliferated after addition of fresh saliva, even after a prolonged stationary phase (106). The streptococcal protein Sic was responsible for this surprising survival capacity by binding to lysozyme and β-defensins in the saliva. Sic is a highly variable bacterial protein that is apparently under selection pressure: 167 Sic variants were identified in one survey, and nearly every base pair change in the *sic* gene resulted in protein sequence changes. Further variation was introduced by deletions (73). During growth in saliva, a number of prophage-encoded proteins are also induced (SpeA, SlaA). Strains from the M28 serotype are also involved in pharyngitis and likewise in invasive infection. M28 strains isolated from both clinical conditions showed a substantial variability in prophage-encoded virulence gene content. From population-based studies, about 30 different phage virulence profiles were identified in each disease category when a total of 560 strains were investigated. Furthermore, significant differences were observed in the frequency of occurrence of prophage toxin gene profiles with infection type (53).

Pharyngitis is the most common clinical manifestation of S. pyogenes. In the United States alone, about 2 million cases are counted annually. James Musser and colleagues performed a systematic bacterial transcriptome analysis during an 86-day follow-up of experimental pharyngitis induced by S. pyogenes in cynomolgus macaques (121). The temporal pattern of S. pyogenes gene expression was tightly linked to the three phases of infection. Notably, prophage genes had the greatest changes in expression throughout the experiment. Carbohydrate metabolism genes were the most abundant transcripts during the colonization phase when cell counts were still low. When the cell counts subsequently increased on day 16 of the infection cycle, the phage gene category had the greatest number of expressed genes. Pyrogenic toxin superantigens, including speA2, were highly expressed during the inflammatory phase. Also, prophage DNase (sdaD2) expression increased during this phase and was correlated with phage replication. However, the most highly expressed virulence genes were the emm1 gene, encoding the M1-specific M protein, and sic, the inhibitor of the complement system. The expression of both genes assists evasion of host defenses, as did SpeB, which degrades most immunoglobulin isotypes.

STAPHYLOCOCCUS AUREUS

Life History between Commensal and Pathogen

S. aureus is a gram-positive coccus that can grow under aerobic and anaerobic conditions. Its natural habitat is the nose and the skin of warm-blooded animals. A large fraction of the human population is colonized with this bacterium. Approximately 20% of healthy people almost always carry an S. aureus strain, whereas 60% harbor it intermittently; only 20% almost never carry S. aureus. On the basis of these data, one could regard S. aureus as a commensal, but physicians see the picture differently. The Centers for Disease Control and Prevention estimates that 16% of hospital-acquired cases of bacteremia in the United States were due to S. aureus. A German study showed even more disturbing correlations: the isolates from the blood were in 80% of the cases molecularly identical to those from the anterior nares of the same patients (122). The commensal S. aureus in the nares is the reservoir not only for other patients in the hospital, but also for the colonized subject. In fact, in a prospective study, it could be demonstrated that 14 of 1,278 patients developed S. aureus bacteremia. Twelve isolates were identical to the isolates of the nares obtained up to 14 months before hospitalization. We have here a paradigm for a facultative pathogen changing smoothly from commensalism to pathogenicity. Indeed, when S. aureus gains access to the bloodstream, it is one of the most lethal human pathogens. In the preantibiotic era, a lethality of 80% was reported, and the rapid increase of antibiotic resistance in S. aureus threatens that we might soon be confronted with a partial return to these conditions. What makes a nasal and skin commensal such a dreadful pathogen? A key to the answer is the clinical observation that S. aureus is, like S. pyogenes, a protean pathogen. It can cause a wide range of diseases, ranging from food poisoning, many skin infections like furunculosis and staphylococcal scalded skin syndrome, to wound infections and systemic infections including a toxic shock syndrome.

PVL-Associated Pneumonia

S. aureus strains encode a large variety of secreted toxins, and these toxins are responsible for most of the clinical symptoms associated with the infections. As in S. pyogenes, prophages play an important role as carriers of virulence genes. I will illustrate this link with two recent clinical reports. Like S. pyogenes, S. aureus is a dynamic pathogen, surprising the clinician with new pathotypes that cause new forms of disease. French clinicians described an "*S. aureus* necrotizing pneumonia" associated with high lethality and dramatic clinical manifestations like hemoptysis and purulent expectorations preceded by an influenza-like syndrome (50). A respiratory distress syndrome developed quickly, and necropsies showed massive ulcerations of the respiratory tract associated with staphylococci. Restriction analysis of the genomes revealed many distinct bacterial types, but a common genetic denominator in the patients was the possession of the PVL virulence genes. The abbreviation stands for Panton-Valentine leukocidin, a group of bacterial toxins that kill leukocytes by creating pores in the cell membrane. The active pore-forming toxin is actually assembled from two components (47). LukS recognizes a specific receptor on the host cell membrane. Subsequent association with the second component, LukM, leads to the formation of a pore resembling that of α-hemolysin. The influenza-*Staphylococcus* link is not new: it was already reported by physicians conducting the necropsies of young U.S. soldiers dying from the Spanish flu in 1918. They reported that half of the cases of this fulminant form of influenza were associated with staphylococci (36).

PVL-Associated Fasciitis

The second new disease entity is necrotizing fasciitis, better known as the flesh-eating infection. This disease made headlines in association with *S. pyogenes*, and a link with *S. aureus* was not observed until 2003. U.S. physicians reported a cluster of 14 cases from Los Angeles; all isolates belonged to the same genotype: the methicillin-resistant USA300 pulsed-field type (79). The strains again carried the PVL toxin genes and, notably, none of the other toxin genes commonly carried by *S. aureus*. This constitutes a circumstantial but compelling link between this new pathology and the PVL genes. The precise clinical manifestation (fasciitis versus pneumonia) might depend on the genetic background of *S. aureus,* concomitant infections, or host disposition (the fasciitis patient group had a high percentage of intravenous drug users).

PVL Epidemiology and Spread

PVL genes showed up in still other epidemiological conditions. Clinicians distinguish between MRSA and MSSA, methicillin-resistant and methicillin-sensitive *S. aureus* strains, respectively. MRSA is a major cause of hospital-acquired infections worldwide. MRSA strains are typically resistant to multiple antibiotics, complicating the choice of an appropriate treatment. It is commonly accepted that MRSA strains flourish in the hospital environment under the selection pressure of massive antibiotic use. Community-associated MRSA have now increasingly been identified. Epidemiologists from the Centers for Disease Control and Prevention identified MRSA strains as the cause for a furunculosis outbreak in Alaska (5). Notably, 97% of the MRSA isolates carried the PVL genes, whereas none of the MSSA strains showed this prophage-encoded virulence factor. Clinicians from California likewise identified community-associated MRSA strains from subjects with skin and soft tissue infections. In contrast to hospital-associated MRSA strains, these isolates carried the type IV staphylococcal chromosomal cassette mec (SCC*mec*) element, which encodes only methicillin resistance but not multiantibiotic resistance, as is commonly seen in hospital strains. Moreover, the microbiologists found a close association of SCC*mec* with the PVL genes (44). The scientists proposed that the PVL/SCC*mec* combination provides a substantial gain in fitness to the strains that have acquired these mobile elements, explaining their recent spread to multiple areas in the United States (92).

PVL Prophage Genomics

PVL genes were detected in different clonal groups of MRSA strains, raising the question of their origin. Several PVL-carrying prophages were described in *S. aureus* genomes: φPVL, φSLT, φPV83, or φMW2. In phage PVL, the two toxin genes, *lukS* and *lukF*, were located between the phage lysin gene and the right attachment site (64). Very similar toxin genes were found at the same location in a morphologically and molecularly distinct *S. aureus* phage, SLT. Sequence comparison suggested that the entire region surrounding the attachment site was the result of a modular exchange between both phages (86). *S. aureus* prophage PV83 also encodes a leukocidin, this time a *lukM* and *lukF* gene combination. As in other staphylococcal phages, these toxin genes were located between the lysin gene and the right attachment site. The toxin genes in PV83 are flanked by a transposase gene, suggesting that this gene cassette derived from a mobile DNA element (134). The different *S. aureus* phages showed a patch-wise pattern of relatedness, as predicted by the modular theory of phage evolution (18).

PVL Cell Biology

Despite the importance of the phage-encoded PVL virulence factors, the sequence of events occurring during its cytotoxic action was only recently defined (49). High concentration (200 nM) of recombinant PVL induced necrosis in the exposed cells, with swelling, loss of cell membrane integrity, and the release of cellular contents. At low (5 nM) concentration, however, apoptosis was observed, with rounding of cells and nuclei and chromatin condensation. In neutrophils (polymorphonuclear leukocytes) the target of PVL action was the mitochondrion: after 5 min exposure, the mitochondrial transmembrane potential decreased. Inhibitor studies revealed that PVL activated the caspase in the mitochondrial pathway. PVL induced the release of proapoptotic proteins from mitochondria as cytochrome c. Hence, bacterial invasion of polymorphonuclear leukocyte is not needed for the effect; soluble PVL can find its way into its mitochondria to induce apoptosis in the neutrophils. This action will decrease neutrophils available for phagocytosis of bacteria and thus weaken the first line of cellular defense against the invading *S. aureus* pathogen.

Further Virulence Genes

Phage-encoded virulence factors were described nearly 50 years ago with a phage that could convert nontoxigenic strains to α-

hemolysin production (15). However, it took more than 20 years after that discovery until a toxin gene was located on a staphylococcal phage by molecular means. The staphylococcal enterotoxin A gene *sea* was mapped near the attachment site of the temperate phage PS42-D (14). Southern hybridization tests revealed that the *sea* genes in staphylococcal strains were associated with a family of phages, rather than with one particular phage.

The exfoliative toxin (ET) is an extracellular protein that is responsible for the scalded skin syndrome in some *S. aureus* infections (69). The causative proteins are ETA and ETB. The former is encoded by prophage ETA, the latter by a large plasmid. Again, the *eta* gene is located downstream of the phage lysin gene. Notably, ETA protein was demonstrated by Western blot analysis in the supernatant from the lysogen. Its toxic activity was demonstrated in mouse experiments (133). Interestingly, ETA production was not stimulated by mitomycin C induction, suggesting that it is constitutively produced from its own promoter and is neither regulated by the preceding lysis gene cassette nor upregulated with phage replication during prophage induction.

Comparative Phage Genomics

The potential medical importance of staphylococcal phages motivated in recent years a number of phage sequencing projects. In one particular effort, 27 *S. aureus* phages obtained from different phage collection centers were sequenced (68). With respect to genome size, three classes could be distinguished, which concurred with their morphological characterization as Podo-, Sipho-, and Myo-Viridae, showing genomes of about 20, 40, and 130 kb in size, respectively. With respect to DNA sequence relatedness, two and three clades could be distinguished in the first two classes, while the *Myoviridae* were quite homogeneous. Two observations were instructive with respect to the sequence space of the phage DNA world. The first was the sheer number of new genes identified in this 1.3-Mb sequencing effort: 522 genes, or 25% of the total, showed no database match, and half of the ORFs have matches restricted to other *S. aureus* phage genes. The cumulative *S. aureus* phage gene pool is likely to exceed that of the *S. aureus* core genome. Second, the microbiologists observed an extensive mosaicism in the construction of the phage genomes. A substantial diversity of phage genomes is thus created by modular exchanges between existing phages creating ever new assemblies of phage genomes.

Comparative *S. aureus* Genomics
MRSA is now the main etiological agent of nosocomial infection, and vancomycin, the only antibiotic effective against it, is no longer effective against all *S. aureus* isolates. The methicillin-resistant strain N315 isolated in 1982 differed from the vancomycin-resistant strain Mu50 isolated in 1997 by less than 4% at the nucleotide level (67). Most of the differences in genome structure were due to the insertion of Mu50-specific mobile genetic elements, including a Mu50-specific prophage. Two phages were shared between both strains: phiN315 and phiMu50A. The two prophages carry known virulence factors: a gene encoding enterotoxin P (*sep* gene), a superantigen involved in the symptoms of food poisoning, and a staphylokinase (*sak* gene). In addition, an M-like protein fragment is encoded by a gene preceding *sep*. The virulence genes flank the phage lysis cassette at both sides. However, the two prophages are not identical. Especially over the lysogeny and early genes, the two prophages differed by numerous small modular exchanges.

One major difference between the two sequenced *S. aureus* strains was prophage phiMu50B. Strain N315 showed no prophage at this position. PhiMu50B shared segments of sequence relatedness with phage ETA especially over the late genes, but lacks the ET toxin gene. In its place, it carries four anonymous genes. A potential virulence gene was identified upstream of the integrase gene; it shared similarity with a gene from the staphylococcal pathogenicity island SaPIn1. Notably, both prophages were flanked near their integrase gene by pathogenicity or genomic islands, suggesting a potential link between acquisition of both types of mobile DNA elements in strain Mu50.

A community-acquired *S. aureus* isolate, strain MW2, was also sequenced (4). It differed from strain N315 by numerous insertions/deletions and gene replacements. The most obvious differences were near the origin of replication, including the DNA element encoding methicillin resistance (SCC*mec*). Further differences between the strains were linked to other mobile DNA elements: prophages, transposons, and a number of small genomic islands. MW2 contains two prophages: phiSa2 and phiSa3. PhiSa2 resembles *S. aureus* phage phi12, but possesses in addition *lukS* and *lukF* genes in a constellation identical to phage SLT. PhiSa3 closely resembles phage PVL over most of their genomes, but both phages differed in the content of virulence genes. PhiSa3 contains, in addition to *sak* and *sea*, located around the lysin gene, two new enterotoxin gene alleles, *seg2* and *sek2*. The latter genes are encoded between repressor and integrase genes in the lysogeny module. Prophage Sa3ms from an *S. aureus* genome sequenced at the Sanger center (58) differed from phiSa3 at only 14 bp positions (115). The extensive patch-wise sequence similarities between *S. aureus* phages suggested that multiple recombination events had occurred between N315-, PVL-, PV83-, Sa3-, and phi13-like phages. In the Sa3ms lysogen, a single 1- and 1.7-kb-long transcript was detected with a *sak* and *sea* probe, respectively.

Previously, a promoter was demonstrated upstream of *sea* (16). Mitomycin C induction led to a marked increase of these mRNAs, and the transcription of two higher-molecular-weight forms of mRNA was observed that covered both genes. If the replication of the prophage was prevented by a mutation, no increase in transcription was observed, suggesting that the augmented transcription was the direct result of the increased phage DNA copy number. Also the transcription of the *seg2* and *sek2* genes was increased by mitomycin C (115). This result was surprising because these genes were cotranscribed from a promoter upstream of the *cI*-like repressor gene. This constellation assures in coliphages constitutive expression during lysogeny, but repression via the Cro repressor after prophage induction.

DNA Mobilization by Prophages

A relationship between different types of mobile DNA elements was demonstrated in *S. aureus* with the 15-kb pathogenicity island SaPI1 and the temperate staphylococcal phage 80α (71, 104). SaPI1 carries the gene for the toxic shock syndrome toxin-1 and an integrase gene. It is flanked by a 17-bp repeat, and it is excised and circularized and replicates autonomously. Notably, SaPI1 interferes with phage 80α growth and is encapsidated into special small phage 80α heads commensurate with its smaller DNA size. Upon phage-mediated transfer to a recipient organism, SaPI1 integrates at a specific attachment site by using the SaPI1-encoded integrase. However, as in the case of the P4 satellite phage and its P2 helper phage, phage 80α provides functions for excision, replication, and encapsidation of this pathogenicity island. This peculiar link with a phage assures mobility to the pathogenicity island and may be responsible for the spread of toxic shock syndrome toxin-1 among *S. aureus* strains.

This elegant model should, however, be interpreted with caution because phage 80α was not a naturally occurring phage, but a laboratory product. Efforts were therefore undertaken to demonstrate that such a prophage-mediated transfer of a pathogenicity island can occur in nature. Recently, SaPIbov1, a member of the growing SaPI family, was excised from the chromosome and replicated after SOS induction of naturally occurring prophages φ11 and φ147 (117). The pathogenicity island is again packaged into phage-like particles with smaller heads and transferred at high frequency. The SOS reaction and the subsequent chain of events were also induced by commonly used antibiotics like ciprofloxacin. Treatment with fluoroquinolone antibiotics might thus have the unintended consequence of promoting the spread of bacterial virulence factors.

Dynamics

Microarray analysis demonstrated an extensive variation in gene content among different strains of *S. aureus*, with 22% of the genome comprising dispensable genetic material (48). Eighteen large regions of difference were identified; 10 encoded virulence factors or antibiotic resistance genes. A comparison of the different *S. aureus* genomes demonstrated that not only prophages but also their associated toxin genes are mobile DNA elements of their own, which are not stably associated with an individual prophage (4). Apparently, lateral gene transfer has played a fundamental role in the evolution of not only *S. aureus*, but also its prophages.

In an interesting study, the dynamic nature of the colonizing *S. aureus* population was followed prospectively (51). The scientists chose two anatomical sites for their investigation: the nares of healthy carriers and the lungs of cystic fibrosis patients. Both were initially colonized with *S. aureus*. Pulsed-field gel electrophoresis was used to monitor genome changes. In total, 19 strain pairs from patients showed genome alterations. These strains were molecularly analyzed to clarify the genetic nature of the changes. In eight cases, genome changes were linked to prophage mobilization. Phage conversion of β toxin production was identified in seven pairs; in one, an ETA-like phage was lost by deletion. Prophage DNA is thus only a temporary passenger of the bacterial genome. Notably, the time between genome alterations was 1 year in the lungs of the cystic fibrosis patients, but 13 years in the nares of the healthy carriers, which the authors explained with the higher antibiotic selection pressure in patients with cystic fibrosis.

The Innate Immune Evasion Cluster

In *S. pyogenes*, one gets the impression that pathogenicity is regulated via polylysogeny, i.e., the permutation of different combinations of prophages, each encoding one or two virulence genes. *S. aureus* genomes tend to contain fewer prophages, but some prophages encode more virulence genes than encountered on *S.*

pyogenes prophages. In the *S. aureus* literature, triple-, quadruple-, and quintuple-converting phages are encountered (38). Particularly impressive combinations of virulence genes were recently reported in β-hemolysin-converting bacteriophages. These phages target and inactivate this host gene during prophage DNA integration, a frequently cited case of negative lysogenic conversion (131). These prophages carry a whole battery of virulence genes in the gene order *sea*—lysis cassette—*sak*—*chp*—*scn* over an 8-kb genome segment near the conserved 3′ ends of several β-hemolysin-converting *S. aureus* phages (118). When the authors investigated these prophages in 80 *S. aureus* strains, they found various combinations of these genes. The *scn* gene was present in all of them, *sak* in most of them, and *chp* in half of them; *sea* was present in only two of seven different combinations; and in two further combinations, the *sea* gene was replaced by *sep*. The various combinations are probably the result of recombinations over this lysogenic conversion cassette during superinfections. Notably, the different genes have a common denominator: all encode proteins that interfere with the first line of defense of the immune system against invading pathogens, and hence the name IEC (immune evasion cluster). The authors observed that the IEC was associated with many different prophages; the common denominators were conserved 5′ and 3′ ends encoding the *attP-int* (attachment site–integrase) directing the phage to the shared bacterial *attB* site in the host β-hemolysin gene and the IEC, respectively.

It should be noted that in the phage particle both prophage ends are adjacent and form one conserved segment of DNA. The authors noted that these prophages were associated with *S. aureus* strains that showed many different pulsed-field patterns. They concluded that the IEC-carrying prophages move around in the population. In fact, these prophages could be induced, and they could transfer IEC to an indicator strain. The fact that IEC were associated with different phages allowed this virulence cassette to cover a huge host range of *S.*

aureus strains. Consequently, IEC prophages are found with up to 90% incidence in human *S. aureus* strains, which is a unique feature among staphylococcal mobile elements carrying virulence factors. The *pvl* and *eta* genes, for example, are found in less than 20% of the strains. The stability of the IEC-containing prophages is also partly explained by the observation that some IEC prophages encode in addition a restriction-modification system that protects the lysogen against superinfection with a set of 23 typing phages from *S. aureus* (e.g., reference 41).

Staphylococcal Enterotoxin A (SEA)

In the following section, I will summarize functional studies published on the proteins of the IEC in the order of their location on this mobile DNA element. The first gene, *sea*, encodes a protein that induces on human monocytes a downmodulation of cell-surface chemokine binding sites (101). Specifically, the binding of monocyte chemotactic proteins (MCP)-1 and the macrophage inflammatory proteins (MIP)-1α and -β was prevented by SEA in a concentration-dependent way. Without their binding, no mobilization of intracellular calcium occurred. MCP and MIP belong to the chemokine family of small proteins that comprise chemoattractant cytokines that play a pivotal role in the extravasation of leukocytes from the circulation and thus represent a first line of defense against invading pathogens. Consequently, SEA antagonized the directed migration of monocytes to MIP and MCP. SEA is a superantigen that activates monocytes by binding directly to MHC class II molecules in a manner independent of TCR coengagement. Superantigen binding to MHC II stimulates a signal transduction pathway in the target cell. Fittingly, MHC class II–specific antibodies simulated the SEA effect and inhibitors of the signal pathway abrogated the SEA effect. Furthermore, addition of serine protease inhibitors also abrogated the effect of SEA on monocyte binding of MCP and MIP, suggesting that their receptors were degraded by a SEA-induced secretion of proteases. SEA thus subverts an important arm of the innate immune system.

Staphylokinase (SAK)

The next phage-encoded gene on the IEC, *sak*, looks like an unlikely *S. aureus* virulence factor because it is widely used in human medicine. It encodes staphylokinase, a potent activator of host plasminogen, the precursor of the fibrinolytic protease plasmin. SAK is used as a bolus infusion for thrombolysis in myocardial infarction. However, staphylokinase is not an enzyme; it binds plasminogen during this action in a 1:1 complex. Microbiologists doubted that this was the main function of staphylokinase during infection. Recently, it was shown that staphylokinase also binds other human proteins called α-defensins, which are more likely candidates for the physiological function of staphylokinase from the viewpoint of the invading pathogen (62).

Defensins are a group of peptides with cationic charge, which are secreted by neutrophils. They bind to the anionic phospholipids on the bacterial surface, leading to the permeabilization of the microbial membrane by pore formation. Host membranes are protected from this action as a result of the high content of cholesterol and neutral phospholipids on the surface. Staphylokinase binds to defensin and thus interferes with their potent killing effect. Notably, an inhibitor of the protease activity of SAK on plasminogen did not inhibit the antidefensin effect of SAK, thus dissociating both physiological effects of this staphylococcal protein. The authors of this study demonstrated that the sensitivity of *S. aureus* strains to α-defensin-mediated killing depended on their SAK production. *S. aureus* caused severe arthritis in 100% of mice upon injection into the joints; the same bacteria injected with α-defensin induced arthritis in only a third of the animals. When SAK was added, this percentage increased to 62%. SAK is not only elaborated during infection, but also 67% of healthy carriers were colonized with SAK-producing strains in the nares,

pointing to a fundamental role of this protein during the commensal life of *S. aureus*.

CHIPS

This acronym stands for chemotaxis inhibitory protein of *Staphylococcus*. This protein was isolated by chromatography and gel filtration by using as a marker the inhibition of formylated peptide binding to neutrophils (40). When *S. aureus* invades the human host, complement is rapidly activated, resulting in the opsonization of the bacteria and the generation of large amounts of C5a. C5a and formylated peptides (the latter are side products of bacterial translation) are classical chemoattractants and constitute the first triggers of the innate immune system. Protein sequencing revealed that CHIPS is the product of the prophage *chp* gene located on IEC. CHIPS binds to human neutrophils, less to monocytes, and not to lymphocytes, and inhibits the C5a-induced, but not the interleukin-8-induced calcium mobilization in neutrophils. CHIPS binding is rather specific to human neutrophils; it is produced in vivo and inhibits in vivo neutrophil influx from the circulation into the peritoneum. CHIPS might be of medical use in the development of new anti-inflammatory compounds for diseases in which damage by neutrophils plays a role. CHIPS binds specifically and with high affinity to the C5a (C5aR) and the formylated peptide receptor (FPR) and thus prevents the binding of the natural ligands (97). The N-terminal 6 aa peptide of CHIPS was sufficient to mimic the anti-FPR activity, and CHIPS mutated at aa position 1 and 3 had lost anti-FPR activity while the anti-C5aR activity was maintained (55).

SCIN

SCIN stands for staphylococcal complement inhibitor. It is the last gene of the pathogenicity island SaPI5 carried on the β-hemolysin-converting phages and fulfills an immune evasion function (103). The C3 convertase mediates almost all biological activities of the complement system. Cleavage of the central compound C3 results in the release of the ana-phylactic agent C3a and in the covalent at-tachment of C3b to the microbial surface. Phagocytes recognize foreign particles most efficiently after opsonization with serum-derived IgG and C3b. Recombinant SCIN prevented deposition of C3b on *S. aureus*, and it thereby reduced the killing of bacteria by neutrophils. SCIN interfered with the activa-tion of all three complement pathways and was also human specific. SCIN is found only in staphylococci isolated from humans, but there were comparable prevalences in patients and healthy carriers, indicating that it is also essen-tial for the survival of *S. aureus* as a commensal in the human host. SCIN seems to protect C3 convertases from decay dissociation. By stabi-lizing C3 convertases, SCIN affects the com-plement cascade at different steps. Because it acts very early in the complement pathways, it can prevent the generation of the membrane attack complex and spares the bacterium this form of killing.

Outlook

The present review demonstrated that prophage-encoded virulence factors play a prominent role in the fight of two important medical pathogens with the human immune system. The reviewed literature seems to con-cur with theoretical predictions, but it is still premature to come to definitive conclusions. The analysis was based on just two pathogens, which both belong to the same phylogenetic group of bacteria, namely, low-GC-content gram-positive bacteria. They might further-more represent extreme cases for the role of phages in the evolution of bacterial pathoge-nicity, and this very fact might have motivated the investigating laboratories to choose just these pathogens for their prophage analysis. It cannot be denied that we know bacterial pathogens that do not contain prophage-encoded virulence factors (e.g., *Helicobacter py-lori*). In fact, half of all sequenced bacterial genomes do not contain prophages. However, virulence-encoding prophages are not re-stricted to low-GC-content gram-positive bacteria, but are also found in high-GC-con-tent gram-positive (*Corynebacterium diphtheriae*) and gram-negative bacteria (*E. coli, Salmonella*).

One might also argue that the peculiar role of prophages in the two investigated pathogens might result from the special nature of these pathogens that are at the same time successful commensals of the human body. However, one might want to add that this commensal-pathogen double nature is a widely distributed situation when, for example, considering food-borne disease (campylobacteriosis, salmonellosis, *E. coli* O157) where human pathogens are de-rived from animal reservoirs. In some of them, e.g., *Salmonella enterica* serovar Typhimurium and *E. coli* O157, prophage-encoded virulence fac-tors likewise play an important role in patho-genesis (Shiga-like toxin in *E. coli* O157). Fur-thermore, both in *E. coli* and in *Salmonella enterica* serovar Typhimurium, many prophages encode virulence factors that also target the first line of the immune defense (for details, see references 19 and 26). These cases clearly show that *S. pyo-genes* and *S. aureus* are not exceptions.

It is, however, still debatable whether bacte-rial pathogenicity evolved from antigrazing strategies developed by bacteria against protist grazing. The formal similarity between both strategies might represent cases of convergent evolution imposed by the superficial similarity between protist amoeba and phagocytes found in animal bodies, which range in morphologi-cal complexity from sponges to humans. The interpretation that bacteria use a "viral" preda-tor against a "grazer" predator when deploying prophage proteins against immune proteins and cells of the host might also be misleading. Actually, when looking at the human situation, one might ask who is the prey and who is the predator. When considering how successfully bacteria have colonized any body surface of humans, one might see us as a food source, if not, frankly, as prey for bacteria. Bacteria, which we call pathogens, might simply use genes trafficked by phages to attack animals. Recent data from *E. coli* O157 showed how many type III secretion effectors are actually elaborated by this lineage and that lambdoid phages took the care to disseminate many of the effector genes (116a). The virulence genes have certainly not evolved in phages but are the result of close bacterial interaction with the eukaryotic cell. Phages are perhaps only the handy gene carriers efficiently shuttling genes around in the bacterial world. In this way, bacteria make the best out of their viral predators in using them in their fight for food against eukaryotic cells.

REFERENCES

1. **Aziz, R. K., R. A. Edwards, W. W. Taylor, D. E. Low, A. McGeer, and M. Kotb.** 2005. Mosaic prophages with horizontally acquired genes account for the emergence and diversification of the globally disseminated M1T1 clone of *Streptococcus pyogenes. J. Bacteriol.* **187:**3311–3318.

2. **Aziz, R. K., S. A. Ismail, H. W. Park, and M. Kotb.** 2004. Post-proteomic identification of a novel phage-encoded streptodornase, Sda1, in invasive M1T1 *Streptococcus pyogenes. Mol. Microbiol.* **54:**184–197.

3. **Aziz, R. K., M. J. Pabst, A. Jeng, R. Kansal, D. E. Low, V. Nizet, and M. Kotb.** 2004. Invasive M1T1 group A *Streptococcus* undergoes a phase-shift in vivo to prevent proteolytic degradation of multiple virulence factors by SpeB. *Mol. Microbiol.* **51:**123–134.

4. **Baba, T., F. Takeuchi, M. Kuroda, H. Yuzawa, K. Aoki, A. Oguchi, Y. Nagai, N. Iwama, K. Asano, T. Naimi, H. Kuroda, L. Cui, K. Yamamoto, and K. Hiramatsu.** 2002. Genome and virulence determinants of high virulence community-acquired MRSA. *Lancet* **359:**1819–1827.

5. **Baggett, H. C., T. W. Hennessy, K. Rudolph, D. Bruden, A. Reasonover, A. Parkinson, R. Sparks, R. M. Donlan, P. Martinez, K. Mongkolrattanothai, and J. C. Butler.** 2004. Community-onset methicillin-resistant *Staphylococcus aureus* associated with antibiotic use and the cytotoxin Panton-Valentine leukocidin during a furunculosis outbreak in rural Alaska. *J. Infect. Dis.* **189:**1565–1573.

6. **Banks, D. J., S. B. Beres, and J. M. Musser.** 2002. The fundamental contribution of phages to GAS evolution, genome diversification and strain emergence. *Trends Microbiol.* **10:**515–521.

7. **Banks, D. J., B. Lei, and J. M. Musser.** 2003. Prophage induction and expression of prophage-encoded virulence factors in group A *Streptococcus* serotype M3 strain MGAS315. *Infect. Immun.* **71:**7079–7086.

8. **Banks, D. J., S. F. Porcella, K. D. Barbian, S. B. Beres, L. E. Philips, J. M. Voyich, F. R. DeLeo, J. M. Martin, G. A. Somerville, and J. M. Musser.** 2004. Progress toward characterization of the group A *Streptococcus* metagenome: complete genome sequence of a macrolide-resistant serotype M6 strain. *J. Infect. Dis.* **190:**727–738.

9. **Banks, D. J., S. F. Porcella, K. D. Barbian, J. M. Martin, and J. M. Musser.** 2003. Structure and distribution of an unusual chimeric genetic element encoding macrolide resistance in phylogenetically diverse clones of group A *Streptococcus. J. Infect. Dis.* **188:**1898–1908.

10. **Barondess, J. J., and J. Beckwith.** 1990. A bacterial virulence determinant encoded by lysogenic coliphage lambda. *Nature* **346:**871–874.

11. **Bensing, B. A., I. R. Siboo, and P. M. Sullam.** 2001. Proteins PblA and PblB of *Streptococcus mitis*, which promote binding to human platelets, are encoded within a lysogenic bacteriophage. *Infect. Immun.* **69:**6186–6192.

12. **Beres, S. B., G. L. Sylva, D. E. Sturdevant, C. N. Granville, M. Liu, S. M. Ricklefs, A. R. Whitney, L. D. Parkins, N. P. Hoe, G. J. Adams, D. E. Low, F. R. DeLeo, A. McGeer, and J. M. Musser.** 2004. Genome-wide molecular dissection of serotype M3 group A *Streptococcus* strains causing two epidemics of invasive infections. *Proc. Natl. Acad. Sci. USA* **101:**11833–11838.

13. **Beres, S. B., G. L. Sylva, K. D. Barbian, B. Lei, J. S. Hoff, N. D. Mammarella, M. Y. Liu, J. C. Smoot, S. F. Porcella, L. D. Parkins, D. S. Campbell, T. M. Smith, J. K. McCormick, D. Y. Leung, P. M. Schlievert, and J. M. Musser.** 2002. Genome sequence of a serotype M3 strain of group A *Streptococcus*: phage-encoded toxins, the high-virulence phenotype, and clone emergence. *Proc. Natl. Acad. Sci. USA* **99:**10078–10083.

14. **Betley, M. J., and J. J. Mekalanos.** 1985. Staphylococcal enterotoxin A is encoded by phage. *Science* **229:**185–187.

15. **Blair, J. E., and M. Carr.** 1961. Lysogeny in staphylococci. *J. Bacteriol.* **82:**984–993.

16. **Borst, D. W., and M. J. Betley.** 1994. Promoter analysis of the staphylococcal enterotoxin A gene. *J. Biol. Chem.* **269:**1883–1888.

17. **Bossi, L., J. A. Fuentes, G. Mora, and N. Figueroa-Bossi.** 2003. Prophage contribution to bacterial population dynamics. *J. Bacteriol.* **185:**6467–6471.

18. **Botstein, D.** 1980. A theory of modular evolution for bacteriophages. *Ann. N.Y. Acad. Sci.* **354:**484–490.

19. **Boyd, E. F., and H. Brüssow.** 2002. Common themes among bacteriophage-encoded virulence factors and diversity among the bacteriophages involved. *Trends Microbiol.* **10:**521–529.

20. **Breitbart, M., P. Salamon, B. Andresen, J. M. Mahaffy, A. M. Segall, D. Mead, F. Azam, and F. Rohwer.** 2002. Genomic analysis of uncultured marine viral communities. *Proc. Natl. Acad. Sci. USA* **99:**14250–14255.

21. **Brinkmann, V., U. Reichard, C. Goosmann, B. Fauler, Y. Uhlemann, D. S. Weiss, Y. Weinrauch, and A. Zychlinsky.** 2004. Neutrophil extracellular traps kill bacteria. *Science* **303:**1532–1535.

22. **Broudy, T. B., and V. A. Fischetti.** 2003. In vivo lysogenic conversion of Tox(-) *Streptococcus pyogenes* to Tox(+) with lysogenic streptococci or free phage. *Infect. Immun.* **71:**3782–3786.

23. **Broudy, T. B., V. Pancholi, and V. A. Fischetti.** 2001. Induction of lysogenic bacteriophage and phage-associated toxin from group A *streptococci* during coculture with human pharyngeal cells. *Infect. Immun.* **69:**1440–1443.

24. **Broudy, T. B., V. Pancholi, and V. A. Fischetti.** 2002. The in vitro interaction of *Streptococcus pyogenes* with human pharyngeal cells induces a phage-encoded extracellular DNase. *Infect. Immun.* **70:**2805–2811.

24a. **Brussow, H.** 2007. *The Quest for Food: a Natural History of Eating.* Springer Publisher, New York, NY.

25. **Brüssow, H., and F. Desiere.** 2001. Comparative phage genomics and the evolution of *Siphoviridae*: insights from dairy phages. *Mol. Microbiol.* **39:**213–222.

26. **Brüssow, H., C. Canchaya, and W.-D. Hardt.** 2004. Phages and the evolution of bacterial pathogens: from genomic rearrangements to lysogenic conversion. *Microbiol. Mol. Biol. Rev.* **68:**560–602.

27. **Brüssow, H., and R. W. Hendrix.** 2002. Phage genomics: small is beautiful. *Cell* **108:**13–16.

28. **Bushman, F.** 2002. *Lateral DNA Transfer.* Cold Spring Harbor Laboratory Press, Cold Spring Harbor, NY.

29. **Canchaya, C., F. Desiere, W. M. McShan, J. J. Ferretti, J. Parkhill, and H. Brussow.** 2002. Genome analysis of an inducible prophage and prophage remnants integrated in the *Streptococcus pyogenes* strain SF370. *Virology* **302:**245–258.

30. **Canchaya, C., G. Fournous, and H. Brussow.** 2004. The impact of prophages on bacterial chromosomes. *Mol. Microbiol.* **53:**9–18.

31. **Canchaya, C., C. Proux, G. Fournous, A. Bruttin, and H. Brussow.** 2003. Prophage genomics. *Microbiol. Mol. Biol. Rev.* **67:**238–276.

32. **Casjens, S.** 2003. Prophages and bacterial genomics: what have we learned so far? *Mol. Microbiol.* **49:**277–300.

33. **Casjens, S., G. F. Hatfull, and R. W. Hendrix.** 1992. Evolution of dsDNA tailed-bacteriophage genomes. *Semin. Virol.* **3:**383–397.

34. **Chatellier, S., N. Ihendyane, R. G. Kansal, F. Khambaty, H. Basma, A. Norrby-Teglund, D. E. Low, A. McGeer, and M. Kotb.** 2000. Genetic relatedness and superantigen expression in group A streptococcus serotype M1 isolates from patients with severe and nonsevere invasive diseases. *Infect. Immun.* **68:**3523–3534.

35. **Chaussee, M. S., G. L. Sylva, D. E. Sturdevant, L. M. Smoot, M. R. Graham, R. O. Watson, and J. M. Musser.** 2002. Rgg influences the expression of multiple regulatory loci to coregulate virulence factor expression in *Streptococcus pyogenes. Infect. Immun.* **70:**762–770.

36. **Chickering, H. T., and J. H. Park.** 1919. *Staphylococcus aureus* pneumonia. *JAMA* **72:**617–626.

37. **Clark, A. J., W. Inwood, T. Cloutier, and T. S. Dhillon.** 2001. Nucleotide sequence of coliphage HK620 and the evolution of lambdoid phages. *J. Mol. Biol.* **311:**657–679.

38. **Coleman, D. C., D. J. Sullivan, R. J. Russell, J. P. Arbuthnott, B. F. Carey, and H. M. Pomeroy.** 1989. *Staphylococcus aureus* bacteriophages mediating the simultaneous lysogenic conversion of beta-lysin, staphylokinase and enterotoxin A: molecular mechanism of triple conversion. *J. Gen. Microbiol.* **135:**1679–1697.

39. **Davies, H. D., A. McGeer, B. Schwartz, K. Green, D. Cann, A. E. Simor, and D. E. Low.** 1996. Invasive group A streptococcal infections in Ontario, Canada. Ontario Group A Streptococcal Study Group. *N. Engl. J. Med.* **335:**547–554.

40. **de Haas, C. J., K. E. Veldkamp, A. Peschel, F. Weerkamp, W. J. van Wamel, E. C. Heezius, M. J. Poppelier, K. P. van Kessel, and J. A. van Strijp.** 2004. Chemotaxis inhibitory protein of *Staphylococcus aureus*, a bacterial antiinflammatory agent. *J. Exp. Med.* **199:**687–695.

41. **Dempsey, R. M., D. Carroll, H. Kong, L. Higgins, C. T. Keane, and D. C. Coleman.** 2005. Sau42I, a BcgI-like restriction-modification system encoded by the *Staphylococcus aureus* quadruple-converting phage Phi42. *Microbiology* **151:**1301–1311.

42. **Desiere, F., W. M. McShan, D. van Sinderen, J. J. Ferretti, and H. Brussow.** 2001. Comparative genomics reveals close genetic relationships

between phages from dairy bacteria and pathogenic streptococci: evolutionary implications for prophage-host interactions. *Virology* **288:**325–341.

43. **Diep, B. A., S. R. Gill, R. F. Chang, T. H. Phan, J. H. Chen, M. G. Davidson, F. Lin, J. Lin, H. A. Carleton, E. F. Mongodin, G. F. Sensabaugh, and F. Perdreau-Remington.** 2006. Complete genome sequence of USA300, an epidemic clone of community-acquired meticillin-resistant *Staphylococcus aureus. Lancet* **367:**731–739.

44. **Diep, B. A., G. F. Sensabaugh, N. S. Somboona, H. A. Carleton, and F. Perdreau-Remington.** 2004. Widespread skin and soft-tissue infections due to two methicillin-resistant *Staphylococcus aureus* strains harboring the genes for Panton-Valentine leucocidin. *J. Clin. Microbiol.* **42:**2080–2084.

45. **Dozois, C. M., F. Daigle, and R. Curtiss III.** 2003. Identification of pathogen-specific and conserved genes expressed in vivo by an avian pathogenic *Escherichia coli* strain. *Proc. Natl. Acad. Sci. USA* **100:**247–252.

46. **Ferretti, J. J., W. M. McShan, D. Ajdic, D. J. Savic, G. Savic, K. Lyon, C. Primeaux, S. Sezate, A. N. Suvorov, S. Kenton, H. S. Lai, S. P. Lin, Y. Qian, H. G. Jia, F. Z. Najar, Q. Ren, H. Zhu, L. Song, J. White, X. Yuan, S. W. Clifton, B. A. Roe, and R. McLaughlin.** 2001. Complete genome sequence of an M1 strain of *Streptococcus pyogenes. Proc. Natl. Acad. Sci. USA* **98:**4658–4663.

47. **Finck-Barbancon, V., G. Duportail, O. Meunier, and D. A. Colin.** 1993. Pore formation by a two-component leukocidin from *Staphylococcus aureus* within the membrane of human polymorphonuclear leukocytes. *Biochim. Biophys. Acta* **1182:**275–282.

48. **Fitzgerald, J. R., D. E. Sturdevant, S. M. Mackie, S. R. Gill, and J. M. Musser.** 2001. Evolutionary genomics of *Staphylococcus aureus*: insights into the origin of methicillin-resistant strains and the toxic shock syndrome epidemic. *Proc. Natl. Acad. Sci. USA* **98:**8821–8826.

49. **Genestier, A. L., M. C. Michallet, G. Prevost, G. Bellot, L. Chalabreysse, S. Peyrol, F. Thivolet, J. Etienne, G. Lina, F. M. Vallette, F. Vandenesch, and L. Genestier.** 2005. *Staphylococcus aureus* Panton-Valentine leukocidin directly targets mitochondria and induces Bax-independent apoptosis of human neutrophils. *J. Clin. Invest.* **115:**3117–3127.

50. **Gillet, Y., B. Issartel, P. Vanhems, J. C. Fournet, G. Lina, M. Bes, F. Vandenesch, Y. Piemont, N. Brousse, D. Floret, and J. Etienne.** 2002. Association between *Staphylococcus aureus* strains carrying gene for Panton-Valentine leukocidin and highly lethal necrotising pneumonia in young immunocompetent patients. *Lancet* **359:**753–759.

51. **Goerke, C., S. Papenberg, S. Dasbach, K. Dietz, R. Ziebach, B. C. Kahl, and C. Wolz.** 2004. Increased frequency of genomic alterations in *Staphylococcus aureus* during chronic infection is in part due to phage mobilization. *J. Infect. Dis.* **189:**724–734.

52. **Graham, M. R., L. M. Smoot, C. A. Migliaccio, K. Virtaneva, D. E. Sturdevant, S. F. Porcella, M. J. Federle, G. J. Adams, J. R. Scott, and J. M. Musser.** 2002. Virulence control in group A *Streptococcus* by a two-component gene regulatory system: global expression profiling and in vivo infection modeling. *Proc. Natl. Acad. Sci. USA* **99:**13855–13860.

53. **Green, N. M., S. B. Beres, E. A. Graviss, J. E. Allison, A. J. McGeer, J. Vuopio-Varkila, R. B. LeFebvre, and J. M. Musser.** 2005. Genetic diversity among type emm28 group A *Streptococcus* strains causing invasive infections and pharyngitis. *J. Clin. Microbiol.* **43:**4083–4091.

54. **Green, N. M., S. Zhang, S. F. Porcella, M. J. Nagiec, K. D. Barbian, S. B. Beres, R. B. LeFebvre, and J. M. Musser.** 2005. Genome sequence of a serotype M28 strain of group a streptococcus: potential new insights into puerperal sepsis and bacterial disease specificity. *J. Infect. Dis.* **192:**760–770.

55. **Haas, P. J., C. J. de Haas, W. Kleibeuker, M. J. Poppelier, K. P. van Kessel, J. A. Kruijtzer, R. M. Liskamp, and J. A. van Strijp.** 2004. N-terminal residues of the chemotaxis inhibitory protein of *Staphylococcus aureus* are essential for blocking formylated peptide receptor but not C5a receptor. *J. Immunol.* **173:**5704–5711.

56. **Halperin, S. A., P. Ferrieri, E. D. Gray, E. L. Kaplan, and L. W. Wannamaker.** 1987. Antibody response to bacteriophage hyaluronidase in acute glomerulonephritis after group A streptococcal infection. *J. Infect. Dis.* **155:**253–261.

57. **Hendrix, R. W., J. G. Lawrence, G. F. Hatfull, and S. Casjens.** 2000. The origins and ongoing evolution of viruses. *Trends Microbiol.* **8:**504–508.

58. **Holden, M. T., E. J. Feil, J. A. Lindsay, S. J. Peacock, N. P. Day, M. C. Enright, T. J. Foster, C. E. Moore, L. Hurst, R. Atkin, A. Barron, N. Bason, S. D. Bentley, C. Chillingworth, T. Chillingworth, C. Churcher, L. Clark, C. Corton, A. Cronin, J. Doggett, L. Dowd, T. Feltwell, Z. Hance, B. Harris, H. Hauser, S. Holroyd, K. Jagels, K. D. James, N. Lennard, A. Line, R. Mayes, S. Moule, K. Mungall, D. Ormond, M. A. Quail, E. Rabbinowitsch, K. Rutherford, M. Sanders, S.**

Sharp, M. Simmonds, K. Stevens, S. White-head, B. G. Barrell, B. G. Spratt, and J. Parkhill. 2004. Complete genomes of two clini-cal *Staphylococcus aureus* strains: evidence for the rapid evolution of virulence and drug resistance. *Proc. Natl. Acad. Sci. USA* **101:**9786–9791.

59. Ikebe, T., A. Wada, Y. Inagaki, K. Sugama, R. Suzuki, D. Tanaka, A. Tamaru, Y. Fujinaga, Y. Abe, Y. Shimizu, and H. Watanabe. 2002. Dis-semination of the phage-associated novel super-antigen gene *speL* in recent invasive and noninva-sive *Streptococcus pyogenes* M3/T3 isolates in Japan. *Infect. Immun.* **70:**3227–3233.

60. Inagaki, Y., F. Myouga, H. Kawabata, S. Yamai, and H. Watanabe. 2000. Genomic dif-ferences in *Streptococcus pyogenes* serotype M3 be-tween recent isolates associated with toxic shock–like syndrome and past clinical isolates. *J. Infect. Dis.* **181:**975–983.

61. Ingrey, K. T., J. Ren, and J. F. Prescott. 2003. A fluoroquinolone induces a novel mitogen-encoding bacteriophage in *Streptococcus canis*. *In-fect. Immun.* **71:**3028–3033.

62. Jin, T., M. Bokarewa, T. Foster, J. Mitchell, J. Higgins, and A. Tarkowski. 2004. *Staph-ylococcus aureus* resists human defensins by produc-tion of staphylokinase, a novel bacterial evasion mechanism. *J. Immunol.* **172:**1169–1176.

63. Juhala, R. J., M. E. Ford, R. L. Duda, A. Youl-ton, G. F. Hatfull, and R. W. Hendrix. 2000. Genomic sequences of bacteriophages HK97 and HK022: pervasive genetic mosaicism in the lamb-doid bacteriophages. *J. Mol. Biol.* **299:**27–51.

64. Kaneko, J., T. Kimura, S. Narita, T. Tomita, and Y. Kamio. 1998. Complete nucleotide se-quence and molecular characterization of the temperate staphylococcal bacteriophage phiPVL carrying Panton-Valentine leukocidin genes. *Gene* **215:**57–67.

65. Kansal, R. G., A. McGeer, D. E. Low, A. Norrby-Teglund, and M. Kotb. 2000. Inverse relation between disease severity and expression of the streptococcal cysteine protease, SpeB, among clonal M1T1 isolates recovered from inva-sive group A streptococcal infection cases. *Infect. Immun.* **68:**6362–6369.

66. Kazmi, S. U., R. Kansal, R. K. Aziz, M. Hooshdaran, A. Norrby-Teglund, D. E. Low, A. B. Halim, and M. Kotb. 2001. Reciprocal, temporal expression of SpeA and SpeB by inva-sive M1T1 group a streptococcal isolates in vivo. *Infect. Immun.* **69:**4988–4995.

67. Kuroda, M., T. Ohta, I. Uchiyama, T. Baba, H. Yuzawa, I. Kobayashi, L. Cui, A. Oguchi, K. Aoki, Y. Nagai, J. Lian, T. Ito, M. Kanamori, H. Matsumaru, A. Maruyama, H. Murakami, A. Hosoyama, Y. Mizutani-Ui,

N. K. Takahashi, T. Sawano, R. Inoue, C. Kaito, K. Sekimizu, H. Hirakawa, S. Kuhara, S. Goto, J. Yabuzaki, M. Kanehisa, A. Ya-mashita, K. Oshima, K. Furuya, C. Yoshino, T. Shiba, M. Hattori, N. Ogasawara, H. Hayashi, and K. Hiramatsu. 2001. Whole genome sequencing of meticillin-resistant *Staph-ylococcus aureus*. *Lancet* **357:**1225–1240.

68. Kwan, T., J. Liu, M. DuBow, P. Gros, and J. Pelletier. 2005. The complete genomes and proteomes of 27 *Staphylococcus aureus* bacterio-phages. *Proc. Natl. Acad. Sci. USA* **102:**5174–5179.

69. Ladhani, S., C. L. Joannou, D. P. Lochrie, R. W. Evans, and S. M. Poston. 1999. Clinical, microbial, and biochemical aspects of the exfolia-tive toxins causing staphylococcal scalded-skin syndrome. *Clin. Microbiol. Rev.* **12:**224–242.

70. Lawrence, J. G., R. W. Hendrix, and S. Cas-jens. 2003. Where are the pseudogenes in bacte-rial genomes? *Trends Microbiol.* **a:**535–540.

71. Lindsay, J. A., A. Ruzin, H. F. Ross, N. Kure-pina, and R. P. Novick. 1998. The gene for toxic shock toxin is carried by a family of mobile pathogenicity islands in *Staphylococcus aureus*. *Mol. Microbiol.* **29:**527–543.

72. Luria, S. E., and M. Delbrück. 1943. Muta-tions of bacteria from virus sensitivity to virus re-sistance. *Genetics* **28:**491–511.

73. Matsumoto, M., N. P. Hoe, M. Liu, S. B. Beres, G. L. Sylva, C. M. Brandt, G. Haase, and J. M. Musser. 2003. Intrahost sequence variation in the streptococcal inhibitor of com-plement gene in patients with human pharyngitis. *J. Infect. Dis.* **187:**604–612.

74. Matz, C., P. Deines, J. Boenigk, H. Arndt, L. Eberl, S. Kjelleberg, and K. Jurgens. 2004. Impact of violacein-producing bacteria on sur-vival and feeding of bacteriovorous nano. *Appl. Environ. Microbiol.* **70:**1593–1599.

75. Matz, C., and S. Kjelleberg. 2005. Off the hook-how bacteria survive protozoan grazing. *Trends Microbiol.* **13:**302–307.

76. McClelland, M., K. E. Sanderson, J. Spieth, S. W. Clifton, P. Latreille, L. Courtney, S. Por-wollik, J. Ali, M. Dante, F. Du, S. Hou, D. Layman, S. Leonard, C. Nguyen, K. Scott, A. Holmes, N. Grewal, E. Mulvaney, E. Ryan, H. Sun, L. Florea, W. Miller, T. Stoneking, M. Nhan, R. Waterston, and R. K. Wilson. 2001. Complete genome sequence of *Salmonella enterica* serovar Typhimurium LT2. *Nature* **413:**852–856.

77. McGrath, S., G. F. Fitzgerald, and D. van Sin-deren. 2002. Identification and characterization of phage-resistance genes in temperate lactococcal bacteriophages. *Mol. Microbiol.* **43:**509–520.

78. Miller, A. A., N. C. Engleberg, and V. J. DiRita. 2001. Repression of virulence genes

by phosphorylation-dependent oligomerization of CsrR at target promoters in *S. pyogenes. Mol. Microbiol.* **40:**976–990.

79. **Miller, L. G., F. Perdreau-Remington, G. Rieg, S. Mehdi, J. Perlroth, A. S. Bayer, A. W. Tang, T. O. Phung, and B. Spellberg.** 2005. Necrotizing fasciitis caused by community-associated methicillin-resistant *Staphylococcus aureus* in Los Angeles. *N. Engl. J. Med.* **352:**1445–1453.

80. **Mirold, S., K. Ehrbar, A. Weissmüller, R. Prager, H. Tschäpe, H. Rüssmann, and W. D. Hardt.** 2001. *Salmonella* host cell invasion emerged by acquisition of a mosaic of separate genetic elements, including *Salmonella* pathogenicity island 1 (SPI1), SPI5, and *sopE2. J. Bacteriol.* **183:**2348–2358.

81. **Mirold, S., W. Rabsch, H. Tschape, and W. D. Hardt.** 2001. Transfer of the *Salmonella* type III effector sopE between unrelated phage families. *J. Mol. Biol.* **312:**7–16.

82. **Morgan, G. J., G. F. Hatfull, S. Casjens, and R. W. Hendrix.** 2002. Bacteriophage Mu genome sequence: analysis and comparison with Mu-like prophages in *Haemophilus, Neisseria* and *Deinococcus. J. Mol. Biol.* **317:**337–359.

83. **Nagiec, M. J., B. Lei, S. K. Parker, M. L. Vasil, M. Matsumoto, R. M. Ireland, S. B. Beres, N. P. Hoe, and J. M. Musser.** 2004. Analysis of a novel prophage-encoded group A Streptococcus extracellular phospholipase A(2). *J. Biol. Chem.* **279:**45909–45918.

84. **Nakagawa, I., K. Kurokawa, A. Yamashita, M. Nakata, Y. Tomiyasu, N. Okahashi, S. Kawabata, K. Yamazaki, T. Shiba, T. Yasunaga, H. Hayashi, M. Hattori, and S. Hamada.** 2003. Genome sequence of an M3 strain of *Streptococcus pyogenes* reveals a large-scale genomic rearrangement in invasive strains and new insights into phage evolution. *Genome Res.* **13:**1042–1055.

85. **Nakayama, K., S. Kanaya, M. Ohnishi, Y. Terawaki, and T. Hayashi.** 1999. The complete nucleotide sequence of phi CTX, a cytotoxin-converting phage of *Pseudomonas aeruginosa*: implications for phage evolution and horizontal gene transfer via bacteriophages. *Mol. Microbiol.* **31:** 399–419.

86. **Narita, S., J. Kaneko, J. Chiba, Y. Piemont, S. Jarraud, J. Etienne, and Y. Kamio.** 2001. Phage conversion of Panton-Valentine leukocidin in *Staphylococcus aureus*: molecular analysis of a PVL-converting phage, phiSLT. *Gene* **268:**195–206.

87. **Nariya, H., A. Nishiyama, and Y. Kamio.** 1997. Identification of the minimum segment in which the threonine246 residue is a potential phosphorylated site by protein kinase A for the LukS-specific function of staphylococcal leukocidin. *FEBS Lett.* **415:**96–100.

88. **Nilsson, A. S., J. L. Karlsson, and E. Haggard-Ljungquist.** 2003. Site-specific recombination links the evolution of P2-like coliphages and pathogenic enterobacteria. *Mol. Biol. Evol.* **21:** 1–13.

89. **Ohnishi, M., K. Kurokawa, and T. Hayashi.** 2001. Diversification of *Escherichia coli* genomes: are bacteriophages the major contributors? *Trends Microbiol.* **9:**481–485.

90. **Ohnishi, M., J. Terajima, K. Kurokawa, K. Nakayama, T. Murata, K. Tamura, Y. Ogura, H. Watanabe, and T. Hayashi.** 2002. Genomic diversity of enterohemorrhagic *Escherichia coli* O157 revealed by whole genome PCR scanning. *Proc. Natl. Acad. Sci. USA* **99:**17043–17048.

91. **Okamoto, K., J. A. Mudd, and J. Marmur.** 1968. Conversion of *Bacillus subtilis* DNA to phage DNA following mitomycin C induction. *J. Mol. Biol.* **34:**429–437.

92. **Pan, E. S., B. A. Diep, E. D. Charlebois, C. Auerswald, H. A. Carleton, G. F. Sensabaugh, and F. Perdreau-Remington.** 2005. Population dynamics of nasal strains of methicillin-resistant *Staphylococcus aureus* and their relation to community-associated disease activity. *J. Infect. Dis.* **192:**811–818.

93. **Pedulla, M. L., M. E. Ford, J. M. Houtz, T. Karthikeyan, C. Wadsworth, J. A. Lewis, D. Jacobs-Sera, J. Falbo, J. Gross, N. R. Pannunzio, W. Brucker, V. Kumar, J. Kandasamy, L. Keenan, S. Bardarov, J. Kriakov, J. G. Lawrence, W. R. Jacobs, Jr., R. W. Hendrix, and G. F. Hatfull.** 2003. Origins of highly mosaic mycobacteriophage genomes. *Cell* **113:**171–182.

94. **Pelludat, C., S. Mirold, and W. D. Hardt.** 2003. The SopEPhi phage integrates into the ssrA gene of *Salmonella enterica* serovar Typhimurium A36 and is closely related to the Fels-2 prophage. *J. Bacteriol.* **185:**5182–5191.

95. **Perna, N. T., G. Plunkett III, V. Burland, B. Mau, J. D. Glasner, D. J. Rose, G. F. Mayhew, P. S. Evans, J. Gregor, H. A. Kirkpatrick, G. Posfai, J. Hackett, S. Klink, A. Boutin, Y. Shao, L. Miller, E. J. Grotbeck, N. W. Davis, A. Lim, E. T. Dimalanta, K. D. Potamousis, J. Apodaca, T. S. Anantharaman, J. Lin, G. Yen, D. C. Schwartz, R. A. Welch, and F. R. Blattner.** 2001. Genome sequence of enterohaemorrhagic *Escherichia coli* O157:H7. *Nature* **409:**529–533.

96. **Porwollik, S., R. M. Wong, and M. McClelland.** 2002. Evolutionary genomics of *Salmonella*: gene acquisitions revealed by microarray analysis. *Proc. Natl. Acad. Sci. USA* **99:**8956–8961.

97. **Postma, B., M. J. Poppelier, J. C. van Galen, E. R. Prossnitz, J. A. van Strijp, C. J. de Haas, and K. P. van Kessel.** 2004. Chemotaxis in-

hibitory protein of *Staphylococcus aureus* binds specifically to the C5a and formylated peptide receptor. *J. Immunol.* **172:**6994–7001.

98. **Prescott, J. F., C. W. Miller, K. A. Mathews, J. A. Yager, and L. DeWinter.** 1997. Update on canine streptococcal toxic shock syndrome and necrotizing fasciitis. *Can. Vet. J.* **38:**241–242.

99. **Proft, T., S. L. Moffatt, C. J. Berkahn, and J. D. Fraser.** 1999. Identification and characterization of novel superantigens from *Streptococcus pyogenes. J. Exp. Med.* **189:**89–102.

100. **Proux, C., D. van Sinderen, J. Suarez, P. Garcia, V. Ladero, G. F. Fitzgerald, F. Desiere, and H. Brussow.** 2002. The dilemma of phage taxonomy illustrated by comparative genomics of Sfi21-like *Siphoviridae* in lactic acid bacteria. *J. Bacteriol.* **184:**6026–6036.

101. **Rahimpour, R., G. Mitchell, M. H. Khandaker, C. Kong, B. Singh, L. Xu, A. Ochi, R. D. Feldman, J. G. Pickering, B. M. Gill, and D. J. Kelvin.** 1999. Bacterial superantigens induce down-modulation of CC chemokine responsiveness in human monocytes via an alternative chemokine ligand-independent mechanism. *J. Immunol.* **162:**2299–2307.

102. **Rohwer, F.** 2003. Global phage diversity. *Cell* **113:**141.

103. **Rooijakkers, S. H., M. Ruyken, A. Roos, M. R. Daha, J. S. Presanis, R. B. Sim, W. J. van Wamel, K. P. van Kessel, and J. A. van Strijp.** 2005. Immune evasion by a staphylococcal complement inhibitor that acts on C3 convertases. *Nat. Immunol.* **6:**920–927.

104. **Ruzin, A., J. Lindsay, and R. P. Novick.** 2001. Molecular genetics of SaPI1—a mobile pathogenicity island in *Staphylococcus aureus. Mol. Microbiol.* **41:**365–377.

105. **Sakaguchi, Y., T. Hayashi, K. Kurokawa, K. Nakayama, K. Oshima, Y. Fujinaga, M. Ohnishi, E. Ohtsubo, M. Hattori, and K. Oguma.** 2005. The genome sequence of *Clostridium botulinum* type C neurotoxin-converting phage and the molecular mechanisms of unstable lysogeny. *Proc. Natl. Acad. Sci. USA* **102:**17472–17477.

106. **Shelburne, S. A., III, C. Granville, M. Tokuyama, I. Sitkiewicz, P. Patel, and J. M. Musser.** 2005. Growth characteristics of and virulence factor production by group A *Streptococcus* during cultivation in human saliva. *Infect. Immun.* **73:**4723–4731.

107. **Smith, N. L., E. J. Taylor, A. M. Lindsay, S. J. Charnock, J. P. Turkenburg, E. J. Dodson, G. J. Davies, and G. W. Black.** 2005. Structure of a group A streptococcal phage-encoded virulence factor reveals a catalytically active triple-stranded beta-helix. *Proc. Natl. Acad. Sci. USA* **102:**17652–17657.

108. **Smoot, J. C., K. D. Barbian, J. J. Van Gompel, L. M. Smoot, M. S. Chaussee, G. L. Sylva, D. E. Sturdevant, S. M. Ricklefs, S. F. Porcella, L. D. Parkins, S. B. Beres, D. S. Campbell, T. M. Smith, Q. Zhang, V. Kapur, J. A. Daly, L. G. Veasy, and J. M. Musser.** 2002. Genome sequence and comparative microarray analysis of serotype M18 group A *Streptococcus* strains associated with acute rheumatic fever outbreaks. *Proc. Natl. Acad. Sci. USA* **99:** 4668–4673.

109. **Smoot, L. M., J. K. McCormick, J. C. Smoot, N. P. Hoe, I. Strickland, R. L. Cole, K. D. Barbian, C. A. Earhart, D. H. Ohlendorf, L. G. Veasy, H. R. Hill, D. Y. Leung, P. M. Schlievert, and J. M. Musser.** 2002. Characterization of two novel pyrogenic toxin superantigens made by an acute rheumatic fever clone of *Streptococcus pyogenes* associated with multiple disease outbreaks. *Infect. Immun.* **70:** 7095–7104.

110. **Smoot, L. M., J. C. Smoot, M. R. Graham, G. A. Somerville, D. E. Sturdevant, C. A. Migliaccio, G. L. Sylva, and J. M. Musser.** 2001. Global differential gene expression in response to growth temperature alteration in group A *Streptococcus. Proc. Natl. Acad. Sci. USA* **98:**10416–10421.

111. **Sriskandan, S., M. Unnikrishnan, T. Krausz, and J. Cohen.** 2000. Mitogenic factor (MF) is the major DNase of serotype M89 *Streptococcus pyogenes. Microbiology* **146:**2785–2792.

112. **Stover, C. K., X. Q. Pham, A. L. Erwin, S. D. Mizoguchi, P. Warrener, M. J. Hickey, F. S. Brinkman, W. O. Hufnagle, D. J. Kowalik, M. Lagrou, R. L. Garber, L. Goltry, E. Tolentino, S. Westbrock-Wadman, Y. Yuan, L. L. Brody, S. N. Coulter, K. R. Folger, A. Kas, K. Larbig, R. Lim, K. Smith, D. Spencer, G. K. Wong, Z. Wu, I. T. Paulsen, J. Reizer, M. H. Saier, R. E. Hancock, S. Lory, and M. V. Olson.** 2000. Complete genome sequence of *Pseudomonas aeruginosa* PA01, an opportunistic pathogen. *Nature* **406:**959–964.

113. **Sumby, P., K. D. Barbian, D. J. Gardner, A. R. Whitney, D. M. Welty, R. D. Long, J. R. Bailey, M. J. Parnell, N. P. Hoe, G. G. Adams, F. R. DeLeo, and J. M. Musser.** 2005. Extracellular deoxyribonuclease made by group A *Streptococcus* assists pathogenesis by enhancing evasion of the innate immune response. *Proc. Natl. Acad. Sci. USA* **102:**1679–1684.

114. **Sumby, P., S. F. Porcella, A. G. Madrigal, K. D. Barbian, K. Virtaneva, S. M. Ricklefs, D. E. Sturdevant, M. R. Graham, J. Vuopio-Varkila, N. P. Hoe, and J. M. Musser.** 2005. Evolutionary origin and emergence of a highly

successful clone of serotype M1 group a *Streptococcus* involved multiple horizontal gene transfer events. *J. Infect. Dis.* **192:**771–782.

115. **Sumby, P., and M. K. Waldor.** 2003. Transcription of the toxin genes present within the Staphylococcal phage phiSa3ms is intimately linked with the phage's life cycle. *J. Bacteriol.* **185:**6841–6851.

116. **Susskind, M. M., and D. Botstein.** 1978. Molecular genetics of bacteriophage P22. *Microbiol. Rev.* **42:**385–413.

116a.**Tobe, T., S. A. Beatson, H. Taniguchi, H. Abe, C. M. Bailey, A. Fivian, R. Younis, S. Matthews, O. Marches, G. Frankel, T. Hayashi, and M. J. Pallen.** 2006. An extensive repertoire of type III secretion effectors in *Escherichia coli* O157 and the role of lambdoid phages in their dissemination. *Proc. Natl. Acad. Sci. USA* **103:**14941–14946.

117. **Ubeda, C., E. Maiques, E. Knecht, I. Lasa, R. P. Novick, and J. R. Penades.** 2005. Antibiotic-induced SOS response promotes horizontal dissemination of pathogenicity island-encoded virulence factors in staphylococci. *Mol. Microbiol.* **56:**836–844.

118. **van Wamel, W. J., S. H. Rooijakkers, M. Ruyken, K. P. van Kessel, and J. A. van Strijp.** 2006. The innate immune modulators staphylococcal complement inhibitor and chemotaxis inhibitory protein of *Staphylococcus aureus* are located on beta-hemolysin-converting bacteriophages. *J. Bacteriol.* **188:**1310–1315.

119. **Ventura, M., A. Bruttin, C. Canchaya, and H. Brüssow.** 2002. Transcription analysis of *Streptococcus thermophilus* phages in the lysogenic state. *Virology* **302:**21–32.

120. **Virtaneva, K., M. R. Graham, S. F. Porcella, N. P. Hoe, H. Su, E. A. Graviss, T. J. Gardner, J. E. Allison, W. J. Lemon, J. R. Bailey, M. J. Parnell, and J. M. Musser.** 2003. Group A *Streptococcus* gene expression in humans and cynomolgus macaques with acute pharyngitis. *Infect. Immun.* **71:**2199–2207.

121. **Virtaneva, K., S. F. Porcella, M. R. Graham, R. M. Ireland, C. A. Johnson, S. M. Ricklefs, I. Babar, L. D. Parkins, R. A. Romero, G. J. Corn, D. J. Gardner, J. R. Bailey, M. J. Parnell, and J. M. Musser.** 2005. Longitudinal analysis of the group A *Streptococcus* transcriptome in experimental pharyngitis in cynomolgus macaques. *Proc. Natl. Acad. Sci. USA* **102:** 9014–9019.

122. **von Eiff, C., K. Becker, K. Machka, H. Stammer, and G. Peters.** 2001. Nasal carriage as a source of *Staphylococcus aureus* bacteremia. Study Group. *N. Engl. J. Med.* **344:**11–16.

123. **Voyich, J. M., K. R. Braughton, D. E. Sturdevant, S. D. Kobayashi, B. Lei, K. Vir-** taneva, D. W. Dorward, J. M. Musser, and F. R. DeLeo. 2003. Genome-wide protective response used by group A *Streptococcus* to evade destruction by human polymorphonuclear leukocytes. *Proc. Natl. Acad. Sci. USA* **100:**1996–2001.

124. **Voyich, J. M., K. R. Braughton, D. E. Sturdevant, A. R. Whitney, B. Said-Salim, S. F. Porcella, R. D. Long, D. W. Dorward, D. J. Gardner, B. N. Kreiswirth, J. M. Musser, and F. R. DeLeo.** 2005. Insights into mechanisms used by *Staphylococcus aureus* to avoid destruction by human neutrophils. *J. Immunol.* **175:**3907–3919.

125. **Wagner, P. L., and M. K. Waldor.** 2002. Bacteriophage control of bacterial virulence. *Infect. Immun.* **70:**3985–3993.

126. **Waldor, M. K.** 1998. Bacteriophage biology and bacterial virulence. *Trends Microbiol.* **6:**295–297.

127. **Waldor, M. K., and J. J. Mekalanos.** 1996. Lysogenic conversion by a filamentous phage encoding cholera toxin. *Science* **272:**1910–1914.

128. **Westmoreland, B. C., W. Szybalski, and H. Ris.** 1969. Mapping of deletions and substitutions in heteroduplex DNA molecules of bacteriophage lambda by electron microscopy. *Science* **163:**1343–1348.

129. **Whiteley, M., M. G. Bangera, R. E. Bumgarner, M. R. Parsek, G. M. Teitzel, S. Lory, and E. P. Greenberg.** 2001. Gene expression in *Pseudomonas aeruginosa* biofilms. *Nature* **413:** 860–864.

130. **Wildschutte, H., D. M. Wolfe, A. Tamewitz, and J. G. Lawrence.** 2004. Protozoan predation, diversifying selection, and the evolution of antigenic diversity in *Salmonella*. *Proc. Natl. Acad. Sci. USA* **101:**10644–10649.

131. **Winkler, K. C., J. de Waart, and C. Grootsen.** 1965. Lysogenic conversion of staphylococci to loss of beta-toxin. *J. Gen. Microbiol.* **39:**321–333.

132. **Wommack, K. E., and R. R. Colwell.** 2000. Virioplankton: viruses in aquatic ecosystems. *Microbiol. Mol. Biol. Rev.* **64:**69–114.

133. **Yamaguchi, T., T. Hayashi, H. Takami, K. Nakasone, M. Ohnishi, K. Nakayama, S. Yamada, H. Komatsuzawa, and M. Sugai.** 2000. Phage conversion of exfoliative toxin A production in *Staphylococcus aureus*. *Mol. Microbiol.* **38:**694–705.

134. **Zou, D., J. Kaneko, S. Narita, and Y. Kamio.** 2000. Prophage, phiPV83-pro, carrying panton-valentine leukocidin genes, on the *Staphylococcus aureus* P83 chromosome: comparative analysis of the genome structures of phiPV83-pro, phiPVL, phi11, and other phages. *Biosci. Biotechnol. Biochem.* **64:**2631–2643.

WHAT GENOMICS HAS TAUGHT US ABOUT GRAM-POSITIVE PROTEIN SECRETION AND TARGETING

Olaf Schneewind and Dominique Missiakas

12

Gram-positive bacteria use several different pathways to transport polypeptides to discrete locations in or across the microbial envelope. Protein secretion beyond the envelope ensures survival of microorganisms because transport substrates provide selective advantages when bacteria are confronted with intensely competitive microbial communities. For example, secreted enzymes degrade complex nutrients and bacteriocins are programmed to destroy competing microbes, whereas other proteins enable symbiotic survival with other organisms. In gram-positive bacteria, most secreted proteins are transported across the plasma membrane via the universally conserved and essential Sec pathway. Proteins carrying a Sec-dependent signal sequence, but lacking any other type of topogenic information, are by default released into the extracellular milieu. Additional sequence motifs within secreted substrates are necessary to target proteins to discrete subcompartments within the envelope. Factors other than the Sec machinery components must therefore be responsible for deciphering such signals and implementing the protein targeting mechanisms within the envelope. Gram-positive bacteria also encode Sec-independent pathways for protein translocation across the membrane, some of which require a complex export apparatus. Here, we will briefly review what is known about the Sec pathway and its modification in gram-positive bacteria and discuss second independent protein targeting pathways.

On the basis of morphological criteria, three distinct cellular compartments can be distinguished in gram-positive bacteria: cytosol, cytoplasmic membrane, and the surrounding cell wall (80). Proteins are found in any one of these three compartments. In addition, some gram-positive bacteria synthesize a crystalline layer of surface proteins (198) deposited around the cell, beyond the cell wall, and are clearly visible by microscopy. A few gram-positive bacteria form spores. Such bacteria are capable of generating morphologically distinct daughter cells via a developmental program of asymmetric cell division (123). The envelope of spores differs from that of mother cells and contains specific sets of proteins (116, 165, 230). Finally, some gram-positive bacteria have evolved protein secretion machines dedicated to the assembly of macromolecular structures such as flagella, type 4 pili, or Flp pili, all of which can be viewed by light or electron microscopy as elongated organelles that protrude from the cell surface.

Olaf Schneewind and Dominique Missiakas   Department of Microbiology, The University of Chicago, Chicago, Illinois 60637.

Bacterial Pathogenomics, Edited by M. J. Pallen et al.
© 2007 ASM Press, Washington, D.C.

As for all proteins that are destined for the bacterial surface or the extracellular media, their travels require passage of an unfolded precursor across the cytoplasmic (plasma) membrane, a process commonly referred to as protein translocation. Only those proteins bearing discrete signal sequences are transported across the membrane. Transport across the membrane is mediated by proteins that not only provide energy and catalysis for transport, but also exert gatekeeper functions, by allowing substrate passage and rejecting other proteins lacking secretion signals. After translocation, some polypeptides are retained in the membrane or envelope. Several conserved mechanisms provide for protein retention in these compartments. Here again, polypeptides residing in the membrane or the envelope are marked with specific "zip codes" that provide for specific molecular interactions and retention.

This brief review describes the general principles of protein secretion in gram-positive bacteria along with a few examples; it is not exhaustive, and we refer readers to other publications for a more comprehensive analysis on this diverse subject. Protein secretion mechanisms are thought to be universally conserved between bacteria. Hence, the factors that catalyze protein secretion can be predicted in silico and identified in genomes of all gram-positive bacteria. Nevertheless, wherever an experimental validation of such predictions seemed available, we sought to describe it.

THE Sec SYSTEM IN ESCHERICHIA COLI

Sec-machinery mediated secretion is an essential pathway that provides for the transport of most proteins across the plasma membrane in addition to supporting biogenesis and assembly of bacterial envelope structures. Sec-mediated protein secretion has been best studied in *Escherichia coli* and, more recently, in *Bacillus subtilis.* Our understanding of the *E. coli* Sec-dependent secretion system serves as a paradigm for all bacterial and eukaryotic cells (56, 164, 168). Although very few studies have been performed by using gram-positive bacte-

ria, genomic analysis of the plethora of genomes available confirms that the genes involved in Sec-dependent secretion are largely conserved, and by extension, the assumption is that the molecular events leading to protein translocation across the plasma membrane must be conserved. But before summarizing what genomics has revealed on protein secretion, we should consider briefly what and how we know about Sec-mediated secretion in *E. coli.*

Translocation through the cytoplasmic membrane is mediated by signal peptides located at the N terminus of a protein (15). Signal peptides are short segments of 15 to 20 hydrophobic amino acids flanked at the N-terminal end by positively charged residues (63, 195). For secreted proteins, signal peptides are proteolytically removed by signal (leader) peptidases upon translocation across the cytoplasmic membrane (31, 32, 43, 44, 50). Upon cleavage, the translocated protein is free of any hydrophobic segment and soluble, and is no longer associated with the membrane (44). The signal peptide–containing precursor is referred to as the preprotein (or precursor), whereas the cleaved and translocated species is its mature form. If a signal peptide is not cleaved off, i.e., it does not harbor a signal peptidase cleavage site, the polypeptide will remain integrated in the membrane as the signal peptide assumes a stop-transfer function (membrane anchor or transmembrane segment). Upon integration into the cytoplasmic membrane, stop transfer signals fold into α-helices and serve as transmembrane segments of integral membrane proteins. Proteins that span the membrane multiple times contain multiple stop transfer signals (the number of transfer signals equals the number of turns in polytopic transmembrane proteins).

Signal peptides are necessary and sufficient for protein translocation across membranes, meaning that any polypeptide fused to the C terminus of a signal peptide is marked for translocation and targeted to the Sec machinery or membrane translocon, a proteinaceous canal though which unfolded proteins must pass. Traveling through this channel requires maintenance of precursor substrates in an export competent state. For example, LacZ fusions to the C terminus of signal peptide–bearing proteins are at first substrates for export, but then arrest at an undefined step along the secretion pathway, blocking the export of other precursors (even those not fused to LacZ) (158). The gene products necessary for the initiation of signal peptide–containing precursors into the secretory pathway were identified as mutants that restored the β-galactosidase activity of these LacZ fusion proteins. The results of the *sec* screens generally matched those of a suppressor screen (*prl*, for "protein localization") that scored for the export of a mutant polypeptide harboring a defective signal peptide (62).

The protein translocation pore of *E. coli* is composed of three membrane proteins, SecYEG, that recognize secretion substrates. SecYEG is also the preprotein translocase channel and requires the SecA ATPase to push polypeptides through the hydrophilic channel (57, 58, 61, 158, 220). SecD, SecF, and YajC, three membrane proteins, associate in a complex (154) and interact with SecYEG to regulate SecA-dependent translocation activity (56) and release translocated proteins from the translocation channel (132). The functionality of the Sec translocon has been demonstrated elegantly in vitro (84, 148).

Signal peptide–bearing precursors are translocated by SecYEG only if they are maintained in a secretion competent state, a function that can be achieved by one of two separate pathways (46, 219). In one pathway, a signal recognition particle (SRP) can bind to the signal peptides of nascent chains and temporally arrest their ribosomal translation (226, 227). The SRP of *E. coli* is a ribonucleoprotein particle consisting of Ffh (also known as P48) and 4.5S RNA (12, 167, 177, 204). Translation resumes after the SRP-ribosome complex docks onto its membrane receptor (FtsY), thereby delivering the nascent polypeptide to the Sec translocation channel (144, 219). Alternatively, signal peptide–bearing precursors may be translocated after their synthesis is completed, and hence translocation is then a post-translational process. Binding of a secretion chaperone such as SecB maintains these precursors in an unfolded, translocation-competent state (113, 171). In *E. coli,* heat shock protein chaperones such as DnaK and DnaJ can also participate in this process, independently of SecB (231) or trigger factor, a cytoplasmic peptidyl prolyl isomerase (89). For all pathways, chaperones bind to the mature part of signal peptide-bearing precursors and SecB may also interact with SecA (21, 47, 172). Once substrate has been initiated into the secretory pathway, chaperones dissociate from the precursor, allowing translocation of the entire polypeptide across the membrane (65).

Although both pathways, SRP-mediated cotranslational secretion and chaperone-implemented posttranslational secretion, penultimately converge at the translocon, discrete substrate features, such as signal peptide hydrophobicity, dictate which of the two roads shall be taken (218, 219). In *E. coli,* the SRP-dependent pathway is essential, and the genes *ffh, ftsY,* or 4.5S RNA (*ffs*) cannot be disrupted (217). Depletion analysis demonstrates that the SRP pathway is required for the secretion of polytopic membrane proteins (217), i.e., proteins harboring hydrophobic and noncleavable stop transfer signals. All other precursors secreted in a posttranslational manner are substrate of signal (leader) peptidase (encoded by *lepB* in *E. coli*), an enzyme that removes the signal peptide and releases mature protein into the periplasm (31, 44).

THE Sec SYSTEM IN GRAM-POSITIVE BACTERIA

Homologs of SecA, SecD, SecE, SecF, SecG, SecY, Ffh, FtsY, and YajC have been identified in the sequenced genome of many gram-positive organisms including *B. subtilis, B. anthracis, Staphylococcus aureus,* and *Listeria monocytogenes* (119). At first approximation, it appears that the translocase (SecYEG) is conserved. A relatively high degree of sequence similarity is shared among SecY proteins, while SecE and SecG, although present in all bacteria, are less well conserved and often shorter than their *E. coli* homologs, in particular in *B. subtilis*, *S. aureus*, and *S. epidermidis*. A *secB* gene could not be found, and it is generally assumed that another chaperone must maintain precursors in an export-competent form. Many gram-positive bacteria genomes carry a second set of secretion genes, *secA-2* and *secY-2*, and often a *secDF* fusion gene, but not single *secD* and *secF* genes (16). In this nomenclature, *secA* and *secY* are referred to as *secA-1* and *secY-1*. Expression of antisense *secY-1* RNA inhibits colony formation of *S. aureus* on agar plates (99), suggesting that, similar to *E. coli* and *B. subtilis*, *secY* expression is essential for staphylococcal growth and survival. In *Streptococcus gordonii*, SecA-2 and SecY-2 are required for export of GspB, a large cell-surface glycoprotein with an LPXTG motif for cell wall attachment (10). GspB contains a very large signal peptide of 90 amino acids. In *Streptococcus parasanguis*, the secretion defect of the *secA-2* mutation is limited to a small group of proteins such as Fap1 and FimA. FimA is a predicted lipoprotein, and Fap1 bears a long signal peptide of 50 amino acids and a C-terminal LPXTG motif. To date, it is not known what determines the specificity of the SecA-1/SecY-1 and SecA-2/SecY-2 translocases.

Ffh, 4.5S RNA, and FtsY are conserved between *E. coli* and gram–positive bacteria. Ffh is essential for growth in *E. coli* and *B. subtilis* but is not essential in *Streptococcus mutans* (42). More experiments need to be performed to understand how membrane proteins are inserted in the plasma membrane of gram-positive bacteria. In *E. coli*, YidC has been identified as a factor that assists in the insertion, folding, and assembly of cytoplasmic membrane proteins within the membrane. YidC interacts with substrates that may or may not be presented to the Sec translocon, and it is also essential for cell viability (181, 190). Indeed, a few proteins are known to insert in the membrane in a Sec-independent manner. The best known example is the M13 procoat protein. Translocation of procoat across the membrane but not procoat targeting to the mem-

brane requires YidC (182). YidC is homologous to *Saccharomyces cerevisiae* Oxa1p, where it was first shown to represent a novel export pathway for the mitochondrial inner membrane (88, 107). This pathway is evolutionarily conserved for the insertion of proteins into the membrane of mitochondria, chloroplasts, and gram-negative bacteria. YidC-like proteins can be found in the genomes of all gram-positive bacteria. In *B. subtilis*, there are two paralogues of YidC: SpoIIIJ and YqjG. Mutations in *spoIIIJ* block sporulation at stage III, whereas *yqjG* is dispensable for this developmental process (149). Together, these genes are indispensable for bacillus growth, suggesting overlapping functions. Most intriguing are the size and localization of YidC paralogues in gram-positive bacteria. These proteins all carry lipoprotein signal sequences, suggesting that they are displayed on the outside of the cell and only tethered to the membrane, instead of being buried within the membrane; furthermore, they are only about half the size of gram-negative YidC.

SUBSTRATE PREDICTIONS FOR Sec TRANSLOCON AND SIGNAL PEPTIDASES

Proteins are targeted for export to the Sec translocon by the presence of a signal sequence at their extreme N terminus. Such signal sequences have a tripartite structure where a short, basic n-region precedes a longer hydrophobic stretch of amino acids (h-region). This in turn is followed by the c-region with the recognition sequence for the enzyme signal peptidase (31, 32). Signal peptidases are anchored in the membrane and function to remove the signal peptides from precursor proteins upon Sec-mediated translocation.

Many methods are available to predict cleavable signal sequences. Although the signal sequence itself is not conserved, the site of cleavage, the c-region, determines which enzyme will cleave the precursor, signal peptidase I (SPI) for secreted proteins and signal peptidase II (SPII) for lipoproteins. SPI often cleaves after an alanine residue, whereas SPII cleaves proteins immediately before the conserved cysteine (222, 223). The activity and substrates of SPI and SPII are conserved between gram-negative and -positive bacteria, but strikingly, gram-positive bacteria often have more than one SPI and SPII, unlike *E. coli*, with only one SPI LepB (233) and one SPII LepA (213). The *lepB* gene in *E. coli* is essential. *B. subtilis* encodes five *lepB*-like genes with the products SipS, SipT, SipU, SipV, and SipW (212) and two *lepA*-like genes, SipS and SipT, are of major importance for the processing of secretory preproteins, growth, and viability (212). Two SPI encoding genes (*spsA* and *spsB*) (41) and only one SPII gene (25, 202) are present in the *S. aureus* chromosome. SpsB can be used to complement an *E. coli* strain that is temperature sensitive for preprotein processing, whereas *spsA* appears to be inactive (41). Indeed, essential catalytic residues found in *E. coli lepB* (112) are missing from the SpsA sequence (41). Nonetheless, *spsA* is a conserved feature of all staphylococci as judged by genome analysis. *S. epidermidis* contains two SpsB homologues, which show the greatest similarity to SipS and SipU of *B. subtilis*. It is not clear what role, if any, SpsA proteins may play in protein secretion.

The specificity of SPI enzymes has been reviewed by van Roosmalen and colleagues (221). *E. coli* SPI recognizes residues at the -1 and -3 positions relative to the cleavage site. Proteomic analysis suggests that all secreted proteins of *B. subtilis* harbor alanine at position -1 and that 71% of these harbor alanine at position -3; in contrast, the nature of amino acid at position -2 varies a lot (210).

Of note, signal peptides of gram-positive bacteria differ from those of their gram-negative counterparts in that they are generally longer and more hydrophobic, and with more charged residues at the N-terminal end (224). Although signal peptides of gram-positive organisms generally function in gram-negative bacteria, the same is not always true when signal peptides of gram-negative bacteria are expressed in gram-positive organisms (189). Could it be that gram-positive bacteria require additional components to recognize signal peptides? This was tested in a biochemical experiment for *B. subtilis* and *S. aureus*. Membranes were purified in the presence or absence of bound ribosomes, and protein contents were compared (1, 2). Although several different polypeptides could be identified as a secretory (S) complex, analysis of the cloned sequences revealed pyruvate dehydrogenase, an enzyme of intermediary metabolism that is presumably not directly involved in protein secretion (86, 87).

PROTEIN TARGETING OR POSTSECRETION (Sec) PROCESSING: AN OVERVIEW

A bacterial secretome, i.e., the sum of all secreted polypeptides, can be gleaned from genome sequences by searching for open reading frames that encode precursor proteins with signal sequences. Because all membrane proteins travel the Sec pathway, the secretome encompasses 40% or more of all proteins synthesized in a bacterial cell. In staphylococci and listerieae, sequence analyses predict that about 8% of all proteins (~2,700) are released from the envelope and hence will diffuse away from the cell. Only a fraction of these proteins will be retained in the envelope after secretion. Several mechanisms can account for retention and anchoring of secreted proteins to the envelope (34, 39). The first example is shared by gram-negative and -positive bacteria and uses an N-terminal diacyl-glycerol modification to retain proteins in the plasma membrane. The products of this reaction are called lipoproteins. However, other mechanisms target proteins to the envelope; those listed are unique to gram-positive bacteria. (i) Sortase-mediated linkage of proteins to peptidoglycan cross bridges provides for protein immobilization at the C-terminal end in the envelope and for protein display on the bacterial surface. (ii) Noncovalent binding of proteins to the cell wall envelope can involve interactions between specific targeting domains and teichoic or lipoteichoic acids. (iii) Alpha-helical membrane anchor structures immobilize surface proteins in the plasma membrane (ActA of

L. monocytogenes). Finally, S-layer proteins are deposited on the surface of gram-positive bacteria to establish specific interactions with cell wall polymers.

Consistent with our earlier prediction, substrates of all pathways carry two signals, an N-terminal signal sequence for secretion by the translocon and a second signal (targeting signal) that dictates the final destiny of the polypeptides.

LIPOPROTEINS

Lipoproteins were first characterized in *E. coli* along with the genetic determinants required for lipoprotein trafficking. Lipoproteins are synthesized in the cytoplasm as precursors with an N-terminal signal peptide for secretion via the Sec pathway (97, 223). Although this signal peptide is rather similar in length and hydrophobicity to that of secreted proteins, it contains a conserved cysteine that becomes the first amino acid upon cleavage of precursor lipoproteins (72, 213). The sulfhydryl moiety of this cysteine residue is modified by lipoprotein diacylglycerol transferase (Lgt). Lgt is a transmembrane protein; its active site is located on the outer leaflet of the plasma membrane and catalyzes the transfer of phosphatidylglycerol to the sulfhydryl group of the conserved cysteine. The substrate of Lgt is the lipoprotein precursor, i.e., the uncleaved precursor engaged in the Sec translocon. The product of this reaction is then cleaved at the modified cysteine by lipoprotein (type II) signal peptidase (Lsp) (35). In gram-negative bacteria, the N-terminal cysteine residue of Braun's murein lipoprotein (19) is modified by N-acyltransferase (Lnt), yielding mature N-acylated lipoprotein (213). Analyses of gram-positive genomes identify the genes *lgt* and *lsp* (encoding SPII); however, bioinformatic analysis failed to reveal the presence of Lnt. Thus, the amine of cysteine-diacylglycerol lipoprotein may not be acylated in gram-positive bacteria.

In gram-negative bacteria, the presence of any amino acid but aspartic acid after the cysteine targets lipoproteins to the outer membrane. This subset of lipoproteins is released from the inner membrane by an ABC transporter complex, LolCDE. Newly released lipoproteins are maintained soluble in the periplasm by interaction with LolA. LolA escorts complexed lipoproteins to a receptor LolB in the outer membrane that ultimately catalyzes the insertion of lipoproteins in the outer membrane. An analysis of signal sequences of gram-positive lipoproteins reveals the presence of aspartic acid as well as of other amino acids after the cysteine. Glycine appears to be the preferred amino acid. Of 66 predicted lipoproteins in the genome of *S. aureus*, 42 carry the motif CG and only 4 carry the motif CD. Although neither LolA nor LolB are found in the genomes of gram-positive bacteria, homologs of LolCDE can be found. Arguably, ABC transporters belong to a conserved family of proteins and are often required for the transport of amino acids and small peptides. Thus, it is not clear whether such genomic analyses truly reveal functional LolCDE homologs in gram-positive bacteria. Perhaps LolCDE, like proteins, are involved in lipoprotein degradation by dislodging the polypeptides from the membrane for degradation. Alternatively, envelope proteases may degrade lipoproteins, and small lipopeptides could be substrates for LolCDE transporter and uptake followed by recycling.

From a functional perspective, lipoproteins serve as receptors and sensors in the envelope (20), and genome analysis suggests that gram-positive bacteria encode between 40 and 70 lipoproteins (48, 193, 202). The pathway for lipoprotein maturation is not essential in gram-positive bacteria (25, 202) because these cells do not need to tether their outer membrane to cell wall peptidoglycan, a role fulfilled by Braun's lipoprotein in *E. coli* (20). Of interest is the role of lipoproteins during host-pathogen interaction. The innate immune system plays an integral role in determining the outcome of bacterial infection. Binding of bacterial molecules called PAMPs, for pathogen-associated molecule patterns, to dedicated Toll-like receptors (TLRs) or Nod proteins triggers specific signaling events and host responses to invading pathogens (139, 140). To date, a dozen different TLRs have been identified in mammals and respond to various PAMPS including lipopolysaccharide, peptidoglycan, lipoteichoic acid, teichoic acid, and DNA (205). Bacterial lipoproteins also function as PAMPs, activating TLR-2 signaling cascades (5, 22, 90, 236). Mutations that block diacyl-glycerol attachment to lipoprotein precursors have recently been shown to increase the virulence of *S. aureus* (25). Staphylococcal variants bearing *apo*-lipoproteins escape immune recognition and fail to elicit an inflammatory response during infection or in a tissue culture assay (25, 202). In short, immune cells are not recruited to sites of infection with these mutant bacteria, presumably because TLR-2-dependent signaling does not occur (25). These results suggest that acylation of lipoproteins is required for initiating and sustaining effective immune responses from infected hosts (25), and lipoproteins might even be the most important PAMP detected by the host during staphylococcal infections (85).

COVALENT ANCHORING OF CELL WALL PROTEINS

Covalent anchoring of secreted proteins to the cell wall is a process that has been best characterized in gram-positive bacteria, presumably because it is required for virulence. This process was first appreciated to occur in group A streptococci and *S. aureus.* The mechanism of anchoring was first solved in *S. aureus* for protein A. Initial experiments focused on solubilizing and purifying streptococcal M protein and staphylococcal protein A from the bacteria. Acid extraction of streptococci caused partial hydrolysis of polypeptides from the cell surface (117). Treatments with detergents, bases, or proteases or boiling were also insufficient to release M protein (67). Solubilization of protein A was overcome by the use of cell wall–hydrolytic enzymes (67, 91). This treatment released intact protein A that could be purified by taking advantage of the ability of protein A to interact with the constant region

of immunoglobulins (69). When purified *S. aureus* cell walls were digested with lysostaphin, a bacteriolytic enzyme secreted by *Staphylococcus simulans*, protein A, was released as a homogeneous population (196), whereas digestion with egg white lysozyme, which does not effectively degrade the staphylococcal cell wall, released small amounts of protein A. This species migrated more slowly on sodium dodecyl sulfate–polyacrylamide gel electrophoresis than the lysostaphin-released counterpart, suggesting an increase in molecular mass (197). The difference in mass could be attributed to the amino sugars *N*-acetylglucosamine and *N*-acetylmuramic acid, suggesting that protein A must be linked to the staphylococcal cell wall (197).

The amino acid sequence of protein A reveals two hydrophobic domains, one at the N terminus and one at the C terminus. Sequence comparison with other cell wall anchored proteins revealed that the C terminus carries an almost invariant LPXTG sequence motif, where X is any amino acid, followed by a C-terminal hydrophobic domain and a tail of mostly positively charged residues (68). Deletion experiments revealed that the complete charged tail and the LPXTG motif are necessary for cell wall anchoring (189). After translocation through the Sec machinery, the N-terminal signal peptide is cleaved, and the C-terminal LPXTG motif sorting signal is then cleaved between the threonine and the glycine residues (150). This cleavage is followed by a transpeptidation reaction between the carboxyl of threonine and the free amino group of pentaglycine cross bridges in the staphylococcal cell wall (187).

Swapping of sorting domains from different bacteria demonstrated the universality and conservation of the cell wall anchoring motif (188). A collection of temperature-sensitive mutants was screened for variants unable to anchor protein A to the cell wall. This experiment combined with a gene complementation approach identified the enzyme sortase A (136). Analysis of mutants lacking *srtA* (the gene encoding sortase A) revealed that sortase

A is responsible for anchoring most cell wall proteins to the cell wall—at least all of those carrying an LPXTG motif (134).

Hence, cell wall–anchored surface proteins of gram-positive bacteria encode two topogenic sequences, an N-terminal signal peptide for secretion of precursor proteins via the Sec pathway and a C-terminal cell wall sorting signal with a conserved LPXTG motif. In staphylococci, sortase anchors proteins to pentaglycine cross bridges that connect peptidoglycan strands. These pentaglycine cross bridges are otherwise covalently linked to the ε-amino group of lysine residues. Gram-positive bacteria do not share the same cell wall structure. For example, two alanine residues are found in place of the pentaglycine cross bridge in *S. pyogenes*, and the T of LPXTG proteins is predicted to be linked to a two-alanine bridge. In *L. monocytogenes*, *meso*-diaminopimelic acid (*m*-Dpm) replaces the lysine residue in peptidoglycan. Here, the T of LPXTG proteins is directly attached to the free ε-amino group of *m*-Dpm (51). Many cell wall–anchored proteins like protein A carry a long signal peptide with the YSIRK-G/S motif (180, 209). Removal of the YSIRK-G/S motif does not abrogate cell wall anchoring of protein A; its role is currently not known (8).

When cell wall sorting signals with LPXTG motif are used as queries in bioinformatic searches, 18 to 22 genes encoding putative sortase A–anchored surface proteins can be identified in the genomes of *S. aureus* or in other related gram-positive pathogens. The function of cell wall–anchored proteins is not exhaustively known; nonetheless, many of them appear to function as microbial surface components recognizing adhesive matrix molecules and thus represent bacterial elements of tissue adhesion and immune evasion (70). Genome analysis suggests that nonpathogenic gram-positive bacteria encode fewer sortase A–anchored surface proteins (48). Some gram-negative bacteria encode *srtA*-like genes, although the physiological requirement of these genes has not been addressed (160). Although most sorting signals carry the LPXTG motif, others harbor variations of this sequence. In this case, it is often

common that the surface protein gene containing such variation resides in the same transcriptional unit with a sortase gene. The assumption is that the two genes encode an enzyme-substrate pair. This conjecture has been experimentally confirmed for *Corynebacterium diphtheriae spa* loci (215), *S. aureus isd-srtB* (137), and *L. monocytogenes svpA-srtB* (152). *srtB* encodes the enzyme sortase B that is required for anchoring of at least one polypeptide needed for heme-iron uptake in staphylococci (137). Enzymes belonging to the sortase B subgroup recognize substrates containing an NPQTN motif rather than the canonical LPXTG (54). A complete classification of sortase enzymes has been reported by Dramsi and colleagues (54).

Very recently, a subgroup of sortases has been found to assemble pili on the surface of *C. diphtheriae* (75, 215). Whereas in gram-negative bacteria, pili are typically formed by noncovalent interactions between pilin subunits, *C. diphtheriae* pili are formed by covalent polymerization of adhesive pilin subunits. In addition to *C. diphtheriae*, genome analysis suggested that *Actinomyces naeslundii*, *Bacillus cereus*, *Clostridium perfringens*, *Enterococcus faecalis*, *Streptococcus agalactiae*, and *Streptococcus pneumoniae* all carry pilin genes and associated sortases (208). Experimental verification of the elaboration of pili on the surfaces of these microbes has now been confirmed for streptococci (53, 118). Why were pili not detected in earlier studies that examined the morphology of gram-positive bacteria by electron microscopy? Presumably, this failure was because the structures were often mistaken as artifacts owing to their small diameter, which is significantly less than that of their gram-negative counterparts (118).

NONCOVALENT ANCHORING OF CELL WALL PROTEINS

Many other proteins decorate the cell wall envelope of gram-positive bacteria but are not covalently attached to the envelope. Internalin B (InlB) of *L. monocytogenes*, a protein involved in bacterial entry into host cells, is one of the best-studied examples. InlB contains an N-terminal signal sequence for Sec-dependent translocation and a C-terminal region with three tandem repeats (about 80 amino acids each), which are highly basic and begin with the dipeptide GW (18). These repeats mediate a loose (noncovalent) association of the protein with lipoteichoic acid, a membrane-anchored polymer present on the surface of gram-positive bacteria (102).

Gram-positive bacteria secrete autolysins via the Sec pathway. These enzymes remain associated with the cell wall after translocation and break covalent bonds in the peptidoglycan of their own cell walls (77). Autolysins have been implicated in various biological functions, such as cell wall turnover, cell separation, cell division, and antibiotic-induced autolysis (192). It has been shown as early as 1975 that *S. pneumoniae* LytA binds to choline-substituted teichoic acids, or lipoteichoic acids by means of 20–amino acid repeats (93, 183). Teichoic acid, like lipoteichoic acid, is a polymer that traverses the cell wall of gram-positive bacteria. The crystal structure of a LytA choline binding domain was solved and revealed a novel solenoid fold made exclusively of β-hairpins stacked in a left-handed superhelix maintained by choline molecules (66). Other autolysins like the amidase Ami of *L. monocytogenes* are retained in the cell wall via a GW module at the C terminus (145, 146) or like staphylococcal amidase Atl through three repeats within the protein that serve as a targeting signal for binding to the cell wall (7). Electron microscopy analysis reveals that staphylococcal amidase and glucosaminidase localize to an equatorial ring on the cell surface that marked the future division site (235). Cell division requires localized hydrolysis as well as de novo biosynthesis of the thick cell wall peptidoglycan layer (80). The mechanisms by which penicillin-binding proteins (which carry out peptidoglycan biogenesis) or cell wall hydrolases are directed to the future cell division sites have thus far remained obscure (191, 192). Presumably, these targeted muralytic enzymes do not hydrolyze the peptidoglycan randomly, but separate dividing cells by cleaving the cell wall at designated sites (79). Although the targeting information of these enzymes can be mapped as repeats within the protein sequence and can be purified as independent domains (7), the specific surface receptor that positions these enzymes at the cell equator remains speculative in nature.

Gram-positive pathogens like *S. pyogenes* and *S. aureus* display fibronectin-binding proteins in their envelope, which enables pathogen binding to epithelial cells via fibronectin (71). In streptococci and staphylococci, fibronectin-binding proteins are secreted by the Sec machinery and contain an LPXTG motif for anchoring by sortase. Intriguingly, fibronectin-binding proteins Fbp54 in *S. pyogenes* (40), PavA in *S. pneumoniae* (92), FbpA in *S. gordonii* (36), and FbpA in *L. monocytogenes* (52) do not contain any conventional secretion signal, anchorage motif, or typical fibronectin-binding sequences but appear to remain tethered to the membrane via a C-terminal anchor (52). It is not clear whether these proteins are secreted in a Sec-dependent manner. For ActA, another virulence factor of *L. monocytogenes* required for bacterial motility within host cells, anchoring to the membrane is achieved by the presence of a hydrophobic stretch of about 20 amino acids followed by positively charged amino acids acting as a stop transfer signal at the C terminus of the protein (109). Although this retention mechanism is not unusual and is likely carried out by the Sec machinery, it does not explain the polar localization of ActA, an intriguing but convenient localization, because ActA polymerizes actin at the pole of invading bacteria as a mean to promote bacterial motility within host cells (110).

S-LAYER PROTEINS

S layers are two-dimensional crystalline protein lattices found on the outermost surface of bacteria (198, 199), although in *B. anthracis* the S layer can coexist with a surface-exposed capsule. Expression and display of the S layer do not interfere with that of the capsule (142). S layers have been identified in a range of species from the *Archaea*, *Bacteria*, and *Eucarya* and diverge greatly in their structures and functions.

Often, S-layer proteins are modified by glyco-sylation (see reference 186 for a review). We make no attempt to describe the hundreds of S layers that have been identified and will in-stead focus on the mechanisms that target S layers on the cell surface of gram-positive bac-teria (see reference 13 for a review). Multiple S layers can be simultaneously expressed on a single cell; in such cases, one S layer often serves as the primary attachment for other lay-ers (60, 108, 143, 216). In *B. stearothermophilus*, the S layer also serves to attach an amylase (59), suggesting that S layers can serve as li-gands for the retention of other secreted pro-teins.

S layers are formed by the entropy-driven aggregation of monomer subunits exported from the cytoplasm via the Sec pathway. S lay-ers are not covalently attached to the cell sur-face and can be extracted, sometimes as sheets, with mild chaotropic agents or conditions. The monomeric subunits will spontaneously re-assemble into a two-dimensional lattice in the geometric crystal pattern of the original cell-associated S layer (9). Structural studies have revealed that S-layer subunits share a region of homology, surface layer homology (SLH), which is responsible for interaction with the cell wall (122, 125, 157). SLH domains can be identified in S-layer subunit protein sequences from several gram-positive species as well as surface proteins of another structure of *Clos-tridium thermocellum* known as the cellulosome (121). SLH domains are found at either the be-ginning or end of mature proteins and are of-ten made of one to three repeating SLH motifs of approximately 50 to 60 residues (124). A comparison of these SLH motifs reveals that they are fairly divergent, with an average iden-tity of 27% and only one universally conserved residue (125).

Cell wall polymers possibly including pep-tidoglycan have been shown to act as the sur-face ligands responsible for the adherence of the S layer to the envelope (157, 176). A two-gene operon *csaAB*, for cell surface anchoring of SLH-containing proteins, was identified in *B. anthracis*. CsaB was shown to add a pyruvyl group to a peptidoglycan-associated polysac-charide fraction in vitro. Further, this modifi-cation was found to be necessary for binding of the SLH domain (141). Genome analysis suggests that many gram-positive bacteria en-code a CsaB ortholog along with predicted S-layer proteins with SLH domains. It is hypoth-esized that these bacteria contain pyruvate in their cell wall (141). Interestingly, pyruvate can also be found bound to teichuronic acid, an-other cell wall polymer of *Geobacillus stearother-mophilus* PV72, where it also serves to anchor SLH domains containing proteins of this or-ganism (129).

SEC-INDEPENDENT SECRETION

Overview

Sec-independent translocation has been ob-served in group A streptococci and other gram-positive species (29). Enzymes of the glycolytic pathway have been isolated from the streptococcal surface and are thought to medi-ate signal transduction and adhesive properties during infection. The mechanism of substrate recognition and transport for these "signal peptide–less" polypeptides has thus far not been revealed (161). Cytolysin-mediated transport of some secreted proteins across the membrane of target cells allows group A strep-tococci to inject polypeptides even into the cytosol of host cells, a mechanism that func-tionally resembles type III secretion of gram-negative bacteria (128). The examples men-tioned above, while intriguing, represent isolated cases and are not conserved secretion systems. We will expand our discussion to de-scribe more general export systems that in-clude the Tat, pseudopilin, flagellin, holin-mediated, and WXG100 transport systems.

The Tat Pathway

The Tat pathway is a conserved pathway that transports proteins across the cytoplasmic membrane of bacteria and the thylakoid mem-brane of plants (11, 147). Unlike the Sec translocon, the Tat machinery does not use ATP and unstructured proteins as substrates, but rather uses the transmembrane proton electrochemical gradient to achieve export of folded (often oligomeric) proteins across the cytoplasmic membrane. In enteric bacteria, many protein substrates that use the Tat pathway bind redox cofactors, and thus the Tat system is essential for energy metabolism in these organisms. Signal sequences that target proteins to the Tat machinery conform to the overall tripartite structure. A striking feature that clearly distinguishes Tat signal sequences is the presence of a consensus motif at the n-region–h-region boundary. This motif can be defined as S-R-R-x-F-L-K. The presence of the almost invariant arginine pair gave rise to the term "twin-arginine" signal peptide and to the name "Tat" (twin-arginine translocation) for the cognate transport system.

In *E. coli*, the *tatA, B, C,* and *E* genes encode membrane-bound components of the Tat translocation machinery (98, 185, 229). The *tatABC* genes are organized in an operon. *tatE* is a cryptic gene duplication of *tatA*, and the proteins encoded by these genes are functionally interchangeable. Under resting (nontranslocating) conditions, TatA and TatBC form separate high-molecular-mass complexes within the membrane. The translocation process is initiated when the twin-arginine signal peptide of a Tat substrate protein binds the TatBC complex in the membrane. The twin-arginine motif is recognized directly by TatC (4). In a proton-motive force (Δp)-dependent manner (234), the TatA complex then associates with the substrate-bound TatBC module, and the substrate protein is translocated across the membrane through a channel formed by multiple TatA protomers (4, 82, 166). Protein transport is probably driven by the transport of protons across the membrane. After transport, the TatA and TatBC complexes dissociate and return to the resting state.

Components of the Tat pathway can be clearly identified in the genomes of gram-positive bacteria but have been poorly characterized, with the exception of *B. subtilis* (101). *B. subtilis* contains two minimal Tat translocases, each composed of a TatA and TatC component, and translocates at least two substrates (101). The TatB-like component is not found in the genome. TatB is missing in most genomes of gram-positive bacteria with the exception of streptomycetes—thus far, the only exception with TatABC translocases. Substrates of Tat translocases in gram-positive bacteria have not been identified to the same extent as in gram-negative bacteria.

Pseudopilin Export (Com) and Tad Pathways

The Com (competence development) pathway is best characterized in *B. subtilis*. This pathway allows for DNA uptake across the membrane (see reference 33 for a review). Three proteins, designated ComEA, ComEC, and ComFA, comprise the bacterial competence-related DNA transformation transporter and assemble to form a translocation apparatus for DNA transport. A second set of genes encoded in the *comG* locus consisting of seven open reading frames (*comGA-GG*) is responsible for secretion and assembly of pilin-like proteins. This process also requires *comC* encoding a type 4 prepilin peptidase and located outside of the *comG* locus. ComGC, ComGD, ComGE, ComGF, and ComGG exhibit similarities with type 4 prepilins, whereas ComGA ATPase and ComGB localize to the cytoplasmic membrane and are postulated to act as energy-transducing protein and exporter of the five prepilins. As structure resembling a type 4 pilus (Tfp) is assembled and thought to function on the outer surface of the plasma membrane by binding DNA and presentation of this molecule to the DNA translocation machinery (see references 37 and 55 for reviews).

A *comG* locus is clearly present in the genome of *Listeria*, where a detailed genomic analysis has been performed (48). *comG* loci are also found in all sequenced staphylococcal genomes (at least ComC, ComGA, ComGB, and ComGC) and other gram-positive bacteria. As in *B. subtilis*, *comC* does not cluster with *comG* genes. This suggests that the Com system is probably not involved in DNA uptake and may catalyze another transport reaction.

In *B. subtilis*, type 4 prepilin-like genes carry an N-terminal signal sequence with the consensus motif (K/R)G(F/Y)BXZE located between the n- and h-domains, with cleavage occurring between G and (F/Y) (see references 55 and 211 for reviews). *comC* and *comG* operon genes of gram-positive bacteria encode proteins that share similarity with proteins found in the type II and type IV secretion systems, as well as the Tfp assembly apparatus of gram-negative bacteria. Hence, along with the Com pathway, these secretion systems are also referred as PSTC, for pilus/secretion/twitching motility/competence (55, 162).

The Tad export apparatus is a newly characterized secretion system involved in the secretion and assembly of Flp (fimbrial low-molecular-weight protein) pili (96). In the gram-negative organism *Actinobacillus actinomycetemcomitans*, this pathway is essential for the ability of the organism to adhere to surfaces and form biofilms (106). Proteins of the Tad apparatus share similarity to those of bacterial type II secretion (100) and type IV pilus (133) systems (163). Recently, the *A. actinomycetemcomitans* TadV protein has been shown to function as a prepilin peptidase of the Flp1 pilin as well as pre-TadE and pre-TadF proteins. TadE and TadF proteins appear to be representatives of a novel subclass of bacterial pseudopilins required for pilus biogenesis (214). Genomic analyses suggest that the Tad system is also present in some gram-positive bacteria, in particular *Clostridium acetobutylicum* (49) and *B. anthracis*, but is absent in *S. aureus* and *L. monocytogenes*. The role of a Tad export apparatus remains to be examined.

Sec-INDEPENDENT SECRETION OF FLAGELLIN

Both gram-negative and -positive bacteria encode genes that are required for flagellum formation. The major subunit of flagella, flagellin, is exported through the hollow conduct made by the basal body across the membrane and hook. Another protein, an anti–sigma factor that negatively controls σ^{28}, is also exported through the hollow conduct, as demonstrated in *Salmonella enterica* serovar Typhimurium (95, 114). The biogenesis of flagella is strictly regulated (115). The assembly of the basal body is followed by formation of the hook and then the flagellar filament. Induction of the flagellar regulon leads to expression of genes required both for the synthesis and assembly of the hook–basal body complex and for the *fliA* and *flgM* regulatory genes. The *fliA* gene encodes an alternative sigma factor, σ^{28}, which is specific for genes required late in assembly, such as flagellin, and genes of the chemosensory system (115, 155). The *flgM* gene encodes an anti-σ^{28} factor that inhibits σ^{28} activity by direct interaction in the absence of a functional hook–basal body intermediate assembly structure (45, 156). Upon completion of the hook–basal body intermediate structure, a signal is transmitted to stop the export of hook subunits and begin export of late assembly structures such as flagellin expressed from σ^{28}-dependent promoters. FlgM is also secreted out of the cell in response to hook–basal body completion (45). Secretion of FlgM through the hook–basal body structure relieves σ^{28}-inhibition, transcription from late promoters ensues, and flagellar assembly is completed. Assembly of the flagellum has been best described by Macnab (127). Sequence and structural similarities exist between the flagellar protein export system and type III secretion systems, responsible for secretion and injection of virulence factors in host cells (see reference 120 for a review). However, such type III–related secretion systems have not been found in the genome of gram-positive bacteria but for the minimal set of genes required for flagellar protein export.

Sec-INDEPENDENT SECRETION

Holins

Bacteriophage lambda of *E. coli* uses the S lysis protein or holin to disrupt the host cell membrane (6, 14, 173). This mechanism allows phage-encoded intrabacterial murein hydrolases to leave the cytoplasm and degrade the peptidoglycan sacculus, which is located in the periplasmic space of gram-negative bacteria. Without murein hydrolysis, phage particles could not be released into the extracellular medium (173). Bacteriophages of gram-positive organisms use similar strategies (178) that allow them to secrete murein hydrolases devoid of an N-terminal signal peptide (228). Once secreted, such enzymes use a cell wall retention signal and remain noncovalently associated with the cell wall, while cleaving the peptidoglycan.

Clostridium difficile, the causative agent of pseudomembranous colitis and antibiotic-associated colitis in humans, releases two toxins, TcdA and TcdB, which are thought to be major virulence factors of this organism (103, 105). TcdA and TcdB are targeted into the cytosol of eukaryotic cells, where they function as glucosidases to modify small GTP binding proteins such as Rho, Rac, and Cdc42 by using UDP-Glc as a cosubstrate (3). This process leads to inactivation of GTPase activity and loss of signaling activity, causing actin rearrangements, fluid loss, and cell death (104). TcdA and TcdB do not contain an N-terminal signal peptide. A small gene *tcdE* is encoded between *tcdA* and *tcdB*. Because genetic analyses are difficult to perform in *C. difficile*, instead, *tcdE* was expressed in *E. coli*, where it caused bacterial cell lysis (206). Analysis of the amino acid sequence of TcdE reveals structural features of bacteriophage holins. It is presumed that TcdE may function to facilitate the release of TcdA and TcdB to the extracellular environment (206, 225).

ESAT-6 Secretion Systems (Ess)

The envelope of mycobacteria is made up of a cytoplasmic (plasma or inner) membrane and a second hydrophobic layer composed of mycolic acid that represents the outer membrane of these microbes (81, 184). The mycolic acid layer has very low permeability; however, small hydrophilic nutrients diffuse through trans-membrane-channel proteins (64), reminiscent of gram-negative outer membrane proteins named porins (153). Passage of proteins across the mycolic acid layer has not been examined,

unlike for gram-negative bacteria (reviewed in reference 120); however, mycobacteria are equipped with the typical Sec machinery for secretion of polypeptides across the plasma membrane. Recently, much attention has been drawn to two secreted virulence factors, ESAT-6 (early secreted antigen target 6 kDa) and CFP-10 (culture filtrate antigen 10 kDa) encoded by the *esxA* (Rv3875) and *esxB* (Rv3874) genes of *Mycobacterium tuberculosis* H37Rv. Secreted ESAT-6 does not contain a cleavable N-terminal signal sequence and is not processed upon secretion (200). Although the exact function of ESAT-6 and CFP-10 remains elusive, *M. tuberculosis* variants lacking the *esxA* gene display defects in the establishment of tuberculosis. Further, the ESAT-6 antigen induces a strong T-cell response during infection (200). The *M. tuberculosis esxA* gene is located within a segment of the genome named region of difference 1 (RD1). Comparative genomics of mycobacteria related to *M. tuberculosis* H37Rv identified multiple regions of difference between virulent, avirulent, and vaccine species of mycobacteria (38, 130, 169). Although RD1 is not the only virulence trait of mycobacteria, it appears to be a prominent factor in attenuating vaccine strains (94, 169, 170).

Although ESAT-6 and CFP-10 were initially thought to be released via mycobacterial lysis, recent work demonstrated that ESAT-6 and CFP-10 are in fact secreted by a specialized secretion system (83, 94, 201). This specialized secretion system has been referred to as Snm for secretion in mycobacteria (201) or ESX-1 (ESAT-6 system-1), to distinguish it from the numerous other loci in the mycobacterial genome bearing ESAT-6-like encoded genes (38). Interestingly, although the Snm/ESX-1 factors were found as mutants with reduced virulence defect, a role in protein secretion had been predicted by bioinformatics analysis (159). Pallen noticed that the 23 ESAT-6 homologues in *M. tuberculosis* were encoded within five gene clusters that also specify large soluble and membrane-bound ATPases with two or more FtsK/SpoIIIE-like domains (FSD) (76, 159). Pallen proposed that FSD ATPases may represent a universal portal for excretion of ESAT-6-like proteins. He also noticed similar cluster arrangements in the genomes of other mycobacteria and related corynebacteria and *Streptomyces coelicolor,* as well as the genomes of gram-positive bacteria such as *S. aureus, L. monocytogenes,* and *B. subtilis* (159). All ESAT-6-like proteins were found to be about 100 amino acids in length, carrying a WXG motif roughly in the middle of the sequence. Pallen referred to these genes as the WXG100 superfamily. Experimental support for this hypothesis was garnered when deletions of FSD-like genes (Rv3870/snm1, Rv3871/snm2) either reduced or prevented the secretion of *M. tuberculosis* ESAT-6 and CFP-10 (83, 94, 170, 201). A third membrane protein encoded by Rv3877/snm4 was also found to be required for secretion of ESAT-6 and CFP-10 (201).

The only gram-positive ESAT-6-like proteins examined thus far are those encoded in the staphylococcal and anthrax genomes (26) (G. Garuti and D. Missiakas, unpublished data). In *S. aureus,* the predicted secretion system has been designated ESAT-6 secretion system (Ess) and is encoded by a cluster of eight genes. The Ess cluster is flanked by two open reading frames, *esxA* and *esxB,* the products of which have been shown to be transported across the envelope of *S. aureus* in a manner requiring the FSD protein EssC as well as two other membrane proteins, EssA and EssB, all of which are encoded within the Ess cluster (26). The importance of EsxA and EsxB secretion during *S. aureus* infection was revealed by analyzing *esxA, esxB,* and *essC* mutants in a murine model of abscess formation. Together these observations point to the Ess locus as a novel secretion pathway in *S. aureus.* The destiny and role of staphylococcal EsxA and EsxB proteins during infection have not been revealed yet. In *M. tuberculosis,* ESAT-6 and CFP-10 are required for cytolysis and bacterial spreading (73). In *M. marinum,* a mutant defective in the secretion of ESAT-6 and CFP-10 is unable to replicate within J774 macrophages because

phagolysosomal fusion cannot be prevented as efficiently as occurs during infection with wild-type mycobacteria (207).

By use of a yeast two-hybrid approach, Cox and colleagues could show that CFP-10, but not ESAT-6, interacts with Snm2, one of the FSD ATPases that promotes CFP-10 and ESAT-6 secretion (201). Further, the last seven amino acids of CFP-10 are necessary and sufficient to mediate interaction with Snm2 as well as secretion of CFP-10 (30). Remarkably, removal of the last seven amino acids also prevented export of ESAT-6. Although it has been shown that ESAT-6 and CFP-10 form a complex (23, 170, 174, 175, 201), the new data reveal that complex formation is necessary for secretion because only CFP-10 carries the secretion signal within its last seven amino acids.

CAN GENOMICS OF GRAM-POSITIVE PROTEIN SECRETION HELP IN VACCINE DESIGN?

Human infections caused by gram-positive bacteria present a serious therapeutic challenge as a result of the emergence of antibiotic-resistant strains (151). Of concern are infections with *S. aureus, S. epidermidis,* and *Enterococcus faecalis,* microorganisms that are the most common cause of bacterial disease in U.S. hospitals (17, 24). These gram-positive human pathogens have acquired resistance mechanisms to virtually all known antibiotics, and the development of novel targets for antimicrobial therapy is therefore urgently needed (194).

Research over the past several decades identified exotoxins, surface proteins, and regulatory molecules as important virulence factors of gram-positive pathogens. For example, during staphylococcal infection, a bacterial census ensures massive secretion of virulence factors when staphylococcal counts are high, presumably increasing the likelihood of spread of bacteria in infected tissues and/or systemic dissemination. Early during infection, secretion of virulence factor is minimal, and on the contrary, surface proteins are produced and displayed in the envelope of the bacterium, presumably favoring adhesion of the pathogen. The unifying theme of all these observations is the multifactorial nature of gram-positive pathogenesis, where many virulence factors contribute overlapping and often redundant roles to the establishment of a wide variety of infectious syndromes and intoxications in humans.

Rappuoli and colleagues exploited information encrypted in bacterial genome sequences to distinguish genes in all strains of a pathogen from conserved genes found in all members of its species (138). Rational vaccine design was achieved by interrogating conserved antigens (secreted or surface displayed) for protective immunity, which led to the identification of multiple surface proteins of group B streptococci as candidates (131). By combining multiple antigens into a single vaccine, broad-spectrum protective immunity against many different clinical isolates was achieved (131).

Recently, a similar approach was used to ask whether a broad-spectrum *S. aureus* vaccine may be derived by testing staphylococcal surface proteins for protective immunity. *S. aureus* sortase A (*srtA*) mutants, which cannot display surface proteins, are unable to establish infections in experimental models, indicating that the sum of all 23 surface proteins is essential for pathogenesis (134). Eight staphylococcal genome sequences in GenBank were examined for the presence of sortase substrate genes by using sorting signals as queries in BLAST searches, and 19 conserved surface protein genes were identified (203). Immunization with four purified surface proteins, IsdA, IsdB, SdrD, and SdrE, generated the highest level of protection in mice as compared with 15 other antigens. When all four antigens were used together, the combined vaccine afforded complete protection against staphylococcal challenge (203). Although the function of all surface proteins is not known, genomic predictions allow their identification, and together, the experimental data suggest that these proteins are good candidates for vaccine studies.

WHAT HAS GENOMICS NOT ANSWERED?

The envelope of gram-positive bacteria represents a formidable barrier for the transport of polypeptides. It is hard to imagine that polypeptides diffuse freely through the thick envelope of staphylococci composed of peptidoglycan, teichoic acid, lipoteichoic acid, and capsular polysaccharides. It is known that deletion of *prsA*, encoding a lipoprotein peptidyl-prolyl isomerase, is not tolerated in *B. subtilis*, presumably because it is required for release of exoproteins into the medium after translocation across the plasma membrane (111). *prsA*-like genes are found in all gram-positive genomes sequenced so far, and we presume these genes will assist secretion of proteins. In group A streptococci and staphylococci, *prsA* mutants did not display a general defect in protein export (126, 232).

Electron microscopic examination of the staphylococcal cell wall envelope has revealed the existence of pore structures in the cell wall envelope that may represent the sites of protein translocation (80). The pores have a dimension of 2 to 3 nm and form a regularly spaced array of holes on both sides of the cell division cleft. The pores are connected with tubular structures that span the entire cell wall envelope (78). Could secretion of signal peptide bearing proteins into the extracellular medium require elements that constitute these pores? If so, how can we predict such genes? Recent advances in knowledge of *S. pyogenes* and *B. subtilis* revealed that the translocon may be part of a larger complex (ExPortal) that organizes transport of signal peptide–bearing precursor proteins to dedicated sites within the bacterial cell wall envelope (27, 179). The genetic requirements that define such secretion sites have not yet been recognized. Although proteins released in the extracellular milieu of *S. pyogenes* travel through the one and only ExPortal channel, the same organism appears to target secretion of sortase substrates at distinct yet defined subcellular regions. The signal sequence of M protein promotes secretion at the division septum, whereas that of PrtF preferentially promotes secretion at the old pole (28). Does the signal sequence contain information that directs precursor proteins to distinct Sec translocons in the membrane? Microscopy permits this new level of complexity in protein translocation to be addressed and revealed. Being able to "genetically" perturb the images would allow us to validate these hypotheses and perhaps elucidate transport across the envelope of gram-positive bacteria, as well as polarity and subcompartmentalization of protein complexes in the envelope.

REFERENCES

1. Adler, L. A., and S. Arvidson. 1988. Cloning and expression in Escherichia coli of genes encoding a multiprotein complex involved in secretion of proteins from Staphylococcus aureus. J. Bacteriol. 170:5337–5343.

2. Adler, L. A., and S. Arvidson. 1987. Correlation between the rate of exoprotein synthesis and the amount of the multiprotein complex on membrane-bound ribosomes (MBRP-complex) in Staphylococcus aureus. J. Gen. Microbiol. 133:803–813.

3. Aktories, K. 1997. Rho proteins: targets for bacterial toxins. Trends Microbiol. 5:282–288.

4. Alami, M., I. Luke, S. Deitermann, G. Eisner, H. G. Koch, J. Brunner, and M. Muller. 2003. Differential interactions between a twin-arginine signal peptide and its translocase in Escherichia coli. Mol. Cell 12:937–946.

5. Aliprantis, A. O., R. B. Yang, M. R. Mark, S. Suggett, B. Devaux, J. D. Radolf, G. R. Klimpel, P. Godowski, and A. Zychlinsky. 1999. Cell activation and apoptosis by bacterial lipoproteins through Toll-like receptor-2. Science 285:736–739.

6. Altman, E., R. K. Altman, J. M. Garrett, R. J. Grimaila, and R. Young. 1983. S gene product: identification and membrane localization of a lysis control protein. J. Bacteriol. 155:1130–1137.

7. Baba, T., and O. Schneewind. 1998. Targeting of muralytic enzymes to the cell division site of gram-positive bacteria: repeat domains direct autolysin to the equatorial surface ring of Staphylococcus aureus. EMBO J. 17:4639–4646.

8. Bae, T., and O. Schneewind. 2003. The YSIRK-G/S motif of staphylococcal protein A and its role in efficiency of signal peptide processing. J. Bacteriol. 185:2910–2919.

9. Baumeister, W., F. Karrenberg, R. Rachel, A. Engel, B. ten Heggeler, and W. O. Saxton. 1982. The major cell envelope protein of Micrococcus radiodurans (R1). Structural and chemical characterization. Eur. J. Biochem. 125:535–544.

10. Bensing, B. A., and P. M. Sullam. 2002. An accessory sec locus of Streptococcus gordonii is required for export of the surface protein GspB and for normal levels of binding to human platelets. Mol. Microbiol. 44:1081–1094.

11. Berks, B. C., T. Palmer, and F. Sargent. 2003. The Tat protein translocation pathway and its role in microbial physiology. Adv. Microb. Physiol. 47:187–254.

12. Bernstein, H. D., M. A. Poritz, K. Strub, P. J. Hoben, S. Brenner, and P. Walter. 1989. Model for signal sequence recognition from amino-acid sequence of 54K subunit of signal recognition particle. Nature 340:482–486.

13. Beveridge, T. J., P. H. Pouwels, M. Sara, A. Kotiranta, K. Lounatmaa, K. Kari, E. Kerosuo, M. Haapasalo, E. M. Egelseer, I. Schocher, U. B. Sleytr, L. Morelli, M. L. Callegari, J. F. Nomellini, W. H. Bingle, J. Smit, E. Leibovitz, M. Lemaire, I. Miras, S. Salamitou, P. Beguin, H. Ohayon, P. Gounon, M. Matuschek, and S. F. Koval. 1997. Functions of S-layers. FEMS. Microbiol. Rev. 20:99–149.

14. Blasi, U., C. Y. Chang, M. T. Zagotta, K. B. Nam, and R. Young. 1990. The lethal lambda S gene encodes its own inhibitor. EMBO J. 9:981–989.

15. Blobel, G. 1980. Intracellular protein topogenesis. Proc. Natl. Acad. Sci. USA 77:1496–1500.

16. Bolhuis, A., C. P. Broekhuizen, A. Sorokin, M. L. van Roosmalen, G. Venema, S. Bron, W. J. Quax, and J. van Dijl. 1998. SecDF of Bacillus subtilis, a molecular Siamese twin required for the efficient secretion of proteins. J. Biol. Chem. 273:21217–21224.

17. Boyce, J. M. 1990. Increasing prevalence of methicillin-resistant Staphylococcus aureus in the United States. Infect. Control Hosp. Epidemiol. 11:639–642.

18. Braun, L., S. Dramsi, P. Dehoux, H. Bierne, G. Lindahl, and P. Cossart. 1997. InlB: an invasion protein of Listeria monocytogenes with a novel type of surface association. Mol. Microbiol. 25:285–294.

19. Braun, V., and V. Bosch. 1972. Sequence of the murein lipoprotein and the attachment site of the lipid. Eur. J. Biochem. 28:51–69.

20. Braun, V., and H. C. Wu. 1994. Lipoproteins: structure, function, biosynthesis and model for protein export, p. 319–337. In J.-M. Ghuysen and R. Hakenbeck (ed.), Bacterial Cell Wall, vol. 27. Elsevier Biomedical Press, Amsterdam, The Netherlands.

21. Breukink, E., N. Nouwen, A. van Raalte, S. Mizushima, J. Tommassen, and B. de Kruijff.

1995. The C terminus of SecA is involved in both lipid binding and SecB binding. *J. Biol. Chem.* **270:**7902–7907.

22. **Brightbill, H. D., D. H. Libraty, S. R. Krutzik, R. B. Yang, J. T. Belisle, J. R. Bleharski, M. Maitland, M. V. Norgard, S. E. Plevy, S. T. Smale, P. J. Brennan, B. R. Bloom, P. J. Godowski, and R. L. Modlin.** 1999. Host defense mechanisms triggered by microbial lipoproteins through Toll-like receptors. *Science* **285:**732–736.

23. **Brodin, P., M. I. de Jonge, L. Majlessi, C. Leclerc, M. Nilges, S. T. Cole, and R. Brosch.** 2005. Functional analysis of early secreted antigenic target-6, the dominant T-cell antigen of *Mycobacterium tuberculosis*, reveals key residues involved in secretion, complex formation, virulence, and immunogenicity. *J. Biol. Chem.* **280:** 33953–33959.

24. **Brumfitt, W., and J. Hamilton-Miller.** 1989. Methicillin-resistant *Staphylococcus aureus*. *N. Engl. J. Med.* **320:**1188–1199.

25. **Bubeck Wardenburg, J., W. A. Williams, and D. Missiakas.** 2006. Host defenses against *Staphylococcus aureus* infection require recognition of bacterial lipoproteins. *Proc. Natl. Acad. Sci. USA* **103:**13831–13836.

26. **Burts, M. L., W. A. Williams, K. DeBord, and D. M. Missiakas.** 2005. EsxA and EsxB are secreted by an ESAT-6-like system that is required for the pathogenesis of *Staphylococcus aureus* infections. *Proc. Natl. Acad. Sci. USA* **102:**1169–1174.

27. **Campo, N., M. Tjalsma, G. Buist, D. Stepniak, M. Meijer, M. Veenhuis, M. Westermann, J. P. Muller, S. Bron, J. Kok, O. P. Kuipers, and J. D. Jongbloed.** 2004. Subcellular sites for bacterial protein export. *Mol. Microbiol.* **53:**1583–1599.

28. **Carlsson, F., M. Stålhammar-Carlemalm, K. Flärdh, C. Sandin, E. Carlemalm, and G. Lindahl.** 2006. Signal sequence directs localized secretion of bacterial surface proteins. *Nature* **442:** 943–946.

29. **Chaatwal, G. S.** 2002. Anchorless adhesins and invasins of gram-positive bacteria: a new class of virulence factors. *Trends Microbiol.* **10:**205–208.

30. **Champion, P. A., S. A. Stanley, M. M. Champion, E. J. Brown, and J. S. Cox.** 2006. C-terminal signal sequence promotes virulence factor secretion in *Mycobacterium tuberculosis*. *Science* **313:** 1632–1636.

31. **Chang, C. N., G. Blobel, and P. Model.** 1978. Detection of prokaryotic signal peptidase in an *Escherichia coli* membrane fraction: endoproteolytic cleavage of nascent f1 pre-coat protein. *Proc. Natl. Acad. Sci. USA* **75:**361–365.

32. **Chang, C. N., J. B. K. Nielsen, K. Izui, G. Blobel, and J. O. Lampen.** 1982. Idenfication of the signal peptidase cleavage site in *Bacillus licheniformis* prepenicillinase. *J. Biol. Chem.* **257:** 4340–4344.

33. **Chen, I., and D. Dubnau.** 2004. DNA uptake during bacterial transformation. *Nat. Rev. Microbiol.* **2:**241–249.

34. **Chhatwal, G. S.** 2002. Anchorless adhesins and invasins of gram-positive bacteria: a new class of virulence factors. *Trends Microbiol.* **10:**205–208.

35. **Choi, D. S., H. Yamada, T. Mizuno, and S. Mizushima.** 1986. Trimeric structure and localization of the major lipoprotein in the cell surface of *Escherichia coli*. *J. Biol. Chem.* **261:**8953–8957.

36. **Christie, J., R. McNab, and H. F. Jenkinson.** 2002. Expression of fibronectin-binding protein FbpA modulates adhesion in *Streptococcus gordonii*. *Microbiology* **148:**1615–1625.

37. **Claverys, J. P., and B. Martin.** 2003. Bacterial "competence" genes: signatures of active transformation, or only remnants? *Trends Microbiol.* **11:** 161–165.

38. **Cole, S. T., R. Brosch, J. Parkhill, T. Garnier, C. Churcher, D. Harris, S. V. Gordon, K. Eiglmeier, S. Gas, C. E. Barry III, F. Tekaia, K. Badcock, D. Basham, D. Brown, T. Chillingworth, R. Connor, R. Davies, K. Devlin, T. Feltwell, S. Gentles, N. Hamlin, S. Holroyd, T. Hornsby, K. Jagels, A. Krogh, J. McLean, S. Moule, L. Murphy, K. Oliver, J. Osborne, M. A. Quail, M. A. Rajandream, J. Rogers, S. Rutter, K. Seeger, J. Skelton, R. Squares, S. Squares, J. E. Sulston, K. Taylor, S. Whitehead, and B. G. Barrell.** 1998. Deciphering the biology of *Mycobacterium tuberculosis* from the complete genome sequence. *Nature* **393:**537–544.

39. **Cossart, P., and R. Jonquieres.** 2000. Sortase, a universal target for therapeutic agents against gram-positive bacteria? *Proc. Natl. Acad. Sci. USA* **97:**5013–5015.

40. **Courtney, H. S., Y. Li, J. B. Dale, and D. L. Hasty.** 1994. Cloning, sequencing, and expression of a fibronectin/fibrinogen-binding protein from group A streptococci. *Infect. Immun.* **62:**3937–3946.

41. **Cregg, K. M., I. Wilding, and M. T. Black.** 1996. Molecular cloning and expression of the spsB gene encoding an essential type I signal peptidase from *Staphylococcus aureus*. *J. Bacteriol.* **178:** 5712–5718.

42. **Crowley, P. J., G. Svensater, J. L. Snoep, A. S. Bleiweis, and L. J. Brady.** 2004. An ffh mutant of *Streptococcus mutans* is viable and able to physiologically adapt to low pH in continuous culture. *FEMS Microbiol. Lett.* **234:**315–324.

43. **Dalbey, R. E., M. O. Lively, S. Bron, and J. M. van Dijl.** 1997. The chemistry and enzymology of the type I signal peptidases. *Protein Sci.* **6:**1129–1138.

44. **Dalbey, R. E., and W. Wickner.** 1985. Leader peptidase catalyzes the release of exported proteins from the outer surface of the *Escherichia coli* plasma membrane. *J. Biol. Chem.* **260:**15925–15931.

45. **Daughdrill, G. W., M. S. Chadsey, J. E. Karlinsey, K. T. Hughes, and F. W. Dahlquist.** 1997. The C-terminal half of the anti-sigma factor, FlgM, becomes structured when bound to its target, sigma 28. *Nat. Struct. Biol.* **4:**285–291.

46. **De Gier, J. W., Q. A. Valent, G. Von Heijne, and J. Luirink.** 1997. The *E. coli* SRP: preferences of a targeting factor. *FEBS Lett.* **408:**1–4.

47. **den Blaauwen, T., E. Terpetschnig, J. R. Lakowicz, and A. J. Driessen.** 1997. Interaction of SecB with soluble SecA. *FEBS Lett.* **416:**35–38.

48. **Desvaux, M., and M. Hebraud.** 2006. The protein secretion systems in *Listeria*: inside out bacterial virulence. *FEMS. Microbiol. Rev.* **30:**774–805.

49. **Desvaux, M., A. Khan, A. Scott-Tucker, R. R. Chaudhuri, M. J. Pallen, and I. R. Henderson.** 2005. Genomic analysis of the protein secretion systems in *Clostridium acetobutylicum* ATCC 824. *Biochim. Biophys. Acta* **1745:**223–253.

50. **Dev, I. K., and P. H. Ray.** 1990. Signal peptidases and signal peptide hydrolases. *J. Bioenergetics Biomem.* **22:**271–289.

51. **Dhar, G., K. F. Faull, and O. Schneewind.** 2000. Anchor structure of cell wall surface proteins in *Listeria monocytogenes. Biochemistry* **39:**3725–3733.

52. **Dramsi, S., F. Bourdichon, D. Cabanes, M. Lecuit, H. Fsihi, and P. Cossart.** 2004. FbpA, a novel multifunctional *Listeria monocytogenes* virulence factor. *Mol. Microbiol.* **53:**639–649.

53. **Dramsi, S., E. Caliot, I. Bonne, S. Guadagnini, M. C. Prevost, M. Kojadinovic, L. Lalioui, C. Poyart, and P. Trieu-Cuot.** 2006. Assembly and role of pili in group B streptococci. *Mol. Microbiol.* **60:**1401–1413.

54. **Dramsi, S., P. Trieu-Cuot, and H. Bierne.** 2005. Sorting sortases: a nomenclature proposal for the various sortases of gram-positive bacteria. *Res. Microbiol.* **156:**289–297.

55. **Dubnau, D.** 1999. DNA uptake in bacteria. *Annu. Rev. Microbiol.* **53:**217–244.

56. **Duong, F., J. Eichler, A. Price, M. R. Leonard, and W. Wickner.** 1997. Biogenesis of the gram-negative bacterial envelope. *Cell* **91:**567–573.

57. **Duong, F., and W. Wickner.** 1997. The SecD-FYajC domain of preprotein translocase controls preprotein movement by regulating SecA membrane cycling. *EMBO J.* **16:**4871–4879.

58. **Economou, A., and W. Wickner.** 1994. SecA promotes preprotein translocation by undergoing ATP-driven cycles of membrane insertion and deinsertion. *Cell* **78:**835–843.

59. **Egelseer, E., I. Schocher, M. Sara, and U. B. Sleytr.** 1995. The S-layer from *Bacillus stearothermophilus* DSM 2358 functions as an adhesion site for a high-molecular-weight amylase. *J. Bacteriol.* **177:**1444–1451.

60. **Egelseer, E. M., K. Leitner, M. Jarosch, C. Hotzy, S. Zayni, U. B. Sleytr, and M. Sara.** 1998. The S-layer proteins of two *Bacillus stearothermophilus* wild-type strains are bound via their N-terminal region to a secondary cell wall polymer of identical chemical composition. *J. Bacteriol.* **180:**1488–1495.

61. **Eichler, J., and W. Wickner.** 1997. Both an N-terminal 65-kDa domain and a C-terminal 30-kDa domain of SecA cycle into the membrane at SecYEG during translocation. *Proc. Natl. Acad. Sci. USA* **94:**5574–5581.

62. **Emr, S. D., S. Hanley-Way, and T. J. Silhavy.** 1981. Suppressor mutations that restore export of a protein with a defective signal sequence. *Cell* **23:**79–88.

63. **Emr, S. D., J. Hedgpeth, J. M. Clement, T. J. Silhavy, and M. Hofnung.** 1980. Sequence analysis of mutations that prevent export of lambda receptor, an *Escherichia coli* outer membrane protein. *Nature* **285:**82–85.

64. **Faller, M., M. Niederweis, and G. E. Schulz.** 2004. The structure of a mycobacterial outer-membrane channel. *Science* **303:**1189–1192.

65. **Fekkes, P., C. van der Does, and A. J. M. Driessen.** 1997. The molecular chaperone SecB is released from the carboxy-terminus of SecA during initiation of precursor protein translocation. *EMBO J.* **16:**6105–6113.

66. **Fernandez-Tornero, C., R. Lopez, E. Garcia, G. Gimenez-Gallego, and A. Romero.** 2001. A novel solenoid fold in the cell wall anchoring domain of the pneumococcal virulence factor LytA. *Nat. Struct. Biol.* **8:**1020–1024.

67. **Fischetti, V. A.** 1989. Streptococcal M protein: molecular design and biological behavior. *Clin. Microbiol. Rev.* **2:**285–314.

68. **Fischetti, V. A., V. Pancholi, and O. Schneewind.** 1990. Conservation of a hexapeptide sequence in the anchor region of surface proteins from gram-positive cocci. *Mol. Microbiol.* **4:**1603–1605.

69. **Forsgren, A., and J. Sjoquist.** 1966. "Protein A" from *S. aureus*. I. Pseudo-immune reaction

with human gamma-globulin. *J. Immunol.* **97:**822–827.

70. **Foster, T. J., and M. Höök.** 1998. Surface protein adhesins of *Staphylococcus aureus. Trends Microbiol.* **6:**484–488.

71. **Fowler, T., E. R. Wann, D. Joh, S. Johansson, T. J. Foster, and M. Hook.** 2000. Cellular invasion by *Staphylococcus aureus* involves a fibronectin bridge between the bacterial fibronectin-binding MSCRAMMs and host cell beta1 integrins. *Eur. J. Cell. Biol.* **79:**672–679.

72. **Gan, K., S. D. Gupta, K. Sankaran, M. B. Schmid, and H. C. Wu.** 1993. Isolation and characterization of a temperature-sensitive mutant of *Salmonella typhimurium* defective in prolipoprotein modification. *J. Biol. Chem.* **268:**16544–16550.

73. **Gao, L. Y., S. Guo, B. McLaughlin, H. Morisaki, J. N. Engel, and E. J. Brown.** 2004. A mycobacterial virulence gene cluster extending RD1 is required for cytolysis, bacterial spreading and ESAT-6 secretion. *Mol. Microbiol.* **53:**1677–1693.

74. Reference deleted.

75. **Gaspar, A. H., and H. Ton-That.** 2006. Assembly of distinct pilus structures on the surface of *Corynebacterium diphtheriae. J. Bacteriol.* **188:**1526–1533.

76. **Gey Van Pittius, N. C., J. Gamieldien, W. Hide, G. D. Brown, R. J. Siezen, and A. D. Beyers.** 2001. The ESAT-6 gene cluster of *Mycobacterium tuberculosis* and other high G+C gram-positive bacteria. *Genome Biol.* **2:**research0044.1–0044.18.

77. **Ghuysen, J.-M., D. J. Tipper, and J. L. Strominger.** 1966. Enzymes that degrade bacterial cell walls. *Methods Enzymol.* **8:**685–699.

78. **Giesbrecht, P., T. Kersten, H. Maidhof, and J. Wecke.** 1998. Staphylococcal cell wall: morphogenesis and fatal variations in the presence of penicillin. *Microbiol. Mol. Biol. Rev.* **62:**1371–1414.

79. **Giesbrecht, P., T. Kersten, and J. Wecke.** 1992. Fan-shaped ejections of regularly arranged murosomes involved in penicillin-induced death of staphylococci. *J. Bacteriol.* **174:**2241–2252.

80. **Giesbrecht, P., J. Wecke, and B. Reinicke.** 1976. On the morphogenesis of the cell wall of staphylococci. *Int. Rev. Cytol.* **44:**225–318.

81. **Glickman, M. S., and W. R. J. Jacobs.** 2001. Microbial pathogenesis of *Mycobacterium tuberculosis* dawn of a discipline. *Cell* **104:**477–485.

82. **Gohlke, U., L. Pullan, C. A. McDevitt, I. Porcelli, E. de Leeuw, T. Palmer, H. R. Saibil, and B. C. Berks.** 2005. The TatA component of the twin-arginine protein transport system forms channel complexes of variable diameter. *Proc. Natl. Acad. Sci. USA* **102:**10482–10486.

83. **Guinn, K. M., M. J. Hickey, S. K. Mathur, K. L. Zakel, J. E. Grotzke, D. M. Lewinsohn, S. Smith, and D. R. Sherman.** 2004. Individual RD1-region genes are required for export of ESAT-6/CFP-10 and for virulence of *Mycobacterium tuberculosis. Mol. Microbiol.* **51:**359–370.

84. **Hartl, F. U., S. Lecker, E. Schiebel, J. P. Hendrick, and W. Wickner.** 1990. The binding cascade of SecB to SecA to SecY/E mediates preprotein targeting to the *E. coli* plasma membrane. *Cell* **63:**269–279.

85. **Hashimoto, M., K. Tawaratsumida, H. Kariya, A. Kiyohara, Y. Suda, F. Krikae, T. Kirikae, and F. Gotz.** 2006. Not lipoteichoic acid but lipoproteins appear to be the dominant immunobiologically active compounds in *Staphylococcus aureus. J. Immunol.* **177:**3162–3169.

86. **Hemila, H., A. Palva, L. Paulin, L. Adler, S. Arvidson, and I. Palva.** 1991. The secretory S complex in *Bacillus subtilis* is identified as pyruvate dehydrogenase. *Res. Microbiol.* **142:**779–785.

87. **Hemila, H., A. Palva, L. Paulin, S. Arvidson, and I. Palva.** 1990. Secretory S complex of *Bacillus subtilis*: sequence analysis and identity to pyruvate dehydrogenase. *J. Bacteriol.* **172:**5052–5063.

88. **Herrmann, J. M., W. Neupert, and R. A. Stuart.** 1997. Insertion into the mitochondrial inner membrane of a polytopic protein, the nuclear-encoded Oxa1p. *EMBO J.* **16:**2217–2226.

89. **Hesterkamp, T., S. Hauser, H. Lutcke, and B. Bukau.** 1996. *Escherichia coli* trigger factor is a prolyl isomerase that associates with nascent polypeptide chains. *Proc. Natl. Acad. Sci. USA* **93:**4437–4441.

90. **Hirschfeld, M., C. J. Kirschning, R. Schwandner, H. Wesche, J. H. Weis, R. M. Wooten, and J. J. Weis.** 1999. Cutting edge: inflammatory signaling by *Borrelia burgdorferi* lipoproteins is mediated by Toll-like receptor 2. *J. Immunol.* **163:**2382–2386.

91. **Hjelm, H., K. Hjelm, and J. Sjoquist.** 1972. Protein A from *Staphylococcus aureus*. Its isolation by affinity chromatography and its use as an immunosorbent for isolation of immunoglobulins. *FEBS Lett.* **28:**73–76.

92. **Holmes, A. R., R. McNab, K. W. Millsap, M. Rohde, S. Hammerschmidt, J. L. Mawdsley, and H. F. Jenkinson.** 2001. The pavA gene of *Streptococcus pneumoniae* encodes a fibronectin-binding protein that is essential for virulence. *Mol. Microbiol.* **41:**1395–1408.

93. **Höltje, J.-V., and A. Tomasz.** 1975. Specific recognition of choline residues in the cell wall teichoic acid by N-acetyl-muramyl-L-alanine amidase of *Pneumococcus. J. Biol. Chem.* **250:**6072–6076.

94. Hsu, T., S. M. Hingley-Wilson, B. Chen, M. Chen, A. Z. Dai, P. M. Morin, C. B. Marks, J. Padiyar, C. Goulding, M. Gingery, D. Eisenberg, R. G. Russell, S. C. Derrick, F. M. Collins, S. L. Morris, C. H. King, and W. R. Jacobs, Jr. 2003. The primary mechanism of attenuation of bacillus Calmette-Guérin is a loss of secreted lytic function required for invasion of lung interstitial tissue. *Proc. Natl. Acad. Sci. USA* **100:**12420–12425.

95. Hughes, K. T., K. L. Gillen, M. J. Semon, and J. E. Karlinsey. 1993. Sensing structural intermediates in bacterial flagellar assembly by export of a negative regulator. *Science* **262:**1277–1280.

96. Inoue, T., I. Tanimoto, H. Ohta, K. Kato, Y. Murayama, and K. Fukui. 1998. Molecular characterization of low-molecular-weight component protein, Flp, in *Actinobacillus actinomycetemcomitans* fimbriae. *Microbiol. Immunol.* **42:**253–258.

97. Inouye, S., S. Wang, J. Sekizawa, S. Halegoua, and M. Inouye. 1977. Amino acid sequence for the peptide extension on the prolipoprotein of the *Escherichia coli* outer membrane. *Proc. Natl. Acad. Sci. USA* **74:**1004–1008.

98. Jack, R. L., F. Sargent, B. C. Berks, G. Sawers, and T. Palmer. 2001. Constitutive expression of *Escherichia coli tat* genes indicates an important role for the twin-arginine translocase during aerobic and anaerobic growth. *J. Bacteriol.* **183:**1801–1804.

99. Ji, Y., B. Zhang, S. F. Van Horn, P. Warren, G. Woodnutt, M. K. Burnham, and M. Rosenberg. 2001. Identification of critical staphylococcal genes using conditional phenotypes generated by antisense RNA. *Science* **293:**2266–2269.

100. Johnson, T. L., J. Abendroth, W. G. Hol, and M. Sandkvist. 2006. Type II secretion: from structure to function. *FEMS Microbiol. Lett.* **255:**175–186.

101. Jongbloed, J. D., U. Grieger, H. Antelmann, M. Hecker, R. Nijland, S. Bron, and J. M. van Dijl. 2004. Two minimal Tat translocases in *Bacillus. Mol. Microbiol.* **54:**1319–1325.

102. Jonquieres, R., H. Bierne, F. Fiedler, P. Gounon, and P. Cossart. 1999. Interaction between the protein InlB of *Listeria monocytogenes* and lipoteichoic acid: a novel mechanism of protein association at the surface of gram-positive bacteria. *Mol. Microbiol.* **34:**902–914.

103. Just, I., G. Fritz, K. Aktories, M. Giry, M. R. Popoff, P. Boquet, S. Hegenbarth, and C. von Eichel-Streiber. 1994. *Clostridium difficile* toxin B acts on the GTP-binding protein Rho. *J. Biol. Chem.* **269:**10706–10712.

104. Just, I., J. Selzer, M. Wilm, C. von Eichel-Streiber, M. Mann, and K. Aktories. 1995. Glucosylation of Rho proteins by *Clostridium difficile* toxin B. *Nature* **375:**500–503.

105. Just, I., M. Wilm, J. Selzer, G. Rex, C. von Eichel-Streiber, M. Mann, and K. Aktories. 1995. The enterotoxin from *Clostridium difficile* (ToxA) monoglucosylates the Rho proteins. *J. Biol. Chem.* **270:**13932–13936.

106. Kachlany, S. C., P. J. Planet, M. K. Bhattacharjee, E. Kollia, R. DeSalle, D. H. Fine, and D. H. Figurski. 2000. Nonspecific adherence by *Actinobacillus actinomycetemcomitans* requires genes widespread in bacteria and archaea. *J. Bacteriol.* **182:**6169–6176.

107. Kermorgant, M., N. Bonnefoy, and G. Dujardin. 1997. Oxa1p, which is required for cytochrome c oxidase and ATP synthase complex formation, is embedded in the mitochondrial inner membrane. *Curr. Genet.* **31:**302–307.

108. Kist, M. L., and R. G. Murray. 1984. Components of the regular surface array of *Aquaspirillum serpens* MW5 and their assembly in vitro. *J. Bacteriol.* **157:**599–606.

109. Kocks, C., E. Gouin, M. Tabouret, P. Berche, H. Ohayon, and P. Cossart. 1992. *L. monocytogenes*–induced actin assembly requires the *actA* gene product, a surface protein. *Cell* **68:**521–531.

110. Kocks, C., R. Hellio, P. Gounon, H. Ohayon, and P. Cossart. 1993. Polarized distribution of *Listeria monocytogenes* surface protein ActA at the site of directional actin assembly. *J. Cell. Sci.* **105**(Pt. 3)**:**699–710.

111. Kontinen, V. P., P. Saris, and M. Sarvas. 1991. A gene (prsA) of *Bacillus subtilis* involved in a novel, late stage of protein export. *Mol. Microbiol.* **5:**1273–1283.

112. Kuhn, A., and W. Wickner. 1985. Conserved residues of the leader peptide are essential for cleavage by leader peptidase. *J. Biol. Chem.* **260:**15914–15918.

113. Kumamoto, C. A., and O. Francetic. 1993. Highly selective binding of nascent polypeptides by an *Escherichia coli* chaperone protein in vivo. *J. Bacteriol.* **175:**2184–2188.

114. Kutsukake, K., S. Iyoda, K. Ohnishi, and T. Iino. 1994. Genetic and molecular analyses of the interaction between the flagellum-specific sigma and anti-sigma factors in *Salmonella typhimurium. EMBO J.* **13:**4568–4576.

115. Kutsukake, K., Y. Ohya, and T. Iino. 1990. Transcriptional analysis of the flagellar regulon of *Salmonella typhimurium. J. Bacteriol.* **172:**741–747.

116. Lai, E. M., N. D. Phadke, M. T. Kachman, R. Giorno, S. Vazquez, J. A. Vazquez, J. R. Maddock, and A. Driks. 2003. Proteomic

analysis of the spore coats of *Bacillus subtilis* and *Bacillus anthracis*. *J. Bacteriol.* **185**:1443–1454.

117. **Lancefield, R.** 1962. Current knowledge of type-specific M antigens of group A streptococci. *J. Immunol.* **89**:307–313.

118. **Lauer, P., C. D. Rinaudo, M. Soriani, I. Margarit, D. Maione, R. Rosini, A. R. Taddei, M. Mora, R. Rappuoli, G. Grandi, and J. L. Telford.** 2005. Genome analysis reveals pili in group B Streptococcus. *Science* **309**:105.

119. **Lee, V. T., S. K. Mazmanian, and O. Schneewind.** 2001. A program of *Yersinia enterocolitica* type III secretion reactions is triggered by specific host signals. *J. Bacteriol.* **183**:4970–4978.

120. **Lee, V. T., and O. Schneewind.** 2001. Protein secretion and the pathogenesis of bacterial infections. *Genes Dev.* **15**:1725–1752.

121. **Leibovitz, E., H. Ohayon, P. Gounon, and P. Beguin.** 1997. Characterization and subcellular localization of the *Clostridium thermocellum* scaffoldin dockerin binding protein SdbA. *J. Bacteriol.* **179**:2519–2523.

122. **Lemaire, M., H. Ohayon, P. Gounon, T. Fujino, and P. Beguin.** 1995. OlpB, a new outer layer protein of *Clostridium thermocellum*, and binding of its S-layer-like domains to components of the cell envelope. *J. Bacteriol.* **177**: 2451–2459.

123. **Losick, R., and P. Stragier.** 1992. Criss-cross regulation of cell-type specific gene expression during development in *B. subtilis*. *Nature* **355**: 601–604.

124. **Lupas, A.** 1996. A circular permutation event in the evolution of the SLH domain? *Mol. Microbiol.* **20**:897–898.

125. **Lupas, A., H. Engelhardt, J. Peters, U. Santarius, S. Volker, and W. Baumeister.** 1994. Domain structure of the *Acetogenium kivui* surface layer revealed by electron crystallography and sequence analysis. *J. Bacteriol.* **176**:1224–1233.

126. **Ma, Y., A. E. Bryant, D. B. Salmi, S. M. Hayes-Schroer, E. McIndoo, M. J. Aldape, and D. L. Stevens.** 2006. Identification and characterization of bicistronic *speB* and *prsA* gene expression in group A *Streptococcus*. *J. Bacteriol.* **188**:7626–7634.

127. **Macnab, R. M.** 2004. Type III flagellar protein export and flagellar assembly. *Biochim. Biophys. Acta* **1694**:207–217.

128. **Madden, J. C., N. Ruiz, and M. Caparon.** 2001. Cytolysin-mediated translocation (CMT): a functional equivalent of type III secretion in gram-positive bacteria. *Cell* **104**:143–152.

129. **Mader, C., C. Huber, D. Moll, U. B. Sleytr, and M. Sara.** 2004. Interaction of the crystalline bacterial cell surface layer protein SbsB

and the secondary cell wall polymer of *Geobacillus stearothermophilus* PV72 assessed by real-time surface plasmon resonance biosensor technology. *J. Bacteriol.* **186**:1758–1768.

130. **Mahairas, G. G., P. J. Sabo, M. J. Hickey, D. C. Singh, and C. K. Stover.** 1996. Molecular analysis of genetic differences between *Mycobacterium bovis* BCG and virulent *M. bovis*. *J. Bacteriol.* **178**:1274–1282.

131. **Maione, D., I. Margarit, C. D. Rinaudo, V. Masignani, M. Mora, M. Scarselli, H. Tettelin, C. Brettoni, E. T. Iacobini, R. Rosini, N. D'Agostino, L. Miorin, S. Buccato, M. Mariani, G. Galli, R. Nogarotto, V. Nardi Dei, F. Vegni, C. Fraser, G. Mancuso, G. Teti, L. C. Madoff, L. C. Paoletti, R. Rappuoli, D. L. Kasper, J. L. Telford, and G. Grandi.** 2005. Identification of a universal group B streptococcus vaccine by multiple genome screen. *Science* **309**:148–150.

132. **Matsuyama, S., Y. Fujita, and S. Mizushima.** 1993. SecD is involved in the release of translocated secretory proteins from the cytoplasmic membrane of *Escherichia coli*. *EMBO J.* **12**:265–270.

133. **Mattick, J. S.** 2002. Type IV pili and twitching motility. *Annu. Rev. Microbiol.* **56**:289–314.

134. **Mazmanian, S. K., G. Liu, E. R. Jensen, E. Lenoy, and O. Schneewind.** 2000. *Staphylococcus aureus* sortase mutants defective in the display of surface proteins and in the pathogenesis of animal infections. *Proc. Natl. Acad. Sci. USA* **97**:5510–5515.

135. Reference deleted.

136. **Mazmanian, S. K., G. Liu, H. Ton-That, and O. Schneewind.** 1999. *Staphylococcus aureus* sortase, an enzyme that anchors surface proteins to the cell wall. *Science* **285**:760–763.

137. **Mazmanian, S. K., E. P. Skaar, A. H. Gaspar, M. Humayun, P. Gornicki, J. Jelenska, A. Joachmiak, D. M. Missiakas, and O. Schneewind.** 2003. Passage of heme-iron across the envelope of *Staphylococcus aureus*. *Science* **299**:906–909.

138. **Medini, D., C. Donati, H. Tettelin, V. Masignani, and R. Rappuoli.** 2005. The microbial pan-genome. *Curr. Opin. Genet. Dev.* **15**:589–594.

139. **Medzhitov, R.** 2001. Toll-like receptors and innate immunity. *Nat. Rev. Immunol.* **1**:135–145.

140. **Medzhitov, R., P. Preston-Hurlburt, and C. A. J. Janeway.** 1997. A human homologue of the *Drosophila* Toll protein signals activation of adaptive immunity. *Nature* **388**:394–397.

141. **Mesnage, S., T. Fontaine, T. Mignot, M. Delepierre, M. Mock, and A. Fouet.** 2000. Bacterial SLH domain proteins are noncovalently anchored to the cell surface via a

conserved mechanism involving wall polysaccharide pyruvylation. *EMBO J.* **19:**4473–4484.

142. **Mesnage, S., E. Tosi-Couture, P. Gounon, M. Mock, and A. Fouet.** 1998. The capsule and S-layer: two independent and yet compatible macromolecular structures in *Bacillus anthracis. J. Bacteriol.* **180:**52–58.

143. **Mesnage, S., E. Tosi-Couture, M. Mock, P. Gounon, and A. Fouet.** 1997. Molecular characterization of the *Bacillus anthracis* main S-layer component: evidence that it is the major cell-associated antigen. *Mol. Microbiol.* **23:**1147–1155.

144. **Miller, J. D., H. D. Bernstein, and P. Walter.** 1994. Interaction of *E. coli* Ffh/4.5S ribonucleoprotein and FtsY mimics that of mammalian signal recognition particle and its receptor. *Nature* **367:**657–659.

145. **Milohanic, E., R. Jonquieres, P. Cossart, P. Berche, and J. L. Gaillard.** 2001. The autolysin Ami contributes to the adhesion of *Listeria monocytogenes* to eukaryotic cells via its cell wall anchor. *Mol. Microbiol.* **39:**1212–1224.

146. **Milohanic, E., R. Jonquieres, P. Glaser, P. Dehoux, C. Jacquet, P. Berche, P. Cossart, and J. L. Gaillard.** 2004. Sequence and binding activity of the autolysin-adhesin Ami from epidemic *Listeria monocytogenes* 4b. *Infect. Immun.* **72:**4401–4409.

147. **Mori, H., and K. Cline.** 2001. Post-translational protein translocation into thylakoids by the Sec and DeltapH-dependent pathways. *Biochim. Biophys. Acta* **1541:**80–90.

148. **Muller, M., and G. Blobel.** 1984. In vitro translocation of bacterial proteins across the plasma membrane of *Escherichia coli. Proc. Natl. Acad. Sci. USA* **81:**7421–7425.

149. **Murakami, T., K. Haga, M. Takeuchi, and T. Sato.** 2002. Analysis of the *Bacillus subtilis spoIIIJ* gene and its paralogue gene, *yqjG. J. Bacteriol.* **184:**1998–2004.

150. **Navarre, W. W., and O. Schneewind.** 1994. Proteolytic cleavage and cell wall anchoring at the LPXTG motif of surface proteins in gram-positive bacteria. *Mol. Microbiol.* **14:**115–121.

151. **Neu, H. C.** 1992. The crisis in antibiotic resistance. *Science* **257:**1064–1073.

152. **Newton, S. M., P. E. Klebba, C. Raynaud, Y. Shao, X. Jiang, I. Dubail, C. Archer, C. Frehel, and A. Charbit.** 2005. The svpA-srtB locus of *Listeria monocytogenes*: fur-mediated iron regulation and effect on virulence. *Mol. Microbiol.* **55:**927–940.

153. **Nikaido, H.** 1992. Porins and specific channels of bacterial outer membranes. *Mol. Microbiol.* **6:**435–442.

154. **Nouwen, N., M. Piwowarek, G. Berrelkamp, and A. J. Driessen.** 2005. The

large first periplasmic loop of SecD and SecF plays an important role in SecDF functioning. *J. Bacteriol.* **187:**5857–5860.

155. **Ohnishi, K., K. Kutsukake, H. Suzuki, and T. Lino.** 1990. Gene fliA encodes an alternative sigma factor specific for flagellar operons in *Salmonella typhimurium. Mol. Gen. Genet.* **221:**139–147.

156. **Ohnishi, K., K. Kutsukake, H. Suzuki, and T. Lino.** 1992. A novel transcriptional regulation mechanism in the flagellar regulon of *Salmonella typhimurium*: an antisigma factor inhibits the activity of the flagellum-specific sigma factor, sigma F. *Mol. Microbiol.* **6:**3149–3157.

157. **Olabarria, G., J. L. Carrascosa, M. A. de Pedro, and J. Berenguer.** 1996. A conserved motif in S-layer proteins is involved in peptidoglycan binding in *Thermus thermophilus. J. Bacteriol.* **178:**4765–4772.

158. **Oliver, D. B., and J. Beckwith.** 1981. *E. coli* mutant pleiotropically defective in the export of secreted proteins. *Cell* **25:**765–772.

159. **Pallen, M. J.** 2002. The ESAT-6/WXG100 superfamily—and a new gram-positive secretion system? *Trends Microbiol.* **10:**209–212.

160. **Pallen, M. J., R. R. Chaudhuri, and I. R. Henderson.** 2003. Genomic analysis of secretion systems. *Curr. Opin. Microbiol.* **6:**519–527.

161. **Pancholi, V., and V. A. Fischetti.** 1992. A major surface protein on group A streptococci is a glyceraldehyde-3-phosphate-dehydrogenase with multiple binding activity. *J. Exp. Med.* **176:**415–426.

162. **Peabody, C. R., Y. J. Chung, M. R. Yen, D. Vidal-Ingigliardi, A. P. Pugsley, and M. H. Saier, Jr.** 2003. Type II protein secretion and its relationship to bacterial type IV pili and archaeal flagella. *Microbiology* **149:**3051–3072.

163. **Planet, P. J., S. C. Kachlany, R. DeSalle, and D. H. Figurski.** 2001. Phylogeny of genes for secretion NTPases: identification of the widespread tadA subfamily and development of a diagnostic key for gene classification. *Proc. Natl. Acad. Sci. USA* **98:**2503–2508.

164. **Pohlschröder, M., W. A. Prinz, E. Hartmann, and J. Beckwith.** 1997. Protein translocation in the three domains of life: variations on a theme. *Cell* **91:**563–566.

165. **Popham, D. L., J. Helin, C. E. Costello, and P. Setlow.** 1996. Muramic acid lactam in peptidoglycan of *Bacillus* spores is required for spore outgrowth but not for spore dehydration or heat resistance. *Proc. Natl. Acad. Sci. USA* **93:**15405–15410.

166. **Porcelli, I., E. de Leeuw, R. Wallis, E. van den Brink-van der Laan, B. de Kruijff, B. A. Wallace, T. Palmer, and B. C. Berks.**

2002. Characterization and membrane assembly of the TatA component of the *Escherichia coli* twin-arginine protein transport system. *Biochemistry* **41:**13690–13697.

167. **Poritz, M. A., K. Strub, and P. Walter.** 1988. Human SRP RNA and *E. coli* 4.5S RNA contain a highly homologous structural domain. *Cell* **55:**4–6.

168. **Pugsley, A. P.** 1993. The complete general secretory pathway in gram-negative bacteria. *Microbiol. Rev.* **57:**50–108.

169. **Pym, A. S., P. Brodin, R. Brosch, M. Huerre, and S. T. Cole.** 2002. Loss of RD1 contributed to the attenuation of the live tuberculosis vaccines *Mycobacterium bovis* BCG and *Mycobacterium microti*. *Mol. Microbiol.* **46:**709–717.

170. **Pym, A. S., P. Brodin, L. Majlessi, R. Brosch, C. Demangel, A. Williams, K. E. Griffiths, G. Marchal, C. Leclerc, and S. T. Cole.** 2003. Recombinant BCG exporting ESAT-6 confers enhanced protection against tuberculosis. *Nat. Med.* **9:**533–539.

171. **Randall, L. L.** 1992. Peptide binding by chaperone SecB: implications for recognition of non-native structure. *Science* **257:**241–245.

172. **Randall, L. L., T. B. Topping, and S. J. Hardy.** 1990. No specific recognition of leader peptide by SecB, a chaperone involved in protein export. *Science* **248:**860–863.

173. **Reader, R. W., and L. Siminovitch.** 1971. Lysis defective mutants of bacteriophage lambda: genetics and physiology of S cistron mutants. *Virology* **43:**607–622.

174. **Renshaw, P. S., K. L. Lightbody, V. Veverka, F. W. Muskett, G. Kelly, T. A. Frenkiel, S. V. Gordon, R. G. Hewinson, B. Burke, J. Norman, R. A. Williamson, and M. D. Carr.** 2005. Structure and function of the complex formed by the tuberculosis virulence factors CFP-10 and ESAT-6. *EMBO J.* **24:**2491–2498.

175. **Renshaw, P. S., P. Panagiotidou, A. Whelan, S. V. Gordon, R. G. Hewinson, R. A. Williamson, and M. D. Carr.** 2002. Conclusive evidence that the major T-cell antigens of the *Mycobacterium tuberculosis* complex ESAT-6 and CFP-10 form a tight, 1:1 complex and characterization of the structural properties of ESAT-6, CFP-10, and the ESAT-6★CFP-10 complex. Implications for pathogenesis and virulence. *J. Biol. Chem.* **277:**21598–21603.

176. **Ries, W., C. Hotzy, I. Schocher, U. B. Sleytr, and M. Sara.** 1997. Evidence that the N-terminal part of the S-layer protein from *Bacillus stearothermophilus* PV72/p2 recognizes a secondary cell wall polymer. *J. Bacteriol.* **179:**3892–3898.

177. **Romisch, K., J. Webb, J. Herz, S. Prehn, R. Frank, M. Vingron, and B. Dobberstein.** 1989. Homology of 54K protein of signal-recognition particle, docking protein and two *E. coli* proteins with putative GTP-binding domains. *Nature* **340:**478–482.

178. **Ronda-Lain, C., R. Lopez, A. Tapia, and A. Tomasz.** 1977. Role of the pneumococcal autolysin (murein hydrolase) in the release of progeny bacteriophage and in the bacteriophage-induced lysis of the host cells. *J. Virol.* **21:**366–374.

179. **Rosch, J., and M. Caparon.** 2004. A microdomain for protein secretion in gram-positive bacteria. *Science* **304:**1513–1515.

180. **Rosenstein, R., and F. Gotz.** 2000. Staphylococcal lipases: biochemical and molecular characterization. *Biochimie* **82:**1005–1014.

181. **Samuelson, J. C., M. Chen, F. Jiang, I. Moller, M. Wiedmann, A. Kuhn, G. J. Phillips, and R. E. Dalbey.** 2000. YidC mediates membrane protein insertion in bacteria. *Nature* **406:**637–641.

182. **Samuelson, J. C., F. Jiang, L. Yi, M. Chen, J. W. de Gier, A. Kuhn, and R. E. Dalbey.** 2001. Function of YidC for the insertion of M13 procoat protein in *Escherichia coli*: translocation of mutants that show differences in their membrane potential dependence and Sec requirement. *J. Biol. Chem.* **276:**34847–34852.

183. **Sánchez-Puelles, J. M., C. Ronda, J.-L. García, R. Lopez, and E. García.** 1986. Searching for autolysin functions: characterization of a pneumococcal mutant deleted in the *lytA* gene. *Eur. J. Biochem.* **158:**289–293.

184. **Sandberg, A. L., S. Ruhl, R. A. Joralmon, M. J. Brennan, M. J. Sutphin, and J. O. Cisar.** 1995. Putative glycoprotein and glycolipid polymorphnuclear leukocyte receptors for the *Actinomyces naeslundii* WVU45 fimbrial lectin. *Infect. Immun.* **63:**2625–2631.

185. **Sargent, F., E. G. Bogsch, N. R. Stanley, M. Wexler, C. Robinson, B. C. Berks, and T. Palmer.** 1998. Overlapping functions of components of a bacterial Sec-independent protein export pathway. *EMBO J.* **17:**3640–3650.

186. **Schaffer, C., and P. Messner.** 2004. Surface-layer glycoproteins: an example for the diversity of bacterial glycosylation with promising impacts on nanobiotechnology. *Glycobiology* **14:**31R–42R.

187. **Schneewind, O., A. Fowler, and K. F. Faull.** 1995. Structure of the cell wall anchor of surface proteins in *Staphylococcus aureus*. *Science* **268:**103–106.

188. **Schneewind, O., D. Mihaylova-Petkov, and P. Model.** 1993. Cell wall sorting signals in

surface protein of gram-positive bacteria. *EMBO J.* **12**:4803–4811.

189. **Schneewind, O., P. Model, and V. A. Fischetti.** 1992. Sorting of protein A to the staphylococcal cell wall. *Cell* **70**:267–281.

190. **Scotti, P. A., M. L. Urbanus, J. Brunner, J. W. de Gier, G. von Heijne, C. van der Does, A. J. Driessen, B. Oudega, and J. Luirink.** 2000. YidC, the *Escherichia coli* homologue of mitochondrial Oxa1p, is a component of the Sec translocase. *EMBO J.* **19**:542–549.

191. **Shockman, G. D., and J. F. Barrett.** 1983. Structure, function, and assembly of cell walls of gram-positive bacteria. *Annu. Rev. Microbiol.* **37**:501–527.

192. **Shockman, G. D., and J.-V. Holtje.** 1994. Microbial peptidoglycan (murein) hydrolases, p. 131–166. *In* J.-M. Ghuysen and R. Hakenbeck (ed.), *Bacterial Cell Wall*, vol. 27. Elsevier Biochemical Press, Amsterdam, The Netherlands.

193. **Sibbald, M. J., A. K. Ziebandt, S. Engelmann, M. Hecker, A. de Jong, H. J. Harmsen, G. C. Raangs, I. Stokroos, J. P. Arends, J. Y. Dubois, and J. M. van Dijl.** 2006. Mapping the pathways to staphylococcal pathogenesis by comparative secretomics. *Microbiol. Mol. Biol. Rev.* **70**:755–788.

194. **Sieradzki, K., R. B. Roberts, S. W. Haber, and A. Tomasz.** 1999. The development of vancomycin resistance in a patient with methicillin-resistant *Staphylococcus aureus* infection. *N. Engl. J. Med.* **340**:517–523.

195. **Silhavy, T. J., S. A. Benson, and S. D. Emr.** 1983. Mechanisms of protein localization. *Microbiol. Rev.* **47**:313–344.

196. **Sjoquist, J., B. Meloun, and H. Hjelm.** 1972. Protein A isolated from *Staphylococcus aureus* after digestion with lysostaphin. *Eur. J. Biochem.* **29**:572–578.

197. **Sjöquist, J., J. Movitz, I.-B. Johansson, and H. Hjelm.** 1972. Localization of protein A in the bacteria. *Eur. J. Biochem.* **30**:190–194.

198. **Sleytr, U. B.** 1997. Basic and applied S-layer research: an overview. *FEMS Microbiol. Rev.* **20**:5–12.

199. **Sleytr, U. B., and A. M. Glauert.** 1976. Ultrastructure of the cell walls of two closely related clostridia that possess different regular arrays of surface subunits. *J. Bacteriol.* **126**:869–882.

200. **Sorensen, A. L., S. Nagai, G. Houen, P. Andersen, and A. B. Andersen.** 1995. Purification and characterization of a low-molecular-mass T-cell antigen secreted by *Mycobacterium tuberculosis*. *Infect. Immun.* **63**:1710–1717.

201. **Stanley, S. A., S. Raghavan, W. W. Hwang, and J. S. Cox.** 2003. Acute infection and macrophage subversion by *Mycobacterium tuberculosis* require a specialized secretion system. *Proc. Natl. Acad. Sci. USA* **100**:13001–13006.

202. **Stoll, H., J. Dengjel, C. Nerz, and F. Gotz.** 2005. *Staphylococcus aureus* deficient in lipidation of prelipoproteins is attenuated in growth and immune activation. *Infect. Immun.* **73**:2411–2423.

203. **Stranger-Jones, Y. K., T. Bae, and O. Schneewind.** 2006. Vaccine assembly from surface proteins of *Staphylococcus aureus*. *Proc. Natl. Acad. Sci. USA* **103**:16942–16947.

204. **Struck, J. C., H. Y. Toschka, T. Specht, and V. A. Erdmann.** 1988. Common structural features between eukaryotic 7SL RNAs, eubacterial 4.5S RNA and scRNA and archaebacterial 7S RNA. *Nucleic Acids Res.* **16**:7740.

205. **Takeda, K., T. Kaisho, and S. Akira.** 2003. Toll-like receptors. *Annu. Rev. Immunol.* **21**:335–376.

206. **Tan, K. S., B. Y. Wee, and K. P. Song.** 2001. Evidence for holin function of tcdE gene in the pathogenicity of *Clostridium difficile*. *J. Med. Microbiol.* **50**:613–619.

207. **Tan, T., W. L. Lee, D. C. Alexander, S. Grinstein, and J. Liu.** 2006. The ESAT-6/CFP-10 secretion system of *Mycobacterium marinum* modulates phagosome maturation. *Cell Microbiol.* **8**:1417–1429.

208. **Telford, J. L., M. A. Barocchi, I. Margarit, R. Rappuoli, and G. Grandi.** 2006. Pili in gram-positive pathogens. *Nat. Rev. Microbiol.* **4**:509–519.

209. **Tettelin, H., K. E. Nelson, I. T. Paulsen, J. A. Eisen, T. D. Read, S. Peterson, J. Heidelberg, R. T. DeBoy, D. H. Haft, R. J. Dodson, A. S. Durkin, M. Gwinn, J. F. Kolonay, W. C. Nelson, J. D. Peterson, L. A. Umayam, O. White, S. L. Salzberg, M. R. Lewis, D. Radune, E. Holtzapple, H. Khouri, A. M. Wolf, T. R. Utterback, C. L. Hansen, L. A. McDonald, T. V. Feldblyum, S. Angiuoli, T. Dickinson, E. K. Hickey, I. E. Holt, B. J. Loftus, F. Yang, H. O. Smith, J. C. Venter, B. A. Dougherty, D. A. Morrison, S. K. Hollingshead, and C. M. Fraser.** 2001. Complete genome sequence of a virulent isolate of *Streptococcus pneumoniae*. *Science* **293**:498–506.

210. **Tjalsma, H., H. Antelmann, J. D. Jongbloed, P. G. Braun, E. Darmon, R. Dorenbos, J. Y. Dubois, H. Westers, G. Zanen, W. J. Quax, O. P. Kuipers, S. Bron, M. Hecker, and J. M. van Dijl.** 2004. Proteomics of protein secretion by *Bacillus subtilis*: separating the "secrets" of the secretome. *Microbiol. Mol. Biol. Rev.* **68**:207–233.

211. **Tjalsma, H., A. Bolhuis, J. D. Jongbloed, S. Bron, and J. M. van Dijl.** 2000. Signal

peptide-dependent protein transport in *Bacillus subtilis*: a genome-based survey of the secretome. *Microbiol. Mol. Biol. Rev.* **64:**515–547.

212. **Tjalsma, H., A. Bolhuis, M. L. van Roosmalen, T. Wiegert, W. Schumann, C. P. Broekhuizen, W. J. Quax, G. Venema, S. Bron, and J. M. van Dijl.** 1998. Functional analysis of the secretory precursor processing machinery of *Bacillus subtilis*: identification of a eubacterial homolog of archaeal and eukaryotic signal peptidases. *Genes Dev.* **12:**2318–2331.

213. **Tokunaga, M., H. Tokunaga, and H. C. Wu.** 1982. Post-translational modification and processing of *Escherichia coli* prolipoprotein in vitro. *Proc. Natl. Acad. Sci. USA* **79:**2255–2259.

214. **Tomich, M., D. H. Fine, and D. H. Figurski.** 2006. The TadV protein of *Actinobacillus actinomycetemcomitans* is a novel aspartic acid prepilin peptidase required for maturation of the Flp1 pilin and TadE and TadF pseudopilins. *J. Bacteriol.* **188:**6899–6914.

215. **Ton-That, H., and O. Schneewind.** 2003. Assembly of pili on the surface of *Corynebacterium diphtheriae*. *Mol. Microbiol.* **50:**1429–1438.

216. **Tsuboi, A., R. Uchihi, R. Tabata, Y. Takahashi, H. Hashiba, T. Sasaki, H. Yamagata, N. Tsukagoshi, and S. Udaka.** 1986. Characterization of the genes coding for two major cell wall proteins from protein-producing *Bacillus brevis* 47: complete nucleotide sequence of the outer wall protein gene. *J. Bacteriol.* **168:**365–373.

217. **Ulbrandt, N. D., J. A. Newitt, and H. D. Bernstein.** 1997. The *E. coli* signal recognition particle is required for the insertion of a subset of inner membrane proteins. *Cell* **88:**187–196.

218. **Valent, Q. A., D. A. Kendall, S. High, R. Kusters, B. Oudega, and J. Luirink.** 1995. Early events in preprotein recognition in *E. coli*: interaction of SRP and trigger factor with nascent polypeptides. *EMBO J.* **14:**5494–5505.

219. **Valent, Q. A., P. A. Scotti, S. High, J. W. de Gier, G. von Heijne, G. Lentzen, W. Wintermeyer, B. Oudega, and J. Luirink.** 1998. The *Escherichia coli* SRP and SecB targeting pathways converge at the translocon. *EMBO J.* **17:**2504–2512.

220. **van der Wolk, J. P., J. G. de Wit, and A. J. Driessen.** 1997. The catalytic cycle of the *Escherichia coli* SecA ATPase comprises two distinct preprotein translocation events. *EMBO J.* **16:**7297–7304.

221. **van Roosmalen, M. L., N. Geukens, J. D. Jongbloed, H. Tjalsma, J. Y. Dubois, S. Bron, J. M. van Dijl, and J. Anné.** 2004. Type I signal peptidases of gram-positive bacteria. *Biochim. Biophys. Acta* **1694:**279–297.

222. **von Heijne, G.** 1983. Patterns of amino acids near signal-sequence cleavage sites. *Eur. J. Biochem.* **133:**17–21.

223. **von Heijne, G.** 1989. The structure of signal peptides from bacterial lipoproteins. *Protein. Eng.* **2:**531–534.

224. **von Heijne, G., and L. Abrahmsen.** 1989. Species-specific variation in signal peptide design. Implications for protein secretion in foreign hosts. *FEBS Lett.* **244:**439–446.

225. **Voth, D. E., and J. D. Ballard.** 2005. *Clostridium difficile* toxins: mechanism of action and role in disease. *Clin. Microbiol. Rev.* **18:**247–263.

226. **Walter, P., and G. Blobel.** 1980. Purification of a membrane-associated protein complex required for protein translocation across the endoplasmic reticulum. *Proc. Natl. Acad. Sci. USA* **77:**7112–7116.

227. **Walter, P., and G. Blobel.** 1981. Translocation of proteins across the endoplasmic reticulum III. Signal recognition protein (SRP) causes signal sequence-dependent and site-specific arrest of chain elongation that is released by microsomal membranes. *J. Cell Biol.* **91:**557–561.

228. **Wang, A., B. J. Wilkinson, and R. K. Jayaswal.** 1991. Sequence analysis of a *Staphylococcus aureus* gene encoding a peptidoglycan hydrolase activity. *Gene* **102:**105–109.

229. **Weiner, J. H., P. T. Bilous, G. M. Shaw, S. P. Lubitz, L. Frost, G. H. Thomas, J. A. Cole, and R. J. Turner.** 1998. A novel and ubiquitous system for membrane targeting and secretion of cofactor-containing proteins. *Cell* **93:**93–101.

230. **Wickus, G. G., A. D. Warth, and J. L. Strominger.** 1972. Appearance of muramic lactam during cortex synthesis in sporulating cultures of *Bacillus cereus* and *Bacillus megaterium*. *J. Bacteriol.* **111:**625–627.

231. **Wild, J., E. Altman, T. Yura, and C. A. Gross.** 1992. DnaK and DnaJ heat shock proteins participate in protein export in *Escherichia coli*. *Genes Dev.* **6:**1165–1172.

232. **Williams, W. A., M. L. Burts, J. Bubeck Wardenburg, and D. Missiakas.** *Staphylococcus aureus* prsA is dispensable for protein secretion. *J. Bacteriol.* Submitted.

233. **Wolfe, P. B., P. Silver, and W. Wickner.** 1982. The isolation of homogeneous leader peptidase from a strain of *Escherichia coli* which overproduces the enzyme. *J. Biol. Chem.* **257:**7898–7902.

234. **Yahr, T. L., and W. T. Wickner.** 2001. Functional reconstitution of bacterial Tat translocation in vitro. *EMBO J.* **20:**2472–2479.

235. **Yamada, S., M. Sugai, H. Komatsuzawa, S. Nakashima, T. Oshida, A. Matsumoto, and H. Suginaka.** 1996. An autolysin ring

associated with cell separation of *Staphylococcus aureus. J. Bacteriol.* **178:**1565–1571.

236. **Yoshimura, A., E. Lien, R. R. Ingalls, E. Tuomanen, R. Dziarski, and D. Golen-** **bock.** 1999. Cutting edge: recognition of gram-positive bacterial cell wall components by the innate immune system occurs via Toll-like receptor 2. *J. Immunol.* **163:**1–5.

WHAT GENOMICS HAS TAUGHT US ABOUT BACTERIAL CELL WALL BIOSYNTHESIS

Lynn G. Dover

13

The bacterial cell wall is a massive, covalently linked structure that confers rigidity by opposing the tremendous turgor pressure developed within the cell and contributes to the shape of the microorganism it bounds. The major structural component is the murein or peptidoglycan (PG) sacculus, a baglike structure (251) assembled from a polymer of two amino sugars and up to four amino acids (204). Almost all bacteria possess a PG-based cell wall (notable exceptions being the mycoplasmas and chlamydiae), and this can be variously modified to produce more elaborate structures as found in the *Corynebacterineae* and gram-negative genera. Both of these groups of bacteria are adorned with an outer asymmetric lipid bilayer tethered to the underlying PG, the outer membranous structures of the *Corynebacterineae* being based on mycolic acids (64) and that of the gram-negative bacteria containing an outer lipooligosaccharide or lipopolysaccharide leaflet (168).

In order to present a coherent commentary on the impact of genomics on our understanding of cell wall biosynthesis, this chapter will focus only on the major structural polysaccharide components of bacterial cell walls. Over 40 years ago, it was recognized that the remodeling of the cell wall, necessary to be compatible with cell growth and division, must include a highly concerted hydrolysis of some of its covalent bonds to allow efficient incorporation of new polymer (251). Postgenomic analyses have afforded insight into the control mechanisms necessary to direct these crucial hydrolyses and also related modifications to the PG of pathogens to protect it from hydrolysis by lytic enzymes such as lysozyme. Likewise, genomics has shed light on the peculiar drug-sensitivity profile and basic cell biology of the chlamydiae and has provided a major impetus to our understanding of the biosynthesis of the particularly complex cell wall of the prevalent bacterial pathogen of our time, *Mycobacterium tuberculosis*.

This chapter is organized to relate a commentary on the basic processes of cell wall biosynthesis derived from decades of chemical analyses and classical genetics studies in the pregenomic era. This seeks to provide the context for a coherent, if not exhaustive, account of the effect of genomics on our understanding of the coordinated biosyntheses, modifications, and hydrolyses that are required to produce these fascinating structures.

Lynn G. Dover   School of Biosciences, University of Birmingham, Birmingham B15 2TT, United Kingdom.

Bacterial Pathogenomics, Edited by M. J. Pallen et al.
© 2007 ASM Press, Washington, D.C.

CHEMICAL STRUCTURE AND MOLECULAR ARCHITECTURE OF BACTERIAL CELL WALLS

PG is composed of an array of glycan chains cross-linked by short peptides (204). The glycan component is constructed from a repeating disaccharide unit of *N*-acetylglucosamine (GlcNAc) and its derivative, *N*-acetylmuramic acid (MurNAc), with each sugar being linked together by β-1→4 glycosidic bonds (87). A short peptide, L-alanyl-D-isoglutamyl-*meso*-diaminopimelyl-D-alanine in *Escherichia coli*, is attached via an amide linkage to the carboxyl group of each MurNAc residue; a small proportion of these peptides contains no alanine, whereas a smaller population contains an extra alaninyl residue. Although this basic composition is common to almost all gram-negative bacteria as well as some gram-positive bacilli, several variations in terms of peptide composition and cross-linking patterns and sugar modification have been described and are useful taxonomically (89, 204).

The disaccharide components of these muropeptide units form relatively short linear chains with reported average lengths of 16 to 30 sugar residues in *E. coli* (37, 90, 98, 172). Adjacent glycan chains are bonded to each other by cross-links in their peptide side chains, with most occurring between the carboxyl group of the fourth residue, D-alanine, with the free amino group of the diaminopimelate (DAP) residue of the adjacent peptide, although some cross-linking between DAP residues also occurs (89). In an early computer-assisted three-dimensional model of PG, the sugars were suggested to be tilted with respect to each other to accommodate the lactoyl moiety of the MurNAC residues, producing a twist in the muropeptides that ultimately results in a spiral arrangement of the glycan chain. Consecutive peptide side chains were then thought to project from this helix at approximately 90° intervals (12). There are currently several models of cell wall architecture and growth that relate two different arrangements of the glycan strands relative to the plane of the cytoplasmic membrane; classical models describe an architecture based on glycan strands lying parallel to the membrane (124), whereas the glycan strands in the new scaffold model are arranged perpendicular to its plane (56, 60). A particularly interesting debate has developed in the literature regarding the relative merits of these models, and the reader is directed to excellent reviews for a full description of these models (56–60, 124, 244, 260). Recently, the 37-step chemical synthesis of GlcNAc-MurNAc(pentapeptide)-GlcNAc-MurNAc(pentapeptide) (105) facilitated the elucidation of the three-dimensional solution structure of a PG fragment by nuclear magnetic resonance analysis (152); progress had previously been limited by the lack of pure and discrete fragments for analyses. This structure is defined as an ordered, right-handed helical saccharide conformation in which three GlcNAc-MurNAc disaccharide pairs would correspond to one turn of the helix. In contrast to the well-defined saccharide conformation, the pentapeptide stems show much greater flexibility in aqueous solution; the D-Lac, L-Ala, and D-Glu moieties adopt a limited number of conformations, while their L-Lys-D-Ala-D-Ala termini are highly disordered (152).

Regarding the orientation of the stem peptides, threefold symmetry with respect to the axis of the oligosaccharide helix is apparent, thereby predisposing each strand for cross-linking to a maximum of three neighboring strands. On the basis of this structure and the expected degree of incomplete cross-linking, in silico modeling of a larger cell wall fragment was carried out. The model describes a honeycomb pattern forming pores from 7 to 12 nm across, large enough to accommodate the enzymes for PG biosynthesis and the PG-interacting region of the TolC outer membrane channel; indeed, the crystal structure of this part of TolC (127) appears perfectly embraced by the smallest pores (152). These authors argue that the scaffold model is very attractive with regard to their observations regarding TolC. Recently, Beveridge and coworkers have made tremendous progress in imaging of bacterial cell walls by cryoelectron microscopy and atomic force microscopy, which has provided new detailed insights into cell wall organization in both gram-positive (140, 141, 233) and gram–negative bacteria (139). Although the variety of pore sizes observed by atomic force microscopy (233) is consistent with those predicted from the scaffold model of Meroueh et al. (152), the resolution of glycan strand orientation remains to be achieved, and the organization of PG is still unclear. Indeed, an argument that elements of both the parallel and scaffold models might coexist in the same wall and that the sacculus may not possess a particularly uniform structure has also recently been made (260).

ENZYMOLOGY OF PG MONOMER BIOSYNTHESIS

The construction of the murein sacculus is essentially a two-stage process. In the first, a disaccharide-peptide monomer unit is assembled by using UDP-linked and then polyprenyl phosphate–linked intermediates (3, 35, 55, 68, 174–177, 238). Initially, in the simplest scenarios, six cytoplasmic reactions catalyzed by MurA to MurF form the UDP-linked MurNAc-pentapeptide moiety, which is then transferred to the undecaprenyl phosphate carrier lipid to form lipid I. Further modification by the addition of a GlcNAc residue forms lipid II, which bears the complete monomer for PG polymerization and is translocated to the outer face of the membrane by an unknown mechanism. Next, transglycosylases (TGs) catalyze the polymerization of the glycan chains and transpeptidases (TPs); the penicillin-binding proteins (PBPs) cross-link the peptide cross-bridges between glycan chains and thus incorporate nascent material into the existing PG sacculus framework. A summary of all these steps is illustrated in Fig. 1.

The initial reactions in the production of the disaccharide-peptide monomer unit concern the conversion of UDP-GlcNAc to UDP-MurNAc. First, MurA, the cellular target of the streptomycete-derived antibiotic fosfomycin (42, 104, 206), transfers enolpyruvate from phosphoenolpyruvate to the hydroxyl group occupying C3 of UDP-GlcNAc (94, 206, 246, 262). In the second MurB-catalyzed reaction, the enolpyruvate moiety is reduced to form UDP-MurNAc (19, 93, 226),

a product that also acts as a feedback inhibitor of MurA (159, 262).

The newly added D-lactoyl group of MurNAc moiety is then modified by the addition of the pentapeptide stem peptide, which cross-links the glycan strands of mature PG. These peptides usually consist of L-Ala, D-Glu, a diamino acid (often *meso*-DAP for gram–negative or L-Lys for gram–positive organisms) and D-Ala, followed by D-Ala, D-Ser, or D-lactate (204). The first three of these individual amino acids are added sequentially by highly specific synthetases (MurC to MurE) via their carboxyl adenylation (68, 217). The transition from MurNAc-tripeptide to MurNAc-pentapeptide is catalyzed by MurF, which sequentially binds ATP, UDP-MurNAc-tripeptide, and then the dipeptide product of the D-Ala-D-Ala-ligase, Ddl (4, 114, 166).

The provision of the D-amino acids required for PG biosynthesis is facilitated by amino acid racemases and epimerases. D-Alanine is formed by alanine racemase in a pyridoxal-5-phosphate (PLP)-dependent reaction (35). The production of *meso*-(DL)DAP and D-glutamate is PLP independent; *meso*-DAP is usually formed from LL-DAP via DAP-epimerase, while glutamate can be transformed by glutamate racemase. However, an alternative route to the latter can occur via the PLP-dependent D-amino acid aminotransferase, which deaminates D-Ala to form pyruvate and then transfers the amino moiety to α-ketoglutarate (78).

At completion of the pentapeptide chain, the modified MurNAc residue can be transferred to the undecaprenyl phosphate carrier lipid via the MraY transferase (32, 113), forming lipid I, and thereafter the glycosyltransferase MurG forms a β-1→4 linkage with GlcNAc to complete the disaccharide-pentapeptide PG monomer unit, lipid II (36, 151, 225).

Several noteworthy variations in stem peptide composition occur in gram-positive genera; variant C-terminal ends occur (204), and the incorporation of a D-lactoyl residue has been shown to be the basis for resistance to the glycopeptide antibiotics vancomycin and teicoplanin in enterococci (8). Furthermore,

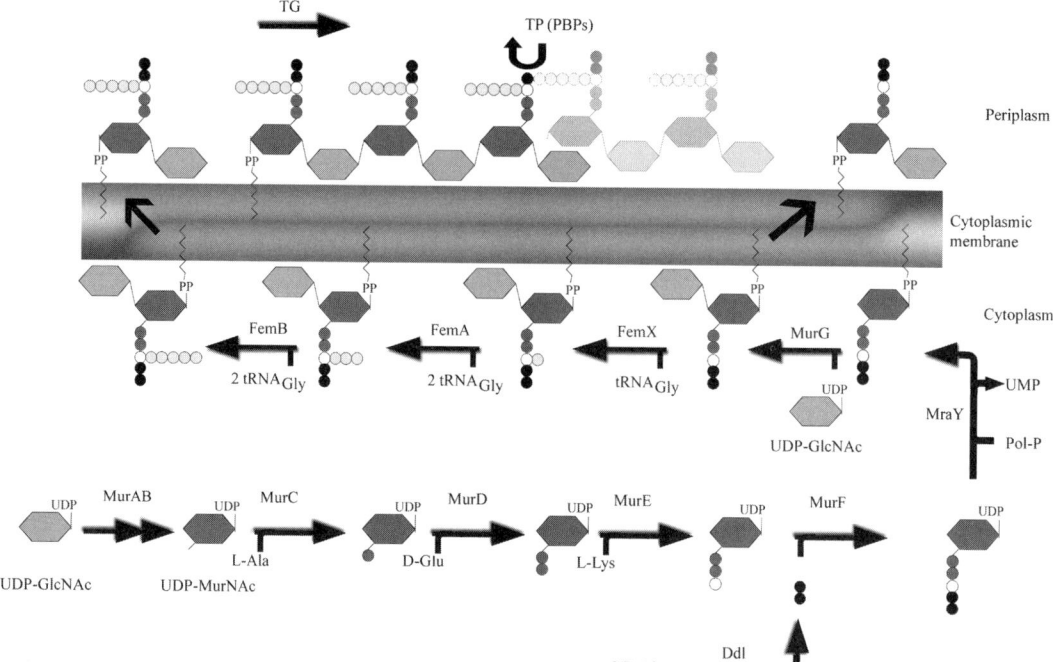

FIGURE 1 Schematic representation of PG biosynthesis. MurA and MurB catalyze the conversion of the UDP-GlcNAc to UDP-MurNAc before MurC and MurD initiate stem peptide synthesis by adding individual amino acid residues. MurE adds the diamino acid that is crucial to stem peptide cross linking. Ddl forms the di-alaninyl peptide that is ligated to the UDP-MurNac-tripeptide. The completed UDP-MurNAc-pentapeptide is transferred via MraY to a polyprenyl monophosphate carrier lipid in the cytoplasmic membrane, represented here by the large gray bar. The introduction of a GlcNAc residue from UDP-GlcNAc via a β-1→4 linkage that completes the basic lipid II PG monomer unit is catalyzed by MurG. At this point in species that do not contain interpeptide bridges, this lipid II is then translocated to the periplasmic face of the membrane, represented here by the arrow within the membrane, to take part in PG polymerization and cross linking. Interpeptide synthesis for *S. aureus* is depicted in the following reactions; FemX adds a single glycine residue to the ε-amino group of the lysine residue of the stem peptide. FemA and FemB then add pairs of Gly residues to the growing interpeptide to complete the Gly$_5$ unit before translocation to the periplasmic face of the membrane, where glycan chains are polymerized via transglycosylases (TGs) and nascent strands are incorporated into existing murein, represented here by a more lightly shaded strand, by the transpeptidase (TP) activity of the PBPs. Both of these reactions likely occur concomitantly.

several species elaborate side chains from the ε-amino group of the L-lysine residue occupying the third position of the stem peptide. These variations include D-Asn or D-Asp in *Enterococcus faecium*, L-Ala-L-Ala in *Enterococcus faecalis*, (L-Ala/L-Ser)-L-Ala in *Streptococcus pneumoniae*, and Gly$_5$ in *Staphylococcus aureus* (204). The L–amino acids and glycine that comprise these interpeptide bridges are not activated by adenylating transferases as with the D-amino acids of the stem peptides, but are

derived from aminoacyl-tRNA (121) by a nonribosomal mechanism that is discussed more fully later, either at the level of the nucleotide-linked intermediates or at the lipid I stage (121, 142, 181, 217).

POLYMERIZATION AND CROSS-LINKING OF PG STRANDS

On the outer face of the cytoplasmic membrane, two types of activity are involved in the polymerization of PG and its insertion into the preexisting murein; TGs catalyze the linear polymerization of the glycan strands and TPs cross-link these strands into the growing sacculus or septum. Whether all TGs involved in the polymerization of the glycan chains incorporate their modified disaccharide substrates at the nonreducing or reducing termini of the growing PG strand remains to be established, but the latter mechanism has been demonstrated in a few gram-positive organisms (80, 247, 256).

In terms of protein architecture, TGs either occur as the N-terminal domains of the so-called high-molecular-mass–penicillin-binding proteins (HMM-PBPs) or as stand-alone monofunctional TGs that do not bind penicillins. In normally growing bacteria, the processes of transglycosylation and transpeptidation are tightly coupled reactions. Transglycosylation can proceed independently of transpeptidation in cell-free systems and in growing cells treated with penicillin and other β-lactam antibiotics, specific inhibitors of transpeptidation. The existence of non-cross-linked nascent PG remains unclear and may be difficult to distinguish from degradative products. In the presence of moenomycin, a specific transglycosylation inhibitor, monomer polymerization via transpeptidation does not occur (222), thus rendering the addition of monomer units to preexisting murein by transpeptidation before linking them together via transglycosylation a highly unlikely scenario. Thus the cross-linking of PG and its insertion into existing murein via transpeptidation likely follows or is concomitant with glycan strand polymerization (239).

GENOME-DRIVEN INSIGHTS INTO PG POLYMERIZATION

By use of the limited sequence data available in the pregenomic era, Ghuysen determined that the HMM-PBPs could be divided into two classes, depending on the function of their non-penicillin-binding (n-PB) modules (86); class A HMM-PBPs contain a TG activity that has been directly demonstrated for PBP1a and PBP1b in *E. coli* (165) and PBP2 of *S. aureus*

(179). The PB domains are active TPs; thus, class A HMM-PBPs are able to perform all the basic functions required for PG polymerization. Class B HMM-PBPs also possess TP activity but contain n-PB modules that have no apparent TG activity. The function of this module is not known, but roles in mediating important protein-protein interactions (109) or folding have been suggested (92). The lack of TG activity suggests that these class B PBPs must cooperate with either a monofunctional TG or the TG module of a class A HMM-PBP in order to polymerize PG. Strong support for this assumption is provided by the observation that the SCC*mec*-encoded class B HMM-PBP of methicillin-resistant *Staphylococcus aureus* (MRSA) PBP2A requires the intact TG module of the native staphylococcal PBP2 to confer methicillin resistance (179). However, the TP activity (180) of PBP2A is relatively poor, and when MRSA is treated with sufficient methicillin to inactivate native staphylococcal TPs, it produces a poorly cross-linked cell wall (52), although this is obviously sturdy enough to survive the methicillin insult. Similarly, PBP5fm of *E. faecium* is another low-affinity PBP that confers resistance to penicillin and other β-lactam antibiotics but possesses relatively poor TP activity (199, 211). It appears that the evolution of these "rescue" TPs (199) has resulted in the sacrifice of catalytic efficiency in favor of low β-lactam affinity. The structures of two low-affinity PBPs have been solved (53, 199), and several structural features have been highlighted that may be implicated in reduced β-lactam affinity.

In a study to determine the TG partner for the *E. faecalis* PBP5, Arbeloa et al. (7) determined that the genes encoding all three known class A HMM-PBPs could be disrupted with only decreases in generation time and PG cross-linking as a consequence. Their observation that the residual TG activity was moenomycin insensitive strongly suggests that *E. faecalis* possesses a novel PG TG (7) that awaits identification. A similar observation regarding a *Bacillus subtilis* mutant containing deletions in all four of its class A PBPs and containing no obviously related glycosyltransferase that is able to synthesize an intact PG with only minor modifications also indicates that a novel enzyme can perform the TG function required for PG synthesis (150)

The considerable expansion of the databases driven by the efforts of the various bacterial genome-sequencing consortia prompted a reevaluation of the HMM-PBPs in 1998 (92) when 63 PBPs were analyzed with the aim of shedding light on their different roles in PG assembly. Hierarchical analyses determined that the 29 class A HMM-PBPs fell into defined clusters, three including gram-negative PBPs, three containing gram-positive PBPs, and a distinct cluster containing mycobacterial PBPs. *E. coli* contains three class A HMM-PBPs (92). PBP1a and PBP1b are members of separate subclusters yet can functionally substitute for each other, but loss of both is fatal (122); PBP1c, which appears to carry an inactive TG module, cannot replace these (92). Similarly, *Bacillus subtilis* possesses three class A HMM-PBPs (PBP1, PBP2c, and PBP4), one from each cluster, that can also substitute for each other but with successive losses exacerbating growth and morphological defects (182). Together these observations suggest that these differently clustered TG-TP hybrid PBPs may be paralogues determining subtle differences in PG cross-linking that have thus far evaded resolution (92).

Similarly, the class B HMM-PBPs formed distinct clusters, two incorporating gram-negative PBPs and three comprising gram-positive PBPs. Further core-based analyses were carried out to identify features that might be responsible for this clustering (92). This analysis highlighted distinctive motifs in the n-PB modules of class A versus class B PBPs, conserved motifs in the PB domains, and a series of adducts that occur at various locations along the peptide chain. Contrary to expectations, however, the PB domains of class A and class B PBPs all clustered according to class, with subclasses again diverging according to phylogeny; almost invariably, the PB and n-PB domains associated within very similar clusters

(92). As seen with their class A counterparts, *E. coli* PBP2 and PBP3, which belong to different subclasses, cannot functionally replace each other. These, however, are certainly paralogues. PBP3 is intimately involved in cell septation, forming complexes with several cell-cycle-related proteins including FtsZ, a tubulin homologue that initiates formation of a midcell multiprotein ringlike structure that mediates the invagination of the septum (215, 253). PBP2, on the other hand, is not required for septation but is involved in lateral wall expansion and also plays a role in shape maintenance (119, 193). Septum complex formation and PBP localization have been recently and extensively reviewed (201, 242).

MAINTENANCE OF CELL DIVISION GENE CLUSTERS AND GENOMIC CHANNELING

Most of the progress toward our current understanding of PG biosynthesis was derived from studies that used *E. coli* in the pregenomic era. Many of the key Mur enzymes (discussed earlier) were cloned and characterized having previously been mapped on the chromosome by using the array of existing temperature-sensitive lethal mutants. A number of these genes were found to cluster at 2 min on the chromosome map along with *fts* genes important to cell division (238, 261) and including *ftsI*, which encodes the class B PBP3. However, not all of the known PG synthesis genes are present in this division/cell wall (DCW) cluster, with others being distributed about the *E. coli* chromosome (33, 62, 63, 186). The 16 genes of the *E. coli* DCW cluster are all orientated in the same direction for transcription and are efficiently arranged with several open reading frames (ORFs) overlapping (Fig. 2), permitting speculation that this region was crucial for the regulation of cellular growth and cell division. This genetic organization was found to be identical in *Haemophilus influenzae* Rd (76), suggesting that it might be highly conserved among gram-negative rods. The discovery that the gram-positive rod *Bacillus subtilis* also possesses a similar cluster that fea-

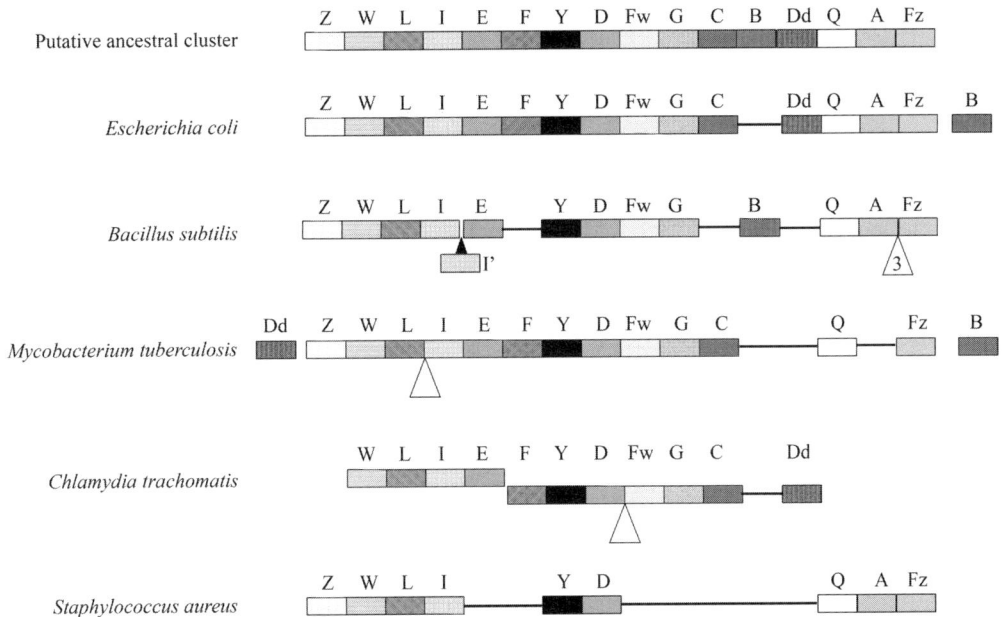

FIGURE 2 Maintenance of DCW gene clusters. The DCW clusters of several of the bacteria discussed herein are schematically represented; coding sequences are not represented to scale in order to facilitate alignment. The triangles denote the positions of single gene insertions within the cluster, apart from in *B. subtilis*, where three sporulation-specific genes are included. Similarly, a sporulation-specific PBP gene I′ SpoVD (48) is indicated above the PBP-encoding *ftsI* column. The discontinuity in the *C. trachomatis* alignment represents the conserved clustering of these genes at another part of the chromosome. When orthologues are removed from their position in the cluster but are retained in the genome in a locus nearby, they are placed to one side. The key to the gene symbols placed above each cluster is as follows: Z, *mraZ*; W, *mraW*; L, *ftsL*; I, *ftsI*; I′, *spoVD*; E, *murE*; Y, *mraY*; D, *murD*; Fw, *ftsW*; G, *murG*; C, *murC*; B, *murB*, Dd, *ddlB*; Q, *ftsQ*; A, *ftsA*; Fz, *ftsZ*.

tured the substitution of several sporulation-specific genes, including an extra sporulation-specific PBP, in place of some of the PG biosynthesis genes supported the suggestion of a possible evolutionary relationship between PG synthesis and cell division (49, 158). The existence of similar but less-well-conserved DCW clusters was subsequently established in the gram-positive cocci *Staphylococcus aureus* and *E. faecalis* (187).

More recent comparisons of DCW clusters from various bacterial genomes show that the order of genes present in DCW clusters is well conserved, and a hypothetical archetypal cluster that contains all of the genes and is similar to that of *E. coli* has been defined (169). Phylogenetic analyses of bacterial morphologies have shown that other bacterial cell shapes probably arose independently and several times from a rod-shaped progenitor (210) (Fig. 2). Presumably this ancestor possessed an ordered DCW cluster that has been broadly maintained in bacilli but has been dispersed or rearranged in other lineages such as cocci (156) or has been modified to accommodate morphological phenomena such as sporulation. In all of these DCW clusters, two homologous ORFs of unknown function precede *ftsL* and a PBP-encoding gene *ftsI*, which in *E. coli* (PBP3) plays a role in septum formation and cell division (215, 253). Because the PBP-encoding genes of DCW clusters show greatest homology with those from other DCW clusters, it was suggested that they all might play a role in cell division (187). Most of the genes that have dispersed from the DCW clusters of gram-positive organisms encode enzymes that supply cytoplasmic PG precursors.

One relationship that appears paramount is the juxtaposition of *ftsA* and *ftsZ*, which is consistent with evidence of their coordinated regulation being important for cell division (47, 54). Because many of the *E. coli* DCW ORFs overlapped, more widespread coordinate regulation of DCW genes was proposed to influence the unavoidably interwoven processes of cell division and PG biosynthesis (61). The *E. coli* cluster possesses one initial promoter, several internal promoters, and a single transcriptional terminator, thus facilitating the generation of partial transcripts, which appear to have differential stability (207), as well as a transcript that might encompass the whole cluster (241). The regulatory mechanisms influencing transcription from these promoters remain unknown, suggesting that the potential for differential and even temporal regulation might, in part, orchestrate the cell cycle (156).

New insights into the significance of gene order and its conservation continue to develop (156, 169, 227, 228). The tendency of proteins encoded by genes whose genetic linkage remains strictly conserved to physically interact (169) was a factor that led Mingorance et al. to suggest "genomic channeling" as a means of orchestrating bacterial cell division, specifically in ensuring provision of sufficient precursors for septum formation during bacillary division (156). Here they develop the idea that the organization and broad conservation of the bacterial DCW cluster over billions of years of evolutionary time against the generalized trend toward genome reorganization (115, 227, 228) is a product of the necessary formation of multiprotein functional complexes that channel (214) precursors at specific sites of PG biosynthesis. Bacillary morphogenesis has been suggested to be the product of a two-site competition process (198), with shape being determined by the relative rates of later wall PG expansion and septum synthesis. Rather than the biosynthetic complex forming septal PG competing for precursors with its counterpart producing PG for wall extension from a limited pool in the bulk of the cytoplasm (198), Mingorance et al. suggest that multienzyme complexes for both lateral wall expansion and septum formation are recruited as necessary, and both effectively produce and channel dedicated pools of precursors (156). Because coccal cell wall growth is essentially related to the septal mode of growth (198), evolutionary transitions to coccal morphology have relaxed the selective pressure to maintain such tight coordination of DCW components and have permitted their dispersal.

An early observation drawn from a comparison of DCW clusters seems to correlate protein architecture with the shape of the organism. The Div1B/FtsQ proteins, implicated in the determination of bacterial shape, from several coccal DCW clusters possessed an extended hydrophilic region preceding the largest hydrophobic region of the protein towards their N termini (187). Whether this N-terminal extension plays any role in coccal morphology remains to be determined.

THE CHLAMYDIAL PARADOX

Chlamydiaceae remain significant and unusual intracellular pathogens causing ocular, sexually transmitted, and respiratory infections. They have an atypical biphasic developmental cycle in which metabolically inert extracellular elementary bodies (EBs) infect host epithelial cells and orchestrate a specialized phagosome termed an inclusion. There, the EBs become metabolically active and differentiate into reticulate bodies (RBs) that are able to divide by binary fission before eventually forming EBs once more. EBs released by cellular lysis can then proliferate the infection (259). Morphologically, both stages resemble gram-negative bacteria in that they possess both a cytoplasmic and an outer membrane in which lipopolysaccharide has been detected (171). However, EBs are rigid and stable while RBs are pleiomorphic and fragile (71, 162). Unlike all other gram-negative bacteria, the *Chlamydiaceae* lack detectable PG (41, 163); the necessary osmotic stability of the extracellular EB phase is provided via a network of disulfide cross-linked envelope proteins (100). However, an intriguing paradox persisted; despite their atypical envelope structure, chlamydiae are sensitive to PG-specific cell wall inhibitors such as penicillins and D-cycloserine (163). Although the initial conversion of EBs to RBs is not hindered by penicillins, it has been suggested that the development of EBs from RBs is affected (11). Specifically, penicillins appear to inhibit RB division (11), causing an accumulation of aberrant structures, which resume normal development on removal of the penicillin (143, 163). The interference of penicillins with the development of *C. trachomatis* had been related to the presence of three PBPs (11, 163); the PBPs are detectable in both developmental forms of the bacterium but appear to be less accessible to penicillin in the EBs, which could explain their resistance to the drugs (11). Similarly, the sensitivity of chlamydiae to D-cycloserine, which inhibits the alanine racemase and D-alanyl-D-alanine ligase (Ddl) of PG-containing bacteria, provided limited evidence of a role for an alanine-containing PG in chlamydial development (41, 110, 163).

The determination of the genome sequence for *Chlamydia trachomatis* (218) highlighted genes that appeared to encode a near-complete pathway for the synthesis of PG (41). Candidates for the structural genes encoding the known PBPs were identified, along with a putative Ddl coding sequence that is fused to a *murC* homologue. Furthermore, homologues of almost all of the genes known to participate in or be associated with PG synthesis in *E. coli* were present in the *C. trachomatis* genome (41, 99, 218). Genomic data from the nine sequenced chlamydiae (representing five species) confirm the conservation of this proposed PG biosynthetic pathway (15, 39, 190, 191, 194, 209, 218, 231).

Transcriptional studies confirmed that at least some of these genes were expressed in chlamydiae; transcripts of *murA*, *murB*, and *pbp2* were detected in the early stages of the EB to RB transition in *C. trachomatis* (147).

However, microarray analyses of *C. trachomatis* transcription revealed a marked increase in expression of PG synthesis genes at a later developmental stage, that of RB division (18, 167). Proteomic studies detected MurG (160) and MurC-Ddl (237) in EBs while a comparison of EB and RB enzyme expression patterns detected MurE exclusively in RBs (212). The latter observation is clearly consistent with both the upregulation of PG biosynthetic enzymes seen in dividing RBs (18, 167) and the morphological aberrations detected on treatment of chlamydiae with cell wall biosynthesis inhibitors (11, 143, 163).

PBP1 and PBP2 were classified as class B HMM-PBPs, bifunctional transglycosylases-transpeptidases, whereas PBP3 resembled a low-molecular-mass class A PBP with a probable carboxypeptidase activity. An almost complete *dcw* cluster (Fig. 2) was also identified in *C. trachomatis*, although occupying two distant loci. The lack of a gene encoding alanine racemase suggested that DdlA might be the cellular target of D-cycloserine and that it may need to acquire D-Ala from its host for PG synthesis. The presence of two homologues of *dagA*, which encodes a D-Ala permease, is wholly consistent with the acquisition of D-Ala from host cells. However, it has been argued that there is little evidence that mammalian cells contain sufficient quantities of D-amino acids (145) to support chlamydial growth, which is consistent with the failure of alanine racemase mutants of both *Listeria monocytogenes* and *Shigella flexneri* to survive in mammalian cells unless exogenous D-Ala is supplied (145). Nevertheless, free D-Ala has been detected in several mammalian tissues including brain, liver, kidney, blood, and urine (97, 161), and its synthesis via intestinal bacteria has been suggested due to the sensitivity of urinary tract D-Ala concentrations to antibiotics (125). McCoy and Maurelli have suggested an equally plausible alternative route to chlamydial D-Ala provision; all of the sequenced chlamydial genomes contain homologues of a *Treponema denticola* cystalysin (145), a member of the α family of PLP-dependent enzymes in which side reactions are commonly observed, whose alanine racemase activity has been demonstrated in vitro (23).

Recently, biochemical characterization of these PG gene products has provided evidence that they possess relevant enzymic activity. The purified product of *C. trachomatis murA* expressed in *E. coli* releases Pi from phosphoenolpyruvate in the presence of UDP-GlcNAc, and the viability of an *E. coli murA* mutant can be supported by expression of *C. trachomatis murA* (147). The MurC enzyme of *C. trachomatis*, which is synthesized as a bifunctional MurC-Ddl product, was expected to possess UDP-MurNAc:L-alanine ligase activity. The MurC domain of the *C. trachomatis* bifunctional protein has been functionally investigated in *E. coli*; it complements a temperature-sensitive, lethal *E. coli* mutant in *murC*. The purified recombinant MurC domain displayed in vitro ATP-dependent UDP-MurNAc:L-alanine ligase activity, although its substrate specificity appears to be relaxed. Studies using L-alanine, L-serine, and glycine revealed comparable V_{max}/K_m values; assuming that a functional PG is elaborated in chlamydiae, it is not clear which amino acid might occupy the first position of the pentapeptide (106). Further genetics-based studies on this bifunctional enzyme showed that the expression of the chlamydial gene can complement a D-Ala-D-Ala ligase *E. coli* auxotroph, demonstrating that it also encodes a functional ligase (146). This ligase activity was also demonstrated in vitro by using purified recombinant enzyme, and inhibition studies confirmed MurC-Ddl as a cellular target for D-cycloserine (146), as anticipated.

To date, no genes encoding the TG activities required to form the linear glycan chains of PG have been recognized in the sequenced chlamydiae. Although no class A HMM-PBPs or homologous monofunctional glycosyltransferase-encoding chlamydial sequences have been identified, a glycanless chlamydial PG (88) need not necessarily be the only conclusion. Because the *B. subtilis* and *E. faecalis* mutants mentioned earlier that are disrupted in all of their class A HMM-PBP genes are viable and synthesize intact PG (7, 150), then at least one novel enzyme possesses a TG activity appropriate for PG polymerization. It therefore remains possible that the chlamydiae also possess novel glycosyltransferases and are able to produce a glycan-containing polymer. These issues, however, will be adequately resolved only after the isolation of chlamydial PG and its complete chemical characterization. However, another and possibly more important issue remains: what is the functional significance of the proposed biosynthesis of undetectable levels of PG?

Chopra et al. argued that continued maintenance of this genetic investment totalling 1.8% of the *C. trachomatis* genome implies that PG plays an important, yet atypical, role in chlamydial cell biology. The lack of the elsewhere highly conserved *ftsZ* in *C. trachomatis* suggested the possibility that its role in cell division might be fulfilled here by a limited amount of PG (41). Interestingly, Brown and Rockey (34) identified a chlamydial antigen, which they termed SEP, that localized in rings within or between RBs (34), reminiscent of the ringlike structures formed by FtsZ in dividing eubacterial cells (28). Through deduction, the antiserum was found to react with the mycobacterial cell wall core component of the adjuvant used. Because chlamydiae lack the arabinogalactan (AG) and mycolic acid components of the mycobacterial material, the antibodies were most likely reacting with its PG component and, consistent with this notion, were used to label the entire periphery of *E. coli* (34). The development of these ring structures might indicate a pivotal role played by PG in RB division, and in agreement with this, treatment of chlamydiae with cell wall biosynthesis inhibitors led to aberrant forms exhibiting marked antigenic redistribution, with SEP occurring mainly at the periphery of the aberrant RBs (34). Whether this rearrangement is symptomatic of, or causal to, the development of aberrant forms is unclear.

MRSA, β-LACTAM RESISTANCE, AND THE FORMATION OF GRAM-POSITIVE INTERPEPTIDE BRIDGES

Recent genome-driven advances in the biosynthesis of the interpeptide bridges elaborated by several important gram-positive organisms can be directly traced to transposon mutagenesis studies focusing on one of the most significant health issues of recent years: the molecular basis for β-lactam tolerance in MRSA (21, 22). The methicillin resistance determinant in *S. aureus* resides on additional DNA, SCC*mec* (staphylococcal cassette chromosome *mec*, nomenclature established in reference 107), not present in β-lactam-sensitive cells. A component gene *mecA* encodes an additional (fifth) low-affinity PBP, PBP2A (PBP2′), known to be essential for cell wall synthesis in the presence of inhibitory concentrations of methicillin and thus likely carrying out PG transpeptidation when the other PBPs are inactivated by the drug (52). However, besides SCC*mec*, other chromosomally determined products were described as factors essential for methicillin resistance, or *fem* factors, whose inactivation lowered the level of resistance (21).

A chromosomal gene encoding one of these factors, *femA,* was mapped to a locus distant from SCC*mec*. Although FemA was needed for cell growth in the presence of β-lactam antibiotics, it had no influence on the synthesis of PBP2A. Molecular analyses suggested that *femA* was transcribed on a polycistronic mRNA containing a homologue, *femB*, presumably the product of gene duplication (22). However, unlike *mec*, the *femAB* operon belongs to the housekeeping genes of several staphylococci (112, 123). Characterization of FemA function via analyses of inactivation mutants revealed a decrease in PG-associated glycine (138), suggesting that its native role is to be involved in the Gly₅ interpeptide side chain formation after addition of the first Gly residue (51). Complementation analyses in a *femAB* mutant determined FemA to form Gly 2–3 and FemB to contribute Gly 4–5 (219).

Subsequently, an unknown factor, FemX, was proposed to be responsible for the incorporation of the first glycine (126). Analysis of the *S. aureus* ATCC 55748 genome for *fem*AB homologues revealed three interesting genes, *fmhA, fmhB,* and *fmhC,* and of these, only *fmhB* was shown to be essential and was postulated to represent *femX* (236). Substitution of its own promoter with an inducible promoter allowed analysis of the essential gene, revealing that reduced expression retarded growth and produced a marked accumulation of non-cross-linked, non-Gly-modified muropeptide, proving that FmhB satisfied the criteria to be the hypothetical FemX (196). Recently in vitro assembly of a complete lipid II–Gly₅ has been achieved using lipid II, staphylococcal tRNA as the source of activated glycine (121), and His₆-tagged FemA, FemB, and FemX (FmhB); biochemical analyses revealed that all Fem proteins are independently active and that FemX acts directly on lipid II, with FemA and FemB adding two glycine residues to lipid II–Gly₁ and lipid II–Gly₃, respectively, without releasing Gly₂ and Gly₄ intermediate products (205).

Genomic analyses of gram-positive bacteria have determined that all species possessing *fem* homologues produce branched-chain PG with glycine and/or L–amino acids in the interpeptide (195). The genes encoding the alanine and/or serine transferases of *S. pneumoniae*, *murMN* (*fibAB*), have been identified (74, 250). Although the interpeptide linkage is dispensable in the pneumococcus (208), in contrast to the staphylococcal situation, where at least one glycine residue is required for viability, it is required for the expression of penicillin resistance, and inactivation of these genes produces similar phenotypes as seen in *S. aureus fem* mutants; the level of total cell wall cross-linking is reduced, and mutants are hypersusceptible to all classes of cell wall synthesis inhibitors (73, 250). Interestingly, many pneumococcal genes, including *murMN* and PBP2x, a low-affinity PBP implicated in penicillin resistance, exhibit mosaic structure (72, 95, 96), offering a molecular basis for the argument that pneumococcal cell walls widely differ in terms of cross-linking even if similar strains both retain *murMN* (74). Variations in the genes are reflected in the catalytic properties of their products (72, 213), with the highly resistant strains containing higher proportions of branched muropeptides, signifying that they possess a *murM* allele that encodes a highly efficient form of the enzyme. Furthermore, only transformation with *murMN* genes from a highly resistant donor strain, along with a low-affinity PBP gene, results in the development of high-level penicillin resistance in a previously susceptible recipient strain (213).

Analysis of a partial *E. faecalis* genome sequence revealed two ORFs encoding products related to *S. aureus* FemX (30). Recombinant expression of these ORFs in *E. coli* revealed that BppA1 possessed UDP-MurNAc-pentapeptide:L-Ala ligase activity but likely acted preferentially on the polyprenyl phosphate–linked intermediate in vivo because it displayed a substantially poorer apparent K_m for the UDP-linked intermediate than did the bona fide UDP-MurNAc-pentapeptide:L-Ala ligase of *Weissella viridescens* (30, 181). This is also consistent with the high degree of identity (39%) shared with *S. pneumoniae* MurM, which fulfills the L-Ala/L-Ser ligase role in that bacterium. The authors also noted that the other FemX homologue, BppA2, shared 37% homology with *S. pneumoniae* MurN and might represent an enterococcal orthologue. A subsequent study demonstrated UDP-MurNAc-hexapeptide:L-Ala ligase activity in vitro and determined that deletion of *bppA2* was associated with production of muropeptides bearing a single L-alanyl substitution that were efficiently cross-linked by the DD-TPs. As in other systems, this deletion impaired the expression of the intrinsic penicillin resistance of the enterococcus, suggesting that the low-affinity PBP requires a dialanyl interpeptide to function adequately (31). Recently Arbeloa et al. were able to engineer mosaic interpeptides in *E. faecalis* by heterospecific expression of *femXAB* from *S. aureus* (6). Interpeptides containing between one and five residues and including Gly instead of L-Ala at the aminoterminus were produced and used as both donors and acceptors in PG cross-linking, indicating that the TPs of *E. faecalis* have a broad specificity. Similarly, heterospecific expression of PBP5m and PBP2a in *E. faecalis* and *E. faecium* conferred β-lactam resistance, suggesting that these low-affinity PBPs also exhibit a broad substrate specificity (6). Together, these observations show that the conservation of interpeptide sequence in members of the same species is the product of the high specificity of the Fem enzymes rather than being imposed by the substrate specificity of the native TPs (6).

ENDOPEPTIDASE-RESISTANT MODIFICATION OF GRAM-POSITIVE INTERPEPTIDE BRIDGES

Several bacterial species produce PG hydrolases that lyse other bacterial cells, and it has been suggested that they might confer a competitive advantage in environments where nutrients are limited or might allow producer organisms to establish an ecological niche within a competitive environment (223). Sequence variations, specifically substitution of Gly2 and Gly4 with L-Ser residues, in the interpeptides of strains of *Staphylococcus simulans*, *Staphylococcus carnosus*, *S. aureus*, and *Staphylococcus capitis* correlate with resistance to staphylolytic glycyl–glycine endopeptidases lysostaphin and ALE-1 (50, 103, 220, 221, 232, 264). The genes for endopeptidase (*end*) and endopeptidase resistance (*epr*) reside on plasmid pACK1 in *Staphylococcus simulans* biovar Staphylolyticus. A pACK1 fragment containing *end* was cloned and introduced into *S. aureus*. The recombinant *S. aureus* cells produced lysostaphin and were resistant to lysis by the enzyme, which indicated that the cloned fragment also contained *epr*. Treatments to remove proteins, teichoic acids, and lipoteichoic acids did not change the lysostaphin sensitivity of walls from strains of *S. simulans* biovar Staphylolyticus or of *S. aureus* with and without *epr*. Treatment of purified PGs with endopeptidase confirmed that resistance or susceptibility was a property of the PG itself. Compositional analyses revealed that cross bridges in the PG of *epr*-positive cells contained more serine and fewer glycine residues than those of cells without *epr* (50). Other authors characterized the *epr* gene, applying the designation *lif* for lysostaphin immunity factor, which occupied a locus adjacent to the *lss* (*end*) gene responsible for production of the prolysosthaphin. The *lif* gene product conferred lysostaphin resistance on *S. carnosus*, increased the Ser content of PG, and shared significant homology with FemAB (232). The plasmid also carried a putative Ser-tRNA synthase gene that appears to be dispensable for Lif-mediated lysostaphin immunity. The three-gene cassette was found to be flanked by the vestiges of two insertion sequences, suggesting that it had been acquired by the plasmid via horizontal gene transfer from another staphylococcus with increased lysostaphin resistance such as *Staphylococcus epidermis* or *Staphylococcus haemolyticus* (232). The similarity of Lif to FemB in particular was confirmed by the ability of Lif to complement FemB, but not FemA, in mutants of *S. aureus*, resulting in L-Ser incorporation in interpeptides (235). Similar studies ascribed very similar properties on the product of *epr* in resistance to ALE-1 of *S. capitas* (221) and in *Streptococcus milleri*, where immunity to the millericin B endopeptidase is mediated through substitution of L-Leu for L-Thr in the interpeptide by the Fem homologue MilF (26).

RECENT GENOME-DRIVEN ADVANCES IN THE ENZYMOLOGY OF PG ACETYLATION AND DEACETYLATION

The biosynthesis, maintenance, and remodeling of PG required to facilitate bacterial growth and cell division are a dynamic process that requires the coordination of both synthetic and lytic enzymes. The introduction of O-acetyl groups into PG has been suggested as a means to exert one level of control in the process as the modification inhibits the lytic transglycosylases, the major endogenous lytic enzymes. This O-acetylation is effective in the

protection of PG against lytic transglycosylases as the modification occurs at the C-6 hydroxyl group of MurNAc residues, thus precluding the formation of their 1,6-anhydroMurNAc lysis product. Similarly, PG degradation by the lysozymes that form components of the innate immune systems of eukaryotes is also inhibited by its O-acetylation. These lytic enzymes destabilize the sacculus by hydrolyzing the β-1→4 glycosidic linkages between GlcNAc and MurNAc residues. Several studies have shown that a variety of both gram-positive and gram-negative bacterial pathogens use O-acetylation to mediate lysozyme resistance.

Two Modes of PG O-Acetylation

Until recently, very little information regarding the incorporation of these acetyl groups into PG and none regarding their subsequent removal was available. Two models were proposed for the transfer of acetate from cytoplasmic CoA carriers to the muramoyl residues of the PG sacculus in their periplasmic location (43). One was modeled on the proposed pathway for acetylation of the pseudomonad exopolysaccharide alginate (79), which composes an integral membrane protein that translocates the acetate to the periplasm, where it is accepted by a second protein that transfers it specifically to the C6-hydroxyl of the muramoyl residues. The alternative model describes a single enzyme similar to NodX of *Rhizobium leguminosarum*, which both translocates the acetate and transfers it to its intended site, in the case of NodX the carbohydrate-based nodulation factor (75). Recently, evidence for PG acetylation systems conforming to both of these mechanistic models has been published (20, 248).

Staphylococcus species are unusual in that they are almost completely insensitive to lysozyme. Several possibilities were suggested to explain the mechanism of lysozyme resistance in *Staphylococcus aureus*, including O-acetylation of muramoyl residues. Scanning of the *S. aureus* N315 genome (129) for genes encoding proteins sharing sequence similarity with acetyltransferases revealed four strong candidate genes that might encode PG acetyl-transferases. Each of the four orthologous genes in *S. aureus* SA113 was interrupted by different antibiotic resistance cassettes, which allowed the construction of strains bearing multiple knockout mutations. A lysozyme sensitivity assay revealed that only mutation of one of these genes, named *oatA* and encoding an integral membrane protein, resulted in an increased sensitivity to lysozyme. Complementation of the mutant strain with a plasmid carrying *oatA* under the control of its own promoter reestablished the resistant phenotype (20). Careful high-performance liquid chromatography and mass spectrometry analyses of purified PG and released monomeric muropeptides from both the parent strain and the *oatA* mutant revealed the loss of acetyl groups in the mutant that could only be linked via the C6-hydroxyl of the muramoyl residues in the parent strain. Structural studies of hen egg white lysozyme (29) suggested that modification of the C6-hydroxyl of muramoyl residues would likely inhibit substrate binding. Because cell wall teichoic acids are also attached via a phosphodiester linkage to this hydroxyl group, the resistance to lysozyme of a *tagO* mutant (252), which is completely devoid of cell wall teichoic acid, was also assessed. The resistance retained by this mutant could possibly be due to frequency of modification with teichoic acid being much lower than that for O-acetylation and that O-acetylation could become more frequent in the absence of teichoic acids. However, it appears clear that OatA is responsible for PG O-acetylation in *S. aureus* and that this modification is indeed responsible for the inherent resistance of this pathogenic bacterium to lysozyme.

Clarke and coworkers (249) identified a homologue of the putative acetate translocator *algI* from *Pseudomonas aeruginosa* (gene products share 33% identity, 51% similarity), in the genome of *Neisseria gonorrhoeae* FA190 (http://www.genome.ou.edu/gono.html), a pathogen that is known to O-acetylate PG and does not produce alginate. Furthermore, the hypothetical protein encoded therein contained appro-

priately positioned signature motifs associated with family 1 acetyltransferases, inviting the proposition that this protein might function in the O-acetylation of PG in *N. gonorrhoeae*, and thus the gene was named *patA* for PG O-acetyltransferase A.

Analyses of sequences downstream of the *patA* locus implicated a further two interesting ORFs in acetate metabolism. The third ORF of this cluster possessed 49% sequence similarity to the catalytic domain of an acetyl xylan esterase from *Ruminococcus flavefaciens*. This enzyme belongs to family 3 carbohydrate esterases, which all facilitate the degradation of xylan by removing the O-linked acetyl groups that otherwise protect the polymer (248). As was the case with PatA, putative key residues were conserved in the gonococcus protein, and the predicted protein fold was also consistent with CE3 family members. Because the gonococcus is not xylanolytic, this hypothetical protein was proposed to catalyze the de-O-acetylation of PG and was named Ape1 for O-acetyl PG esterase 1. The predicted product of the second ORF in the cluster weakly resembles lysophospholipase and related esterases but only bears 18.6% similarity to Ape1. However, because it also contained potentially significant conserved residues, the ORF was tentatively assigned *ape2*.

Subsequent probing of the NCIB database, using PatA, Ape1, and Ape2 as queries in BLASTP, identified the existence of similar gene clusters, termed O-acetyl PG (OAP) clusters, in a further 18 genomes from both gram-negative and gram-positive bacteria, including the important human pathogens *Neisseria meningitidis*, *Campylobacter jejuni*, *Helicobacter pylori*, and *Bacillus anthracis*, none of which produces alginate. Sequence alignment analyses of the hypothetical Ape proteins identified unique signature motifs that permitted their subdivision into three Ape families, with family 1 being further subdivided into three subfamilies. OAPs exist in four configurations, each of which is characteristic of the component Ape (sub)families present (249). Interestingly, the OAP clusters of gram-positive organisms possess only one *ape* gene, *ape3*, whereas all the gram-negative species carried two genes, including at least one *ape1* subfamily component, and in some cases *ape2*.

The significant *O*-acetylation (14 to 69%) of muramoyl residues in PG purified from late exponential phase cultures of 9 of a panel of 10 OAP cluster-containing organisms provided compelling circumstantial evidence that these clusters could be involved in *O*-acetylation of PG; of those tested, only the xylanolytic *Agrobacterium tumefaciens* exhibited a poor PG modification, suggesting that the genes of its putative OAP cluster might actually function in xylan degradation.

Subsequently, molecular cloning and characterization of Ape1a from *N. gonorrhoeae* revealed that the enzyme was an esterase capable of releasing ester-linked acetate from a variety of substrates and had no de-*N*-acetylase activity (248). Although the purified recombinant enzyme de-*O*-acetylated xylan, it displayed a higher specific activity with *O*-acetylated PG, which is less frequently modified than xylan. Because all authentic acetyl xylan esterases studied to date have been shown to be devoid of activity on PG, these studies provide convincing evidence that Ape1a is indeed an *O*-acetyl PG esterase. Consistent with this role, failure to achieve functional inactivation of *ape1* in *N. gonorrhoeae* by generating an insertion mutant strongly indicates that Ape1a is essential for viability (248).

PG Deacetylation Protects Pathogens from Lytic Enzymes

Another PG modification strategy to avoid lysis via mammalian hydrolases that is relevant to pathogenicity is the de-*N*-acetylation of PG (2, 245, 254). Over 80% of the glucosamine and 10% of the muramic acid residues in *S. pneumoniae* were found to be non-*N*-acetylated, explaining the extraordinary resistance of the PG to the hydrolytic action of lysozyme. Indeed *N*-acetylation of this PG promoted its degradation by lysozyme, a property previously noted in PG from *Bacillus cereus* (2, 254).

Polysaccharide deacetylases belong to carbohydrate esterase family 4 (CE4), which includes chitin deacetylases, acetylxylan esterases, xylanases, rhizobial NodB chitooligosaccharide deacetylases, and PG deacetylases (45). All share a universal conserved polysaccharide deacetylase domain, and all members of this family catalyze the hydrolysis of either *N*-linked acetyl groups from GlcNAc residues (chitin deacetylase, NodB, and PG GlcNAc deacetylase), or *O*-linked acetyl groups from *O*-acetylxylose residues (acetylxylan esterase, xylanase). A large number of ORFs encoding putative polysaccharide deacetylases have been identified in genomes of gram-positive bacteria (45).

Only one gene, *pgdA*, of the *S. pneumoniae* genome was identified as potentially encoding a PG GlcNAc deacetylase through sequence similarity to fungal chitin deacetylases and rhizobial NodB chitooligosaccharide deacetylases. Inactivation of *pgdA* in *S. pneumoniae* led to the production of fully *N*-acetylated glycan and became hypersensitive to exogenous lysozyme in the stationary phase of growth (245).

The *pgdA* gene product is a putative secreted protein, which is likely translocated across the cytoplasmic membrane by components of the general secretory pathway and remains anchored to the cytoplasmic membrane by its *N*-terminal membrane domain from where it could access its substrate, PG (245). This supposed location of the enzyme suggests the deacetylation reaction is a secondary modification, consistent with the known pathway of PG synthesis that leads to a fully acetylated glycan backbone. The presence of nonacetylated amino sugars in PG is not limited to pneumococci (5, 9, 101, 102, 263) and might be related to the presence of PgdA-like enzymes in other bacteria.

The sequencing of the genomes of *Bacillus cereus* (116) and *Bacillus anthracis* (192) identified in each 10 ORFs encoding putative polysaccharide deacetylases, six of which were proposed to encode PG deacetylases. Psylinakis et al. cloned and characterized two of these from *B. cereus*, BC1960 and BC3618 (185). Both enzymes deacetylated only the GlcNAc residue of the synthetic muropeptide *N*-acetyl-D-glucosamine-(β-1,4)-*N*-acetylmuramyl-L-alanine-D-isoglutamine and similarly, analysis of the constituent muropeptides of PG from *B. subtilis* and *H. pylori* resulting from incubation of the enzymes BC1960 and BC3618 confirmed that both enzymes deacetylate GlcNAc residues of PG. Additionally, enzymatic deacetylation of chemically acetylated vegetative PG from *B. cereus* by BC1960 and BC3618 resulted in increased resistance to lysozyme digestion.

The unusual occurrence of multiple putative deacetylases in these *Bacillus* genomes has led to speculation that these enzymes may play different roles in cellular, developmental, or environmental biology of *B. cereus and B. anthracis*. These genes may represent paralogues that are involved in cell wall biosynthesis/modification in diverse events such as sporulation, spore germination, vegetative growth, and, of course, host-microbe interactions in the case of *B. anthracis* (185).

IMPACT OF GENOMICS ON CELL WALL SYNTHESIS IN *CORYNEBACTERINEAE*

Arguably the greatest impact of genomic technologies in this field has been on our understanding of cell wall biosynthesis in the *Corynebacterineae*. This taxon includes both *Corynebacterium* and *Mycobacterium*, genera that possess members of unsurpassed clinical importance. The species comprising the *Mycobacterium tuberculosis* complex (*Mycobacterium africanum*, *M. tuberculosis*, and *Mycobacterium bovis*) represent the single most frequent cause of mortality from infections with a single organism. Their ability to cause tuberculosis decades after initial infection and the deadly synergy that has developed with human immunodeficiency virus and AIDS are exacerbated by emerging multidrug resistance and represents an enormous public health problem worldwide (65, 67, 69, 164, 258). Nontuberculous opportunistic mycobacterial pathogens such as *Mycobacterium*

avium also commonly infect people with acquired immunodeficiency syndrome (118), *Mycobacterium leprae* maintains the global burden of leprosy (197), and other pathogenic members of the *Corynebacterineae*, such as *Corynebacterium diphtheriae*, remain prevalent (108, 144).

Many of the drugs used to treat tuberculosis inhibit biosynthesis of components of the extremely complex mycobacterial cell wall (Fig. 3A), a structure that is considered essential to its survival and pathogenicity (83). Briefly, a layer of mycolic acid, long-chain, branched β-hydroxy fatty acid residues, form the inner leaflet of an outer membrane. These are covalently tethered to the underlying PG via their esterification to an AG polysaccharide (Fig. 3B). The outer leaflet of the outer membrane is composed of a complex array of unusual lipids as well as glycoconjugates of mycolic acids. This membrane provides a highly effective permeability barrier to many drugs, phages, and components of the immune response. An architecturally similar but less complex cell wall core structure is conserved across the *Corynebacterineae*, which presented the possibility that comparative genomics might scout new routes toward our understanding of the construction of these fascinating structures (64).

To date, the genomes of nine members of *Corynebacterineae* have been annotated and published (40, 44, 66, 77, 84, 120, 133, 170, 229), and it is the comparison of *M. tuberculosis* with the corynebacteria and *M. leprae* that has proven most useful in this respect. *M. leprae* is an intracellular pathogen that has yet to be grown in axenic laboratory culture. The examination of its genome highlighted the high degree of genetic decay that has occurred with many genes being annotated as nonfunctional pseudogenes, and thus *M. leprae* has been thought of as maintaining a minimal gene set for mycobacterial pathogenesis (66). Despite this massive genetic decay, *M. leprae* possesses an intact cell wall and represents a useful reference when considering candidate genes for essential cell wall–related enzymes (243). The corynebacteria possess a similar cell wall core

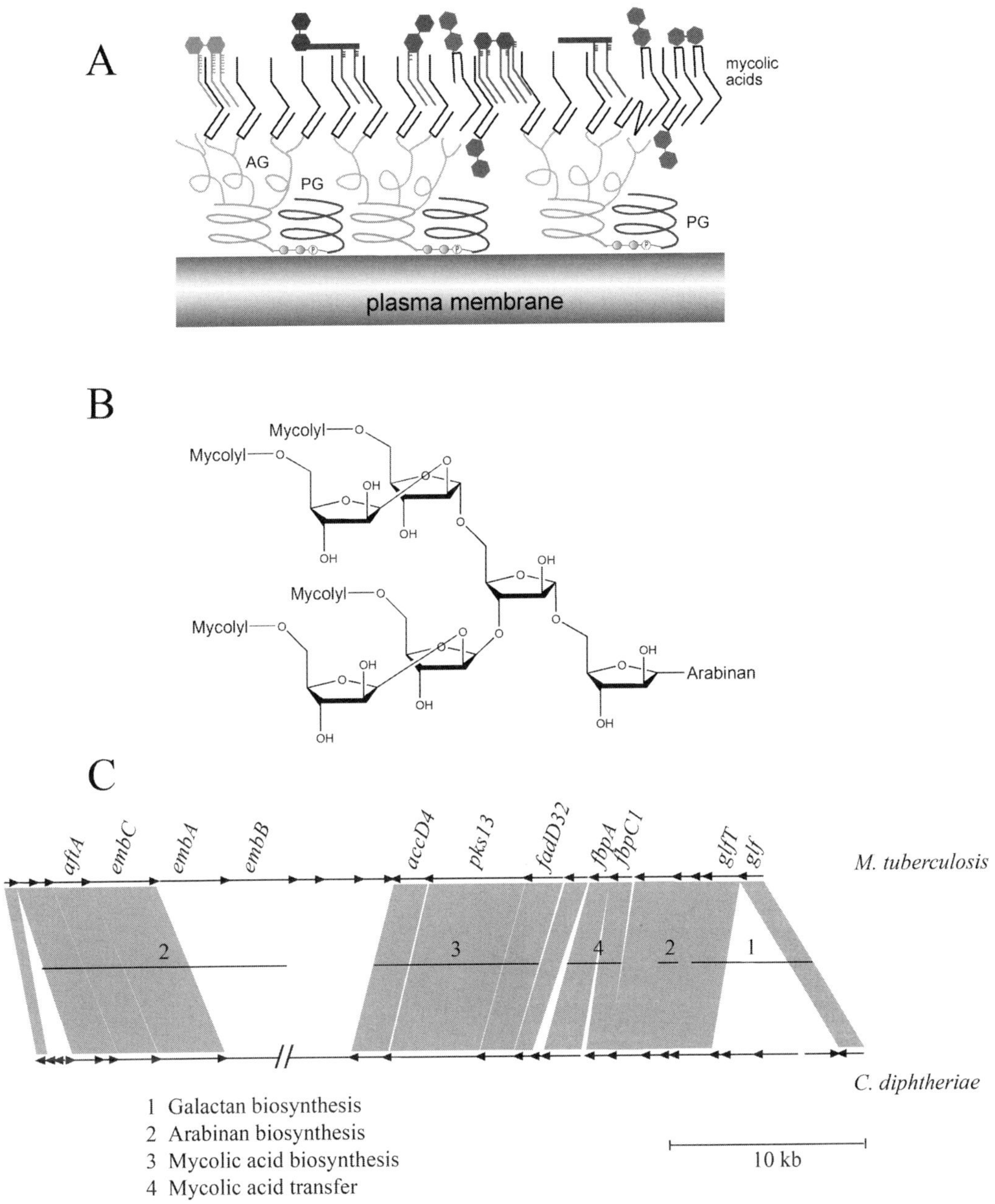

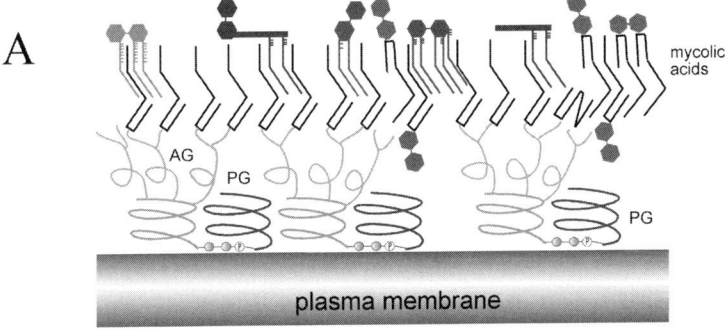

A

mycolic
acids

AG

PG

PG

plasma membrane

B

Mycolyl—O

Mycolyl—O

OH

OH

OH

Mycolyl—O

Mycolyl—O

OH

OH

OH

OH

OH

Arabinan

OH

C

aftA   embC   embA   embB   accD4   pks13   fadD32   fbpA fbpC1   glfT glf

M. tuberculosis

2   3   4   2   1

C. diphtheriae

1 Galactan biosynthesis
2 Arabinan biosynthesis
3 Mycolic acid biosynthesis
4 Mycolic acid transfer

10 kb

FIGURE 3 Architecture and genetic organization of *Mycobacterium tuberculosis* cell wall. (A) Adaptation of current model of cell wall (157), which favors the scaffold model of PG organization. The *M. tuberculosis* wall possesses a high proportion of covalently attached mycolic acid residues (black), which form the inner leaflet of an outer cell wall permeability barrier. The membranous structure is completed by intercalating trehalose-based mycolate containing glycolipids along with a diverse repertoire of complex lipids. The membrane structure is tethered to the peptidogly-can (PG, dark gray) of the murein sacculus via arabinogalactan (AG, light gray). The arabinan domain of the my-colylarabinogalactan (mAGP) is branched and linked to the coiled (in some recent models [58]) galactan domains that intercalate with the coiled glycan domains of PG; these polysaccharide chains are linked by the Rha-GlcNAc-phosphate linker unit shown. (B) Hexarabinofuranosyl motives representing sites for mycoalte deposition in the cell wall. (C) Conserved genetic organization of cell wall biosynthesis in *Corynebacterineae*. Probable orthologues are linked by gray bars. The central lines crossing these bars are numbered with respect to gene product function. In *C. diphtheriae*, a homologous continuum encompassing orthologues of *glf* through to *accD4* is disrupted by the insertion of four genes that appear to encode glycine-betaine production. A smaller cluster with similar genetic organization to the region Rv3789 to *embC* also occurs in *C. diphtheriae*, although some 480 kb distant.

to that of the mycobacteria, the principal apparent differences being the size of the mycolic acid residues and the reduced complexity of the free cell wall lipids (64).

Mycobacterial PG Synthesis

The PG found within the mycobacteria and other mycolata is unremarkable and similar to that of *E. coli* (117, 178, 257), although, in *M. tuberculosis*, the muramic acid residues can either be glycolylated or acetylated (10, 224) and the D-isoglutamate residue of the stem peptide can be amidated at its carboxyl group (91). The galactan domain of AG is linked via a disaccharide phosphate [α-*L*-Rha*p*-(1→3)-D-GlcNAc-(1→P)] linker unit (LU) to the C-6 of MurNGly residues. The galactan domain consists of a linear polymer of around 30 β-D-galactofuranose (β-D-Gal*f*) residues linked via alternating (1→5) and (1→6) glycosidic bonds (46). Periodically arabinan chains (three per galactan chain) are linked to the 5 position of some 6-linked Gal*f* residues (46), probably nearer to the reducing end of the galactan chain (25). These arabinan chains are almost entirely composed of α-D-Ara*f* residues. The bulk of the arabinan domain is polymerized via (1→5) glycosidic linkages, with branching being introduced by 3,5-α-D-Ara*f* residues (46). The nonreducing termini are decorated with a [β-D-Ara*f*-(1→2)-α-D-Ara*f*]₂-3,5-α-D-Ara*f*-(1→5)-α-D-Ara*f* motif (Fig. 3B), and within a particular arabinan chain, two-thirds of these terminal arabinofuranoside motifs are esterified with mycolates (46, 149).

Recent studies of the AG of *C. diphtheriae* showed that the linkage profile of its AG was broadly similar to that of *M. tuberculosis* (188). A unique feature of the *C. diphtheriae* AG was the absence of 3,5-linked Ara*f* residues, suggesting the lack of an α-D-(1→3)-arabinofuranosyltransferase activity and, as a consequence, a less elaborate linear terminal Ara*f* motif for corynomycolate deposition (188). The motif is, however, conserved in *C. glutamicum* (1).

Several PBPs have now been identified in mycobacteria (13, 14, 27, 91, 132, 137). Among these are two putative class A HMM-PBP of

M. tuberculosis, encoded by Rv0050 (*ponA*) and Rv3682. A soluble version of the *ponA* product (s-PBP1★) has been constructed and shown to bind benzylpenicillin (27), whereas Rv3682 has been proposed to be a penicillin-resistant PBP (91). Like in *E. coli*, during exponential growth, the cross-links between the stem peptides of mycobacteria and corynebacteria occur between the fourth residue (D-Ala) and the third (*meso*-diaminopimelate) of another. During stationary phase, several bacteria, including *M. tuberculosis*, alter their PG cross-linking pattern, incorporating increasing amounts of (3→3)-linked peptides along with the (4→3)-linked variety. This increasing frequency of (3→3) interpeptide bridges has been associated with penicillin resistance, and it is likely that these linkages are created by the penicillin-resistant HMM-PBPs such as Rv3682 (91).

Like many other bacteria (91, 92), both *M. tuberculosis* and *C. diphtheriae* appear to possess a single representative of three subclasses of class B HMM-PBPs. Rv2163*c* and DIP1604 both appear to encode class B3 HMM-PBPs (91), and their loci are associated with the *dcw* clusters of the respective organisms (Fig. 2). Similarly, *DIP55* and Rv0016 have been assigned to subclass B–like I (91), and these are clustered with orthologues encoding cell division and cell shape–determining factors such as FtsW and PknAB. Another class B HMM-PBP is apparently encoded by Rv2864c and DIP1497, and it is likely that these do not bind penicillins. Consequently, these have been assigned to the penicillin-resistant subclass B–like II (91). *M. tuberculosis* contains three genes encoding free-standing PBPs that may be involved as auxiliary cell cycle proteins in (4→3)-linked PG synthesis (91). These comprise genes encoding putative D-alanyl-D-alanine carboxypeptidases *dacB1* and *dacB2* and Rv3627c (91).

Clustering of AG Biosynthesis Genes

Belanger and Inamine suggested the possibility of a large AG biosynthetic gene cluster comprising 31 genes and stretching from Rv3779 to Rv3809c, covering 48.5 kb of the chromosome (Fig. 3C) (17). Among these genes are *glf* and *glfT* (galactan polymerization), *aftA* and *embCAB* (arabinan deposition), and *fbpA* (mycolyltransfer). Comparison of this region with the equivalent from *C. diphtheriae* showed that part of the arrangement is well conserved but with two major differences: a probable glycine-betaine synthesis cassette is inserted between the *glf* and *glfT* homologues, and the whole conserved region is split into two discontinuous segments resulting in the *emb* homologue of *C. diphtheriae* lying over 460 Kb away from the *glfT* homologue. One cluster consists of homologues of *embC*-Rv3789 (DIP0159 to DIP0166), which includes a three-gene insertion. The other cluster encodes orthologues of Glf, GlfT, Rv3805 to Rv3807, mycolyltransferases, and three genes that encode a fatty acid CoA ligase, a non-biotinylated CoA carboxylase subunit and a polyketide synthase (Pks13) recently implicated in mycolic acid formation (82, 183, 184, 234) (Fig. 3C). Analysis of the genes of this cluster in particular, as well as the genomes of these related organisms as a whole, via heterospecific cloning experiments and reverse-genetic strategies has vastly accelerated research toward a more complete understanding of the enzymology of mycobacterial cell wall biosynthesis, both from an intellectual standpoint and with regard to the urgent requirement for new antimycobacterial agents.

This initial step in the synthesis of AG involves the formation of a polyprenyl-P-P-GlcNAc unit (GL1), followed by the incorporation of Rha from dTDP-Rha (154). Subsequent addition of Gal*f* residues derived from UDP-Gal*p* via a UDP-galactopyranose mutase results in a highly polymerized lipid-linked (24, 155) polymer found to consist of 35 to 50 residues, and glycosidic linkage analysis indicated the presence of alternating 5- and 6-linked linear galactan components of mycobacterial cell walls with a small amount of branching. The incorporation of Ara from synthetic decaprenyl-P-Ara*f* (130) into the same polymer (155) has been achieved in vitro, sug-

gesting that the total synthesis of the arabinan domain of AG may occur while still linked to the polyprenyl-P (Pol-P) carrier.

The initial reaction in the glycosylation of the DPP carrier is the transfer of GlcNAc-1-phosphate from its UDP donor; the enzyme responsible remains unidentified. The complementation of a *wbbL* mutant of *E. coli,* deficient in Rha transfer in lipopolysaccharide biosynthesis, with Rv3265c implicates its product as the probable rhamnosyltransferase involved in linker unit synthesis (148). The enzymes providing the dTDP-Rha donor substrate, RmlA to RmlD, have also been identified, expressed in *E. coli* and identified as essential in mycobacteria (134–136).

Similarly, the provision of the nucleotide sugar donor for galactan deposition has also been defined in *M. tuberculosis*. Two reactions are required to reach UDP-Gal*f* from UDP-Glc*p;* the first is catalyzed by UDP-Glc*p* epimerase to form UDP-Gal*p*, which is then converted to the furanose form by UDP-Gal*p* mutase; both enzymes have been cloned and characterized (255). Several glycosyltransferase activities may be required to carry out the synthesis of the large Pol-P-LU-galactan polymers; a UDP-Gal*f*:Rha-GlcNAc-P-Pol Gal*f* transferase activity, β-D-Gal*f* transferases forming the alternating (1→5) and (1→6) internal glycosidic linkages of the galactan, and possibly a further transferase that deposits the terminal Gal*f* residue. Several candidate genes from *M. tuberculosis* were highlighted as potential Gal*f* transferases by using hydrophobic cluster analysis, which uses a plot of neighboring hydrophobic residues along a theoretical α-helix to suggest secondary structure features (38, 81, 131). Because sequence homology between glycosyltransferases is often very low, hydrophobic cluster analysis is particularly useful in categorizing these enzymes into functional classes (85). Genes from *M. tuberculosis* H37Rv showing homology to glycosyltransferases were analyzed by hydrophobic cluster analysis and the plots analyzed for recognized features of inverting glycosyltransferases (200), i.e., those that carry out sugar transfers in which the configuration of the anomeric carbon is inverted. These features include a domain A with alternating hydrophobic α-helices and β-strands in which the DXD motifs, commonly found in glycosyltransferases, are positioned close to the C-terminal end of strong vertical hydrophobic clusters (200). Several candidates were identified within the *M. tuberculosis* H37Rv genome satisfying these criteria. One in particular, Rv3808c, appeared to be a good candidate in that it occupied the locus adjacent to *glf*, encoding the UDP-galactopyranose mutase. Because these genes overlapped, it was considered likely that they comprised part of an operon. The putative transferase also contained the signature QXXRW motif of the less-well-defined domain B, which is found only in processive enzymes, i.e., those that carry out multiple sugar transfers (200). Molecular cloning and overexpression of Rv3808c were achieved by two groups (128, 155). Mikusova et al. (155) successfully overexpressed Rv3808c in *M. smegmatis*, observing an increased yield of a bona fide galactofuran in the overproducing strain. Kremer et al. (128) designed neoglycolipid acceptor substrates incorporating a Gal*f* disaccharide moiety, the sugars either being linked β-(1→5) or β-(1→6). Glycosidic linkage analysis of trisaccharide and tetrasaccharide products showed that the new linkages emulated the repeating nature of native galactan. Thus the product of Rv3808c was a processive enzyme capable of depositing alternating (1→5) and (1→6)-linked Gal*f* residues from UDP-galactose. This enzyme, now designated GlfT, is likely to be responsible for the bulk of galactan deposition in the cell wall of *M. tuberculosis*, but it still remains unclear whether it directs the deposition of the initial galactofuranose residue on the rhamnose residue of the linker unit. The importance of galactan synthesis to the growth of mycobacteria was demonstrated by the disruption of *glf* in *M. smegmatis* by allele replacement. Two recovery plasmids bearing the *M. tuberculosis glf* and *glfT* genes were required to isolate the mutant, suggesting a polar effect on *glfT* expression. The requirement for func-

tional copies of both genes for mycobacterial growth thus emphasizes the importance of galactan biosynthesis as a target for novel chemotherapeutics (173).

The origins of AG arabinfuranosyl residues is suggested to proceed via the nonoxidative pentose shunt route (202). Scherman et al. have shown that both Pol-P-Ara and Pol-P-Rib are derived from 5-phosphoribose pyrophosphate (PRPP) with a 2′ epimerase mediating the Rib→Ara conversion at an intermediate stage (203). The enzymes that direct the formation of Pol-P-Ara in *M. tuberculosis* have recently been identified. On the basis of similarity to proteins implicated in the arabinosylation of the *Azorhizobium caulidans* nodulation factor, three genes, Rv3790, Rv3791, and Rv3806c, were cloned from the *Mycobacterium tuberculosis* genome and their products characterized. The latter encodes an integral membrane protein that can form Pol-P-5-phosphoribose from PRPP and Pol-P (1, 111). Together, Rv3790 and Rv3791 are able to catalyze the 2′ epimerization of the ribosyl moiety, which occurs through oxidation of the C2-hydroxyl group to likely form decaprenylphosphoryl-2-keto-β-D-erythropentofuranose, which is then reduced to form Pol-P-β-D-Ara*f*, although neither protein individually is sufficient to support the activity (153).

The products of the *emb* locus of *M. avium* were identified as the targets for the antimycobacterial drug ethambutol. The inhibitory effect of ethambutol on the incorporation of Ara*f* from Pol-P-Ara into arabinan in vitro was partially recovered in extracts from cells overexpressing *embB* (16). In accord with these observations, many ethambutol resistance–associated mutations occur within *embB* and in particular within a predicted "periplasmic" region of the protein (189, 216, 230). *M. tuberculosis* possesses three closely related *emb* genes, clustered *embCAB*. A similar genetic organization occurs in *M. smegmatis*, and recent gene knockout studies in this organism have shed light on this apparent redundancy (70). Individual mutants inactivated in *embC*, *embA*, and *embB* were characterized. All three strains were viable, but of them, the *embB*-negative mutant was most profoundly affected. Compromised cell wall integrity was indicated as morphological changes, and the cells displayed increased sensitivity to hydrophobic drugs and detergents. The arabinose content of the AG was diminished for both the *embA* and *embB* strains, with considerable effects on the formation of the terminal hexaarabinofuranosyl motives, specifically the addition of the β-D-Araf-1→2-β-D-Araf disaccharide to the 3 position of the 3,5-linked Araf residue, resulting in a linear terminal motif. The AG formation of the *embC* strain, however, seemed unaffected, yet arabinan deposition in the lipoglycan lipoarabinomannan was abolished. These data confirmed that the mycobacterial Emb proteins are intimately involved in the process of cell wall arabinan deposition and that EmbA and EmbB are crucial to the formation of the hexaarabinofuranosyl motives of AG that are crucial for the esterification of mycolic acids. Because the expression of active Emb proteins in a heterologous system is still to be achieved, their role as arabinosyltransferases remains unconfirmed. Because these proteins contain 11 to 13 predicted transmembrane helices, the possibility arises that they could play a part in arabinan export rather than polymerization, or even combine both functions as a bifunctional protein. Another scenario is that EmbA and EmbB could combine an export role with the recruitment of appropriate, and as yet undefined, arabinosyltransferase activities specifical to AG synthesis. A similar role for EmbC could be envisaged in lipoarabinomannan biosynthesis.

The scenario in *C. diphtheriae* appeared vastly simplified by its possession of a single Emb homologue (DIP0159), which is most closely related to EmbC of the mycobacteria. Because *C. diphtheriae* AG contains no 3,5-linked Araf residues, this appears entirely consistent with the lack of EmbAB functions, i.e., the deposition of the β-D-Araf-1→2-β-D-Araf disaccharide to the C-3 position of the 3,5-linked Araf residue as in mycobacteria. Because *C. diphtheriae* does contain arabinan-containing lipoglycans, it is possible that this single Emb protein possesses broader acceptor specificity than its mycobacterial counterparts and can direct the arabinan deposition in both molecules but is unable to produce the more elaborate structures that gene duplications and the accumulation of point mutations may have allowed in the mycobacteria. However, *C. glutamicum* also carries a single *embC*-like gene and possesses a terminally branched arabinan, suggesting that the true scenario is more complex (1). Deletion of this gene has recently shed light on the priming of the arabinan domain of AG. Alderwick et al. established that, in contrast to mycobacterial *embA* and *embB* mutants, deletion of *C. glutamicum emb* led to a highly truncated AG possessing only t-Araf residues that indicated for the first time the exact site of attachment of arabinan chains in a corynebacterial AG and suggesting the presence of a novel enzyme responsible for "priming" the galactan domain for further elaboration by Emb (1)

SUMMARY

The use of genome-derived information has undoubtedly accelerated research into bacterial cell wall biosynthesis, providing extra impetus to research in the less-well-characterized bacteria. For instance, in the mycobacterial field, progress toward the definition of the biosynthesis of the essential AG polysaccharide has been achieved with great alacrity since the determination of the *M. tuberculosis* genome sequence. Similarly, recognition of new enzyme families in interpeptide synthesis and PG modification has been greatly facilitated by the use of genomic data. One of the most fascinating genome-driven findings sheds new light on the previously puzzling chlamydial paradox. The demonstration that these intriguing pathogens have maintained much of the genetic wherewithal to synthesize PG poses several questions. The discovery that antigenic material that is likely to be PG forms an interesting ring structure at midcell during division inspires the quest to define the presumably crucial role a vanishingly small amount of PG can play in the division process of a natural *ftsZ* mutant. In this era of rapidly emerging multidrug resistance, our efforts to understand bacterial pathogens through study of their cell wall biosynthesis, in identification of novel targets, by defining modes of action of current drugs, and by investigating the development of resistance must keep pace with our rapidly evolving adversaries. We must continue to make good use of genomic information to find and exploit the chinks in their armor that might push to the limit their notorious ability to bounce back.

ACKNOWLEDGMENT

L.G.D. thanks Prof. G. S. Besra for his continuing support through Medical Research Council (United Kingdom) funding.

REFERENCES

1. **Alderwick, L. J., E. Radmacher, M. Seidel, R. Gande, P. G. Hitchen, H. R. Morris, A. Dell, H. Sahm, L. Eggeling, and G. S. Besra.** 2005. Deletion of Cg-emb in *Corynebacterianeae* leads to a novel truncated cell wall arabinogalactan, whereas inactivation of Cg-ubiA results in an arabinan-deficient mutant with a cell wall galactan core. *J. Biol. Chem.* **280:**32362–32371.
2. **Amano, K., H. Hayashi, Y. Araki, and E. Ito.** 1977. The action of lysozyme on peptidoglycan with N-unsubstituted glucosamine residues. Isolation of glycan fragments and their susceptibility to lysozyme. *Eur. J. Biochem.* **76:**299–307.
3. **Anderson, J. S., M. Matsuhashi, M. A. Haskin, and J. L. Strominger.** 1965. Lipid-phosphoacetylmuramyl-pentapeptide and lipid-phosphodisaccharide-pentapeptide: presumed membrane transport intermediates in cell wall synthesis. *Proc. Natl. Acad. Sci. USA* **53:**881–889.
4. **Anderson, M. S., S. S. Eveland, H. R. Onishi, and D. L. Pompliano.** 1996. Kinetic mechanism of the *Escherichia coli* UDPMurNAc-tripeptide D-alanyl-D-alanine-adding enzyme: use of a glutathione S-transferase fusion. *Biochemistry* **35:**16264–16269.
5. **Araki, Y., S. Fukuoka, S. Oba, and E. Ito.** 1971. Enzymatic deacetylation of N-acetylglucosamine residues in peptidoglycan from *Bacillus cereus* cell walls. *Biochem. Biophys. Res. Commun.* **45:**751–758.
6. **Arbeloa, A., J. E. Hugonnet, A. C. Sentilhes, N. Josseaume, L. Dubost, C. Monsempes, D. Blanot, J. P. Brouard, and M. Arthur.** 2004. Synthesis of mosaic peptidoglycan cross-bridges by hybrid peptidoglycan assembly pathways in gram-positive bacteria. *J. Biol. Chem.* **279:**41546–41556.
7. **Arbeloa, A., H. Segal, J. E. Hugonnet, N. Josseaume, L. Dubost, J. P. Brouard, L. Gutmann, D. Mengin-Lecreulx, and M. Arthur.** 2004. Role of class A penicillin-binding proteins in PBP5-mediated beta-lactam resistance in *Enterococcus faecalis. J. Bacteriol.* **186:**1221–1228.
8. **Arthur, M., and P. Courvalin.** 1993. Genetics and mechanisms of glycopeptide resistance in enterococci. *Antimicrob. Agents Chemother.* **37:**1563–1571.
9. **Atrih, A., G. Bacher, G. Allmaier, M. P. Williamson, and S. J. Foster.** 1999. Analysis of peptidoglycan structure from vegetative cells of *Bacillus subtilis* 168 and role of PBP 5 in peptidoglycan maturation. *J. Bacteriol.* **181:**3956–3966.
10. **Azuma, I., D. W. Thomas, A. Adam, J. M. Ghuysen, R. Bonaly, J. F. Petit, and E. Lederer.** 1970. Occurrence of N-glycolylmuramic acid in bacterial cell walls. A preliminary survey. *Biochim. Biophys. Acta* **208:**444–451.
11. **Barbour, A. G., K. Amano, T. Hackstadt, L. Perry, and H. D. Caldwell.** 1982. *Chlamydia trachomatis* has penicillin-binding proteins but not detectable muramic acid. *J. Bacteriol.* **151:**420–428.
12. **Barnickel, G., D. Naumann, H. Bradaczek, H. Labischinski, and P. Glesbrecht.** 1983. Computer aided molecular modeling of the three-dimensional structure of bacterial peptidoglycan, p. 61–66. *In* R. Hakenbeck, J.-V. Holtje, and H. Labischinski (ed.), *The Target of Penicillin.* Walter de Gruyter & Co., Berlin, Germany.
13. **Basu, J., R. Chattopadhyay, M. Kundu, and P. Chakrabarti.** 1992. Purification and partial characterization of a penicillin-binding protein from *Mycobacterium smegmatis. J. Bacteriol.* **174:**4829–4832.
14. **Basu, J., S. Mahapatra, M. Kundu, S. Mukhopadhyay, M. Nguyen-Disteche, P. Dubois, B. Joris, J. Van Beeumen, S. T. Cole, P. Chakrabarti, and J. M. Ghuysen.** 1996. Identification and overexpression in *Escherichia coli* of a *Mycobacterium leprae* gene, *pon1*, encoding a high-molecular-mass class A penicillin-binding protein, PBP1. *J. Bacteriol.* **178:**1707–1711.
15. **Bavoil, P. M., R. Hsia, and D. M. Ojcius.** 2000. Closing in on *Chlamydia* and its intracellular bag of tricks. *Microbiology* **146:**2723–2731.
16. **Belanger, A. E., G. S. Besra, M. E. Ford, K. Mikusova, J. T. Belisle, P. J. Brennan, and J. M. Inamine.** 1996. The *embAB* genes of *Mycobacterium avium* encode an arabinosyl transferase involved in cell wall arabinan biosynthesis that is

the target for the antimycobacterial drug ethambutol. *Proc. Natl. Acad. Sci. USA* **93:**11919–11924.

17. **Belanger, A. E., and J. M. Inamine.** 2000. Genetics of cell wall biosnthesis. *In* G. F. Hatfull and
W. R. Jacobs, Jr. (ed.), *Molecular Genetics of Mycobacteria.* ASM Press, Washington, DC.

18. **Belland, R. J., G. Zhong, D. D. Crane, D.
Hogan, D. Sturdevant, J. Sharma, W. L.
Beatty, and H. D. Caldwell.** 2003. Genomic
transcriptional profiling of the developmental cy-
cle of *Chlamydia trachomatis. Proc. Natl. Acad. Sci.
USA* **100:**8478–8483.

19. **Benson, T. E., C. T. Walsh, and J. M. Hogle.**
1996. The structure of the substrate-free form of
MurB, an essential enzyme for the synthesis of
bacterial cell walls. *Structure* **4:**47–54.

20. **Bera, A., S. Herbert, A. Jakob, W. Vollmer,
and F. Gotz.** 2005. Why are pathogenic staphylococci so lysozyme resistant? The peptidoglycan
O-acetyltransferase OatA is the major determinant for lysozyme resistance of *Staphylococcus aureus. Mol. Microbiol.* **55:**778–787.

21. **Berger-Bachi, B.** 1983. Insertional inactivation
of staphylococcal methicillin resistance by Tn551.
J. Bacteriol. **154:**479–487.

22. **Berger-Bachi, B., L. Barberis-Maino, A.
Strassle, and F. H. Kayser.** 1989. FemA, a host-
mediated factor essential for methicillin resistance
in *Staphylococcus aureus*: molecular cloning and
characterization. *Mol. Gen. Genet.* **219:**263–269.

23. **Bertoldi, M., B. Cellini, A. Paiardini, M. Di
Salvo, and C. Borri Voltattorni.** 2003. *Treponema denticola* cystalysin exhibits significant alanine racemase activity accompanied by transami-
nation: mechanistic implications. *Biochem. J.* **371:**
473–483.

24. **Besra, G. S., and P. J. Brennan.** 1997. The
mycobacterial cell wall: biosynthesis of arabinogalactan and lipoarabinomannan. *Biochem. Soc.
Trans.* **25:**845–850.

25. **Besra, G. S., K. H. Khoo, M. R. McNeil, A.
Dell, H. R. Morris, and P. J. Brennan.** 1995. A
new interpretation of the structure of the my-
colyl-arabinogalactan complex of *Mycobacterium
tuberculosis* as revealed through characterization of
oligoglycosylalditol fragments by fast-atom bombardment mass spectrometry and 1H nuclear
magnetic resonance spectroscopy. *Biochemistry* **34:**
4257–4266.

26. **Beukes, M., and J. W. Hastings.** 2001. Self-pro-
tection against cell wall hydrolysis in *Streptococcus
milleri* NMSCC 061 and analysis of the millericin
B operon. *Appl. Environ. Microbiol.* **67:**3888–3896.

27. **Bhakta, S., and J. Basu.** 2002. Overexpression,
purification and biochemical characterization of a
class A high-molecular-mass penicillin-binding
protein (PBP), PBP1★ and its soluble derivative

from *Mycobacterium tuberculosis. Biochem. J.* **361:**
635–639.

28. **Bi, E. F., and J. Lutkenhaus.** 1991. FtsZ ring
structure associated with division in *Escherichia
coli. Nature* **354:**161–164.

29. **Blake, C. C., D. F. Koenig, G. A. Mair, A. C.
North, D. C. Phillips, and V. R. Sarma.** 1965.
Structure of hen egg-white lysozyme. A three-
dimensional Fourier synthesis at 2 Angstrom resolution. *Nature* **206:**757–761.

30. **Bouhss, A., N. Josseaume, D. Allanic, M.
Crouvoisier, L. Gutmann, J. L. Mainardi,
D. Mengin-Lecreulx, J. van Heijenoort,
and M. Arthur.** 2001. Identification of the
UDP-MurNAc-pentapeptide:L-alanine ligase for
synthesis of branched peptidoglycan precursors
in *Enterococcus faecalis. J. Bacteriol.* **183:**5122–
5127.

31. **Bouhss, A., N. Josseaume, A. Severin, K.
Tabei, J. E. Hugonnet, D. Shlaes, D. Mengin-
Lecreulx, J. Van Heijenoort, and M. Arthur.**
2002. Synthesis of the L-alanyl-L-alanine cross-
bridge of *Enterococcus faecalis* peptidoglycan. *J. Biol.
Chem.* **277:**45935–45941.

32. **Brandish, P. E., K. I. Kimura, M. Inukai, R.
Southgate, J. T. Lonsdale, and T. D. Bugg.**
1996. Modes of action of tunicamycin, li-
posidomycin B, and mureidomycin A: inhibition
of phospho-N-acetylmuramyl-pentapeptide translocase from *Escherichia coli. Antimicrob. Agents
Chemother.* **40:**1640–1644.

33. **Brown, E. D., J. L. Marquardt, J. P. Lee, C. T.
Walsh, and K. S. Anderson.** 1994. Detection
and characterization of a phospholactoyl-enzyme
adduct in the reaction catalyzed by UDP-N-
acetylglucosamine enolpyruvoyl transferase, MurZ.
Biochemistry **33:**10638–10645.

34. **Brown, W. J., and D. D. Rockey.** 2000. Identification of an antigen localized to an apparent
septum within dividing chlamydiae. *Infect. Immun.*
68:708–715.

35. **Bugg, T. D., and C. T. Walsh.** 1992. Intracellular steps of bacterial cell wall peptidoglycan
biosynthesis: enzymology, antibiotics, and antibiotic resistance. *Nat. Prod. Rep.* **9:**199–215.

36. **Bupp, K., and J. van Heijenoort.** 1993. The final step of peptidoglycan subunit assembly in
Escherichia coli occurs in the cytoplasm. *J. Bacteriol.*
175:1841–1843.

37. **Burman, L. G., and J. T. Park.** 1983. Changes
in the composition of *Escherichia coli* murein as it
ages during exponential growth. *J. Bacteriol.* **155:**
447–453.

38. **Callebaut, I., G. Labesse, P. Durand, A.
Poupon, L. Canard, J. Chomilier, B. Henrissat, and J. P. Mornon.** 1997. Deciphering protein sequence information through hydrophobic

cluster analysis (HCA): current status and perspectives. *Cell. Mol. Life Sci.* **53:**621–645.

39. **Carlson, J. H., S. F. Porcella, G. McClarty, and H. D. Caldwell.** 2005. Comparative genomic analysis of *Chlamydia trachomatis* oculotropic and genitotropic strains. *Infect. Immun.* **73:** 6407–6418.

40. **Cerdeno-Tarraga, A. M., A. Efstratiou, L. G. Dover, M. T. Holden, M. Pallen, S. D. Bentley, G. S. Besra, C. Churcher, K. D. James, A. De Zoysa, T. Chillingworth, A. Cronin, L. Dowd, T. Feltwell, N. Hamlin, S. Holroyd, K. Jagels, S. Moule, M. A. Quail, E. Rabbinowitsch, K. M. Rutherford, N. R. Thomson, L. Unwin, S. Whitehead, B. G. Barrell, and J. Parkhill.** 2003. The complete genome sequence and analysis of *Corynebacterium diphtheriae* NCTC13129. *Nucleic Acids Res.* **31:**6516–6523.

41. **Chopra, I., C. Storey, T. J. Falla, and J. H. Pearce.** 1998. Antibiotics, peptidoglycan synthesis and genomics: the chlamydial anomaly revisited. *Microbiology* **144**(Pt. 10):2673–2678.

42. **Christensen, B. G., W. J. Leanza, T. R. Beattie, A. A. Patchett, B. H. Arison, R. E. Ormond, F. A. Kuehl, Jr., G. Albers-Schonberg, and O. Jardetzky.** 1969. Phosphonomycin: structure and synthesis. *Science* **166:**123–125.

43. **Clarke, A. J., H. Strating, and N. T. Blackburn.** 2000. Pathways for O-acetylation of bacterial cell wall polymers, p. 187–212. *In* R. J. Doyle (ed.), *Glycomicrobiology.* Plenum Publishing Co. Ltd., New York.

44. **Cole, S. T., R. Brosch, J. Parkhill, T. Garnier, C. Churcher, D. Harris, S. V. Gordon, K. Eiglmeier, S. Gas, C. E. Barry III, F. Tekaia, K. Badcock, D. Basham, D. Brown, T. Chillingworth, R. Connor, R. Davies, K. Devlin, T. Feltwell, S. Gentles, N. Hamlin, S. Holroyd, T. Hornsby, K. Jagels, A. Krogh, J. McLean, S. Moule, L. Murphy, K. Oliver, J. Osborne, M. A. Quail, M. A. Rajandream, J. Rogers, S. Rutter, K. Seeger, J. Skelton, R. Squares, S. Squares, J. E. Sulston, K. Taylor, S. Whitehead, and B. G. Barrell.** 1998. Deciphering the biology of *Mycobacterium tuberculosis* from the complete genome sequence. *Nature* **393:**537–544.

45. **Coutinho, P. M., and B. Henrissat.** 1999. Carbohydrate-active enzymes: an integrated database approach, p. 3–12. *In* H. J. Gilbert, G. Davies, B. Henrissat, and B. Svensson (ed.), *Carbohydrate-active Enzymes: An Integrated Database Approach.* Royal Society of Chemistry, Cambridge, United Kingdom.

46. **Daffe, M., P. J. Brennan, and M. McNeil.** 1990. Predominant structural features of the cell wall arabinogalactan of *Mycobacterium tuberculosis* as revealed through characterization of oligoglycosyl alditol fragments by gas chromatography/mass spectrometry and by ^{1}H and ^{13}C NMR analyses. *J. Biol. Chem.* **265:**6734–6743.

47. **Dai, K., and J. Lutkenhaus.** 1991. ftsZ is an essential cell division gene in *Escherichia coli. J. Bacteriol.* **173:**3500–3506.

48. **Daniel, R. A., S. Drake, C. E. Buchanan, R. Scholle, and J. Errington.** 1994. The *Bacillus subtilis* spoVD gene encodes a mother-cell-specific penicillin-binding protein required for spore morphogenesis. *J. Mol. Biol.* **235:**209–220.

49. **Daniel, R. A., A. M. Williams, and J. Errington.** 1996. A complex four-gene operon containing essential cell division gene pbpB in *Bacillus subtilis. J. Bacteriol.* **178:**2343–2350.

50. **DeHart, H. P., H. E. Heath, L. S. Heath, P. A. LeBlanc, and G. L. Sloan.** 1995. The lysostaphin endopeptidase resistance gene (*epr*) specifies modification of peptidoglycan cross bridges in *Staphylococcus simulans* and *Staphylococcus aureus. Appl. Environ. Microbiol.* **61:**1475–1479.

51. **de Jonge, B. L., T. Sidow, Y. S. Chang, H. Labischinski, B. Berger-Bachi, D. A. Gage, and A. Tomasz.** 1993. Altered muropeptide composition in *Staphylococcus aureus* strains with an inactivated femA locus. *J. Bacteriol.* **175:**2779–2782.

52. **de Jonge, B. L., and A. Tomasz.** 1993. Abnormal peptidoglycan produced in a methicillin-resistant strain of *Staphylococcus aureus* grown in the presence of methicillin: functional role for penicillin-binding protein 2A in cell wall synthesis. *Antimicrob. Agents Chemother.* **37:**342–346.

53. **Dessen, A., N. Mouz, E. Gordon, J. Hopkins, and O. Dideberg.** 2001. Crystal structure of PBP2x from a highly penicillin-resistant *Streptococcus pneumoniae* clinical isolate: a mosaic framework containing 83 mutations. *J. Biol. Chem.* **276:**45106–45112.

54. **Dewar, S. J., K. J. Begg, and W. D. Donachie.** 1992. Inhibition of cell division initiation by an imbalance in the ratio of FtsA to FtsZ. *J. Bacteriol.* **174:**6314–6316.

55. **Dietrich, C. P., A. V. Colucci, and J. L. Strominger.** 1967. Biosynthesis of the peptidoglycan of bacterial cell walls. V. Separation of protein and lipid components of the particulate enzyme from Micrococcus lysodeikticus and purification of the endogenous lipid acceptors. *J. Biol. Chem.* **242:** 3218–3225.

56. **Dmitriev, B., F. Toukach, and S. Ehlers.** 2005. Towards a comprehensive view of the bacterial cell wall. *Trends Microbiol.* **13:**569–574.

57. **Dmitriev, B. A., S. Ehlers, and E. T. Rietschel.** 1999. Layered murein revisited: a fundamentally new concept of bacterial cell wall

structure, biogenesis and function. *Med. Microbiol. Immunol. (Berl.)* **187:**173–181.

58. **Dmitriev, B. A., S. Ehlers, E. T. Rietschel, and P. J. Brennan.** 2000. Molecular mechanics of the mycobacterial cell wall: from horizontal layers to vertical scaffolds. *Int. J. Med. Microbiol.* **290:**251–258.

59. **Dmitriev, B. A., F. V. Toukach, O. Holst, E. T. Rietschel, and S. Ehlers.** 2004. Tertiary structure of *Staphylococcus aureus* cell wall murein. *J. Bacteriol.* **186:**7141–7148.

60. **Dmitriev, B. A., F. V. Toukach, K. J. Schaper, O. Holst, E. T. Rietschel, and S. Ehlers.** 2003. Tertiary structure of bacterial murein: the scaffold model. *J. Bacteriol.* **185:**3458–3468.

61. **Donachie, W. D.** 1993. The cell cycle of *Escherichia coli. Annu. Rev. Microbiol.* **47:**199–230.

62. **Doublet, P., J. van Heijenoort, and D. Mengin-Lecreulx.** 1992. Identification of the *Escherichia coli murI* gene, which is required for the biosynthesis of D-glutamic acid, a specific component of bacterial peptidoglycan. *J. Bacteriol.* **174:** 5772–5779.

63. **Dougherty, T. J., J. A. Thanassi, and M. J. Pucci.** 1993. The *Escherichia coli* mutant requiring D-glutamic acid is the result of mutations in two distinct genetic loci. *J. Bacteriol.* **175:**111–116.

64. **Dover, L. G., A. M. Cerdeno-Tarraga, M. J. Pallen, J. Parkhill, and G. S. Besra.** 2004. Comparative cell wall core biosynthesis in the mycolated pathogens, *Mycobacterium tuberculosis* and *Corynebacterium diphtheriae. FEMS Microbiol. Rev.* **28:**225–250.

65. **Dye, C., S. Scheele, P. Dolin, V. Pathania, and M. C. Raviglione.** 1999. Consensus statement. Global burden of tuberculosis: estimated incidence, prevalence, and mortality by country. WHO Global Surveillance and Monitoring Project. *JAMA* **282:**677–686.

66. **Eiglmeier, K., J. Parkhill, N. Honore, T. Garnier, F. Tekaia, A. Telenti, P. Klatser, K. D. James, N. R. Thomson, P. R. Wheeler, C. Churcher, D. Harris, K. Mungall, B. G. Barrell, and S. T. Cole.** 2001. The decaying genome of *Mycobacterium leprae. Lepr. Rev.* **72:**387–398.

67. **Ellner, J. J., A. R. Hinman, S. W. Dooley, M. A. Fischl, K. A. Sepkowitz, M. J. Goldberger, T. M. Shinnick, M. D. Iseman, and W. R. Jacobs, Jr.** 1993. Tuberculosis symposium: emerging problems and promise. *J. Infect. Dis.* **168:** 537–551.

68. **El Zoeiby, A., F. Sanschagrin, and R. C. Levesque.** 2003. Structure and function of the Mur enzymes: development of novel inhibitors. *Mol. Microbiol.* **47:**1–12.

69. **Enarson, D. A., and J. F. Murray.** 1996. Global epidemiology of tuberculosis, p. 3–11. *In* B. R.

Bloom (ed.), *Tuberculosis: Pathogenesis, Protection and Control.* ASM Press, Washington, DC.

70. **Escuyer, V. E., M. A. Lety, J. B. Torrelles, K. H. Khoo, J. B. Tang, C. D. Rithner, C. Frehel, M. R. McNeil, P. J. Brennan, and D. Chatterjee.** 2001. The role of the *embA* and *embB* gene products in the biosynthesis of the terminal hexaarabinofuranosyl motif of *Mycobacterium smegmatis arabinogalactan. J. Biol. Chem.* **276:**48854–48862.

71. **Everett, K. D., and T. P. Hatch.** 1995. Architecture of the cell envelope of *Chlamydia psittaci* 6BC. *J. Bacteriol.* **177:**877–882.

72. **Filipe, S. R., E. Severina, and A. Tomasz.** 2000. Distribution of the mosaic structured *murM* genes among natural populations of *Streptococcus pneumoniae. J. Bacteriol.* **182:**6798–6805.

73. **Filipe, S. R., E. Severina, and A. Tomasz.** 2001. The role of murMN operon in penicillin resistance and antibiotic tolerance of *Streptococcus pneumoniae. Microb. Drug Resist.* **7:**303–316.

74. **Filipe, S. R., and A. Tomasz.** 2000. Inhibition of the expression of penicillin resistance in *Streptococcus pneumoniae* by inactivation of cell wall muropeptide branching genes. *Proc. Natl. Acad. Sci. USA* **97:**4891–4896.

75. **Firmin, J. L., K. E. Wilson, R. W. Carlson, A. E. Davies, and J. A. Downie.** 1993. Resistance to nodulation of cv. Afghanistan peas is overcome by nodX, which mediates an O-acetylation of the *Rhizobium leguminosarum* lipo-oligosaccharide nodulation factor. *Mol. Microbiol.* **10:**351–360.

76. **Fleischmann, R. D., M. D. Adams, O. White, R. A. Clayton, E. F. Kirkness, A. R. Kerlavage, C. J. Bult, J. F. Tomb, B. A. Dougherty, J. M. Merrick, et al.** 1995. Whole-genome random sequencing and assembly of *Haemophilus influenzae* Rd. *Science* **269:**496–512.

77. **Fleischmann, R. D., D. Alland, J. A. Eisen, L. Carpenter, O. White, J. Peterson, R. DeBoy, R. Dodson, M. Gwinn, D. Haft, E. Hickey, J. F. Kolonay, W. C. Nelson, L. A. Umayam, M. Ermolaeva, S. L. Salzberg, A. Delcher, T. Utterback, J. Weidman, H. Khouri, J. Gill, A. Mikula, W. Bishai, W. R. Jacobs, Jr., J. C. Venter, and C. M. Fraser.** 2002. Whole-genome comparison of *Mycobacterium tuberculosis* clinical and laboratory strains. *J. Bacteriol.* **184:**5479–5490.

78. **Fotheringham, I. G., S. A. Bledig, and P. P. Taylor.** 1998. Characterization of the genes encoding D-amino acid transaminase and glutamate racemase, two D-glutamate biosynthetic enzymes of *Bacillus sphaericus* ATCC 10208. *J. Bacteriol.* **180:**4319–4323.

79. **Franklin, M. J., and D. E. Ohman.** 1996. Identification of algI and algJ in the *Pseudomonas aeruginosa* alginate biosynthetic gene cluster which are

required for alginate *O* acetylation. *J. Bacteriol.* **178:**2186–2195.

80. **Fuchs-Cleveland, E., and C. Gilvarg.** 1976. Oligomeric intermediate in peptidoglycan biosynthesis in *Bacillus megaterium. Proc. Natl. Acad. Sci. USA* **73:**4200–4204.

81. **Gaboriaud, C., V. Bissery, T. Benchetrit, and J. P. Mornon.** 1987. Hydrophobic cluster analysis: an efficient new way to compare and analyse amino acid sequences. *FEBS Lett.* **224:**149–155.

82. **Gande, R., K. J. Gibson, A. K. Brown, K. Krumbach, L. G. Dover, H. Sahm, S. Shioyama, T. Oikawa, G. S. Besra, and L. Eggeling.** 2004. Acyl-CoA carboxylases (accD2 and accD3), together with a unique polyketide synthase (Cg-pks), are key to mycolic acid biosynthesis in *Corynebacterianeae* such as *Corynebacterium glutamicum* and *Mycobacterium tuberculosis. J. Biol. Chem.* **279:**44847–44857.

83. **Gao, L. Y., F. Laval, E. H. Lawson, R. K. Groger, A. Woodruff, J. H. Morisaki, J. S. Cox, M. Daffe, and E. J. Brown.** 2003. Requirement for kasB in *Mycobacterium* mycolic acid biosynthesis, cell wall impermeability and intracellular survival: implications for therapy. *Mol. Microbiol.* **49:**1547–1563.

84. **Garnier, T., K. Eiglmeier, J. C. Camus, N. Medina, H. Mansoor, M. Pryor, S. Duthoy, S. Grondin, C. Lacroix, C. Monsempe, S. Simon, B. Harris, R. Atkin, J. Doggett, R. Mayes, L. Keating, P. R. Wheeler, J. Parkhill, B. G. Barrell, S. T. Cole, S. V. Gordon, and R. G. Hewinson.** 2003. The complete genome sequence of *Mycobacterium bovis. Proc. Natl. Acad. Sci. USA* **100:**7877–7882.

85. **Geremia, R. A., E. A. Petroni, L. Ielpi, and B. Henrissat.** 1996. Towards a classification of glycosyltransferases based on amino acid sequence similarities: prokaryotic alpha-mannosyltransferases. *Biochem. J.* **318**(Pt. 1)**:**133–138.

86. **Ghuysen, J. M.** 1991. Serine beta-lactamases and penicillin-binding proteins. *Annu. Rev. Microbiol.* **45:**37-67.

87. **Ghuysen, J. M.** 1968. Use of bacteriolytic enzymes in determination of wall structure and their role in cell metabolism. *Bacteriol. Rev.* **32:**425–464.

88. **Ghuysen, J. M., and C. Goffin.** 1999. Lack of cell wall peptidoglycan versus penicillin sensitivity: new insights into the chlamydial anomaly. *Antimicrob. Agents Chemother.* **43:**2339–2344.

89. **Glauner, B., and U. Schwarz.** 1983. The analysis of murein composition with high-pressure-liquid chromatography, p. 29–34. *In* R. Hakenbeck, J.-V. Holtje, and H. Labischinski (ed.), *The Target of Penicillin.* Walter de Gruyter & Co., Berlin, Germany.

90. **Gmeiner, J., P. Essig, and H. H. Martin.** 1982. Characterization of minor fragments after digestion of *Escherichia coli* murein with endo-*N,O*-diacetylmuramidase from *Chalaropsis*, and determination of glycan chain length. *FEBS Lett.* **138:**109–112.

91. **Goffin, C., and J. M. Ghuysen.** 2002. Biochemistry and comparative genomics of SxxK superfamily acyltransferases offer a clue to the mycobacterial paradox: presence of penicillin-susceptible target proteins versus lack of efficiency of penicillin as therapeutic agent. *Microbiol. Mol. Biol. Rev.* **66:**702–738.

92. **Goffin, C., and J. M. Ghuysen.** 1998. Multimodular penicillin-binding proteins: an enigmatic family of orthologs and paralogs. *Microbiol. Mol. Biol. Rev.* **62:**1079–1093.

93. **Gunetileke, K. G., and R. A. Anwar.** 1966. Biosynthesis of uridine diphospho-*N*-acetyl muramic acid. *J. Biol. Chem.* **241:**5740–5743.

94. **Gunetileke, K. G., and R. A. Anwar.** 1968. Biosynthesis of uridine diphospho-*N*-acetylmuramic acid. II. Purification and properties of pyruvate-uridine diphospho-*N*-acetylglucosamine transferase and characterization of uridine diphospho-*N*-acetylenopyruvylglucosamine. *J. Biol. Chem.* **243:**5770–5778.

95. **Hakenbeck, R., N. Balmelle, B. Weber, C. Gardes, W. Keck, and A. de Saizieu.** 2001. Mosaic genes and mosaic chromosomes: intra- and interspecies genomic variation of *Streptococcus pneumoniae. Infect. Immun.* **69:**2477–2486.

96. **Hakenbeck, R., and J. Coyette.** 1998. Resistant penicillin-binding proteins. *Cell. Mol. Life Sci.* **54:**332–340.

97. **Hamase, K., R. Konno, A. Morikawa, and K. Zaitsu.** 2005. Sensitive determination of D–amino acids in mammals and the effect of D–amino-acid oxidase activity on their amounts. *Biol. Pharm. Bull.* **28:**1578–1584.

98. **Harz, H., K. Burgdorf, and J. V. Holtje.** 1990. Isolation and separation of the glycan strands from murein of *Escherichia coli* by reversed-phase high-performance liquid chromatography. *Anal. Biochem.* **190:**120–128.

99. **Hatch, T.** 1998. Chlamydia: old ideas crushed, new mysteries bared. *Science* **282:**638–639.

100. **Hatch, T. P.** 1996. Disulfide cross-linked envelope proteins: the functional equivalent of peptidoglycan in chlamydiae? *J. Bacteriol.* **178:**1–5.

101. **Hayashi, H., K. Amano, Y. Araki, and E. Ito.** 1973. Action of lysozyme on oligosaccharides from peptidoglycan *N*-unacetylated at glucosamine residues. *Biochem. Biophys. Res. Commun.* **50:**641–648.

102. **Hayashi, H., Y. Araki, and E. Ito.** 1973. Occurrence of glucosamine residues with free amino groups in cell wall peptidoglycan from bacilli as a factor responsible for resistance to lysozyme. *J. Bacteriol.* **113**:592–598.

103. **Heath, H. E., L. S. Heath, J. D. Nitterauer, K. E. Rose, and G. L. Sloan.** 1989. Plasmid-encoded lysostaphin endopeptidase resistance of *Staphylococcus simulans* biovar staphylolyticus. *Biochem. Biophys. Res. Commun.* **160**:1106–1109.

104. **Hendlin, D., B. M. Frost, E. Thiele, H. Kropp, M. E. Valiant, B. Pelak, B. Weissberger, C. Cornin, and A. K. Miller.** 1969. Phosphonomycin. 3. Evaluation in vitro. *Antimicrob. Agents Chemother.* **9**:297–302.

105. **Hesek, D., M. Lee, K. Morio, and S. Mobashery.** 2004. Synthesis of a fragment of bacterial cell wall. *J. Org. Chem.* **69**:2137–2146.

106. **Hesse, L., J. Bostock, S. Dementin, D. Blanot, D. Mengin-Lecreulx, and I. Chopra.** 2003. Functional and biochemical analysis of *Chlamydia trachomatis* MurC, an enzyme displaying UDP-*N*-acetylmuramate:amino acid ligase activity. *J. Bacteriol.* **185**:6507–6512.

107. **Hiramatsu, K., L. Cui, M. Kuroda, and T. Ito.** 2001. The emergence and evolution of methicillin-resistant *Staphylococcus aureus. Trends Microbiol.* **9**:486–493.

108. **Holmes, R. K.** 2000. Biology and molecular epidemiology of diphtheria toxin and the *tox* gene. *J. Infect. Dis.* **181**(Suppl. 1):S156–S167.

109. **Holtje, J. V.** 1996. A hypothetical holoenzyme involved in the replication of the murein sacculus of *Escherichia coli. Microbiology* **142**(Pt. 8): 1911–1918.

110. **How, S. J., D. Hobson, and C. A. Hart.** 1984. Studies in vitro of the nature and synthesis of the cell wall of *Chlamydia trachomatis. Curr. Microbiol.* **10**:269–274.

111. **Huang, H., M. S. Scherman, W. D'Haeze, D. Vereecke, M. Holsters, D. C. Crick, and M. R. McNeil.** 2005. Identification and active expression of the *Mycobacterium tuberculosis* gene encoding 5-phospho-{alpha}-D-ribose-1-diphosphate: decaprenyl-phosphate 5-phospho-ribosyltransferase, the first enzyme committed to decaprenylphosphoryl-D-arabinose synthesis. *J. Biol. Chem.* **280**:24539–24543.

112. **Hurlimann-Dalel, R. L., C. Ryffel, F. H. Kayser, and B. Berger-Bachi.** 1992. Survey of the methicillin resistance–associated genes mecA, mecR1-mecI, and femA-femB in clinical isolates of methicillin-resistant *Staphylococcus aureus. Antimicrob. Agents Chemother.* **36**:2617–2621.

113. **Ikeda, M., M. Wachi, H. K. Jung, F. Ishino, and M. Matsuhashi.** 1991. The *Escherichia coli* *mraY* gene encoding UDP-*N*-acetylmuramoyl-pentapeptide: undecaprenyl-phosphate phospho-*N*-acetylmuramoyl-pentapeptide transferase. *J. Bacteriol.* **173**:1021–1026.

114. **Ito, E., and J. L. Strominger.** 1960. Enzymatic synthesis of the peptide in a uridine nucleotide from *Staphylococcus aureus. J. Biol. Chem.* **235**:PC5–PC6.

115. **Itoh, T., K. Takemoto, H. Mori, and T. Gojobori.** 1999. Evolutionary instability of operon structures disclosed by sequence comparisons of complete microbial genomes. *Mol. Biol. Evol.* **16**:332–346.

116. **Ivanova, N., A. Sorokin, I. Anderson, N. Galleron, B. Candelon, V. Kapatral, A. Bhattacharyya, G. Reznik, N. Mikhailova, A. Lapidus, L. Chu, M. Mazur, E. Goltsman, N. Larsen, M. D'Souza, T. Walunas, Y. Grechkin, G. Pusch, R. Haselkorn, M. Fonstein, S. D. Ehrlich, R. Overbeek, and N. Kyrpides.** 2003. Genome sequence of *Bacillus cereus* and comparative analysis with *Bacillus anthracis. Nature* **423**:87–91.

117. **Janczura, E., M. Leyh-Bouille, C. Cocito, and J. M. Ghuysen.** 1981. Primary structure of the wall peptidoglycan of leprosy-derived corynebacteria. *J. Bacteriol.* **145**:775–779.

118. **Jones, D., and D. V. Havlir.** 2002. Nontuberculous mycobacteria in the HIV infected patient. *Clin. Chest Med.* **23**:665–674.

119. **Joseleau-Petit, D., D. Thevenet, and R. D'Ari.** 1994. ppGpp concentration, growth without PBP2 activity, and growth-rate control in *Escherichia coli. Mol. Microbiol.* **13**:911–917.

120. **Kalinowski, J., B. Bathe, D. Bartels, N. Bischoff, M. Bott, A. Burkovski, N. Dusch, L. Eggeling, B. J. Eikmanns, L. Gaigalat, A. Goesmann, M. Hartmann, K. Huthmacher, R. Kramer, B. Linke, A. C. McHardy, F. Meyer, B. Mockel, W. Pfefferle, A. Puhler, D. A. Rey, C. Ruckert, O. Rupp, H. Sahm, V. F. Wendisch, I. Wiegrabe, and A. Tauch.** 2003. The complete *Corynebacterium glutamicum* ATCC 13032 genome sequence and its impact on the production of L-aspartate-derived amino acids and vitamins. *J. Biotechnol.* **104**:5–25.

121. **Kamiryo, T., and M. Matsuhashi.** 1969. Sequential addition of glycine from glycyl-tRNA to the lipid-linked precursors of cell wall peptidoglycan in *Staphylococcus aureus. Biochem. Biophys. Res. Commun.* **36**:215–222.

122. **Kato, J., H. Suzuki, and Y. Hirota.** 1985. Dispensability of either penicillin-binding protein-1a or -1b involved in the essential process for cell elongation in *Escherichia coli. Mol. Gen. Genet.* **200**:272–277.

123. Kobayashi, N., H. Wu, K. Kojima, K. Taniguchi, S. Urasawa, N. Uehara, Y. Omizu, Y. Kishi, A. Yagihashi, and I. Kurokawa. 1994. Detection of *mecA*, *femA*, and *femB* genes in clinical strains of staphylococci using polymerase chain reaction. *Epidemiol. Infect.* **113:** 259–266.

124. Koch, A. L. 1998. Orientation of the peptidoglycan chains in the sacculus of *Escherichia coli*. *Res. Microbiol.* **149:**689–701.

125. Konno, R., A. Niwa, and Y. Yasumura. 1990. Intestinal bacterial origin of D-alanine in urine of mutant mice lacking D-amino-acid oxidase. *Biochem. J.* **268:**263–265.

126. Kopp, U., M. Roos, J. Wecke, and H. Labischinski. 1996. Staphylococcal peptidoglycan interpeptide bridge biosynthesis: a novel antistaphylococcal target? *Microb. Drug Resist.* **2:** 29–41.

127. Koronakis, V., J. Eswaran, and C. Hughes. 2004. Structure and function of TolC: the bacterial exit duct for proteins and drugs. *Annu. Rev. Biochem.* **73:**467–489.

128. Kremer, L., L. G. Dover, C. Morehouse, P. Hitchin, M. Everett, H. R. Morris, A. Dell, P. J. Brennan, M. R. McNeil, C. Flaherty, K. Duncan, and G. S. Besra. 2001. Galactan biosynthesis in *Mycobacterium tuberculosis*. Identification of a bifunctional UDP-galactofuranosyltransferase. *J. Biol. Chem.* **276:**26430–26440.

129. Kuroda, M., T. Ohta, I. Uchiyama, T. Baba, H. Yuzawa, I. Kobayashi, L. Cui, A. Oguchi, K. Aoki, Y. Nagai, J. Lian, T. Ito, M. Kanamori, H. Matsumaru, A. Maruyama, H. Murakami, A. Hosoyama, Y. Mizutani-Ui, N. K. Takahashi, T. Sawano, R. Inoue, C. Kaito, K. Sekimizu, H. Hirakawa, S. Kuhara, S. Goto, J. Yabuzaki, M. Kanehisa, A. Yamashita, K. Oshima, K. Furuya, C. Yoshino, T. Shiba, M. Hattori, N. Ogasawara, H. Hayashi, and K. Hiramatsu. 2001. Whole genome sequencing of meticillin-resistant *Staphylococcus aureus*. *Lancet* **357:**1225–1240.

130. Lee, R. E., K. Mikusova, P. J. Brennan, and G. S. Besra. 1995. Synthesis of the mycobacterial arabinose donor beta-D-arabinofuranosyl-1-monophosphoryldecaprenol, development of a basic arabinosyltransferase assy, and identification of ethambutol as an arabinosyltransferase inhibitor. *J. Am. Chem. Soc.* **117:**11829–11832.

131. Lemesle-Varloot, L., B. Henrissat, C. Gaboriaud, V. Bissery, A. Morgat, and J. P. Mornon. 1990. Hydrophobic cluster analysis: procedures to derive structural and functional information from 2-D-representation of protein sequences. *Biochimie* **72:**555–574.

132. Lepage, S., P. Dubois, T. K. Ghosh, B. Joris, S. Mahapatra, M. Kundu, J. Basu, P. Chakrabarti, S. T. Cole, M. Nguyen-Disteche, and J. M. Ghuysen. 1997. Dual multimodular class A penicillin-binding proteins in *Mycobacterium leprae*. *J. Bacteriol.* **179:**4627–4630.

133. Li, L., J. P. Bannantine, Q. Zhang, A. Amonsin, B. J. May, D. Alt, N. Banerji, S. Kanjilal, and V. Kapur. 2005. The complete genome sequence of *Mycobacterium avium* subspecies paratuberculosis. *Proc. Natl. Acad. Sci. USA* **102:**12344–12349.

134. Ma, Y., J. A. Mills, J. T. Belisle, V. Vissa, M. Howell, K. Bowlin, M. S. Scherman, and M. McNeil. 1997. Determination of the pathway for rhamnose biosynthesis in mycobacteria: cloning, sequencing and expression of the *Mycobacterium tuberculosis* gene encoding alpha-D-glucose-1-phosphate thymidylyltransferase. *Microbiology* **143**(Pt. 3):937–945.

135. Ma, Y., F. Pan, and M. McNeil. 2002. Formation of dTDP-rhamnose is essential for growth of mycobacteria. *J. Bacteriol.* **184:**3392-3395.

136. Ma, Y., R. J. Stern, M. S. Scherman, V. D. Vissa, W. Yan, V. C. Jones, F. Zhang, S. G. Franzblau, W. H. Lewis, and M. R. McNeil. 2001. Drug targeting *Mycobacterium tuberculosis* cell wall synthesis: genetics of dTDP-rhamnose synthetic enzymes and development of a microtiter plate-based screen for inhibitors of conversion of dTDP-glucose to dTDP-rhamnose. *Antimicrob. Agents Chemother.* **45:**1407–1416.

137. Mahapatra, S., S. Bhakta, J. Ahamed, and J. Basu. 2000. Characterization of derivatives of the high-molecular-mass penicillin-binding protein (PBP) 1 of *Mycobacterium leprae*. *Biochem. J.* **350**(Pt. 1):75–80.

138. Maidhof, H., B. Reinicke, P. Blumel, B. Berger-Bachi, and H. Labischinski. 1991. *femA*, which encodes a factor essential for expression of methicillin resistance, affects glycine content of peptidoglycan in methicillin-resistant and methicillin-susceptible *Staphylococcus aureus* strains. *J. Bacteriol.* **173:**3507–3513.

139. Matias, V. R., A. Al-Amoudi, J. Dubochet, and T. J. Beveridge. 2003. Cryo-transmission electron microscopy of frozen-hydrated sections of *Escherichia coli* and *Pseudomonas aeruginosa*. *J. Bacteriol.* **185:**6112–6118.

140. Matias, V. R., and T. J. Beveridge. 2005. Cryo-electron microscopy reveals native polymeric cell wall structure in *Bacillus subtilis* 168 and the existence of a periplasmic space. *Mol. Microbiol.* **56:**240–251.

141. Matias, V. R., and T. J. Beveridge. 2006. Native cell wall organization shown by cryo-electron microscopy confirms the existence of a

periplasmic space in *Staphylococcus aureus. J. Bacteriol.* **188:**1011–1021.

142. **Matsuhashi, M., C. P. Dietrich, and J. L. Strominger.** 1965. Incorporation of glycine into the cell wall glycopeptide in *Staphylococcus aureus*: role of sRNA and lipid intermediates. *Proc. Natl. Acad. Sci. USA* **54:**587–594.

143. **Matsumoto, A., and G. P. Manire.** 1970. Electron microscopic observations on the effects of penicillin on the morphology of *Chlamydia psittaci. J. Bacteriol.* **101:**278–285.

144. **Mattos-Guaraldi, A. L., L. O. Moreira, P. V. Damasco, and R. Hirata Junior.** 2003. Diphtheria remains a threat to health in the developing world—an overview. *Mem. Inst. Oswaldo Cruz* **98:**987–993.

145. **McCoy, A. J., and A. T. Maurelli.** 2006. Building the invisible wall: updating the chlamydial peptidoglycan anomaly. *Trends Microbiol.* **14:**70–77.

146. **McCoy, A. J., and A. T. Maurelli.** 2005. Characterization of *Chlamydia* MurC-Ddl, a fusion protein exhibiting D-alanyl-D-alanine ligase activity involved in peptidoglycan synthesis and D-cycloserine sensitivity. *Mol. Microbiol.* **57:**41–52.

147. **McCoy, A. J., R. C. Sandlin, and A. T. Maurelli.** 2003. In vitro and in vivo functional activity of *Chlamydia* MurA, a UDP-*N*-acetylglucosamine enolpyruvyl transferase involved in peptidoglycan synthesis and fosfomycin resistance. *J. Bacteriol.* **185:**1218–1228.

148. **McNeil, M.** 1999. Arabinogalactan in mycobacteria: structure, biosynthesis and genetics, p. 207–223. *In* J. B. Goldberg (ed.), *Genetics of Bacterial Polysaccharides.* CRC Press, Washington, DC.

149. **McNeil, M. R., K. G. Robuck, M. Harter, and P. J. Brennan.** 1994. Enzymatic evidence for the presence of a critical terminal hexa-arabinoside in the cell walls of *Mycobacterium tuberculosis. Glycobiology* **4:**165–173.

150. **McPherson, D. C., and D. L. Popham.** 2003. Peptidoglycan synthesis in the absence of class A penicillin-binding proteins in *Bacillus subtilis. J. Bacteriol.* **185:**1423–1431.

151. **Mengin-Lecreulx, D., L. Texier, M. Rousseau, and J. van Heijenoort.** 1991. The *murG* gene of *Escherichia coli* codes for the UDP-*N*-acetylglucosamine: *N*-acetylmuramyl-(pentapeptide) pyrophosphoryl-undecaprenol *N*-acetylglucosamine transferase involved in the membrane steps of peptidoglycan synthesis. *J. Bacteriol.* **173:**4625–4636.

152. **Meroueh, S. O., K. Z. Bencze, D. Hesek, M. Lee, J. Fisher, T. L. Stemmier, and S. Mobashery.** 2006. Three-dimensional structure of the bacterial cell wall peptidoglycan. *Proc. Natl. Acad. Sci. USA* **103:**4404–4409.

153. **Mikusova, K., H. Huang, T. Yagi, M. Holsters, D. Vereecke, W. D'Haeze, M. S. Scherman, P. J. Brennan, M. R. McNeil, and D. C. Crick.** 2005. Decaprenylphosphoryl arabinofuranose, the donor of the D-arabinofuranosyl residues of mycobacterial arabinan, is formed via a two-step epimerization of decaprenylphosphoryl ribose. *J. Bacteriol.* **187:**8020–8025.

154. **Mikusova, K., M. Mikus, G. S. Besra, I. Hancock, and P. J. Brennan.** 1996. Biosynthesis of the linkage region of the mycobacterial cell wall. *J. Biol. Chem.* **271:**7820-7828.

155. **Mikusova, K., T. Yagi, R. Stern, M. R. McNeil, G. S. Besra, D. C. Crick, and P. J. Brennan.** 2000. Biosynthesis of the galactan component of the mycobacterial cell wall. *J. Biol. Chem.* **275:**33890–33897.

156. **Mingorance, J., J. Tamames, and M. Vicente.** 2004. Genomic channeling in bacterial cell division. *J. Mol. Recognit.* **17:**481–487.

157. **Minnikin, D. E., L. Kremer, L. G. Dover, and G. S. Besra.** 2002. The methyl-branched fortifications of *Mycobacterium tuberculosis. Chem. Biol.* **9:**545–553.

158. **Miyao, A., A. Yoshimura, T. Sato, T. Yamamoto, G. Theeragool, and Y. Kobayashi.** 1992. Sequence of the *Bacillus subtilis* homolog of the *Escherichia coli* cell-division gene *murG. Gene* **118:**147–148.

159. **Mizyed, S., A. Oddone, B. Byczynski, D. W. Hughes, and P. J. Berti.** 2005. UDP-*N*-acetylmuramic acid (UDP-MurNAc) is a potent inhibitor of MurA (enolpyruvyl-UDP-GlcNAc synthase). *Biochemistry* **44:**4011–4017.

160. **Montigiani, S., F. Falugi, M. Scarselli, O. Finco, R. Petracca, G. Galli, M. Mariani, R. Manetti, M. Agnusdei, R. Cevenini, M. Donati, R. Nogarotto, N. Norais, I. Garaguso, S. Nuti, G. Saletti, D. Rosa, G. Ratti, and G. Grandi.** 2002. Genomic approach for analysis of surface proteins in *Chlamydia pneumoniae. Infect. Immun.* **70:**368–379.

161. **Morikawa, A., K. Hamase, and K. Zaitsu.** 2003. Determination of D-alanine in the rat central nervous system and periphery using column-switching high-performance liquid chromatography. *Anal. Biochem.* **312:**66–72.

162. **Moulder, J. W.** 1991. Interaction of chlamydiae and host cells in vitro. *Microbiol. Rev.* **55:**143–190.

163. **Moulder, J. W.** 1993. Why is *Chlamydia* sensitive to penicillin in the absence of peptidoglycan? *Infect. Agents Dis.* **2:**87–99.

164. **Nachega, J. B., and R. E. Chaisson.** 2003. Tuberculosis drug resistance: a global threat. *Clin. Infect. Dis.* **36:**S24–S30.

165. Nakagawa, J., S. Tamaki, S. Tomioka, and M. Matsuhashi. 1984. Functional biosynthesis of cell wall peptidoglycan by polymorphic bifunctional polypeptides. Penicillin-binding protein 1Bs of *Escherichia coli* with activities of transglycosylase and transpeptidase. *J. Biol. Chem.* **259:**13937–13946.

166. Neuhaus, F. C. 1962. The enzymatic synthesis of D-alanyl-D-alanine. I. Purification and properties of D-alanyl-D-alanine synthetase. *J. Biol. Chem.* **237:**778–786.

167. Nicholson, T. L., L. Olinger, K. Chong, G. Schoolnik, and R. S. Stephens. 2003. Global stage-specific gene regulation during the developmental cycle of *Chlamydia trachomatis. J. Bacteriol.* **185:**3179–3189.

168. Nikaido, H., and M. Vaara. 1987. Outer membrane, p. 3–6. *In* J. L. Ingraham, K. B. Low, B. Magasanik, M. Schaechter, H. E. Umbarger, and F. C. Neidhardt (ed.), Escherichia coli *and* Salmonella typhimurium: *Cellular and Molecular Biology*, vol. 1. ASM Press, Washington, DC.

169. Nikolaichik, Y. A., and W. D. Donachie. 2000. Conservation of gene order amongst cell wall and cell division genes in *Eubacteria*, and ribosomal genes in *Eubacteria* and Eukaryotic organelles. *Genetica* **108:**1–7.

170. Nishio, Y., Y. Nakamura, Y. Kawarabayasi, Y. Usuda, E. Kimura, S. Sugimoto, K. Matsui, A. Yamagishi, H. Kikuchi, K. Ikeo, and T. Gojobori. 2003. Comparative complete genome sequence analysis of the amino acid replacements responsible for the thermostability of *Corynebacterium efficiens. Genome Res.* **13:**1572–1579.

171. Nurminen, M., M. Leinonen, P. Saikku, and P. H. Makela. 1983. The genus-specific antigen of *Chlamydia*: resemblance to the lipopolysaccharide of enteric bacteria. *Science* **220:**1279–1281.

172. Obermann, W., and J. V. Holtje. 1994. Alterations of murein structure and of penicillin-binding proteins in minicells from *Escherichia coli. Microbiology* **140**(Pt. 1):79–87.

173. Pan, F., M. Jackson, Y. Ma, and M. McNeil. 2001. Cell wall core galactofuran synthesis is essential for growth of mycobacteria. *J. Bacteriol.* **183:**3991–3998.

174. Park, J. T. 1952. Uridine-5′-pyrophosphate derivatives. I. Isolation from Staphylococcus aureus. *J. Biol. Chem.* **194:**877-884.

175. Park, J. T. 1952. Uridine-5′-pyrophosphate derivatives. II. A structure common to three derivatives. *J. Biol. Chem.* **194:**885–895.

176. Park, J. T. 1952. Uridine-5′-pyrophosphate derivatives. III. Amino acid–containing derivatives. *J. Biol. Chem.* **194:**897–904.

177. Park, J. T., and M. J. Johnson. 1949. Accumulation of labile phosphate in *Staphylococcus aureus* grown in the presence of penicillin. Biochemical evidence for the formation of a covalent acylphosphate linkage between UDP-*N*-acetylmuramate and ATP. *J. Biol. Chem.* **179:**585–592.

178. Petit, J. F., A. Adam, J. Wietzerbin-Falszpan, E. Lederer, and J. M. Ghuysen. 1969. Chemical structure of the cell wall of *Mycobacterium smegmatis*. I. Isolation and partial characterization of the peptidoglycan. *Biochem. Biophys. Res. Commun.* **35:**478–485.

179. Pinho, M. G., H. de Lencastre, and A. Tomasz. 2001. An acquired and a native penicillin-binding protein cooperate in building the cell wall of drug-resistant staphylococci. *Proc. Natl. Acad. Sci. USA* **98:**10886–10891.

180. Pinho, M. G., S. R. Filipe, H. de Lencastre, and A. Tomasz. 2001. Complementation of the essential peptidoglycan transpeptidase function of penicillin-binding protein 2 (PBP2) by the drug resistance protein PBP2A in *Staphylococcus aureus. J. Bacteriol.* **183:**6525–6531.

181. Plapp, R., and J. L. Strominger. 1970. Biosynthesis of the peptidoglycan of bacterial cell walls. XVII. Biosynthesis of peptidoglycan and of interpeptide bridges in *Lactobacillus viridescens. J. Biol. Chem.* **245:**3667–3674.

182. Popham, D. L., and P. Setlow. 1996. Phenotypes of *Bacillus subtilis* mutants lacking multiple class A high-molecular-weight penicillin-binding proteins. *J. Bacteriol.* **178:**2079–2085.

183. Portevin, D., C. de Sousa-D'Auria, C. Houssin, C. Grimaldi, M. Chami, M. Daffe, and C. Guilhot. 2004. A polyketide synthase catalyzes the last condensation step of mycolic acid biosynthesis in mycobacteria and related organisms. *Proc. Natl. Acad. Sci. USA* **101:**314–319.

184. Portevin, D., C. de Sousa-D'Auria, H. Montrozier, C. Houssin, A. Stella, M. A. Laneelle, F. Bardou, C. Guilhot, and M. Daffe. 2005. The acyl-AMP ligase FadD32 and AccD4-containing acyl–CoA carboxylase are required for the synthesis of mycolic acids and essential for mycobacterial growth: identification of the carboxylation product and determination of the acyl-CoA carboxylase components. *J. Biol. Chem.* **280:**8862–8874.

185. Psylinakis, E., I. G. Boneca, K. Mavromatis, A. Deli, E. Hayhurst, S. J. Foster, K. M. Varum, and V. Bouriotis. 2005. Peptidoglycan *N*-acetylglucosamine deacetylases from *Bacillus cereus*, highly conserved proteins in *Bacillus anthracis. J. Biol. Chem.* **280:**30856–30863.

186. Pucci, M. J., L. F. Discotto, and T. J. Dougherty. 1992. Cloning and identification of the *Escherichia coli* murB DNA sequence, which

encodes UDP-*N*-acetylenolpyruvoylglucosamine reductase. *J. Bacteriol.* **174:**1690–1693.

187. **Pucci, M. J., J. A. Thanassi, L. F. Discotto, R. E. Kessler, and T. J. Dougherty.** 1997. Identification and characterization of cell wall-cell division gene clusters in pathogenic gram-positive cocci. *J. Bacteriol.* **179:**5632–5635.

188. **Puech, V., M. Chami, A. Lemassu, M. A. Laneelle, B. Schiffler, P. Gounon, N. Bayan, R. Benz, and M. Daffe.** 2001. Structure of the cell envelope of corynebacteria: importance of the non-covalently bound lipids in the formation of the cell wall permeability barrier and fracture plane. *Microbiology* **147:**1365–1382.

189. **Ramaswamy, S. V., A. G. Amin, S. Goksel, C. E. Stager, S. J. Dou, H. El Sahly, S. L. Moghazeh, B. N. Kreiswirth, and J. M. Musser.** 2000. Molecular genetic analysis of nucleotide polymorphisms associated with ethambutol resistance in human isolates of *Mycobacterium tuberculosis. Antimicrob. Agents Chemother.* **44:**326–336.

190. **Read, T. D., R. C. Brunham, C. Shen, S. R. Gill, J. F. Heidelberg, O. White, E. K. Hickey, J. Peterson, T. Utterback, K. Berry, S. Bass, K. Linher, J. Weidman, H. Khouri, B. Craven, C. Bowman, R. Dodson, M. Gwinn, W. Nelson, R. DeBoy, J. Kolonay, G. McClarty, S. L. Salzberg, J. Eisen, and C. M. Fraser.** 2000. Genome sequences of *Chlamydia trachomatis* MoPn and *Chlamydia pneumoniae* AR39. *Nucleic Acids Res.* **28:**1397–1406.

191. **Read, T. D., G. S. Myers, R. C. Brunham, W. C. Nelson, I. T. Paulsen, J. Heidelberg, E. Holtzapple, H. Khouri, N. B. Federova, H. A. Carty, L. A. Umayam, D. H. Haft, J. Peterson, M. J. Beanan, O. White, S. L. Salzberg, R. C. Hsia, G. McClarty, R. G. Rank, P. M. Bavoil, and C. M. Fraser.** 2003. Genome sequence of *Chlamydophila caviae* (*Chlamydia psittaci* GPIC): examining the role of niche-specific genes in the evolution of the *Chlamydiaceae. Nucleic Acids Res.* **31:**2134–-2147.

192. **Read, T. D., S. N. Peterson, N. Tourasse, L. W. Baillie, I. T. Paulsen, K. E. Nelson, H. Tettelin, D. E. Fouts, J. A. Eisen, S. R. Gill, E. K. Holtzapple, O. A. Okstad, E. Helgason, J. Rilstone, M. Wu, J. F. Kolonay, M. J. Beanan, R. J. Dodson, L. M. Brinkac, M. Gwinn, R. T. DeBoy, R. Madpu, S. C. Daugherty, A. S. Durkin, D. H. Haft, W. C. Nelson, J. D. Peterson, M. Pop, H. M. Khouri, D. Radune, J. L. Benton, Y. Mahamoud, L. Jiang, I. R. Hance, J. F. Weidman, K. J. Berry, R. D. Plaut, A. M. Wolf, K. L. Watkins, W. C. Nierman, A. Hazen, R. Cline, C. Redmond, J. E. Thwaite, O. White,**

S. L. Salzberg, B. Thomason, A. M. Friedlander, T. M. Koehler, P. C. Hanna, A. B. Kolsto, and C. M. Fraser. 2003. The genome sequence of *Bacillus anthracis* Ames and comparison to closely related bacteria. *Nature* **423:**81–86.

193. **Reddy, P. S., A. Raghavan, and D. Chatterji.** 1995. Evidence for a ppGpp-binding site on Escherichia coli RNA polymerase: proximity relationship with the rifampicin-binding domain. *Mol. Microbiol.* **15:**255–265.

194. **Rockey, D. D., J. Lenart, and R. S. Stephens.** 2000. Genome sequencing and our understanding of chlamydiae. *Infect. Immun.* **68:**5473–5479.

195. **Rohrer, S., and B. Berger-Bachi.** 2003. FemABX peptidyl transferases: a link between branched-chain cell wall peptide formation and beta-lactam resistance in gram-positive cocci. *Antimicrob. Agents Chemother.* **47:**837–846.

196. **Rohrer, S., K. Ehlert, M. Tschierske, H. Labischinski, and B. Berger-Bachi.** 1999. The essential *Staphylococcus aureus* gene fmhB is involved in the first step of peptidoglycan pentaglycine interpeptide formation. *Proc. Natl. Acad. Sci. USA* **96:**9351–9356.

197. **Sasaki, S., F. Takeshita, K. Okuda, and N. Ishii.** 2001. *Mycobacterium leprae* and leprosy: a compendium. *Microbiol. Immunol.* **45:**729–736.

198. **Satta, G., R. Fontana, and P. Canepari.** 1994. The two-competing site (TCS) model for cell shape regulation in bacteria: the envelope as an integration point for the regulatory circuits of essential physiological events. *Adv. Microb. Physiol.* **36:**181–245.

199. **Sauvage, E., F. Kerff, E. Fonze, R. Herman, B. Schoot, J. P. Marquette, Y. Taburet, D. Prevost, J. Dumas, G. Leonard, P. Stefanic, J. Coyette, and P. Charlier.** 2002. The 2.4-A crystal structure of the penicillin-resistant penicillin-binding protein PBP5fm from *Enterococcus faecium* in complex with benzylpenicillin. *Cell. Mol. Life Sci.* **59:**1223–1232.

200. **Saxena, I. M., R. M. Brown, Jr., M. Fevre, R. A. Geremia, and B. Henrissat.** 1995. Multidomain architecture of beta-glycosyl transferases: implications for mechanism of action. *J. Bacteriol.* **177:**1419–1424.

201. **Scheffers, D. J., and M. G. Pinho.** 2005. Bacterial cell wall synthesis: new insights from localization studies. *Microbiol. Mol. Biol. Rev.* **69:**585–607.

202. **Scherman, M., A. Weston, K. Duncan, A. Whittington, R. Upton, L. Deng, R. Comber, J. D. Friedrich, and M. McNeil.** 1995. Biosynthetic origin of mycobacterial cell wall arabinosyl residues. *J. Bacteriol.* **177:**7125–7130.

203. **Scherman, M. S., L. Kalbe-Bournonville, D. Bush, Y. Xin, L. Deng, and M. McNeil.** 1996. Polyprenylphosphate-pentoses in mycobacteria are synthesized from 5- phosphoribose pyrophosphate. *J. Biol. Chem.* **271:**29652–29658.

204. **Schleifer, K. H., and O. Kandler.** 1972. Peptidoglycan types of bacterial cell walls and their taxonomic implications. *Bacteriol. Rev.* **36:**407–477.

205. **Schneider, T., M. M. Senn, B. Berger-Bachi, A. Tossi, H. G. Sahl, and I. Wiedemann.** 2004. In vitro assembly of a complete, pentaglycine interpeptide bridge containing cell wall precursor (lipid II-Gly5) of *Staphylococcus aureus*. *Mol. Microbiol.* **53:**675–685.

206. **Schonbrunn, E., S. Sack, S. Eschenburg, A. Perrakis, F. Krekel, N. Amrhein, and E. Mandelkow.** 1996. Crystal structure of UDP-N-acetylglucosamine enolpyruvyltransferase, the target of the antibiotic fosfomycin. *Structure* **4:**1065–1075.

207. **Selinger, D. W., R. M. Saxena, K. J. Cheung, G. M. Church, and C. Rosenow.** 2003. Global RNA half-life analysis in *Escherichia coli* reveals positional patterns of transcript degradation. *Genome Res.*. **13:**216–223.

208. **Severin, A., and A. Tomasz.** 1996. Naturally occurring peptidoglycan variants of *Streptococcus pneumoniae*. *J. Bacteriol.* **178:**168–174.

209. **Shirai, M., H. Hirakawa, M. Kimoto, M. Tabuchi, F. Kishi, K. Ouchi, T. Shiba, K. Ishii, M. Hattori, S. Kuhara, and T. Nakazawa.** 2000. Comparison of whole genome sequences of *Chlamydia pneumoniae* J138 from Japan and CWL029 from USA. *Nucleic Acids Res.* **28:**2311–2314.

210. **Siefert, J. L., and G. E. Fox.** 1998. Phylogenetic mapping of bacterial morphology. *Microbiology* **144**(Pt. 10):2803–2808.

211. **Signoretto, C., M. Boaretti, and P. Canepari.** 1998. Peptidoglycan synthesis by *Enterococcus faecalis* penicillin binding protein 5. *Arch. Microbiol.* **170:**185–190.

212. **Skipp, P., J. Robinson, C. D. O'Connor, and I. N. Clarke.** 2005. Shotgun proteomic analysis of *Chlamydia trachomatis*. *Proteomics* **5:**1558–1573.

213. **Smith, A. M., and K. P. Klugman.** 2001. Alterations in MurM, a cell wall muropeptide branching enzyme, increase high-level penicillin and cephalosporin resistance in Streptococcus pneumoniae. *Antimicrob. Agents Chemother.* **45:**2393–2396.

214. **Spivey, H. O., and J. Ovadi.** 1999. Substrate channeling. *Methods* **19:**306–321.

215. **Spratt, B. G.** 1977. Properties of the penicillin-binding proteins of *Escherichia coli* K12. *Eur. J. Biochem.* **72:**341–352.

216. **Sreevatsan, S., K. E. Stockbauer, X. Pan, B. N. Kreiswirth, S. L. Moghazeh, W. R. Jacobs, Jr., A. Telenti, and J. M. Musser.** 1997. Ethambutol resistance in *Mycobacterium tuberculosis*: critical role of embB mutations. *Antimicrob. Agents Chemother.* **41:**1677–1681.

217. **Staudenbauer, W., and J. L. Strominger.** 1972. Activation of D-aspartic acid for incorporation into peptidoglycan. *J. Biol. Chem.* **247:**5095–5102.

218. **Stephens, R. S., S. Kalman, C. Lammel, J. Fan, R. Marathe, L. Aravind, W. Mitchell, L. Olinger, R. L. Tatusov, Q. Zhao, E. V. Koonin, and R. W. Davis.** 1998. Genome sequence of an obligate intracellular pathogen of humans: *Chlamydia trachomatis*. *Science* **282:**754–759.

219. **Stranden, A. M., K. Ehlert, H. Labischinski, and B. Berger-Bachi.** 1997. Cell wall monoglycine cross-bridges and methicillin hypersusceptibility in a femAB null mutant of methicillin-resistant *Staphylococcus aureus*. *J. Bacteriol.* **179:**9–16.

220. **Sugai, M., T. Fujiwara, T. Akiyama, M. Ohara, H. Komatsuzawa, S. Inoue, and H. Suginaka.** 1997. Purification and molecular characterization of glycylglycine endopeptidase produced by *Staphylococcus capitis* EPK1. *J. Bacteriol.* **179:**1193–1202.

221. **Sugai, M., T. Fujiwara, K. Ohta, H. Komatsuzawa, M. Ohara, and H. Suginaka.** 1997. epr, which encodes glycylglycine endopeptidase resistance, is homologous to femAB and affects serine content of peptidoglycan cross bridges in *Staphylococcus capitis* and *Staphylococcus aureus*. *J. Bacteriol.* **179:**4311–4318.

222. **Suzuki, H., Y. van Heijenoort, T. Tamura, J. Mizoguchi, Y. Hirota, and J. van Heijenoort.** 1980. In vitro peptidoglycan polymerization catalysed by penicillin binding protein 1b of *Escherichia coli* K-12. *FEBS Lett.* **110:**245–249.

223. **Tagg, J. R., A. S. Dajani, and L. W. Wannamaker.** 1976. Bacteriocins of gram-positive bacteria. *Bacteriol. Rev.* **40:**722–756.

224. **Takayama, K., H. L. David, L. Wang, and D. S. Goldman.** 1970. Isolation and characterization of uridine diphosphate-N-glycolylmuramyl-L-alanyl-gamma-D-glutamyl-meso-alpha, alpha'-diaminopimelic acid from *Mycobacterium tuberculosis*. *Biochem. Biophys. Res. Commun.* **39:**7–12.

225. **Taku, A., and D. P. Fan.** 1976. Identification of an isolated protein essential for peptidoglycan synthesis as the N-acetylglucosaminyltransferase. *J. Biol. Chem.* **251:**6154–6156.

226. **Taku, A., K. G. Gunetileke, and R. A. Anwar.** 1970. Biosynthesis of uridine diphospho-N-

acetylmuramic acid. 3. Purification and properties of uridine diphospho-*N*-acetylenolpyruvyl-glucosamine reductase. *J. Biol. Chem.* **245:**5012–5016.

227. **Tamames, J.** 2001. Evolution of gene order conservation in prokaryotes. *Genome Biol.* **2:**RESEARCH0020.

228. **Tamames, J., G. Casari, C. Ouzounis, and A. Valencia.** 1997. Conserved clusters of functionally related genes in two bacterial genomes. *J. Mol. Evol.* **44:**66–73.

229. **Tauch, A., O. Kaiser, T. Hain, A. Goesmann, B. Weisshaar, A. Albersmeier, T. Bekel, N. Bischoff, I. Brune, T. Chakraborty, J. Kalinowski, F. Meyer, O. Rupp, S. Schneiker, P. Viehoever, and A. Puhler.** 2005. Complete genome sequence and analysis of the multiresistant nosocomial pathogen *Corynebacterium jeikeium* K411, a lipid-requiring bacterium of the human skin flora. *J. Bacteriol.* **187:**4671–4682.

230. **Telenti, A., W. J. Philipp, S. Sreevatsan, C. Bernasconi, K. E. Stockbauer, B. Wieles, J. M. Musser, and W. R. Jacobs, Jr.** 1997. The emb operon, a gene cluster of *Mycobacterium tuberculosis* involved in resistance to ethambutol. *Nat. Med.* **3:**567–570.

231. **Thomson, N. R., C. Yeats, K. Bell, M. T. Holden, S. D. Bentley, M. Livingstone, A. M. Cerdeno-Tarraga, B. Harris, J. Doggett, D. Ormond, K. Mungall, K. Clarke, T. Feltwell, Z. Hance, M. Sanders, M. A. Quail, C. Price, B. G. Barrell, J. Parkhill, and D. Longbottom.** 2005. The *Chlamydophila abortus* genome sequence reveals an array of variable proteins that contribute to interspecies variation. *Genome Res.* **15:**629–640.

232. **Thumm, G., and F. Gotz.** 1997. Studies on prolysostaphin processing and characterization of the lysostaphin immunity factor (Lif) of *Staphylococcus simulans* biovar *staphylolyticus*. *Mol. Microbiol.* **23:**1251–1265.

233. **Touhami, A., M. H. Jericho, and T. J. Beveridge.** 2004. Atomic force microscopy of cell growth and division in *Staphylococcus aureus*. *J. Bacteriol.* **186:**3286–3295.

234. **Trivedi, O. A., P. Arora, V. Sridharan, R. Tickoo, D. Mohanty, and R. S. Gokhale.** 2004. Enzymic activation and transfer of fatty acids as acyl-adenylates in mycobacteria. *Nature* **428:**441–445.

235. **Tschierske, M., K. Ehlert, A. M. Stranden, and B. Berger-Bachi.** 1997. Lif, the lysostaphin immunity factor, complements FemB in staphylococcal peptidoglycan interpeptide bridge formation. *FEMS Microbiol. Lett.* **153:**261–264.

236. **Tschierske, M., C. Mori, S. Rohrer, K. Ehlert, K. J. Shaw, and B. Berger-Bachi.** 1999. Iden-

tification of three additional femAB-like open reading frames in *Staphylococcus aureus*. *FEMS Microbiol. Lett.* **171:**97–102.

237. **Vandahl, B. B., S. Birkelund, H. Demol, B. Hoorelbeke, G. Christiansen, J. Vandekerckhove, and K. Gevaert.** 2001. Proteome analysis of the *Chlamydia pneumoniae* elementary body. *Electrophoresis* **22:**1204–1223.

238. **van Heijenoort, J.** 1994. Biosynthesis of the bacterial peptidoglycan unit, p. 39–54. *In* J.-M. Ghuysen and R. Hakenbeck (ed.), *Bacterial Cell Wall*. Elsevier, Dordrecht, The Netherlands.

239. **van Heijenoort, J.** 2001. Formation of the glycan chains in the synthesis of bacterial peptidoglycan. *Glycobiology* **11:**25R–36R.

240. **van Heijenoort, J.** 2001. Recent advances in the formation of the bacterial peptidoglycan monomer unit. *Nat. Prod. Rep.* **18:**503–519.

241. **Vicente, M., M. J. Gomez, and J. A. Ayala.** 1998. Regulation of transcription of cell division genes in the *Escherichia coli dcw* cluster. *Cell. Mol. Life Sci.* **54:**317–324.

242. **Vicente, M., A. I. Rico, R. Martinez-Arteaga, and J. Mingorance.** 2006. Septum enlightenment: assembly of bacterial division proteins. *J. Bacteriol.* **188:**19–27.

243. **Vissa, V. D., and P. J. Brennan.** 2001. The genome of *Mycobacterium leprae*: a minimal mycobacterial gene set. *Genome Biol.* **2:**REVIEWS 1023.

244. **Vollmer, W., and J. V. Holtje.** 2004. The architecture of the murein (peptidoglycan) in gram-negative bacteria: vertical scaffold or horizontal layer(s)? *J. Bacteriol.* **186:**5978–5987.

245. **Vollmer, W., and A. Tomasz.** 2000. The pgdA gene encodes for a peptidoglycan N-acetylglucosamine deacetylase in *Streptococcus pneumoniae*. *J. Biol. Chem.* **275:**20496–20501.

246. **Wanke, C., R. Falchetto, and N. Amrhein.** 1992. The UDP-*N*-acetylglucosamine 1-carboxyvinyl-transferase of *Enterobacter cloacae*. Molecular cloning, sequencing of the gene and overexpression of the enzyme. *FEBS Lett.* **301:**271–276.

247. **Ward, J. B., and H. R. Perkins.** 1973. The direction of glycan synthesis in a bacterial peptidoglycan. *Biochem. J.* **135:**721–728.

248. **Weadge, J. T., and A. J. Clarke.** 2006. Identification and characterization of O-acetylpeptidoglycan esterase: a novel enzyme discovered in *Neisseria gonorrhoeae*. *Biochemistry* **45:**839–851.

249. **Weadge, J. T., J. M. Pfeffer, and A. J. Clarke.** 2005. Identification of a new family of enzymes with potential O-acetylpeptidoglycan esterase activity in gram-positive and gram-negative bacteria. *BMC Microbiol.* **5:**49.

250. **Weber, B., K. Ehlert, A. Diehl, P. Reichmann, H. Labischinski, and R. Hakenbeck.**

2000. The fib locus in Streptococcus pneumoniae is required for peptidoglycan crosslinking and PBP-mediated beta-lactam resistance. *FEMS Microbiol. Lett.* **188**:81–85.

251. **Weidel, W., and H. Pelzer.** 1964. Bagshaped macromolecules—a new outlook on bacterial cell walls. *Adv. Enzymol. Relat. Areas Mol. Biol.* **26**:193–232.

252. **Weidenmaier, C., A. Peschel, Y. Q. Xiong, S. A. Kristian, K. Dietz, M. R. Yeaman, and A. S. Bayer.** 2005. Lack of wall teichoic acids in *Staphylococcus aureus* leads to reduced interactions with endothelial cells and to attenuated virulence in a rabbit model of endocarditis. *J. Infect. Dis.* **191**:1771–1777.

253. **Weiss, D. S., K. Pogliano, M. Carson, L. M. Guzman, C. Fraipont, M. Nguyen-Disteche, R. Losick, and J. Beckwith.** 1997. Localization of the *Escherichia coli* cell division protein FtsI (PBP3) to the division site and cell pole. *Mol. Microbiol.*. **25**:671–681.

254. **Westmacott, D., and H. R. Perkins.** 1979. Effects of lysozyme on *Bacillus cereus* 569: rupture of chains of bacteria and enhancement of sensitivity to autolysins. *J. Gen. Microbiol.* **115**: 1–11.

255. **Weston, A., R. J. Stern, R. E. Lee, P. M. Nassau, D. Monsey, S. L. Martin, M. S. Scherman, G. S. Besra, K. Duncan, and M. R. McNeil.** 1997. Biosynthetic origin of mycobacterial cell wall galactofuranosyl residues. *Tuber. Lung Dis.* **78**:123–131.

256. **Weston, A., J. B. Ward, and H. R. Perkins.** 1977. Biosynthesis of peptidoglycan in wall plus membrane preparations from micrococcus-luteus—direction of chain extension, length of chains and effect of penicillin on cross-linking. *J. Gen. Microbiol.* **99**:171–181.

257. **Wietzerbin, J., B. C. Das, J. F. Petit, E. Lederer, M. Leyh-Bouille, and J. M. Ghuysen.** 1974. Occurrence of D-alanyl-(D)-meso-diaminopimelic acid and meso-diaminopimelyl-meso-diaminopimelic acid interpeptide linkages in the peptidoglycan of *Mycobacteria. Biochemistry* **13**:3471–3476.

258. **World Health Organization.** 2002 report on infectious diseases 2002. Scaling up the response to infectious diseases: a way out of poverty. Available at: http://www.who.int/infectious-disease-report/2002/. Accessed 26 January 2007.

259. **Wyrick, P. B.** 2000. Intracellular survival by *Chlamydia. Cell. Microbiol.* **2**:275–282.

260. **Young, K. D.** 2006. Too many strictures on structure. *Trends Microbiol.* **14**:155–156.

261. **Yura, T., H. Mori, H. Nagai, T. Nagata, A. Ishihama, N. Fujita, K. Isono, K. Mizobuchi, and A. Nakata.** 1992. Systematic sequencing of the *Escherichia coli* genome: analysis of the 0–2.4 min region. *Nucleic Acids Res.* **20**:3305–3308.

262. **Zemell, R. I., and R. A. Anwar.** 1975. Pyruvate-uridine diphospho-N-acetylglucosamine transferase. Purification to homogeneity and feedback inhibition. *J. Biol. Chem.* **250**:3185–3192.

263. **Zipperle, G. F., Jr., J. W. Ezzell, Jr., and R. J. Doyle.** 1984. Glucosamine substitution and muramidase susceptibility in *Bacillus anthracis. Can. J. Microbiol.* **30**:553–559.

264. **Zygmunt, W. A., H. P. Browder, and P. A. Tavormina.** 1967. Lytic action of lysostaphin on susceptible and resistant strains of *Staphylococcus aureus. Can. J. Microbiol.* **13**:845–853.

WHAT GENOMICS HAS TAUGHT US ABOUT INTRACELLULAR PATHOGENS: THE EXAMPLE OF *LISTERIA MONOCYTOGENES*

Carmen Buchrieser and Pascale Cossart

14

Pathogens that are able to enter and multiply within human cells are responsible for multiple diseases and millions of deaths worldwide. Thus, the challenge is to elucidate these pathogen-specific and cell biological mechanisms involved in intracellular growth and spread. Many different techniques, such as molecular genetics, tissue culture systems, high-resolution microscopy, in vivo infection models, and, recently, in vivo imaging techniques have been applied to the study of the mechanisms of intracellular pathogenesis. Since the publication of the first bacterial genome sequence in 1995 (47), a tremendous increase in genomic information has substantially altered our view on bacterial pathogenesis and has led to the application of many different genomic and postgenomic approaches in microbial research. In this chapter, we will discuss the insights gained from genomics of intracellular pathogens using *Listeria monocytogenes* as an example.

LISTERIA MONOCYTOGENES: AN INTRACELLULAR OPPORTUNISTIC PATHOGEN

Listeria monocytogenes belongs to the genus *Listeria*, a group of a gram-positive, low G+C content bacteria closely related to *Bacillus subtilis*. The genus comprises six species, two of which are pathogenic, *L. monocytogenes* for humans and animals, and *L. ivanovii*, mainly for ruminants. The remaining four species, *L. welshimeri, L. seeligeri, L. innocua*, and *L. grayi*, are apathogenic environmental saprophytes. *L. monocytogenes*, a facultative intracellular parasite that invades and multiplies within diverse eukaryotic cell types, like epithelial cells and macrophages, has become a model organism for deciphering the molecular determinants of intracellular pathogenesis, but it is also a serious human pathogen. It is the causative agent of listeriosis, a foodborne infection with a mortality rate up to 30% (102, 112). The two main clinical manifestations are sepsis and meningitis. Meningitis is often complicated by encephalitis, a pathology that is unusual for bacterial infections (123). Gastroenteritis due to *L. monocytogenes* has also been reported (2, 28).

Listeria monocytogenes is widely present in nature, and it has been isolated from animals such as cattle, sheep, goats, poultry, and, infrequently, wild animals (43). *L. monocytogenes* has

Carmen Buchrieser Unité de Génomique des Microorganismes Pathogènes, Institut Pasteur, 25, rue du Dr. Roux, and CNRS URA 2171, F-75015 France. *Pascale Cossart* Unité des Interactions Bactéries-Cellules, Institut Pasteur, 28, rue du Dr. Roux; INSERM, U604; and INRA, USC 2020, Paris, F-75015 France.

the important capacity to adapt and survive in extreme environments, such as high salt concentration (10% NaCl), a broad pH range (from 4.5 to 9.0), and a wide temperature range. The ability to grow between −1 and 45°C increases the contamination risk in dairy products, meats, seafood, and other processed food products via selective enrichment during refrigeration. *Listeria* can also survive long periods of drying and freezing with subsequent thawing (87). *Listeria monocytogenes* is an environmental bacterium that lives, for example, on decomposing plants, but the acquisition of virulence factors, most probably by horizontal gene transfer, allows *L. monocytogenes* to also infect humans and many other mammalian hosts.

An important feature of the virulence of *L. monocytogenes* is its capacity to cross three barriers: the intestinal barrier, the blood-brain barrier, and the placental barrier. In addition, *L. monocytogenes* is able to promote its own uptake into host cells through pathogen-induced phagocytosis. Figure 1 shows a schematic representation of the intracellular life cycle and the proteins known to be implicated in the different steps. After internalization, which is medi-

ated mainly by the two invasion proteins InlA and InlB (34, 48), *L. monocytogenes* resides in a vacuole. The bacterium lyses the vacuolar membrane by a combined action of the pore-forming protein listeriolysin (50, 71, 91) and two phospholipases (PI-PlcA, PI-Plc) and consequently escapes into the cytosol, where it is able to multiply while making use of specific transporters like the hexosephosphate transporter Hpt (21) to gain carbohydrates from the host cell. Concomitant with intracellular replication, *L. monocytogenes* expresses the surface protein ActA to polymerize the host actin to induce its own movement (29, 72). This ability to move in the host cell cytoplasm and to spread within tissues by directly passing from one cell to another is an essential determinant of pathogenicity (24, 25). The propulsive force for intracellular movement is thought to be generated by continuous actin assembly at the rear end of the bacterium. Moving bacteria that reach the plasma membrane induce the formation of long membranous protrusions that are internalized by neighboring cells, thus mediating the spread of infection (Fig. 1). The unrelated pathogens *Shigella* and *Rickettsia* use a similar process of actin-based motility to dis-

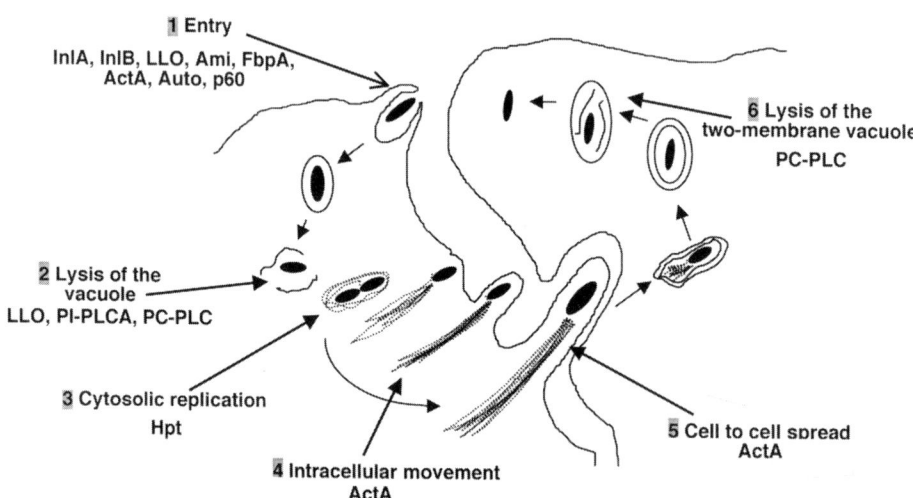

FIGURE 1 Schematic representation of the infection cycle of *Listeria monocytogenes*. The successive steps are: entry (1), lysis of the vacuole (2), intracellular replication (3), intracellular movements (4), cell-to-cell spread (5), formation, and lysis (6) of two-membrane vacuole. Virulence factors involved at the different steps are indicated (from reference 121).

seminate in infected tissues (54, 56). The cell biology of this infectious process has been investigated in detail (103). The regulatory factor, necessary for the regulation of the expression of all of these virulence genes, is PrfA (82, 92). PrfA activates the genes necessary for intracellular replication that are located on the so-called virulence gene cluster (*prfA, plcA, hly, mpl, actA,* and *plcB*), as well as the expression of *inlA* and *inlB*, encoding the invasion proteins (InlA and InlB) (37), and *hpt*, important for intracellular proliferation (21).

Despite the elucidation of the important players necessary for entry and intracellular replication of *L. monocytogenes* during the pregenomic area (for reviews, see references 103 and 123), many open questions remained to be answered: How does *L. monocytogenes* cross the blood-brain barrier? What is the genetic basis of virulence differences among different *L. monocytogenes* strains? How does *L. monocytogenes* switch between life as an environmental bacterium and an intracellular, human pathogen?

An important step forward in *Listeria* research—and a step that helped partly answer these questions—was the establishment and publication of the first complete *Listeria* genome sequences, those of *L. monocytogenes* EGDe and its apathogenic relative *Listeria innocua* in 2001 (51) (http://genolist.pasteur.fr/ListiList/). Two years later, three additional *L. monocytogenes* sequences were published (96). The availability of these complete sequences paved the way for major breakthroughs in understanding the biodiversity and biology of *L. monocytogenes* in particular and *Listeria* in general.

LISTERIA GENOMES SHOW HIGHLY CONSERVED ORGANIZATION, BUT EACH HAS UNIQUE INTERSPERSED REGIONS

The *L. monocytogenes* EGDe (serovar 1/2a) genome comprises 2,944,528 bp with an average G+C content of 39% and 2,853 predicted protein-coding genes. The *L. innocua* CLIP11626 (serovar 6a) chromosome is 3,011,209 bp long with an average G+C content of 37% and 2,981 predicted protein-coding genes (Color Plate 4, Table 1) (51). Analysis of the two *Listeria* genomes allowed the identification of common and specific features highlighting the differences between pathogenic and nonpathogenic *Listeria* strains. One interesting common feature is the finding that 2,587 of the 2,853 *L. monocytogenes* genes have an orthologous gene in the *L. innocua* genome (51). Furthermore, a perfect conservation of the order and the relative orientation of these orthologous genes was identified, indicating a high stability in the genome organization of *Listeria* and a close phylogenetic relationship of the two *Listeria* genomes (13). However, despite this high number of common genes, considerable differences in gene content are also present, some of which are without doubt related to the ability of *L. monocytogenes* to cause disease in humans and animals.

The *L. monocytogenes* EGDe-specific genes are present in 100 DNA fragments scattered throughout the entire chromosome. Interestingly, their G+C content varies from 24 to 46%. The average G+C content of all predicted coding regions is 38%, whereas that of the genes specific to *L. monocytogenes* is only 34%. The G+C content of the genes compared with that of their flanking regions has been used as a marker for phylogenetic origin. In *L. monocytogenes*, 54 of the 100 specific regions had a significantly lower G+C content than the flanking regions, and six had a significantly higher G+C content, suggesting recent acquisition by horizontal gene transfer. The *L. innocua* CLIP11626 specific genes are clustered in 63 regions containing one to seven genes. Analysis of the G+C content of these regions gave similar results. Thirty have a lower and three a higher GC content than their flanking regions (13). Interestingly, many of the *L. innocua*–specific surface proteins are located in such regions with distinct G+C content, suggesting that they were acquired by horizontal gene transfer, probably from other bacteria with lower G+C content, such as, for example, *Streptococcus*. This particular organization of a number of small regions within the *Listeria* genomes suggested that many acquisition and deletion events have led to the present genome content.

TABLE 1 General features of published *Listeria* genome sequences

Parameter	L. monocytogenes EGDe (1/2a)	L. monocytogenes F6854 (1/2a)[c]	L. monocytogenes F2365 (4b)	L. monocytogenes H7858 (4b)[c]	L. innocua CLIP11262 (6a)
Size of the chromosome (bp)	2,944,528	~ 2,953,211	2,905,310	~ 2,893,921	3,011,209
G+C content (%)	38	37.8	38	38	37.4
G+C content of protein-coding genes (%)	38.4	38.5	38.5	38.4	38.0
Total no. of CDS[a]	2,853	2,973	2,847	3,024	2,973
Percentage coding	89.2	90.3	88.4	89.5	89.1
No. of prophage regions	1	3	2	2	5
Monocins	1	1	1	1	1
Plasmid	—	—	—	1 (94 CDS)	1 (79 CDS)
No. of strain-specific genes[b]	61	97	51	69	78
No. of transposons	1 (Tn916 like)	—	—	—	—
No. of rRNA operons	6	6	6	6	6
No. of tRNA genes	67	67[c]	67	65[c]	66

[a]CDS, coding sequence.
[b]Except prophage genes.
[c]Draft genome sequence (eightfold coverage without gap closure).

In 2004, the complete genome sequence of an additional *L. monocytogenes* isolate (strain F2365, serovar 4b) and two almost complete genome sequences of *L. monocytogenes* strains (strain F6854, serovar 1/2a, and H7858, serovar 4b) (8 × coverage) (96) were published. Thus, four *L. monocytogenes* (EGDe, F6854, F2365, H7858) and one *L. innocua* genome sequence are available today. Comparison of all five sequences further underlined the marked genome conservation observed from the first analysis. The *L. monocytogenes* genomes have a very similar size, with 2,893,921 to 2,953,211 bp; the same average G+C content of 38%; and about 2,900 predicted protein-coding genes (Table 1) (51, 96). No inversions or shifts of large genome segments are observed, and the conservation of the order and of the relative orientation of the orthologous genes is nearly perfect. This conserved genome organization may be related to the low occurrence of insertion sequence (IS) elements, suggesting that IS transposition or IS-mediated deletions are not key evolutionary mechanisms in *Listeria*. The chromosome of the serotype 4b strains (F2365 and H7858) lacks intact ISs completely but does contain four transposases of the IS3 family that are present in homologous locations in both strains. The serotype 1/2a strains (F6854 and EGDe) each contain three and *L. innocua* four of these elements. Furthermore, the serotype 1/2a strains contain an intact IS element named ISLmo1; two copies are present in strain F6854, and three, one of which is not intact, are present in EGDe. ISLmo1 is missing in the serotype 4b and the *L. innocua* strains.

Similar to the comparison of *L. monocytogenes* EGDe and *L. innocua* CLIP11262 (13, 51), the comparison of all five genomes now available showed that each strain contains between 50 to 100 strain-specific genes (Table 1) scattered around the chromosome in 60 to 100 regions of one to several kilobase pairs. In addition, the *Listeria* genomes contain one to five (*L. innocua*) prophage regions. *Listeria innocua* and all *L. monocytogenes* except the serotype 4b strain (F2365) contain a phage of the A118 family (51, 85) inserted in the *comK* gene, and each sequenced strain carries a monocin region (encoding an incompletely assembled phage particle derived from a cryptic prophage). Furthermore, analysis of the genome sequence identified a nearly complete set of putative DNA uptake genes homologous to *B. subtilis* competence genes, although *Listeria* are not known to be competent. However, ComQ, ComS, ComX, and ComFB, which are involved in regulation of competence in *B. subtilis*, are missing. Even if *Listeria* may in some conditions be naturally competent, the signals inducing competence seem to be different from those of *B. subtilis*. However, this finding indicates that competence may play a role in the acquisition of genes and the evolution of *Listeria*.

SPECIFIC FEATURES OF *LISTERIA* AS DEDUCED FROM WHOLE-GENOME ANALYSIS AND COMPARISON

The virulence gene cluster (*prfA-plcA-hly-mpl-actA-plcB*) and the invasion locus (*inlAB*), which are the two loci of *L. monocytogenes* that are probably most important, allowing this bacterium the transition from the environment to the life inside mammalian cells, were identified by transposon mutagenesis in the pregenomic era. However, genome sequencing and analysis identified several new features of *Listeria*, some of which are undoubtedly related to their intracellular parasitism, but others are probably involved in adaptation and survival in the environment. These particular features are an exceptionally large number and a wide variety of different surface protein coding genes, an abundance of transport proteins (in particular proteins dedicated to carbohydrate transport), and an extensive regulatory repertoire.

Surface Proteins

Several listerial surface proteins play key roles in the interaction of *L. monocytogenes* with mammalian host cells, like the major virulence factors of *L. monocytogenes* Internalin (InlA)

(48, 80) and InlB (9). These two proteins, which are necessary and sufficient for entry into eukaryotic cells, belong to an important protein family in *Listeria*, the internalin family. Internalin-family proteins include an N-terminal signal–peptide sequence, followed by several leucine-rich tandem repeats (LRRs). Several internalins contain an interrepeat region, which is structurally related to an immunoglobulin-like domain. The internalin family can be divided into three subfamilies on the basis of their association with bacteria. The first subfamily includes internalins that contain a cell wall anchor in the C-terminal part, comprising the sorting motif LPXTG, by which they are covalently anchored to the cell wall. The second subfamily is represented by InlB, which is loosely associated with the bacterial surface through its terminal, so-called GW modules, because these repeats start with the GW dipeptide. The third subfamily represented by InlC contains internalins that are predicted to be secreted because they do not have any surface-anchoring domains (61). In addition to the *inlAB* locus, five other internalin loci had been found in *L. monocytogenes* (*inlC, inlE, inlF, inlG,* and *inlH*) (36, 40, 106) before its genome was sequenced. The identification of seven different internalins, proteins rarely found in other bacterial species, had already suggested their importance for *L. monocytogenes*, although for many (*inlC, inlE, inlF, inlG,* and *inlH*), the precise role still remains largely unknown.

The real importance of the internalin protein family in *Listeria* was revealed by the analysis of the complete genome sequence of strain *L. monocytogenes* EGDe (51): 25 internalin coding genes were identified (51). Nineteen of these contain an LPXTG anchoring motif, one possesses GW modules, and five are secreted internalins (51, 61). In addition to the internalins, the search for surface protein coding genes in the complete genome sequence identified 22 other genes that code for proteins containing an LPXTG motif (14, 51) (Table 2). With 41 LPXTG anchored proteins, 19 of which belong to the internalin/LRR

family, *L. monocytogenes* contains more LPXTG proteins than any other gram-positive bacterium (13 LPXTG proteins in *Streptococcus pyogenes* [45]; 18 in *Staphylococcus aureus* N315 [78]).

In addition to the internalin protein, InlB, which is attached to the lipoteichoic acids via GW modules, eight members of this family are present in the *L. monocytogenes* EGDe genome (Table 2). One of those is Ami, a surface protein that was shown to contribute to the adhesion of *L. monocytogenes* to eukaryotic cells (95); another is auto (*lmo1076*), a recently described autolysin (15).

Analysis of the distribution of these surface protein coding genes in the five sequenced genomes showed that seven internalin/LPXTG proteins, 11 LPXTG, seven GW module-containing proteins, and three secreted internalins are the core set of this class of surface/secreted proteins because they are present in all four sequenced *L. monocytogenes* and the sequenced *L. innocua* strain (Table 2). Six internalins, seven LPXTG, one GW module containing protein (InlB), and two secreted internalins are species specific; they are present in all *L. monocytogenes* strains sequenced, but are absent from *L. innocua* CLIP11262. In addition, each strain encodes specific ones like one internalin protein in *L. monocytogenes* EGDe (*lmo2026*), two in strain F6854 (*LMOf6854_0338, LMOf6854_0365*), one in strain F2365 (*LMOf2365_0282*), and six in *L. innocua* (*Lin0559, Lin0661, Lin0739, Lin0740, Lin0803, Lin2724*) (Table 2).

Another surface protein of *L. monocytogenes* involved in virulence is the actin-polymerizing protein ActA (72) (30). This protein has a signal sequence and a hydrophobic C-terminal region that anchors it to the cell membrane. In strain EGDe, 11 such hydrophobic tail proteins are present. Their distribution among the sequenced genomes is conserved: only one (Lmo0082) is missing in one of the sequenced *L. monocytogenes* strains (H6854). Furthermore, the p60-like protein family, named after p60, a major surface protein from *L. monocytogenes*, which has a murein hydrolase and contributes to cell division and uptake in some mammalian

TABLE 2 Presence and distribution of internalins, LPXTG, and GW module containing surface proteins in sequenced *L. monocytogenes* and *L. innocua*[a]

EGDE-1/2a	Internalin	F6854-1/2a	F2365-4b	H7858-4b	CLIP11262-6a
LPXTG/LRR containing proteins					
Lmo0262	*inlG*	LMOf6854_0275	—	LMOh7858_0292	—
Lmo0264	*inlE*	**LMOf6854_0277**	**LMOf2365_0283**	**LMOh7858_0295**	—
Lmo1136		LMOf6854_1175	LMOf2365_1144	LMOh7858_1209	Lin1101
Lmo0263	*inlH*	**LMOf6854_0276**	**LMOf2365_0281**	**LMOh7858_0293**	—
Lmo0610		LMOf6854_0650	LMOf2365_0639	LMOh7858_0671	Lin0619
Lmo1289		LMOf6854_1331	—	LMOh7858_1375	Lin1328
Lmo1290		**LMOf6854_1332**	**LMOf2365_1307**	**LMOh7858_1376**	—
Lmo0514		LMOf6854_0547	LMOf2365_0543	LMOh7858_0574	Lin0514
Lmo2026		—	—	—	—
Lmo0331		LMOf6854_0363	LMOf2365_0347	LMOh7858_0367	Lin0354
Lmo0732		LMOf6854_0778	LMOf2365_0768	LMOh7858_0798	Lin0741
Lmo0801		LMOf6854_0846	—	—	—
Lmo0433	*inlA*	**LMOf6854_0469**	**LMOf2365_0471**	**LMOh7858_0498**	—
Lmo0409	*inlF*	—	LMOf2365_0429	LMOh7858_0457	—
Lmo0171		LMOf6854_0180	—	—	—
Lmo2821	*inlJ*	**LMOf6854_2939**	**LMOf2365_2812**	**LMOh7858_3085**	—
Lmo2396		LMOf6854_2455	LMOf2365_2370	LMOh7858_2546	Lin2495
Lmo0327		LMOf6854_0336	LMOf2365_0345	LMOh7858_0365	Lin0352
Lmo0333	*inlI*	**LMOf6854_0368**	**LMOf2365_0350**	**LMOh7858_0370**	—
—		LMOf6854_0338	—	—	—
—		LMOf6854_0365	—	—	—
—		—	—	LMOh7858_0291	Lin0290
—		—	—	—	Lin0661
—		—	—	—	Lin0739
—		—	—	—	Lin2724
—		LMOf6854_0367	LMOf2365_0349	LMOh7858_0369	—
—	*inlD*	—	LMOf2365_0282	—	—
—		LMOf6854_0364	LMOf2365_1254	LMOh7858_1321	Lin1204
—		LMOf6854_0833	LMOf2365_0805	LMOh7858_0843	—
—		—	LMOf2365_2416	LMOh7858_2593	Lin2537
LPXTG-motif containing proteins					
Lmo0550		LMOf6854_0591	LMOf2365_0579	LMOh7858_0608	Lin0554
Lmo2714		LMOf6854_2833	LMOf2365_2694	LMOh7858_2978	Lin2862
Lmo0842		**LMOf6854_0887**	**LMOf2365_0859**	**LMOh7858_0898**	—
Lmo0835		**LMOf6854_0880**	**LMOf2365_0852**	**LMOh7858_0891**	—
Lmo0880		LMOf6854_0926	LMOf2365_0899	LMOh7858_0940	Lin0879
Lmo0627		LMOf6854_0667	LMOf2365_0656	LMOh7858_0691	Lin0636
Lmo0435		LMOf6854_0472	—	—	Lin0457
Lmo0463		**LMOf6854_0500**	**LMOf2365_2052**	**LMOh7858_0527**	—
Lmo1413		**LMOf6854_1456**	**LMOf2365_1432**	**LMOh7858_1508**	—
Lmo0320		**LMOf6854_0328**	**LMOf2365_0338**	**LMOh7858_0356**	—
Lmo0160		LMOf6854_0169	LMOf2365_0175	LMOh7858_0183	Lin0203

(Continued)

TABLE 2 *Continued*

EGDE-1/2a	Internalin	F6854-1/2a	F2365-4b	H7858-4b	CLIP11262-6a
Lmo0159		LMOf6854_0168	LMOf2365_0174	LMOh7858_0182	Lin0202
Lmo1115		—	—	—	—
Lmo1799		LMOf6854_1857	—	LMOh7858_1923	Lin1913
Lmo1666		**LMOf6854_1722**	**LMOf2365_1690**	**LMOh7858_1785**	—
Lmo2085		**LMOf6854_2146**	**LMOf2365_2117**	**LMOh7858_2214**	—
Lmo2576		LMOf6854_2637	—		—
Lmo0725		LMOf6854_0772	LMOf2365_0761	LMOh7858_0791	Lin0733
Lmo2179		LMOf6854_2243	LMOf2365_2212	LMOh7858_2313	Lin2283
Lmo2178		LMOf6854_2242	LMOf2365_2211	LMOh7858_2312	Lin2282
Lmo0175		LMOf6854_0184	LMOf2365_0186	LMOh7858_0194	Lin0214
Lmo0130		LMOf6854_0143	LMOf2365_0148	LMOh7858_0155	Lin0177
—		—	LMOf2365_2210	LMOh7858_2312	Lin2281
—		—	LMOf2365_1974	LMOh7858_2075	Lin0141
—		—	LMOf2365_0694	LMOh7858_0723	—
—		—	LMOf2365_0413	LMOh7858_0439	Lin0415
—		—	LMOf2365_0693	LMOh7858_0721	Lin0665
—		—	LMOf2365_0498	LMOh7858_2150	—
—		—	LMOf2365_0374	LMOh7858_0394	Lin0372
—		—	—	—	Lin0559
—		—	—	—	Lin0740
—		—	—	—	Lin0803
GW-module containing proteins					
Lmo0434	*inlB*	**LMOf6854_0470**	**pseudogene**	**LMOh7858_0498**	—
Lmo1215		LMOf6854_1254	LMOf2365_1224	LMOh7858_1289	Lin1178
Lmo1216		LMOf6854_1255	LMOf2365_1225	LMOh7858_1290	Lin1179
Lmo1521		LMOf6854_1568	LMOf2365_1540	LMOh7858_1621	Lin1556
Lmo2203		LMOf6854_2268	LMOf2365_2236	LMOh7858_2337	Lin2306
Lmo2558	*ami*	LMOf6854_2618	LMOf2365_2530	LMOh7858_2709	Lin2703
Lmo2591		LMOf6854_2707	LMOf2365_2564	LMOh7858_2748	Lin2738
Lmo2713		LMOf6854_2832	LMOf2365_2693	LMOh7858_2977	Lin2861
Lmo1076	*auto*	—	—	—	—
—		LMOf6854_1129	LMOf2365_1093	LMOh7858_1143	Lin1064
Secreted internalins					
Lmo1786	*inlC*	**LMOf6854_1845**	**LMOf2365_1812**		—
Lmo2445		LMOf6854_2507	LMOf2365_2418	LMOh7858_2595	Lin2539
Lmo2027		—	—	—	—
Lmo2470		**LMOf6854_2531**	**LMOf2365_2443**	**LMOh7858_2619**	—
Lmo0549		LMOf6854_0590	LMOf2365_0578	LMOh7858_0607	Lin0553
—		LMOf6854_0284	LMOf2365_0289	LMOh7858_0301	Lin0295

[a]EGDe: *Listeria monocytogenes* EGDe (serovar 1/2a); F6854: *L. monocytogenes* F6854 (serovar 1/2a); F2365: *L. monocytogenes* F2365 (serovar 4b); H7858: *L. monocytogenes* H7858 (serovar 4b); CLIP11262: *Listeria innocua* (serovar 6a). Bold type indicates proteins present in all *L. monocytogenes* but not in *L. innocua*.

cells (77), contains three additional members in *L. monocytogenes* as deduced from genome analysis. p60 and two of the three p60 like proteins identified in the EGDe genome (Lmo0394, Spl) are present in all sequenced genomes; the fourth (Lmo1104) is specific for strain *L. monocytogenes* EGDe.

The most abundant class of surface proteins are lipoproteins, a class of bacterial surface proteins that may be implicated in adherence to different substrates, to host tissues, or to other bacteria, as well as in conjugation, signaling, or metabolic functions. Seventy-one lipoproteins, identified by their characteristic signal sequence, are predicted in the *L. monocytogenes* EGDe genome, and 69 are predicted in *L. innocua*. Taken together, 4.7% of the predicted genes of *L. monocytogenes* EGDe are dedicated to surface proteins.

Although the major known virulence factors among the surface proteins like InlA, InlB, or ActA are conserved, there is a pronounced diversity within the surface proteins of the different strains of *L. monocytogenes*, as described above and in Table 2. Except surface proteins present in strain EGDe but missing in the other sequenced *L. monocytogenes* genomes, several surface proteins specific to each of these strains were identified. This indicates that this protein family might be strongly implicated in strain-specific and species-specific features of *Listeria* and that differences in the surface protein repertoire might be related to strain differences in virulence as well as to niche adaptation.

Transporters

Another surprising feature of the *Listeria* genomes is the abundance of transport proteins (e.g., 11.6% of all predicted genes of *L. monocytogenes* EGDe), a feature probably related to its property to colonize a broad range of ecosystems, including eukaryotic cells. These comprise in particular proteins that are dedicated to carbohydrate transport, which probably confers on *Listeria* its ability to colonize a broad range of ecosystems, at least in part. As in most bacterial genomes, the predominant class corresponds to ABC transporters. Interestingly, most of the carbohydrate transport proteins belong to phosphoenolpyruvate-dependent phosphotransferase system (PTS)-mediated carbohydrate transport genes. The PTS allows the use of different carbon sources, and in many bacteria studied so far, the PTS is a crucial link between metabolism and regulation of catabolic operons (4, 75). The *Listeria* genomes contain an unusually large number of PTS loci (e.g., 92 in *L. monocytogenes* EGDe, which is nearly twice as many as *Escherichia coli* and nearly three times as many as *B. subtilis*). Many of these PTS systems are conserved in the different sequenced genomes; however, differences can be observed. An example is the family of β-glucoside-specific PTSs, of which eight are present in *L. monocytogenes* serotype 1/2a, two of those are missing in the *L. monocytogenes* serotype 4b strains, and five are missing from *L. innocua*.

Regulators

Given the fact that *L. monocytogenes* is a ubiquitous, opportunistic pathogen that needs a variety of combinatorial pathways to adapt its metabolism to a given niche, an extensive regulatory repertoire is needed. Indeed, a little more than 7% of the *Listeria* genes predicted in the genomes are dedicated to regulatory proteins. *Listeria* have almost twice as many regulators as *Staphylococcus aureus* despite the similar genome size. Only *Pseudomonas aeruginosa* (117), another ubiquitous, opportunistic pathogen, encodes with over 8% of its predicted genes a higher proportion of regulatory proteins. Interestingly, diversity among the regulatory genes is not very pronounced, neither among the different *L. monocytogenes* genomes nor with respect to *L. innocua*, suggesting their implication primarily in features common to the lifestyle of *Listeria* outside a mammalian host. The most studied regulatory gene of *L. monocytogenes* is *prfA*, which encodes the master regulator of virulence. In line with its function in regulating the expression of genes coding proteins necessary for the entry and for intracellular multiplication of *L. monocytogenes*, PrfA is absent from *L. innocua* but conserved in all *L. monocytogenes* strains.

NEW VIRULENCE FACTORS IDENTIFIED FROM GENOME SEQUENCE

After the determination of the first *Listeria* genome sequences in 2001, genome analysis and comparative genomics have been undertaken to identify new virulence genes. Subsequent functional analyses of many of these genes have led to the identification of several proteins implicated in the establishment and maintenance of infection by *L. monocytogenes*.

Surface Proteins

Among possible bacterial factors interacting with host tissues and involved in virulence, surface proteins are privileged candidates. Many of the listerial surface proteins contain the LPXTG cell wall anchoring motif, and/or belong to the internalin family. Some have already been shown to be important for *Listeria* virulence and host-pathogen interactions. Thus, much research focus was put on this class of proteins. Primarily, the mechanism of LPXTG anchoring was studied. As was first shown for protein A of *Staphylococcus aureus*, the anchoring of LPXTG proteins to the cell wall is mediated by a transpeptidase, called sortase (90). Comparative sequence analysis between *L. monocytogenes* and *Staphylococcus aureus* identified two genes in the *L. monocytogenes* genome that encode putative sortases. Inactivation of *srtA* inhibits anchoring of LPXTG proteins, including InlA, to the peptidoglycan and attenuates virulence after intravenous or oral infection of mice (8, 49). Because InlA does not contribute significantly to virulence of *Listeria* in normal mice after oral infection (79, 81), the attenuation of a sortase A mutant in orally infected mice suggests that it anchors other LPXTG proteins to the peptidoglycan, which are implicated in infection and which remain to be identified.

The second sortase gene is *srtB*. SrtB seems to have only two substrates carrying an NXZTN motif and not an LPXTG motif, and it does not seem to be required for the infectious process (7). The *svpA-srtB* locus contains a well-conserved Fur box, indicating iron reg-

ulation of this locus: real-time polymerase chain reaction analyses and anti-SvpA immunoblot testing showed that *svpA* transcription and SvpA production markedly increased in response to iron deprivation. Thus, iron availability controls through Fur-mediated regulation transcription of the *svpA-srtB* locus and attachment of SvpA to the cell wall (98). Recently, a nongel proteomic analysis of the substrates of SrtA and SrtB identified both substrates of SrtB, Lmo2185, and Lmo2186 (SvpA and SvpB) that recognize the NAKTN sorting motifs and the NKVTN motif, respectively (105). Furthermore, 13 LPXTG-containing proteins were identified as substrates of SrtA (105).

In addition to the identification of the proteins necessary for cell wall anchoring of the *Listeria* LPXTG proteins, the function of several surface proteins identified from the genome was investigated. On the basis of genome analysis and comparison, a GW module protein that is absent from the *L. innocua* and *L. monocytogenes* sv4b genomes was further characterized. Functional analysis revealed that this protein, named Auto, is a novel autolysin that expresses autolytic activity. It is involved in invasion of eukaryotic cells and in virulence (15). Inactivation of the *aut* gene leads to decreased invasiveness of *L. monocytogenes* into several epithelial and fibroblastic cell lines. However, entry of *L. innocua* expressing Auto is similar to that of *L. innocua*, indicating that Auto is necessary but not sufficient for entry (15).

Another surface protein showing autolytic activity was identified through genome analysis, the peptidoglycan hydrolase called MurA. About 15 years ago, the so-called p60 surface protein, which has been shown to possess cell wall hydrolase activity, was identified in *L. monocytogenes* (77). Spontaneous mutants exhibiting deficient expression of p60 show rough colony morphology and form long chains separated by long septa and are reduced in virulence in the mouse model of infection (77). It was initially thought that p60 is essential for cell viability; however, a mutant harboring a transposon in the p60 coding gene has been isolated, suggesting that other proteins are able to compensate for loss of p60 (6, 130). In order to identify this protein, the genome sequence was searched for a candidate identifying *lmo2691* (*murA*) predicted to encode a cell wall hydrolase with similarity to p60 (18). Functional characterization of this protein revealed that it shows murein hydrolase activity and suggests that MurA is involved in generalized autolysis of *L. monocytogenes* (18).

lmo0320 was selected for further functional studies on the basis of genome analysis and comparisons because it is predicted to code an LPXTG protein and because it is absent from the apathogenic *L. innocua* strain. As a result of its implication in virulence, this protein was named Vip, for virulence protein. Vip is positively regulated by PrfA, it is anchored to the bacterial cell wall by sortase A, and it is required for entry in some mammalian cells. Vip is important for infection in vivo and seems to play a role in the crossing of the intestinal barrier, as well as in later stages of the infectious process. The endoplasmic reticulum resident chaperone Gp96 was identified as cellular receptor. Vip thus appears as a new virulence factor exploiting Gp96 as a receptor for cell invasion and/or signaling events that may interfere with the host immune response in the course of infection (16).

A surface protein lacking a classical signal sequence has been identified through signature-tagged mutagenesis and subsequent mapping of the inactivated gene on the *L. monocytogenes* EGDe genome. This protein shows strong homologies to the fibronectin-binding proteins PavA of *Streptococcus pneumoniae*, Fbp54 of *S. pyogenes*, and FbpA of *S. gordonii* and was thus named FbpA for fibronectin-binding protein A (35). It is a substrate for the SecA2 pathway (35), a pathway involved in the secretion of proteins lacking a classical signal sequence (83, 84). Purified FbpA binds to human fibronectin immobilized onto microtiter plate wells, and it contributes to cell adherence. Interestingly, FbpA seems also to modulate the levels of listeriolysin O and InlB at a posttranscriptional stage (35), suggesting that

this molecule could also function as a molecular chaperone preventing degradation of some virulence proteins.

A large-scale comparative genomics approach implicating complete genome sequence comparison and strain comparisons by macroarray hybridization (discussed below) pointed to two surface protein coding genes, *lmo0333* and *lmo2821*, now named *inlI* and *inlJ*, respectively, that are present in all *L. monocytogenes* (113 strains tested) but are absent from all nonpathogenic *Listeria* spp. (32). The *inlJ* deletion mutant is significantly attenuated in virulence after intravenous infection of mice or oral inoculation of hEcad mice (110). *inlJ* encodes an LRR protein that is structurally related to the listerial invasion factor internalin. The consensus sequence of the LRR defined a novel subfamily of cysteine-containing proteins belonging to the internalin family in *L. monocytogenes*. The newly studied surface proteins described here are present in all *L. monocytogenes* strains investigated. Further studies should focus on strain-specific surface proteins because diversity among these may account for strain differences in virulence and in niche adaptation.

Transporters

The importance of transporters for environmental survival and growth was demonstrated with the identification of *hpt*, a gene encoding a hexose phosphate transporter (21). For its identification, the knowledge of the genome sequence was of utmost importance. It was previously observed that *L. monocytogenes* efficiently utilizes, in a PrfA-dependent manner, not only glucose-1-phosphate but also hexose phosphates, including glucose-6-phosphate, fructose-6-phosphate, and mannose-6-phosphate (108). Thus, the uptake of hexose phosphates by *L. monocytogenes* in a PrfA-dependent manner leads to the idea that a PrfA-regulated gene is involved. The genome sequence was screened for yet-unknown PrfA-regulated genes by searching for the known PrfA box sequence, the DNA-binding motif of PrfA (93, 114). One of the genes identified through this screen was predicted to encode a protein similar to UhpT, a hexose phosphate permease found in enteric bacteria. Interestingly, *hpt* is absent from the nonpathogenic *L. innocua*. Functional analysis showed that Hpt allows *L. monocytogenes* to utilize phosphorylated sugars such as glucose-1-phosphate within the host cytosol and thus contributes to the bacterial virulence within the mammalian host cell. Deletion of the *hpt* gene resulted in impaired listerial intracytosolic proliferation and attenuated virulence in mice. Thus, Hpt is involved in the replicative phase of the intracellular parasitism of *L. monocytogenes* and is a clear example that adaptation to intracellular parasitism involves mimicry of physiological mechanisms of the eukaryotic host cell (21). It has been shown that this transporter can mediate the in vivo uptake of the antibiotic fosfomycin, thus leading to fosfomycin sensitivity of intracellular *L. monocytogenes*, despite the fact that *L. monocytogenes* is resistant to fosfomycin when tested in vitro with conventional methods (113).

Regulators

Bacterial survival requires sophisticated and sensitive signal transduction pathways to continuously monitor external conditions and rapidly respond to extracellular and cytoplasmic signals reflecting changes in the environment. Major players among these are the so-called two-component systems, where signal acquisition involves autophosphorylation of a sensor histidine kinase and transduction takes place when the kinase phosphorylates its cognate response regulator protein, leading in turn to a specific alteration of gene expression and providing bacteria with the flexibility to adapt to a wide variety of stimuli. Signal-transducing histidine kinase/response regulator pairs control functions ranging from bacterial development to pathogenicity (for a review, see reference 62). These paired families of highly conserved proteins appear to be ubiquitous throughout the bacterial world. The information provided by the genome sequence allowed the identification of 15 histidine kinases and 16 response regulators in the *L. monocytogenes* genome. Some have

already been studied in more detail: *lisR*/*lisK* (26, 27); *cheY*/*cheA* (46), *agrA*/*agrC* (3), and *cesR*/*cesK* (69); and *virR*/*virS* (89). These different two-component systems have been shown to be implicated in acid, ethanol, and oxidative stress; osmosensing and osmoregulation; and swarming and virulence, respectively.

The two-component systems *virR*/*virS* had been identified through signature-tagged mutagenesis (STM) (89). Deletion of *virR* greatly decreased virulence in mice and its invasion in cell-culture experiments. By means of a transcriptomic approach, 12 genes regulated by VirR, including the *dlt* operon, were identified. Another VirR-regulated gene is homologous to *mprF*, which encodes a protein that modifies membrane phosphatidyl glycerol with L-lysine and which is involved in resistance to human defensins in *Staphylococcus aureus*. VirR thus appears to control virulence by a global regulation of surface-component modifications.

Recently a systematic approach to study the role of these putative two-component systems was undertaken. Williams and colleagues (128) introduced in-frame deletions into 15 of 16 response regulator genes and characterized the resulting mutants. In this study, the deletion of the individual response regulator genes had only minor effects on in vitro and in vivo growth of the bacteria, except for DegU. The mutant carrying a deletion in the ortholog of the *Bacillus subtilis* response regulator gene *degU* showed a clearly reduced virulence in mice, indicating that DegU is involved in the regulation of virulence-associated genes. This regulator is particularly interesting because its kinase, DegS, is missing in *L. monocytogenes*. Two-dimensional gel electrophoresis, mass spectrometry, and microarray analyses identified that the two *L. monocytogenes* operons encoding flagellum-specific genes and the monocistronically transcribed *flaA* gene are positively regulated by DegU at 24°C, but are not expressed at 37°C. In accordance with these results, an isogenic mutant of *L. monocytogenes* EGDe with a deletion of the response regulator gene *degU* lacked motility as a result of the absence of flagella (129).

Another regulator implicated in flagellar regulation and virulence of *L. monocytogenes* was identified recently by using affinity purification of proteins bound to the *flaA* promoter, identification by mass spectrometry, and subsequent search of the genome sequence. This approach identified *lmo0674*, which was not previously predicted to encode a regulatory protein. This protein was designated MogR, for motility gene repressor, because it was shown to be a transcriptional repressor of flagellar motility and required for virulence (59). Deletion of *mogR* reduced the capacity of *L. monocytogenes* for cell-to-cell spread, and a 250-fold decrease of in vivo virulence in a mouse infection model was observed (59). Most interestingly, MogR is the first known example of a transcriptional repressor functioning as a master regulator controlling nonhierarchical expression of flagellar motility genes (116). It tightly represses expression of flagellin (FlaA) during extracellular growth at 37°C and during intracellular infection. MogR is also required for full virulence in a murine model of infection. The severe virulence defect of MogR-negative bacteria is due to overexpression of FlaA. Specifically, overproduction of FlaA in MogR-negative bacteria causes pleiotropic defects in bacterial division (chaining phenotype), intracellular spread, and virulence in mice. MogR represses transcription of all known flagellar motility genes by binding directly to a DNA recognition site. MogR repression of transcription is antagonized in a temperature-dependent manner by the DegU response regulator, and DegU further regulates FlaA levels through a posttranscriptional mechanism (116).

Gene regulation by small RNAs has been recognized as an important posttranscriptional regulatory mechanism for several years. A variety of mechanisms utilized by small RNAs to control gene expression, usually at the level of mRNA translation and/or degradation, has been described, mainly in *Escherichia coli* (53). In most bacteria, these RNAs control stress response and translation of secreted proteins; therefore, they are key regulators of environmental adaptation, particularly during host infection. Small RNA regulators are proving to be multifunctional and have provided explanations for a number of previously mysterious regulatory effects. The availability of complete genome sequences and powerful bioinformatics tools allows the search for small regulatory RNAs at the genome level. Thus, in recent years, genome-wide screens for novel small noncoding RNAs have been performed in various bacteria by using both computational and experimental approaches (for a review, see reference 124). Recently, Christiansen and colleagues identified the first three sRNA molecules in *L. monocytogenes* by coimmunoprecipitation with Hfq, followed by enzymatic sequencing of the Hfq-binding RNA molecules (22). These small RNAs were named Lhr for Listeria Hfq-binding RNA. LhrA has a size of 268 bp, LrhB of 150 bp, and LrhC of 100 bp. Their characterization revealed a number of unique features displayed by each individual sRNA. LrhA is encoded from within an annotated gene (*lmo2257*) in the *L. monocytogenes* EGDe genome. Analogous to most regulatory sRNAs in *Escherichia coli*, the stability of this sRNA is highly dependent on the presence of Hfq. LrhB appears to be produced by a transcription attenuation mechanism, and LrhC is present in five copies at two different locations within the *L. monocytogenes* EGDe genome. The cellular levels of the sRNAs are growth phase dependent and vary in response to growth medium. All three sRNAs are expressed when *L. monocytogenes* multiplies within mammalian cells (22).

Genes may also be controlled by a posttranscriptional mechanism involving 5′-untranslated region (5′-UTR). Recently it has been shown that the translation of three *L. monocytogenes* virulence determinants (*inlA*, *hly*, and *actA*) is controlled by their 5′-UTR (115, 118, 131). Furthermore, the 5′-UTR preceding the virulence regulator PrfA was shown to act as a thermosensor in which the UTR controls regulation (67). A genome-wide search for 5′-UTR located in front of genes that encode known virulence factors and surface proteins in *L. monocytogenes* was conducted. This study identified 43 putative UTRs preceding genes that encode surface proteins, suggesting that the expression of InlC, Mpl, UhpT, BsH, BilE, FbpA, InlJ, and Vip is controlled by a posttranscriptional mechanism (86).

Physiology and Metabolism

It is generally accepted that *L. monocytogenes* is an aerobically growing microaerophilic (carbon dioxydophilic) organism. However, *L. monocytogenes* is able to survive and to colonize the mammalian gut where it encounters anaerobic conditions. Analysis of the genome sequence identified a single continuous locus that coded for the proteins necessary for vitamin B_{12} synthesis in *L. monocytogenes* and in *L. innocua*. Furthermore, the proteins necessary for degradation of the carbon sources ethanolamine and propanediol in a coenzyme B_{12}-dependent manner are encoded within the same region. Vitamin B_{12} synthesis and degradation of ethanolamine and propanediol may be important for anaerobic growth of *Listeria*.

The ability to synthesize vitamin B_{12} is unevenly distributed in living organisms. A search in currently available genome sequences reveals that vitamin B_{12} biosynthesis genes are found in just over one-third of the bacteria sequenced so far (for a review, see reference 107). The proteins deduced from the genome sequence of *L. monocytogenes* and *L. innocua* share the highest homology with those of *S. enterica* serovar Typhimurium (38 to 65% protein identity). Genes coding for CbiD, CbiG, and CbiK, specifically associated with the anaerobic pathway, are present in *Listeria*. This suggests that the two *Listeria* species also contain the oxygen-independent pathway, like *Salmonella*. Downstream of the cobalamin biosynthesis genes, *Listeria* contain orthologs of genes necessary in *Salmonella* for the coenzyme B_{12}-dependent degradation of ethanolamine and propanediol. In *Listeria*, all three gene clusters seem to have been acquired en bloc by horizontal gene transfer because they are organized contiguously around the terminus of replication (13). *S. enterica* serovar Ty-

phimurium synthesizes vitamin B_{12} anaerobically (66) and can use ethanolamine and 1,2-propanediol as the sole carbon and energy source for growth (104). Vitamin B_{12}-dependent anaerobic degradation of ethanolamine and propanediol could enable *L. monocytogenes* to use ethanolamine and 1,2-propanediol as a carbon and energy source for growth under the anaerobic conditions encountered in the mammalian gut, where both substances are believed to be abundant. Taken together, it is tempting to speculate that the vitamin B_{12} synthesis genes together with the *pdu* and *eut* operons play a role during listerial infection.

Genome analysis also allowed the identification of a previously unknown and alternative route of glutathione (GSH) biosynthesis in *Listeria*, which probably plays an important role in protection against oxidative stress (52). GSH is the predominant low-molecular-weight peptide thiol present in living organisms. In bacteria, it plays a pivotal role in many metabolic processes, such as thiol redox homeostasis, protection against reactive oxygen species, protein folding, and provision of electrons via NADPH to reductive enzymes such as ribonucleotide reductase. A broad survey of the distribution of thiols in microorganisms revealed that several species of gram-positive bacteria, including *Listeria*, streptococci, and enterococci, produce significant amounts of GSH (97). However, the source of GSH in these bacteria has remained a puzzle because their genomes do not contain a canonical *gshB* gene (42). Copley and Dhillon (23) identified, among others in the *L. monocytogenes* genome, a gene containing an N-terminal domain that encodes a molecule significantly related to bacterial γ-glutamylcysteine ligases (GshA) and a C-terminal domain that encodes a molecule that bears little resemblance to typical bacterial GSH synthetases (GshB) but is clearly related to the ATP-grasp superfamily of proteins. Gopal and colleagues (52) demonstrated that this gene encodes a multidomain protein (termed GshF) that carries out complete synthesis of GSH. Furthermore, GSH biosynthesis seems to be involved in virulence of *L. monocy-*

togenes as a *gsf* deletion mutant is not only defective in GSH synthesis but also impaired in growth and survival in mouse macrophages and Caco-2 enterocyte-like cells (52).

Another protein identified from analysis and comparison of the genome sequences of *L. monocytogenes* and *L. innocua* is a bile salt hydrolase, which is absent from *L. innocua* and other nonpathogenic *Listeria* spp. Bile salt hydrolase is an enzyme that deconjugates conjugated bile salts (38). Bile salt hydrolysis (BSH) is believed to be a mechanism to protect bacteria from bile salt toxicity (60). The *bsh* gene (*lmo2067*) of *L. monocytogenes* is controlled by sigma B (94, 119) and by PrfA (38). Moreover, bile salt hydrolysis activity increases at low oxygen concentration. Deletion of *bsh* results in decreased resistance to bile in vitro, reduced bacterial fecal carriage after oral infection, reduced virulence after intravenous infection, and a reduced level of liver infection, demonstrating that bile salt hydrolase is a novel *L. monocytogenes* virulence factor involved in the intestinal and hepatic phases of listeriosis.

The ability of microorganisms to adapt their metabolism to a restrictive environment has undoubtedly contributed to their success as pathogens. An example is the above-mentioned hexose phosphate transporter (Hpt), which allows pathogenic *Listeria* spp. to use hexose phosphates in the host cytosol as a carbon and energy source (21). Another example is the lipoate protein ligase LplA1, which seems to allow *L. monocytogenes* the use of host-derived lipoic acid. LplA1 is important for in vivo replication of *L. monocytogenes*, as shown in the mouse model. An *L. monocytogenes* strain lacking the lipoate protein ligase LplA1 is defective for growth specifically in the host cytosol and is less virulent in animals by a factor of 300 (100).

GLOBAL GENE EXPRESSION PROFILING IDENTIFIES VIRULENCE GENES AND INTRACELLULAR GROWTH REQUIREMENTS

Knowledge of the *L. monocytogenes* genome sequence has allowed researchers to use global gene expression analysis in diverse conditions by means of microarrays. To date, several analyses of the regulons of some *L. monocytogenes* regulators and the study of gene expression in the infected host have been undertaken in order to identify new virulence genes and to better understand the intracellular life of *L. monocytogenes*.

The most important virulence genes of *L. monocytogenes* are regulated by the transcriptional activator PrfA (82, 92). With the aim of elucidating the complete set of genes regulated by PrfA, a systematic approach was undertaken by Milohanic and colleagues, who used whole-genome macroarrays to compare the expression profiles of a wild-type and a *prfA* deletion mutant in three different conditions (94). By this approach, three groups of differently regulated genes were identified. Group I comprises, in addition to the 10 previously known (*prfA, plcA, hly, mpl, actA, plcB, inlA, inlB, inlC,* and *hpt*), 2 new PrfA-regulated genes (*lmo2219* and *lmo0788*); all of them are preceded by a PrfA binding site. Gene *lmo2219* encodes a protein similar to PrsA of *B. subtilis,* a posttranslation molecular chaperon (73, 74) and gene *lmo0788* code probably for a membrane protein of unknown function. Group II comprises 8 negatively regulated genes, and group III contains 53 genes, of which only 2 are preceded by a PrfA-binding site. However, a binding site for the alternative sigma factor B was identified for most of the group III genes. These results suggested that PrfA can act either as an activator or as a repressor, and that PrfA may activate or repress different sets of genes in association with different sigma factors. Furthermore, this study suggests for the first time that an interplay between PrfA and sigma B exists (94).

A coupled bioinformatics/microarray approach applied to identify sigma B–regulated genes confirmed the overlap between the PrfA and the sigma B regulon (70). In this study, SigB-dependent promoter sequences were searched in the *L. monocytogenes* EGDe genome sequence. This search led to the identification of 170 candidate sigma B–dependent promoter sequences. On the basis of this bioinformatics result, a specialized 208-gene microarray, which included 166 genes downstream of the predicted SigB-dependent promoters as well as selected virulence and stress response genes, was developed (70). Hybridization of this array with RNA from wild-type and a *sigB* deletion mutant identified 55 clearly SigB-dependent genes, including both stress response genes (e.g., *gadB, ctc,* and the GSH reductase gene *lmo1433*) and virulence genes (e.g., *inlA, inlB,* and *bsh*). These data suggest that SigB is not only important for survival under environmental stress conditions, but also contributes to regulation of virulence gene expression in *L. monocytogenes* (70). Another approach using in silico *sigB* promoter search and proteomics of the compared wild-type and a *sigB* deletion mutant in response to acid stress, high hydrostatic pressure, and freeze survival in *L. monocytogenes* EGDe identified many of the same genes, thus corroborating the microarray results (126).

Another alternative sigma factor, σ^{54}, encoded by the *rpoN* gene, was investigated by comparing the global gene expression of the wild-type *L. monocytogenes* EGDe and an *rpoN* mutant (1). In the conditions used, most of the genes whose expression was modulated belonged to carbohydrate metabolism, in particular to pyruvate metabolism. However, only one operon, *mptACD*, seems to be directly controlled by σ^{54}, whereas the change in gene expression for the remaining over 70 genes seems to be indirectly controlled. RpoN seems thus mainly involved in carbohydrate metabolism (1).

A challenging question to answer is, what is the gene expression and the growth requirement of intracellular *L. monocytogenes*? Two different studies tried to answer this question by using global expression profiling of intracellular grown bacteria. Although many key steps in the intracellular lifestyle of this gram-positive pathogen are well characterized, our knowledge about the factors required for cytosolic growth is still rather limited. Comparison of the transcriptional profile of intracellular *L. monocytogenes* in epithelial cells and bacteria grown in brain heart infusion medium identi-

fied approximately 19% of the genes as differentially expressed. These included genes encoding transporter proteins essential for the uptake of carbon and nitrogen sources, factors involved in anabolic pathways, stress proteins, transcriptional regulators, and proteins of unknown function. Interfacing the results obtained from screening a random mutant library and expression profiling provided evidence that *L. monocytogenes* can use alternative carbon sources like phosphorylated glucose and glycerol and nitrogen sources like ethanolamine during replication in epithelial cells and that the pentose phosphate cycle, but not glycolysis, is the predominant pathway of sugar metabolism in the host environment. Furthermore, the synthesis of arginine, isoleucine, leucine, and valine, as well as a species-specific PTS, plays a major role in the intracellular growth of *L. monocytogenes* (68).

A second study used the same approach but investigated the intracellular gene expression profile of *L. monocytogenes* EGDe in the murine macrophage cell line P388D1 inside the vacuolar and cytosolic environments (20). Similar to the results obtained by Joseph and colleagues (68), 17% of the total genome was mobilized to enable adaptation for intracellular growth. Intracellularly expressed genes showed responses typical of glucose limitation within bacteria, with a decrease in the amount of mRNA encoding enzymes in the central metabolism and a temporal induction of genes involved in alternative-carbon-source utilization pathways and their regulation. Adaptive intracellular gene expression involved genes that are associated with virulence, the general stress response, cell division, and changes in cell wall structure and included many genes with unknown functions. A total of 41 genes were species specific, being absent from the genome of the nonpathogenic *L. innocua* CLIP 11262 strain, and 25 genes were strain specific, i.e., absent from the genome of the previously sequenced *L. monocytogenes* F2365 serotype 4b strain, suggesting heterogeneity in the gene pool required for intracellular survival of *L. monocytogenes* in host cells (20).

Further expression profiling studies using microarrays and different mutants and growth conditions of this intracellular pathogen, as well as refined functional analysis of the differentially expressed genes, will lead to a more in-depth understanding of the virulence mechanisms of *L. monocytogenes* and of intracellular pathogens in general.

COMPARATIVE GENOMICS REVEALS IMPORTANT DIVERSITY AMONG *L. MONOCYTOGENES* STRAINS AND *LISTERIA* SPECIES

Within the species *L. monocytogenes*, differences in virulence among strains seem to exist. Epidemiological data indicate that not all strains of *L. monocytogenes* are equally capable of causing disease in humans. Isolates from only four (1/2a, 1/2c, 1/2b, 4b) of the 13 serovars identified within this species are responsible for over 98% of the human listeriosis cases reported (64). Furthermore, all major foodborne outbreaks of listeriosis, as well as the majority of sporadic cases, have been caused by serovar 4b strains, suggesting that strains of this serovar may possess unique virulence properties. Heterogeneity in virulence has also been observed in the mouse infection model (11). A number of different typing and population genetic studies suggested that different genetic divisions/lineages exist within the species *L. monocytogenes*, which correlate with serovars (1/2a, 1/2c, and 3c strains as lineage I; 4b, 1/2b, and 3b strains as lineage II; and 4a and 4c strains as lineage III) (5, 12, 57, 101). Genetic analyses by multilocus sequence typing of virulence-associated genes, restriction fragment length polymorphism analysis, and ribotyping suggested that epidemic strains are mostly found in lineage II and sporadic strains in lineage I and II, whereas lineage III strains are extremely rare and are mostly animal pathogens (65, 127). Sequencing of housekeeping and virulence genes suggested that serovars 4b, 1/2b, and 3b are highly clonal, while serovars 1/2a, 1/2c, and 3c show greater diversity and more evidence for horizontal gene transfer (99).

The availability of genome sequences now allows the use of other techniques, such as DNA arrays, to investigate and describe diversity among *Listeria* strains belonging to different lineages or different populations or showing different epidemiological characteristics in order to explore whether genetic differences among and within the previously described lineages exist and, if so, whether they can be attributed to virulence differences and different niche adaptation. Different studies have been undertaken with the aim of answering these questions.

DNA arrays carrying the specific gene pool of three sequenced *Listeria* genomes (*L. monocytogenes* EGDe serovar 1/2a, *L. monocytogenes* CLIP80459, *L. innocua* CLIP11262) as well as genes coding known virulence and surface proteins were used to evaluate the variability in gene content within a *Listeria* strain collection (93 *L. monocytogenes* and 20 *Listeria* spp.) covering all serovars, species, and epidemiological characteristics of the genus *Listeria*. The large data set obtained from 113 *Listeria* strains allowed grouping of strains according to shared genetic profiles. The strains clustered according to their species definition and the *L. monocytogenes* strains were subgrouped into the previously defined three lineages (I, II, and III). Within each lineage, two subdivisions were distinguished, and specific markers were identified (Fig. 2; Table 3).

Nineteen genes were associated specifically with lineage I (Table 3, group A). The *bvr* locus (*bvrABC*), a β-glucoside-specific PTS system (10), was present only in isolates of lineage I and in the two 4c strains. This finding was surprising because this locus was previously described as being implicated in virulence gene expression (10). Eight genes allowed the subdivision of lineage I. They were present in lineage I.2 (serovars 1/2c and 3c) but generally absent from lineage I.1 (serovars 1/2a and 3a) (Table 2, group B). Five of the 53 serovar 4b–specific genes were identified as markers for lineage II (Table 2, group C). Two code for transcriptional regulators, and three code for surface proteins containing an LPXTG anchor. Because

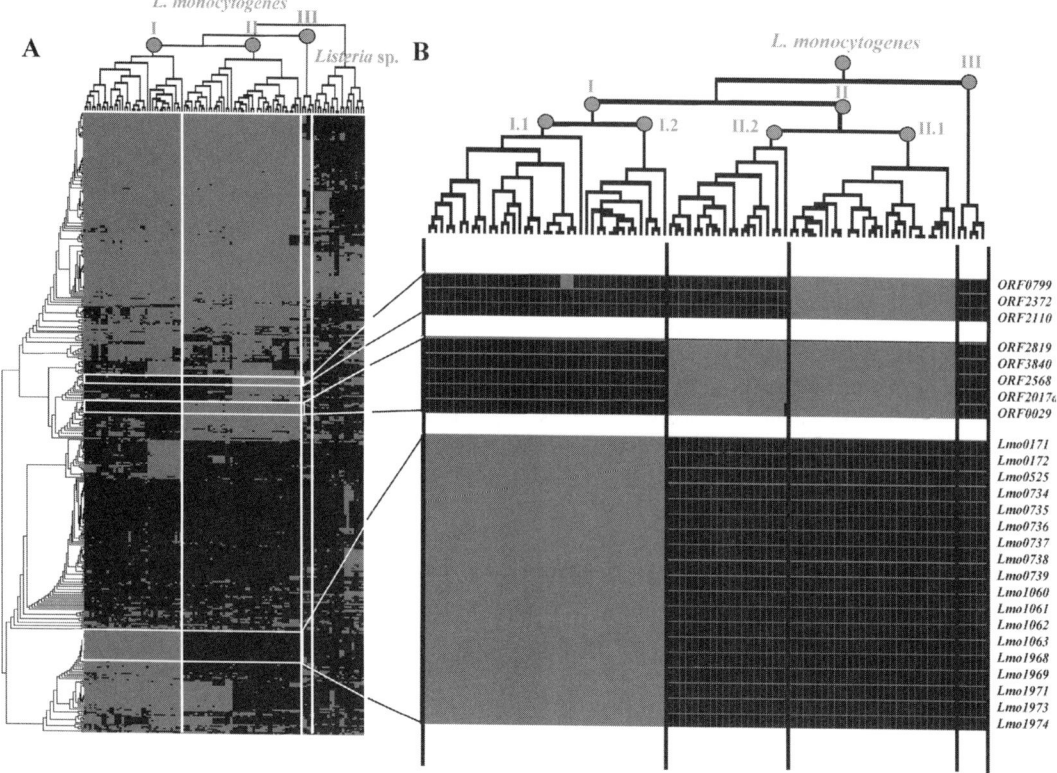

FIGURE 2 *Listeria* genetic diversity. Gray and black denote presence and absence of genes, respectively. (A) The dendrogram shows estimates of genomic relationships of the 113 strains constructed by hierarchical cluster analysis with the program J-Express. Phylogenetic lineages and subgroups are indicated. (B) Enlargements represent the blocs of lineage-specific genes whose numbers are indicated on the right-hand side. I: lineage I (serovars 1/2a; 1/2c, 3a, 3c); II: lineage II (serovars 4b, 4d, 4e,1/2b, 3b); III: lineage III (serovars 4a, 4c); I.1: serovars 1/2a, 3a; I.2: serovars 1/2c, 3c; II.1: serovars 4b, 4d, 4e; II.2: serovars 1/2b, 3b.

serovar 4b strains are mainly responsible for human listeriosis, it is of particular interest to identify markers for 4b strains. Interestingly, 35 of the 53 serovar 4b genes spotted on the array were conserved in all 4b strains, suggesting their implication in characteristic features of 4b strains. Taken together, the gene content comparisons showed that the previously defined lineages are also reflected in specific gene content (32). The identification of selective markers for the different subpopulations are an essential contribution for the construction of rapid, accurate identification and subtyping tools, and led to the development of molecular serotyping by multiplex PCR (31, 33), which

should now be a powerful tool applicable in health institutions and the food industry. Finally, the identification of genes consistently absent or present in epidemic-associated *L. monocytogenes* strains opens the way for mutational and functional analysis of these genes in order to decipher the molecular basis for the increased pathogenic potential of certain *L. monocytogenes* strains.

Further analysis of the gene content of these strains with respect to virulence genes revealed that all known virulence factors (*inlAB, prfA, plcA, hly, mpl, actA, plcB, uhpT,* and *bsh*) are present in all *L. monocytogenes* strains tested. Thus, other as yet uncharacterized factors must be

TABLE 3 *L. monocytogenes* lineage-specific marker genes

Gene name	No. of strains with marker genes						Functional category
	Lineage I		Lineage II		Lineage III		
	I.1 (1/2a, 3a) (27 strains)	I.2 (1/2c, 3c) (12 strains)	II.1 (4b, 4d, 4e) (27 strains)	II.2 (1/2b, 3b) (20 strains)	III.1 (4a) (3 strains)	III.2 (4c) (2 strains)	
A Lmo 0171	27	12	0	0	0	0	Cell surface proteins
Lmo 0172	27	12	0	0	0	0	Transposon and IS
Lmo 0525	27	12	0	0	0	0	Unknown
Lmo 0734	27	12	0	0	0	0	Regulation
Lmo 0735	27	12	0	0	0	0	Specific pathways
Lmo 0736	27	12	0	0	0	0	Specific pathways
Lmo 0737	27	12	0	0	0	0	Unknown
Lmo 0738	27	12	0	0	0	0	Transport/binding proteins and lipoproteins
Lmo 0739	27	12	0	0	0	0	Specific pathways
Lmo 1060	27	12	0	0	0	0	Regulation
Lmo 1061	27	12	0	0	0	0	Sensors
Lmo 1062	27	12	0	0	0	0	Transport/binding proteins and lipoproteins
Lmo 1063	27	12	0	0	0	0	Transport/binding proteins and lipoproteins
Lmo 1968	27	12	0	0	0	0	Metabolism of amino acids
Lmo 1969	27	12	0	0	0	0	Specific pathways
Lmo 1970	27	12	0	0	0	0	Metabolism of lipids
Lmo 1971	27	12	0	0	0	0	Transport/binding proteins and lipoproteins
Lmo 1973	27	12	0	0	0	0	Transport/binding proteins and lipoproteins
Lmo 1974	27	12	0	0	0	0	Regulation
bvrC	27	12	0	0	0	**2**	Unknown
bvrB	27	12	0	0	0	**2**	Transport/binding proteins and lipoproteins
B Lmo 0151	3	12	0	0	0	0	Unknown
Lmo 0466	2	12	0	0	0	0	Unknown
Lmo 0467	2	12	0	0	0	0	Unknown
Lmo 0469	2	12	0	0	0	0	Unknown
Lmo 0470	2	12	0	0	0	0	DNA restrictions and modifications

(Continued)

TABLE 3 *Continued*

Gene name	Lineage I		Lineage II		Lineage III		Functional category
	I.1 (1/2a, 3a) (27strains)	I.2 (1/2c, 3c) (12 strains)	II.1 (4b, 4d, 4e) (27 strains)	II.2 (1/2b, 3b) (20 strains)	III.1 (4a) (3 strains)	III.2 (4c) (2 strains)	
Lmo 0471	2	12	0	0	0	0	Unknown
Lmo 1118	1	12	0	0	0	0	Unknown
Lmo 1119	1	12	0	0	0	0	DNA restrictions and modifications
C *ORF2819*	0	0	27	20	0	0	Unknown, similar to hypothetical transcriptional regulator
ORF3840	0	0	27	20	0	0	Unknown, similar to transcriptional regulator
ORF2568	0	0	27	20	0	0	Unknown, similar to internalin proteins, putative peptidoglycan bound protein (LPXTG)
ORF2017	0	0	27	20	0	0	Unknown, similar to internalin proteins, putative peptidoglycan bound protein (LPXTG)
ORF0029	0	0	27	19	0	0	Unknown, similar to internalin proteins, putative peptidoglycan bound protein (LPXTG)
D *ORF0799*	0	0	27	0	0	0	Unknown
ORF2372	0	0	27	0	0	0	Unknown, similar to teichoic acid protein precurser C
ORF2110	0	0	27	0	0	0	Unknown, putative secreted protein

No. of strains with marker genes

responsible for these differences (32). Probably surface proteins play a role. The pronounced diversity among LPXTG proteins, identified by whole-genome comparisons, was further substantiated by this comparative genomics study using DNA/DNA array hybridization. The distribution of 55 genes coding for putative surface proteins belonging to three sequenced *Listeria* genomes (*L. monocytogenes* EGDe serovar 1/2a, *L. monocytogenes* CLIP80459) was investigated. This identified 25 surface protein-coding genes of the internalin/LPXTG/GW motif containing family as specific for the species *L. monocytogenes,* including *inlAB* (32). On the basis of these results, two internalin protein-encoding genes present in all genomes of *L. monocytogenes* strains tested but absent from all other *Listeria* species were selected for functional characterization. They were named *inlI* (*lmo0333*) and *inlJ* (*lmo2821*) (110).

The distribution of surface proteins among the *L. monocytogenes* strains also mirrored the three lineages as each lineage, and each subgroup within a lineage, is characterized by a specific surface-protein combination (32). One of the important questions is whether epidemic *L. monocytogenes* grouped in lineage II are characterized by a specific gene content, which may explain its higher potential to cause human listeriosis. Indeed, the DNA/DNA array hybridization, as well as previous studies, attributes specific gene content to these strains (32, 41, 132). Epidemic *L. monocytogenes* strains isolated and investigated in the United States are defined by specific markers distinguishing them from other serovar 4b strains, and the array-based study identified 35 serovar 4b–specific genes, some of which are coding for surface proteins. However, most of these markers identified to date are not functionally characterized, or their characterization did not yet allow the reason for the higher prevalence of serovar 4b strains in human listeriosis to be defined. In contrast, the fact that some strains seem less virulent for humans may be related to missing genes. Most interestingly, in the rarely isolated *L. monocytogenes* serovar 4a

strains (lineage III), which are mostly animal pathogens, 13 of the 25 *L. monocytogenes* specific surface proteins, including all known internalins except *inlAB,* were missing. The lack of these surface proteins as well as the lack of additional genes of as yet unknown function may explain why lineage III strains are mainly found in animals but not in human listeriosis. This is in line with a recent study that investigated *L. monocytogenes* populations present at the farm, in food processing plants, at retail, and in the human population. Analyses of over 400 strains suggested that *L. monocytogenes* populations adapted to different niches exist. This study identified one dominant strain among the strains collected from the food processing plants. This major pork product strain was not identified among human isolates, indicating its adaptation to its particular niche. DNA array characterization of these strains and its comparison to a gene content database of 295 *L. monocytogenes* strains identified a specific genetic profile and a specific pattern of presence and absence of 15 genes. Interestingly, five of the genes specifically missing in this group are predicted to encode internalins and cell surface proteins. Thus, it seems again that surface protein diversity contributes to the different potentials to multiply in different niches (63). The elucidation of the functions of the different surface proteins and the putative strain specific characters they confer will be one of the challenging questions for the future and may provide additional insights into the tropism of *L. monocytogenes* toward different cell types.

When these results are taken together, the species *L. monocytogenes* and even the genus *Listeria* show a well-conserved genome organization and a high number of orthologous genes, but also a considerable number of strain-specific traits, most of them organized in many small plasticity zones. Furthermore, the three subgroups present within the species *L. monocytogenes* clearly correspond to distinct evolutionary lineages, with specific histories of gene gain and loss by which each lineage has adapted to a primary niche and probably also to a different virulence potential for humans.

FROM ENVIRONMENT TO LIFE AS HUMAN INTRACELLULAR PATHOGEN

The natural habitat of *L. monocytogenes* is thought to be a surface layer of soil rich in decaying vegetation (44). In a study investigating the incidence of *Listeria* in nature, 20% were isolated from soil and plant material, 15% from feces of deer and stag, 27% from moldy fodder and wildlife feeding ground, and 17% from birds (125). Thus, it was suggested that *L. monocytogenes* is a saprophytic organism, which lives in plant-soil environments (125). However, *L. monocytogenes* is also an intracellular pathogen, capable of causing serious infection in humans and many animal species (39, 109, 123). How did *L. monocytogenes* acquire the capacity to infect humans and animals, to be able to promote its uptake into eukaryotic cells, to resist the immune defense, and to be able to replicate in the eukaryotic cell?

The capacity to resist eukaryotic cells may have evolved for the purposes of avoiding predation by, or predating upon, environmental microscopic eukaryotes, such as amoebae and nematodes (58, 88, 120). However, the origin of the known virulence genes is still unclear, although comparative sequence analysis provides insight into the possible evolution of pathogenesis in *Listeria*. There are two major virulence loci in *L. monocytogenes*, the *inlAB* locus coding for the major invasion proteins and the virulence gene cluster coding for the main proteins for intracellular replication, together with their positive regulator PrfA (*prfA*, *plcA*, *hly*, *mpl*, *actA*, and *plcB*). Many different studies have investigated the virulence gene cluster and its flanking regions in all *Listeria* species by sequencing and subsequent comparison (17, 19, 55, 76, 111). A fully functional virulence gene cluster is present only in *L. monocytogenes* and *L. ivanovii*, and a similar cluster with additional genes is present in *L. seeligeri* (Fig. 3). Comparison of the genomic region in *L. monocytogenes*, *L. innocua*, and *B. subtilis* indicated that this virulence gene cluster was acquired by a common ancestor of *Listeria* and that *L. innocua* subsequently lost most of it (51) (Fig. 3).

Analysis of the sequence of the virulence gene cluster locus and its flanking regions and phylogenetic analyses on the sequences of *prs*, *ldh*, *orfA*, and *orfB* (all directly flanking the virulence gene cluster), as well as the *iap* gene and the 16S and 23S rRNA coding genes, which are located at different sites in the listerial chromosomes, indicated that *L. grayi* represents the deepest branch within the genus. The remaining five species form two groups. One lineage represents *L. monocytogenes* and *L. innocua*; the other contains *L. welshimeri*, *L. ivanovii*, and *L. seeligeri*. The most likely scenario is that the virulence gene cluster was acquired by the common ancestor of *L. monocytogenes*, *L. innocua*, *L. ivanovii*, *L. seeligeri*, and *L. welshimeri* and that the pathogenic capability has been lost in two separate events represented by *L. innocua* and *L. welshimeri*. This hypothesis is also supported by the location of the putative deletion breakpoints of the virulence gene cluster within *L. innocua* and *L. welshimeri* (111) (Fig. 3).

The origin of the large internalins like InlA and InlB is thought to have evolved in *Listeria* after initial combination of the LPXTG membrane anchor motif with an LRR motif to a protointernalin. This protointernalin was then probably duplicated several times and evolved by further intragenic recombination, mutations, and positive selection (51, 122). However, it is also possible that the *inlAB* locus evolved in a similar way as the virulence gene cluster. Comparison of the organization of the *inlAB* locus and its flanking regions in different *L. monocytogenes* and *L. innocua* strains indicates that it is present at the same chromosomal location in all strains (Fig. 4). Furthermore, this chromosomal region might be a plasticity region because it has undergone several insertion and deletion events during evolution. Interestingly, the organization of the *inlAB* chromosomal region shows lineage-specific organization. As shown in Fig. 4, the comparison of the genomic region where the *inlAB* genes are pres-

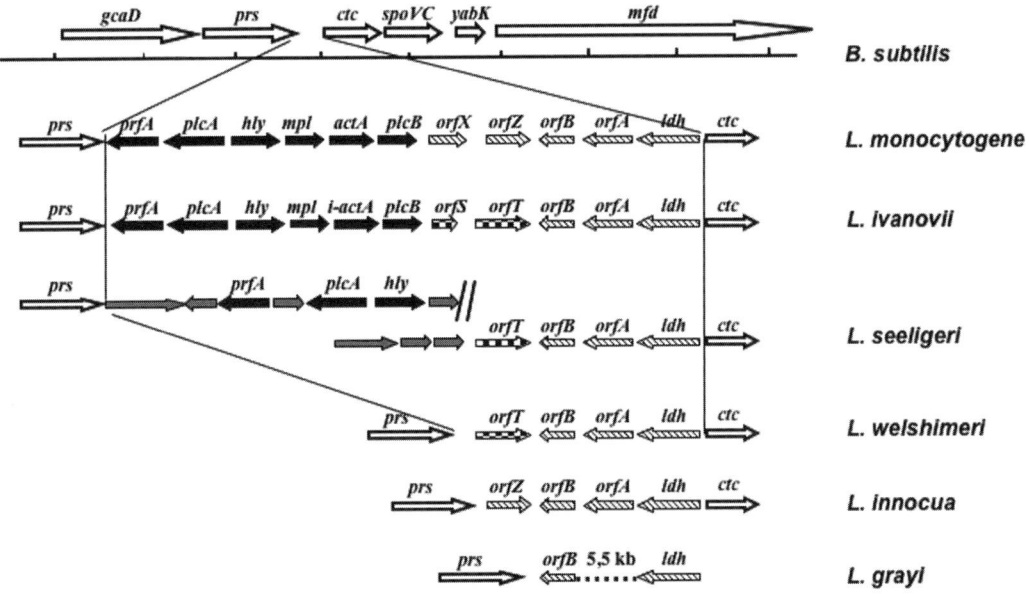

FIGURE 3 Schematic presentation of the virulence gene cluster in *Listeria* and its comparison with the orthologous region in *Bacillus subtilis*. Orthologous genes among the different *Listeria* sp. are depicted in the same color or shading pattern. The gene cluster is flanked by the housekeeping genes (light blue arrows) *prs* and *ldh* in all six species of *Listeria* and are also present in *B. subtilis*. Known virulence genes de facto or potentially regulated by PrfA are depicted in red (freely adapted from references 19, 51, and 111).

ent among the five sequenced *L. monocytogenes* genomes and the one *L. innocua* genome is clearly lineage dependent, with lineage I (serovar 1/2a strains) and linage II (serovar 4b strains) each having a similar organization, but one different from each other.

Comparison of the *inlAB* locus region with the orthologous part of the *L. innocua* genome allows an evolutionary scenario for how this region might have evolved to be proposed. A common ancestor of *L. monocytogenes* and *L. innocua* might have acquired a DNA region carrying the *inlAB* locus as well as the downstream LPXTG protein and the upstream *wapA*-like protein. Later in evolution, lineage I (serovar 1/2a strains) lost the downstream region carrying the wapA-like protein, and lineage II (serovar 4b strains) lost the downstream part carrying the LPXTG protein. Subsequent insertions in this region then led to the present structure of the *L. monocytogenes* lineage II (serovar 4b strains) *inlAB* region. In contrast, in

L. innocua, as seen for the virulence locus, only the virulence genes *inlAB* were lost (Fig. 4).

Further analysis, in particular the availability of the genome sequences of additional *L. monocytogenes* strains and of each member of the genus *Listeria*, will soon shed more light on the question of how *Listeria* became a human pathogen.

CONCLUDING REMARKS AND FUTURE DIRECTIONS

Sequencing the genomes of four *L. monocytogenes* strains and one *L. innocua* strain has brought a wealth of new data and insight into this fascinating pathogen. At present, different genomics approaches are applied and provide for the first time a global view and a more complete knowledge of gene distribution and the genetic content present in the gene pool of the genus *Listeria*. This information represents a fundamental basis for functional studies to better understand phenotypic and virulence

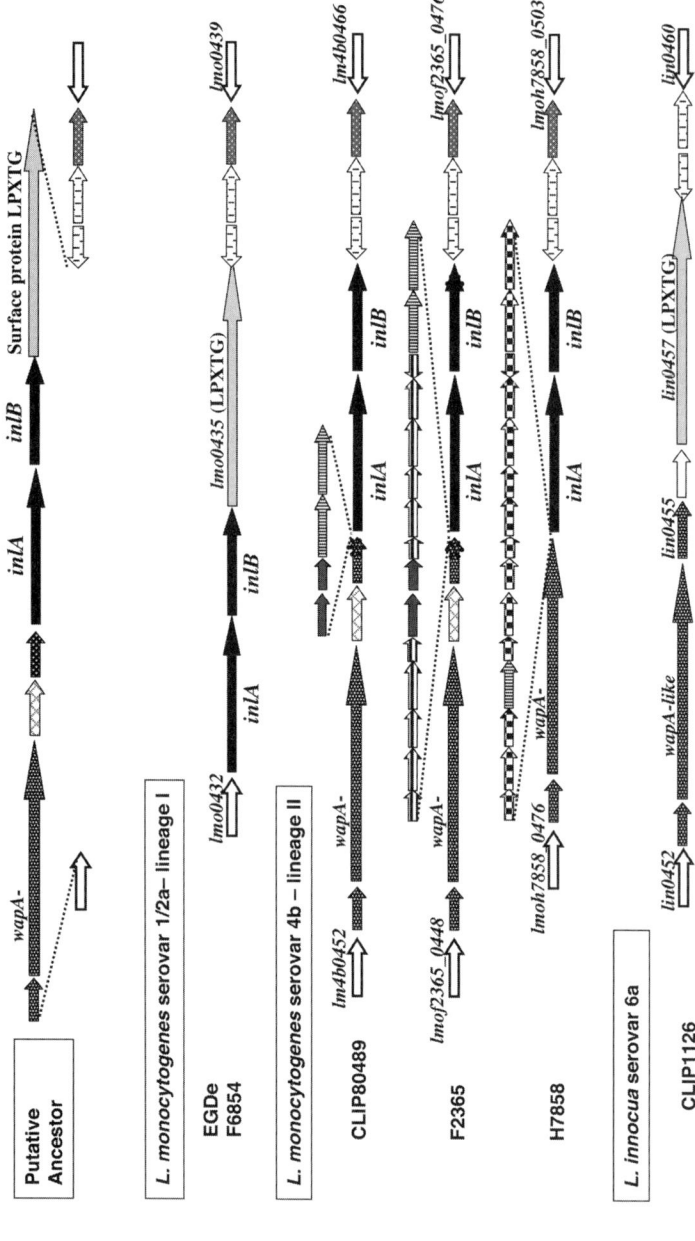

FIGURE 4 Schematic presentation of the *inlAB* locus and the flanking regions of *L. monocytogenes* and *L. innocua* and its hypothetical ancestral organization. EGD, F6854, CLIP80459, F2365, H7858: *L. monocytogenes* strain designations. Clip 11262: *L. innocua* strain designation. Orthologous genes are depicted in the same shading pattern. Black indicates *inlAB* locus; dotted lines, specific regions with respect to the other genomes; *lmo*, gene names of *L. monocytogenes* EGDe, *lm4b*, gene names of *L. monocytogenes* CLIP80459, *lmof2365*, gene names of *L. monocytogenes* F2365, *lmoh7858*, gene names of *L. monocytogenes* H7858; *lin*, gene names of *L. innocua* CLIP11262. A star designates pseudogenes.

differences between *L. monocytogenes* strains and to gain knowledge on the evolution of the pathogen. The ongoing sequencing projects aimed at determining the complete genome sequence of one representative of each species of the genus *Listeria* by the Institut Pasteur and the German PathoGenomiK network (http://www.pasteur.fr/recherche/unites/gmp/; http://www.genomik.uniwuerzburg.de/ seq.htm) and the determination of the complete genome sequence of an additional 19 *Listeria* strains by the Broad Institute (http://www.broad.mit.edu/seq/msc/) will be the driving force for understanding the function of the many factors encoded by the genome, whether involved in virulence or not, and to understand strain-specific differences in niche adaptation and virulence.

ACKNOWLEDGMENTS
We thank many of the colleagues who have contributed in different ways to this research. This work received financial support from the Institut Pasteur (GPH 9). Pascale Cossart is an international research scholar from the Howard Hughes Medical Institute.

REFERENCES
1. **Arous, S., C. Buchrieser, P. Folio, P. Glaser, A. Namane, M. Hebraud, and Y. Hechard.** 2004. Global analysis of gene expression in an *rpoN* mutant of *Listeria monocytogenes*. *Microbiology* **150:**1581–1590.
2. **Aureli, P., G. C. Fiorucci, D. Caroli, G. Marchiaro, O. Novara, L. Leone, and S. Salmaso.** 2000. An outbreak of febrile gastroenteritis associated with corn contaminated by *Listeria monocytogenes*. *N. Engl. J. Med.* **342:**1236–1241.
3. **Autret, N., C. Raynaud, I. Dubail, P. Berche, and A Charbit.** 2003. Identification of the *agr* locus of *Listeria monocytogenes*: role in bacterial virulence. *Infect. Immun.* **71:**4463–4471.
4. **Barabote, R. D., and M. H. Saier.** 2005. Comparative genomic analyses of the bacterial phosphotransferase system. *Microbiol. Mol. Biol. Rev.* **69:**608–634.
5. **Bibb, W. F., B. Schwartz, B. G. Gellin, B. D. Plikaytis, and R. E. Weaver.** 1989. Analysis of *Listeria monocytogenes* by multilocus enzyme electrophoresis and application of the method to epidemiologic investigations. *Int. J. Food Microbiol.* **8:**233–239.
6. **Bielecki, J.** 1994. Insertions within *iap* gene of *Listeria monocytogenes* generated by plasmid pLIV are not lethal. *Acta Microbiol. Pol.* **43:**133–143.
7. **Bierne, H., C. Garandeau, M. G. Pucciarelli, S. Sabet, F. Newton, F. Garcia-del Portillo, P. Cossart, and A. Charbit.** 2004. Sortase B, a new class of sortase in *Listeria monocytogenes*. *J. Bacteriol.* **186:**1972–1982.
8. **Bierne, H., S. K. Mazmanian, M. Trost, M. G. Pucciarelli, G. Liu., P. Dehoux, L. Jansch, F. Garcia-del Portillo, O. Schneewind, and P. Cossart.** 2002. Inactivation of the *srtA* gene in *Listeria monocytogenes* inhibits anchoring of surface proteins and affects virulence. *Mol. Microbiol.* **43:**869–881.
9. **Braun, L., S. Dramsi, P. Dehoux, H. Bierne, G. Lindahl, and P. Cossart.** 1997. InlB: an invasion protein of *Listeria monocytogenes* with a novel type of surface association. *Mol. Microbiol.* **25:**285–294.
10. **Brehm, K., M. T. Ripio, J. Kreft, and J. A. Vazquez-Boland.** 1999. The *bvr* locus of *Listeria monocytogenes* mediates virulence gene repression by beta-glucosides. *J. Bacteriol.* **181:**5024–5032.
11. **Brosch, R., B. Catimel, G. Milon, C. Buchrieser, E. Vindel, and J. Rocourt.** 1993. Virulence heterogeneity of *Listeria monocytogenes* strains from various sources (food, human, animal) in immunocompetent mice and its association with typing characteristics. *J. Food Prot.* **56:**296–301.
12. **Brosch, R., J. Chen, and J. B. Luchansky.** 1994. Pulsed-field fingerprinting of listeriae: identification of genomic divisions for *Listeria monocytogenes* and their correlation with serovar. *Appl. Environ. Microbiol.* **60:**2584–2592.
13. **Buchrieser, C., C. Rusniok, F. Kunst, P. Cossart, and P. Glaser.** 2003. Comparison of the genome sequences of *Listeria monocytogenes* and *Listeria innocua*: clues for evolution and pathogenicity. *FEMS Immunol. Med. Microbiol.* **35:**207–213.
14. **Cabanes, D., P. Dehoux, O. Dussurget, L. Frangeul, and P. Cossart.** 2002. Surface proteins and the pathogenic potential of *Listeria monocytogenes*. *Trends Microbiol.* **5:**238–245.
15. **Cabanes, D., O. Dussurget, P. Dehoux, and P. Cossart.** 2004. Auto, a surface associated autolysin of *Listeria monocytogenes* required for entry in eukaryotic cells and virulence. *Mol. Microbiol.* **51:**1601–1614.
16. **Cabanes, D., S. Sousa, A. Cebria, M. Lecuit, F. Garcia-del Portillo, and P. Cossart.** 2005. Gp96 is a receptor for a novel *Listeria monocytogenes* virulence factor, Vip, a surface protein. *EMBO J.* **24:**2827–2838.
17. **Cai, S., and M. Wiedmann.** 2001. Characterization of the *prfA* virulence gene cluster insertion site in non-hemolytic *Listeria* spp.: probing the evolution of the *Listeria* virulence gene island. *Curr. Microbiol.* **43:**271–277.

18. **Carroll, S. A., T. Hain, U. Technow, A. Darji, P. Pashalidis, S. W. Joseph, and T. Chakraborty.** 2003. Identification and characterization of a peptidoglycan hydrolase, MurA, of *Listeria monocytogenes*, a muramidase needed for cell separation. *J. Bacteriol.* **185:**6801–6808.

19. **Chakraborty, T., T. Hain, and E. Domann.** 2000. Genome organization and the evolution of the virulence gene locus in *Listeria* species. *Int. J. Med. Microbiol.* **2:**167–174.

20. **Chatterjee, S. S., H. Hossain, S. Otten, C. Kuenne, K. Kuchmina, S. Machata, E. Domann, T. Chakraborty, and T. Hain.** 2006. Intracellular gene expression profile of *Listeria monocytogenes*. *Infect. Immun.* **74:**1323–1338.

21. **Chico-Calero, I., M. Suarez, B. Gonzalez-Zorn, M. Scortti, J. Slaghuis, W. Goebel, and J. A. Vazquez-Boland.** 2002. Hpt, a bacterial homolog of the microsomal glucose- 6-phosphate translocase, mediates rapid intracellular proliferation in *Listeria*. *Proc. Natl. Acad. Sci. USA* **99:**431–436.

22. **Christiansen, J. K., J. S. Nielsen, T. Ebersbach, P. Valentine-Hansen, L. Sogaard-Andersen, and B. H. Kallipolitis.** 2006. Identification of small Hfq-binding RNAs in *Listeria monocytogenes*. *RNA* **12:**1383–1396.

23. **Copley, S. D., and J. K. Dhillon.** 2002. Lateral gene transfer and parallel evolution in the history of glutathione biosynthesis genes. *Genome Biol.* **3:**research0025.

24. **Cossart, P., and C. Kocks.** 1994. The actin based motility of the intracellular pathogen *Listeria monocytogenes*. *Mol. Microbiol.* **13:**395–402.

25. **Cossart, P., and M. Lecuit.** 1998. Interactions of *Listeria monocytogenes* with mammalian cells during entry and actin-based movement: bacterial factors, cellular ligands and signaling. *EMBO J.* **17:**3797–3806.

26. **Cotter, P. D., N. Emerson, C. G. M. Gahan, and C. Hill.** 1999. Identification and disruption of *lisRK*, a genetic locus encoding a two-component signal transduction system involved in stress tolerance and virulence in *Listeria monocytogenes*. *J. Bacteriol.* **181:**6840–6843.

27. **Cotter, P. D., C. M. Guinane, and C. Hill.** 2002. The LisRK signal transduction system determines the sensitivity of *Listeria monocytogenes* to nisin and cephalosporins. *Antimicrob. Agents Chemother.* **46:**2784–2790.

28. **Dalton, C. B., C. C. Austyin, J. Sobel, P. S. Hayes, W. F. Bibb, L. M. Graves, B. Swaminathan, M. E. Proctor, and P. M. Griffin.** 1997. An outbreak of gastroenteritis and fever due to *Listeria monocytogenes* in milk. *N. Engl. J. Med.* **336:**100–105.

29. **Domann, E., J. Wehland, M. Rohde, S. Pistor, M. Hartl, W. Goebel, M. Leimeister-Wächter, M. Wuenscher, and T. Chakraborty.** 1992. A novel bacterial gene in *Listeria monocytogenes* required for host cell microfilament interaction with homology to the proline-rich region of vinculin. *EMBO J.* **11:**1981–1990.

30. **Domann, E., S. Zechel, A. Lingnau, R. Hain, A. Darji, T. Nichterlein, J. Wehland, and T. Chakraborty.** 1997. Identification and characterization of a novel PrfA-regulated gene in *Listeria monocytogenes* whose product, IrpA, is highly homologous to Internalin proteins, which contain leucine-rich repeats. *Infect. Immun.* **65:**101–109.

31. **Doumith, M., C. Buchrieser, P. Glaser, C. Jacquet, and P. Martin.** 2004. Differentiation of the major *Listeria monocytogenes* serovars by multiplex PCR. *J. Clin. Microbiol.* **42:**3819–3822.

32. **Doumith, M., C. Cazalet, N. Simoes, L. Frangeul, C. Jaquet, F. Kunst, P. Martin, P. Cossart, P. Glaser, and C. Buchrieser.** 2004. New aspects regarding evolution and virulence of *Listeria monocytogenes* revealed by comparative genomics. *Infect. Immun.* **72:**1072–1083.

33. **Doumith, M., C. Jacquet, P. Gerner-Smidt, L. M. Graves, S. Loncarevic, T. Mathisen, A. Morvan, C. Salcedo, M. Torpdahl, J. A. Vazquez, and P. Martin.** 2005. Multicenter validation of a multiplex PCR assay for differentiating the major *Listeria monocytogenes* serovars 1/2a, 1/2b, 1/2c, and 4b: toward an international standard. *J. Food Prot.* **68:**2648–2650.

34. **Dramsi, S., I. Biswas, E. Maguin, L. Braun, P. Mastroeni, and P. Cossart.** 1995. Entry of *Listeria monocytogenes* into hepatocytes requires expression of InlB, a surface protein of the internalin multigene family. *Mol. Microbiol.* **16:**251–261.

35. **Dramsi, S., F. Bourdichon, D. Cabanes, M. Lecuit, H. Fsihi, and P. Cossart.** 2004. FbpA, a novel multifunctional *Listeria monocytogenes* virulence factor. *Mol. Microbiol.* **53:**639–649.

36. **Dramsi, S., P. Dehoux, M. Lebrun, P. L. Goossens, and P. Cossart.** 1997. Identification of four new members of the internalin multigene family of *Listeria monocytogenes* EGD. *Infect. Immun.* **65:**1615–1625.

37. **Dramsi, S., M. Lebrun, and P. Cossart.** 1996. Molecular and genetic determinants involved in invasion of mammalian cells by *Listeria monocytogenes*. *Curr. Topics Microbiol. Immunol.* **209:**61–77.

38. **Dussurget, O., D. Cabanes, P. Dehoux, M. Lecuit, C. Buchrieser, P. Glaser, and P. Cossart.** 2002. *Listeria monocytogenes* bile salt hydrolase is a PrfA-regulated virulence factor involved in the intestinal and hepatic phases of listeriosis. *Mol. Microbiol.* **45:**1095–1106.

39. **Dussurget, O., J. Pizarro-Cerda, and P. Cossart.** 2004. Molecular determinants of *Listeria mo-*

nocytogenes virulence. *Annu. Rev. Microbiol.* **58:** 587–610.

40. **Engelbrecht, F., S.-K. Chun, C. Ochs, J. Hess, F. Lottspeich, W. Goebel, and Z. Sokolovic.** 1996. A new PrfA-regulated gene of *Listeria monocytogenes* encoding a small, secreted protein which belongs to the family of internalins. *Mol. Microbiol.* **21:**823–837.

41. **Evans, M. R., B. Swaminathan, L. M. Graves, E. Altermann, T. R. Klaenhammer, R. C. Fink, S. Kernodle, and S. Kathariou.** 2004. Genetic markers unique to *Listeria monocytogenes* serotype 4b differentiate epidemic clone II (hot dog outbreak strains) from other lineages. *Appl. Environ. Microbiol.* **70:**2383–2390.

42. **Fahey, R. C., W. C. Brown, W. B. Adams, and M. B. Worsham.** 1978. Occurrence of glutathione in bacteria. *J. Bacteriol.* **133:**1126–1129.

43. **Farber, J. M., and P. I. Peterkin.** 1991. *Listeria monocytogenes*, a food-borne pathogen. *Microbiol. Rev.* **55:**476–511.

44. **Fenlon, D. R.** 1999. Listeria monocytogenes *in the Natural Environment*, 2nd ed. Marcel Dekker, New York, NY.

45. **Feretti, J., et. al.** 2001. Complete genome sequence of an M1 strain of *Streptococcus pyogenes*. *Proc. Natl. Acad. Sci. USA* **98:**4658–5663.

46. **Flanary, P. L., R. D. Allen, L. Dons, and S. Kathariou.** 1999. Insertional inactivation of the *Listeria monocytogenes cheYA* operon abolishes response to oxygen gradients and reduces the number of flagella. *Can. J. Microbiol.* **45:**646–652.

47. **Fleischmann, R. D., M. D. Adams, O. White, R. A. Clayton, E. F. Kirkness, A. R. Kerlavage, C. J. Bult, J. F. Tomb, B. A. Dougherty, J. M. Merrick, et al.** 1995. Whole-genome random sequencing and assembly of *Haemophilus influenzae* Rd. *Science* **269:**496–512.

48. **Gaillard, J.-L., P. Berche, C. Frehel, E. Gouin, and P. Cossart.** 1991. Entry of *Listeria monocytogenes* into cells is mediated by internalin, a repeat protein reminiscent of surface antigens from gram-positive cocci. *Cell* **65:**1127–1141.

49. **Garandeau, C., H. Reglier-Poupet, I. Dubail, J. L. Beretti, P. Berche, and A. Charbit.** 2002. The sortase A of *Listeria monocytogenes* is involved in processing of internalin and in virulence. *Infect. Immun.* **70:**1382–1390.

50. **Geoffroy, C., A. M. Gilles, and J. E. Alouf.** 1981. The sulfhydryl groups of the thiol-dependent cytolytic toxin from *B. alvei*: evidence for one essential sulfhydryl group. *Biochem. Biophys. Res. Commun.* **99:**781–788.

51. **Glaser, P., L. Frangeul, C. Buchrieser, C. Rusniok, A. Amend, F. Baquero, P. Berche, H. Bloecker, P. Brandt, T. Chakraborty, A. Charbit, F. Chetouani, E. Couve, A. de Daruvar, P. Dehoux, E. Domann, G. Dominguez-Bernal, E. Duchaud, L. Durant, O. Dussurget, K. D. Entian, H. Fsihi, F. G. Portillo, P. Garrido, L. Gautier, W. Goebel, N. Gomez-Lopez, T. Hain, J. Hauf, D. Jackson, L. M. Jones, U. Kaerst, J. Kreft, M. Kuhn, F. Kunst, G. Kurapkat, E. Madueno, A. Maitournam, J. M. Vicente, E. Ng, H. Nedjari, G. Nordsiek, S. Novella, B. de Pablos, J. C. Perez-Diaz, R. Purcell, B. Remmel, M. Rose, T. Schlueter, N. Simoes, A. Tierrez, J. A. Vazquez-Boland, H. Voss, J. Wehland, and P. Cossart.** 2001. Comparative genomics of *Listeria* species. *Science* **294:**849–852.

52. **Gopal, S., I. Borovok, A. Ofer, M. Yanku, G. Cohen, W. Goebel, J. Kreft, and Y. Aharonowitz.** 2005. A multidomain fusion protein in *Listeria monocytogenes* catalyzes the two primary activities for glutathione biosynthesis. *J. Bacteriol.* **187:**3839–3847.

53. **Gottesman, S.** 2004. The small RNA regulators of *Escherichia coli*: roles and mechanisms. *Annu. Rev. Microbiol.* **58:**303–328.

54. **Gouin, E., H. Gantelet, C. Egile, I. Lasa, H. Ohayon, V. Villiers, P. Gounon, P. J. Sansonetti, and P. Cossart.** 1999. A comparative study of the actin-based motilities of the pathogenic bacteria *Listeria monocytogenes*, *Shigella flexneri* and *Rickettsia conorii*. *J. Cell Sci.* **112:** 1697–1708.

55. **Gouin, E., J. Mengaud, and P. Cossart.** 1994. The virulence gene cluster of *Listeria monocytogenes* is also present in *Listeria ivanovii*, an animal pathogen, and *Listeria seeligeri*, a nonpathogenic species. *Infect. Immun.* **62:**3550–3553.

56. **Gouin, E., M. D. Welch, and P. Cossart.** 2005. Actin-based motility of intracellular pathogens. *Curr. Opin. Microbiol.* **8:**35–45.

57. **Graves, L., B. Swaminathan, M. Reeves, S. B. Hunter, R. E. Weaver, B. D. Plikaytis, and A. Schuchat.** 1994. Comparison of ribotyping and multilocus enzyme electrophoresis for subtyping of *Listeria monocytogenes* isolates. *J. Clin. Microbiol.* **32:**2936–2943.

58. **Greub, G., and D. Raoult.** 2004. Microorganisms resistant to free-living amoebae. *Clin. Microbiol. Rev.* **17:**413–433.

59. **Grundling, A., L. S. Burrack, H. G. Bouwer, and D. E. Higgins.** 2004. *Listeria monocytogenes* regulates flagellar motility gene expression through MogR, a transcriptional repressor required for virulence. *Proc. Natl. Acad. Sci. USA* **101:**12318–12323.

60. **Gunn, J. S.** 2000. Mechanisms of bacterial resistance and response to bile. *Microb. Infect.* **2:**907–913.

61. **Hamon, M., H. Bierne, and P. Cossart.** 2006. *Listeria monocytogenes*: a multifaceted model. *Nat. Rev. Microbiol.* **4:**423–434.

62. **Hoch, J. A., and T. J. Silhavy (ed.).** 1995. *Two-Component Signal Transduction.* ASM Press, Washington, DC.

63. **Hong, E., M. Doumith, S. Duperrier, I. Giovannacci, A. Morvan, P. Glaser, C. Buchrieser, C. Jacquet, and P. Martin.** 2006. Genetic diversity of *Listeria monocytogenes* recovered from infected persons and pork, seafood and dairy products on retail sale in France during 2000 and 2001. *Int. J. Food Microbiol.* 22 Dec [Epub ahead of print].

64. **Jacquet, C., E. Gouin, D. Jeannel, P. Cossart, and J. Rocourt.** 2002. Expression of ActA, Ami, InlB, and listeriolysin O in *Listeria monocytogenes* of human and food origin. *Appl. Environ. Microbiol.* **68:**616–622.

65. **Jeffers, G. T., J. L. Bruce, P. L. McDonough, J. Scarlett, K. J. Boor, and M. Wiedmann.** 2001. Comparative genetic characterization of *Listeria monocytogenes* isolates from human and animal listeriosis cases. *Microbiology* **147:**1095–1104

66. **Jeter, R. M., B. M. Olivera, and J. R. Roth.** 1984. *Salmonella typhimurium* synthesizes cobalamin (vitamin B12) de novo under anaerobic growth conditions. *J. Bacteriol.* **159:**206–213.

67. **Johansson, J., P. Mandin, A. Renzoni, C. Chiaruttini, M. Springer, and P. Cossart.** 2002. An RNA thermosensor controls expression of virulence genes in *Listeria monocytogenes. Cell* **110:**551–561.

68. **Joseph, B., K. Przybilla, C. Stuhler, K. Schauer, J. Slaghuis, T. M. Fuchs, and W. Goebel.** 2006. Identification of *Listeria monocytogenes* genes contributing to intracellular replication by expression profiling and mutant screening. *J. Bacteriol.* **188:**556–568.

69. **Kallipolitis, B. H., H. Ingmer, C. G. Gahan, C. Hill, and L. Sogaard-Andersen.** 2003. CesRK, a two-component signal transduction system in *Listeria monocytogenes,* responds to the presence of cell wall-acting antibiotics and affects beta-lactam resistance. *Antimicrob. Agents Chemother.* **47:**3421–3429.

70. **Kazmierczak, M. J., S. C. Mithoe, K. J. Boor, and M. Wiedmann.** 2003. *Listeria monocytogenes* sigma B regulates stress response and virulence functions. *J. Bacteriol.* **185:**5722–5734.

71. **Kingdon, G. C., and C. P. Sword.** 1970. Effects of *Listeria monocytogenes* hemolysin on phagocytic cells and lysosmes. *Infect. Immun.* **1:**356–362.

72. **Kocks, C., E. Gouin, M. Tabouret, P. Berche, H. Ohayon, and P. Cossart.** 1992. *Listeria monocytogenes*–induced actin assembly requires the *actA* gene product, a surface protein. *Cell* **68:**521–531.

73. **Kontinen, V. P., P. Saris, and M. Sarvas.** 1991. A gene (*prsA*) of *Bacillus subtilis* involved in a novel, late stage of protein export. *Mol. Microbiol.* **5:**1273–1283.

74. **Kontinen, V. P., and M. Sarvas.** 1993. The PrsA lipoprotein is essential for protein secretion in *Bacillus subtilis* and sets a limit for high-level secretion. *Mol. Microbiol.* **4:**727–737.

75. **Kotrba, P., M. Inui, and H. Yukawa.** 2001. Bacterial phosphotransferase system (PTS) in carbohydrate uptake and control of carbon metabolism. *J. Biosci. Bioeng.* **92:**502–517.

76. **Kreft, J., J.-A. Vazquez-Boland, S. D.-B. Altrock, G., and W. Goebel.** 2002. Pathogenicity islands and other virulence elements in *Listeria. Curr. Top. Microbiol. Immunol.* **264:**109–125.

77. **Kuhn, M., and W. Goebel.** 1989. Identification of an extracellular protein of *Listeria monocytogenes* possibly involved in intracellular uptake by mammalian cells. *Infect. Immun.* **57:**55–61.

78. **Kuroda, M., T. Ohta, I. Uchiyama, T. Baba, H. Yuzawa, I. Kobayashi, L. Cui, A. Oguchi, K. Aoki, Y. Nagai, J. Lian, T. Ito, M. Kanamori, H. Matsumaru, A. Maruyama, H. Murakami, A. Hosoyama, Y. Mizutani-Ui, N. K. Takahashi, T. Sawano, R. Inoue, C. Kaito, K. Sekimizu, H. Hirakawa, S. Kuhara, S. Goto, J. Yabuzaki, M. Kanehisa, A. Yamashita, K. Oshima, K. Furuya, C. Yoshino, T. Shiba, M. Hattori, N. Ogasawara, H. Hayashi, and K. Hiramatsu.** 2001. Whole genome sequencing of methicillin-resistant *Staphylococcus aureus. Lancet* **357:**1225–1240.

79. **Lecuit, M., S. Dramsi, C. Gottardi, M. Fredor-Chaiken, B. Gumbiner, and P. Cossart.** 1999. A single amino acid in E-cadherin responsible for host specificity toward the human pathogen *Listeria monocytogenes. EMBO J.* **18:**3956–3963.

80. **Lecuit, M., R. Hurme, J. Pizarro-Cerda, H. Ohayon, B. Geiger, and P. Cossart.** 2000. A role for a- and b-catenins in bacterial uptake. *Proc. Natl. Acad. Sci. USA* **97:**10008–10013.

81. **Lecuit, M., S. Vandormael-Pournin, J. Lefort, M. Huerre, P. Gounon, C. Dupuy, C. Babinet, and P. Cossart.** 2001. A transgenesis model for listeriosis: role of internalin in crossing the intestinal barrier. *Science* **5522:**1722–1725.

82. **Leimeister-Wächter, M., C. Haffner, E. Domann, W. Goebel, and T. Chakraborty.** 1990. Identification of a gene that positively regulates expression of listeriolysin, the major virulence factor of *Listeria monocytogenes. Proc. Natl. Acad. Sci. USA* **87:**8336–8340.

83. **Lenz, L. L., S. Mohammadi, A. Geissler, and D. A. Portnoy.** 2003. SecA2-dependent secretion of autolytic enzymes promotes *Listeria monocytogenes* pathogenesis. *Proc. Natl. Acad. Sci. USA* **100:**12432–12437.

84. **Lenz, L. L., and D. A. Portnoy.** 2002. Identification of a second *Listeria secA* gene associated with protein secretion and the rough phenotype. *Mol. Microbiol.* **45:**1043–1056.

85. **Loessner, M. J., R. B. Inman, P. Lauer, and R. Calendar.** 2000. Complete nucleotide sequence, molecular analysis and genome structure of bacteriophage A118 of *Listeria monocytogenes*: implications for phage evolution. *Mol. Microbiol.* **2:**324–340.

86. **Loh, E., J. Gripenland, and J. Johansson.** 2006. Control of *Listeria monocytogenes* virulence by 5'-untranslated RNA. *Trends Microbiol.* **14:**294–298.

87. **Lou, Y., and A. E. Yousef.** 1998. Characteristics of *Listeria monocytogenes* important to food processors, p. 00–00. *In* E. T. Ryser and E. H. Marth (ed.), Listeria, *Listeriosis and Food Safety*. Marcel Dekker Inc., New York, NY.

88. **Ly, T. M., and H. E. Muller.** 1990. Ingested *Listeria monocytogenes* survive and multiply in protozoa. *J. Med. Microbiol.* **33:**51–54.

89. **Mandin, P., H. Fsihi, O. Dussurget, M. Vergassola, E. Milohanic, A. Toledo-Arana, I. Lasa, J. Johansson, and P. Cossart.** 2005. VirR, a response regulator critical for *Listeria monocytogenes* virulence. *Mol. Microbiol..* **57:**1367–1380.

90. **Mazmanian, S. K., G. Liu, H. Ton-That, and O. Schneewind.** 1999. *Staphylococcus aureus* sortase, an enzyme that anchors surface proteins to the cell wall. *Science* **285:**760–763.

91. **Mengaud, J., J. Chenevert, C. Geoffroy, J. L. Gaillard, and P. Cossart.** 1987. Identification of the structural gene encoding the SH-activated hemolysin of *Listeria monocytogenes*: listeriolysin O is homologous to streptolysin O and pneumolysin. *Infect. Immun.* **55:**3225–3227.

92. **Mengaud, J., S. Dramsi, E. Gouin, J. A. Vazquez-Boland, G. Milon, and P. Cossart.** 1991. Pleiotropic control of *Listeria monocytogenes* virulence factors by a gene which is autoregulated. *Mol. Microbiol.* **5:**2273–2283.

93. **Mengaud, J., M. F. Vicente, and P. Cossart.** 1989. Transcriptional mapping and nucleotide sequence of the *Listeria monocytogenes hlyA* region reveal structural features that may be involved in regulation. *Infect. Immun.* **57:**3695–3701.

94. **Milohanic, E., P. Glaser, J. Y. Coppee, L. Frangeul, Y. Vega, J. A. Vazquez-Boland, F. Kunst, P. Cossart, and C. Buchrieser.** 2003. Transcriptome analysis of *Listeria monocytogenes* identifies three groups of genes differently regulated by PrfA. *Mol. Microbiol.* **47:**1613–1625.

95. **Milohanic, E., R. Jonquicres, P. Cossart, P. Berche, and J. L. Gaillard.** 2001. The autolysin Ami contributes to the adhesion of *Listeria monocytogenes* to eukaryotic cells via its cell wall anchor. *Mol. Microbiol.* **39:**1212–1234.

96. **Nelson, K. E., D. E. Fouts, E. F. Mongodin, J. Ravel, R. T. DeBoy, J. F. Kolonay, D. A. Rasko, S. V. Angiuoli, S. R. Gill, I. T. Paulsen, J. Peterson, O. White, W. C. Nelson, W. Nierman, M. J. Beanan, L. M. Brinkac, S. C. Daugherty, R. J. Dodson, A. S. Durkin, R. Madupu, D. H. Haft, J. Selengut, S. Van Aken, H. Khouri, N. Fedorova, H. Forberger, B. Tran, S. Kathariou, L. D. Wonderling, G. A. Uhlich, D. O. Bayles, J. B. Luchansky, and C. M. Fraser.** 2004. Whole genome comparisons of serotype 4b and 1/2a strains of the food-borne pathogen *Listeria monocytogenes* reveal new insights into the core genome components of this species. *Nucleic Acids Res.* **32:**2386–2395.

97. **Newton, G. L., K. Arnold, M. S. Price, C. Sherrill, S. B. Delcardayre, Y. Aharonowitz, G. Cohen, J. Davies, R. C. Fahey, and C. Davis.** 1996. Distribution of thiols in microorganisms: mycothiol is a major thiol in most actinomycetes. *J. Bacteriol.* **178:**1990–1995.

98. **Newton, S. M., P. E. Klebba, C. Raynaud, Y. Shao, X. Jiang, I. Dubail, C. Archer, C. Frehel, and A. Charbit.** 2005. The svpA-srtB locus of *Listeria monocytogenes*: fur-mediated iron regulation and effect on virulence. *Mol. Microbiol.* **55:**927–940.

99. **Nightingale, K. K., K. Windham, and M. Wiedmann.** 2005. Evolution and molecular phylogeny of *Listeria monocytogenes* isolated from human and animal listeriosis cases and foods. *J. Bacteriol.* **187:**5537–5551.

100. **O'Riordan, M., M. A. Moors, and D. A. Portnoy.** 2003. *Listeria* intracellular growth and virulence require host-derived lipoic acid. *Science* **302:**462–464.

101. **Piffaretti, J. C., H. Kressebuch, M. Aeschbacher, J. Bille, E. Bannerman, J. M. Musser, R. K. Selander, and J. Rocourt.** 1989. Genetic characterization of clones of the bacterium *Listeria monocytogenes* causing epidemic disease. *Proc. Natl. Acad. Sci. USA* **86:**3818–3822.

102. **Pinner, R. W., A. Schuchat, B. Swaminathan, P. S. Hayes, K. A. Deaver, R. E. Weaver, B. D. Plikaytis, M. Reeves, C. V. Broome, and J. D. Wenger.** 1992. Role of foods in sporadic listeriosis. II. Microbiologic and epidemiologic investigation. *JAMA* **267:**2081–2082.

103. **Pizarro-Cerda, J., and P. Cossart.** 2006. Subversion of cellular functions by *Listeria monocytogenes*. *J. Pathol.* **208:**215–223.

104. **Price-Carter, M., J. Tingey, T. A. Bobik, and J. R. Roth.** 2001. The alternative electron acceptor tetrathionate supports B12-dependent anaerobic growth of *Salmonella enterica* serovar

Typhimurium on ethanolamine or 1,2-propanediol. *J. Bacteriol.* **183:**2463–2475.

105. **Pucciarelli, M. G., E. Calvo, C. Sabet, H. Bierne, P. Cossart, and F. Garcia-del Portillo.** 2005. Identification of substrates of the *Listeria monocytogenes* sortases A and B by a nongel proteomic analysis. *Proteomics* **5:**4808–4817.

106. **Raffelsbauer, D., A. Bubert, F. Engelbrecht, J. Scheinpflug, A. Simm, J. Hess, S. H. Kaufmann, and W. Goebel.** 1998. The gene cluster inlC2DE of *Listeria monocytogenes* contains additional new internalin genes and is important for virulence in mice. *Mol. Gen. Genet.* **260:**144–158.

107. **Raux, E., H. L. Schubert, and M. J. Warren.** 2000. Biosynthesis of cobalamin (vitamin B12): a bacterial conundrum. *Cell. Mol. Life Sci.* **57:** 1880–1893.

108. **Ripio, M., K. Brehm, M. Lara, M. Suarez, and J. A. Vazquez-Boland.** 1997. Glucose-1-phosphate utilization by *Listeria monocytogenes* is PrfA dependent and coordinately expressed with virulence factors. *J. Bacteriol.* **179:**7174–7180.

109. **Roberts, A. J., and M. Wiedmann.** 2003. Pathogen, host, and environmental factors contributing to the pathogenesis of listeriosis. *Cell. Mol. Life Sci.* **60:**1–15.

110. **Sabet, C., M. Lecuit, D. Cabanes, P. Cossart, and H. Bierne.** 2005. LPXTG protein InlJ, a newly identified internalin involved in *Listeria monocytogenes* virulence. *Infect. Immun.* **73:**6912–6922.

111. **Schmid, M. W., E. Y. Ng, R. Lampidis, M. Emmerth, M. Walcher, J. Kreft, W. Goebel, M. Wagner, and K. H. Schleifer.** 2005. Evolutionary history of the genus *Listeria* and its virulence genes. *Syst. Appl. Microbiol.* **28:**1–18.

112. **Schuchat, A., K. A. Deaver, J. D. Wenger, B. D. Plikaitis, L. Mascola, R. W. Pinner, A. L. Reingold, and C. V. Broome.** 1992. Role of foods in sporadic listeriosis. I. A case control study of dietary risk factors. *JAMA* **267:**2041–2045.

113. **Scortti, M., L. Lacharme-Lora, M. Wagner, I. Chico-Calero, P. Losito, and J. A. Vazquez-Boland.** 2006. Coexpression of virulence and fosfomycin susceptibility in *Listeria*: molecular basis of an antimicrobial in vitro-in vivo paradox. *Nat. Med.* **12:**515–517.

114. **Sheehan, B., A. Klarsfeld, R. Ebright, and P. Cossart.** 1996. A single substitution in the putative helix-turn-helix motif of the pleiotropic activator PrfA attenuates *Listeria monocytogenes* virulence. *Mol. Microbiol.* **20:**785–797.

115. **Shen, A., and D. E. Higgins.** 2005. The 5′ untranslated region-mediated enhancement of in-

tracellular listeriolysin O production is required for *Listeria monocytogenes* pathogenicity. *Mol. Microbiol.* **57:**1460–1473.

116. **Shen, A., and D. E. Higgins.** 2006. The MogR transcriptional repressor regulates nonhierarchical expression of flagellar motility genes and virulence in *Listeria monocytogenes. PLoS Pathog.* **2:**e30.

117. **Stover, C. K., X. Q. Pham, A. L. Erwin, S. D. Mizoguchi, P. Warrener, M. J. Hickey, F. S. Brinkman, W. O. Hufnagle, D. J. Kowalik, M. Lagrou, R. L. Garber, L. Goltry, E. Tolentino, S. Westbrock-Wadman, Y. Yuan, L. L. Brody, S. N. Coulter, K. R. Folger, A. Kas, K. Larbig, R. Lim, K. Smith, D. Spencer, G. K. Wong, Z. Wu, I. T. Paulsen, J. Reizer, M. H. Saier, R. E. Hancock, S. Lory, and M. V. Olson.** 2000. Complete genome sequence of *Pseudomonas aeruginosa* PA01, an opportunistic pathogen. *Nature* **406:**959–964.

118. **Stritzker, J., C. Schoen, and W. Goebel.** 2005. Enhanced synthesis of internalin A in aro mutants of *Listeria monocytogenes* indicates posttranscriptional control of the *inlAB* mRNA. *J. Bacteriol.* **187:**2836–2845.

119. **Sue, D., K. J. Boor, and M. Wiedmann.** 2003. Sigma(B)-dependent expression patterns of compatible solute transporter genes *opuCA* and *lmo1421* and the conjugated bile salt hydrolase gene *bsh* in *Listeria monocytogenes. Microbiology* **149:**3247–3256.

120. **Thomsen, L. E., S. S. Slutz, M. W. Tan, and H. Ingmer.** 2006. *Caenorhabditis elegans* is a model host for *Listeria monocytogenes. Appl. Environ. Microbiol.* **72:**1700–1701.

121. **Tilney, L. G., and D. A. Portnoy.** 1989. Actin filaments and the growth, movement, and spread of the intracellular bacterial parasite *Listeria monocytogenes. J. Cell Biol.* **109:**1597–1608.

122. **Tsai, Y. H., R. H. Orsi, K. K. Nightingale, and M. Wiedmann.** 2006. *Listeria monocytogenes* internalins are highly diverse and evolved by recombination and positive selection. *Infect. Genet. Evol.* **9:**9.

123. **Vazquez-Boland, J.-A., M. Kuhn, P. Berche, T. Chakraborty, G. Dominguez-Bernal, W. Goebel, B. Gonzalez-Zorn, J. Wehland, and J. Kreft.** 2001. *Listeria* pathogenesis and molecular virulence determinants. *Clin. Microbiol. Rev.* **14:**1–57.

124. **Vogel, J., and C. M. Sharma.** 2005. How to find small non-coding RNAs in bacteria. *Biol. Chem.* **386:**1219–1238.

125. **Weis, J., and H. P. R. Seeliger.** 1975. Incidence of *Listeria monocytogenes* in nature. *Appl. Microbiol.* **30:**29-32.

126. **Wemekamp-Kamphuis, H. H., J. A. Wouters, P. P. de Leeuw, T. Hain, T. Chakraborty, and T. Abee.** 2004. Identification of sigma factor sigma B–controlled genes and their impact on acid stress, high hydrostatic pressure, and freeze survival in *Listeria monocytogenes* EGD-e. *Appl. Environ. Microbiol.* **70:**3457–3466.

127. **Wiedmann, M., J. L. Bruce, C. Keating, A. E. Johnson, P. L. McDonough, and C. A. Batt.** 1997. Ribotypes and virulence gene polymorphisms suggest three distinct *Listeria monocytogenes* lineages with differences in pathogenic potential. *Infect. Immun.* **65:**2707–2716.

128. **Williams, T., S. Bauer, D. Beier, and M. Kuhn.** 2005. Construction and characterization of *Listeria monocytogenes* mutants with in-frame deletions in the response regulator genes identified in the genome sequence. *Infect. Immun.* **73:**3152–3159.

129. **Williams, T., B. Joseph, D. Beier, W. Goebel, and M. Kuhn.** 2005. Response regulator DegU of *Listeria monocytogenes* regulates the expression of flagella-specific genes. *FEMS Microbiol. Lett.* **252:**287–298.

130. **Wisniewski, J. M., and J. E. Bielecki.** 1999. Intracellular growth of *Listeria monocytogenes* insertional mutant deprived of protein p60. *Acta Microbiol. Pol.* **48:**317–329.

131. **Wong, K. K., H. G. Bouwer, and N. E. Freitag.** 2004. Evidence implicating the 5′ untranslated region of *Listeria monocytogenes actA* in the regulation of bacterial actin-based motility. *Cell. Microbiol.* **6:**155–166.

132. **Yildirim, S., W. Lin, A. D. Hitchins, L. A. Jaykus, E. Altermann, T. R. Klaenhammer, and S. Kathariou.** 2004. Epidemic clone I–specific genetic markers in strains of *Listeria monocytogenes* serotype 4b from foods. *Appl. Environ. Microbiol.* **70:**4158–4164.

GENOMIC ANALYSIS OF PLANT PATHOGENIC BACTERIA

Gail M. Preston, David S. Guttman, and Ian Toth

15

Plant pathogenic lifestyles have evolved independently in many of the major groups of bacteria, including *Proteobacteria, Actinobacteria, Firmicutes,* and *Mollicutes.* Soft rots may be caused by *Erwinia carotovora* (γ-*Proteobacteria*) or *Clostridium* (*Firmicute*). Dystrophies such as galls and witches' brooms can be caused by *Phytoplasma* (*Mollicute*), *Pantoea agglomerans,* and *Agrobacterium tumefaciens* (γ- and α-*Proteobacteria*), or by *Rhodococcus fascians* (*Actinobacteria*). Wilting, stunting, yellowing, necrosis, and decline may similarly be attributed to any one of a wide range of bacteria. Many plant pathogenic bacteria colonize the same natural habitats as benign endophytic, epiphytic, and rhizosphere-colonizing microorganisms and may behave in a similar way until disease-inducing conditions ensue. This provides ample opportunity for convergent evolution of physiological adaptations and for bidirectional lateral gene transfer and innovation between bacteria and between kingdoms. Some bacteria, such as *Pseudomonas aeruginosa* and *Burkholderia cepacia,* are able to cause opportunistic infections in both plants and animals, in addition to infecting soil organisms such as nematodes and amoebae, raising questions about the origin and evolution of bacterial pathogenicity mechanisms.

Plant pathogenic bacteria are relative latecomers to the genome-sequencing boom, with the first plant pathogen genome, that of *Xylella fastidiosa,* being released in 2000 (146). However, a large number of plant pathogen genomes have been completed in the past 5 years (Table 1), and subsequent analyses of these genomes have provided significant insights into the biology of these bacteria, their relationship to animal pathogens, and the mechanisms and evolution of plant pathogenesis. Comparative analyses of genome data from related organisms that cause distinctly different disease symptoms, such as *Erwinia amylovora* (vascular wilt, blight) and *Erwinia chrysanthemi* (soft rot), or *Xanthomonas campestris* pv. *campestris* (necrosis, rot) and *Xanthomonas axonopodis* pv. *citri* (canker), will prove increasingly valuable for understanding the etiology of plant diseases.

Genome analyses of plant pathogenic bacteria have not only brought new questions and new technology to the field of plant pathology, but have also brought new ideas and approaches to pathogenomics. The relative lack

Gail M. Preston   Department of Plant Sciences, University of Oxford, South Parks Road, Oxford OX1 3RB, United Kingdom.   *David S. Guttman*   Department of Botany, Centre for the Analysis of Genome Evolution and Function, University of Toronto, 25 Willcocks Street, Toronto, ON M5S 3B3, Canada.   *Ian Toth*   The Scottish Crop Research Institute, Invergowrie, Dundee DD2 5DA, United Kingdom.

Bacterial Pathogenomics, Edited by M. J. Pallen et al.
© 2007 ASM Press, Washington, D.C.

TABLE 1 Completed genome sequences of plant pathogenic bacteria

Pathogen	Taxonomy	Pathology	Host range	Genome sequence (size, GenBank ID/website, year)
Xylella fastidiosa 9a5c	γ-*Proteobacteria*	Chlorosis, leaf scorch, dystrophy	Broad (insect vector)	2.7 Mb AE003849 (2000)
Pseudomonas aeruginosa PAO1	γ-*Proteobacteria*	Watersoaking, necrosis	Broad	6.3 Mb AE004091 (2000)
Agrobacterium tumefaciens C58	α-*Proteobacteria*	Galls	Broad	5.7 Mb AE008688 (2001) 5.7 Mb AE007869 (2001)
Ralstonia solanacearum GMI1000	β-*Proteobacteria*	Wilt, chlorosis, necrosis	Broad	5.8 Mb AL646052 (2002)
Xanthomonas campestris pv. *campestris* ATCC33913	γ-*Proteobacteria*	Necrosis, chlorosis, black rot, wilt	Crucifers	5.1 Mb AE008922 (2002)
Xanthomonas axonopodis pv. *citri* 306	γ-*Proteobacteria*	Canker, necrosis, dieback	Citrus	5.27 Mb AE008923 (2002)
Xylella fastidiosa Temecula1	γ-*Proteobacteria*	Chlorosis, leaf scorch, dystrophy	Broad (insect vector)	2.5 Mb AE009442 (2003)
Pseudomonas syringae pv. *tomato* DC3000	γ-*Proteobacteria*	Chlorosis and necrosis	Tomato, *Arabidopsis*	6.5 Mb AE016853 (2003)
Erwinia carotovora subsp. *atroseptica* SCRI1043	γ-*Proteobacteria*	Soft rot, black rot of stems, wilt	Potato	5.1 Mb BX950851 (2004)
Onion yellows *Phytoplasma* OY-M	*Mollicute*	Dystrophy—witches' broom	Broad (onion, chrysanthemum) (insect vector)	0.9 Mb AP006628 (2004)
Leifsonia xyli subsp. *xyli* CTCB07	*Actinobacteria*	Stunting, discoloration	Sugarcane	2.6 Mb, AE016822 (2004)
Pseudomonas syringae pv. *syringae* B728a	γ-*Proteobacteria*	Necrosis	Broad (bean)	6.1 Mb CP000075 (2005)
Pseudomonas syringae pv. *phaseolicola* 1448a	γ-*Proteobacteria*	Chlorosis, necrosis	Bean	5.9 Mb CP000058 (2005)
Xanthomonas campestris pv. *campestris* 8004	γ-*Proteobacteria*	Necrosis, chlorosis, black rot, wilt	Crucifers	5.1 Mb CP000050 (2005)
Xanthomonas campestris pv. *vesicatoria* 85-10	γ-*Proteobacteria*	Necrosis, chlorosis	Crucifers, Solanaceae	5.4 Mb AM039952 (2005)
Xanthomonas oryzae pv. *oryzae* KAXCC10331	γ-*Proteobacteria*	Watersoaking, chlorosis, wilt	Rice, Graminaceae	4.5 Mb AE013598 (2005), NIAB, Macrogen
Xanthomonas oryzae pv. *oryzae* MAFF 311018	γ-*Proteobacteria*	Watersoaking, chlorosis, wilt	Rice, Graminaceae	4.8 Mb AP008229 (2006)
Aster yellows, witches' broom, *Phytoplasma*	*Mollicute*	Dystrophy, yellowing, witches' broom	Broad (insect vector)	0.7 Mb CP000061 (2006)
Streptomyces scabies 87–22	*Actinobacteria*	Necrosis, dystrophy	Potato	10.1 Mb, http://www.sanger.ac.uk /Projects/S_scabies/
Clavibacter michiganensis subsp. *sepedonicus* ATCC33113	*Actinobacteria*	Wilt, ring rot of tubers	Potato	3.4 Mb, http://www.sanger.ac .uk/Projects/C_michiganensis/
Erwinia amylovora Ea283	γ-*Proteobacteria*	Necrosis, dieback	Apple, pear, Maloideae	3.8 Mb, http://www.sanger .ac.uk/Projects/E_amylovora/

of knowledge of plant cell biology compared with that of animals has fueled the development of bioinformatic and experimental techniques for investigating the breadth and diversity of bacterial pathogenicity mechanisms. Analyses of pathogen-induced dysfunction in plant cells have provided significant insights into plant cell biology, most notably through the identification and characterization of bacterial "effectors," as discussed below. Current and future investigations will continue to take advantage of the fact that plants share many commonalities in signal transduction with animal cells and are well suited to rapid, high-throughput analyses of host-microbe interactions, with fewer regulatory and ethical restrictions than work on animal hosts. The infection courts of many plant diseases are within the plant leaf or root, which are well suited to in vivo imaging by fluorescent and confocal microscopy. In this chapter, we provide an overview of how genome sequence data are changing our understanding of the mechanisms and evolution of plant pathogenesis, and we discuss the opportunities and challenges genome data present for future research. By necessity, we largely focus on gram-negative proteobacterial pathogens, for which large amounts of genome sequence data and experimental data are available. However, it is important to note that genome sequences have recently been published for two phytoplasmas and for *Leifsonia xyli* subsp. *xyli* (7, 108, 116), and that *Streptomyces scabies* and *Clavibacter michiganensis* subsp. *sepedonicus* genomes have recently been completed by the Wellcome Trust Sanger Institute (http://www.sanger.ac.uk) (Table 1). Analyses of these genomes are likely to generate many insights into other plant pathogenic taxa in the near future.

GENOME ORGANIZATION AND VARIATION

Comparative genomic analyses of bacterial genomes permit us to identify common features that may be required for their ecological and evolutionary success. Even more importantly, comparative genomics allows us to identify features that vary among different bacteria, and that explain the genetic bases of their different lifestyles or evolutionary makeup. Such flexible or accessory components of the genome may contain genes with a putative role in disease or in niche/host specialization, and therefore may provide clear candidates for future study.

Horizontal gene transfer is crucial to the evolution of bacteria, and plant pathogens are no exception. These pathogens commonly acquire foreign DNA through the uptake of plasmids, which often encode essential pathogenic traits (87, 166). The Ti (tumor-inducing) plasmid of the crown gall pathogen *Agrobacterium tumefaciens* is solely responsible for the symptoms of this disease, supported by chromosomal traits that allow *Agrobacterium* to survive on the plant surface (161). Ti plasmid functions can be transferred to non-*Agrobacterium* species, such as *Rhizobium,* rendering them capable of transforming plants (19, 34). In addition, foreign DNA uptake via phage transduction, conjugation, or transformation may lead to the integration of foreign DNA into the chromosome or existing plasmids to form genomic islands (6). These islands often differ in their GC content from the rest of the genome and are rich in transposons, integrases, and phage-related genes (6, 29, 87). Where they contain pathogenesis-related genes, they are termed pathogenicity islands (PIs) (70).

Some pathogens show only limited gene acquisition—for example, the functionally restricted pathogens *Xylella fastidiosa* Temecula (Pierce's disease of grapevines) and *X. fastidiosa* 9a5c (citrus variegated chlorosis) have only one obvious genomic island each (162). The majority of plant pathogens, on the other hand, contain large numbers of genomic islands, and it is clear that in many cases they play an essential role in disease (6). For example, the type III secretion system (T3SS) and/or many of the genes encoding T3SS-secreted effector proteins (T3SE) in *Pseudomonas, Ralstonia, Erwinia*, and *Xanthomonas* spp., appear to have been acquired through horizontal gene transfer (4, 10, 58, 112). Genes encoding the *P. syringae* phytotoxins corona-tine and syringomycin reside on PIs, as do their recently discovered counterparts in *E. carotovora* subsp. *atroseptica* SCR1043 (Eca1043) (2, 10, 144). Other factors that may have been subject to horizontal transfer include adhesins, polysaccharides, various secretion systems, and a large number of conserved hypothetical or unknown proteins that may play roles in pathogenesis.

PIs have also been identified in other plant pathogenic taxa. The thaxtomin biosynthesis genes of the potato scab pathogen *S. scabies* are found in a large (320 to 660 kb) PI, along with a gene encoding the necrosis-inducing protein *nec1* (86). Phytoplasmas have small genomes that contain large clusters of potential mobile units (PMU). These PMU appear to be unique to phytoplasmas and contain *tra5* insertion se-quences and genes for specialized sigma factors and membrane proteins. Phytoplasmas lack some recombination functions found in re-lated mycoplasmas, and it is possible that these PMU have an important role in genome plas-ticity (7).

COMPARATIVE GENOMICS
Comparative genomic data from closely related plant pathogens are available for *P. syringae* and *Xanthomonas* spp., and they can be used to de-velop hypotheses about the molecular basis of host specificity and of differences in symptoms and disease progression. Six agronomically im-portant strains of *Xanthomonas* have been fully sequenced: two strains of *Xanthomonas cam-pestris* pv. *campestris* (NCPPB528 [Xcc] and 8004 [Xcc8004]), the causal agent of black rot in brassicas, and a pathogen of the model plant *Arabidopsis thaliana* (39, 127); *X. campestris* pv. *vesicatoria* 85-10 (Xcv), the causal agent of bac-terial spot disease of peppers and tomatoes (155); two strains of *Xanthomonas oryzae* pv. *oryzae* (KACC10331 [Xoo] and MAFF 311018), the causal agent of bacterial blight of rice (92, 114); and *Xanthomonas axonopodis* pv. *citri* XV101/Str306 (Xac), the causal agent of citrus canker (39). The evolutionary relation-ships among these strains have not yet been studied. However, a BLAST-based analysis of these genomes identified 2,999 shared coding sequences (66.8%) making up the *Xanthomonas* core genome (155). A total of 184 (4.1%), 87 (1.9%), 45 (1.0%), and 548 (12.2%) of coding sequences were unique to Xac, Xcc, Xoo, and Xcv, respectively, with most of these unique sequences encoding hypothetical and con-served hypothetical proteins (155).

The three sequenced *P. syringae* strains indi-vidually represent three of the largest phyloge-netic groups in the *P. syringae* species complex, and therefore they provide a robust picture of the extent of genetic diversity found in this species (76, 140). *P. syringae* pv. *tomato* DC3000 (PtoDC3000) is the causal agent of bacterial speck of tomato and of the model plant *A. thaliana* (21); *P. syringae* pv. *phaseolicola* 1448A (Pph1448A) is the causal agent of halo blight of bean (80); and *P. syringae* pv. *syringae* B728A (PsyB728A) is the causal agent of bacterial brown spot of bean (50). A reciprocal smallest distance analysis, which relies on maximum likelihood estimations of evolutionary distances and best reciprocal global alignments rather than the local alignments used in BLAST (167), was used to identify the core and flex-ible genome of *P. syringae*. The three sequenced pathovars share 3,779 coding sequences (~71.7%) in their core genome. A total of 1,296 (23.1%), 825 (16.1%), and 735 (14.4%) cod-ing sequences are unique for PtoDC3000, Pph1448A, and PsyB728A, respectively (H. Nahal and D. Guttman, in preparation).

Neither *P. syringae* nor *Xanthomonas* sp. are closely related to animal pathogens; however the genome sequence of the enteric plant pathogen Eca1043, the causal agent of blackleg and soft rot diseases of potato, offers a unique opportunity to compare an enterobacterial plant pathogen with animal pathogenic cousins such as *Escherichia coli*, *Salmonella*, and *Yersinia* species (10, 121). A reciprocal best-hit analysis and GenomeDiagram visualization software were used to compare the Eca1043 genome with 229 available bacterial genome sequences, including 13 from animal pathogenic entero-bacteria (APE) (126, 157). Eca1043 shares over

75% of its coding sequences with other enterobacteria, most of which are present on the chromosomal backbone and represent core functions such as metabolism, cell division, motility, and chemotaxis. Seventeen major genomic islands, on the other hand, share only half of their coding sequences with enterobacteria, and they show a large increase in the proportion of coding sequences that make no reciprocal best hit or that show a greater similarity to other plant-associated bacteria (PAB). The genomic islands also contain a larger proportion of coding sequences associated with cell surface proteins, plant-associated functions, transport, regulation, and pathogenicity, which may account for the plant-associated lifestyle of this *E. coli*–like plant pathogen. Color Plate 5 shows a circular image generated by Genome-Diagram comparing the Eca1043 genome sequence to those of other PAB and APE. In general, Eca1043 is clearly more related to the APE than the PAB in terms of amino acid sequence similarity. However, some horizontally acquired islands show an increase in the number of reciprocal best hits and amino acid sequence similarity to PAB compared with APE, e.g., coding sequences associated with coronofacic acid phytotoxin production (A), T3SS (B), and nitrogen fixation (C).

PATHOGENICITY AND VIRULENCE FACTORS

Pathogenicity and virulence factors have received the greatest focus in the genomic analysis of plant pathogens. Pathogenicity factors are those systems that are necessary for either the initiation or maintenance of disease. They are generally core components of the genome of a pathogenic species, meaning that they are found in all isolates of that species, because any strain that lacks them will not be able to compete in that niche. Virulence factors can be considered facultative components of virulence. They may be necessary in certain niches (e.g., specific hosts), but not in others. As such, they are commonly components of the flexible or accessory genome, which is that portion of the genome that varies among strains

within a species. Once a virulence factor does not positively contribute to bacterial fitness, it may be lost via mutational degeneration or by excision, which is often mediated by mobile genetic elements. Host specificity factors are a special subclass of virulence factors that contribute to virulence in a host-specific manner. Each type of factor may act during initial colonization of the host, during invasion of the host tissue, during the maintenance of infection, or even during transmission to a new susceptible host. Figure 1 provides an overview of some of the key pathogenicity and virulence factors commonly associated with plant pathogenic bacteria.

Attachment

Host attachment is a critical first step in pathogen-host interactions. When a phytopathogen initially comes into contact with a potential host, it must adhere to the plant surface or risk being dislodged by rain and wind. Attachment is mediated by components in the bacterial envelope such as extracellular polysaccharides (EPS), lipopolysaccharide (LPS), and pili (137).

EPS are adhesive polysaccharide polymers that are either excreted into the environment or bound to the cell as a loose slime capsule. In addition to facilitating attachment, the EPS layer prevents desiccation, aids in the evasion of host detection, and plays a central role in biofilm formation and the occlusion of water-transporting xylem vessels, leading to wilt symptoms (46, 110). The most important EPS in *P. syringae* is alginate, which is important for both epiphytic fitness and virulence (179). Alginate production is under the control of the *gacS/gacA* global activator two-component system (27, 175), which also regulates homoserine lactone production, swarming, toxin production, and other systems important for plant-pathogen interactions (72, 88, 89, 128). Alginate synthesis genes appear to be part of the core virulence repertoire of *P. syringae* because they are found in all three genomes.

Xanthomonads secrete the characteristic EPS xantham gum, which is synthesized from the *gum* operon. This molecule is essential for

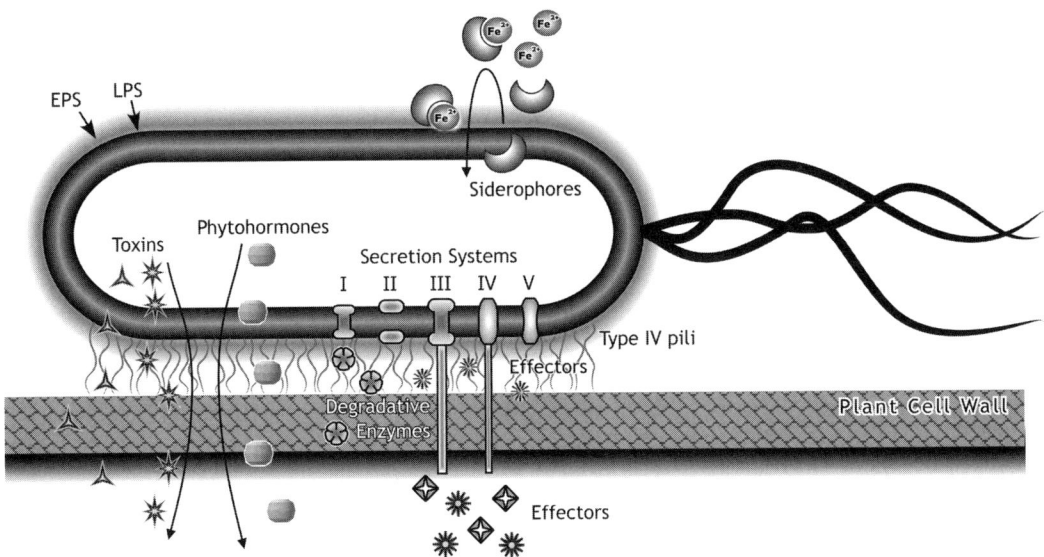

FIGURE 1 Pathogenicity and virulence factors produced by proteobacterial plant pathogens.

Xoo virulence in rice (42), although in Xcc, it positively contributes to aggressiveness but is not required for virulence (84). Xcc also carries a gene for cyclic β-1,2-glucan synthetase, which is associated with virulence and colonization in members of the *Rhizobiaceae* family and which is not found in Xac (39, 40). Interestingly, the *gum* regulon is found in a 15-kb region that has an atypical G+C content and which is flanked by an insertion sequence element in Xcv, perhaps indicating that it was brought into this species by horizontal gene transfer.

In the enterobacterial phytopathogens *E. amylovora* and *Pantoea stewartii*, EPS is a major virulence determinant, assisting in the colonization of plant hosts by blocking xylem vessels (95). Although genome sequencing has revealed that Eca1043 possesses EPS synthesis genes similar to these pathogens, no association with virulence has yet been found; however, mutations in LPS structure in *E. carotovora* have been associated with reduced virulence (158).

LPS is an essential component of the bacterial outer membrane and is found in all gram-negative bacteria. LPS is considered a pathogen-associated molecular pattern because it stimulates the host innate immune response (107). One component of the LPS layer is the O antigen, which forms a hydrophilic surface layer. This region is often variable both in sequence and in genomic location among species and pathovars, and it is likely to play an important role in host specificity (9, 39, 43, 118). *A. thaliana* plants respond to bacterial LPS with a rapid burst of nitric oxide and the activation of defense genes (181). This response is mediated by a nitric oxide synthase that is required for resistance to *P. syringae*. LPS has also been shown to act as a potent endotoxin (173), suppress host immunity, and promote virulence in animals (64, 174).

Bacteria produce many types of hairlike surface appendages called pili or fimbriae, often with adhesins at their tips, which play important roles in attachment to host surfaces. *P. syringae* type IV pili influence leaf surface adherence and epiphytic growth. Nonpiliated strains adhere poorly and are uniformly distributed on leaf surfaces, whereas piliated strains are localized to stomata (138, 152). Xcv also produces type IV pili, with mutants of this system exhibiting reduced aggregation in the laboratory but still retaining wild-type aggressiveness toward their tomato host (115). Eca1043 has two gene clusters representing type IV pili, although neither has yet been implicated in virulence (10). However, the genome sequence has shown that one of these clusters is found on what appears to be a mobile genetic element similar to the SPI-7 PI of *Salmonella enterica* serovar Typhi, which contains a similar gene cluster. Identifying the biological role of adhesins can be challenging because phenotypes can be subtle and specific to cell type or host. The *hecA* adhesin of *E. chrysanthemi* was identified through its proximity to the T3SS PI; however, phenotypes associated with this protein were evident only in bacterial interactions with *Nicotiana clevelandii*, out of several different host plants tested (135).

The availability of genome sequence data will greatly facilitate proteomic analyses of adhesive structures. Proteomic analysis of extracellular proteins produced by *X. fastidiosa* showed that this bacterium expresses a wide range of adhesive structures, including type IV fimbriae, mrk pili, and hsf surface fibrils (147). The same study noted an unusual codon bias in highly expressed extracellular proteins, which may be related to the slow growth of this specialized bacterium.

Penetration

Most bacterial pathogens lack the ability to penetrate the waxy cuticle that covers plant surfaces. They must enter through natural openings and wounds, or be transmitted by insect vectors such as xylem-feeding sharpshooters and phloem-feeding aphids (Fig. 2) (132, 153, 172). Some insect-transmitted bacteria, such as the xylem pathogen *X. fastidiosa,* are able to replicate inside their insect vectors. The mechanisms underpinning *X. fastidiosa*–insect interactions are poorly understood, but it is clear that *X. fastidiosa* is specifically adapted to colonize the insect foregut, and that the transmission efficiency of *X. fastidiosa* (ability of a vector to successfully acquire and transmit the bacteria) varies widely, depending on the combination of vector species and host plant (132). Comparative genome analyses have identified *X. fastidiosa*–specific adhesins and EPS biosynthesis genes similar to *Xanthomonas gum* genes that may have a key role in enabling bacteria to establish biofilms in insect hosts (40, 146, 149, 151). The *X. fastidiosa* genome also provides a potential insight into another aspect of *X. fastidiosa* penetration: movement across the sieve plates that separate each xylem cell. The genome encodes cell wall degrading enzymes that may be used to degrade this barrier (146).

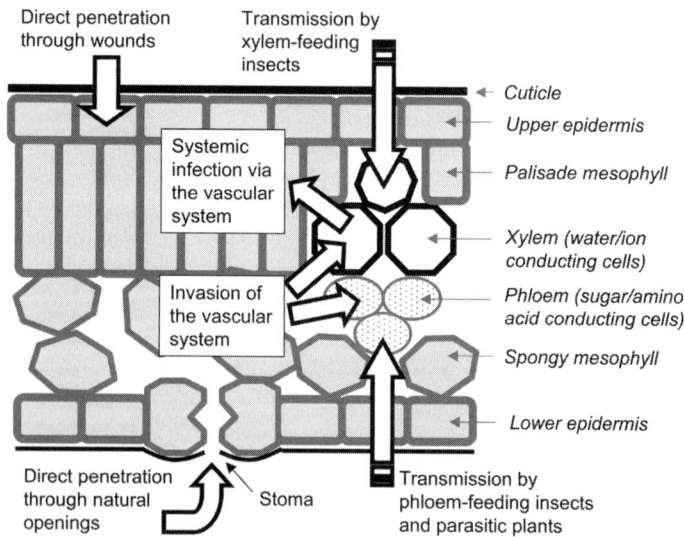

FIGURE 2 Main routes of bacterial penetration and infection.

Ice-nucleating *P. syringae* and *Erwinia* strains are psychrophiles that use extracellular ice-nucleating proteins to raise the temperature at which plants suffer frost damage and thereby promote wounding of plant surfaces (65, 68, 177). The ice-nucleation proteins of *P. syringae* have proven to be useful tools for cell surface expression of a wide range of proteins (94). Genome analyses show that ice-nucleation genes are present in PsyB728a, but not in the other two sequenced *P. syringae* strains (50). Although ice nucleation has been most extensively studied in *Pseudomonas* and *Erwinia*, it has also been reported in *Xanthomonas* (183), and a putative ice-nucleating protein is present in the genome of Xcc8004.

Siderophores

Siderophores are low-molecular-weight molecules that permit the chelation and then import of iron from the environment. They play an important function during pathogenesis because iron is often a limiting resource in the extracellular environment (168). Pyoverdine is perhaps the most important *P. syringae* siderophore, but it is not present in all three sequenced genomes. Siderophores are responsible for the fluorescence seen among most pseudomonads (38). An alternative polyketide synthase (nonribosomal peptide) siderophore called yersiniabactin is found in PtoDC3000 and Pph1448, but not in PsyB728A. The large polyketide synthase gene believed to be important for yersiniabactin synthesis shares a high degree of similarity with the *Yersinia pestis* homolog (21). Although PsyB728A does not carry the yersiniabactin siderophore, it does contain a second siderophore gene cluster that is similar to the achromobactin hydroxy/carboxylate class of siderophores produced by *E. chrysanthemi* and *E. carotovora* (50, 54). Eca1043, in addition to achromobactin, also contains the *E. coli*–like siderophore enterobactin. However, unlike *E. chrysanthemi*, where there is now an extensive body of evidence relating siderophore production to virulence, the importance of siderophores in Eca1043 remains to be investigated (10).

Phytotoxins and Phytohormones

PHASEOLOTOXIN

Phaseolotoxin is a chlorosis-inducing phytotoxin, which inhibits ornithine carbamoyltransferase (OCTase), a central enzyme in the urea cycle, thereby resulting in arginine deficiency (11). Phaseolotoxin-producing strains make a phaseolotoxin-insensitive OCTase encoded by the *argK* locus. Pph1448A produces phaseolotoxin and carries two OCTases, one of which is encoded by *argK* (PSPPH_4319) clustered with other loci associated with toxin production (80). PtoDC3000 does not produce phaseolotoxin and carries only the ortholog of the Pph1448A phaseolotoxin-sensitive OCTase gene (PSPPH_3895, PSPTO_4164). PsyB728A also does not produce phaseolotoxin, yet it carries two genes encoding OCTase. The first is a highly similar ortholog of the phaseolotoxin-sensitive enzyme (Psyr_3901); the second is a highly divergent homolog of the resistant enzyme (Psyr_2685). PsyB728A also carries homologs to six genes of unknown function found in one of the phaseolotoxin biosynthetic clusters.

CORONATINE

Coronatine is a *P. syringae* polyketide phytotoxin that induces chlorosis and lesions in host cells (11). Coronatine is formed by the conjugation of coronafacic acid (CFA) to coronamic acid (CMA) by coronafacic ligase (Cfl). The structure of coronatine mimics that of methyl jasmonate, an important growth regulator and signaling molecule that is synthesized by plants under biological stress. Coronatine enhances bacterial virulence, but it is not required for disease. Its production is associated with the induction of jasmonate and wound-responsive genes, and the suppression of pathogenesis-related genes during *P. syringae* infection of *A. thaliana* (184). The coronatine biosynthetic genes are encoded in the PtoDC3000 genome, although they are more frequently plasmid localized (12, 13). None of the other sequenced pathovars of *P. syringae* carries these genes or produces coronatine (76). Surprisingly, a gene cluster highly similar in both structure and organization to the *cfa* cluster of *P. syringae* is also found in Eca1043, together with the gene encoding Cfl. However, it lacks genes encoding CMA. Nevertheless, mutations in the Eca1043 *cfa* cluster lead to reduced virulence, suggesting that Cfl may conjugate other amino acids to CFA to produce a range of coronafacoyl compounds (10). The presence and role in virulence of genes encoding such compounds, together with an active T3SS, suggest that the archetypical necrotroph *E. carotovora* may undergo a biotrophic phase during part of its infection process (157).

SYRINGOMYCINS
Lipodepsinonapeptides are a group of phytotoxins that induce necrosis by forming pores in the plasma membrane of the host plant cell (11). Two major classes of lipodepsinonapeptides have been identified in *P. syringae*, including the syringomycin group, which include syringotoxin, syringostatin, and pseudomycin, all of which differ only in their 3-hydroxy fatty acid moiety, and the syringopeptin group. The secretion of these phytotoxins promotes passive transmembrane influx of H^+ and Ca^{2+} ions, acidifying the cytoplasm, resulting in cell death and the induction of a calcium-related cellular signaling cascade (11). The sequencing of the PsyB728A genome identified genes encoding one of each class of lipodepsinonapeptides, which were similar to toxin biosynthesis genes from *P. syringae* pv. *syringae* B301D (50). Bioinformatic analyses, molecular genetics, and a subgenomic oligonucleotide microarray have been used to define and characterize the toxin-associated genes of *P. syringae* pv. *syringae* B301D (98, 169, 170). No lipodepsinonapeptides have been identified in either PtoDC3000 or Pph1448A. A *syrE* homolog, potentially coding for a syringomycin-like toxin, was found in Xac, but not Xcc (39). Two copies in tandem are also present within the Eca1043 genome, although as with Xac, their role in virulence remains to be established (10). Pore-forming lipodepsinonapeptides such as tolaasin are also key pathogenicity factors in

many mushroom pathogenic *Pseudomonas* (63), but no mushroom pathogenic bacteria have yet been sequenced.

PHYTOHORMONES
The plant hormone indole-3-acetic acid (IAA or auxin) is produced by many phytopathogenic bacteria. IAA production has been shown to contribute to in planta and epiphytic growth, virulence, and the regulation of syringomycin in *P. syringae* (62, 104). The two enzymes required to produce IAA are present in all three sequenced *P. syringae* genomes, although only PtoDC3000 carries the gene for IaaL, which converts IAA to the less active IAA-lysine (21, 80). This locus is intriguing because it is driven by a functional Hrp (T3SS) promoter (21). PsyB728A produces greater amounts of IAA than PtoDC3000, and it has been speculated that PsyB728A might produce IAA through an alternative pathway that permits the pathogen to redirect the flux of the phytoalexin and glucosinolate plant defense pathways in such a way as to suppress the hypersensitive response (50).

Secretion Systems
Protein secretion systems are essential for pathogenesis in most plant pathogenic bacteria and have been extensively studied, particularly in proteobacterial pathogens. The protein secretion systems of plant pathogenic Proteobacteria have recently been reviewed by Preston et al. (124) and are illustrated in Figure 1.

TYPE I SECRETION SYSTEM
The type I secretion system (T1SS) consists of three major proteins that form a continuous channel spanning the bacterial inner membrane, the periplasmic space, and the bacterial outer membrane (14, 154). These ABC protein-mediated transport systems secrete a variety of toxins, proteases, and lipases into the environment. It is exemplified by *E. coli* hemolysin secretion (59), a toxin that causes the lysis of red blood cells. *E. amylovora* and *E. chrysanthemi* use a T1SS to secrete metalloproteases, which contribute to leaf colonization

(182). In addition to the metalloprotease system, the Eca1043 genome sequence reveals four other putative T1SSs that appear to be associated with the secretion of a large repetitive protein (ECA1099), an iron scavenging hemophore (similar to the HasA system in *Serratia marcescens* (15), and a hemagglutinin/hemolysin-like protein (although there is some similarity to a repeat in toxin [RTX]-like protein [10]). The T1SS associated with this latter protein has an outer membrane component similar to the root adhesion protein AggA of *Pseudomonas putida,* suggesting a potential role in attachment to plant surfaces (20).

Both PtoDC3000 and Pph1448A encode multiple T1SS pathway components, although the secreted substrates and their role in virulence have not been shown (80). PsyB728A has a T1SS that is found on the opposite strand from a putative antifreeze gene and two glycosyltransferase genes, which may play a role in the glycosylation and secretion of the antifreeze protein (50). Secreted antifreeze proteins are thought to inhibit the formation of ice crystals, and they may play a role in modulating the ice nucleation activity of these bacteria. PtoDC3000 does not carry any orthologs of the ice nucleation or antifreeze proteins found in PsyB728A (50).

Xac carries an RTX-like toxin (39, 92) that is likely to be secreted through the T1SS. RTX toxins are pore-forming proteins which exhibit cytotoxic or hemolytic activity in a broad range of pathogens (55). These proteins are believed to function as important host-specificity factors because their activity is often limited to specific cell types or hosts (55). Xac carries one and Xoo carries two apparent RTX-like gene clusters. Neither of these loci is found in Xcc.

TYPE II SECRETION SYSTEM

The type II secretion system (T2SS) is an extension of the general secretory pathway delivering proteins from the cytoplasm to the extracellular environment as part of a two-step process (25, 35, 51, 97, 139). It is common among gram-negative bacteria, and in phytopathogens, it is best known in the soft rot *Erwinia* for its role in the secretion of plant cell wall–degrading enzymes, including pectate lyases and cellulases (156). Mutations in the *out* gene cluster, which encodes the T2SS in soft rot *Erwinia*, generally lead to a dramatic reduction in virulence (83). However, recent data have shown that other virulence-associated proteins are also secreted by this system, including Svx, a homologue of the *X. campestris* avirulence protein AvrXca (37).

The T2SS is also found in other phytopathogens including pseudomonads, xanthomonads, *X. fastidiosa*, and *Ralstonia solanacearum* (124). The *R. solanacearum* T2SS has a different gene organization to other T2SSs, secretes at least six major exoproteins, and is required for infection of host plants (58). Pph1448A contains two distinct T2SS pathways. Possible substrates for these pathways include two cellulases, two pectate lyases, a pectin lyase, and a polygalacturonase (80). PtoDC3000 only carries the loci for one T2SS (21), and the potential set of substrates is limited to two cellulases and a pectin lyase. Pph1448A also carries a gene encoding ExeA, which, along with ExeB, is required for T2SS-mediated aerolysin toxin secretion in *Aeromonas hydrophila*. This locus is plasmid localized in both Pph1448A and the radish pathogen *P. syringae* pv. *maculicola* ES4326 (PmaES4326) (150). No homologs of *exeB* are found in either strain.

Both Xcc and Xac have two T2SSs; Xoo has one (155), although Xoo carries more genes predicted to encode proteins that degrade plant cell wall polymers than either Xcc or Xac (92). Xylanase and pectinase function as Xoo virulence factors, and they may facilitate the dissemination of Xoo through plant xylem vessels (131). A secreted lipase also contributes to virulence in Xoo, and a xylanase/lipase double mutant is strongly attenuated in virulence (129). Neither Xcc nor Xac carries a xylanase gene (92). Xoo is also unique among the xanthomonads in that it carries a homolog of a serine protease that enhances wilting in the tomato pathogen *Clavibacter michiganensis* (45).

Although Xcc carries fewer degradative enzymes than Xoo, it has more pectin and cellulose degradation genes than Xac (39). Furthermore, Xac lacks two genes from the *rpf* gene cluster (*rpfH* and *rpfI*) that regulate the synthesis of extracellular degrading enzymes (39). Mutations in *rpfI* have been shown to dramatically reduce the levels of the secreted enzymes in Xcc (44, 176). These differences may in part explain why Xac does not cause as severe tissue degradation as Xcc during infection (39).

TYPE III SECRETION SYSTEM

The type III secretion system (T3SS) is a specialized protein secretion system that enables bacteria to inject T3SE directly from the bacterial cell into the cytoplasm of its host (5, 71, 79, 180). The T3SS apparatus spans both bacterial membranes and the periplasmic space, and it makes direct contact with the host via the T3SS pilus. The T3SS is often, but not always (8), essential for plant pathogenesis in the strains in which it is present, and it can also play an important role in establishing mutualistic bacteria-plant relationships (102). The T3SSs of phytopathogens are encoded by 20 or more genes, many of which are highly conserved among both plant and animal pathogens. The T3SSs of phytopathogens fall into two of the five known T3SS groups, with *Erwinia* and *Pseudomonas syringae* belonging to the Hrp1 group, and *Xanthomonas* spp. and *Ralstonia* spp. belonging to the Hrp2 group (53). T3SS gene clusters often show different GC content to the rest of the genome and do not always correlate with rRNA-based phylogenies, indicating their acquisition via horizontal gene transfer (70, 71). One possible exception is the *R. solanacearum hrp* gene cluster, which is not flanked by mobile elements, shows no variation in GC content, and is strongly conserved in terms of colinearity and sequence similarity with clusters from xanthomonads and *Burkholderia pseudomallei* (which do show PI features) (58).

T3SEs interact directly with host proteins and are thought to primarily function by suppressing the host defense response (1, 5, 18, 28, 48, 69, 71, 111, 113, 160). In animal pathogens, they are also known to cause cytoskeletal changes and may have a direct cytotoxic effect (23, 48, 49, 67, 75, 159). Given their intimate interactions inside the host plant cell, it is not surprising that they may also inadvertently alert the host of a pathogen attack. Plants have evolved large suites of resistance proteins that monitor T3SE targets. When these targets are altered by T3SE, the resistance proteins trigger a massive defense response, called the hypersensitive response (HR), which is usually effective in stopping pathogen proliferation. Intriguingly, although the *R. solanacearum* genome contains several genes encoding proteins homologous to HR–eliciting T3SE identified in other genera, only one, PopP2, has ever been shown to elicit the HR in a classical gene for-gene manner (58).

PtoDC3000 appears to have the largest suite of T3SEs of any plant or animal pathogen, including other *P. syringae* strains. This strain carries at least 33 functional T3SEs and another 20 T3SEs that are either disrupted or of questionable status (A. Collmer, personal communication). PsyB728A has 27 putative T3SEs, and Pph1448A has at least 22 (50, 80, 164). Multiple T3SEs are encoded in conserved and variable T3SE loci flanking the T3SS gene cluster; others are scattered singly or in clusters around the genome (4, 6, 81).

The Pph1448A genome sequence delivered one very unexpected T3SS-related finding. In addition to the Hrp1 T3SS, this strain contains a cluster of 12 genes that are homologous to the T3SS genes found in *Rhizobium*, *Photorhabdus*, *Aeromonas* spp., and *P. aeruginosa*. It appears that this system is not functional, at least as a T3SS, because it cannot complement mutations in the primary T3SS (80).

All of the sequenced xanthomonads use a T3SS during pathogenesis. The *hrp* genes that encode the T3SS apparatus are similar in all four sequenced strains, except for minor changes in gene order, the insertion of transposase genes in the *hrp* cluster of Xoo, and variation in genome location (39, 92). As in *P. syringae,* regions flanking the xanthomonad *hrp* cluster carry variable suites of T3SEs and accessory elements (24, 69, 112, 155). Xcv carries 20 confirmed or putative T3SEs (69, 155). Of these 15 clear homologs or similar partial sequences that are found in Xcc, 10 are in Xac, 8 in Xoo, 8 in PtoDC3000, and 8 in *R. solanacearum* (155). There is no association, with respect to their joint presence or absence, among the Xcv T3SE found in PtoDC3000 and *R. solanacearum* ($P = 0.65$, Fisher's exact test, two tailed). PtoDC3000, *R. solanacearum*, and *Xanthomonas* genomes all carry *xopQ*, a *P. syringae hopQ1-1* homolog with putative inosine-uridine nucleoside *N*-ribohydrolase activity, while all three species of *Xanthomonas* carry *avrBs2, xopF1, xopF2, xopN, xopP*, and *xopX*. Of these six T3SEs, only *avrBs2*, a putative glycerophosphoryl-diester phosphodiesterase, has a predicted function (69, 155). Twelve of 17 or 20 T3SEs or putative T3SEs are shared among Xcc and Xac, respectively (39). Neither Xcv nor Xcc carries the well-studied *avrBs3* T3SE, whereas four homologues were found in both Xoo and Xac (39, 69, 92, 155).

An interesting difference between *Xanthomonas* and other plant pathogens is the relative deficiency of T3SEs with protease activity (74). YopT homologs, such as the *avrPphB* T3SE from *P. syringae*, are absent from *Xanthomonas*, and YopJ homologs, such as the large HopZ family from *P. syringae* are quite rare. Xcv carries four YopJ family members, AvrXv4, AvrBsT, AvrRxv, and XopJ (74, 134), and although the latter two T3SEs have low-similarity homologs in Xcc, none is found in either Xac or Xoo. The only other known *Xanthomonas* protease T3SE is XopD, which is found in Xcv but no other strain.

Genome sequence data and associated comparative and functional analyses provide the only solution to the challenge posed by functional redundancy among T3SE and other virulence factors in *P. syringae* and *Xanthomonas*. Mutations in single T3SE frequently have no effect on pathogenicity or virulence. One study identified and deleted eight candidate T3SEs in Xcc and observed no significant change in pathogenicity, even when all eight were deleted, although one mutation caused a change from virulence to avirulence in one host cultivar (26). Research into T3SE function is therefore currently proceeding from bioinformatic and functional identification of candidate T3SE to analyses of secretion competence, systemic mutation of single and multiple T3SEs, transgenic and heterologous expression of single and multiple T3SEs, and characterization of host targets and cellular localization (1, 5, 18, 28, 48, 66, 69, 71, 111,113).

T3SE redundancy is a less significant problem in *Erwinia*, which possess T3SSs with *hrp* clusters similar to each other and to *P. syringae* in terms of both sequence similarity and gene order (73). The T3SS has been most thoroughly studied in *E. amylovora*, with only three proteins, DspE, HrpN, and HrpW, currently known to be secreted through this system, DspE and HrpN are both required for *E. amylovora* pathogenesis (16, 17, 57, 171). DspE is widely conserved among phytopathogens and is similar to AvrE of *P. syringae* and to RopE of *Pseudomonas fluorescens* (123). DspE has been shown to interact with a receptor-like kinase in its host apple (41,105). T3SSs are also present in soft rot *Erwinia*, with DspE and HrpN both playing important roles in pathogenesis (8, 73, 130, 136, 178). No other T3SE has been identified in *Erwinia* to date.

TYPE IV SECRETION SYSTEM

The type IV secretion system (T4SS) encodes the conjugative sex pilus used by bacteria to exchange plasmids. It has also been intensively investigated for its ability to move DNA-protein complexes and toxins into eukaryotic cells (32, 33, 52, 145). The plant pathogen *Agrobacterium tumefaciens* has one of the best-studied T4SSs. It delivers gall-forming (oncogenic) DNA-protein complexes into plant cells through its VirB T4SS (91, 161). T4SS have also been found in Pph1448A, PsyB728A and PmaES4326 (50, 80, 145, 150), but not PtoDC3000. The T4SS is plasmid borne in Pph1448A and PmaES4326, but localized to a genomic island in Psy B728A. This island is inserted at tRNA-Lys-2, and it also encodes copper and arsenic resistance genes, none of which have homologs in PtoDC3000 (50).

T4SSs were not known to occur in *Xanthomonas*, *R. solanacearum*, or *Erwinia* until the release of the corresponding genome sequences. Eca1043 has what appears to be a horizontally acquired island containing a *virB*-like operon encoding a putative T4SS and, like Psy B728A, genes encoding arsenic resistance, suggesting a possible common origin for these islands in the two pathogens (10). Mutational analysis of genes within the *virB*-like operon reduced virulence in Eca1043, although the nature of its potential role in disease remains to be determined. Xac has both a chromosomally encoded T4SS and a second plasmid-borne system. Xcc has only the single chromosomally encoded system (3, 39). Xcv is interesting in that the largest plasmid carries a T4SS similar to the Lcm/Dot system found in human pathogens. This is the first report of this system in a plant pathogen (155). There is no report of a T4SS in Xoo. *R. solanacearum* carries a T4SS system related to systems found on IncP or Ti plasmids as part of a conjugative transposon (58).

Necrosis-Inducing Proteins

Although relatively little is known about secretion systems and secreted proteins in nonproteobacterial plant pathogens, it is worth noting that Nep1-like (NLP)-secreted proteins have been identified in a wide range of plant-associated fungi, oomycetes, and bacteria (*Streptomyces*, *Bacillus*, *Frankia*, *Vibrio*, *Erwinia*) (60, 120). Large families of NLP proteins are present in the genomes of plant pathogenic fungi, and many of these proteins have been shown to cause necrosis or cell death in plant cells. The expression of the *E. carotovora* NLP Nip promotes larger lesions and more rot, and the wide distribution of these proteins suggests extensive horizontal transfer and effective secretion by many taxa. The gram-positive scab pathogen *S. scabies* secretes another type of necrogenic protein, Nec1, encoded by a gene with atypical GC content that may have been acquired through horizontal gene transfer (22).

PHYSIOLOGICAL ADAPTATION TO LIFE ON PLANTS

Most bacterial plant pathogens initiate plant colonization as biotrophs, deriving nutrients from living plant cells. Biotrophic plant pathogens occupy four main niches: the plant surface, intercellular spaces, cytoplasm, and xylem (Fig. 2). Epiphytic and rhizosphere colonizing pathogens such as *Agrobacterium tumefaciens* colonize plant surfaces, absorbing plant exudates and gaining direct access to plant cells via wounds (47, 103). *P. syringae* is well adapted for life as a foliar epiphyte, but can also colonize the intercellular spaces between plant cells and multiply inside susceptible plants (122). Intracellular endophytes such as *Phytoplasma* live inside plant cells, separated from the cytoplasm by a single unit membrane, and are frequently found in sugar-transporting phloem cells (31). Some xylem-colonizing pathogens, such as *X. fastidiosa* and *L. xyli* subsp. *xyli*, are xylem specialists (106, 108); others, such as *R. solanacearum*, colonize a range of tissues (142). The composition of xylem sap is similar to that of intercellular fluid, so we might expect to find nutritional similarities between intercellular and xylem-colonizing pathogens. The biotrophic niches occupied by plant pathogens are physiologically distinct from the necrotrophic niches that are formed in the later stages of many bacterial infections, when plant cells lyse and die.

Genome analyses of plant pathogens have highlighted three aspects of physiological adaptation to life on plants: specialization, innovation, and flexibility.

Specialization and Genome Reduction

The most dramatic adaptations to life on plants are found in obligate and fastidious pathogens such as *X. fastidiosa*, *Phytoplasma* spp., and *L. xyli* subsp. *xyli*. These bacteria have small genomes and restricted functionality (31, 108, 109, 116). The chromosome of Bermuda grass white leaf phytoplasma is only 0.53 Mb (101). In comparison, strains of the aphid symbiont *Buchnera aphidicola*, which is frequently cited as having the smallest bacterial genome (61), have genomes ranging from 0.45 to 0.64 Mb. The 0.87-Mb genome of *Phytoplasma asteris* OY strain encodes fewer metabolic functions than *Mycoplasma* genomes, and it even lacks all eight ATP synthase subunits (116). This may mean that it imports host ATP by an as yet uncharacterized mechanism, or that it is dependent on host glycolysis. Although it has lost many biosynthetic functions, it does retain 27 different transporter genes, presumably to take in host metabolites (82, 116). Despite their small genome size, some phytoplasmas have been shown to infect a wide range of host plants, and mixed infections have been reported in a number of independent studies (7, 31, 93). However, in nature, phytoplasma dissemination is limited by interactions with insect vectors (153, 172).

Some strains of the xylem-restricted pathogen *X. fastidiosa* also show a broad host range in experimental studies, but others are host specific. Evidence that host restriction is associated with genome reduction in *X. fastidiosa* can be seen in DNA microarray-based comparative analyses of the narrow host-range *X. fastidiosa* CVC strain J1a12 with the broad host range sequenced strain 9a5c. This study demonstrated that 14 coding sequences of strain 9a5c are absent or highly divergent in strain J1a12, including an arginase and an adhesin (90).

The genome of the xylem-colonizing sugar cane pathogen *L. xyli* subsp. *xyli* provides a fascinating case study of the process of genome reduction in action. Thirteen percent of the genes in this bacterium are pseudogenes, and it lacks genes for vitamin biosynthesis, oxidative phophorylation, and gluconeogenesis (108).

A significant problem in characterizing genome-reduced bacteria is the challenge of growing these highly specialized symbionts in pure culture. Researchers have used the genome of *X. fastidiosa* to design synthetic media that support increased axenic growth of this bacterium. The most effective media contained amino acids such as serine and methionine, for which biosynthetic enzymes are absent, plus amino acids that are abundant in xylem sap (96). Only 2.7% of all open reading frames in *X. fastidiosa* are identifiable as major transporters, which may be indicative of a host niche with predictable and relatively invariant nutrient sources and stressors.

One interesting feature in the *X. fastidiosa* genome is the presence of a candidate ATP:ADP transporter, a gene typically found in intracellular animal parasites such as chlamydiae and rickettsiae (143). Recent studies have shown that ATP is present in the apoplastic fluid of plant cells and has a potential signaling role in plant defenses (148). ATP depletion promotes plant cell death (30). ATP utilization may be a useful feature for a bacterium that colonizes the "dead" cells that act as water conduits in plants, but it could hinder an intercellular biotroph living in the apoplast.

Although genome data provide a useful starting point, genome predictions of reduced functionality can be misleading and need to be verified experimentally. Bioinformatic analyses suggested that two enzymes essential for fatty acid synthesis were absent in *X. fastidiosa*, one of which was holo-acyl-carrier-protein synthase. However, further research showed that *X. fastidiosa* is able to synthesize fatty acids from acetate, and identified a gene (*hetI*) that could fulfill this function (117).

Gene deletion and gene inactivation also provide an effective means of modifying gene expression and evading plant recognition. The *P. syringae* and *R. solanacearum* genomes provide examples of potential virulence genes that have been transiently or irreversibly inactivated (21, 50, 58, 80). For example, *R. solanacearum* homologues of *P. syringae* avrRpm1 and avrD have frameshifts, while two other probable T3SEs have been disrupted by transposable elements (58). *R. solanacearum* also contains candidate contingency loci in the promoter regions or coding sequences of T3SE that could facilitate rapid inactivation or variation of T3SE (58).

Innovation
Although obligate and fastidious pathogens such as *X. fastidiosa* appear to have a relatively restricted diet, experimental and genomic analyses

of facultative plant pathogens such as *P. syringae,* *E. carotovora,* and *R. solanacearum* suggest that they are able to use a wide range of plant metabolites, from simple sugars and amino acids to complex secondary metabolites (10, 39, 50, 58).

Studholme et al. (151) analyzed the Pfam protein domain database to show that the genomes of *P. syringae, E. carotovora,* and other facultative plant-pathogenic bacteria contain a large number of "overrepresented" protein domains when compared with the genomes of bacteria that colonize other niches. These domains included transporter domains such as BPD_transp_1, BPD_transp_2, and ABC_tran, chemotaxis domains such as MCP_signal, and genes associated with protein secretion. Genes containing these domains may give these bacteria the ability to migrate toward, take up, and use a wide variety of plant and microbial-derived carbon, nitrogen, sulfur, and phosphorus sources. Plants produce an extraordinary variety of secondary metabolites that could act as nutrient sources for plant-colonizing bacteria. Degradation of plant secondary metabolites could also serve a second function: to detoxify and deplete toxic metabolites such as phytoalexins and cyanogenic glucosides. Degradation of some plant metabolites can confer an additional benefit by generating products that suppress plant defenses (77, 163). However, although detoxification of plant metabolites has been shown to play an important role in fungal colonization of plants, it has been little studied with respect to bacterial pathogens (99, 119). Studholme et al. (151) observed that the genomes of plant pathogens such as PtoDC3000 contain an unusually large number of putative glycosyl hydrolases, which could act to detoxify antimicrobial chemicals and suppress plant defenses (151). Further support for this hypothesis comes from the observation that a protein belonging to glycosyl hydrolase family 10 from *Clavibacter michiganensis* subsp. *michiganensis* is able to degrade the tomato saponin α-tomatine. However, mutants lacking this protein were not affected in virulence on tomato (85).

Studholme et al. (151) also used the Pfam database to examine the presence and distribution of unique domain architectures in plant pathogens—combinations of protein domains that are only found in PAB. These included a significant number of functionally uncharacterized enzymes that could be associated with nutrient utilization or detoxification of plant metabolites. However, one surprising conclusion from these domain architecture analyses was the observation that there were relatively few plant-specific architectures distributed across multiple plant-associated strains, even when comparing multiple species within a genus. The greatest degree of overlap was seen in *Xanthomonas* genomes. This suggests that many features associated with plant pathogenesis are also shared with animal pathogens and environmental bacteria or have been newly generated within each lineage (151).

Flexibility

Most facultative plant pathogens have a relatively high "bacterial IQ," which is to say they contain a large number of membrane bound and intracellular signaling proteins. *P. syringae* was singled out by Galperin (56) for containing an exceptionally high number of monocyte chemoattractant proteins, with Eca1043 close behind (0.9 and 0.8% of all proteins, respectively [10]) (Table 2). Surprisingly, although soil-borne pathogens such as *A. tumefaciens, R. solanacearum,* and *E. carotovora* might be predicted to encounter a greater diversity of environmental conditions and niches, the plant pathogens with the highest proportion and number of regulatory domains in this analysis are *P. syringae, X. campestris,* and *X. axonopodis,* which live as epiphytes and endophytes on plant foliage and display a high degree of host specificity (56) (Table 2). This could indicate that these pathogens experience a greater degree of environmental heterogeneity in their natural habitats in terms of both abiotic (light, temperature, water availability) and biotic signals (nutrient gradients, variation in host physiology, host defenses).

PLANT PATHOGENS IN THE ENVIRONMENT

A largely uninvestigated element in understanding the content and function of plant pathogen genomes, particularly the genomes

TABLE 2 Regulatory domains of representative plant pathogens[a]

Organism name	Phylum	GenBank accession no.	All proteins	HisKin[b]	MCP[b]	STYK[b]	GGDEF[b]	EAL[b]	HD-GYP[b]	AC1[b]	AC2[b]	AC3[b]	Total[c]	%
Agrobacterium tumefaciens	Alpha	AE008688	5,402	53 (32)	20 (14)	2 (1)	27 (17)	14 (10)	1 (0)	0	0	3 (2)	109	2
Ralstonia solanacearum	Beta	AL646052	5,116	44 (37)	22 (22)	3 (1)	22 (16)	14 (6)	2 (0)	0	1 (0)	0	98	1.9
Pseudomonas syringae	Gamma	AE016853	5,608	69 (42)	48 (41)	8 (2)	35 (18)	21 (14)	1 (0)	1 (0)	0	0	164	2.9
Pseudomonas aeruginosa	Gamma	AE004091	5,567	63 (45)	26 (23)	8 (2)	33 (19)	21 (14)	3 (0)	1 (0)	0	1 (1)	140	2.5
Erwinia carotovora	Gamma	BX950851	4,472	28 (23)	36 (36)	2 (1)	16 (11)	11 (5)	0	1 (0)	1 (0)	0	92	2.1
Xanthomonas axonopodis	Gamma	AE008923	4,427	61 (38)	21 (18)	10 (3)	30 (14)	14 (5)	3 (0)	0	0	1 (1)	130	2.9
Xanthomonas campestris	Gamma	AE008922	4,181	55 (31)	20 (16)	6 (3)	31 (15)	14 (5)	3 (0)	0	0	1 (1)	120	2.9
Xylella fastidiosa 9a5c	Gamma	AE003849	2,832	14 (9)	1 (1)	2 (1)	3 (1)	3 (1)	1 (0)	0	0	0	23	0.8
Onion yellows phytoplasma		AP006628	754	0	0	0	0	0	0	0	0	0	0	0
Caulobacter crescentus[1]	Alpha	AE005673	3,737	62 (39)	18 (12)	2 (1)	10 (4)	10 (5)	0	0	0	3 (1)	98	2.6
Chromobacterium violaceum[d]	Beta	AE016825	4,407	50 (32)	42 (33)	4 (3)	43 (20)	26 (7)	11 (0)	0	0	0	160	3.6
Pseudomonas putida[d]	Gamma	AE015451	5,350	67 (44)	27 (22)	5 (1)	36 (21)	21 (11)	2 (1)	1 (0)	0	0	142	2.6
Bacillus thuringiensis[d]	Firmi	AE017355	5,117	58 (50)	12 (10)	4 (2)	9 (6)	7 (4)	1 (1)	0	0	0	85	1.7

[a]After Galperin (56). The analysis only lists dedicated signaling systems that consist of more than two individual components and excludes single-domain regulators such as the AraC family.
[b]Numbers in bold indicate the total number of proteins of each kind; numbers in parentheses indicate the number with at least one predicted transmembrane segment.
[c]Numbers may be smaller than the sum of all types of proteins because GGDEF and EAL domains are often fused on a single polypeptide chain.
[d]Representative of nonpathogenic environmental bacteria from each phylum.

of facultative plant pathogens, rests on understanding what plant pathogenic bacteria do when not causing disease. The most extensive literature in this area focuses on epiphytic leaf colonization by *Pseudomonas syringae*. Phyllosphere microbiology has been studied for three practical reasons: pathogenesis, biocontrol, and ice nucleation.

Of the three *P. syringae* strains for which genome sequences are available, *P. syringae* pv. *syringae* PsyB278a is thought to be the most effective epiphyte. It is much less sensitive to UV irradiation and to osmotic and oxidative stress, and has more catalase and *rulAB* genes than PtoDC3000 (50). One explanation for higher osmotolerance in PsyB728a, suggested by genome analyses, could be the presence of genes for ectoine synthesis. PsyB728a also contains additional antibacterial S2 pyocin, a distinctive set of toxin biosynthesis genes, and more genes for antibiotic resistance than PtoDC3000, which could help it to survive in a more competitive environment. Despite these adaptations, PsyB728a and other *P. syringae* strains are not generally efficient rhizosphere-colonizing bacteria. The reasons for this are poorly defined but could include an inability to tolerate micro- and anaerobic conditions, to compete for nutrients, and to resist antimicrobial factors and predators. Comparisons of *P. syringae* genomes with the genomes of soil *Pseudomonas* will help us to explore these factors further in the future.

Genome sequence data provide an invaluable scaffold on which to build transposon mutagenesis and promoter-trapping analyses, in order to build a more detailed picture of niche colonization. Marco et al. (100) surveyed leaf surface-induced genes in PsyB728a by using in vivo expression technology, screening for promoters that induced expression of a promoterless *metXW* operon in planta, but failed to rescue methionine auxotrophy in vitro. They identified 45 genes, most of which were similar to genes from *P. syringae* or other *Pseudomonas* species. These included genes for nutrient acquisition, such as the *ssuE* gene, which is involved in sulfur acquisition. Sur-

prisingly, they also included known and candidate virulence genes. This suggests either that genes involved in virulence and pathogenicity have as yet unknown roles in colonization of leaf surfaces, or that the regulatory mechanisms governing virulence gene expression are unable to discriminate between some environments. Unfortunately, we do not yet have a nonpathogenic epiphyte genome for comparison, but it will be interesting to uncover the extent to which "virulence" genes are conserved in these commensal bacteria.

Similar questions about the distribution and function of plant pathogenicity genes, and the bias imposed by pathogen-focused genome sequencing, have been raised in studies of rhizosphere-colonizing *Pseudomonas fluorescens*. Preston et al. (123) reported that the nonpathogenic plant growth–promoting bacterium *P. fluorescens* SBW25 contained a plant-induced T3SS very similar to that of *P. syringae*. Subsequent genome analyses have identified additional effectors, including homologues of AvrA (*P. syringae*) and ExoY (*P. aeruginosa*) that have been confirmed to be secreted (165), and shown that a homologue of the *P. syringae* effector AvrE is coregulated with the T3SS (78). A T3SS has also been reported in *P. fluorescens* KD, although this system does not appear to be plant induced and is proposed to act in bacteria-oomycete interactions (133).

Perhaps the greatest surprise to arise from the Eca1043 genome sequence was the large number of genes that appeared to be associated with a life on plants but not directly related to pathogenesis (10, 157). For example, the genome contains a (*nif*) gene cluster with a putative role in nitrogen fixation—a function often associated with rhizosphere colonization, and one that may provide benefit to the colonized plant host (36). It also contains genes or gene clusters similar to *aggA* of *Pseudomonas putida* involved in attachment to roots (20), phenazine antibiotic production that may limit the growth of competing microbes and contribute to ecological competence (125), and the ability to take up and catabolize opine/rhizopine-like compounds produced by the plant during infection by members of the *Rhizobiaceae* (141). Of those genes associated with pathogenesis in *E. carotovora*, several could also equally be involved in plant colonization when not causing disease. Such factors include the T3SS, regulation (including quorum sensing), polysaccharides, motility, and iron acquisition, as well as, unlike *P. syringae*, the ability to thrive in microaerophilic environments. *R. solanacearum* also appears to be well adapted for prolonged periods of saprotrophic survival in soil and water and carries a large repertoire of catabolic genes, hydrolytic enzymes, and metal and drug resistance mechanisms (58).

CONCLUSION
The diversity of plant-bacteria interactions and the relatively small size of the global community of phytobacteriologists compared with the number of medical microbiologists mean that researchers must face the challenge of deciding whether future investigations should embrace this diversity or focus on fundamental investigations into a limited number of model bacteria and host plants. There is also the key question of whether we should prioritize investigations that are likely to support the development of new technologies for pathogen detection and disease control. Many plant pathogens are facultative pathogens, able to persist in the environment in the absence of an appropriate host or favorable environmental conditions, forming part of the anonymous and largely unstudied masses of environmental bacteria that colonize root and leaf surfaces. They share the same genetic reservoirs, and in some instances the same pathogenicity factors or even hosts as opportunistic human pathogens such as *Pseudomonas aeruginosa* or the *Burkholderia cepacia* complex. Perhaps, then, we should be more careful about investigating pathogens only on the basis of their role in disease, and perhaps we should be cautious about labeling genes as "pathogenicity genes." Instead, we should begin to consider these organisms in a wider environmental context. A broader perspective of pathogen-plant associations may offer new opportunities for disease control, and genomics is the ideal tool with which to begin such investigations.

REFERENCES
1. **Abramovitch, R. B., and G. B. Martin.** 2004. Strategies used by bacterial pathogens to suppress plant defenses. *Cur. Opin. Plant Biol.* **7:**356–364.
2. **Alarcon-Chaidez, F. J., A. Penaloza-Vazquez, M. Ullrich, and C. L. Bender.** 1999. Characterization of plasmids encoding the phytotoxin coronatine in *Pseudomonas syringae*. *Plasmid* **42:**210–220.
3. **Alegria, M. C., D. P. Souza, M. O. Andrade, C. Docena, L. Khater, C. H. I. Ramos, A. C. R. da Silva, and C. S. Farah.** 2005. Identification of new protein-protein interactions involving the products of the chromosome- and plasmid-encoded type IV secretion loci of the phytopathogen *Xanthomonas axonopodis* pv. *citri*. *J. Bacteriol.* **187:**2315–2325.
4. **Alfano, J. R., A. O. Charkowski, W.-L. Deng, J. L. Badel, T. Petnicki-Ocwieja, K. van Dijk, and A. Collmer.** 2000. The *Pseudomonas syringae* Hrp pathogenicity island has a tripartite mosaic structure composed of a cluster of type III secretion genes bounded by exchangeable effector and conserved effector loci that contribute to parasitic fitness and pathogenicity in plants. *Proc. Natl. Acad. Sci. USA* **97:**4856–4861.
5. **Alfano, J. R., and A. Collmer.** 2004. Type III secretion system effector proteins: double agents in bacterial disease and plant defense. *Ann. Rev. of Phytopathol.* **42:**385–414.
6. **Arnold, D. L., A. Pitman, and R. W. Jackson.** 2003. Pathogenicity and other genomic islands in plant pathogenic bacteria. *Mol. Plant Pathol.* **4:**407–420.
7. **Bai, X., J. Zhang, A. Ewing, S. A. Miller, A. Jancso Radek, D. V. Shevchenko, K. Tsukerman, T. Walunas, A. Lapidus, J. W. Campbell, and S. A. Hogenhout.** 2006. Living with genome instability: the adaptation of phytoplasmas to diverse environments of their insect and plant hosts. *J. Bacteriol.* **188:**3682–3696.
8. **Bauer, D. W., A. J. Bogdanove, S. V. Beer, and A. Collmer.** 1994. *Erwinia chrysanthemi hrp* genes and their involvement in soft rot pathogenesis and elicitation of the hypersensitive response. *Mol. Plant Microbe Interact.* **7:**573–581.
9. **Bedini, E., C. DeCastro, G. Erbs, L. Mangoni, J. M. Dow, M.-A. Newman, M. Parrilli, and C. Unverzagt.** 2005. Structure-dependent modulation of a pathogen response in plants by synthetic O-antigen polysaccharides. *J. Am. Chem. Soc.* **127:**2414–2416.

10. **Bell, K. S., M. Sebaihia, L. Pritchard, M. T. G. Holden, L. J. Hyman, M. C. Holeva, N. R. Thomson, S. D. Bentley, L. J. C. Churcher, K. Mungall, R. Atkin, N. Bason, K. Brooks, T. Chillingworth, K. Clark, J. Doggett, A. Fraser, Z. Hance, H. Hauser, K. Jagels, S. Moule, H. Norbertczak, D. Ormond, C. Price, M. A. Quail, M. Sanders, D. Walker, S. Whitehead, G. P. C. Salmond, P. R. J. Birch, J. Parkhill, and I. K. Toth.** 2004. Genome sequence of the enterobacterial phytopathogen *Erwinia carotovora* subsp. *atroseptica* and characterization of virulence factors. *Proc. Natl. Acad. Sci. USA* **101:**11105–11110.

11. **Bender, C. L., F. Alarcon-Chaidez, and D. C. Gross.** 1999. *Pseudomonas syringae* phytotoxins: mode of action, regulation, and biosynthesis by peptide and polyketide synthetases. *Microbiol. Mol. Biol. Rev.* **63:**266–292.

12. **Bender, C. L., D. K. Malvick, and R. E. Mitchell.** 1989. Plasmid-mediated production of the phytotoxin coronatine in *Pseudomonas syringae* pv. *tomato. J. Bacteriol.* **171:**807–812.

13. **Bender, C. L., S. A. Young, and R. E. Mitchell.** 1991. Conservation of plasmid DNA sequences in coronatine-producing pathovars of *Pseudomonas syringae. Appl. Environ. Microbiol.* **57:** 993–999.

14. **Binet, R., S. Letoffe, J. M. Ghigo P. Delepelaire, and C. Wandersman.** 1997. Protein secretion by gram-negative bacterial ABC exporters—a review. *Gene* **192:**7–11.

15. **Binet, R., and C. Wandersman.** 1996. Cloning of the *Serratia marcescens hasF* gene encoding the Has ABC exporter outer membrane component: a TolC analogue. *Mol. Microbiol.* **22:**265–273.

16. **Bogdanove, A. J., D. W. Bauer, and S. V. Beer.** 1998. *Erwinia amylovora* secretes DspE, a pathogenicity factor and functional AvrE homolog, through the Hrp (type III secretion) pathway. *J. Bacteriol.* **180:**2244–2247.

17. **Boureau, T., H. El Maarouf-Bouteau, A. Garnier, M.-N. Brisset, C. Perino, I. Pucheu, and M.-A. Barny.** 2006. DspA/E, a type III effector essential for *Erwinia amylovora* pathogenicity and growth in planta, induces cell death in host apple and nonhost tobacco plants. *Mol. Plant Microbe. Interact.* **19:**16–24.

18. **Bretz, J. R., and S. W. Hutcheson.** 2004. Role of type III effector secretion during bacterial pathogenesis in another kingdom. *Infect. Immun.* **72:**3697–3705.

19. **Broothaerts, W., H. J. Mitchell, B. Weir, S. Kaines, L. M. A. Smith, W. Yang, J. E. Mayer, C. Roa-Rodriguez, and R. A. Jefferson.** 2005. Gene transfer to plants by diverse species of bacteria. *Nature* **433:**629–633.

20. **Buell, C. R., and A. J. Anderson.** 1992. Genetic analysis of the *aggA* locus involved in agglutination and adherence of *Pseudomonas putida*, a beneficial pseudomonad. *Mol. Plant Microbe Interact.* **5:**154–162.

21. **Buell, C. R., V. Joardar, M. Lindeberg, J. Selengut, I. T. Paulsen, M. L. Gwinn, R. J. Dodson, R. T. Deboy, A. S. Durkin, J. F. Kolonay, R. Madupu, S. Daugherty, L. Brinkac, M. J. Beanan, D. H. Haft, W. C. Nelson, T. Davidsen, N. Zafar, L. Zhou, J. Liu, Q. Yuan, H. Khouri, N. Fedorova, B. Tran, D. Russell, K. Berry, T. Utterback, S. E. Van Aken, T. V. Feldblyum, M. D'Ascenzo, W.-L. Deng, A. R. Ramos, J. R. Alfano, S. Cartinhour, A. K. Chatterjee, T. P. Delaney, S. G. Lazarowitz, G. B. Martin, D. J. Schneider, X. Tang, C. L. Bender, O. White, C. M. Fraser, and A. Collmer.** 2003. The complete genome sequence of the Arabidopsis and tomato pathogen *Pseudomonas syringae* pv. *tomato* DC3000. *Proc. Natl. Acad. Sci. USA* **100:**10181–10186.

22. **Bukhalid, R. A., S. Y. Chung, and R. Loria.** 1998. *nec1*, a gene conferring a necrogenic phenotype, is conserved in plant-pathogenic *Streptomyces* spp. and linked to a transposase pseudogene. *Mol. Plant Microbe Interact.* **11:**960–967.

23. **Buttner, D., and U. Bonas.** 2003. Common infection strategies of plant and animal pathogenic bacteria. *Curr. Opin. Plant Biol.* **6:**312–319.

24. **Buttner, D., L. Noel, F. Thieme, and U. Bonas.** 2003. Genomic approaches in *Xanthomonas campestris* pv. *vesicatoria* allow fishing for virulence genes. *J. Biotechnol.* **106:**203–214.

25. **Cao, T. B., J. Saier, and H. Milton.** 2003. The general protein secretory pathway: phylogenetic analyses leading to evolutionary conclusions. *Biochim. Biophys. Acta* **1609:**115–125.

26. **Castaneda, A., J. D. Reddy, B. El-Yacoubi, and D. W. Gabriel.** 2005. Mutagenesis of all eight *avr* genes in *Xanthomonas campestris* pv. *campestris* had no detected effect on pathogenicity, but one *avr* gene affected race specificity. *Mol. Plant Microbe. Interact.* **18:**1306–1317.

27. **Castaneda, M., J. Guzman, S. Moreno, and G. Espin.** 2000. The GacS sensor kinase regulates alginate and poly-beta-hydroxybutyrate production in *Azotobacter vinelandii. J. Bacteriol.* **182:** 2624–2628.

28. **Chang, J. H., A. K. Goel, S. R. Grant, and J. L. Dangl.** 2004. Wake of the flood: ascribing functions to the wave of type III effector proteins of phytopathogenic bacteria. *Curr. Opin. Microbiol.* **7:**11–18.

29. **Chen, L.-L.** 2006. Identification of genomic islands in six plant pathogens. *Gene* **374:**134–141.

30. Chivasa, S., B. K. Ndimba, W. J. Simon, K. Lindsey, and A. R. Slabas. 2005. Extracellular ATP functions as an endogenous external metabolite regulating plant cell viability. *Plant Cell* **17:**3019–3034.

31. Christensen, N. M., K. B. Axelsen, M. Nicolaisen, and A. Schulz. 2005. Phytoplasmas and their interactions with hosts. *Trends Plant Sci.* **10:**526–535.

32. Christie, P. J., K. Atmakuri, V. Krishnamoorthy, S. Jakubowski, and E. Cascales. 2005. Biogenesis, architecture, and function of bacterial type IV secretion systems. *Annu. Rev. Microbiol.* **59:**451–485.

33. Christie, P. J., and J. P. Vogel. 2000. Bacterial type IV secretion: conjugation systems adapted to deliver effector molecules to host cells. *Trends Microbiol.* **8:**354–360.

34. Chung, S.-M., M. Vaidya, and T. Tzfira. 2006. *Agrobacterium* is not alone: gene transfer to plants by viruses and other bacteria. *Trends Plant Sci.* **11:**1–4.

35. Cianciotto, N. P. 2005. Type II secretion: a protein secretion system for all seasons. *Trends Microbiol.* **13:**581–588.

36. Cocking, E. C. 2003. Endophytic colonisation of plant roots by nitrogen-fixing bacteria. *Plant Soil* **252:**169–175.

37. Corbett, M., S. Virtue, K. Bell, P. Birch, T. Burr, L. Hyman, K. Lilley, S. Poock, I. Toth, and G. Salmond. 2005. Identification of a new quorum-sensing controlled virulence factor in *Erwinia carotovora* subsp. *atroseptica* secreted via the type II targeting pathway. *Mol. Plant Microbe Interact.* **18:**334–342.

38. Cornelis, P., and S. Matthijs. 2002. Diversity of siderophore-mediated iron uptake systems in fluorescent pseudomonads: not only pyoverdines. *Environ. Microbiol.* **4:**767–798.

39. da Silva, A. C. R., J. A. Ferro, F. C. Reinach, C. S. Farah, L. R. Furlan, R. B. Quaggio, C. B. Monteiro-Vitorello, M. A. V. Sluys, N. F. Almeida, L. M. C. Alves, A. M. do Amaral, M. C. Bertolini, L. E. A. Camargo, G. Camarotte, F. Cannavan, J. Cardozo, F. Chambergo, L. P. Ciapina, R. M. B. Cicarelli, L. L. Coutinho, J. R. Cursino-Santos, H. El-Dorry, J. B. Faria, A. J. S. Ferreira, R. C. C. Ferreira, M. I. T. Ferro, E. F. Formighieri, M. C. Franco, C. C. Greggio, A. Gruber, A. M. Katsuyama, L. T. Kishi, R. P. Leite, E. G. M. Lemos, M. V. F. Lemos, E. C. Locali, M. A. Machado, A. M. B. N. Madeira, N. M. Martinez-Rossi, E. C. Martins, J. Meidanis, C. F. M. Menck, C. Y. Miyaki, D. H. Moon, L. M. Moreira, M. T. M. Novo, V. K. Okura, M. C. Oliveira, V. R. Oliveira, H. A. Pereira, A. Rossi, J. A. D. Sena, C. Silva, R. F. de Souza, L. A. F. Spinola, M. A. Takita, R. E. Tamura, E. C. Teixeira, R. I. D. Tezza, M. Trindade dos Santos, D. Truffi, S. M. Tsai, F. F. White, J. C. Setubal, and J. P. Kitajima. 2002. Comparison of the genomes of two *Xanthomonas* pathogens with differing host specificities. *Nature* **417:**459–463.

40. da Silva, F. R., A. L. Vettore, E. L. Kemper, A. Leite, and P. Arruda. 2001. Fastidian gum: the *Xylella fastidiosa* exopolysaccharide possibly involved in bacterial pathogenicity. *FEMS Microbiol. Lett.* **203:**165–171.

41. DebRoy, S., R. Thilmony, Y.-B. Kwack, K. Nomura, and S. Y. He. 2004. A family of conserved bacterial effectors inhibits salicylic acid-mediated basal immunity and promotes disease necrosis in plants. *Proc. Natl. Acad. Sci. USA* **101:**9927–9932.

42. Dharmapuri, S., and R. V. Sonti. 1999. A transposon insertion in the gumG homologue of *Xanthomonas oryzae* pv. *oryzae* causes loss of extracellular polysaccharide production and virulence. *FEMS Microbiol. Lett.* **179:**53–59.

43. Dow, J. M., A. E. Osbourn, T. J. Wilson, and M. J. Daniels. 1995. A locus determining pathogenicity of *Xanthomonas campestris* is involved in lipopolysaccharide biosynthesis. *Mol. Plant Microbe Interact.* **8:**768–777.

44. Dow, J. M., J.-X. Feng, C. E. Barber, J.-L. Tang, and M. J. Daniels. 2000. Novel genes involved in the regulation of pathogenicity factor production within the *rpf* gene cluster of *Xanthomonas campestris*. *Microbiology* **146:**885–891.

45. Dreier, J., D. Meletzus, and R. Eichenlaub. 1997. Characterization of the plasmid encoded virulence region pat-1 of phytopathogenic *Clavibacter michiganensis* subsp. *michiganensis*. *Mol. Plant Microbe Interact.* **10:**195–206.

46. Dunne, W. M. J. 2002. Bacterial adhesion: seen any good biofilms lately? *Clin. Microbiol. Rev.* **15:**155–166.

47. Escobar, M. A., and A. M. Dandekar. 2003. *Agrobacterium tumefaciens* as an agent of disease. *Trends Plant Sci.* **8:**380–386.

48. Espinosa, A., and J. R. Alfano. 2004. Disabling surveillance: bacterial type III secretion system effectors that suppress innate immunity. *Cell. Microbiol.* **6:**1027–1040.

49. Fallman, M., and A. Gustavsson. 2005. Cellular mechanisms of bacterial internalization counteracted by *Yersinia*, p. 135–188. *In* K. W. Jeon (ed.), *A Survey of Cell Biology*. Academic Press, San Diego, CA.

50. Feil, H., W. S. Feil, P. Chain, F. Larimer, G. DiBartolo, A. Copeland, A. Lykidis, S. Trong, M. Nolan, E. Goltsman, J. Thiel, S. Malfatti,

J. E. Loper, A. Lapidus, J. C. Detter, M. Land, P. M. Richardson, N. C. Kyrpides, N. Ivanova, and S. E. Lindow. 2005. Comparison of the complete genome sequences of *Pseudomonas syringae* pv. *syringae* B728a and pv. *tomato* DC3000. *Proc. Natl. Acad. Sci. USA* **102:**11064–11069.

51. **Filloux, A.** 2004. The underlying mechanisms of type II protein secretion. *Biochim. Biophys. Acta* **1694:**163–179.

52. **Foulongne, V., S. Michaux-Charachon, D. O'Callaghan, and M. Ramuz.** 2002. Type IV secretion systems and bacterial virulence. *MS Med. Sci.* **18:**439–447.

53. **Foultier, B., P. Troisfontaines, S. Muller, F. R. Opperdoes, and G. R. Cornelis.** 2002. Characterization of the *ysa* pathogenicity locus in the chromosome of *Yersinia enterocolitica* and phylogeny analysis of type III secretion systems. *J. Mol. Evol.* **55:**37–51.

54. **Franza, T., B. Mahe, and D. Expert.** 2005. *Erwinia chrysanthemi* requires a second iron transport route dependent of the siderophore achromobactin for extracellular growth and plant infection. *Mol. Microbiol.* **55:**261–275.

55. **Frey, J., and P. Kuhnert.** 2002. RTX toxins in *Pasteurellaceae. Int. J. Med. Microbiol.* **292:**149–158.

56. **Galperin, M.** 2005. A census of membrane-bound and intracellular signal transduction proteins in bacteria: bacterial IQ, extroverts and introverts. *BMC Microbiol.* **5:**35.

57. **Gaudriault, S., L. Malandrin, J. P. Paulin, and M. A. Barny.** 1997. DspA, an essential pathogenicity factor of *Erwinia amylovora* showing homology with AvrE of *Pseudomonas syringae*, is secreted via the Hrp secretion pathway in a DspB-dependent way. *Mol. Microbiol.* **26:**1057–1069.

58. **Genin, S., and C. Boucher.** 2004. Lessons learned from the genome analysis of *Ralstonia solanacearum. Annu. Rev. Phytopathol.* **42:**107–134.

59. **Gentschev, I., G. Dietrich, and W. Goebel.** 2002. The *E. coli* (alpha)-hemolysin secretion system and its use in vaccine development. *Trends Microbiol.* **10:**39–45.

60. **Gijzen, M., and T. Nurnberger.** 2006. Nep1-like proteins from plant pathogens: recruitment and diversification of the NPP1 domain across taxa. *Phytochemistry* **67:**1800–1807.

61. **Gil, R., B. Sabater-Munoz, V. Perez-Brocal, F. J. Silva, and A. Latorre.** 2006. Plasmids in the aphid endosymbiont *Buchnera aphidicola* with the smallest genomes. A puzzling evolutionary story. *Gene* **370:**17–25.

62. **Glickmann, E., L. Gardan, S. Jacquet, S. Hussain, M. Elasri, A. Petit, and Y. Dessaux.** 1998. Auxin production is a common feature of most pathovars of *Pseudomonas syringae. Mol. Plant Microbe Interact.* **11:**156–162.

63. **Godfrey, S. A. C., S. A. Harrow, J. W. Marshall, and J. D. Klena.** 2001. Characterization by 16S rRNA sequence analysis of pseudomonads causing blotch disease of cultivated *Agaricus bisporus. Appl. Env. Microbiol.* **67:**4316–4323.

64. **Goldberg, J. B., and G. B. Pier.** 1996. *Pseudomonas aeruginosa* lipopolysaccharides and pathogenesis. *Trends Microbiol.* **4:**490–494.

65. **Green, R. L., and G. J. Warren.** 1985. Physical and functional repetition in a bacterial ice nucleation gene. *Nature* **317:**645–648.

66. **Greenberg, J. T., and B. A. Vinatzer.** 2003. Identifying type III effectors of plant pathogens and analyzing their interaction with plant cells. *Curr. Opin. Microbiol.* **6:**20–28.

67. **Guiney, D. G., and M. Lesnick.** 2005. Targeting of the actin cytoskeleton during infection by *Salmonella* strains. *Clin. Immunol.* **114:**248–255.

68. **Gurian-Sherman, D., and S. Lindow.** 1993. Bacterial ice nucleation: significance and molecular basis. *FASEB J.* **7:**1338–1343.

69. **Gurlebeck, D., F. Thieme, and U. Bonas.** 2006. Type III effector proteins from the plant pathogen *Xanthomonas* and their role in the interaction with the host plant. *J. Plant Physiol.* **163:**233–255.

70. **Hacker, J., and J. B. Kaper.** 2000. Pathogenicity islands and the evolution of microbes. *Annu. Rev. Microbiol.* **54:**641–679.

71. **He, S. Y., K. Nomura, and T. S. Whittam.** 2004. Type III protein secretion mechanism in mammalian and plant pathogens. *Biochim. Biophys. Acta* **1694:**181–206.

72. **Heeb, S., and D. Haas.** 2001. Regulatory roles of the GacS/GacA two-component system in plant-associated and other gram-negative bacteria. *Mol. Plant Microbe Interact.* **14:**1351–1363.

73. **Holeva, M. C., K. Bell, L. Hyman, A. O. Avrova, S. C. Whisson, P. R. Birch, and I. K. Toth.** 2004. Use of a pooled transposon mutation grid to demonstrate roles in disease development for *Erwinia carotovora* subsp. *atroseptica* putative type III secreted effector (DspE/A) and helper (HrpN) proteins. *Mol. Plant Microbe Interact.* **17:**943–950.

74. **Hotson, A., and M. B. Mudgett.** 2004. Cysteine proteases in phytopathogenic bacteria: identification of plant targets and activation of innate immunity. *Curr. Opin. Plant Biol.* **7:**384–390.

75. **Hueck, C. J.** 1998. Type III protein secretion systems in bacterial pathogens of animals and plants. *Microbiol. Mol. Biol. Rev.* **62:**379–433.

76. **Hwang, M. S. H., R. L. Morgan, S. F. Sarkar, P. W. Wang, and D. S. Guttman.** 2005. Phylogenetic characterization of virulence and resistance phenotypes of *Pseudomonas syringae. Appl. Environ. Microbiol.* **71:**5182–5191.

77. **Ito, S.-i., T. Eto, S. Tanaka, N. Yamauchi, H. Takahara, and T. Ikeda.** 2004. Tomatidine

and lycotetraose, hydrolysis products of (alpha)-tomatine by *Fusarium oxysporum* tomatinase, suppress induced defense responses in tomato cells. *FEBS Lett.* **571:**31–34.

78. **Jackson, R. W., G. M. Preston, and P. B. Rainey.** 2005. Genetic characterisation of *Pseudomonas fluorescens* SBW25 *rsp* gene expression in the phytosphere and in vitro. *J. Bacteriol.* **187:** 8477–8488.

79. **Jin, Q., R. Thilmony, J. Zwiesler-Vollick, and S.-Y. He.** 2003. Type III protein secretion in *Pseudomonas syringae*. *Microbes Infect.* **5:**301–310.

80. **Joardar, V., M. Lindeberg, R. W. Jackson, J. Selengut, R. Dodson, L. M. Brinkac, S. C. Daugherty, R. DeBoy, A. S. Durkin, M. G. Giglio, R. Madupu, W. C. Nelson, M. J. Rosovitz, S. Sullivan, J. Crabtree, T. Creasy, T. Davidsen, D. H. Haft, N. Zafar, L. Zhou, R. Halpin, T. Holley, H. Khouri, T. Feldblyum, O. White, C. M. Fraser, A. K. Chatterjee, S. Cartinhour, D. J. Schneider, J. Mansfield, A. Collmer, and C. R. Buell.** 2005. Whole-genome sequence analysis of *Pseudomonas syringae* pv. *phaseolicola* 1448A reveals divergence among pathovars in genes involved in virulence and transposition. *J. Bacteriol.* **187:**6488–6498.

81. **Joardar, V., M. Lindeberg, D. J. Schneider, A. Collmer, and C. R. Buell.** 2005. Lineage-specific regions in *Pseudomonas syringae* pv. *tomato* DC3000. *Mol. Plant Pathol.* **6:**53–64.

82. **Kakizawa, S., K. Oshima, and S. Namba.** 2006. Diversity and functional importance of phytoplasma membrane proteins. *Trends Microbiol.* **14:**254–256.

83. **Kang, Y., J. Huang, M. Guozhang, L.-Y. He, and M. A. Schell.** 1994. Dramatically reduced virulence of mutants of *Pseudomonas solanacearum* defective in export of extracellular proteins across the outer membrane. *Mol. Plant Microbe Interact.* **7:**370–377.

84. **Katzen, F., D. U. Ferreiro, C. G. Oddo, M. V. Ielmini, A. Becker, A. Puhler, and L. Ielpi.** 1998. *Xanthomonas campestris* pv. *campestris* gum mutants: effects on xanthan biosynthesis and plant virulence. *J. Bacteriol.* **180:**1607–1617.

85. **Kaup, O., I. Grafen, E. M. Zellermann, R. Eichenlaub, and K. H. Gartemann.** 2005. Identification of a tomatinase in the tomato-pathogenic actinomycete *Clavibacter michiganensis* subsp. *michiganensis* NCPPB382. *Mol. Plant Microbe. Interact.* **18:**1090–1098.

86. **Kers, J. A., K. D. Cameron, M. V. Joshi, R. A. Bukhalid, J. E. Morello, M. J. Wach, D. M. Gibson, and R. Loria.** 2005. A large, mobile pathogenicity island confers plant pathogenicity on *Streptomyces* species. *Mol. Microbiol.* **55:**1025–1033.

87. **Kim, J. F., and J. R. Alfano.** 2002. Pathogenicity islands and virulence plasmids of bacterial plant pathogens. *Curr. Top. Microbiol. Immunol.* **264:**127–147.

88. **Kinscherf, T. G., and D. K. Willis.** 1999. Swarming by *Pseudomonas syringae* B728a requires *gacS* (*lemA*) and *gacA* but not the acyl-homoserine lactone biosynthetic gene *ahlI*. *J. Bacteriol.* **181:** 4133–4136.

89. **Kitten, T., T. G. Kinscherf, J. L. McEvoy, and D. K. Willis.** 1998. A newly identified regulator is required for virulence and toxin production in *Pseudomonas syringae*. *Mol. Microbiol.* **28:**917–929.

90. **Koide, T., P. A. Zaini, L. M. Moreira, R. Z. N. Vencio, A. Y. Matsukuma, A. M. Durham, D. C. Teixeira, H. El-Dorry, P. B. Monteiro, A. C. R. da Silva, S. Verjovski-Almeida, A. M. da Silva, and S. L. Gomes.** 2004. DNA microarray-based genome comparison of a pathogenic and a nonpathogenic strain of *Xylella fastidiosa* delineates genes important for bacterial virulence. *J. Bacteriol.* **186:**5442–5449.

91. **Lai, E.-M., and C. I. Kado.** 2000. The T-pilus of *Agrobacterium tumefaciens*. *Trends Microbiol.* **8:** 361–369.

92. **Lee, B.-M., Y.-J. Park, D.-S. Park, H.-W. Kang, J.-G. Kim, E.-S. Song, I.-C. Park, U.-H. Yoon, J.-H. Hahn, B.-S. Koo, G.-B. Lee, H. Kim, H.-S. Park, K.-O. Yoon, J.-H. Kim, C.-H. Jung, N.-H. Koh, J.-S. Seo, and S.-J. Go.** 2005. The genome sequence of *Xanthomonas oryzae* pathovar oryzae KACC10331, the bacterial blight pathogen of rice. *Nucleic Acids Res.* **33:**577–586.

93. **Lee, I.-M., R. E. Davis, and D. E. Gundersen-Rindal.** 2000. Phytoplasma: phytopathogenic mollicutes. *Ann. Rev. Microbiol.* **54:**221–255.

94. **Lee, J.-S., K.-S. Shin, J.-G. Pan, and C.-J. Kim.** 2000. Surface-displayed viral antigens on *Salmonella* carrier vaccine. *Nat. Biotechnol.* **18:** 645–648.

95. **Leigh, J. A., and D. L. Coplin.** 1992. Exopolysaccharides in plant-bacterial interactions. *Annu. Rev. Microbiol.* **46:**307–346.

96. **Lemos, E. G. d. M., L. M. C. Alves, and J. C. Campanharo.** 2003. Genomics-based design of defined growth media for the plant pathogen *Xylella fastidiosa*. *FEMS Microbiol. Lett.* **219:**39–45.

97. **Lory, S.** 1998. Secretion of proteins and assembly of bacterial surface organelles: shared pathways of extracellular protein targeting. *Curr. Opin. Microbiol.* **1:**27–35.

98. **Lu, S. E., N. Wang, J. Wang, Z. J. Chen, and D. C. Gross.** 2005. Oligonucleotide microarray analysis of the *salA* regulon controlling phytotoxin production by *Pseudomonas syringae* pv. *syringae*. *Mol. Plant Microbe Interact.* **18:**324–333.

99. **Maor, R., and K. Shirasu.** 2005. The arms race continues: battle strategies between plants and fungal pathogens. *Curr. Opin. Microbiol.* **8:**399–404.

100. **Marco, M. L., J. Legac, and S. E. Lindow.** 2005. *Pseudomonas syringae* genes induced during colonization of leaf surfaces. *Environ. Microbiol.* **7:**1379–1391.

101. **Marcone, C., H. Neimark, A. Ragozzino, U. Lauer, and E. Seemuller.** 1999. Chromosome sizes of phytoplasmas composing major phylogenetic groups and subgroups. *Phytopathology* **89:**805–810.

102. **Marie, C., W. J. Broughton, and W. J. Deakin.** 2001. *Rhizobium* type III secretion systems: legume charmers or alarmers? *Curr. Opin. Plant Biol.* **4:**336–342.

103. **Matthysse, A. G., M. Marry, L. Krall, M. Kaye, B. E. Ramey, C. Fuqua, and A. R. White.** 2005. The effect of cellulose overproduction on binding and biofilm formation on roots by *Agrobacterium tumefaciens. Mol. Plant Microbe. Interact.* **18:**1002–1010.

104. **Mazzola, M., and F. F. White.** 1994. A mutation in the indole-3-acetic-acid biosynthesis pathway of *Pseudomonas syringae* pv. *syringae* affects growth in *Phaseolus vulgaris* and syringomycin production. *J. Bacteriol.* **176:**1374–1382.

105. **Meng, X., J. M. Bonasera, J. F. Kim, R. M. Nissinen, and S. V. Beer.** 2006. Apple proteins that interact with DspA/E, a pathogenicity effector of *Erwinia amylovora*, the fire blight pathogen. *Mol. Plant Microbe Interact.* **19:**53–61.

106. **Meng, Y., Y. Li, C. D. Galvani, G. Hao, J. N. Turner, T. J. Burr, and H. C. Hoch.** 2005. Upstream migration of *Xylella fastidiosa* via pilus-driven twitching motility. *J. Bacteriol.* **187:**5560–5567.

107. **Miller, S. I., R. K. Ernst, and M. W. Bader.** 2005. LPS, TLR4 and infectious disease diversity. *Nat. Rev. Microbiol.* **3:**36–46.

108. **Monteiro-Vitorello, C. B., L. E. A. Camargo, M. A. Van Sluys, J. P. Kitajima, D. Truffi, A. M. do Amaral, R. Harakava, J. C. F. de Oliveira, D. Wood, M. C. de Oliveira, C. Miyaki, M. A. Takita, A. C. R. da Silva, L. R. Furlan, D. M. Carraro, G. Camarotte, N. F. Almeida, H. Carrer, L. L. Coutinho, H. A. El-Dorry, M. I. T. Ferro, P. R. Gagliardi, E. Giglioti, M. H. S. Goldman, G. H. Goldman, E. T. Kimura, E. S. Ferro, E. E. Kuramae, E. G. M. Lemos, M. V. F. Lemos, S. M. Z. Mauro, M. A. Machado, C. L. Marino, C. F. Menck, L. R. Nunes, R. C. Oliveira, G. G. Pereira, W. Siqueira, A. A. de Souza, S. M. Tsai, A. S. Zanca, A. J. G. Simp-** son, S. M. Brumbley, and J. C. Setúbal. 2004. The genome sequence of the gram-positive sugarcane pathogen *Leifsonia xyli* subsp. *xyli. Mol. Plant Microbe Interact.* **17:**827–836.

109. **Moreira, L. M., R. F. De Souza, L. A. Digiampietri, A. C. R. Da Silva, and J. C. Setubal.** 2005. Comparative analyses of *Xanthomonas* and *Xylella* complete genomes. *OMICS* **9:**43–76.

110. **Morris, C. E., and J.-M. Monier.** 2003. The ecological significance of biofilm formation by plant-associated bacteria. *Ann. Rev. Phytopathol.* **41:**429–453.

111. **Mudgett, M. B.** 2005. New insights to the function of phytopathogenic bacterial type III effectors in plants. *Ann. Rev. Plant Biol.* **56:**509–531.

112. **Noel, L., F. Thieme, D. Nennstiel, and U. Bonas.** 2002. Two novel type III–secreted proteins of *Xanthomonas campestris* pv. *vesicatoria* are encoded within the *hrp* pathogenicity island. *J. Bacteriol.* **184:**1340–1348.

113. **Nomura, K., M. Melotto, and S.-Y. He.** 2005. Suppression of host defense in compatible plant-*Pseudomonas syringae* interactions. *Curr. Opin. Plant Biol.* **8:**361–368.

114. **Ochiai, H., Y. Inoue, M. Takeya, A. Sasaki, and H. Kaku.** 2005. Genome sequence of *Xanthomonas oryzae* pv. *oryzae* suggests contribution of large numbers of effector genes and insertion sequences to its race diversity. *JARQ* **39:**275–287.

115. **Ojanen-Reuhs, T., N. Kalkkinen, B. Westerlund-Wikstrom, J. van Doorn, K. Haahtela, E. Nurmiaho-Lassila, K. Wengelnik, U. Bonas, and T. Korhonen.** 1997. Characterization of the *fimA* gene encoding bundle-forming fimbriae of the plant pathogen *Xanthomonas campestris* pv. *vesicatoria. J. Bacteriol.* **179:**1280–1290.

116. **Oshima, K., S. Kakizawa, H. Nishigawa, J. Hee-Young, W. Wei, S. Suzuki, R. Nakata, D. Arashida, S.-I. Miyata, M. Ugaki, and N. Shigetou.** 2004. Reductive evolution suggested from the complete genome sequence of a plant-pathogenic phytoplasma. *Nat. Genet.* **36:**27–29.

117. **Osiro, D., J. R. C. Muniz, H. D. Coleta Filho, A. A. de Sousa, M. A. Machado, R. C. Garratt, and L. A. Colnago.** 2004. Fatty acid synthesis in *Xylella fastidiosa*: correlations between genome studies, 13C NMR data, and molecular models. *Biochem. Biophys. Res. Commun.* **323:**987–995.

118. **Patil, P. B., and R. V. Sonti.** 2004. Variation suggestive of horizontal gene transfer at a lipopolysaccharide (lps) biosynthetic locus in

Xanthomonas oryzae pv. *oryzae*, the bacterial leaf blight pathogen of rice. *BMC Microbiol.* **4:**40.

119. **Pedras, M. S. C., and P. W. K. Ahiahonu.** 2005. Metabolism and detoxification of phytoalexins and analogs by phytopathogenic fungi. *Phytochemistry* **66:**391–411.

120. **Pemberton, C. L., and G. P. C. Salmond.** 2004. The Nep1-like proteins—a growing family of microbial elicitors of plant necrosis. *Mol. Plant Pathol.* **5:**353–359.

121. **Perombelon, M. C.** 2002. Potato diseases caused by soft rot erwinias: an overview of pathogenesis. *Plant Pathol.* **51:**1–12.

122. **Preston, G.** 2000. *Pseudomonas syringae* pv. *tomato*: the right pathogen, of the right plant, at the right time. *Mol. Plant Pathol.* **1:**263–275.

123. **Preston, G. M., N. Bertrand, and P. B. Rainey.** 2001. Type III secretion in plant growth-promoting *Pseudomonas fluorescens* SBW25. *Mol. Microbiol.* **41:**999–1014.

124. **Preston, G. M., D. Studholme, and I. Caldelari.** 2005. Profiling the secretomes of plant pathogenic Proteobacteria. *FEMS Microbiol. Rev.* **29:**331–360.

125. **Price-Whelan, A., L. E. P. Dietrich, and D. K. Newman.** 2006. Rethinking "secondary" metabolism: physiological roles for phenazine antibiotics. *Nat. Chem. Biol.* **2:**71–78.

126. **Pritchard, L., J. White, P. R. J. Birch, and I. K. Toth.** 2006. GenomeDiagram: a python package for the visualisation of large-scale genomic data. *Bioinformatics* **22:**616–617.

127. **Qian, W., Y. Jia, S.-X. Ren, Y.-Q. He, J.-X. Feng, L.-F. Lu, Q. Sun, G. Ying, D.-J. Tang, H. Tang, W. Wu, P. Hao, L. Wang, B.-L. Jiang, S. Zeng, W.-Y. Gu, G. Lu, L. Rong, Y. Tian, Z. Yao, G. Fu, B. Chen, R. Fang, B. Qiang, Z. Chen, G.-P. Zhao, J.-L. Tang, and C. He.** 2005. Comparative and functional genomic analyses of the pathogenicity of phytopathogen *Xanthomonas campestris* pv. *campestris*. *Genome Res.* **15:**757–767.

128. **Quinones, B., C. J. Pujol, and S. E. Lindow.** 2004. Regulation of AHL production and its contribution to epiphytic fitness in *Pseudomonas syringae*. *Mol. Plant Microbe Interact.* **17:**521–531.

129. **Rajeshwari, R., G. Jha, and R. V. Sonti.** 2005. Role of an in planta-expressed xylanase of *Xanthomonas oryzae* pv. *oryzae* in promoting virulence on rice. *Mol. Plant Microbe Interact.* **18:**830–837.

130. **Rantakari, A., O. Virtaharju, S. Vahamiko, S. Taira, E. T. Palva, H. T. Saarilahti, and M. Romantschuk.** 2001. Type III secretion contributes to the pathogenesis of the soft-rot pathogen *Erwinia carotovora*: partial characteriza-

tion of the *hrp* gene cluster. *Mol. Plant Microbe Interact.* **14:**962–968.

131. **Ray, S. K., R. Rajeshwari, and R. V. Sonti.** 2000. Mutants of *Xanthomonas oryzae* pv. *oryzae* deficient in general secretory pathway are virulence deficient and unable to secrete xylanase. *Mol. Plant Microbe. Interact.* **13:**394–401.

132. **Redak, R. A., A. H. Purcell, J. R. S. Lopes, M. J. Blua, R. F. Mizell III, and P. C. Andersen.** 2004. The biology of xylem fluid-feeding insect vectors of *Xylella fastidiosa* and their relation to disease epidemiology. *An. Rev. Entomol.* **49:**243–270.

133. **Rezzonico, F., C. Binder, G. Defago, and Y. Moenne-Loccoz.** 2005. The type III secretion system of biocontrol *Pseudomonas fluorescens* KD targets the phytopathogenic Chromista *Pythium ultimum* and promotes cucumber protection. *Mol. Plant Microbe Interact.* **18:**991–1001.

134. **Roden, J. A., B. Belt, J. B. Ross, T. Tachibana, J. Vargas, and M. B. Mudgett.** 2004. A genetic screen to isolate type III effectors translocated into pepper cells during *Xanthomonas* infection. *Proc. Natl. Acad. Sci. USA* **101:**16624–16629.

135. **Rojas, C. M., J. H. Ham, W. L. Deng, J. J. Doyle, and A. Collmer.** 2002. HecA, a member of a class of adhesins produced by diverse pathogenic bacteria, contributes to the attachment, aggregation, epidermal cell killing, and virulence phenotypes of *Erwinia chrysanthemi* EC16 on *Nicotiana clevelandii* seedlings. *Proc. Natl. Acad. Sci. USA* **99:**13142–13147.

136. **Rojas, C. M., J. H. Ham, L. M. Schecter, J. F. Kim, S. V. Beer, and A. Collmer.** 2004. The *Erwinia chrysanthemi* EC16 *hrp/hrc* gene cluster encodes an active Hrp type III secretion system that is flanked by virulence genes functionally unrelated to the Hrp system. *Mol. Plant Microbe Interact.* **17:**644–653.

137. **Romantschuk, M.** 1992. Attachment of plant pathogenic bacteria to plant surfaces. *Annu. Rev. Phytopathol.* **30:**225–243.

138. **Romantschuk, M., and D. H. Bamford.** 1986. The causal agent of halo blight in bean, *Pseudomonas syringae* pv. *phaseolicola*, attaches to stomato via its pili. *Microb. Pathog.* **1:**139–148.

139. **Russel, M.** 1998. Macromolecular assembly and secretion across the bacterial cell envelope: type II protein secretion systems. *J. Mol. Biol.* **279:**485–499.

140. **Sarkar, S. F., and D. S. Guttman.** 2004. Evolution of the core genome of *Pseudomonas syringae*, a highly clonal, endemic plant pathogen. *Appl. Environ. Microbiol.* **70:**1999–2012.

141. **Savka, M. A., Y. Dessaux, P. Oger, and S. Rossbach.** 2002. Engineering bacterial

competitiveness and persistence in the phytosphere. *Mol. Plant Microbe. Interact.* **15**:866–874.

142. **Schell, M. A.** 2000. Control of virulence and pathogenicity genes of *Ralstonia solanacearum* by an elaborate sensory network. *Annu. Rev. Phytopathol.* **38**:263–292.

143. **Schmitz-Esser, S., N. Linka, A. Collingro, C. L. Beier, H. E. Neuhaus, M. Wagner, and M. Horn.** 2004. ATP/ADP translocases: a common feature of obligate intracellular amoebal symbionts related to chlamydiae and rickettsiae. *J. Bacteriol.* **186**:683–691.

144. **Scholz-Schroeder, B. K., J. D. Soule, S. E. Lu, I. Grgurina, and D. C. Gross.** 2002. A physical map of the syringomycin and syringopeptin gene clusters localized to an approximately 145-kb DNA region of *Pseudomonas syringae* pv. *syringae* strain B301D. *Mol. Plant Microbe Interact.* **14**:1426–1435.

145. **Sexton, J., and J. P. Vogel.** 2002. Type IVB secretion by intracellular pathogens. *Traffic* **3**: 178–195.

146. **Simpson, A. J. G., F. C. Reinach, P. Arruda, F. A. Abreu, M. Acencio, R. Alvarenga, L. M. C. Alves, J. E. Araya, G. S. Baia, C. S. Baptista, M. H. Barros, E. D. Bonaccorsi, S. Bordin, J. M. Bove, M. R. S. Briones, M. R. P. Bueno, A. A. Camargo, L. E. A. Camargo, D. M. Carraro, H. Carrer, N. B. Colauto, C. Colombo, F. F. Costa, M. C. R. Costa, C. M. Costa-Neto, L. L. Coutinho, M. Cristofani, E. Dias-Neto, C. Docena, H. El-Dorry, A. P. Facincani, A. J. S. Ferreira, V. C. A. Ferreira, J. A. Ferro, J. S. Fraga, S. C. Franca, M. C. Franco, M. Frohme, L. R. Furlan, M. Garnier, G. H. Goldman, M. H. S., Gomes, S. L. Goldman, A. Gruber, P. L. Ho, J. D. Hoheisel, M. L. Junqueira, E. L. Kemper, J. P. Kitajima, J. E. Krieger, E. E. Kuramae, F. Laigret, M. R. Lambais, L. C. C. Leite, E. G. M. Lemos, M. V. F. Lemos, S. A. Lopes, C. R. Lopes, J. A. Machado, M. A. Machado, A. M. B. N. Madeira, H. M. F. Madeira, C. L. Marino, M. V. Marques, E. A. L. Martins, E. M. F. Martins, A. Y. Matsukuma, C. F. M. Menck, E. C. Miracca, C. Y. Miyaki, C. B. Monteiro-Vitorello, D. H. Moon, M. A. Nagai, A. L. T. O. Nascimento, L. E. S. Netto, A. Nhani, F. G. Nobrega, L. R. Nunes, M. A. Oliveira, M. C. de Oliveira, R. C. de Oliveira, D. A. Palmieri, A. Paris, B. R. Peixoto, G. A. G. Pereira, H. A. Pereira, J. B. Pesquero, R. B. Quaggio, P. G. Roberto, V. Rodrigues, A. J. de M. Rosa, V. E. de Rosa, R. G. de Sa, R. V. Santelli, H. E. Sawasaki, A. C. R. da Silva, A. M. da Silva, F. R. da Silva, W. A. Silva, J. F. da Sil-

veira, M. L. Z. Silvestri, W. J. Siqueira, A. A. de Souza, A. P. de Souza, M. F. Terenzi, D. Truffi, S. M. Tsai, M. H. Tsuhako, H. Vallada, M. A. Van Sluys, S. Verjovski-Almeida, A. L. Vettore, M. A. Zago, M. Zatz, J. Meidanis, and J. C. Setubal.** 2000. The genome sequence of the plant pathogen *Xylella fastidiosa*. *Nature* **406**:151–157.

147. **Smolka, M. B., D. Martins, F. V. Winck, C. E. Santoro, R. R. Castellari, F. Ferrari, I. J. Brum, E. Galembeck, H. D. C. Filho, M. A. Machado, S. Marangoni, and J. C. Novello.** 2003. Proteome analysis of the plant pathogen *Xylella fastidiosa* reveals major cellular and extracellular proteins and a peculiar codon bias distribution. *Proteomics* **3**:224–237.

148. **Song, C. J., I. Steinebrunner, X. Wang, S. C. Stout, and S. J. Roux.** 2006. Extracellular ATP induces the accumulation of superoxide via NADPH oxidases in *Arabidopsis*. *Plant Physiol.* **140**:1222–1232.

149. **Souza, L. C. A., N. A. Wulff, P. Gaurivaud, A. G. Mariano, A. C. D. Virgilio, J. L. Azevedo, and P. B. Monteiro.** 2006. Disruption of *Xylella fastidiosa* CVC *gumB* and *gumF* genes affects biofilm formation without a detectable influence on exopolysaccharide production. *FEMS Microbiol. Lett.* **257**:236–242.

150. **Stavrinides, J., and D. S. Guttman.** 2004. Nucleotide sequence and evolution of the five-plasmid complement of the phytopathogen *Pseudomonas syringae* pv. *maculicola* ES4326. *J. Bacteriol.* **186**:5101–5115.

151. **Studholme, D., A. Downie, and G. Preston.** 2005. Unique domain architectures in plant pathogenic Proteobacteria. *BMC Genomics* **6**:17.

152. **Suoniemi, A., K. Bjorklof, K. Haahtela, and M. Romantschuk.** 1995. Pili of *Pseudomonas syringae* pathovar *syringae* enhance initiation of bacterial epiphytic colonization of bean. *Microbiology* **141**:497–503.

153. **Suzuki, S., K. Oshima, S. Kakizawa, R. Arashida, H.-Y. Jung, Y. Yamaji, H. Nishigawa, M. Ugaki, and S. Namba.** 2006. Interaction between the membrane protein of a pathogen and insect microfilament complex determines insect-vector specificity. *Proc. Natl. Acad. Sci. USA* **103**:4252–4257.

154. **Thanassi, D. G., and S. J. Hultgren.** 2000. Multiple pathways allow protein secretion across the bacterial outer membrane. *Curr. Opin. Cell Biol.* **12**:420–430.

155. **Thieme, F., R. Koebnik, T. Bekel, C. Berger, J. Boch, D. Buttner, C. Caldana, L. Gaigalat, A. Goesmann, S. Kay, O. Kirchner, C. Lanz, B. Linke, A. C. McHardy, F. Meyer, G. Mittenhuber, D. H. Nies, U. Nies-**

bach-Klosgen, T. Patschkowski, C. Ruckert, O. Rupp, S. Schneiker, S. C. Schuster, F.-J. Vorholter, E. Weber, A. Puhler, U. Bonas, D. Bartels, and O. Kaiser. 2005. Insights into genome plasticity and pathogenicity of the plant pathogenic bacterium *Xanthomonas campestris* pv. *vesicatoria* revealed by the complete genome sequence. *J. Bacteriol.* **187:**7254–7266.

156. Thomson, N. R., J. D. Thomas, and G. P. C. Salmond. 1999. *Virulence Determinants in the Bacterial Phytopathogen* Erwinia. Academic Press Inc., San Diego, CA.

157. Toth, I. K., L. Pritchard, and P. R. J. Birch. 2006. Comparative genomics reveals what makes an enterobacterial plant pathogen. *Annu. Rev. Phytopathol.* **44:**305–336.

158. Toth, I. K., C. J. Thorpe, S. D. Bentley, V. Mulholland, L. J. Hyman, M. C. Perombelon, and G. P. C. Salmond. 1999. Mutation in a gene required for lipopolysaccharide and enterobacterial common antigen biosynthesis affects virulence in the plant pathogen *Erwinia carotovora* subsp. *atroseptica*. *Mol. Plant Microbe Interact.* **12:**499–507.

159. Tran Van Nhieu, G., J. Enninga, P. Sansonetti, and G. Grompone. 2005. Tyrosine kinase signaling and type III effectors orchestrating *Shigella* invasion. *Curr. Opin. Microbiol.* **8:**16–20.

160. Truman, W., M. T. Zabala, and M. Grant. 2006. Type III effectors orchestrate a complex interplay between transcriptional networks to modify basal defence responses during pathogenesis and resistance. *Plant J.* **46:**14–33.

161. Tzfira, T., and V. Citovsky. 2006. *Agrobacterium*-mediated genetic transformation of plants: biology and biotechnology. *Curr. Opin. Biotechnol.* **17:**147–154.

162. Van Sluys, M. A., M. C. de Oliveira, C. B. Monteiro-Vitorello, C. Y. Miyaki, L. R. Furlan, L. E. A. Camargo, A. C. R. da Silva, D. H. Moon, M. A. Takita, E. G. M. Lemos, M. A. Machado, M. I. T. Ferro, F. R. da Silva, M. H. S. Goldman, G. H. Goldman, M. V. F. Lemos, H. El-Dorry, S. M. Tsai, H. Carrer, D. M. Carraro, R. C. de Oliveira, L. R. Nunes, W. J. Siqueira, L. L. Coutinho, E. T. Kimura, E. S. Ferro, R. Harakava, E. E. Kuramae, C. L. Marino, E. Giglioti, I. L. Abreu, L. M. C. Alves, A. M. do Amaral, G. S. Baia, S. R. Blanco, M. S. Brito, F. S. Cannavan, A. V. Celestino, A. F. da Cunha, R. C. Fenille, J. A. Ferro, E. F. Formighieri, L. T. Kishi, S. G. Leoni, A. R. Oliveira, V. E. Rosa, Jr., F. T. Sassaki, J. A. D. Sena, A. A. de Souza, D. Truffi, F. Tsukumo, G. M. Yanai, L. G. Zaros, E. L. Civerolo, A. J. G. Simpson, N. F. Almeida, Jr., J. C. Se-

tubal, and J. P. Kitajima. 2003. Comparative analyses of the complete genome sequences of Pierce's Disease and Citrus Variegated Chlorosis strains of *Xylella fastidiosa*. *J. Bacteriol.* **185:**1018–1026.

163. van Wees, S. C. M., and J. Glazebrook. 2003. Loss of non-host resistance of *Arabidopsis* NahG to *Pseudomonas syringae* pv. *phaseolicola* is due to degradation products of salicylic acid. *Plant J.* **33:**733–742.

164. Vencato, M., F. Tian, J. R. Alfano, R. Buell, S. Cartinhour, G. DeClerck, D. S. Guttman, J. Stavrinides, V. Joardar, M. Lindeberg, P. Bronstein, J. Mansfield, C. R. Myers, A. Collmer, and D. J. Schneider. 2006. Bioinformatics-enabled inventory of the Hrp regulon and type III secretion system effector proteins of *Pseudomonas syringae* pv. *phaseolicola* 1448a. *Mol. Plant Microbe Interact.* **19:**1193–1206.

165. Vinatzer, B. A., J. Jelenska, and J. T. Greenberg. 2005. Bioinformatics correctly identifies many type III secretion substrates in the plant pathogen *Pseudomonas syringae* and the biocontrol isolate *P. fluorescens* SBW25. *Mol. Plant Microbe Interact.* **18:**877–888.

166. Vivian, A., J. Murillo, and R. W. Jackson. 2001. The roles of plasmids in phytopathogenic bacteria: mobile arsenals? *Microbiol.* **147:**763–780.

167. Wall, D. P., H. B. Fraser, and A. E. Hirsh. 2003. Detecting putative orthologs. *Bioinformatics* **19:**1710–1711.

168. Wandersman, C., and P. Delepelaire. 2004. Bacterial iron sources: from siderophores to hemophores. *Ann. Rev. Microbiol.* **58:**611–647.

169. Wang, N., S.-E. Lu, A. R. Records, and D. C. Gross. 2006. Characterization of the transcriptional activators SalA and SyrF, which are required for syringomycin and syringopeptin production by *Pseudomonas syringae* pv. *syringae*. *J. Bacteriol.* **188:**3290–3298.

170. Wang, N., S.-E. Lu, Q. Yang, S.-H. Sze, and D. C. Gross. 2006. Identification of the *syr-syp* box in the promoter regions of genes dedicated to syringomycin and syringopeptin production by *Pseudomonas syringae* pv. *syringae* B301D. *J. Bacteriol.* **188:**160–168.

171. Wei, Z.-M., R. J. Laby, C. H. Zumoff, D. W. Bauer, S. Y. He, A. Collmer, and S. V. Beer. 1992. Harpin, elicitor of the hypersensitive response produced by the plant pathogen *Erwinia amylovora*. *Science* **257:**85–88.

172. Weintraub, P. G., and L. Beanland. 2006. Insect vectors of phytoplasmas. *Annu. Rev. Entomol.* **51:**91–111.

173. West, M. A., and W. Heagy. 2002. Endotoxin tolerance: a review. *Crit. Care Med.* **30:**S64–S73.

174. **West, N. P., P. Sansonetti, J. Mounier, R. M. Exley, C. Parsot, S. Guadagnini, M.-C. Prevost, A. Prochnicka-Chalufour, M. Delepierre, M. Tanguy, and C. M. Tang.** 2005. Optimization of virulence functions through glucosylation of *Shigella* LPS. *Science* **307:**1313–1317.

175. **Willis, D. K., J. J. Holmstadt, and T. G. Kinscherf.** 2001. Genetic evidence that loss of virulence associated with *gacS* or *gacA* mutations in *Pseudomonas syringae* B728a does not result from effects on alginate production. *Appl. Environ. Microbiol.* **67:**1400–1403.

176. **Wilson, T. J. G., N. Bertrand, J.-L. Tang, J.-X. Feng, M.-Q. Pan, C. E. Barber, J. M. Dow, and M. J. Daniels.** 1998. The *rpfA* gene of *Xanthomonas campestris* pathovar campestris, which is involved in the regulation of pathogenicity factor production, encodes an aconitase. *Mol. Microbiol.* **28:**961–970.

177. **Wisniewski, M., S. E. Lindow, and E. N. Ashworth.** 1997. Observations of ice nucleation and propagation in plants using infrared video thermography. *Plant Physiol.* **113:**327–334.

178. **Yang C. H., M. Gavilanes-Ruiz, Y. Okinaka, R. Vedel, I. Berthuy, M. Boccara, J. W. Chen, N. T. Perna, and N. T. Kee.** 2002. *hrp* genes of *Erwinia chrysanthemi* 3937 are important virulence factors. *Mol. Plant Microbe Interact.* **15:**472–480.

179. **Yu, J., A. Penaloza-Vazquez, A. M. Chakrabarty, and C. L. Bender.** 1999. Involvement of the exopolysaccharide alginate in the virulence and epiphytic fitness of *Pseudomonas syringae* pv. *syringae*. *Mol. Microbiol.* **33:**712–720.

180. **Zaharik, M. L., S. Gruenheid, A. J. Perrin, and B. B. Finlay.** 2002. Delivery of dangerous goods: type III secretion in enteric pathogens. *Int. J. Med. Microbiol.* **291:**593–603.

181. **Zeidler, D., U. Zahringer, I. Gerber, I. Dubery, T. Hartung, W. Bors, P. Hutzler, and J. Durner.** 2004. Innate immunity in *Arabidopsis thaliana*: lipopolysaccharides activate nitric oxide synthase (NOS) and induce defense genes. *Proc. Natl. Acad. Sci. USA* **101:**15811–15816.

182. **Zhang, Y., D. D. Bak, H. Heid, and K. Geider.** 1999. Molecular characterization of a protease secreted by *Erwinia amylovora*. *J. Mol. Biol.* **289:**1239–1251.

183. **Zhao, J. L., and C. S. Orser.** 1990. Conserved repetition in the ice nucleation gene inaX from *Xanthomonas campestris* pv. *translucens*. *Mol. Gen. Genet.* **223:**163–166.

184. **Zhao, Y., R. Thilmony, C. L. Bender, A. Schaller, S. Y. He, and G. A. Howe.** 2003. Virulence systems of *Pseudomonas syringae* pv. *tomato* promote bacterial speck disease in tomato by targeting the jasmonate signaling pathway. *Plant J.* **36:**485–499.

PHOTORHABDUS: GENOMICS OF A PATHOGEN AND SYMBIONT

Richard H. ffrench-Constant, Andrea Dowling, Michelle Hares, Guowei Yang, and Nicholas Waterfield

16

Photorhabdus bacteria live in tight association with nematodes of the genus *Heterorhabditis* that infect insects (20, 21). *Photorhabdus* are gramnegative enteropathogens that live in the gut of the infective juvenile (IJ) nematode (20). These IJs actively seek out and penetrate insects. Once inside the open blood system (hemocoel) of the insect, the nematodes then regurgitate 50 to 200 bacterial cells directly into the insect blood or hemolymph (9). After their appearance in the insect blood system, we know that the bacteria are recognized by the immune system and that an immune defense is also launched (16). Thus, immune recognition genes are upregulated and antibacterial peptides are synthesized. This immune response must be at least partially effective, because RNAi-mediated knockdown of recognition protein encoding genes increases host susceptibility to *Photorhabdus* infection (16). However, the bacteria are able to overcome this immune response and subsequently kill the insect host, presumably via the release of insecticidal toxins and proteases (3, 19). Once dead, the insect is then bioconverted by the bacteria, and in turn, the nematodes then feed and reproduce on

the bacteria themselves (Fig. 1). After several rounds of nematode replication, a new generation of IJs are produced, which re–take up *Photorhabdus* bacteria and then emerge out of the insect cadaver to colonize new hosts (20).

This amazing lifestyle raises several interesting questions for the genomics of this bacterium. How can one bacterium be both a symbiont with the vector nematode and a pathogen to its insect host? What are the switch genes that control the switch from symbiosis to pathogenicity? Can we identify islands in the genome involved in pathogenicity, symbiosis, or both? In this review, we will attempt to address these questions by looking at the genomics and functional genomics of *Photorhabdus*. Initially we will compare the *Photorhabdus* genome with *Escherichia coli* to identify putative pathogenicity islands. We will then move on to comparisons between different *Photorhabdus* species in order to identify islands that are putatively specific to different species with different life cycles. At present, we have full genome sequences from both an exclusive insect pathogen, *P. luminescens* TT01, and an insect pathogen that has also been recovered from human wounds, *P. asymbiotica*. Comparisons between these two very different species should therefore also shed light on the evolution of vertebrate pathogenicity in the *Photorhabdus* genus.

Richard H. ffrench-Constant, Andrea Dowling, and Michelle Hares Department of Biology, University of Exeter in Cornwall, Penryn TR10 9EZ, United Kingdom. *Guowei Yang and Nicholas Waterfield* Department of Biology, University of Bath, Bath BA2 7AY, United Kingdom.

Bacterial Pathogenomics, Edited by M. J. Pallen et al.
© 2007 ASM Press, Washington, D.C.

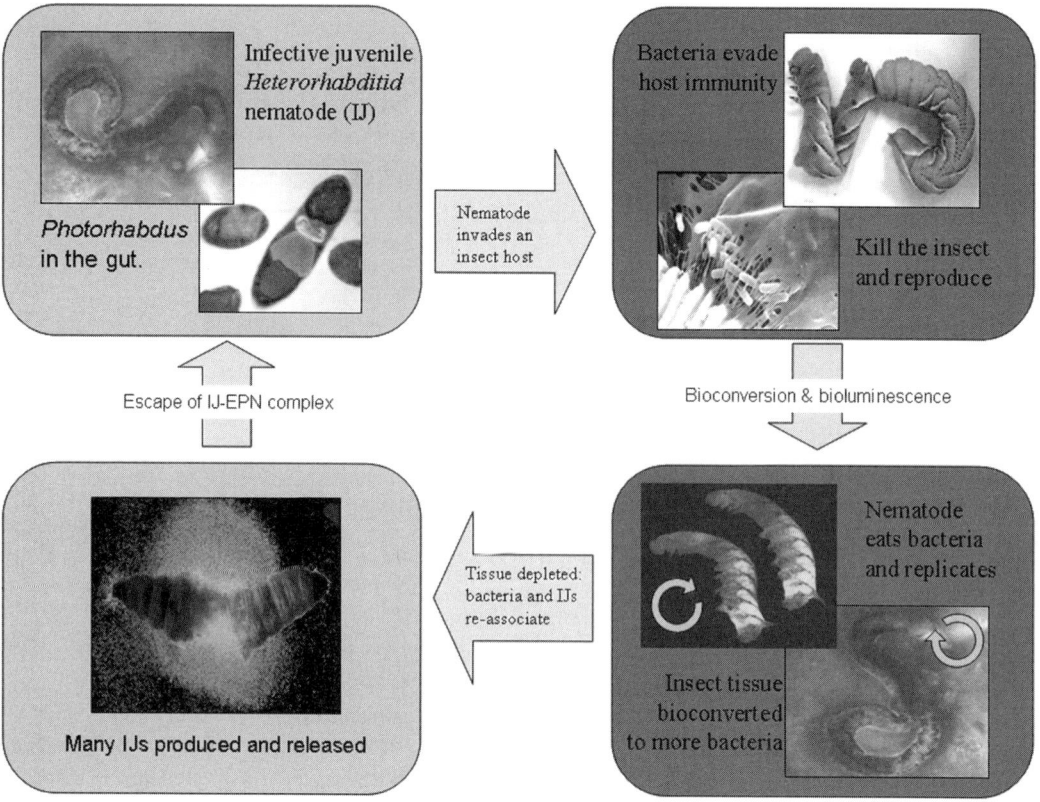

FIGURE 1 Life cycle of *Photorhabdus*. The bacteria are carried by a vector nematode that seeks out and penetrates insects. The bacteria are then regurgitated from the nematode gut and are used to kill the insect and bioconvert the corpse. In the meantime, while feeding off the bacteria, the nematodes undergo several rounds of reproduction. When the cadaver is exhausted, the nematodes produce a new generation of infective juveniles, which reuptake the bacteria and exit the cadaver to seek new hosts.

GENOMIC COMPARISONS WITH *E. COLI*

Genomic Islands within an *E. coli*–Like Backbone

Comparison of *Photorhabdus* with the backbone of genes found in *E. coli* and other *Enterobacteriaceae* can help us understand which genes are required simply to occupy the gut and which genes are involved in the specialized insect- and nematode-associated parts of the *Photorhabdus* lifestyle. Initial sample sequencing of *P. luminescens* W14 led to the discovery of an amazing range of new toxins and proteases apparently produced by this insect pathogen (18). The first full genomic sequence of *P. luminescens* TT01 then confirmed that the *Photorhabdus*

genome contains more toxins and proteases than any other bacterial genome sequenced to date (15). The TT01 genome is 5,688,987 bp long and contains 4,839 predicted protein-coding genes. There are many genomic islands (GIs) relative to the common enterobacterial backbone. As expected, these are located near tRNA genes, Rhs elements, flagella operons, phage integrase genes, IS elements, and the ubiquitous enterobacterial repetitive intergenic consensus sequences. Many of these horizontally acquired genes are predicted to encode a large number of adhesins, toxins, hemolysins, proteases, lipases, and antibiotic synthesizing genes. Although some are clear homologues of GIs from other species, others appear unique to *Photorhabdus* and/or other entomopathogens. It is not our intention here to list all of these genes, which has been published elsewhere (15), but here we will review what we know of the pathogenicity islands revealed by comparisons between *Photorhabdus* and *E. coli* and describe what we have learned from their functional genomic analysis.

Islands Encoding Toxin Complexes

The first genes encoding insecticidal toxin complexes (Tc's) were cloned using antibodies raised against purified proteins (2). The oral activity of these toxins (42) is surprising because the bacteria are released directly into the insect blood system, and therefore any toxin action on the gut would be expected to be from the blood side rather than from the gut lumen. After the initial description of four loci each encoding a different toxin complex, and termed *tca, tcb, tcc,* and *tcd,* large islands encoding these toxins have come to light (44, 46). In these islands, the *tc* genes are often found as tandem arrays of similar genes (e.g., *tcdA1, tcdA2,* and *tcdA3*), leading to a highly complicated no-menclature. Comparisons between *tcd*-GIs of different strains of *Photorhabdus* revealed that gain/loss of *tc* genes are probably associated with linked *tccC* genes that are not only essential for the production of active toxin complex but are also homologous to the recombination hot spot (*rhs*) element core genes found in most bacteria. Functional studies have now confirmed that three components are necessary for full toxicity; these have been termed A, B, and C components for simplicity (17). In the *tcd*-like operons, the A component represents a single large toxin gene, exemplified by *tcdA,* whereas in the *tca*-like operons, the A component is genetically split into two smaller genes, *tcaA* and *tcaB.* Interestingly, the *tca*-like loci have only been seen in the chromosomes of *Photorhabdus* and the *Yersinia pseudotuberculosis* clonal group (which includes *Y. pestis*). Alternatively, the *tcd*-like operons are seen on both chromosomes and plasmids and in a far wider range of species, including *Photorhabdus, Xenorhabdus, Yersinia enterocolitica, Y. frederiksenii, Serratia entomophila, Burkholderia pseudo-*

mallei, and even the gram-positive *Paenobacillus,* suggesting a more mobile natural history.

The toxin ABC arrangement is most clearly seen in the pADAP plasmid of *Serratia entomophila. S. entomophila* is a free-living bacterium that causes amber disease in the New Zealand grass grub (27). In *Serratia,* the three ABC genes (*tcd* type) have been shown to be necessary for the clearance of the insect gut seen in the disease after ingestion of the bacteria (25, 27). This gut clearance is very similar to the phenotype of caterpillars that have ingested Tc toxins. In *S. entomophila,* the three ABC genes are found on a 120-kb self-conjugating plasmid (Fig. 2). In *P. luminescens,* the equivalent genes are found at a number of different *tc* loci (Fig. 2). Our most recent functional work on these toxins suggests that both A and BC constitute toxins in their own right (unpublished data). Therefore, these pathogenicity islands may in fact be encoding a mixture of toxins with different activities. For example, different A components (e.g., *tcdA1* and *tcdA2*) may have different activities against different insects, as shown for A components of the other nematode vectored insect pathogen *Xenorhabdus* (32, 35) and therefore effectively represent toxin mixtures. Given that the bacterium cannot predict which type of insect it will be vectored into by the nematode (because the nematodes are capable of penetrating any larval or adult insect), this range of toxin activities may make sense in terms of killing a broad range of hosts. This potentially large host range probably represents the selective force that has driven the unusually large number and redundancy of virulence factors in the *Photorhabdus* genome.

Although we still have little idea of the mode of action of the Tc toxins, the genomic structure of *tc* loci can be compared with the *Salmonella* plasmid-borne virulence (*spv*) factor locus. Specifically, both the predicted amino acid sequences of *spvA* and *spvB* show similarity to elements of the *tca* and *tcc* locus. This similarity in amino acid sequences and the elements of conserved genomic organization between the *spv-ABC* and *tccABC/tcaABC* loci (Fig. 3) suggests that it may possess a conserved function against the insect immune system. In this respect, it is

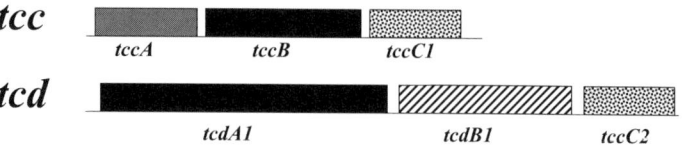

FIGURE 2 Diagram showing the relationship between *tc* genes found on the *pADAP* plasmid of the free-living *Serratia entomophila* and the nematode-associated *Photorhabdus luminescens*. Note that the *S. entomophila sepABC* form A, B, and C components of the toxin complex proteins and that these then correspond to elements of the *tca*, *tcc*, and *tcd* locus of *P. luminescens*, as indicated by shading.

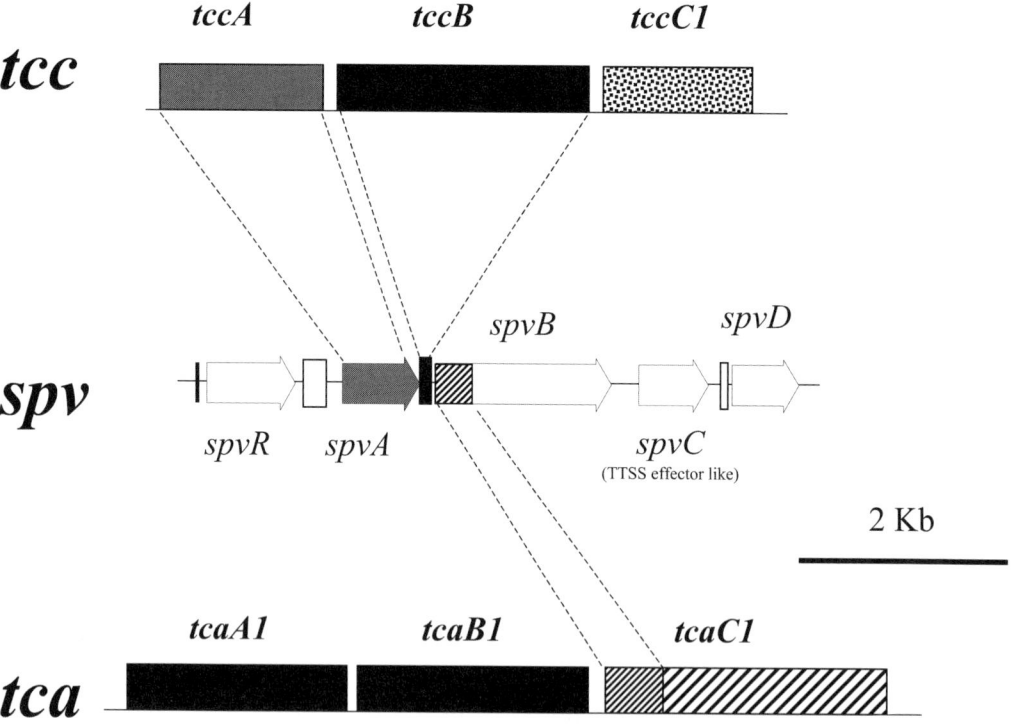

FIGURE 3 Relationship between the *Salmonella* plasmid-borne virulence (spv) locus and *tca* and *tcc* from *P. luminescens*. Note that *spvA*, is predicted to be *tccA*-like and that following *spvA*, there is a fragment of a *tccB*-like gene. The predicted amino acid sequence of the N terminus of *spvB* is also similar to *tcaC1*. These organizational relationships suggest that the similarity of these three loci could relate to a common plasmid-borne ancestor (see text).

interesting to note that in *Salmonella*, *spvB* has been implicated in persistence in macrophages (31), and SpvB has recently been shown to be a mono(ADP-ribosyl)-transferase (33). Structure predictions of SpvB indicate that it is composed of a C-terminal ADP-ribosyl transferase domain fused via a poly proline stretch to an N domain resembling the N domain of TcaC (33). This N terminus is now known to allow SpvB secretion from the bacteria, with the aid of a type II sec-dependent signal leader. TcaC lacks such a leader, and evidence suggests that it may be involved in host cell nuclear import. This suggests that the Tc toxins may have a similar modular structure, with the rest of *tcaC* encoding an as yet unknown effector.

Islands Encoding Mcf Toxins

The "Makes caterpillars floppy" (Mcf) toxin is the dominant insecticidal toxin in *Photorhabdus* (11). Mcf was discovered because expression of this gene alone in *E. coli* is sufficient both to promote survival of the recombinant bacteria in the face of the insect immune system and also to kill the host (11). Mcf therefore destroys the insect phagocytes that are sent to destroy the invading bacteria and the insect gut, leading to the floppy appearance of the caterpillar (11). Mcf1 causes apoptosis in both insect and mammalian cells and appears to act by mimicking a BH3 domain-only protein, which promotes apoptosis in the host mitochondrion (11, 13). This mechanism is unique in bacterial toxins. To date, we have demonstrated that Mcf1-treated cells show nuclear fragmentation, DNA laddering, and caspase-3 activation, which are all consistent with the promotion of apoptosis (13). More recently, we have also shown that Mcf1-treated cells show release of cytochrome C from their mitochondria, a finding consistent with this being the site of action of Mcf1 (unpublished data). The gene encoding Mcf1 is found at different genomic locations in different *Photorhabdus* strains (Fig. 4), suggesting either

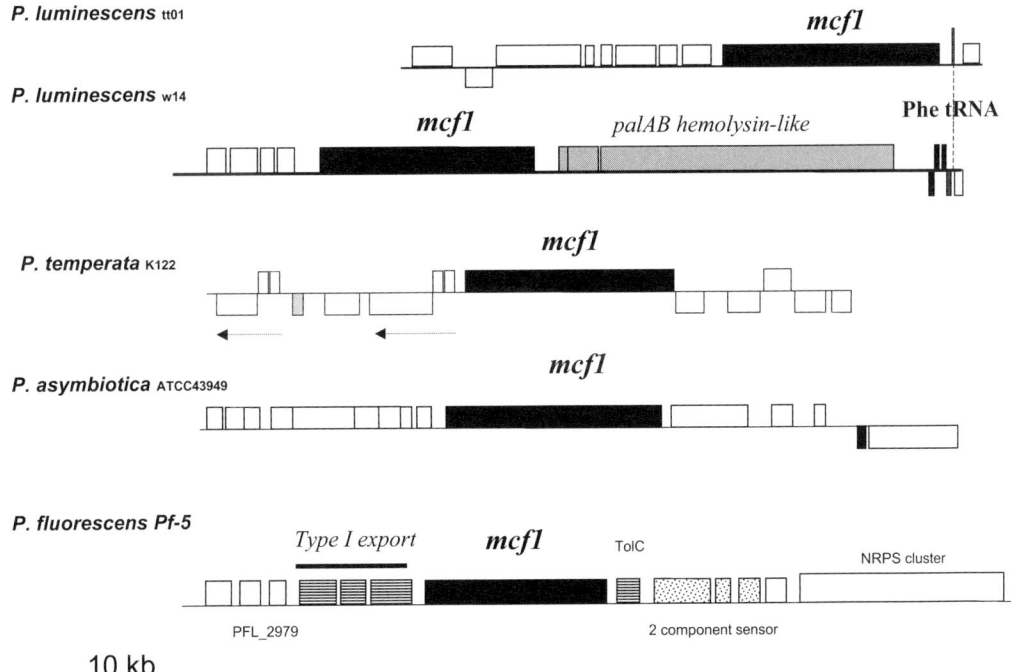

FIGURE 4 Gene encoding Mcf1 toxin, found in different genomic locations in different *Photorhabdus* strains. In *P. luminescens* TT01 and W14, the *mcf1* gene is located near a Phe tRNA in a classic PAI organization. In *P. temperata* and *P. asymbiotica*, mcf1 is found at different genomic locations away from a predicted tRNA. Note that in *Pseudomonas syringae* pf5 that *mcf1* is encoded next to type I export machinery (see text for discussion).

that it is mobile or that it has been acquired independently. However, all *Photorhabdus* strains, and indeed *Xenorhabdus* and *Pseudomonas fluorescens* Pf-5, also carry a copy of *mcf1*, suggesting it is an essential virulence factor. The presence of this insecticidal toxin in *P. fluorescens* Pf-5 is interesting because this strain is not typically regarded as an insect pathogen but is a biocontrol strain used to suppress the growth of plant pathogenic fungi and oomycetes in soil. Sequence comparisons of the Mcf proteins suggest a tripartite structure with toxic N termini and middle domain and C-terminal domains potentially involved in membrane translocation and bacterial export, respectively.

After the discovery of Mcf1, a second similar gene, Mcf2, was discovered in the same *P. luminescens* W14 genome (45). In contrast to Mcf1, the N terminus of Mcf2 has similarity to

the type III–delivered HrmA avirulence gene from *Pseudomonas syringae* (45). The *hrmA* gene from *P. syringae* pv. *syringae* has been shown to confer avirulence on the virulent bacterium *P. syringae* pv. *tabaci* in all examined tobacco cultivars. This suggests that Mcf-like toxins may have a similar central and C-terminal section but differ in their N termini, which appear to have homology to different known bacterial effectors (Fig. 5). The comparative genomic organization of Mcf1- and Mcf2-encoding loci is also interesting because they often contain type I export machinery, providing a clue as to how such large proteins may exit the bacterial cell (Fig. 5). This is supported by the observation that the C-terminal portion of all the Mcf proteins resembles a domain from repeat in toxin–like exported proteins known to be secreted via type I systems.

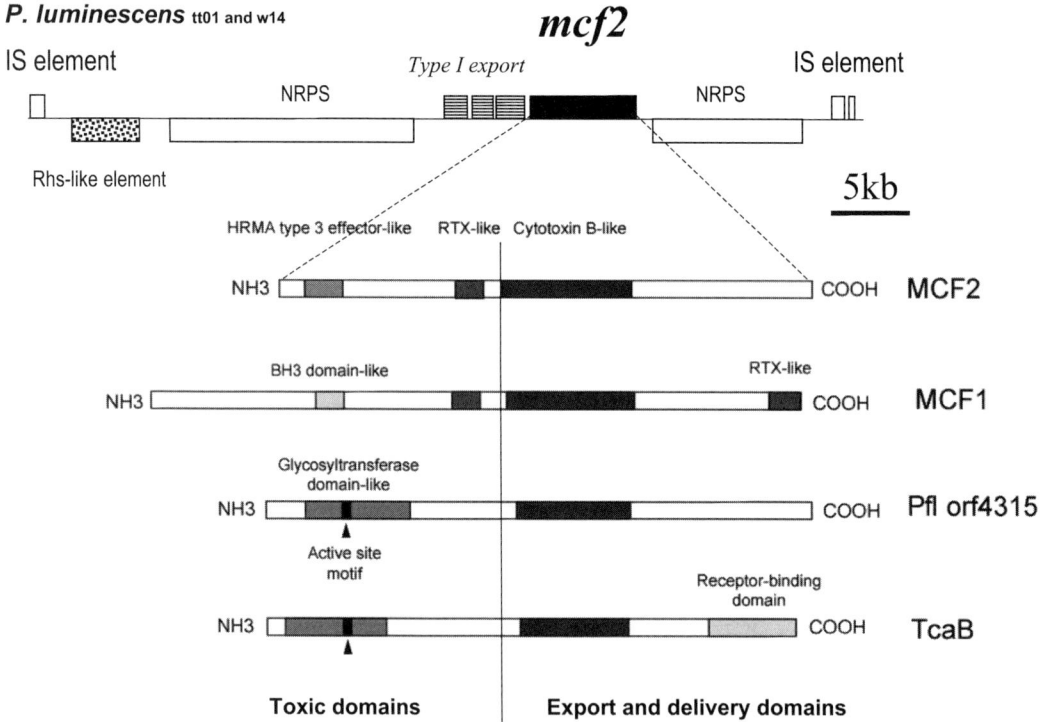

FIGURE 5 Genomic location of *mcf2* and its relationship to the rest of the group of Mcf-like toxins. The gene encoding Mcf2 in both *P. luminescens* TT01 and W14 is encoded next to type I export machinery, suggesting type I secretion for this toxin. Below this diagram is shown the modular nature of this group of toxins. In each case, the N terminus of the Mcf-like toxin encodes a putative effector domain, whereas the rest of the toxin is inferred to be involved in toxin export (C terminus) and translocation (central domain).

Photorhabdus Virulence Cassettes

Other toxin-encoding islands in *Photorhabdus* genomes include the *Photorhabdus* virulence cassettes, or PVCs. These islands are found as tandem repeats of prophage-like loci (47, 50). Each PVC unit is divided into conserved PVC elements that encode components of the prophage and a payload region that carries open reading frames encoding proteins with predicted similarity to known bacterial effectors (Fig. 6). In *P. luminescens* TT01, four PVC loci are repeated in tandem array between a *muk*-replicon stability loci and a type IV DNA conjugation pilus operon. Comparison of PVC loci of different *Photorhabdus* strains shows that different strains possess different complements of units (Fig. 7).

This suggests that the PVCs are mobile between strains and that certain loci represent integration hot spots in which they can insert end to end, such as in TT01. The basic organization of a single PVC is similar to the antifeeding prophage from *Serratia entomophila*. In *S. entomophila*, the antifeeding prophage, which is also found on the pADAP plasmid, alongside the *sepABC* genes, causes larvae of the New Zealand grass grub to cease feeding (26). Interestingly, this plasmid also utilizes a type IV DNA conjugation pilus, similar to that located in the PVC repeat region of *Photorhabdus*, suggesting a common mechanism of horizontal transfer.

When spun down from recombinant *E. coli*, the PVCs show an unusual structure with visible

The *Serratia* anti-feeding prophage

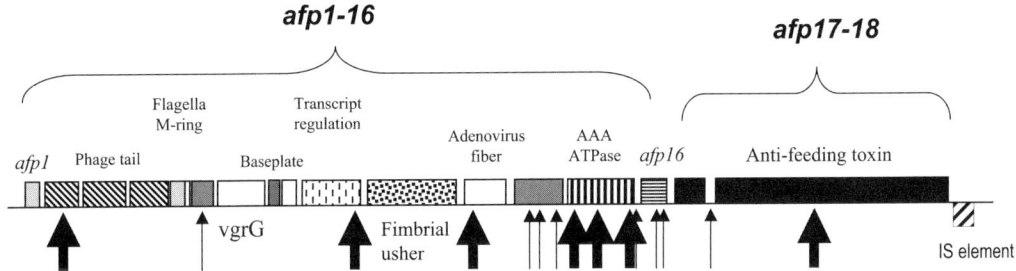

A Photorhabdus Virulence Cassette: PVCasy_lopT

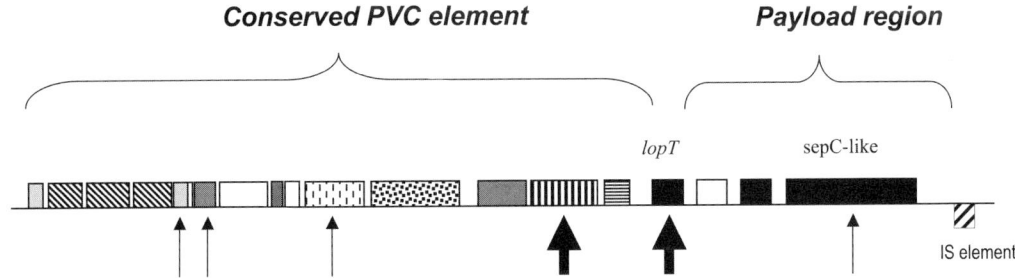

FIGURE 6 Basic organization of a *Photorhabdus* virulence cassette (PVC) and comparison with the antifeeding prophage from *S. entomophila*. Note that the PVC locus is composed of conserved PVC elements encoding phage components such as phage tails and baseplates, and a "payload" region that contains open reading frames with predicted similarity to known bacterial effectors. The organization of the antifeeding prophage from *S. entomophila*, which stops feeding of New Zealand grass grub, is shown for comparison (see text).

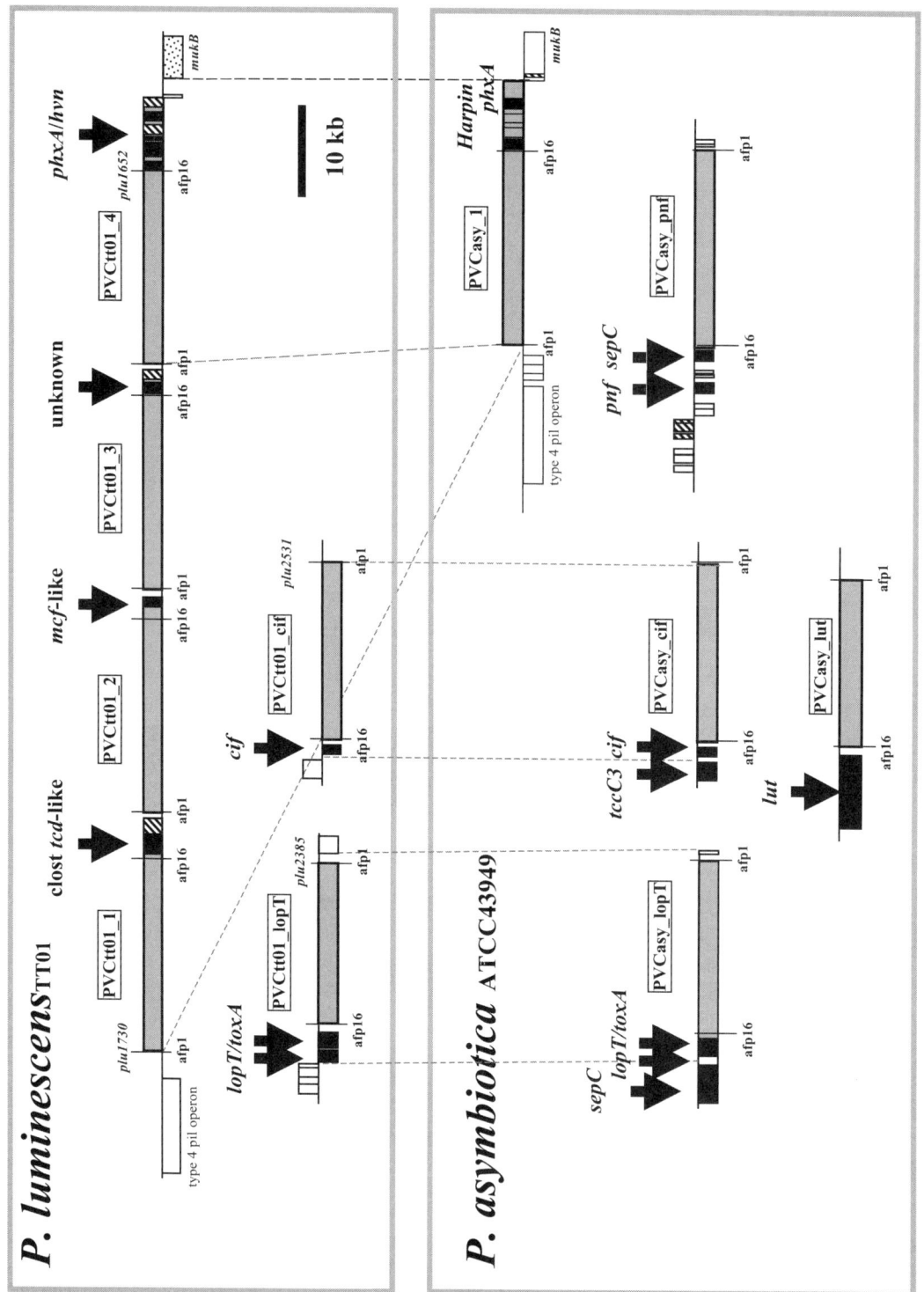

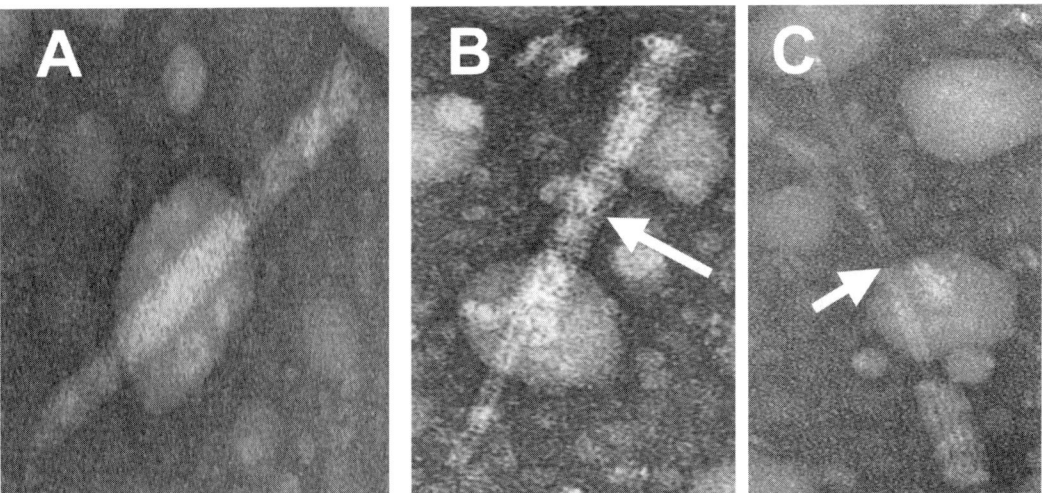

FIGURE 8 Plates showing the structure of a *Photorhabdus* virulence cassette (PVC) as seen under transmission electron microscopy. Note the inner and outer sheath of the PVC, which is reminiscent of an R-type pyocin, which is a bacteriocin. Note that the outer sheath (arrow in B) can be contracted, revealing the long inner sheath (arrow in C).

similarity to an R-type pyocin (Fig. 8). However, unlike these bacteriocins, the PVCs have no antibacterial activity but instead have potent injectable toxicity to larvae of the wax moth *Galleria* (50). Larvae injected with PVC extract blacken and die quickly, and they also show a reduction in circulating hemocytes, which die after actin rearrangement. Similarly, expression of putative PVC effectors within mammalian tissue culture cells also causes actin rearrangement, suggesting that this may be their mode of action (50). It remains to be seen if the antifeeding prophage from *Serratia* has a similar mode of action or structure. However, the similarity of the structure of the PVCs to bacteriocins suggests the fascinating hypothesis that these previously antibacterial structures may have been modified to confer insecticidal activity. It is not yet clear whether the pyocin-like structure is responsible for delivering the payload-region-

encoded toxin into the host cell, but if this were the case, it would represent a novel system, reminiscent of a secreted type III molecular syringe. Interestingly, despite the availability of a large number of genome sequences, to date, PVC elements have been seen only in the genomes of *Photorhabdus* and pADAP, suggesting a limited host applicability.

PirAB Binary Toxins

The *Photorhabdus* insect-related, or Pir, toxins were first recognized from the genome of *P. luminescens* TT01 (15). In the TT01 genome, two open reading frames, *plu4092* and *plu4436*, were described as encoding proteins similar to a protein from the Colorado potato beetle (*Leptinotarsa decemlineata*) that was described as being a juvenile hormone esterase or JHE (38–41). These genes are located near the highly related *plu4093* and *plu4437*, which

FIGURE 7 Comparison of genomic organization of *Photorhabdus* virulence cassette (PVC) loci in *P. luminescens* TT01 and *P. asymbiotica* ATCC43949. Note that multiple PVC units in TT01 (units 1–4) correspond to single PVC units (unit pil) in *P. asymbiotica*. The boxes represent predicted open reading frames (ORFs). ORFs shaded gray correspond to PVC cassettes, and those in black (highlighted with arrows) show similarity to known bacterial effectors.

were therefore assumed to be orphans derived via duplication (15). These loci are similar in their organization in the different species *P. asymbiotica*, and the presence of several enterobacterial repetitive intergenic consensus sequences around the loci in each species suggests that they may have been horizontally acquired (43). These genes were shown to encode orally active toxins via the testing of a single TT01 bacterial artificial chromosome clone that contained the locus *plu4093–plu4092*. This clone showed oral toxicity to three different species of mosquito, *Aedes aegypti*, *Culex pipiens*, and *Anopheles gambiae*. Toxicity to the caterpillar pest *Plutella xylostella* was also shown for a bacterial artificial chromosome containing *plu4437–plu4436*, suggesting that both loci encode orally active insecticidal toxins (15). When *plu4093* and *plu4092* were expressed together in *E. coli*, high levels of both caterpillar and mosquito toxicity were seen (15). Although the study demonstrated novel toxins with oral activity, it was confusing because it assumed that the similar protein in Colorado potato beetle was indeed a JHE. This therefore implied that the PirAB toxins may have their mode of action by interfering with insect development, as juvenile hormone is required to maintain the insect in the larval stage each time it sheds its skin.

In order to clarify the relationship between the PirAB toxins and the protein labeled as a JHE in the Colorado potato beetle, we cloned *plu4093* and *plu4092* from TT01 and tested them for injectable activity against *Galleria* (43). We confirmed that both toxins, termed PirA and PirB, must be together to confer insecticidal activity. However, we also tested PirAB for any JHE activity, which it would have to possess in order to interfere with insect development in the anticipated fashion. We confirmed that PirAB lacked any JHE activity at all (43). Further, we examined the literature on the original description of the JHE-like protein in the beetle, and we found no direct evidence for JHE metabolism. Rather, we found simply that the protein is differentially expressed throughout insect development in a manner consistent with its being under hormonal regulation (38–41). Most recently, similarity between the developmentally regulated beetle protein and β-leptinotarsin-h has been noted, which provides an interesting alternative hypothesis (10). This protein, purified from the hemolymph of a different beetle species, *L. haldemani*, is a potent neuroactive toxin that stimulates Ca^{2+} influx and neurotransmitter release (10), suggesting that it may be a pore-forming toxin. This similarity with β-leptinotarsin-h therefore provides an alternative, but not tested, mode of action for PirAB as pore-forming toxins with an unknown site of action. Given the level of oral toxicity induced by these binary toxins, and their small size, the PirAB toxins warrant further investigation as alternatives to Bt toxins in insect control.

Type III Secretion Systems

Like many pathogenic bacteria, *Photorhabdus* carry islands encoding type III secretion systems, or TTSSs. In *Photorhabdus*, these have been implicated in attaching to the insect phagocytes and in delivering effectors into these cells in order to prevent phagocytosis. Comparisons of the TTSS islands found in the insect pathogens *P. luminescens* TT01 and W14 and in *P. asymbiotica* are interesting. In all *Photorhabdus* species studied to date (*P. luminescens*, *P. temperata*, and *P. asymbiotica*), the TTSS is encoded at the same location in the genome, suggesting that they were acquired as a block before speciation of the genus (5). The different TTSS loci show a conserved structure of genes encoding the molecular syringe, but they differ in the linked effector genes that are encoded within these islands (Fig. 9). This flexibility in the effector genes has led to the interesting hypothesis that their variation may reflect differences in *Photorhabdus* ecology (5). However, one *Photorhabdus* TTSS effector, termed LopT, is a homologue of YopT from *Yersinia pestis*. In *Y. pestis*, YopT inhibits phagocytosis by macrophages; similarly, LopT has been shown to inhibit the phagocytosis of *Photorhabdus* by insect phagocytes (4). This

P. luminescens tt01 and w14

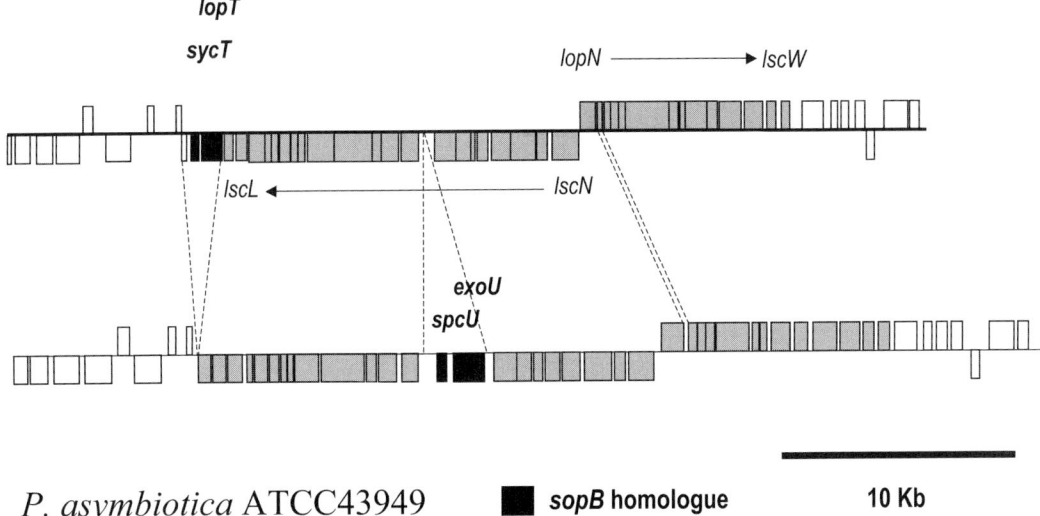

P. asymbiotica ATCC43949   ■ **sopB** homologue   **10 Kb**

FIGURE 9   Comparison between the type III secretion systems (TTSS) in *P. luminescens* and *P. asymbiotica*. Note the similar organization of the genes encoding the TTSS but the variable presence of different effector genes with similarity to lopT, sycT, exoU, and spcU TTSS delivered effectors.

shows the striking similarity in the mechanisms by which both mammalian and insect pathogens avoid phagocytosis (29).

GENOMIC COMPARISONS BETWEEN *PHOTORHABDUS* SPECIES

Tc Loss and Gain

Comparisons between the genomes of different *Photorhabdus* strains have also allowed us to understand different phenotypes associated with specific toxins such as the Tc's. For example, in strain W14, both the cells and the supernatants of the bacterial broth are orally toxic to caterpillars, whereas in TT01, only the cells are toxic (30). Genomic comparisons of these strains by DNA hybridization to a microarray confirmed that although both strains had *tcd* components, TT01 lacked *tcaAB* (30). This led to the discovery that *tcd* encodes an oral toxin associated with the cell (presumably displayed in the outer membrane of the bacterium) but that *tca* encodes a toxin that is released into the

bacterial supernatant (30). In this way, genomics can directly inform biology. Subsequent analysis of the genomic events leading to these differences can be plotted on a family tree of all *Photorhabdus* species (Fig. 10).

Looking at this tree, we can now see that the common ancestor to all *Photorhabdus* likely had a *tca*-like locus. The *tcd* locus was then acquired only by the branch containing strains W14 and TT01, conferring oral toxicity to the bacterial cell. Interestingly, in the remaining branches of the tree, such as that carrying *P. temperata* K122, the *tca* locus appears to have been lost at least twice independently, suggesting why oral toxicity to *Manduca* has not been documented in these branches of the tree. Interestingly, however, the K122 strain has recently been shown to have oral activity against a different lepidopteran, *Prays oleae* (36), which is a pest of olives. However, the individual *tc* genes conferring this activity have not been identified. This suggests that the *tc* genes in the different branches of the *Photorhabdus* tree may

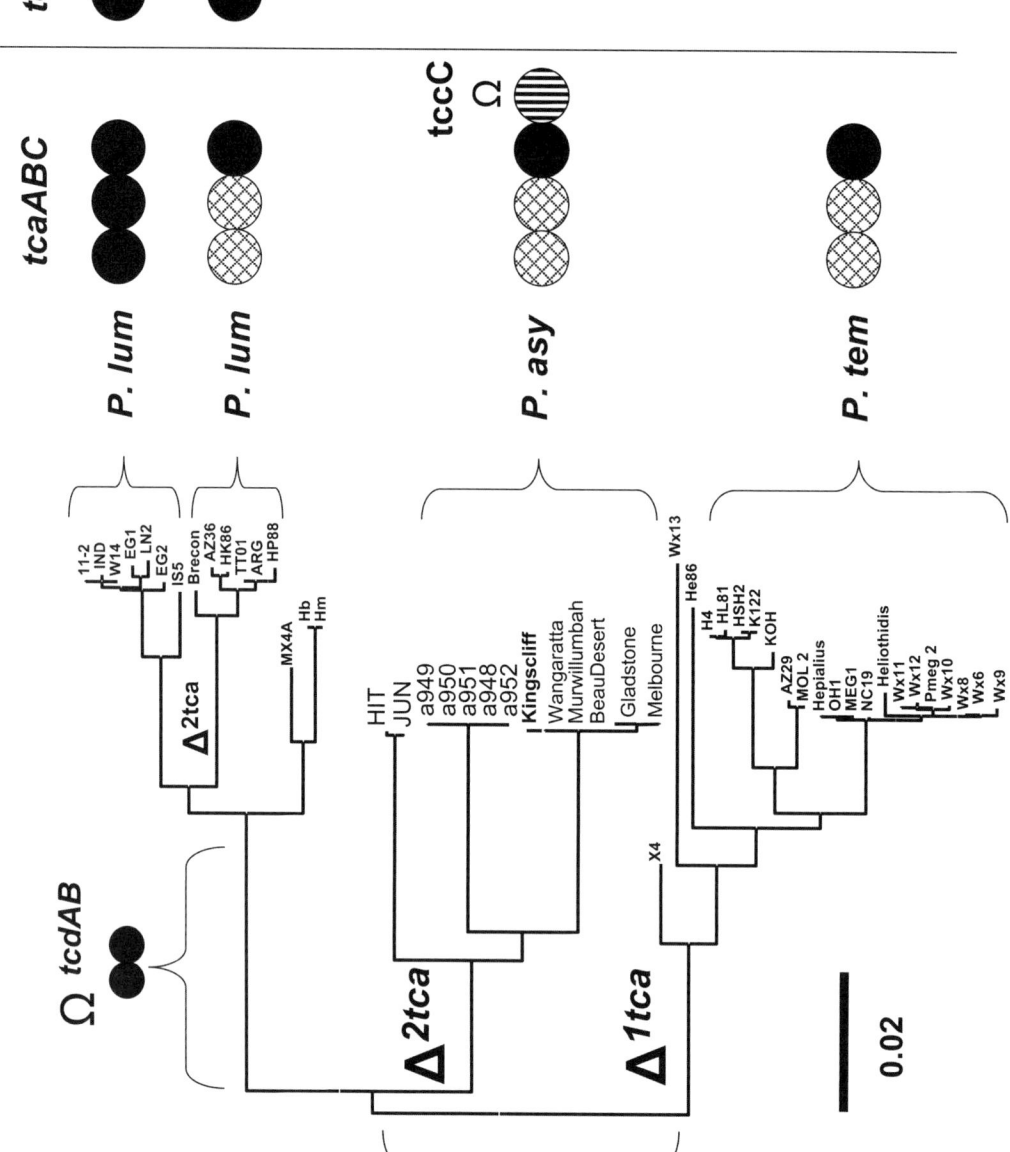

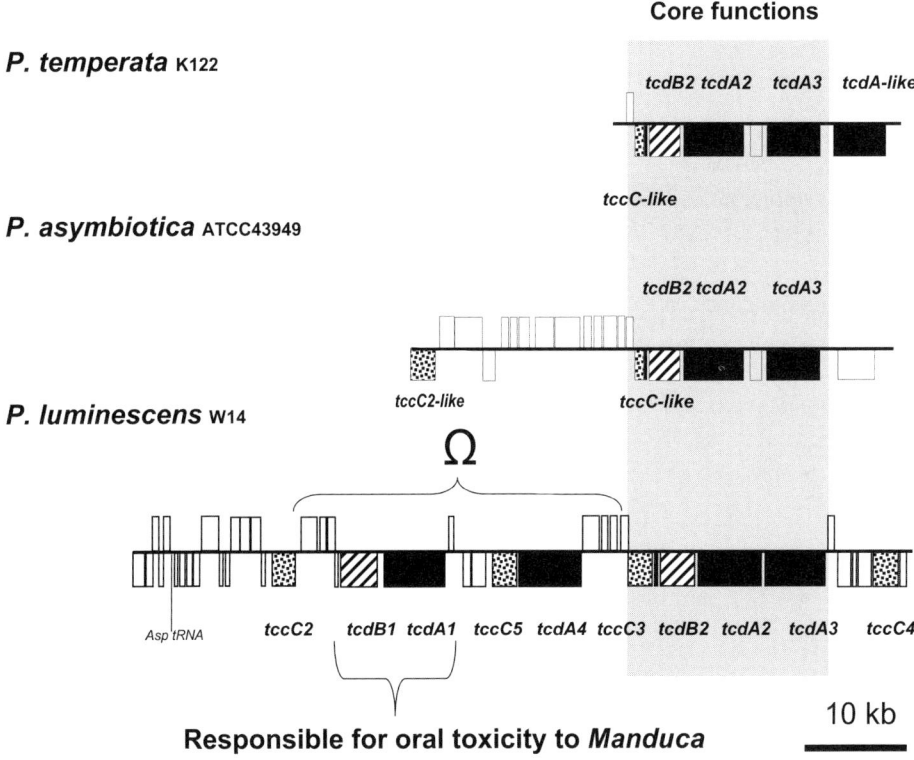

FIGURE 11 Diagram showing the gain of part of the *tcd* island that confers cell-associated oral toxicity to *Manduca*. Note that in the other species, this part of the *tcd* PAI has never been present.

have activities against different groups of caterpillars. Comparative genomics of the *tcd* locus (Fig. 11) confirms that the part of the *tcd* PAI responsible for conferring oral toxicity to *Manduca* has only been acquired by members of the *P. luminescens* W14 group. Similarly, comparisons of the *tca* loci of different *Photorhabdus* species show that three different deletions have led to the three independent losses of *tca* components in *P. temperata*, *P. asymbiotica*, and *P. luminescens* TT01 (Fig. 12). Finally, and for reasons we do not yet under-

stand, these losses mimic the loss of the *tca*-like locus also seen in different *Yersinia*.

Nematode Association

DNA-based microarrays have been used in the comparative genomics of different *Photorhabdus* species and strains in order to attempt to identify regions involved in the specificity of the *Photorhabdus* nematode interaction (22). By comparing strains isolated from different *Heterorhabditis* hosts, researchers hoped to identify the factors responsible for the specific association with one

FIGURE 10 Family tree of *Photorhabdus* strains and species showing the relationship between the presence of different *tc* loci and the resulting insecticidal phenotype. Note that the common ancestor for all *Photorhabdus* had a *tcaABC* locus. The *tcd* locus was then acquired by the *P. luminescens* (*P. lum*) group containing W14 and TT01. Subsequently, *tca* was independently lost at least three times, once in the TT01 group (Δ*1tca*), and then twice in the *P. asymbiotica* (*P. asy*, Δ*2tca*) or *P. temperata* (*P. temp*, Δ*3tca*) groups. This accounts for the differences in the oral toxicity of the cells (Tcd associated) and supernatants (Tca associated) seen between the strains. Solid black circles indicate the presence of a gene; shaded circles indicate its absence.

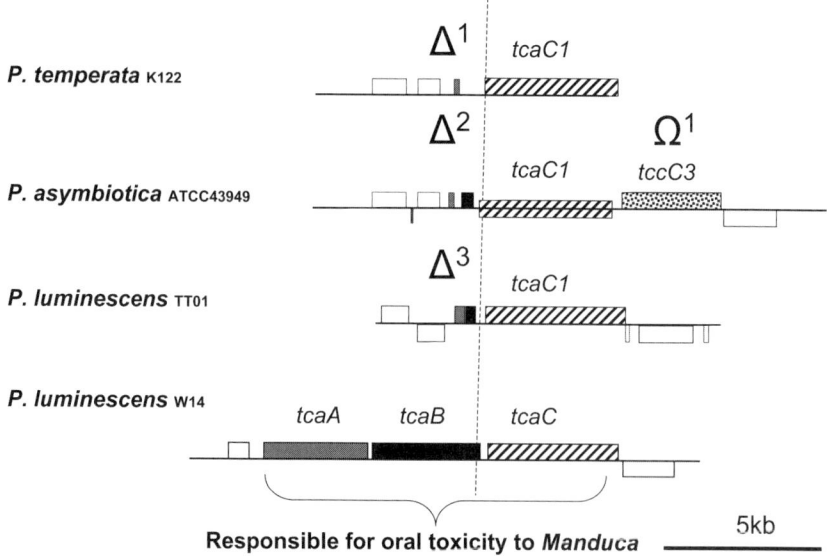

FIGURE 12 Independent losses of *tca* locus elements in different *Photorhabdus* strains and species. Note the complete *tcaABC* locus in *P. luminescens* W14 that confers oral toxicity of bacterial supernatants to *Manduca*. Note also that the *tcaAB* locus has been independently deleted in each of the other three strains (see text).

particular species, *H. bacteriophora*. In this way, 31 regions of the *Photorhabdus* genome were identified that were flexible—that is, differing between strains associated with different nematode hosts. These regions included loci encoding different pili, siderophores, Tc toxins, lipopolysaccharides, and antibiotics (22). These candidate loci form useful targets for testing via gene knockout and subsequent tests for nematode reassociation. This identification of regions producing siderophores and antibiotics as being potentially important in the bacterial association with the nematode was also supported by the results of a screen of transposon-induced mutants of *P. luminescens* for the loss of the ability to support growth and reproduction of *H. bacteriophora* nematodes (7). One mutant, NGR209, consistently failed to support both nematode growth and reproduction, and it was also defective in the production of siderophore and antibiotic activities. The transposon was inserted into an open reading frame homologous to *E. coli EntD*, a 4′-phosphopantetheinyl (Ppant) transferase, which is required for the biosynthesis of the catechol siderophore enter-

obactin. Ppant transferases catalyze the transfer of the Ppant moiety from coenzyme A to holo-acyl, -aryl, or -peptidyl carrier proteins required for the biosynthesis of fatty acids, polyketides (including polyketide antibiotics), or nonribosomal peptides (7).

It is therefore possible that disruption of this gene leads to a failure to make both a siderophore and polyketide antibiotics and that one, the other, or both of these bacterial gene products are required for nematode growth and reproduction. More recently, the lack of siderophore activity as the cause for the inadequacy of the *ngrA* mutant in supporting nematode growth and reproduction has been eliminated by disruption of the genes encoding the siderophore (8). However, the role of polyketide antibiotics in the nematode association remains to be tested. Iron is important in the relationship between the bacterium and nematode, as supported by the observation that a mutation in *exbD* of *P. temperata* K122 can no longer support nematode growth and development (49). The *exbD* gene encodes a component of the ExbB-ExbD-TonB complex that uptakes siderophores, suggesting that the mutant bacteria were unable to sequester enough iron (29). Consistent with this hypothesis, defects in both pathogenicity and symbiosis could be rescued by the addition of $FeCl_3$ (49).

Insect versus Human Pathogens

Given the significant differences between mammalian and insect hosts in number, distribution, longevity, and complexity of immune responses, we can expect obligate pathogens of invertebrates to use very different strategies of virulence to those used by human pathogens (48). For example, insect pathogens would not require a means of evading the rapidly adapting acquired immunity afforded by somatic cell recombination of the mammalian response. Nor would there be much advantage to establishing long-term chronic infections in hosts that have only short lifespans. Despite this, pathogens of both types must combat similar germ-line immune responses, arguably the most important aspect of immunity. These include almost identical phagocytic cellular responses and antibacterial peptide humoral responses. This means that virulence factors evolved to combat such host responses are likely to be equally effective in both mammals and insects. The strategies by which these factors are deployed, however, are likely to differ. Although most *Photorhabdus* strains that have been isolated represent obligate insect pathogens, only one of the three species, *P. asymbiotica*, has been isolated as the causative agent of human infections (24). This means that *Photorhabdus* is ideal for elucidating the molecular adaptations in invertebrate and mammalian pathogenicity. After the sequencing of the insect pathogen *P. luminescens* TT01, we have been sequencing the genome of *P. asymbiotica* ATCC43949. These strains are capable of killing insects and of forming serious infections in humans (24). Most recently, we have discovered that they are also vectored by nematodes, just like the strict entomopathogens *P. luminescens* and *P. temperata* (23). We have therefore been comparing the genome of the insect pathogen with that of the human patho-

gen in an attempt to understand what makes *P. asymbiotica* so pathogenic. Preliminary analysis has revealed a number of striking similarities and differences between the two genomes. The genome of the insect pathogen *P. luminescens* is 5.6 Mb, whereas that of *P. asymbiotica* is 5.0 Mb. Nevertheless, there is still 0.88 Mb that is unique to the human pathogen, suggesting that extensive adaptation has occurred.

As discussed earlier, we can see that *P. asymbiotica* possesses different complements of type III– and PVC-encoded effector genes. Type III effectors include the replacement of the *lopT* gene, which is responsible for antiphagocytosis in the insect pathogen with a homologue of *exoU*, a known cytotoxin from *Pseudomonas*. In addition, the human pathogen has acquired a homologue of SopB, a type III effector gene essential for intracellular survival in *Salmonella*. We believe that this reflects a difference in virulence strategies of the two *Photorhabdus* species, with the *P. asymbiotica* possibly evading adaptive immunity by entering host immune cells. Clinical evidence and preliminary tissue culture experiments support this. In addition to differences in type III effectors, a second type III–like locus similar to one found in *Vibrio parahaemolyticus* is seen in the human pathogen. In *Vibrio*, this locus is believed to be crucial to its particular strategy of human infection.

Further comparative genomics has allowed us to identify many GIs, some common to both *P. luminescens* and *P. asymbiotica* but absent from *E. coli*, and others unique to either type. One of the largest classes of genes present in the genomes of both strains is the nonribosomal-peptide synthase (NRPS) gene clusters. The functions of the secondary metabolites produced by these clusters cannot be predicted from the sequences, but we can see that eight are common to the insect and human pathogen, and at least six are unique to ATCC43949 and five to strain TT01. We have recently shown that at least four of these clusters from ATCC43949 can produce toxins that kill insects when heterologously expressed in *E. coli* on cosmid clones. These toxic NRPS clusters were identified by using powerful complementary approaches to screen the whole *P. asymbiotica* genome for gene products toxic to insects, nematodes, and amoebae. These approaches include screening cosmid libraries for resistance to predation by nematodes or amoebae and the ability to survive the immune response in insects. In addition to the NRPS clusters, these screens have revealed novel protein toxins and secretion systems and have also allowed us to ascribe specific functions to otherwise unknown gene sequences.

GENOMIC COMPARISONS WITH *YERSINIA*

After the complete sequencing of the *Y. pestis* CO92 genome and the documentation of a *tca*-like locus that was apparently disrupted, it was speculated that *Y. pestis* may have lost this putatively insecticidal toxin on its way to acquiring an insect vector, the rat flea (34). In other words, the *Y. pestis* ancestor needed to lose some of its putative Tc-associated insecticidal activity be-

fore it could be successfully vectored by fleas. However, our recent detailed examination of the biology of the Tc's in different *Yersinia* raises an interesting alternative hypothesis. *Y. pseudotuberculosis* and some strains of *Y. pestis* have an intact copy of a *tca*-like locus (Fig. 13). Thus, the disruption of *tca* in the original *Y. pestis* strain may not be representative of the group as a whole. Further, we have recently shown, in collaboration with David Erickson and Jim Hinnebusch, that BC toxins from *Y. pseudotuberculosis* are in fact toxic to rat fleas when administered via blood feeding. This suggests, contrary to the original hypothesis, that the *tc* genes of both *Y. pestis* and its immediate ancestor *P. pseudotuberculosis* may be toxic to insects.

Photorhabdus Switches between Symbiosis and Pathogenicity

These descriptions of the comparative genomics of *Photorhabdus* have identified several

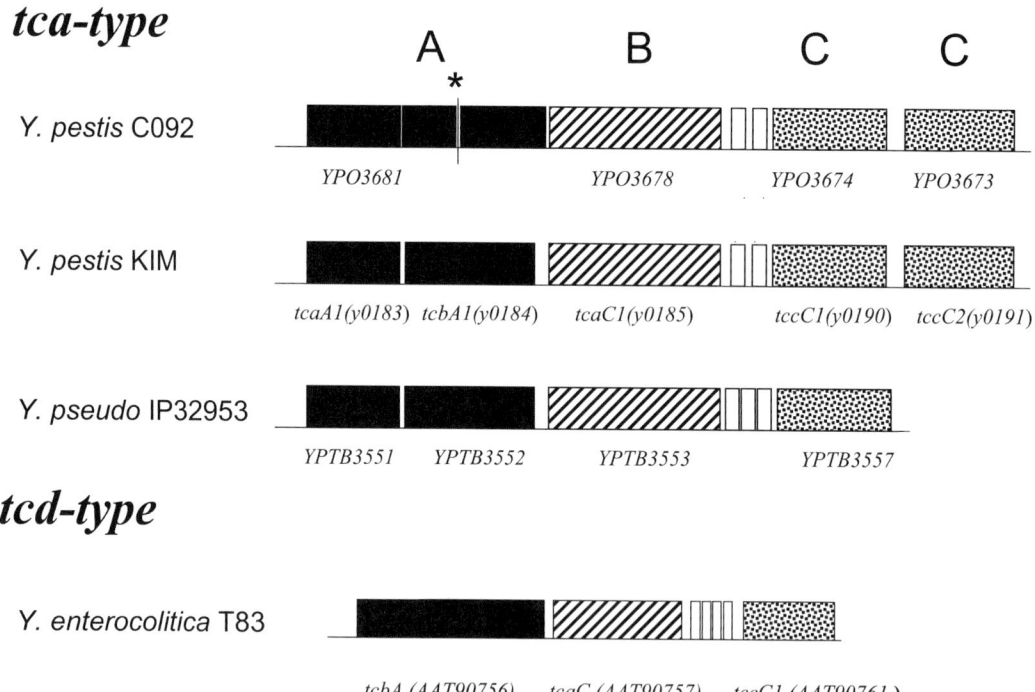

FIGURE 13 Diagram showing *tca*-like loci in different strains and species of *Yersinia*. In *Y. pestis* CO92, *tcaA* is disrupted (asterisk), but in *Y. pestis* KIM, this ORF is intact. Note that similar loci are found in the close relatives *Y. pseudotuberculosis* and *Y. enterocolitica*, which may have implications for the evolution of the flea association of *Y. pestis* (see text).

virulence factors that help the pathogen overcome its insect host, but they have not illustrated how the nematode symbiont switches into an insect pathogen. To recognize that *Photorhabdus* has been regurgitated out of the nematode gut, it must require sensors capable of detecting the insect hemolymph. To this end, it is interesting that a mutation in the *phoP* gene has been shown to render *Photorhabdus* avirulent to insects (12). The *phoP* gene encodes the response regulator of a two-component pathway capable of detecting low levels of Mg^{2+} (6). However, the precise mechanism whereby the *Photorhabdus* PhoPQ two-component pathway recognizes that the bacteria are within the insect is not currently clear (29). The demonstration that a second mutant, in the *pgbPE* operon, renders *Photorhabdus* avirulent and unable to recolonize the IJ nematode (1), suggests an interesting, and related, link between pathogenicity and symbiosis. The *pbgPE* operon is homologous to *pmrHFIJKLM* in *Salmonella enterica*, which encodes proteins responsible for the modification of lipid A of lipopolysaccharide with L-amino-arabinose. In *Salmonella*, the expression of this operon is controlled by Mg^{2+} and the PhoPQ two-component sensor; this is also the case in *Photorhabdus* (29). It therefore seems likely that *Photorhabdus* PhoPQ controls the expression of genes required both for insect infection and nematode reassociation, as suggested by Joyce and others (29).

As an alternative approach to study the genetic regulation of the switch between mutualism and pathogenicity, the laboratory of David Clarke has used the primary and secondary variants of *Photorhabdus* as a model of symbiosis and pathogenicity (see review by Joyce et al. [29]). When isolated from their vector nematodes, *Photorhabdus* colonies typically have a primary, highly pigmented morphology. However, after prolonged passaging on culture plates, secondary, unpigmented, variant colonies arise. This switch between primary and secondary variants appears to be nonreversible, and the genetic basis of this switch remains unclear. Both variants are equally virulent when injected into insect larvae, but only the primary variant can colonize the intestinal

tract of the IJ nematode and support nematode growth and development (28). The primary variant expresses several phenotypes that are absent from the secondary variant, including the production of extracellular enzymes, pigments, antibiotics, and light that are required for nematode growth and development (28). A more recent proteomic analysis of the two variants has confirmed that few proteins are secreted by the secondary variants and that proteins upregulated in the primary variants are involved in oxidative stress, energy metabolism, and translation (37).

Clarke and colleagues argue that the differences in pigment, protease, and antibiotic production shown by these forms mimic the switch between symbiosis and pathogenicity (28, 29). They have therefore screened for mutants that affect the switch between primary and secondary variants. They found that the production of these symbiosis factors is repressed in the secondary variant by the protein encoded by a gene with homology to *hexA* from *Erwinia*. Moreover, removing the repression of the symbiosis factors in the secondary variant results in a significant attenuation of virulence to larvae of the greater wax moth (28). This suggests that during a normal infection, pathogenicity and symbiosis must be temporally separated and that HexA is involved in the regulation of this pathogen-symbiont transition (28). These findings give us a clue as to the regulatory mechanisms responsible for the switch between symbiont and pathogen.

Functional Genomics of *Photorhabdus*

As a result of the relative closeness of *Photorhabdus* and *E. coli*, we have had considerable success in isolating *Photorhabdus* virulence factors via the screening of individual clones from cosmid libraries expressed in *E. coli*. However, we recently have modified our screens of *Photorhabdus* cosmid libraries to switch from screens looking for simple gain of function in individual recombinant *E. coli* clones to screens in which pools of clones are screened for different phenotypes—for example, ability to persist within an insect. In the latter screen, we took a complete *P. asymbiotica* cosmid library and

Gain of persistence clusters

Gain of toxicity clusters

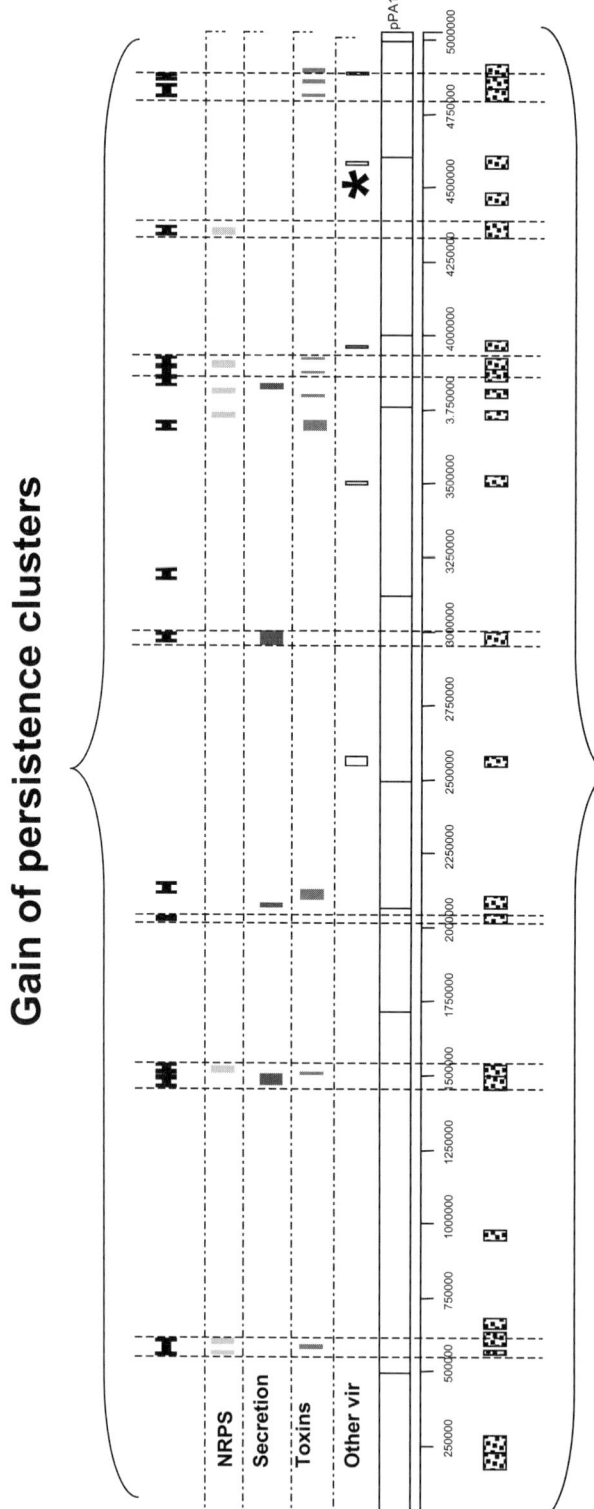

FIGURE 14 Diagram showing the relative location of cosmid clones from *P. asymbiotica* either causing toxicity in individual insects (below the line) or allowing for the persistence of recombinant *E. coli* injected into insects (above the line). The line represents the complete genome sequence of *P. asymbiotica*. Note that clusters of clones causing either toxicity or persistence are often the same (vertical dotted lines). A general classification of these clusters is given to the left of the diagram. The star represents the location of the toxic PVC*pnf* cluster (see text).

injected it into a single late-instar larva of *Manduca*. We then recovered clones that persisted in the face of the insect immune system, plated them, and end-sequenced them. This allowed us to plot the location of *E. coli* clones capable of persisting in the insect as a mixture.

Strikingly, in many cases, the same clones that allowed persistence within an insect correlated with those that caused toxicity (Fig. 14). This screen not only verified the presence of known toxins such as *mcf*, but also gave new insights into potential alternative functions for these loci. For example, the non–orally toxic "core" region of the tcd pathogenicity island facilitates persistence in the face of the insect immune system, suggesting it may play an unexpected role in overcoming the insect immune system. We also noted some loci that conferred toxicity but not persistence, and vice versa. This includes the PVC*pnf* element, which is highly toxic to insects but not stable in *E. coli*. This novel approach means that we can now screen for different phenotypes, such as cell attachment, by screening pools of cosmid clones or, indeed, complete libraries. This streamlining of the screening process will not only allow us to cross-correlate the effect of single clones in several screens, but will also allow us to screen for a range of more subtle phenotypes beyond simple insect death.

CONCLUSIONS

We started by answering three fundamental questions about the biology of *Photorhabdus*. How can one bacterium be both a symbiont with the vector nematode and a pathogen to its insect host? What genes control the switch from symbiosis to pathogenicity? Can we identify islands in the genome involved in pathogenicity, symbiosis, or both? Here we have gained some clues into the regions of the genome that may be involved in either insect pathogenicity or nematode symbiosis. We have also seen some tantalizing glimpses of the types of genes that may be involved in switching between these two very different lifestyles. Most importantly, this review raises an interesting alternative hypothesis: that the expression of the same genes may be responsible for both pathogenic and symbiotic interactions. Most studies to date have focused on anti-insect virulence and have ignored the possibility that the same genes could also be used to interact with the nematode host. Clearly, we still have a long way to go in understanding how this amazing bacterium can switch between being a benign symbiont and a virulent pathogen.

ACKNOWLEDGMENTS

This work was supported by a Royal Society Merit Award to R.ff.-C. and an Exploiting Genomics grant from the BBSRC to R.ff.-C. and N.W.

REFERENCES

1. **Bennett, H. P., and D. J. Clarke.** 2005. The *pbgPE* operon in *Photorhabdus luminescens* is required for pathogenicity and symbiosis. *J. Bacteriol.* **187:**77–84.

2. **Bowen, D., T. A. Rocheleau, M. Blackburn, O. Andreev, E. Golubeva, R. Bhartia, and R. H. ffrench-Constant.** 1998. Insecticidal toxins from the bacterium *Photorhabdus luminescens*. *Science* **280:**2129–2132.

3. **Bowen, D. J., T. A. Rocheleau, C. K. Grutzmacher, L. Meslet, M. Valens, D. Marble, A. Dowling, R. Ffrench-Constant, and M. A. Blight.** 2003. Genetic and biochemical characterization of PrtA, an RTX-like metalloprotease from *Photorhabdus*. *Microbiology* **149:**1581–1591.

4. **Brugirard-Ricaud, K., E. Duchaud, A. Givaudan, P. A. Girard, F. Kunst, N. Boemare, M. Brehelin, and R. Zumbihl.** 2005. Site-specific antiphagocytic function of the *Photorhabdus luminescens* type III secretion system during insect colonization. *Cell. Microbiol.* **7:**363–371.

5. **Brugirard-Ricaud, K., A. Givaudan, J. Parkhill, N. Boemare, F. Kunst, R. Zumbihl, and E. Duchaud.** 2004. Variation in the effectors of the type III secretion system among *Photorhabdus* species as revealed by genomic analysis. *J. Bacteriol.* **186:**4376–4381.

6. **Charnnongpol, S., M. Cromie, and E. A. Groisman.** 2003. Mg^{2+} sensing by the Mg^{2+} sensor PhoQ of *Salmonella enterica*. *J. Mol. Biol.* **325:**795–807.

7. **Ciche, T. A., S. B. Bintrim, A. R. Horswill, and J. C. Ensign.** 2001. A phosphopantetheinyl transferase homolog is essential for *Photorhabdus luminescens* to support growth and reproduction of the entomopathogenic nematode *Heterorhabditis bacteriophora*. *J. Bacteriol.* **183:**3117–3126.

8. **Ciche, T. A., M. Blackburn, J. R. Carney, and J. C. Ensign.** 2003. Photobactin: a catechol siderophore produced by *Photorhabdus luminescens*,

an entomopathogen mutually associated with *Heterorhabditis bacteriophora* NC1 nematodes. *Appl. Environ. Microbiol.* **69:**4706–4713.

9. **Ciche, T. A., and J. C. Ensign.** 2003. For the insect pathogen *Photorhabdus luminescens*, which end of a nematode is out? *Appl. Environ. Microbiol.* **69:**1890–1897.

10. **Crosland, R. D., R. W. Fitch, and H. B. Hines.** 2005. Characterization of beta-leptinotarsin-h and the effects of calcium flux antagonists on its activity. *Toxicon* **45:**829–841.

11. **Daborn, P. J., N. Waterfield, C. P. Silva, C. P. Y. Au, S. Sharma, and R. H. ffrench-Constant.** 2002. A single *Photorhabdus* gene *makes caterpillars floppy* (*mcf*) allows *Escherichia coli* to persist within and kill insects. *Proc. Natl. Acad. Sci. USA* **99:**10742–10747.

12. **Derzelle, S., E. Turlin, E. Duchaud, S. Pages, F. Kunst, A. Givaudan, and A. Danchin.** 2004. The PhoP-PhoQ two-component regulatory system of *Photorhabdus luminescens* is essential for virulence in insects. *J. Bacteriol.* **186:**1270–1279.

13. **Dowling, A., P. Daborn, N. Waterfield, P. Wang, C. Streuli, and R. ffrench-Constant.** 2004. The insecticidal toxin makes caterpillars floppy (Mcf) promotes apoptosis in mammalian cells. *Cell. Microbiol.* **6:**345–353.

14. Reference deleted.

15. **Duchaud, E., C. Rusniok, L. Frangeul, C. Buchrieser, A. Givaudan, S. Taourit, S. Bocs, C. Boursaux-Eude, M. Chandler, J. F. Charles, E. Dassa, R. Derose, S. Derzelle, G. Freyssinet, S. Gaudriault, C. Medigue, A. Lanois, K. Powell, P. Siguier, R. Vincent, V. Wingate, M. Zouine, P. Glaser, N. Boemare, A. Danchin, and F. Kunst.** 2003. The genome sequence of the entomopathogenic bacterium *Photorhabdus luminescens*. *Nat. Biotechnol.* **21:**1307–1313.

16. **Eleftherianos, I., P. J. Millichap, R. H. ffrench-Constant, and S. E. Reynolds.** 2006. RNAi suppression of recognition protein mediated immune responses in the tobacco hornworm *Manduca sexta* causes increased susceptibility to the insect pathogen *Photorhabdus*. *Dev. Comp. Immunol.* **30:**1099–1107.

17. **ffrench-Constant, R., and N. Waterfield.** 2005. An ABC guide to the bacterial toxin complexes. *Adv. Appl. Microbiol.* **58C:**169–183.

18. **ffrench-Constant, R. H., N. Waterfield, V. Burland, N. T. Perna, P. J. Daborn, D. Bowen, and F. R. Blattner.** 2000. A genomic sample sequence of the entomopathogenic bacterium *Photorhabdus luminescens* W14: potential implications for virulence. *Appl. Environ. Microbiol.* **66:**3310–3329.

19. **ffrench-Constant, R., N. Waterfield, P. Daborn, S. Joyce, H. Bennett, C. Au, A.**

Dowling, S. Boundy, S. Reynolds, and D. Clarke. 2003. *Photorhabdus*: towards a functional genomic analysis of a symbiont and pathogen. *FEMS Microbiol. Rev.* **26:**433–456.

20. **Forst, S., and D. Clarke.** 2002. Bacteria-nematode symbiosis, p. 57–77. *In* R. Gaugler (ed.), *Entomopathogenic Nematology.* CAB International, London.

21. **Forst, S., B. Dowds, N. Boemare, and E. Stackebrandt.** 1997. *Xenorhabdus* and *Photorhabdus* spp.: bugs that kill bugs. *Ann. Rev. Microbiol.* **51:**47–72.

22. **Gaudriault, S., E. Duchaud, A. Lanois, A. S. Canoy, S. Bourot, R. Derose, F. Kunst, N. Boemare, and A. Givaudan.** 2006. Whole-genome comparison between *Photorhabdus* strains to identify genomic regions involved in the specificity of nematode interaction. *J. Bacteriol.* **188:**809–814.

23. **Gerrard, J. G., S. A. Joyce, D. J. Clarke, R. H. ffrench-Constant, G. R. Nimmo, D. F. Looke, E. J. Feil, L. Pearce, and N. R. Waterfield.** 2006. Nematode symbiont for *Photorhabdus asymbiotica*. *Emerg. Infect. Dis.* **12:**1562–1564.

24. **Gerrard, J., N. Waterfield, R. Vohra, and R. ffrench-Constant.** 2004. Human infection with *Photorhabdus asymbiotica*: an emerging bacterial pathogen. *Microb. Infect.* **6:**229–237.

25. **Glare, T. R., G. E. Corbett, and A. J. Sadler.** 1993. Association of a large plasmid with amber disease of the New Zealand grass grub, *Costelytra zealandica*, caused by *Serratia entomophila* and *Serratia proteamaculans*. *J. Invertebr. Pathol.* **62:**165–170.

26. **Hurst, M. R., T. R. Glare, and T. A. Jackson.** 2004. Cloning *Serratia entomophila* antifeeding genes—a putative defective prophage active against the grass grub *Costelytra zealandica*. *J. Bacteriol.* **186:**5116–5128.

27. **Hurst, M. R., T. R. Glare, T. A. Jackson, and C. W. Ronson.** 2000. Plasmid-located pathogenicity determinants of *Serratia entomophila*, the causal agent of amber disease of grass grub, show similarity to the insecticidal toxins of *Photorhabdus luminescens*. *J. Bacteriol.* **182:**5127–5138.

28. **Joyce, S. A., and D. J. Clarke.** 2003. A *hexA* homologue from *Photorhabdus* regulates pathogenicity, symbiosis and phenotypic variation. *Mol. Microbiol.* **47:**1445–1457.

29. **Joyce, S. A., R. J. Watson, and D. J. Clarke.** 2006. The regulation of pathogenicity and mutualism in *Photorhabdus*. *Curr. Opin. Microbiol.* **9:**127–132.

30. **Marokhazi, J., N. Waterfield, G. LeGoff, E. Feil, R. Stabler, J. Hinds, A. Fodor, and R. H. ffrench-Constant.** 2003. Using a DNA microarray to investigate the distribution of insect

virulence factors in strains of *photorhabdus* bacteria. *J. Bacteriol.* **185**:4648–4656.

31. **Matsui, H., C. M. Bacot, W. A. Garlington, T. J. Doyle, S. Roberts, and P. A. Gulig.** 2001. Virulence plasmid-borne *spvB* and *spvC* genes can replace the 90-kilobase plasmid in conferring virulence to *Salmonella enterica* serovar Typhimurium in subcutaneously inoculated mice. *J. Bacteriol.* **183**:4652–4658.

32. **Morgan, J. A., M. Sergeant, D. Ellis, M. Ousley, and P. Jarrett.** 2001. Sequence analysis of insecticidal genes from *Xenorhabdus nematophilus* PMFI296. *Appl. Environ. Microbiol.* **67**:2062–2069.

33. **Otto, H., D. Tezcan-Merdol, R. Girisch, F. Haag, M. Rhen, and F. Koch-Nolte.** 2000. The *spvB* gene-product of the *Salmonella enterica* virulence plasmid is a mono(ADP-ribosyl)transferase. *Mol. Microbiol.* **37**:1106–1115.

34. **Parkhill, J., B. W. Wren, N. R. Thomson, R. W. Titball, M. T. Holden, M. B. Prentice, M. Sebaihia, K. D. James, C. Churcher, K. L. Mungall, S. Baker, D. Basham, S. D. Bentley, K. Brooks, A. M. Cerdeno-Tarraga, T. Chillingworth, A. Cronin, R. M. Davies, P. Davis, G. Dougan, T. Feltwell, N. Hamlin, S. Holroyd, K. Jagels, A. V. Karlyshev, S. Leather, S. Moule, P. C. Oyston, M. Quail, K. Rutherford, M. Simmonds, J. Skelton, K. Stevens, S. Whitehead, and B. G. Barrell.** 2001. Genome sequence of *Yersinia pestis*, the causative agent of plague. *Nature* **413**:523–527.

35. **Sergeant, M., P. Jarrett, M. Ousley, and J. A. Morgan.** 2003. Interactions of insecticidal toxin gene products from *Xenorhabdus nematophilus* PMFI296. *Appl. Environ. Microbiol.* **69**:3344–3349.

36. **Tounsi, S., A. E. Aoun, M. Blight, A. Rebai, and S. Jaoua.** 2006. Evidence of oral toxicity of *Photorhabdus temperata* strain K122 against *Prays oleae* and its improvement by heterologous expression of *Bacillus thuringiensis* cry1Aa and cry1Ia genes. *J. Invertebr. Pathol.* **91**:131–135.

37. **Turlin, E., G. Pascal, J. C. Rousselle, P. Lenormand, S. Ngo, A. Danchin, and S. Derzelle.** 2006. Proteome analysis of the phenotypic variation process in *Photorhabdus luminescens*. *Proteomics* **6**:2705–2725.

38. **Vermunt, A. M., A. B. Koopmanschap, J. M. Vlak, and C. A. de Kort.** 1997. Cloning and sequence analysis of cDNA encoding a putative juvenile hormone esterase from the Colorado potato beetle. *Insect Biochem. Mol. Biol.* **27**:919–928.

39. **Vermunt, A. M., A. B. Koopmanschap, J. M. Vlak, and C. A. de Kort.** 1998. Evidence for two juvenile hormone esterase-related genes in the Colorado potato beetle. *Insect Mol. Biol.* **7**:327–336.

40. **Vermunt, A. M., A. B. Koopmanschap, J. M. Vlak, and C. A. de Kort.** 1999. Expression of the juvenile hormone esterase gene in the Colorado potato beetle, *Leptinotarsa decemlineata*: photoperiodic and juvenile hormone analog response. *J. Insect Physiol.* **45**:135–142.

41. **Vermunt, A. M., A. M. Vermeesch, and C. A. de Kort.** 1997. Purification and characterization of juvenile hormone esterase from hemolymph of the Colorado potato beetle. *Arch. Insect Biochem. Physiol.* **35**:261–277.

42. **Waterfield, N., A. Dowling, S. Sharma, P. J. Daborn, U. Potter, and R. H. ffrench-Constant.** 2001. Oral toxicity of *Photorhabdus luminescens* W14 toxin complexes in *Escherichia coli. Appl. Environ. Microbiol.* **67**:5017–5024.

43. **Waterfield, N., S. G. Kamita, B. D. Hammock, and R. ffrench-Constant.** 2005. The *Photorhabdus* Pir toxins are similar to a developmentally regulated insect protein but show no juvenile hormone esterase activity. *FEMS Microbiol. Lett.* **245**:47–52.

44. **Waterfield, N. R., D. J. Bowen, J. D. Fetherston, R. D. Perry, and R. H. ffrench-Constant.** 2001. The *tc* genes of *Photorhabdus*: a growing family. *Trends Microbiol.* **9**:185–191.

45. **Waterfield, N. R., P. J. Daborn, A. J. Dowling, G. Yang, M. Hares, and R. H. ffrench-Constant.** 2003. The insecticidal toxin Makes caterpillars floppy 2 (Mcf2) shows similarity to HrmA, an avirulence protein from a plant pathogen. *FEMS Microbiol. Lett.* **229**:265–270.

46. **Waterfield, N. R., P. J. Daborn, and R. H. ffrench-Constant.** 2002. Genomic islands in *Photorhabdus*. *Trends Microbiol.* **10**:541–545.

47. **Waterfield, N. R., P. J. Daborn, and R. H. ffrench-Constant.** 2004. Insect pathogenicity islands in the insect pathogenic bacterium *Photorhabdus*. *Physiol. Entomol.* **29**:240–250.

48. **Waterfield, N. R., B. W. Wren, and R. H. ffrench-Constant.** 2004. Invertebrates as a source of emerging human pathogens. *Nat. Rev. Microbiol.* **2**:833–841.

49. **Watson, R. J., S. A. Joyce, G. V. Spencer, and D. J. Clarke.** 2005. The *exbD* gene of *Photorhabdus temperata* is required for full virulence in insects and symbiosis with the nematode *Heterorhabditis. Mol. Microbiol.* **56**:763–773.

50. **Yang, G., A. J. Dowling, U. Gerike, R. H. ffrench-Constant, and N. R. Waterfield.** 2006. *Photorhabdus* virulence cassettes confer injectable insecticidal activity against the wax moth. *J. Bacteriol.* **188**:2254–2261.

INDEX